QUANTUM ELECTRONICS

QUANTUM ELECTRONICS

Third Edition

Amnon Yariv

California Institute of Technology

WILEY

JOHN WILEY & SONS

New York • Chichester • Brisbane • Toronto • Singapore

LIBRARY OF CONGRESS

Library of Congress Cataloging-in-Publication Data:

Yariv, Amnon.
Quantum electronics / Amnon Yariv.—3rd ed.
p. cm.
Bibliography: p.
Includes index.
ISBN 0-471-60997-8
1. Quantum electronics. I. Title.
QC688.Y37 1988
537.5—dc19

Printed in the United States of America

10 9 8 7 6 5 4 3 2

To **The Memory of My Father**

Preface

The thirteen years that have intervened since the appearance of the second edition of this book have witnessed some important developments in the field of lasers and quantum electronics. Foremost among them are: phase-conjugate optics and its myriad applications, the long wavelength quaternary semiconductor laser, and the deepened understanding of the physics of semiconductor lasers—especially that applying to their current modulation and limiting linewidth, laser arrays and the related concept of supermodes, quantum well semiconductor lasers, the role of phase amplitude coupling in laser noise, and free-electron lasers. The present edition retains nearly all the material of the second edition. There are four new chapters on semiconductor lasers, quantum well lasers, free-electron lasers, and phase-conjugate optics. In addition, the chapters on laser noise and third-order nonlinear effects have been extensively revised.

I benefited from teaching the material and from feedback at Caltech by colleagues and students who aided me in tightening and improving numerous points throughout the book. To these individuals special thanks are due. I am grateful to my administrative assistant, Jana Mercado, for her assistance in preparing this edition. I also take pleasure in thanking the talented group of students and ex-students, especially K. Lau, K. Vahala, C. Harder, T. Koch, M. Cronin-Golomb, S. Kwong, and the late C. Lindsey, whose research efforts are responsible for much of the new material added to this edition. The material on quantum well lasers benefited greatly from my association with Y. Arakawa while David Crouch contributed a major part of the treatment of squeezed states.

Pasadena, California
September 1987

Amnon Yariv

Preface to the Second Edition

This textbook introduces the main principles involved in the study and practice of quantum electronics, which include the theory of laser oscillators, a wide range of optical phenomena, and devices that owe their existence to the intense and coherent optical fields made possible by the laser.

The emphasis is almost exclusively on fundamental principles. An attempt is made, however, to bridge the gap between theory and practice through the use of numerical examples based on real situations.

Approximately one-half of this edition is new. In addition, a number of topics related to microwave phenomena and magnetic resonance were omitted. The major changes are as follows.

1. The addition of treatments on Gaussian beam propagation in lenslike media, optical resonators, density matrix formulation of the interaction of light and matter, theory of laser oscillation, Van der Pol noise analysis of lasers, dye lasers, amplification in vibrational-rotational transitions, double heterojunction lasers, mode locking in homogeneously broadened lasers, Q-switching, saturated amplifiers and amplification of spontaneous emission, acoustooptic interactions, self-induced transparency, photon echoes, spontaneous parametric fluorescence, distributed feedback lasers, and mode coupling in dielectric waveguides.
2. The deletion of chapters dealing with microwave masers, magnetism, magnetic resonance, and microwave parametric oscillators.
3. An exclusive use of the meter-kilogram-second (MKS) unit system.

This text is primarily for the graduate student in physics and applied physics. The latter category often includes students in departments of electrical engineering and material science.

The typical Caltech student taking the course from which this book was developed has a background of a one-year rigorous course in quantum mechanics and one course in electromagnetic theory. These are courses taken by the more advanced students in their senior year but often in the first year of graduate school. A good familiarity with these two topics is assumed, although most of the prerequisite background material is included here.

The book can be used as a basis for a one-year course in quantum electronics or, alternatively, for these one-semester courses:

1. Lasers: Chapters 5–13.
2. Nonlinear Optical Effects and Stimulated Scattering Phenomena: Chapters 14 (part dealing with acoustooptics), 15–18.
3. Optical Modes and Propagation Phenomena: Chapters 5–7, 14, 19.

Course 1 makes heavy use of quantum mechanics. In course 2 quantum mechanics is needed only in Chapter 15, while in course 3 it is not used at all. An electromagnetic background is needed in all three courses.

I apologize to any of my colleagues whose work has not been acknowledged or adequately represented in this book. Since this is primarily a textbook, the material was chosen mainly because of pedagogic considerations rather than chronological precedence.

I thank Ruth Stratton and Dian Rapchak for typing and proofreading the original manuscript and Paula Samazan for assisting with the references.

Thanks are due to Dr. Jack Comly who made important contributions to Chapters 15 and 18 and to Mr. H. W. Yen who has gone over the whole manuscript rederiving the results and checking the internal consistency of the text.

Amnon Yariv

About the Author

Amnon Yariv, a native of Israel, received his higher education at the University of California at Berkeley. After four years at the Bell Telephone Laboratories, he joined the California Institute of Technology (Caltech) where he is the Thomas G. Myers Professor of Electrical Engineering and Applied Physics. At Caltech he studies with a team of doctoral students and postdoctoral fellows, a number of research topics in laser physics, nonlinear optics, and optoelectronics. Some of his major contributions include the invention and co-invention of the fields of integrated optoelectronic circuits, phase conjugate optics, and the authoring of the first papers on the theory of mode locking and nonlinear quantum optics. Dr. Yariv is also the founder and chairman of the board of ORTEL Corporation, a semiconductor laser company in Alhambra, California.

When not working, which is rare, he spends time with his family (wife—Fran and three daughters) as well as with a piano, a tennis racket, and a windsurfer.

Contents

CHAPTER 1

Basic Theorems and Postulates of Quantum Mechanics

1.0 INTRODUCTION

In this chapter we shall consider some of the basic postulates and theorems of quantum mechanics. These are general and independent of the specific system studied. The application of these results to special problems will be the concern of Chapter 2 and, to a lesser extent, of the rest of the book.

1.1 THE SCHRÖDINGER WAVE EQUATION

According to quantum mechanics, the behavior of a particle is described by the wavefunction $\psi(\mathbf{r}, t)$, which is a solution of the Schrödinger wave equation

$$\left[-\frac{\hbar^2}{2m}\nabla^2 + V(\mathbf{r}, t)\right]\psi = i\hbar\frac{\partial\psi}{\partial t} \tag{1.1-1}$$

$V(\mathbf{r}, t)$ is the potential energy function of the particle and $\hbar = h/2\pi$ where h is Planck's universal constant.

By associating the differential operator $-i\hbar\,\nabla$ with particle linear momentum $\mathbf{p}$, that is,

$$\mathbf{p} \to -i\hbar\,\nabla \tag{1.1-2}$$

the operator on the left side of (1.1-1) can be associated with the sum of the kinetic and potential energy of the particle.

$$E \to -\frac{\hbar^2}{2m}\nabla^2 + V(\mathbf{r}, t) = i\hbar\frac{\partial}{\partial t} \tag{1.1-3}$$

Statistical Interpretation of the Wavefunction

Consider a very large number of independent spaces with identical potential functions $V(\mathbf{r}, t)$. The motion of a particle in each one of these spaces is described by the same $\psi(\mathbf{r}, t)$. The a-priori probability $P(\mathbf{r}, t)\,dv$ of finding any given particle inside a volume dv (centered about $\mathbf{r}$) is taken as the fraction of the particles found, by measurement, to be inside dv at time t. According to quantum mechanics, the probability density $P(\mathbf{r}, t)$ is given by

$$P(\mathbf{r}, t) = \psi^*(\mathbf{r}, t)\psi(\mathbf{r}, t) \tag{1.1-4}$$

where the asterisk superscript stands for the complex conjugate of the quantity in question.

The statistical interpretation of $(\psi^*\psi)$ is made plausible by showing, as will be done in Section 1.2, that the average motion of a particle as determined from the statistical point of view agrees with its classical counterpart. The final arbiter of the validity of this statement is, of course, the agreement of the results derived using it with experiment.

The first condition resulting from the statistical interpretation is that the total probability of finding the particle somewhere in space is finite and is a constant, that is,

$$\int_{\text{all space}} P(\mathbf{r}, t)\, dv = \int_{\text{all space}} \psi^*(\mathbf{r}, t)\psi(\mathbf{r}, t)\, dv = \text{constant} \tag{1.1-5}$$

or that

$$\frac{d}{dt}\int_{\text{all space}} \psi^*(\mathbf{r}, t)\psi(\mathbf{r}, t)\, dv = 0 \tag{1.1-6}$$

The proof proceeds as follows

$$\frac{d}{dt}\int_V \psi^*\psi\, dv = \int_V \frac{\partial}{\partial t}(\psi^*\psi)\, dv = \int_V \left(\psi\frac{\partial\psi^*}{\partial t} + \frac{\partial\psi}{\partial t}\psi^*\right) dv$$

substituting for $\partial\psi/\partial t$ and $\partial\psi^*/\partial t$ from (1.1-1), the terms involving $V(\mathbf{r}, t)$ drop out, and the result is

$$\frac{i\hbar}{2m}\int_V (\psi^*\, \nabla^2\psi - \psi\, \nabla^2\psi^*)\, dv$$

Use is now made of Green's theorem

$$\int_V (f\,\nabla^2 g - g\,\nabla^2 f)\, dv = \int_A (f\,\nabla g - g\,\nabla f)\cdot \mathbf{n}\, da \tag{1.1-7}$$

where f and g are two arbitrary scalar functions, A is the surface bounding V, and $\mathbf{n}$ is the unit outward normal vector. This leads to

$$\frac{d}{dt}\int_V (\psi^*\psi)\, dv = \frac{i\hbar}{2m}\int_A (\psi^*\,\nabla\psi - \psi\,\nabla\psi^*)\cdot\mathbf{n}\, da \tag{1.1-8}$$

If the volume V is that of all space, the admissible solutions of (1.1-1) are indeed those in which the behavior of ψ as $r \to \infty$ is such that the integration specified by (1.1-8) leads to a zero result. If the volume V is finite, the same result is obtained by choosing ψ so that its value on the bounding surface A leads to a zero result in (1.1-8). This point is discussed further in Problem 1.1.

Having proven (1.1-5), we are free to normalize ψ so that the constant appearing in (1.1-5) is unity. This is consistent with the probabilistic interpretation, since the probability of finding the particle somewhere in all of space is unity, that is,

$$\int_{V=\text{all space}} P(\mathbf{r}, t)\, dv = \int_V \psi^*(\mathbf{r}, t)\psi(\mathbf{r}, t)\, dv = 1 \tag{1.1-9}$$

Particle Density Current

We start by rewriting (1.1-8)

$$\frac{d}{dt}\int_V (\psi^*\psi)\, dv = \frac{i\hbar}{2m}\int_A (\psi^* \nabla\psi - \psi \nabla\psi^*) \cdot \mathbf{n}\, da$$

where V is any arbitrary volume.

The use of Gauss's theorem

$$\int_V (\nabla \cdot \mathbf{B})\, dv = \int_A \mathbf{B} \cdot \mathbf{n}\, da \tag{1.1-10}$$

for any arbitrary vector function $\mathbf{B}$ leads to

$$\frac{\partial}{\partial t}(\psi^*\psi) = -\nabla \cdot \left[\frac{i\hbar}{2m}(\psi \nabla\psi^* - \psi^* \nabla\psi)\right] \tag{1.1-11}$$

in direct analogy with the charge conservation condition in electricity

$$\nabla \cdot \mathbf{i}_e = -\frac{\partial\rho}{\partial t}$$

where ρ is the charge density and $\mathbf{i}_e$ is the electric current density, we define

$$\mathbf{i} = \frac{i\hbar}{2m}(\psi \nabla\psi^* - \psi^* \nabla\psi) \tag{1.1-12}$$

as the particle probability current density. Equation (1.1-11) is thus a statement of the conservation of particle probability. The quantum mechanical counterpart of the classical motion of a particle can be viewed as the specification of the particle probability current density as a function of space and time.

Expectation Value

The expectation value of a physical observable is the ensemble average, in the sense described in the paragraph preceding (1.1-4), of the measurements of this observable. Alternatively, it can be defined as the mathematical expectation for the result of a single measurement of this observable. The expectation value for the position radius vector $\mathbf{r}$ is

$$\langle\mathbf{r}\rangle = \int \mathbf{r}\psi^*(\mathbf{r}, t)\psi(\mathbf{r}, t)\, dv = \int \psi^*\mathbf{r}\psi\, dv \tag{1.1-13}$$

So far we have no reason to prefer one or the other of the two expressions for $\langle\mathbf{r}\rangle$ given in (1.1-13). When the observable whose expectation value is to be evaluated is a function of the coordinates only, the two methods lead, obviously, to the same result. This is not necessarily true when the observable is a function of momenta and/or of energy. This point can be illustrated if we write the expectation value for the momentum component p_x. If we use (1.1-2), the contenders for $\langle p_x\rangle$ become

$$\langle p_x\rangle = -i\hbar \int \frac{\partial}{\partial x}(\psi^*\psi)\, dv$$

and

$$\langle p_x \rangle = -i\hbar \int \psi^* \frac{\partial \psi}{\partial x}\, dv$$

It is clear that these two procedures will lead, in general, to different results. The issue can be settled by making the reasonable requirement that when calculating the expectation value of the particle energy $\langle E \rangle$, we obtain the same result by using either of the two energy operators given in (1.1-3), that is,

$$\left\langle -\frac{\hbar^2}{2m} \nabla^2 + V(\mathbf{r}, t) \right\rangle = \left\langle i\hbar \frac{\partial}{\partial t} \right\rangle \tag{1.1-14}$$

From Schrödinger's equation (1.1-1), it follows that

$$\psi^* \left(-\frac{\hbar^2}{2m} \nabla^2 + V \right) \psi = i\hbar \psi^* \frac{\partial \psi}{\partial t}$$

so that the first procedure for finding the expectation value of an operator applies, since

$$\langle E \rangle = \int \psi^* \left(-\frac{\hbar^2}{2m} \nabla^2 + V \right) \psi \, dv = i\hbar \int \psi^* \frac{\partial \psi}{\partial t}\, dv \tag{1.1-15}$$

The second procedure, on the other hand, requires that

$$\left(-\frac{\hbar^2}{2m} \nabla^2 + V \right) (\psi^* \psi) = i\hbar \frac{\partial}{\partial t} (\psi^* \psi)$$

Carrying out the indicated operations of the last equation shows that it cannot be reconciled with the Schrödinger equation. The procedure for finding the expectation value of an operation A, which depends explicitly on the coordinates, momenta, energy, and time, is thus

$$\langle A(\mathbf{r}, \mathbf{p}, E, t) \rangle = \int \psi^* A \left(\mathbf{r}, -i\hbar \nabla, i\hbar \frac{\partial}{\partial t}, t \right) \psi \, dv \tag{1.1-16}$$

We next show that our choice of (1.1-16) for calculation $\langle A \rangle$ leads to expectation values of physical observables that agree with their classical counterparts. The proofs, which tend to be rather tedious, are greatly simplified here because all the operators considered are Hermitian.

Hermitian Operators. The Hermitian adjoint $A^\dagger$ of an operator A is defined as the operator satisfying the relation[1]

$$\int f^*(Ag)\, dv = \int (A^\dagger f)^* g \, dv \tag{1.1-17}$$

where f and g are two arbitrary scalar functions. If $A^\dagger = A$, that is, when

[1] See, for example, N. Dunford and J. T. Schwartz, *Linear Operators* (New York: Interscience, 1958), Part I, p. 350.

$$\int f^*(Ag)\, dv = \int (Af)^* g\, dv \tag{1.1-18}$$

the operator A is called a Hermitian operator.

It will be left for Chapter 3 to prove that operators representing physical observables are indeed Hermitian. In the meantime, we shall use this fact in some of the following proofs.

The Time Rate of Change of the Expectation Value

In addition to the procedure for obtaining the expectation value of an operator $\langle A \rangle$, we can derive a very general, and most useful, result involving the time rate of change of the expectation value.

$$\begin{aligned}
\frac{d\langle A \rangle}{dt} &= \frac{d}{dt}\int \psi^* A\psi\, dv = \int \frac{\partial}{\partial t}(\psi^* A\psi)\, dv \\
&= \int \left(\psi^* \frac{\partial}{\partial t} A\psi + \frac{\partial \psi^*}{\partial t} A\psi\right) dv \\
&= \int \left[\psi^*\left(\frac{\partial A}{\partial t}\psi + A\frac{\partial \psi}{\partial t}\right) + \frac{\partial \psi^*}{\partial t}(A\psi)\right] dv
\end{aligned}$$

Defining the Hamiltonian operator H by

$$H = -\frac{\hbar^2}{2m}\nabla^2 + V \tag{1.1-19}$$

we have, according to (1.1-1),

$$H\psi = i\hbar \frac{\partial \psi}{\partial t}$$

which, substituted for $\partial\psi/\partial t$ above, results in

$$\begin{aligned}
\frac{d\langle A \rangle}{dt} &= \int \left[\psi^* \frac{\partial A}{\partial t}\psi - \frac{i}{\hbar}\psi^* AH\psi + \frac{i}{\hbar}(H\psi)^* A\psi\right] dv \\
&= \left\langle \frac{\partial A}{\partial t} \right\rangle + \frac{i}{\hbar}\int (\psi^* HA\psi - \psi^* AH\psi)\, dv \qquad (1.1\text{-}20) \\
&= \left\langle \frac{\partial A}{\partial t} \right\rangle + \frac{i}{\hbar}\langle [H, A] \rangle
\end{aligned}$$

where, owing to the Hermiticity of H, we put

$$\int (H\psi)^* A\psi\, dv = \int \psi^* HA\psi\, dv$$

and the commutator of H and A is defined by

$$[H, A] = HA - AH$$

This definition will be used for any pair of operators.

Ehrenfest's Theorem

The theorem shows that the classical equations of motion for a single particle

$$m \frac{d\mathbf{r}}{dt} = \mathbf{p}, \qquad \frac{d\mathbf{p}}{dt} = -\nabla V \tag{1.1-21}$$

are obeyed, according to quantum mechanics, when all the vectors appearing in (1.1-21) are replaced by the expectation values of the corresponding quantum mechanical operators.

Part 1:

$$\frac{d\langle x\rangle}{dt} = \frac{d}{dt}\int \psi^* x\psi\, dv = \int \left(\psi^* x \frac{\partial\psi}{\partial t} + \frac{\partial\psi^*}{\partial t} x\psi\right) dv$$

$$= \int \left[\psi^* x \left(\frac{i\hbar}{2m}\nabla^2\psi - \frac{i}{\hbar} V\psi\right) + x\psi\left(-\frac{i\hbar}{2m}\nabla^2\psi^* + \frac{i}{\hbar}V\psi^*\right)\right] dv$$

$$= \frac{i\hbar}{2m}\int \left(\psi^* x\, \nabla^2\psi - \nabla^2\psi^*(x\psi)\right) dv$$

If we use the Hermiticity of ∇^2, the last expression becomes

$$\frac{i\hbar}{2m}\int \left[\psi^* x\, \nabla^2\psi - \psi^*\, \nabla^2(x\psi)\right] dv$$

but

$$\nabla^2(x\psi) = x\,\nabla^2\psi + 2\frac{\partial\psi}{\partial x}$$

so that

$$\frac{d\langle x\rangle}{dt} = -\frac{i\hbar}{m}\int \psi^* \frac{\partial\psi}{\partial x}\, dv = \frac{\langle p_x\rangle}{m} \tag{1.1-22}$$

Part 2:

We use Eq. (1.1-20).

$$\frac{d\langle p_x\rangle}{dt} = \frac{i}{\hbar}\langle[H, p_x]\rangle = \frac{i}{\hbar}\int \psi^*\left[\left(-\frac{\hbar^2}{2m}\nabla^2 + V\right), -i\hbar\frac{\partial}{\partial x}\right]\psi\, dv$$

$$= \int \left[\psi^* V \frac{\partial\psi}{\partial x} - \psi^*\frac{\partial}{\partial x}(V\psi)\right) dv$$

$$= -\int \left(\psi^* \frac{\partial V}{\partial x}\psi\right) dv \tag{1.1-23}$$

$$= -\left\langle\frac{\partial V}{\partial x}\right\rangle$$

Use has been made of the fact that the operators ∇^2 and $\partial/\partial x$ commute.

The Momentum Wavefunction. Consider the Fourier transform of the wavefunction $\psi(\mathbf{r}, t)$ defined by[2]

$$\Phi(\mathbf{p}, t) = \left(\frac{1}{2\pi\hbar}\right)^{3/2} \int_{-\infty}^{+\infty} e^{-i(\mathbf{p}\cdot\mathbf{r})/\hbar}\psi(\mathbf{r}, t)\, dv \tag{1.1-24}$$

so that

$$\psi(\mathbf{r}, t) = \left(\frac{1}{2\pi\hbar}\right)^{3/2} \int_{-\infty}^{+\infty} e^{i(\mathbf{p}\cdot\mathbf{r})/\hbar}\Phi(\mathbf{p}, t)\, d^3\mathbf{p}$$

where $d^3\mathbf{p} = dp_x\, dp_y\, dp_z$.

Using the last equation in Schrödinger's equation (1.1-1), we find—the actual derivation being assigned as a problem—that $\Phi(\mathbf{p}, t)$ satisfies

$$\left[\frac{p^2}{2m} + V(\mathbf{r} \to i\hbar\, \nabla_{\mathbf{p}}, \mathbf{p}, t)\right]\Phi(\mathbf{p}, t) = i\hbar\, \frac{\partial\Phi(\mathbf{p}, t)}{\partial t} \tag{1.1-25}$$

where $\mathbf{r} \to i\hbar\, \nabla_{\mathrm{p}}$ signifies that wherever $x_i (i = 1, 2, 3)$ appears in V, it is to be replaced by $i\hbar\, \partial/\partial p_i$.

Equation (1.1-25) is the wave equation in $\mathbf{p}$ space, the expression inside the square brackets being in energy operator. Since $p^2/2m$ is the kinetic energy term, we identify $\mathbf{p}$ in (1.1-24) as the momentum vector.

The statistical interpretation of $\Phi(\mathbf{p}, t)$ is similar to that of $\psi(\mathbf{r}, t)$, and $\Phi(\mathbf{p}, t)\, dp_x\, dp_y\, dp_z$ is the probability that the particle momentum at time t is within the differential volume $dp_x\, dp_y\, dp_z$ centered on $\mathbf{p}$ in momentum space.

The expectation value of any operator A is calculated accordingly by

$$\langle A \rangle = \int \Phi^*(\mathbf{p}, t) A\left(\mathbf{p}, i\hbar\, \nabla_{\mathrm{p}}, i\hbar\, \frac{\partial}{\partial t}, t\right) \Phi(\mathbf{p}, t)\, dp_x, dp_y\, dp_z \tag{1.1-26}$$

The normalization of $\Phi(\mathbf{p}, t)$, $\int \Phi\Phi^*\, d^3\mathbf{p} = 1$, follows from Parseval's theorem, according to which

$$\int_{-\infty}^{+\infty} |F(\mathbf{k})|^2\, d^3\mathbf{k} = \int_{-\infty}^{+\infty} |f(\mathbf{r})|^2\, dv \tag{1.1-27}$$

where $F(\mathbf{k})$ and $f(\mathbf{r})$ are a Fourier transform pair. It will be left as a (nontrivial) exercise (Problems 1.4 and 1.5) to show that $\langle A \rangle$ calculated by (1.1-26) is identical to that obtained from (1.1-16).

1.2 THE TIME-INDEPENDENT SCHRÖDINGER WAVE EQUATION

Starting with the time-dependent wave equation

$$-\frac{\hbar^2}{2m}\, \nabla^2\psi + V\psi = i\hbar\, \frac{\partial\psi}{\partial t} \tag{1.2-1}$$

[2] Notice that we choose the coordinate of the conjugate space as $\mathbf{p}/\hbar$ rather than the conventional $\mathbf{k}$.

let $\psi(\mathbf{r}, t) = u(\mathbf{r})g(t)$. Equation (1.2-1) becomes

$$\frac{1}{u}\left(-\frac{\hbar^2}{2m}\nabla^2 u + Vu\right) = \frac{i\hbar}{g}\frac{dg}{dt} \tag{1.2-2}$$

For a time-independent potential function $V(\mathbf{r})$, we can separate (1.2-2) by means of a separation constant E

$$-\frac{\hbar^2}{2m}\nabla^2 u(\mathbf{r}) + V(\mathbf{r})u = Eu \tag{1.2-3}$$

and

$$\frac{i\hbar}{g(t)}\frac{dg(t)}{dt} = E \tag{1.2-4}$$

The solution of (1.2-4) is

$$g(t) = g(0)e^{-i(E/\hbar)t}$$

Equation (1.2-3) is referred to as the time-independent Schrödinger equation, and its solutions $u_E(\mathbf{r})$ as the time-independent energy wavefunctions. The set (discrete or continuous) of allowed values of E are the eigenvalues and are determined (as will be shown in Chapter 2) by the boundary conditions of $u(\mathbf{r})$ or by the necessary behavior at infinity. We will assume that the set $u_E(\mathbf{r})$ generated by (1.2-3) is a complete set and will proceed to derive some general results concerning these wavefunctions.

The Orthonormality of the Wavefunctions

We will show first that any two members of the set $u_E(\mathbf{r})$ with different E's are orthogonal in the sense

$$\int u_{E'}^*(\mathbf{r})u_E(\mathbf{r})\,dv = 0 \qquad E' \neq E$$

The starting point is the time-independent wave equation (1.2-3)

$$-\frac{\hbar^2}{2m}\nabla^2 u_E + Vu_E = Eu_E$$

where we dropped the functional notation. We rewrite the equation for $u_{E'}^*$, multiply the first equation by $u_{E'}^*$ the second by u_E, and subtract one equation from the other. The result is

$$-\frac{\hbar^2}{2m}(u_{E'}^*\,\nabla^2 u_E - u_E\,\nabla^2 u_{E'}^*) + V(u_E u_{E'}^* - u_E u_{E'}^*) = (E - E')u_E u_{E'}^*$$

Integrating the last equation over a volume V and invoking Green's theorem (1.1-7) give

$$-\frac{\hbar^2}{2m}\int_A (u_{E'}^*\,\nabla u_E - u_E\,\nabla u_{E'}^*)\cdot\mathbf{n}\,da = (E - E')\int_V u_{E'}^*\,u_E\,dv$$

The integral on the left side vanishes, as discussed in connection with Equation 1.1-8, with the result that

$$\int u_{E'}^*(\mathbf{r}) u_E(\mathbf{r})\, dv = 0, \qquad E' \neq E \tag{1.2-5}$$

We have used the fact that, since the E's are the eigenvalues of an Hermitian operator, they are real and $(E')^* = E'$. The normalization of $\psi(\mathbf{r}, t)$ is achieved by requiring that $g(0)g^*(0) = 1$ and by the condition

$$\int |u_E|^2\, dv = 1 \tag{1.2-6}$$

so that $u_E(\mathbf{r})\exp[-(iE/\hbar)t]$ is a particular (normalized) solution of the time-dependent wave equation.

The Significance of *E*

The separation constant E has, up to this point, a mere mathematical significance as the eigenvalue of the time-independent wave equation (1.2-3). Consider the case when the wavefunction is given by the particular solution

$$\psi(\mathbf{r}, t) = u_E(\mathbf{r}) e^{-(iE/\hbar)t} \tag{1.2-7}$$

The expectation value of the total energy operator, which from now on will be called the Hamiltonian operator, is

$$\langle H \rangle = \left\langle -\frac{\hbar^2}{2m} \nabla^2 + V \right\rangle = \left\langle i\hbar \frac{\partial}{\partial t} \right\rangle = E \tag{1.2-8}$$

where use has been made of (1.2-3) and (1.2-6). E emerges as the expectation value of the total energy operator H for the time-independent potential function $V(\mathbf{r})$.

Some Mathematical Properties of the Wavefunctions

Since the $u_E(\mathbf{r})$ constitute a complete orthonormal set, they can be used for the expansion of any arbitrary function $G(\mathbf{r})$

$$G(\mathbf{r}) = \sum_E g_E u_E(\mathbf{r}) \tag{1.2-9}$$

and using the orthonormality conditions (1.2-5, 6), we obtain

$$g_E = \int G(\mathbf{r}') u_E^*(\mathbf{r}')\, dv'$$

$$G(\mathbf{r}) = \sum_E \left[\int u_E^*(\mathbf{r}') G(\mathbf{r}')\, dv' \right] u_E(\mathbf{r}) \tag{1.2-10}$$

$$= \int dv' G(\mathbf{r}') \sum_E u_E^*(\mathbf{r}') u_E(\mathbf{r})$$

where the integration extends over all space. We can write in general

$$G(\mathbf{r}) = \int dv' G(\mathbf{r}')\, \delta(\mathbf{r}' - \mathbf{r})$$

from which we conclude that

$$\sum_E u_E^*(\mathbf{r}')u_E(\mathbf{r}) = \delta(\mathbf{r}' - \mathbf{r}) \tag{1.2-11}$$

and that

$$\int_{V'} \sum_E u_E^*(\mathbf{r}')u_E(\mathbf{r})\, dv' \begin{cases} = 1 & \text{if } \mathbf{r} \text{ is in } V' \\ = 0 & \text{otherwise} \end{cases} \tag{1.2-12}$$

The Complex Vector Space

The orthonormal and complete set of eigenfunctions generated by a linear operator can be viewed as a set of *mutually orthogonal unit vectors* in an abstract (Hilbert) vector space. The advantage of this point of view is that many of the mathematical properties and manipulations involving these functions become analogous to familiar properties and operations involving vectors in ordinary space. By an extension of this idea we may regard an arbitrary function $G(\mathbf{r})$ as a vector in the complex vector space.

To illustrate this point of view, consider the energy eigenfunctions u_E. Defining the operation $(u_{E'}, u_E)$ by

$$(u_{E'}, u_E) = \int u_{E'}^* u_E\, dv \tag{1.2-13}$$

and calling it the scalar (dot) product of the "vectors" $u_{E'}$ and u_E, we can write the orthonormality condition (1.2-5) as

$$(u_{E'}, u_E) = \delta_{E'E}$$

The expansion of an arbitrary function $G(\mathbf{r})$ in terms of the u_E's, Eq. (1.2-10), takes the form

$$G = \sum_E (u_E, G)u_E \tag{1.2-14}$$

and is interpreted as expressing the vector, that is, function, G in terms of the unit vectors u_E, so that (u_E, G) is the *projection of G along u_E*.

The effect of linear operators in this space is to alter, in general, the directions and magnitudes of the vectors on which they operate. The eigenvectors of a given operator, say A, are those vectors whose direction in space is not altered when operated on by A.

Fundamental Postulates of Quantum Mechanics

According to the abstract formulation of quantum mechanics, a physical state of a system corresponds to a complex state vector. The set u_E, for example, constitutes all the possible energy states of the system. A physical measurement on the system constitutes a disturbance and, in general, alters the state of the system. Quantum mechanically, *the process of measuring a physical quantity (energy, momentum, etc.) corresponds to operation on the state vector by the operator corresponding to the physically observable.*

Assume, for example, that the energy state of the system is u_E, that is, the system is known, with certainty, to have a total energy E. If a second measurement of a physical observable A is undertaken, this corresponds to operating on the state vector u_E with the linear operator A. If u_E is not an eigenvector of A, the operation alters, by definition, the state u_E so that one is no longer certain about the energy of the system. If, on the other hand, u_E is also an eigenvector of A, the energy state is unaltered and we can state precisely both the energy and the value of the physical quantity that corresponds to the operator A.

The necessary and sufficient condition that any two linear operators, in this case H and A, have the same eigenvectors is that they commute. To prove the sufficiency part, assume $HA = AH$. It follows that $AHu_E = HAu_E = AEu_E = EAu_E$. The vector Au_E is thus an eigenvector of H with an eigenvalue E. If there is only one eigenvector associated with E, it follows that $Au_E = Cu_E$, where C is any constant. If there is more than one eigenvector u_E with an eigenvalue E (i.e., the set E is degenerate), it is possible to take linear combinations of the u_E's of this subspace that are orthonormal and also eigenvectors of A. The proof of the necessity condition is left as an exercise.

Let us summarize: *The necessary and sufficient condition that any two physical observables be simultaneously and precisely measurable is that their respective operators commute.*

When the operators A and H do not commute, the interpretation of the measurement process, due to Born, is the following.

The wavefunction $\psi(\mathbf{r}, t)$ is expanded in terms of the eigenfunctions (here we revert to talking about functions rather than vectors) v_n of the operator A, that is,

$$\psi(\mathbf{r}, t) = \sum c_n(t) v_n(\mathbf{r})$$

where

$$A v_n(\mathbf{r}) = a_n v_n(\mathbf{r})$$

The eigenvalues a_n constitute the set of all possible measurements of the observable A. The probability that any given value a_n is found by the measurement at time t is $|c_n(t)|^2$. According to this interpretation,

$$\langle A(t) \rangle = \sum_n a_n \, |c_n(t)|^2 \tag{1.2-15}$$

using

$$c_n = \int v_n^*(\mathbf{r}) \psi(\mathbf{r}, t) \, dv$$

leads to

$$\begin{aligned}
\langle A(t) \rangle &= \sum_n a_n \int \psi^*(\mathbf{r}', t) v_n(\mathbf{r}') \, dv' \int \psi(\mathbf{r}, t) v_n^*(\mathbf{r}) \, dv \\
&= \int dv' \psi^*(\mathbf{r}', t) A' \int dv \psi(\mathbf{r}, t) \sum_n v_n^*(\mathbf{r}) v_n(\mathbf{r}') \\
&= \int \psi^*(\mathbf{r}', t) A' \psi(\mathbf{r}', t) \, dv'
\end{aligned}$$

where A' operates on functions of $\mathbf{r}'$ and use has been made of the closure property, $\sum_n v_n^*(\mathbf{r})v_n(\mathbf{r}') = \delta(\mathbf{r} - \mathbf{r}')$.

The value $\langle A(t)\rangle$ obtained according to (1.2-15) is thus consistent with the previous procedure, Eq. (1.1-16).

The Uncertainty Principle

Having established that members of certain pairs of physical observables cannot be measured simultaneously with certainty, it is of interest to discover the magnitude of the uncertainties involved.

One such pair of variables is the position x and momentum p_x of a particle, because the operator $(\hbar/i)(\partial/\partial x)$ corresponding to p_x does not commute with x. We have already shown that the wavefunctions $\Phi(\mathbf{p}/\hbar, t)$ and $\psi(\mathbf{r}, t)$ form a Fourier transform pair. It follows directly from the mathematical properties of such pairs that reasonably defined (this will be done in the proof that follows) widths of these functions denoted as $\Delta(p_x/\hbar)$ and Δx will satisfy the relation $\Delta(p_x/\hbar)\ \Delta x \sim 1$. It should be possible, however, to extract this information from the wave equation since the behavior of the particle is completely specified by it.

The uncertainties in coordinate and in momentum are characterized by the respective mean-square deviations from the average

$$\langle\Delta x^2\rangle = \int \psi^*(x - \langle x\rangle)^2\psi\, dv$$
$$\langle\Delta p_x^2\rangle = \int \psi^*\left(-i\hbar\frac{\partial}{\partial x} - \langle p_x\rangle\right)^2 \psi\, dv \tag{1.2-16}$$

If we take $\langle x\rangle = \langle p_x\rangle = 0$ [this will be true if $\psi(x)$ is an even function of x], the product of uncertainties becomes

$$\langle\Delta p_x^2\rangle\langle\Delta x^2\rangle = -\hbar^2 \int \psi^* \frac{\partial^2\psi}{\partial x^2}\, dv \int \psi^* x^2 \psi\, dv$$

From the Hermiticity of $i\,\partial/\partial x$ (or from simple integration by parts), it follows that

$$\int \psi^* \frac{\partial^2\psi}{\partial x^2}\, dv = -\int \frac{\partial\psi^*}{\partial x}\frac{\partial\psi}{\partial x}\, dv$$

and in consequence

$$\langle\Delta p_x^2\rangle\langle\Delta x^2\rangle = \hbar^2 \int \frac{\partial\psi^*}{\partial x}\frac{\partial\psi}{\partial x}\, dv \int \psi^* x^2 \psi\, dv \tag{1.2-17}$$

By the Schwarz inequality[3]

$$\int ff^*\, dv \int g^*g\, dv \geq \left[\frac{1}{2}\left(\int fg^*\, dv + \int gf^*\, dv\right)\right]^2 \tag{1.2-18}$$

[3] The Schwarz inequality states that the sum of the lengths of two sides of a triangle exceeds or equals the third. That is, $|\mathbf{A}|\,|\mathbf{B}| \geq \mathbf{A}\cdot\mathbf{B}$ where $\mathbf{A}$ and $\mathbf{B}$ are any two vectors. When extended to complex vector space, it becomes $(\mathbf{f}, \mathbf{f})(\mathbf{g}, \mathbf{g}) \geq \{\frac{1}{2}[(\mathbf{f}, \mathbf{g}) + (\mathbf{g}, \mathbf{f})]\}^2$ for two arbitrary functions (vectors) $\mathbf{f}$ and $\mathbf{g}$. This form is identical to (1.2-18).

if we let $f = \partial\psi/\partial x$, $g = x\psi$ gives

$$\langle\Delta p_x^2\rangle\langle\Delta x^2\rangle \geq \frac{\hbar^2}{4}\left(\int \frac{\partial\psi}{\partial x} x\psi^* \, dv + \int x\psi \frac{\partial\psi^*}{\partial x} \, dv\right)^2$$

$$= \frac{\hbar^2}{4}\left[\int x \frac{\partial}{\partial x}(\psi\psi^*) \, dv\right]^2 = \frac{\hbar^2}{4}$$

The term within the last brackets is equal to -1, as can be shown by integration by parts (or the Hermiticity of $i\,\partial/\partial x$).

Defining Δx as $\langle\Delta x^2\rangle^{1/2}$ and similarly for Δp_x, we write

$$\Delta p_x \, \Delta x \geq \frac{\hbar}{2} \tag{1.2-19}$$

Minimum Uncertainty Wavepacket

It is of considerable interest to discover when the equality sign of (1.2-19) applies, that is, under what conditions is the uncertainty product $\Delta p_x \, \Delta x$ a minimum. It is clear from the geometrical interpretation that the equality in Eq. (1.2-18) applies when f and g are equal within a multiplicative constant. If we use again the substitutions $f = \partial\psi/\partial x$, $g = x\psi$, the equality sign is satisfied when

$$\frac{\partial\psi}{\partial x} = -ax\psi, \qquad \psi = Ne^{-(ax^2/2)}$$

where a is a constant.

The normalization constant N is determined by requiring that $\int \psi^*\psi \, dx = 1$. The result is $N = (a/\pi)^{1/4}$. In a similar fashion, we can identify the constant a by

$$\langle\Delta x^2\rangle = \int_{-\infty}^{-\infty} \psi^2 x^2 \, dx = \left(\frac{a}{\pi}\right)^{1/2} \int_{-\infty}^{+\infty} x^2 e^{-ax^2} \, dx = \frac{1}{2a}$$

The normalized expression for the minimum uncertainty $\psi(x)$ becomes

$$\psi(x, 0) = \left(\frac{1}{2\pi\langle\Delta x^2\rangle}\right)^{1/4} e^{-(x^2/4\langle\Delta x^2\rangle)} \tag{1.2-20}$$

where the time coordinate, chosen arbitrarily at $t = 0$, was restored. An important result is the fact that the minimum uncertainty wavepacket has a Gaussian form. Since the Fourier transform of one Gaussian is another one, it follows that $\Phi(p, 0)$, the corresponding particle momentum wavepacket, is also a Gaussian. Since, according to (1.1-27), $\Phi(p, 0)$ is normalized, we can write it directly as

$$\Phi(p, 0) = \left(\frac{1}{2\pi\langle\Delta p^2\rangle}\right)^{1/4} e^{-(p^2/4\langle\Delta p^2\rangle)} \tag{1.2-21}$$

It will be left as an exercise to show that $\langle\Delta p^2\rangle$ in (1.2-21) satisfies the uncertainty principle (1.2-19) with an equality sign.

Other pairs of conjugate observables obeying an uncertainty relation are

$$\Delta E \, \Delta t \geqslant \frac{\hbar}{2}$$

$$\Delta\phi_z \, \Delta J_z \geqslant \frac{\hbar}{2} \tag{1.2-22}$$

$$\Delta n \, \Delta\Phi \geqslant 1$$

where E and t are the energy and time of a system, ϕ_z and J_z are the azimuthal angle and the angular momentum of a system about an arbitrary (z) axis, and n and Φ refer to the number of quanta and oscillation phase of an harmonic oscillator.[4]

The Uncertainty Principle in Communication

In this section we shall illustrate how the uncertainty principle can be used to give, in a quick fashion, certain answers concerning communication problems.

The Spread of an Antenna Beam. Consider first the problem of designing a large microwave antenna for communication purposes. One of the most important problems in antenna design is that of concentrating the radiated power into the smallest possible divergence angle θ. The beam spread can be considered to be due to the uncertainty Δp_x in the transverse momentum p_x of photons that are emitted by the antenna of diameter $2R$. Taking the uncertainty $\Delta x = R$ leads to

$$\theta \approx \frac{\Delta p_x}{p_z} = \frac{\lambda}{2\pi R} \tag{1.2-23}$$

where the relation $p_z = h/\lambda$ was used.

This is, of course, a very old problem and the result (1.2-23) is well known. Approached as a problem in electromagnetic theory, Eq. (1.2-21) results from the Fourier transform relationship that exists between the field intensity distribution in the antenna plane and the far-field angular distribution of the radiated field. This Fourier transform relationship is also the basis for the uncertainty relations in quantum mechanics, as shown in the discussion surrounding Eq. (1.2-20).

Information Capacity of a Communication Link. Having just become experts in antenna design, we may contemplate, for a moment, the information capacity of the beam radiated by the antenna. It is known from the work of Shanon that the capacity of a communication channel in bits per second is

[4] R. Serber and C. H. Townes, *Symposium on Quantum Electronics* (New York: Columbia University Press, 1960), p. 233.

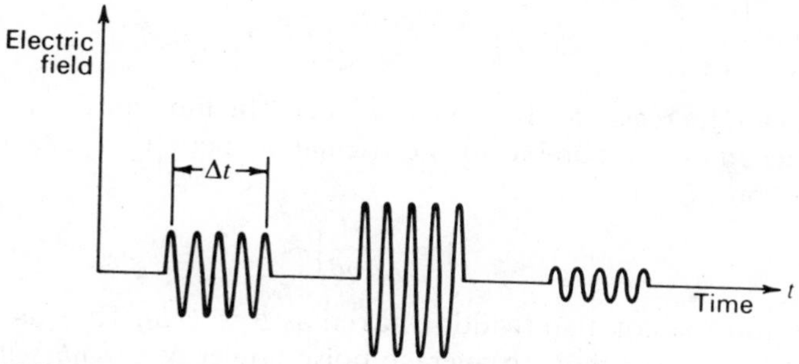

FIGURE 1.1 A sequence of radiation pulses with the information coded into the amplitudes.

given by[5]

$$C = \left[\log_2\left(1 + \frac{S}{N}\right)\right] B \tag{1.2-24}$$

where S and N are the average signal and noise powers, respectively, and B is the transmission bandwidth in cycles per second.

There are numerous sources of noise in communication links and their study is of no concern here. Even if all other noise sources are "cleaned up," there remains, however, an effective noise source that we may consider as being of quantum mechanical origin. To be specific, consider the information carried by a sequence of pulses as shown in Figure 1.1. The average power is taken as $\bar{p}$, so that the average energy per pulse is $\bar{p}\ \Delta t$. The transmitted information is coded into the amplitude of the pulses so that each predetermined value of pulse energy is associated with a message. According to information theory,[6] the information contents, in bits, of each pulse is given by the logarithm (to the base 2) of the number of distinguishable values (in this case energies) it may possess. The average information contents per pulse is thus

$$\bar{I}_{\text{pulse}} = \log_2\left(\frac{\bar{p}\ \Delta t}{\Delta E}\right) \tag{1.2-25}$$

where the energy resolution ΔE is the smallest energy difference between pulses that can be measured. Since the time available for measuring the energy of a given pulse is $2\ \Delta t$, the energy resolution ΔE is given according to the uncertainty principle (1.2-22) by $\Delta E \sim \hbar/2\ \Delta t$. Substitution in (1.2-25) gives

$$\bar{I}_{\text{pulse}} \sim \log_2\left[\frac{2\bar{p}(\Delta t)^2}{\hbar}\right]$$

The average information transmission rate is equal to the average information contents per pulse multiplied by the number of pulses per second $1/(2\ \Delta t)$.

[5] C. E. Shanon and W. Weaver, *The Mathematical Theory of Communication* (Urbana, Ill.: University of Illinois Press, 1949), p. 67.

[6] Ibid.

$$C \sim \log_2 \left[\frac{2\bar{p}(\Delta t)^2}{\hbar}\right]\left(\frac{1}{2\,\Delta t}\right) \tag{1.2-26}$$

It is clear that C increases as Δt is made smaller. The limit for this procedure is when Δt becomes comparable to the oscillation period f^{-1}.[7] At this point, (1.2-26) becomes

$$C \sim \log_2 \left[\frac{\bar{p}}{\hbar\omega B}\right] B \tag{1.2-27}$$

where the transmission bandwidth is taken as $B = 1/\Delta t$. Thus, according to (1.2-24), we may associate an average noise power $N \sim \hbar\omega B$ with what is classically a noise-free channel. The origin of this noise is, as shown above, quantum mechanical, and it is indeed the limiting factor on channel capacity when the condition $\hbar\omega > kT$ is satisfied. A rigorous treatment of this problem has been given by Gordon.[8] We shall return to this subject in Chapter 13, where the noise properties of some specific systems are described.

Supplementary References

1. Leighton, R. B., *Principles of Modern Physics* (New York: McGraw-Hill, 1959).
2. Messiah, A., *Quantum Mechanics* (New York: Interscience, 1961).
3. Schiff, L. I., *Quantum Mechanics* (New York: McGraw-Hill, 1959).

Problems

1.1 If the behavior of $\psi(\mathbf{r}, t)$ as $r \to \infty$ is dominated by r^{-n}, what values can n assume if the integral (1.1-8)

$$\int_A (\psi^* \nabla\psi - \psi\, \nabla\psi^*) \cdot \mathbf{n}\, da$$

taken over the surface at infinity is to vanish.

1.2 Show that $(d\langle p_x\rangle)/dt = -\langle \partial V/\partial x\rangle$ without using Hermiticity.

1.3 What is the value of $[p_x, x]$? Using this result and (1.1-20), derive

$$\frac{d\langle x\rangle}{dt} = \frac{\langle p_x\rangle}{m}$$

1.4 Show that if $\psi(\mathbf{r}, t)$ defined by

$$\psi(\mathbf{r}, t) = \left(\frac{1}{2\pi\hbar}\right)^{3/2} \int_{-\infty}^{+\infty} e^{i\mathbf{p}\cdot\mathbf{r}/\hbar}\Phi(\mathbf{p}, t)\, d^3\mathbf{p}$$

is to satisfy Schrödinger's equation, $\Phi(\mathbf{p}, t)$ satisfies the equation

$$\left[\frac{p^2}{2m} + V(\mathbf{r} \to i\hbar\, \nabla_p, \mathbf{p}, t)\right] \Phi(\mathbf{p}, t) = i\hbar \frac{\partial \Phi(\mathbf{p}, t)}{\partial t}$$

where $\mathbf{r} \to i\hbar\, \nabla_p$ means that x_i is to be replaced by $i\hbar\, \partial/\partial p_i$

[7] For shorter Δt the concept of frequency becomes meaningless.

[8] J. P. Gordon, in *Advances in Quantum Electronics*, J. R. Singer, ed. (New York: Columbia University Press, 1961), p. 509. Also in *Proc. IEEE* **50,** 1 (1962).

Hint:
Show that

$$\int_{-\infty}^{+\infty} \frac{\partial \Phi}{\partial p_x} e^{ip_x x/\hbar}\, dp_x = -\frac{ix}{\hbar} \int_{-\infty}^{+\infty} \Phi e^{ip_x x/\hbar}\, dp_x$$

for $\Phi(-\infty) = \Phi(+\infty) = 0$.

1.5 Show that the expectation value $\langle A \rangle$ is the same whether calculated by (1.1-26) or by (1.1-16).

1.6 Prove that operators A and B commute if they have the same eigenfunctions.

1.7 Prove that

$$\frac{\hbar^2}{m} \int_{-\infty}^{\infty} u_j^* \frac{\partial u_i}{\partial x}\, dv = (E_i - E_j) \int_{-\infty}^{\infty} u_j^* x u_i\, dv$$

where $Hu_i = E_i u_i$.

Hint:
Calculate first the commutator $[H, x]$.

1.8 Show that if the uncertainty product $\Delta p_x\, \Delta x$ of a free particle obeys

$$\Delta p_x\, \Delta x \geqslant \frac{\hbar}{2}$$

the uncertainty product for the time t and energy $E = p_x^2/2m$ obeys $\Delta E\, \Delta t \geqslant \hbar/2$.

1.9 Show that $[x, p_x^n] = i\hbar n p_x^{n-1}$.

Hint:
What is $(\partial^n/\partial x^n)(xu)$?

1.10 Using (1.1-17), show that the eigenfunctions of a Hermitian operator are orthogonal.

C H A P T E R 2

Some Solutions of the Time-Independent Schrödinger Equation

2.0 INTRODUCTION

In this chapter we shall solve two eigenvalue problems. This will provide concrete demonstrations for the abstract material in Chapter 1, as well as working formulas for subsequent chapters. Specifically, we shall find the energy eigenvalues and corresponding eigenfunctions for the harmonic oscillator and for angular momentum operators.

2.1 PARITY

Consider the wave equation

$$-\frac{\hbar^2}{2m}\nabla^2 u(\mathbf{r}) + V(\mathbf{r})u(\mathbf{r}) = Eu(\mathbf{r})$$

replacing $\mathbf{r}$ by $-\mathbf{r}$ gives

$$-\frac{\hbar^2}{2m}\nabla^2 u(-\mathbf{r}) + V(-\mathbf{r})u(-\mathbf{r}) = Eu(-\mathbf{r})$$

It follows that if $V(-\mathbf{r}) = V(\mathbf{r})$, $u(-\mathbf{r})$ is also a solution of the wave equation with an eigenvalue E. We can now construct two new solutions

$$\begin{aligned} u_e(\mathbf{r}) &= u(\mathbf{r}) + u(-\mathbf{r}) \\ u_o(\mathbf{r}) &= u(\mathbf{r}) - u(-\mathbf{r}) \end{aligned} \tag{2.1-1}$$

where u_e is an even function and u_o is odd. The functions u_e and u_o are also solutions of Schrödinger's equation with the same eigenvalue E. If the set E is nondegenerate, all four functions must be multiples of the same function. Two cases are possible: (1) $u(\mathbf{r})$ is a multiple of $u_e(\mathbf{r})$ and $u_o(\mathbf{r})$ is zero. (2) $u(\mathbf{r})$ is a multiple of $u_o(\mathbf{r})$ and $u_e(\mathbf{r})$ is zero. The eigenfunction $u(\mathbf{r})$ is thus seen to possess even or odd parity, that is,

$$u(-\mathbf{r}) = \pm u(\mathbf{r})$$

If E belongs to a degenerate set containing n eigenfunctions, it is possible to construct n linearly independent superpositions of these functions that have definite parity. It should be noted that the eigenfunctions possess a definite parity only when $V(\mathbf{r}) = V(-\mathbf{r})$, that is, when the potential field has inversion symmetry. There are cases, such as the potential in noncentrosymmetric crystals, when this condition is not satisfied so that the functions are neither odd nor even.

2.2 THE HARMONIC OSCILLATOR

The harmonic oscillator consists of a mass m acted on by a restoring force that is proportional to its displacement from some point (which is taken as the origin). The solution of this problem provides an illustration of the manner in which the necessary behavior of the wavefunction at infinity is used to determine the eigenvalues. Other situations in quantum mechanics, such as electromagnetic radiation in an enclosure and the propagation of sound waves in solids and liquids, can be shown, as will be done in Chapter 4, to be formally equivalent to that of the harmonic oscillator and their treatment in the following chapters will rely on the material developed here.

The wave equation of the harmonic oscillator in one dimension becomes

$$\left(\frac{p_x^2}{2m} + \tfrac{1}{2}Kx^2\right)u = -\frac{\hbar^2}{2m}\frac{d^2u}{dx^2} + \tfrac{1}{2}Kx^2u = Eu \qquad (2.2\text{-}1)$$

using the following substitutions:

$$\xi = \alpha x \qquad \alpha^4 = \frac{mK}{\hbar^2} = \left(\frac{m\omega}{\hbar}\right)^2 \qquad \omega^2 = \frac{K}{m} \qquad \lambda = \frac{2E}{\hbar\omega}$$

Equation (2.2-1) becomes

$$\frac{d^2u}{d\xi^2} + (\lambda - \xi^2)u = 0 \qquad (2.2\text{-}2)$$

for $\xi^2 \gg \lambda$ the behavior of u is dominated by the $e^{-(1/2)\xi^2}$ term, so that it is natural to assume a solution of the form

$$u(\xi) = H(\xi)e^{-(1/2)\xi^2} \qquad (2.2\text{-}3)$$

where $H(\xi)$ is a polynomial of a finite order.

Substituting $u(\xi)$ from (2.2-3) in (2.2-2) leads to the equation

$$\frac{d^2H}{d\xi^2} - 2\xi\frac{dH}{d\xi} + (\lambda - 1)H = 0 \qquad (2.2\text{-}4)$$

In line with the comment following (2.2-3), we assume a power series expansion

$$H(\xi) = \xi^s(a_0 + a_1\xi + a_2\xi^2 + \cdots) \qquad (2.2\text{-}5)$$

where, for definiteness, $a_0 \neq 0$.

Substituting (2.2-5) in (2.2-4) and equating separately the coefficients of the various powers of ξ to zero yield

$$
\begin{aligned}
&s(s-1)a_0 = 0 \\
&(s+1)sa_1 = 0 \\
&(s+2)(s+1)a_2 - (2s+1-\lambda)a_0 = 0 \qquad (2.2\text{-}6)\\
&\quad\vdots \\
&(s+v+2)(s+v+1)a_{v+2} - (2s+2v+1-\lambda)a_v = 0
\end{aligned}
$$

Since $a_0 \neq 0$, it follows from the first of Eqs. (2.2-6) that $s = 0$ or $s = 1$, from the second that $s = 0$ or $a_1 = 0$ or both. The last equation shows how the general coefficient a_{v+2} can be determined from a_v.

The first case to be considered is that of $s = 0$. Since $a_0 \neq 0$, the only way to terminate the sequence of a_v with v even is to have

$$\lambda = 2v + 1 \qquad (2.2\text{-}7)$$

for some v. Since v is even, λ can take on the values $1, 5, 9, \ldots$. This choice of λ will not terminate, as is made clear from the last of (2.2-6), the odd a_v sequence. The only way to guarantee a finite number of terms in $H(\xi)$ is to put $a_1 = 0$. This prevents the odd a_v series from ever "getting off the ground." The same argument is repeated with $s = 1$; again, only even v terms are allowed and $a_1 = 0$. λ takes on the sequence of values

$$\lambda = 2v + 3 = 3, 7, 11, \ldots \qquad (2.2\text{-}8)$$

and the resultant polynomial $H_n(\xi)$ is odd (since the even polynomial is now multiplied by ξ). If we combine (2.2-7) and (2.2-8), the allowed values for λ become

$$\lambda = 2n + 1 \qquad n = 1, 2, 3, \ldots \qquad (2.2\text{-}8a)$$

which when using $\lambda = 2E/\hbar\omega$ gives

$$E_n = \hbar\omega(n + \tfrac{1}{2}) \qquad (2.2\text{-}9)$$

for the energy of the nth eigenstate.

According to (2.2-9), the harmonic oscillator, even in its lowest energy state, $n = 0$, has a finite amount of energy, $\frac{1}{2}\hbar\omega$. The lowest energy of a classical harmonic oscillator is zero. This essential difference is a manifestation of the uncertainty principle. For the classical harmonic oscillator to have zero energy, both its momentum p_x and position x must be simultaneously zero. This, according to the uncertainty principle, is impossible. The division of uncertainty between p_x and x, which minimizes the total energy E_0 while satisfying the uncertainty principle, gives $E_0 \approx \hbar\omega$. To show this, we take the total energy as

$$E = \left(\frac{1}{2} K(\Delta x)^2 + \frac{(\Delta p_x)^2}{2m}\right)$$

letting $\Delta p_x \, \Delta x = \hbar/2$

$$E = \frac{1}{2}\left(\frac{K\hbar^2}{4(\Delta p_x)^2} + \frac{(\Delta p_x)^2}{m}\right)$$

which when minimized with respect to Δp_x gives

$$E_{\min} = \tfrac{1}{2}\hbar\omega$$

This result suggests that the lowest energy wavefunction $u_0(x)$ is a minimum uncertainty wavepacket, that is, a Gaussian. This indeed is the case since, according to (2.2-3), $u_0(x) \propto H_0(\alpha x)e^{-(1/2)\alpha^2 x^2}$ where $H_0(\alpha x) = a_0$.

Hermite Polynomials

The solutions of Eq. (2.2-4) that correspond to the different values of $\lambda = 2n + 1$ are seen to be polynomials of order n. The polynomials are even when n is an even integer or odd when n is odd. If we put $\lambda = 2n + 1$, the differential equation for the polynomials $H_n(\xi)$, which are known as the Hermite polynomials, becomes

$$\frac{d^2H_n}{d\xi^2} - 2\xi\frac{dH_n}{d\xi} + 2nH_n = 0 \tag{2.2-10}$$

These polynomials are conveniently derived by means of the power series expansion of the function $e^{-s^2+2s\xi}$ according to

$$G(\xi, s) = e^{\xi^2-(s-\xi)^2} = e^{-s^2+2s\xi} = \sum_{n=0}^{\infty} \frac{H_n(\xi)}{n!} s^n \tag{2.2-11}$$

It is a simple matter, left as a problem, to show that $H_n(\xi)$ defined by (2.2-11) satisfies the differential equation (2.2-10).

To generate the Hermite polynomials, we notice that, according to (2.2-11),

$$\begin{aligned} H_n(\xi) &= \left\{\frac{\partial^n}{\partial s^n}\left[e^{\xi^2-(s-\xi)^2}\right]\right\}_{s=o} \\ &= e^{\xi^2}(-1)^n\left[\frac{\partial^n}{\partial \xi^n} e^{-(s-\xi)^2}\right]_{s=0} = e^{\xi^2}(-1)^n \frac{d^n}{d\xi^n} e^{-\xi^2} \end{aligned} \tag{2.2-12}$$

Applying (2.2-12) to generate, as an example, the first three H_n's gives

$$H_0(\xi) = 1, \qquad H_1(\xi) = 2\xi, \qquad H_2(\xi) = 4\xi^2 - 2 \tag{2.2-13}$$

This particular sequence of $H_n(\xi)$ corresponds to choosing $a_0 = 1$ in Eq. (2.2-5) and is common practice.

The Harmonic Oscillator—Creation and Annihilation Operators

The harmonic oscillator wavefunction is, according to (2.2-3),

$$u_n(x) = N_n e^{-(1/2)\alpha^2x^2} H_n(\alpha x) \tag{2.2-14}$$

The normalization constant N_n is determined by requiring that $\int_{-\infty}^{+\infty} u_n^* u_n \, dx = 1$.

$$\int_{-\infty}^{+\infty} u_n^* u_n \, dx = N_n^2 \int_{-\infty}^{+\infty} e^{-\alpha^2 x^2} H_n^2(\alpha x) \, dx = \frac{N_n^2}{\alpha} \int_{-\infty}^{+\infty} e^{-\xi^2} H_n^2(\xi) \, d\xi = 1$$

The evaluation of N_n is most conveniently performed with the aid of the generating function. Consider the integral

$$\int_{-\infty}^{+\infty} e^{-s^2+2s\xi-t^2+2t\xi-\xi^2} \, d\xi = \sum_{n=0}^{\infty} \sum_{m=0}^{\infty} \frac{s^n t^m}{n!m!} \int_{-\infty}^{+\infty} H_n(\xi) H_m(\xi) e^{-\xi^2} \, d\xi \tag{2.2-15}$$

Replacing the definite integral on the left side of (2.2-15) with its value $\pi^{1/2} e^{2st}$ gives

$$\pi^{1/2} e^{2st} = \pi^{1/2} \sum_{0}^{\infty} \frac{2^n s^n t^n}{n!} = \sum_{n} \sum_{m} \frac{s^n t^m}{n!m!} \int_{-\infty}^{+\infty} H_n(\xi) H_m(\xi) e^{-\xi^2} \, d\xi$$

Equating the coefficients of equal powers of $s^n t^m$ on both sides results in

$$\int_{-\infty}^{+\infty} H_n(\xi) H_m(\xi) e^{-\xi^2} \, d\xi = \begin{cases} 0 & m \neq n \\ \pi^{1/2} n! 2^n & m = n \end{cases} \tag{2.2-16}$$

Substitution in (2.2-14) identifies N_n as

$$N_n = \left(\frac{\alpha}{\pi^{1/2} n! 2^n} \right)^{1/2} \tag{2.2-17}$$

The normalized wavefunction is thus

$$u_n(x) = \left(\frac{\alpha}{\pi^{1/2} n! 2^n} \right)^{1/2} H_n(\alpha x) e^{-(1/2)\alpha^2 x^2} \tag{2.2-18}$$

A plot of some of the low-order $u_n(x)$ is given in Figure 2.1.

We will now use the generating function to evaluate the integral

$$\int_{-\infty}^{+\infty} u_n^* \frac{\partial u_m}{\partial x} \, dx$$

Consider first the integral

$$\int_{-\infty}^{+\infty} G(\xi, s) e^{-(\xi^2/2)} \frac{\partial}{\partial \xi} [G(\xi, t) e^{-(\xi^2/2)}] \, d\xi \tag{2.2-19}$$

which with (2.2-11) becomes

$$\begin{aligned} &\int_{-\infty}^{+\infty} e^{-s^2+2s\xi} e^{-(\xi^2/2)} \frac{\partial}{\partial \xi} [e^{-(\xi^2/2)+2t\xi-t^2}] \, d\xi \\ &= \sum_{n} \sum_{m} \frac{s^n t^m}{n!m!} \int_{-\infty}^{+\infty} H_n(\xi) e^{-(\xi^2/2)} \frac{\partial}{\partial \xi} [H_m(\xi) e^{-(\xi^2/2)}] \, d\xi \qquad (2.2\text{-}20) \\ &= \int_{-\infty}^{+\infty} e^{-(s^2+t^2)} e^{+2\xi(s+t)} e^{-\xi^2} (-\xi + 2t) \, d\xi \end{aligned}$$

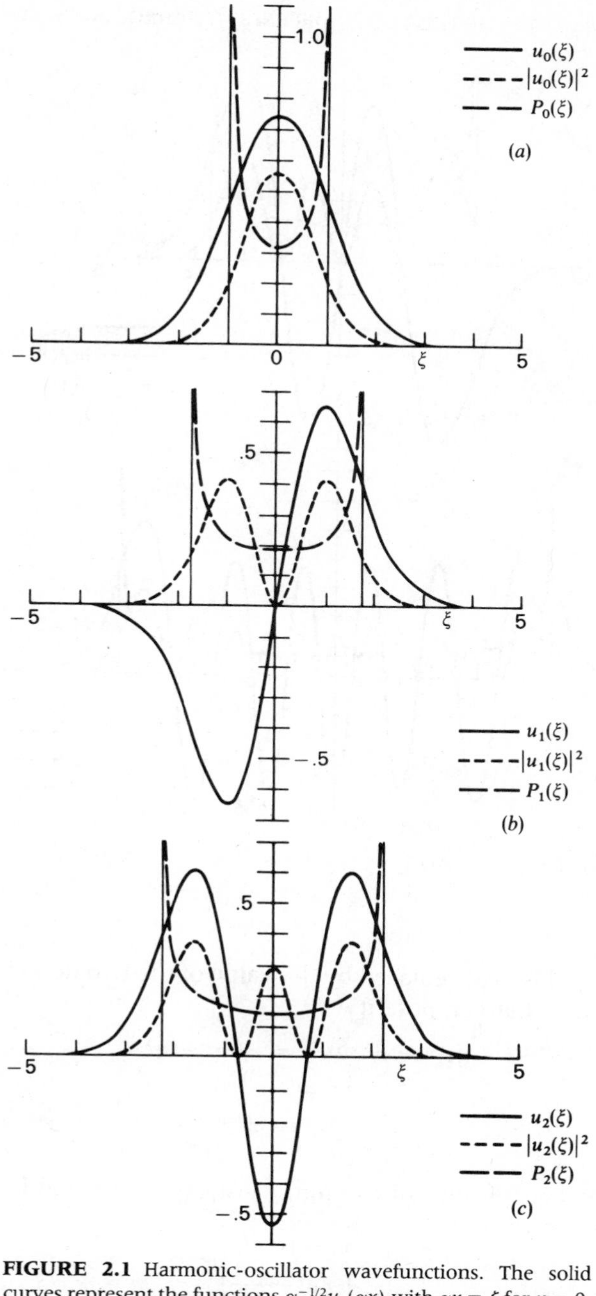

FIGURE 2.1 Harmonic-oscillator wavefunctions. The solid curves represent the functions $\alpha^{-1/2}u_n(\alpha x)$ with $\alpha x = \xi$ for $n = 0$, 1, 2, 3, and 10. The dotted curves represent $\alpha^{-1}u_n u_n$ for the same values of n. The dashed curves represent the probability distribution for a classical oscillator having the same energy as the corresponding quantum-mechanical oscillator. The vertical lines define the limits of the classical motion. *Source:* R. B. Leighton, *Principles of Modern Physics* (New York, McGraw-Hill, 1959).

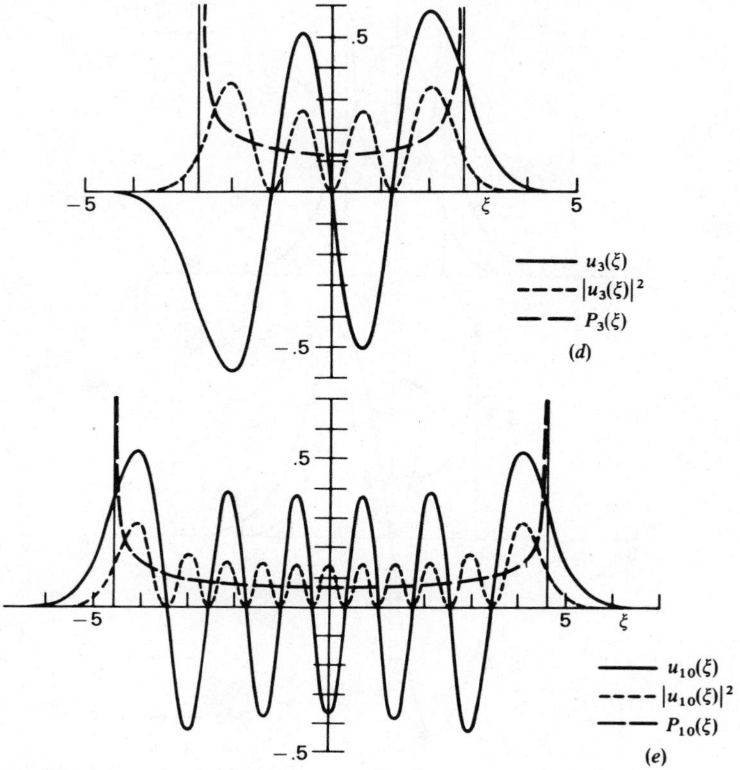

FIGURE 2.1 *(Continued)*

This last integral is next replaced by the value of the two definite integrals (in the same order) that compose it

$$-\pi^{1/2}(s + t)e^{2st} + \pi^{1/2}2te^{2st} = \pi^{1/2}(t - s)e^{2st}$$

$$= \pi^{1/2} \sum_{n=0}^{\infty} 2^n \frac{s^n t^{n+1} - s^{n+1}t^n}{n!}$$

Equating equal coefficients of $s^n t^m$ in the last expression and in (2.2-20), and using (2.2-18) give

$$\int_{-\infty}^{+\infty} u_n \frac{du_m}{dx}\,dx = \begin{cases} \alpha\left(\dfrac{n+1}{2}\right)^{1/2} & m = n + 1 \\ -\alpha\left(\dfrac{n}{2}\right)^{1/2} & m = n - 1 \\ 0 & \text{otherwise} \end{cases} \qquad (2.2\text{-}21)$$

We can use (2.2-21) to obtain another useful integral involving u_n and u_m.

We make use of the following relation that holds (see Problem 1.8) for any pair of one-particle wavefunctions:

$$\int_{-\infty}^{+\infty} u_j^*(\mathbf{r}) x u_i(\mathbf{r})\, dv = \frac{\hbar^2}{m(E_i - E_j)} \int_{-\infty}^{+\infty} u_j^* \frac{\partial u_i}{\partial x}\, dv \tag{2.2-22}$$

A direct substitution of (2.2-21) gives

$$\int_{-\infty}^{+\infty} u_n(x) x u_m(x)\, dx = \begin{cases} \dfrac{1}{\alpha}\left(\dfrac{n+1}{2}\right)^{1/2} = \sqrt{\dfrac{\hbar}{2m\omega}}\,(n+1)^{1/2} & m = n+1 \\ \dfrac{1}{\alpha}\left(\dfrac{n}{2}\right)^{1/2} = \sqrt{\dfrac{\hbar}{2m\omega}}\, n^{1/2} & m = n-1 \\ 0 & \text{otherwise} \end{cases} \tag{2.2-23}$$

where the relation $E_{n+1} - E_n = \hbar\omega$ was used. Another important integral, whose proof is assigned as Problem 2.3, is

$$\int_{-\infty}^{+\infty} u_n(x) x^2 u_m(x)\, dx = \begin{cases} \dfrac{2n+1}{2\alpha^2} & m = n \\ \dfrac{\sqrt{(n+1)(n+2)}}{2\alpha^2} & m = n+2 \\ 0 & n \neq m \neq n \pm 2 \end{cases} \tag{2.2-24}$$

Consider next the operators a and a^+ defined by

$$a = \frac{\alpha}{\sqrt{2}} x + \frac{i}{\sqrt{2}\hbar\alpha} p_x = \frac{\alpha}{\sqrt{2}} x + \frac{1}{\sqrt{2}\alpha}\frac{\partial}{\partial x}$$

and

$$a^\dagger = \frac{\alpha}{\sqrt{2}} x - \frac{i}{\sqrt{2}\hbar\alpha} p_x = \frac{\alpha}{\sqrt{2}} x - \frac{1}{\sqrt{2}\alpha}\frac{\partial}{\partial x} \tag{2.2-25}$$

With these definitions and with the aid of Eqs. (2.2-21) and (2.2-24), it follows directly that

$$\int_{-\infty}^{+\infty} u_n a u_m\, dx = \begin{cases} \sqrt{n+1} & m = n+1 \\ 0 & \text{otherwise} \end{cases}$$

$$\int_{-\infty}^{+\infty} u_n a^\dagger u_m\, dx = \begin{cases} \sqrt{n} & m = n-1 \\ 0 & \text{otherwise} \end{cases} \tag{2.2-26}$$

To learn more about the operators $a^\dagger$ and a, we notice that Eqs. (2.2-26) are consistent with the relations

$$a^\dagger u_n = \sqrt{n+1}\, u_{n+1}$$

$$a u_n = \sqrt{n}\, u_{n-1} \tag{2.2-27}$$

To show that Eqs. (2.2-27) are indeed correct, we need only prove that the functions $a^\dagger u_n$ and $a u_n$ satisfy the Schrödinger equation with the eigenvalues E_{n+1} and E_{n-1}, respectively. This is left as an exercise. An alternative approach uses the recursion formulas for $H_n(\xi)$ as discussed in Problem 2.8.

The operators $a^\dagger$ and a usually are referred to as the creation and annihilation operators, respectively. These names are used because, according to (2.2-27), the operation with $a^\dagger$ (or a) on the wavefunction u_n, corresponding to a state with n quanta of energy $\hbar\omega$, leads to a new wavefunction with $n + 1$ (or $n - 1$) quanta, thus "creating" (or "annihilating") one unit (quantum) $\hbar\omega$ of excitation.

The commutator of a and $a^\dagger$ is

$$[a, a^\dagger] = \left[\left(\frac{\alpha}{\sqrt{2}}x + \frac{i}{\sqrt{2}\hbar\alpha}p_x\right), \left(\frac{\alpha}{\sqrt{2}}x - \frac{i}{\sqrt{2}\hbar\alpha}p_x\right)\right]$$

$$= -\frac{i}{2\hbar}[x, p_x] + \frac{i}{2\hbar}[p_x, x]$$

which using the commutator relation

$$[p_x, x] = -i\hbar$$

becomes

$$[a, a^\dagger] = 1 \tag{2.2-28}$$

Using (2.2-25), we can express x and p_x in terms of a and $a^\dagger$ as

$$\begin{aligned} x &= \frac{1}{\sqrt{2}\alpha}(a^\dagger + a) \\ p_x &= \frac{i\hbar\alpha}{\sqrt{2}}(a^\dagger - a) \end{aligned} \tag{2.2-29}$$

Substituting (2.2-29) in the Hamiltonian

$$H = \frac{p_x^2}{2m} + \frac{1}{2}Kx^2$$

and using the relations $\alpha^4 = mK/\hbar^2$ and $\omega = \sqrt{K/m}$ yield

$$H = \frac{\hbar\omega}{2}(aa^\dagger + a^\dagger a)$$

Using $[a, a^\dagger] = 1$, we obtain

$$H = \hbar\omega(a^\dagger a + \tfrac{1}{2}) \tag{2.2-30}$$

This is a most useful form of the harmonic oscillator Hamiltonian and it will be encountered in several subsequent developments.

The operator $a^\dagger a$ commutes with H and has the number of quanta n for its eigenvalue. To show this, we use Eq. (2.2-27)

$$a^\dagger a u_n = a^\dagger \sqrt{n}\, u_{n-1} = n u_n$$

so that $Hu_n = \hbar\omega(n + \frac{1}{2})u_n$ in agreement with (2.2-9).

A far simpler procedure for evaluating integrals involving u_n, such as those of Eqs. (2.2-21), and (2.2-23), is by use of the recursion formulas for the Hermite polynomials. This is discussed in Problem 2.8.

2.3 THE SCHRÖDINGER EQUATION IN SPHERICALLY SYMMETRIC POTENTIAL FIELDS

A spherically symmetric potential is one in which the potential depends only on the distance from the origin, that is, $V(\mathbf{r}) = V(r)$.

In classical dynamics, a particle moving under the influence of such a force field has a constant angular momentum $\mathbf{L}$, where $\mathbf{L} = \mathbf{r} \times \mathbf{p}$.

$$\frac{d\mathbf{L}}{dt} = \frac{d\mathbf{r}}{dt} \times \mathbf{p} + \mathbf{r} \times \frac{d\mathbf{p}}{dt} = \mathbf{r} \times \mathbf{F}$$

where $\mathbf{F}$ is the force acting on the particle $\mathbf{F} = -\nabla V(\mathbf{r})$. For $V(\mathbf{r}) = V(r)$, we have

$$\frac{d\mathbf{L}}{dt} = -\mathbf{r} \times \mathbf{a}_r \frac{dV}{dr} = 0$$

where $\mathbf{a}_r$ is a unit vector in the $\mathbf{r}$ direction. Classically, both E and $\mathbf{L}$ are conserved in a spherically symmetric potential field. The corresponding statement in quantum mechanics is that the three operators H, $\mathbf{L}^2$, and L_z (where z is any arbitrary direction) commute, so that the energy, the magnitude of the angular momentum, and its projection along any arbitrary axis (z) can be measured simultaneously with precision.

Schrödinger Equation and Its Solutions in a Spherically Symmetric Potential Field

The Schrödinger equation in spherical coordinates is

$$-\frac{\hbar^2}{2m}\left[\frac{1}{r^2}\frac{\partial}{\partial r}\left(r^2\frac{\partial}{\partial r}\right) + \frac{1}{r^2 \sin\theta}\frac{\partial}{\partial\theta}\left(\sin\theta\frac{\partial}{\partial\theta}\right) + \frac{1}{r^2\sin^2\theta}\frac{\partial^2}{\partial\phi^2}\right]u + V(r)u = Eu \quad (2.3\text{-}1)$$

For a spherically symmetric potential, u can be separated into a product of functions

$$u(\mathbf{r}) = R(r)Y(\theta, \phi)$$

Substitution in (2.3-1) and introducing the separation constant λ yield

$$\frac{1}{R}\frac{d}{dr}\left(r^2\frac{dR}{dr}\right) + \frac{2mr^2}{\hbar^2}[E - V(r)] = \lambda$$

and (2.3-2)

$$-\left[\frac{1}{\sin\theta}\frac{\partial}{\partial\theta}\left(\sin\theta\frac{\partial}{\partial\theta}\right) + \frac{1}{\sin^2\theta}\frac{\partial^2}{\partial\phi^2}\right]Y = \lambda\, Y$$

The radial differential equation can be rewritten as

$$\frac{1}{r^2}\frac{d}{dr}\left(r^2\frac{dR}{dr}\right) + \left\{\frac{2m}{\hbar^2}[E - V(r)] - \frac{\lambda}{r^2}\right\}R = 0 \quad (2.3\text{-}3)$$

whereas the angular equation is

$$\frac{1}{\sin\theta}\frac{\partial}{\partial\theta}\left(\sin\theta\,\frac{\partial Y}{\partial\theta}\right) + \frac{1}{\sin^2\theta}\frac{\partial^2 Y}{\partial\phi^2} + \lambda Y = 0 \tag{2.3-4}$$

This last equation can be separated further into the product $Y = \Theta(\theta)\Phi(\phi)$ by the introduction of the separation constant m^2.[1] The two equations that result are

$$\frac{1}{\Phi}\frac{d^2\Phi}{d\phi^2} = -m^2 \tag{2.3-5}$$

and

$$\frac{1}{\sin\theta}\frac{d}{d\theta}\left(\sin\theta\,\frac{d\Theta}{d\theta}\right) + \left(\lambda - \frac{m^2}{\sin^2\theta}\right)\Theta = 0 \tag{2.3-6}$$

The solution for Φ becomes

$$\Phi = Ae^{im\phi} + Be^{-im\phi} \qquad m \neq o$$

$$\Phi = A' + B'\phi \qquad m = 0$$

Requiring that $u(r, \theta, \phi + 2\pi) = u(r, \theta, \phi)$, that is, that u be a single-valued function of the real space coordinates, limits us to integral values of m and makes B' zero. Both of these conditions are satisfied by writing

$$\Phi = \frac{1}{\sqrt{2\pi}}\,e^{im\phi} \qquad m = 0, \pm 1, \pm 2, \pm 3 \tag{2.3-7}$$

where the $(2\pi)^{-1/2}$ factor was introduced for normalization purposes so that $\int_0^{2\pi} \Phi^*\Phi\, d\phi = 1$.

Introducing the variable $\omega = \cos\theta$ and taking $\Theta(\theta) = P(\omega)$, we obtain

$$\frac{d}{d\omega}\left[(1 - \omega^2)\frac{dP}{d\omega}\right] + \left(\lambda - \frac{m^2}{1 - \omega^2}\right)P = 0 \tag{2.3-8}$$

Using the series substitution method in a manner similar to that employed in the harmonic oscillator, we find there is a physically acceptable solution for $P(\omega)$ that remains finite for $\omega = \pm 1$ ($\theta = 0$ and π). This occurs only when $\lambda = l(l+1)$ and $|m| \leq l$. This solution $P^l_m(\omega)$ is called the associated Legendre function and it is a polynomial of order $l - |m|$ multiplied by $(1 - \omega^2)^{|m|/2}$.

The complete solution to (2.3-1) can be written as

$$u(r, \theta, \phi) = \frac{1}{\sqrt{2\pi}}\,R_{nl}(r)e^{im\phi}N_{lm}P^l_m(\cos\theta) \qquad |m| \leq l \tag{2.3-9}$$

n is the added quantum number that is introduced by solving the radial equation (2.3-3). The normalization constant N_{lm} is determined from the

[1] This m should not be confused with the mass m appearing in Schrödinger equation.

requirement

$$|N_{lm}|^2 \int_0^{\pi} [P_m^l(\cos\theta)]^2 \sin\theta \, d\theta = 1$$

Using the equality, we obtain

$$\int_{-1}^{+1} P_m^l(\omega) P_m^{l'}(\omega)\, d\omega \begin{cases} = \left(\dfrac{2}{2l+1}\right) \dfrac{(l+|m|)!}{(l-|m|)!} & l = l' \\ = 0 & l \neq l' \end{cases} \qquad (2.3\text{-}10)$$

N_{lm} is found to be

$$N_{lm} = \left[\frac{2l+1}{2} \frac{(l-|m|)!}{(l+|m|)!}\right]^{1/2} \qquad (2.3\text{-}11)$$

and the normalized angular part of the wavefunction becomes

$$Y_m^l(\theta, \phi) = i^{m+|m|} \left[\frac{2l+1}{4\pi} \frac{(l-|m|)!}{(l+|m|)!}\right]^{1/2} P_m^l(\cos\theta) e^{im\phi} \qquad (2.3\text{-}12)$$

Some of the low-order $Y_m^l(\theta, \phi)$ functions are

$$Y_0^0 = \left(\frac{1}{4\pi}\right)^{1/2} \qquad Y_1^1 = -\left(\frac{3}{8\pi}\right)^{1/2} \sin\theta e^{i\phi}$$

$$Y_0^1 = \left(\frac{3}{4\pi}\right)^{1/2} \cos\theta \qquad Y_{-1}^1 = \left(\frac{3}{8\pi}\right)^{1/2} \sin\theta e^{-i\phi}$$

The radial part of the wavefunction, $R(r)$, is derived from a solution of (2.3-3). *Unlike the angular part* $Y_m^l(\theta, \phi)$, it depends on the specific form of $V(r)$. Since λ in (2.3-3) has been shown to be equal to $l(l+1)$, the solutions of Eq. (2.3-3) will be characterized by l and the additional quantum number n introduced in the process of solving it. For the same reason, the energy eigenvalues will depend, in general, on n and l, but not on m, since m does not appear in (2.3-3). It follows that since $|m| \leq l$, there will be, in general, at least $2l + 1$ eigenfunctions (each with a different m) for each energy E_{nl}.

The Parity of the Wavefunction Y_m^l

The parity of the eigenfunction is found by determining what happens to the wavefunction when $\mathbf{r}$ is replaced by $-\mathbf{r}$. In spherical coordinates, this corresponds to the transformation

$$r \to r \qquad \theta \to \pi - \theta \qquad \phi \to \phi + \pi$$

Since $P_m^l(\omega)$, where $\omega = \cos\theta$, is equal to $(1-\omega^2)^{|m|/2}$ times a polynomial (in ω) of order $l - |m|$, we have

$$\begin{aligned} Y_m^l(\pi - \theta, \phi + \pi) &= (-1)^{l-|m|} P_m^l(\theta) e^{im\phi} e^{im\pi} \\ &= P_m^l(\theta) e^{im\phi} (-1)^l = Y_m^l(\theta, \phi)(-1)^l \end{aligned}$$

so that the parity is even when l is even and odd when l is odd. Otherwise stated, Y^l_m has the parity $(-1)^l$.

2.4 THE ANGULAR MOMENTUM OPERATORS AND THEIR EIGENFUNCTIONS

The classical angular momentum $\mathbf{L}$ of a particle is $\mathbf{L} = \mathbf{r} \times \mathbf{p}$ where $\mathbf{L}$ is measured with respect to the origin. Using the operator equivalent $\mathbf{p} \to -i\hbar\boldsymbol{\nabla}$, we can write the operators L_x, L_y, and L_z as

$$\begin{aligned} L_x &= yp_z - zp_y = -i\hbar\left(y\frac{\partial}{\partial z} - z\frac{\partial}{\partial y}\right) \\ L_y &= zp_x - xp_z = -i\hbar\left(z\frac{\partial}{\partial x} - x\frac{\partial}{\partial z}\right) \\ L_z &= xp_y - yp_x = -i\hbar\left(x\frac{\partial}{\partial y} - y\frac{\partial}{\partial x}\right) \end{aligned} \tag{2.4-1}$$

In spherical coordinates, these operators become

$$\begin{aligned} L_x &= i\hbar\left(\sin\phi\frac{\partial}{\partial\theta} + \cot\theta\cos\phi\frac{\partial}{\partial\phi}\right) \\ L_y &= i\hbar\left(-\cos\phi\frac{\partial}{\partial\theta} + \cot\theta\sin\phi\frac{\partial}{\partial\phi}\right) \\ L_z &= -i\hbar\frac{\partial}{\partial\phi} \end{aligned} \tag{2.4-2}$$

By using (2.4-2), the operator $\mathbf{L}^2 = \mathbf{L}\cdot\mathbf{L}$ can be written as

$$\mathbf{L}^2 = -\hbar^2\left[\frac{1}{\sin\theta}\frac{\partial}{\partial\theta}\left(\sin\theta\frac{\partial}{\partial\theta}\right) + \frac{1}{\sin^2\theta}\frac{\partial}{\partial\phi^2}\right] \tag{2.4-3}$$

This operator is equal to $\hbar^2$ times the operator appearing on the left side in the last of Eqs. (2.3-2). This operator was found to have as its (normalized) eigenfunctions the set $Y^l_m(\theta, \phi)$, as given by (2.3-12), with eigenvalues $\lambda = l(l+1)$. Consequently, we can write directly the eigenfunctions and eigenvalues of $\mathbf{L}^2$

$$\mathbf{L}^2 Y^l_m(\theta, \phi) = \hbar^2 l(l+1) Y^l_m(\theta, \phi) \tag{2.4-4}$$

The possible values of the magnitude of the angular momentum are thus $\hbar\sqrt{l(l+1)}$. Since $\mathbf{L}^2$ commutes with the Hamiltonian, being proportional to the angular part of the latter, the magnitude of the angular momentum can be measured simultaneously with the total energy.

We consider next the operator $L_z = -i\hbar\,\partial/\partial\phi$. L_z also commutes with the Hamiltonian operator (2.3-1) and consequently has the same eigenfunctions. From (2.3-12), it follows that

$$L_z Y^l_m(\theta, \phi) = m\hbar Y^l_m(\theta, \phi) \tag{2.4-5}$$

so that the result of measuring the projection of **L** along z is one of the values $m\hbar$. It can be verified straightforwardly that both L_x and L_y, as well as L_z, commute with $\mathbf{L}^2$ (and H) so that the component of **L** along any arbitrary direction commutes with $\mathbf{L}^2$.

We consider next the possibility of simultaneous measurement of any two components, say L_x and L_y, of **L**. We must consider the commutator $[L_x, L_y]$

$$\begin{aligned}[L_x, L_y] &= [(yp_z - zp_y), (zp_x - xp_z)] \\ &= [yp_z, zp_x] + [zp_y, xp_z] \\ &= [yp_z zp_x - zp_x yp_x + zp_y xp_z - xp_z zp_y]\end{aligned}$$

Using $[z, p_z] = i\hbar$, so that $zp_z = p_z z + i\hbar$, gives

$$[L_x, L_y] = i\hbar(xp_y - yp_x) = i\hbar L_z$$

The other two commutator relations are obtained by cyclic permutation. The result is

$$\begin{aligned}[L_x, L_y] &= i\hbar L_z \\ [L_y, L_z] &= i\hbar L_x \\ [L_z, L_x] &= i\hbar L_y\end{aligned} \tag{2.4-6}$$

Consequently, no two components of **L** can be measured simultaneously.

Supplementary References

1. Leighton, R. B., *Principles of Modern Physics* (New York: McGraw-Hill, 1959).
2. Messiah, A., *Quantum Mechanics* (New York: Interscience, 1961).
3. Schiff, L. I., *Quantum Mechanics* (New York: McGraw-Hill, 1959).

Problems

2.1 Find the eigenvalues and eigenfunctions of the three-dimensional harmonic oscillator where $V(\mathbf{r}) = \frac{1}{2}K(x^2 + y^2 + z^2)$.

2.2 (a) Derive the second of Eqs. (1.2-22) using the relation $\Delta p_x \, \Delta x \geqslant \hbar/2$. (b) Use the reference mentioned in connection with the last of Eqs. (1.2-22) and outline the proof for $\Delta N \, \Delta\Phi \geqslant 1$.

2.3 Prove that

$$\int_{-\infty}^{+\infty} u_n^2(x) x^2 \, dx = \frac{2n+1}{2\alpha^2}$$

$$\int_{-\infty}^{\infty} x^2 u_{n+2} u_n \, dx = \frac{1}{2\alpha^2} \sqrt{(n+1)(n+2)}$$

where u_n is the harmonic oscillator wavefunction.

2.4 Calculate the expectation value of the kinetic energy $\langle p^2/2m \rangle$ of a harmonic oscillator in the state u_n.

Hint:
Can you make use of Problem 2.3?

2.5 Calculate $\Delta p_x \, \Delta x$ for the state u_n of the harmonic oscillator.

Answer:

$$\Delta p_x \, \Delta x = (n + \tfrac{1}{2})\hbar$$

2.6 Prove that

$$a^\dagger u_n = \sqrt{n+1}\; u_{n+1}$$

$$a u_n = \sqrt{n}\; u_{n-1}$$

There is no loss of generality in taking the operators as

$$a^\dagger = \xi - \frac{\partial}{\partial \xi}$$

$$a = \xi + \frac{\partial}{\partial \xi}$$

where $\xi = \alpha x$.

2.7 Starting with the generating function $G(s, \xi) = e^{-s^2+2s\xi} = \sum_{n=0}^{\infty} [H_n(\xi)/n!]s^n$, show that $H_n(\xi)$ obeys the following recursion formulas:

$$\frac{dH_n}{d\xi} = 2nH_{n-1}$$

$$\xi H_n = \tfrac{1}{2}H_{n+1} + nH_{n-1}$$

2.8 With the aid of the recursion formulas of Problem 2.7, show that for $m = K = \hbar = 1$

$$\int_{-\infty}^{+\infty} u_m^* p_x u_n \, dv = -i\sqrt{\frac{n}{2}}\, \delta_{m,n-1} + i\sqrt{\frac{n+1}{2}}\, \delta_{m,n+1}$$

and that

$$a u_n = \sqrt{u}\; u_{n-1}$$

$$a^\dagger u_n = \sqrt{n+1}\; u_{n+1}$$

where the u_n's are the harmonic oscillator wavefunctions.

2.9 Express the total energy of a three-dimensional harmonic oscillator in terms of annihilation and creation operators.

2.10 Show that L_x and L_y commute with $\mathbf{L} \cdot \mathbf{L}$ and H in a spherically symmetric $V(r)$.

2.11 Show that if

$$[q, p] = i\hbar$$

then

$$[q, p^l] = i\hbar l p^{l-1} \tag{1}$$

$$[p, q^l] = -i\hbar l q^{l-1} \tag{2}$$

Hint:
Use an induction proof.

2.12 Show, using (1) and (2) in Problem 2.11, that

$$[q_i, F(\vec{p}, \vec{q})] = i\hbar \frac{\partial F}{\partial p_i}$$

$$[p_i, F(\vec{q}, t)] = -i\hbar \frac{\partial F}{\partial q_i}$$

C H A P T E R 3

Matrix Formulation of Quantum Mechanics

3.0 INTRODUCTION

In Chapters 1 and 2 the solutions and eigenvalues of the Schrödinger wave equation were obtained from a conventional solution of the differential equation. Eigenfunctions and eigenvalues of Hermitian operators can also be obtained by matrix methods. This formulation is equivalent, formally, to the differential equation approach, and this equivalence will be brought out in this chapter.

3.1 SOME BASIC MATRIX PROPERTIES

It will be assumed that the student is familiar with the basic definitions and operations involving matrices. Consequently, the review of the necessary background of matrix algebra is very sketchy.

The kl element of the product of the matrices A and B is

$$(AB)_{kl} = \sum_m A_{km}B_{ml}$$

and by extension

$$(ABC)_{kl} = \sum_m \sum_n A_{km}B_{mn}C_{nl} \tag{3.1-1}$$

The unit matrix I is defined as the matrix that, when multiplying a matrix B, leaves the latter unchanged.

$$\begin{aligned} BI &= B \\ IB &= B \end{aligned} \tag{3.1-2}$$

It follows that $I_{kl} = \delta_{kl}$.

The inverse matrix B^{-1} of B is the matrix satisfying the conditions

$$\begin{aligned} B^{-1}B &= I \\ BB^{-1} &= I \end{aligned} \tag{3.1-3}$$

The inverse of a product of matrices is equal to the product of the inverse matrices taken in a reverse order.

$$(AB)^{-1} = B^{-1}A^{-1} \tag{3.1-4}$$

The elements of A^{-1} are related to those of A by the relation

$$A_{kl}^{-1} = \frac{\text{cofactor of } A_{lk}}{\text{determinant of } A} \tag{3.1-5}$$

Notice the reverse order of k and l on both sides of the equality sign.

The *Hermitian adjoint* of a matrix A, denoted as $A^\dagger$, is defined by

$$A_{kl}^\dagger = A_{lk}^* \tag{3.1-6}$$

A matrix is *Hermitian* when it is its own Hermitian adjoint, that is, when

$$A^\dagger = A \tag{3.1-7}$$

Equation (3.1-7) can be written, if we use (3.1-6), as

$$A_{kl}^\dagger = A_{kl} = A_{lk}^* \tag{3.1-8}$$

so that in a Hermitian matrix the interchange of rows and columns is equivalent to replacing each element by its complex conjugate. This applies, consequently, only to square matrices. It also follows that the matrix elements A_{kk} along the main diagonal are real.

A *unitary* matrix is defined as the matrix whose Hermitian adjoint is equal to its inverse

$$A^\dagger = A^{-1} \tag{3.1-9}$$

From (3.1-9), it follows that $(AA^\dagger) = I$ for a unitary matrix A that when written in a component form becomes

$$(AA^\dagger)_{kl} = \sum_n A_{kn}A_{nl}^\dagger = \sum_n A_{kn}A_{ln}^* = \delta_{kl} \tag{3.1-10}$$

where use was made of (3.1-6).

3.2 TRANSFORMATION OF A SQUARE MATRIX

A matrix A' derived from a square matrix A by the operation

$$A' = SAS^{-1} \tag{3.2-1}$$

is called the *transformation* of A by (the square matrix) S. Matrix equations are left invariant under a transformation of the individual matrices. A typical equation containing matrix products and sums such as

$$AB + CDE = F$$

can be written as

$$SAS^{-1}SBS^{-1} + SCS^{-1}SDS^{-1}SES^{-1} = SFS^{-1}$$

since $S^{-1}S = I$. The last equality, according to (3.2-1), is

$$A'B' + C'D'E' = F' \tag{3.2-2}$$

3.3 MATRIX DIAGONALIZATION

Of special interest in quantum mechanics is the transformation A (of a square matrix A') that is diagonal. We will illustrate in Section 3.6 that finding the elements A_{kk} and those of S takes the place, in the matrix formulation of quantum mechanics, of solving the wave equation. A is assumed to be diagonal with (unknown) elements $A_{kk} \equiv A_k$ so that

$$(SA'S^{-1})_{kl} = A_k \, \delta_{kl}$$

Postmultiplying the relation $SA'S^{-1} = A$ by S, we obtain

$$SA' = AS$$

and taking the kl element yields

$$\sum_m S_{km} A'_{ml} = \sum_m A_{km} S_{ml} = A_k S_{kl}$$

or

$$\sum_m S_{km} (A'_{ml} - A_k \, \delta_{ml}) = 0 \qquad (3.3\text{-}1)$$

If S is an N-dimensional matrix, we obtain N equations by fixing k and writing (3.3-1) for each value of l between 1 and N. These N simultaneous and homogeneous equations for the N unknowns $S_{k1}, \ldots, S_{kN}$ have nontrivial solutions only when the determinant of the matrix made up of the coefficients vanishes

$$\det \, [A'_{ml} - A_k \, \delta_{ml}] = 0 \qquad (3.3\text{-}2)$$

The N solutions $A_1, \ldots, A_N$ of A_k that result from solving (3.3-2) are the diagonal matrix elements sought. They are also referred to as the N *eigenvalues* of the matrix A'.

The remainder of the problem is to find the elements S_{km} of the transformation matrix. A given A_k is substituted in (3.3-1). The N homogeneous equations can be solved to yield $S_{k1}, S_{k2}, \ldots, S_{kN}$ to within an arbitrary multiplying constant. This procedure is repeated with each of the N A_k's, thus generating the matrix S_{km}. The additional condition necessary to determine S_{km} uniquely is, as will be shown in Section 3.4, that S be unitary, so that, according to (3.1-10),

$$\sum_n S_{kn} S^*_{ln} = \delta_{kl}$$

3.4 REPRESENTATIONS OF OPERATORS AS MATRICES

In the matrix formulation of quantum mechanics, an arbitrary operator A is *represented* by a matrix A. (We will depend on the context to avoid confusion between a matrix A and the operator A.) A particular matrix representation A

is derived through the relation

$$A_{km} = \int u_k^*(\mathbf{r}) A u_m(\mathbf{r}) \, dv \tag{3.4-1}$$

where $u_m(\mathbf{r})$ is any *arbitrary complete orthonormal set of functions*. The representation A_{km} of an operator A is, consequently, not unique and depends on the (arbitrary) choice of the set $u_m(\mathbf{r})$. The operator A can have other representations as well. Let one such representation, obtained in the space v_n, be A'. It follows that

$$A'_{km} = \int v_k^* A v_m \, dv \tag{3.4-2}$$

A Unitary Transformation Matrix

We can expand an arbitrary member of the set v_n in terms of the set u_k

$$v_n(\mathbf{r}) = \sum_k S_{kn} u_k(\mathbf{r}) \tag{3.4-3}$$

The reverse expansion becomes

$$u_k = \sum_n S_{kn}^* v_n \tag{3.4-4}$$

It follows from (3.4-3) or (3.4-4) and the orthonormality of the sets u_k and v_n that

$$S_{kn} = \int u_k^* v_n \, dv \tag{3.4-5}$$

The set of numbers S_{kn} can be regarded as a matrix. We will refer to it as the *transformation* matrix from the space v_n to u_k. The matrix S defined by (3.4-5) is a *unitary* matrix. As a proof, we may show that $SS^\dagger = I$ where I is the unity matrix

$$\begin{aligned}
(SS^\dagger)_{kl} &= \sum_n S_{kn} S_{nl}^\dagger = \sum_n S_{kn} S_{ln}^* \\
&= \sum_n \int u_k^*(\mathbf{r}) v_n(\mathbf{r}) \, dv \int u_l(\mathbf{r}') v_n^*(\mathbf{r}') \, dv' \\
&= \int dv u_k^*(\mathbf{r}) \int u_l(\mathbf{r}') \sum_n v_n(\mathbf{r}) v_n^*(\mathbf{r}') \, dv' \\
&= \int dv u_k^*(\mathbf{r}) \int u_l(\mathbf{r}') \, \delta(\mathbf{r} - \mathbf{r}') \, dv' \\
&= \int u_k^*(\mathbf{r}) u_l(\mathbf{r}) \, dv = \delta_{kl}
\end{aligned}$$

An alternative proof, and one that may shed some more physical insight into the nature of unitary transformations, is to show that a unitary transformation is required so that the (squared) magnitude of an arbitrary vector f (in abstract vector space)

$$(\mathbf{f}, \mathbf{f}) = \int f^* f \, dv$$

remains the same when f is expanded in terms of v_n or u_k. For further discussion, see Problem 3.1.

3.5 TRANSFORMATION OF OPERATOR REPRESENTATIONS

We have considered in Section 3.4 two arbitrary representations of an operator A, one in the space v_n, the other in u_n.

$$A_{kl} = \int u_k^* A u_l \, dv$$

$$A'_{kl} = \int v_k^* A v_l \, dv \tag{3.5-1}$$

$$u_k = \sum_n S_{kn}^* v_n$$

We can derive the matrix A from A' and vice versa by the transformation

$$A = SA'S^{-1} = SA'S^{\dagger} \tag{3.5-2}$$

where S is the unitary transformation matrix defined by the last of Eqs. (3.5-1).

The proof of Eq. (3.5-2) consists of replacing u_k^* and u_l in the first Eqs. (3.5-1) by their expansion according to (3.4-4).

$$\begin{aligned} A_{kl} &= \int u_k^* A u_l \, dv \\ &= \int \left(\sum_n S_{kn} v_n^*\right) A \sum_m S_{lm}^* v_m \, dv \\ &= \sum_n \sum_m S_{kn} \int v_n^* A v_m \, dv S_{lm}^* \\ &= \sum_n \sum_m S_{kn} A'_{nm} S_{ml}^{\dagger} = (SA'S^{\dagger})_{kl} \end{aligned}$$

If the matrix A' is Hermitian, so is the matrix A. In other words, the Hermiticity of a matrix is invariant under unitary transformations. The proof consists of showing that $A_{kl} = A_{lk}^*$ where A is given by (3.5-2)

$$A_{kl} = \sum_{mn} S_{km} A'_{mn} S_{nl}^{\dagger} = \sum_{mn} S_{km} (A'_{nm})^* S_{ln}^*$$

$$A_{lk} = \sum_{mn} S_{lm} A'_{mn} S_{kn}^*$$

where the relations $A'_{mn} = (A'_{nm})^*$ were used. Taking the complex conjugate of the last equation and interchanging m and n lead to

$$(A_{lk})^* = \sum_{mn} S_{ln}^* (A'_{nm})^* S_{km} = A_{kl}$$

which completes the proof.

A corollary of this result is the following: *The eigenvalues of an Hermitian matrix are real.* This result follows from the fact that any unitary transformation of an Hermitian matrix is Hermitian and therefore has real diagonal elements. This applies also to the diagonal transformation whose elements are the eigenvalues.

3.6 DERIVING THE EIGENFUNCTIONS AND EIGENVALUES OF AN OPERATOR BY THE MATRIX METHOD

Finding the eigenfunctions and eigenvalues of an arbitrary Hermitian operator A consists of solving the equation

$$Au_n = A_n u_n \tag{3.6-1}$$

The set of functions u_n is referred to as the eigenfunctions of A, and the (real) numbers A_n are its eigenvalues. As an example of an eigenvalue problem which we have already solved, we may take Eq. (2.2-1). The eigenfunction u_n is given by (2.2-18), while the eigenvalues are, according to (2.2-9), $E_n = \hbar\omega(n + \frac{1}{2})$. A second example is provided by the operator $\mathbf{L}^2$ in a spherically symmetric potential field. The eigenfunctions and eigenvalues are given by Eq. (2.4-4) as

$$\mathbf{L}^2 Y_m^l(\theta, \phi) = \hbar^2 l(l + 1) Y_m^l(\theta, \phi)$$

An alternative approach to solving an eigenvalue problem, such as (3.6-1), is by use of the matrix methods developed above. The matrix representation of A in the Hilbert space u_n (where the u_n's are the eigenfunctions of A) gives a diagonal matrix with the eigenvalues A_n as the elements

$$A_{kl} = \int u_k^* A u_l \, dv = E_l \int u_k^* u_l \, dv = E_l \, \delta_{kl}$$

It follows immediately that if the representation A' of an operator A in any complete and orthonormal set v_n is known, we can obtain directly its eigenvalues and eigenfunctions. The procedure consists of finding, by the method discussed in Section 3.3, the matrix S that diagonalizes A'

$$A_{kl} = (SA'S^\dagger)_{kl} = A_k \, \delta_{kl}$$

The diagonal elements A_k are the eigenvalues of A. Since, according to Section 3.5, the same matrix S is used in transforming from the v_n to the u_n space, the eigenfunctions are given by

$$u_k(\mathbf{r}) = \sum_n S_{kn}^* v_n(\mathbf{r}) \tag{3.6-2}$$

If only the eigenvalues of A and not the eigenfunctions are desired, it is not necessary, according to Section 3.3, to obtain the transformation matrix S. The eigenvalues A_k are the solutions of the determinantal equation

$$\det\,[A'_{ml} - A_k \, \delta_{ml}] = 0 \tag{3.6-3}$$

To convince ourselves that the eigenvalues of A obtained by matrix diagonalization are the same as those obtained by solving (3.6-1), we consider the matrix elements of the operator A in the function space u_n. Using (3.6-1), we obtain

$$A_{nm} = \int u_n^* A u_m \, dv = A_n \, \delta_{nm}$$

so that the matrix A_{nm} is diagonal with its elements being equal to the eigenvalues A_n. Since the two matrices A and A' are derivable from one another by a unitary transformation, they possess the same eigenvalues A_n.

The procedure described above applies to the eigenvalues and eigenfunctions of any Hermitian operator. As an illustration, let us consider formally the Hamiltonian operator H.

$$H = -\frac{\hbar^2}{2m} \boldsymbol{\nabla}^2 + V(\mathbf{r})$$

whose eigenfunctions and eigenvalues are u_k and E_k, respectively, so that

$$Hu_k = E_k u_k \tag{3.6-4}$$

Let the Hamiltonian matrix in the (arbitrary) space v_n be denoted as H'

$$H'_{mn} = \int v_m^* H v_n \, dv \tag{3.6-5}$$

The transformation coefficient S is defined by

$$u_k = \sum S_{kn}^* v_n$$

We will show below that the transformed matrix

$$H = SH'S^\dagger$$

is diagonal with its elements equal to the eigenvalues E_k.

$$\begin{aligned}
(SH'S^\dagger)_{kl} &= \sum_m \sum_n S_{km} H'_{mn} S_{ln}^* \\
&= \sum_m \sum_n \int u_k^*(\mathbf{r}) v_m(\mathbf{r}) \, dv \int v_m^*(\mathbf{r}') H v_n(\mathbf{r}') \, dv' \int u_l(\mathbf{r}'') v_n^*(\mathbf{r}'') \, dv'' \\
&= \int dv u_k^*(\mathbf{r}) \int dv' \sum_m v_m(\mathbf{r}) v_m^*(\mathbf{r}') H \int u_l(\mathbf{r}'') \sum_n v_n(\mathbf{r}') v_n^*(\mathbf{r}'') \, dv'' \\
&= \int dv u_k^*(\mathbf{r}) \int dv' \, \delta(\mathbf{r} - \mathbf{r}') H \int u_l(\mathbf{r}'') \, \delta(\mathbf{r} - \mathbf{r}'') \, dv'' \\
&= \int dv u_k^*(\mathbf{r}) \int \delta(\mathbf{r} - \mathbf{r}') H u_l(\mathbf{r}') \, dv' = \int u_k^*(\mathbf{r}) H u_l(\mathbf{r}) \, dv \\
&= E_l \, \delta_{kl}
\end{aligned}$$

where use has been made of (1.2-11) and of (3.6-4). This completes the proof.

We are now finally in a position to justify the assumption that operators that correspond to physical observables are Hermitian. The necessary and sufficient condition for an operator to have Hermitian matrix representations is that it be Hermitian (see Problem 3.5). Since an Hermitian matrix has real eigenvalues (see Problem 3.4) and these correspond to the possible results of physical measurements, it follows that the corresponding operator must be Hermitian.

3.7 THE HEISENBERG EQUATIONS OF MOTION

In Section 1.1 we derived the equation of motion for an arbitrary operator B as

$$\frac{d\langle B\rangle}{dt} = \left\langle\frac{\partial B}{\partial t}\right\rangle + \frac{i}{\hbar}\langle[H, B]\rangle \tag{3.7-1}$$

where $\langle B\rangle = \int \psi^* B\psi \, dv$ and ψ satisfies the time-dependent Schrödinger equation

$$\frac{\partial\psi}{\partial t} = -\frac{i}{\hbar} H\psi \tag{3.7-2}$$

From the last equation, it follows that[1]

$$\psi(\mathbf{r}, t) = e^{-iHt/\hbar}\psi(\mathbf{r}, 0)$$

so that $\langle B\rangle$ can be written as

$$\begin{aligned}\langle B\rangle &= \int [e^{-iHt/\hbar}\psi(\mathbf{r}, 0)]^* B e^{-iHt/\hbar}\psi(\mathbf{r}, 0)\, dv \\ &= \int \psi(\mathbf{r}, 0)^* e^{iHt/\hbar} B e^{-iHt/\hbar}\psi(\mathbf{r}, 0)\, dv \qquad (3.7\text{-}3) \\ &= \int \psi(\mathbf{r}, 0)^* B_H(t)\psi(\mathbf{r}, 0)\, dv \qquad (3.7\text{-}4)\end{aligned}$$

Moving the operator $e^{-iHt/\hbar}$ to the right of $\psi(\mathbf{r}, 0)$ in (3.7-3) can be justified by expanding it in a power series and treating each power of H as an Hermitian operator. $B_H(t)$ is defined by the last equation as

$$B_H(t) = e^{iHt/\hbar} B e^{-iHt/\hbar} \tag{3.7-5}$$

and is referred to as the Heisenberg form of B.

According to (3.7-4), in the Heisenberg formulation the expectation values of physical observables are evaluated using the wavefunctions at a fixed time ($t = 0$) so that the wavefunctions are stationary. The operators, however, evolve in time in accordance with (3.7-5). To evaluate $\langle B\rangle$, we thus need to know $B_H(t)$. The differential equation describing the evolution of $B_H(t)$ can be derived straightforwardly from (3.7-5) as

$$\begin{aligned}\frac{dB_H(t)}{dt} &= \frac{i}{\hbar}(He^{iHt/\hbar}Be^{-iHt/\hbar} - e^{iHt/\hbar}Be^{-iHt/\hbar}H) + \left(\frac{\partial B}{\partial t}\right)_H \\ &= \frac{i}{\hbar}[H, B_H(t)] + \left(\frac{\partial B}{\partial t}\right)_H\end{aligned} \tag{3.7-6}$$

[1] The operator $e^{-iHt/\hbar}$ is equal to $\sum_{n=0}^{\infty} (-iHt/\hbar)^n/n!$.

This formulation is especially useful in problems involving interactions with quantized boson fields. It will be used in Chapter 17 to describe energy exchange in nonlinear optical processes.

3.8 MATRIX ELEMENTS OF THE ANGULAR MOMENTUM OPERATORS

As an illustration of the ideas developed in this chapter, we shall derive next some matrix elements involving the angular momentum operator. These relationships will play a central role in later developments concerned with the addition of angular momenta and with the orbital and spin magnetic moments.

The raising and lowering operators L^+ and L^-, respectively, are defined by

$$\begin{aligned} L^+ &= L_x + iL_y \\ L^- &= L_x - iL_y \end{aligned} \tag{3.8-1}$$

These operators obey the commutation relations

$$\begin{aligned} [L^\pm, L_z] &= \pm\hbar L^\pm \\ [\mathbf{L}^2, L^\pm] &= 0 \\ [L^+, L^-] &= 2\hbar L_z \end{aligned} \tag{3.8-2}$$

These relations can be proved with the aid of Eqs. (2.4-6). As an illustration, consider the first of Eqs. (3.8-2)

$$[L^\pm, L_z] = [(L_x \pm iL_y), L_z] = [L_x, L_z] \pm i[L_y, L_z] = -i\hbar L_y \pm \hbar L_x = \pm\hbar L^\pm$$

The result of operating with $L^\pm$ on the eigenfunctions $Y^j_m(\theta, \phi)$ of $\mathbf{L}^2$ (here we replaced l by j) can be studied by the following development:

$$L_z(L^\pm Y^j_m) = (L^\pm L_z \pm \hbar L^\pm)Y^j_m = (m\hbar L^\pm \pm \hbar L^\pm)Y^j_m = (m \pm 1)\hbar(L^\pm Y^j_m)$$

where we used the relation $L_z Y^j_m = m\hbar Y^j_m$ and the first of Eqs. (3.8-2). The last equality states that $L^\pm Y^j_m$ is an eigenfunction of L_z with eigenvalues $(m \pm 1)\hbar$. These have been found to be the functions $Y^j_{m\pm1}$. We can thus write directly

$$L^\pm Y^j_m(\theta, \phi) = \hbar C^\pm_m Y^j_{m\pm1}(\theta, \phi) \tag{3.8-3}$$

where the constants $C^\pm_m$ remain to be evaluated.

Consider next the matrix elements of $\mathbf{L}^2$, L_z, and $L^\pm$ in the function space $Y^j_m(\theta, \phi)$. Using Eqs. (2.4-4) and (2.4-5) results in

$$\begin{aligned} (\mathbf{L}^2)_{j,m;j',m'} &= \int_0^\pi \int_0^{2\pi} [Y^j_m(\theta, \phi)]^* \mathbf{L}^2 Y^{j'}_{m'}(\theta, \phi) \sin\theta \, d\theta \, d\phi \\ &= j(j+1)\hbar^2 \delta_{j,j'} \delta_{m,m'} \\ (L_z)_{j,m;j',m'} &= m\hbar \, \delta_{j,j'} \delta_{m,m'} \\ (L^+)_{j,m+1;j,m} &= C^+_m \hbar = C_m \hbar \qquad \text{where } C_m = C^+_m \text{ by definition} \\ (L^-)_{j,m;j,m+1} &= C^-_{m+1}\hbar = (C^+_m)^*\hbar = C^*_m \hbar \end{aligned} \tag{3.8-4}$$

The last two equations involving $(L^{\pm})$ were derived using Eq. (3.8-3). The proof of the relation $C^{-}_{m+1} = (C^{+}_{m})^{*}$ is left as an exercise (Problem 3.9).

In order to derive the constant C_m, consider the mth diagonal element of the third of Eqs. (3.8-2)

$$[L^{+}L^{-} - L^{-}L^{+}]_{m,m} = 2\hbar(L_z)_{m,m} = 2m\hbar^2$$

where the subscript j is omitted since matrix elements involving states with a different j have been shown, in (3.8-4), to be zero. All the relations developed below consequently will assume a constant j. Expanding the matrix elements gives

$$\sum_{m'} L^{+}_{m,m'}L^{-}_{m',m} - L^{-}_{m,m'}L^{+}_{m',m} = L^{+}_{m,m-1}L^{-}_{m-1,m} - L^{-}_{m,m+1}L^{+}_{m+1,m} = 2m\hbar^2$$

or

$$C_{m-1}C^{*}_{m-1} - C^{*}_{m}C_m = 2m$$

where use has been made of the last two equations of (3.8-4). The last relation is rewritten as

$$|C_{m-1}|^2 - |C_m|^2 = 2m \tag{3.8-5}$$

To evaluate $|C_m|$, we use the relation

$$\mathbf{L}^2 = L_z^2 + \tfrac{1}{2}(L^{+}L^{-} + L^{-}L^{+}) \tag{3.8-6}$$

and take the m, m diagonal matrix elements of both sides.

The matrix elements of the first two terms are taken directly from (3.8-4), the result being

$$\begin{aligned} j(j+1)\hbar^2 &= m^2\hbar^2 + \tfrac{1}{2}\sum_{s}(L^{+}_{m,s}L^{-}_{s,m} + L^{-}_{m,s}L^{+}_{s,m}) \\ &= m^2\hbar^2 + \tfrac{1}{2}\hbar^2(C_{m-1}C^{*}_{m-1} + C^{*}_{m}C_m) \\ &= m^2\hbar^2 + \frac{\hbar^2}{2}(|C_{m-1}|^2 + |C_m|^2) \end{aligned} \tag{3.8-7}$$

Combining this result with (3.8-5) gives

$$C_m = \sqrt{j(j+1) - m(m+1)}$$

where the phase of the wavefunction is assumed to be such that C_m is real and positive. Substituting C_m in Eqs. (3.8-3) gives

$$\begin{aligned} L^{+}Y^{j}_{m} &= \hbar\sqrt{j(j+1) - m(m+1)}\; Y^{j}_{m+1} \\ L^{-}Y^{j}_{m} &= \hbar\sqrt{j(j+1) - m(m-1)}\; Y^{j}_{m-1} \end{aligned} \tag{3.8-8}$$

or

$$(L^{-})_{j,m;j,m+1} = (L^{+})_{j,m+1;j,m} = \hbar\sqrt{j(j+1) - m(m+1)} \tag{3.8-9}$$

These are the desired results.

Using Eqs. (3.8-4) and (3.8-9), we can construct the matrices corresponding to the various angular momentum operators in the space $Y^j_m(\theta, \phi)$. Since the matrix elements involving states with different j values are zero, we may limit ourselves to the submatrix within a constant j manifold. (The term manifold will be applied often in this context to describe the subspace made up of the $2j + 1$ eigenfunctions $Y^j_m(\theta, \phi)$ with $-j \leq m \leq j$.) Choosing $j = 1$, as an example, results in the following set of 3×3 matrices

$$\begin{array}{r|rrr|l} & m=1 & 0 & -1 & \\ \hline 1 & 1 & 0 & 0 & \\ \tilde{L}_z = \quad 0 & 0 & 0 & 0 & \hbar \\ -1 & 0 & 0 & -1 & \\ \hline \end{array} \tag{3.8-10}$$

$$\begin{array}{r|rrr|l} & 1 & 0 & -1 & \\ \hline 1 & 1 & 0 & 0 & \\ \tilde{\mathbf{L}}^2 = \quad 0 & 0 & 1 & 0 & 2\hbar^2 \\ -1 & 0 & 0 & 1 & \\ \hline \end{array} \tag{3.8-11}$$

$$\begin{array}{r|rrr|l} & 1 & 0 & -1 & \\ \hline 1 & 0 & 1 & 0 & \\ \tilde{L}^+ = \quad 0 & 0 & 0 & 1 & \sqrt{2}\hbar \\ -1 & 0 & 0 & 0 & \\ \hline \end{array} \tag{3.8-12}$$

$$\begin{array}{r|rrr|l} & 1 & 0 & -1 & \\ \hline 1 & 0 & 0 & 0 & \\ \tilde{L}^- = \quad 0 & 1 & 0 & 0 & \sqrt{2}\hbar \\ -1 & 0 & 1 & 0 & \\ \hline \end{array} \tag{3.8-13}$$

$$\begin{array}{r|rrr|l} & 1 & 0 & -1 & \\ \hline 1 & 0 & 1 & 0 & \\ \tilde{L}_x = \quad 0 & 1 & 0 & 1 & \dfrac{\hbar}{\sqrt{2}} \\ -1 & 0 & 1 & 0 & \\ \hline \end{array} \tag{3.8-14}$$

$$\widetilde{L}_y = \begin{array}{c|ccc|} & 1 & 0 & -1 \\ \hline 1 & 0 & -i & 0 \\ 0 & i & 0 & -i \\ -1 & 0 & i & 0 \\ \hline \end{array} \frac{\hbar}{\sqrt{2}} \tag{3.8-15}$$

3.9 SPIN ANGULAR MOMENTUM

Up to this point, the treatment of the angular momentum operators and their eigenfunctions was based on solving the eigenvalue problem (2.4-4)

$$\mathbf{L}^2(\theta, \phi) Y_m^l(\theta, \phi) = \hbar^2(l + 1)lY_m^l(\theta, \phi)$$

Since the operators were functions of the spatial variables (θ, ϕ), the solutions $Y_m^l(\theta, \phi)$ had to be single valued in real space. This, as shown in Section 2.3, forced m and l to assume integral values.

If the starting point for evolving the theory is taken as the commutation relationship $\mathbf{L} \times \mathbf{L} = i\hbar\mathbf{L}$ [Eq. (2.4-6)], the restriction on l and m mentioned above does not hold. This would be true if particles had, in addition to orbital angular momentum $\mathbf{L} = \mathbf{r} \times \mathbf{p}$, intrinsic angular momentum that does not depend on the spatial coordinates. Such angular momentum operators would commute with any Hamiltonian that depends only on $\mathbf{r}$ and $\mathbf{p}$ and would, consequently, be constants of the motion. This intrinsic angular momentum $\mathbf{S}$ is called spin, as distinguished from orbital, angular momentum $\mathbf{L}$ and indeed is found experimentally. The electron and proton both have a spin of $S_z = \hbar/2$, that is, a total spin angular momentum of $\sqrt{\frac{3}{4}}\hbar$. In treating spin angular momentum operators, we assume that they obey Eq. (2.4-6) so that the matrix representations and the operator manipulations are identical to those of the orbital angular momentum operators except that S can assume half odd integer values as well as integer values. The total eigenfunction specifying the state of a free electron is thus a function of $\mathbf{r}$ and an additional coordinate specifying the projection of the spin angular momentum along any arbitrary (z') direction

$$\psi = \psi(r, \theta, \phi, S_{z'})$$

whereas the total angular momentum operator is given by

$$\mathbf{J} = \mathbf{L} + \mathbf{S} \tag{3.9-1}$$

3.10 ADDITION OF ANGULAR MOMENTA

Consider two particles with total angular momentum quantum numbers j_1 and j_2. The sets of operators describing these particles commute as would be true, for example, with the orbital angular momenta of two different particles, or with the orbital and spin angular momenta of the same particle. Let the eigenfunctions of $\mathbf{L}_1^2$ and $\mathbf{L}_2^2$ be α_{mi}^{ji}. We can form a representation by taking all the possible product functions $\alpha_{m1}^{j1}\beta_{m2}^{j2}$. There are $(2j_1 + 1)(2j_2 + 1)$ such functions. These functions are simultaneous eigenfunctions of $\mathbf{L}_1^2$, $\mathbf{L}_2^2$, L_{1z}, and L_{2z}.

We can form another representation in which $\mathbf{L}_1^2$, $\mathbf{L}_2^2$, $\mathbf{L}^2$, and L_z are diagonal, where $\mathbf{L}$ is the sum angular momentum operator

$$\mathbf{L} = \mathbf{L}_1 + \mathbf{L}_2 \tag{3.10-1}$$

The eigenvalues of L_z and $\mathbf{L}^2$ will correspond, according to the basic postulates of quantum mechanics, to the possible values of the z component and the squared magnitude, respectively, of the angular momentum operator $\mathbf{L} = \mathbf{L}_1 + \mathbf{L}_2$. This procedure amounts to the addition of angular momenta. Since both sets of eigenfunctions span the same Hilbert space, they must be connected by a unitary transformation. The eigenfunctions ψ_m^j of $\mathbf{L}^2$ and L_z consequently can be expanded as a linear superposition of the $\alpha_{m_1}^{j_1}\beta_{m_2}^{j_2}$ functions.

$$\psi_m^j = \sum_{m_1}\sum_{m_2} (j_1 j_2 m_1 m_2 | j_1 j_2 j m)\alpha_{m_1}^{j_1}\beta_{m_2}^{j_2} \tag{3.10-2}$$

where the $(|)$ symbol represents the elements of the unitary transformation matrix connecting the two sets of eigenfunctions. It is clear that only the product wave functions $\alpha_{m_1}^{j_1}\beta_{m_2}^{j_2}$ where $m_1 + m_2 = m$ are to be included in the summation on the right side of (3.10-2). This can be verified by operating on both sides with $L_z = L_{1z} + L_{2z}$. In consequence, we have

$$\psi_m^j = \sum_{m_1} (j_1 j_2 m_1 (m - m_1) | j_1 j_2 j m)\alpha_{m_1}^{j_1}\beta_{m-m_1}^{j_2} \tag{3.10-3}$$

Another property of the $(|)$ coefficients is

$$(j_1 j_2 j_1 j_2 | j_1 j_2 (j_1 + j_2)(j_1 + j_2)) = 1 \tag{3.10-4}$$

so that

$$\psi_{j_1+j_2}^{j_1+j_2} = \alpha_{j_1}^{j_1}\beta_{j_2}^{j_2} \tag{3.10-5}$$

This results from the fact that according to (3.10-3), $m = m_1 + m_2 = j_1 + j_2$ and there is only one product function, given by (3.10-5), where $m_1 + m_2 = j_1 + j_2$. Since $j \geqslant m$, it follows that $j = j_1 + j_2$.

It follows from the above discussion that the maximum value of j is equal to $j_1 + j_2$. With each value of j, there are $2j + 1$ wavefunctions with $-j \leqslant m \leqslant j$. Since the total number of wavefunctions. ψ_m^j must be equal to that of the function space $\alpha_{m_1}^{j_1}\beta_{m_2}^{j_2}$, namely $(2j_1 + 1)(2j_2 + 1)$, it follows that the smallest value of j is $|j_1 - j_2|$. This can be verified from the summation

$$\sum_{j=|j_1-j_2|}^{j_1+j_2} (2j + 1) = (2j_1 + 1)(2j_2 + 1)$$

The coefficients $(|)$ are known as the Clebsch–Gordan coefficients. Condon and Shortley (Reference 1), as an example, tabulate these coefficients for a number of j_1 and j_2.

For simple cases, it is quite easy to derive the eigenfunctions of $\mathbf{L} = \mathbf{L}_1 + \mathbf{L}_2$ starting with (3.10-5). Consider, for example, "adding" the angular momenta of two particles with $j_1 = j_2 = 1$. We have from (3.10-5)

$$\psi_2^2 = \alpha_1^1\beta_1^1$$

applying $L^- = L_1^- + L_2^-$ to both sides and using (3.8-8) result in

$$L^-\psi_2^2 = 2\psi_1^2 = \sqrt{2}(\alpha_0^1\beta_1^1 + \alpha_1^1\beta_0^1)$$

so that

$$\psi_1^2 = \frac{1}{\sqrt{2}}(\alpha_0^1\beta_1^1 + \alpha_1^1\beta_0^1)$$

The next function ψ_1^1 can be written according to (3.10-3) as

$$\psi_1^1 = a\alpha_0^1\beta_1^1 + b\alpha_1^1\beta_0^1$$

Requiring that (ψ_1^1, ψ_1^1) be unity gives (for real a and b)

$$a^2 + b^2 = 1$$

whereas the condition $(\psi_1^1, \psi_1^2) = 0$ gives

$$a + b = 0$$

The last two equations are satisfied by

$$a = \pm\frac{1}{\sqrt{2}}$$

$$b = \mp\frac{1}{\sqrt{2}}$$

that, if we choose arbitrarily the upper sign, gives

$$\psi_1^1 = \frac{1}{\sqrt{2}}(\alpha_0^1\beta_1^1 - \alpha_1^1\beta_0^1) \qquad (3.10\text{-}6)$$

This procedure can be used to generate the remaining eigenfunctions.

3.11 TIME-INDEPENDENT PERTURBATION THEORY

In this section we pose the following problem: Given a Hamiltonian H_0 and its spectrum of eigenfunctions u_m and eigenvalues E_m

$$H_0u_m = E_mu_m \qquad (3.11\text{-}1)$$

What are the new eigenfunctions and eigenvalues when the Hamiltonian is changed, adiabatically, from H_0 to $H_0 + H'$? One method of solution would be to diagonalize the matrix $H_0 + H'$ in the u_n representation as discussed in Section 3.6. This method is often used in practice. If $H_0 \gg H'$, we can employ perturbation theory and obtain analytic expressions for the perturbation of u_n and E_n to any desired order. This is the concern of this section.

The Hamiltonian operator is taken as $H_0 + \lambda H'$ where $0 < \lambda < 1$ is a parameter that "turns the perturbation on" ($\lambda = 1$) or "off" ($\lambda = 0$). We are looking for the energies W and functions ψ satisfying

$$(H_0 + \lambda H')\psi = W\psi \tag{3.11-2}$$

Expanding ψ and W in a power series in λ

$$\begin{aligned} \psi &= \psi_0 + \lambda\psi_1 + \lambda^2\psi_2 + \cdots \\ W &= W_0 + \lambda W_1 + \lambda^2 W_2 + \cdots \end{aligned} \tag{3.11-3}$$

and substituting in (3.11-2) give

$$(H_0 + \lambda H')(\psi_0 + \lambda\psi_1 + \lambda^2\psi_2 + \cdots)$$
$$= (W_0 + \lambda W_1 + \lambda^2 W_2 \cdots)(\psi_0 + \lambda\psi_1 + \lambda^2\psi_2 + \cdots)$$

Equating the coefficients for λ^0, λ^1, and λ^2 on both sides of the last equation gives

$$\begin{aligned} H_0\psi_0 &= W_0\psi_0 \\ H_0\psi_1 + H'\psi_0 &= W_0\psi_1 + W_1\psi_0 \\ H_0\psi_2 + H'\psi_1 &= W_0\psi_2 + W_1\psi_1 + W_2\psi_0 \end{aligned} \tag{3.11-4}$$

respectively. Comparing the first of Eqs. (3.11-4) with (3.11-1) identifies the zero-order solutions as

$$\begin{aligned} \psi_0 &= u_m \\ W_0 &= E_m \end{aligned}$$

where u_m and E_m are the eigenfunction and eigenvalue at the absence of perturbation. Next, we expand ψ_1 in terms of u_n as

$$\psi_1 = \sum_n a_n^{(1)} u_n \tag{3.11-5}$$

and substitute it in the second of Eqs. (3.11-4). The result is

$$\sum_n a_n^{(1)} E_n u_n + H' u_m = E_m \sum_n a_n^{(1)} u_n + W_1 u_m$$

Multiplying by u_k^* and integrating give

$$E_k a_k^{(1)} + H'_{km} = E_m a_k^{(1)} + W_1 \delta_{km} \tag{3.11-6}$$

that for $k \neq m$ yields

$$a_k^{(1)} = \frac{H'_{km}}{E_m - E_k} \qquad k \neq m \tag{3.11-7}$$

Putting $k = m$ in (3.11-6) gives

$$W_1 = H'_{mm} \tag{3.11-8}$$

We still need to evaluate $a_m^{(1)}$. This is done by requiring that $\psi = u_m + \psi_1$ be normalized to unity

$$\int \left[u_m + \sum_n a_n^{(1)} u_n\right]^* \left[u_m + \sum_s a_s^{(1)} u_s\right] dv$$
$$= 1 + a_m^{(1)} + a_m^{*(1)} + \sum_n a_n^{(1)} a_n^{*(1)} = 1 \tag{3.11-9}$$

that, if we neglect the second-order term, gives $a_m^{(1)} = 0$ for a choice of phases that renders $a_m^{(1)}$ real.

The eigenfunction and eigenvalue to first-order perturbation are thus given as

$$\psi = u_m + \sum_{k \neq m} \frac{H'_{km}}{E_m - E_k} u_k$$
$$W = E_m + H'_{mm} \tag{3.11-10}$$

Second-Order Perturbation

The starting point is the third of Eqs. (3.11-4). Expanding ψ_2 according to

$$\psi_2 = \sum_n a_n^{(2)} u_n$$

and substituting in (3.11-4) give

$$\sum_n a_n^{(2)} E_n u_n + H' \sum_n a_n^{(1)} u_n = \sum_n a_n^{(2)} E_m u_n + W_1 \psi_2 + W_2 u_m$$

Substituting for ψ its expansion according to (3.11-5), and then multiplying by u_k^* and integrating result in

$$a_k^{(2)} E_k + \sum_n a_n^{(1)} H'_{kn} = a_k^{(2)} E_m + W_1 a_k^{(1)} + W_2 \delta_{mk} \tag{3.11-11}$$

Setting $k = m$ gives

$$W_2 = \sum_n a_n^{(1)} H'_{mn} - W_1 a_m^{(1)}$$
$$= \sum_{n \neq m} a_n^{(1)} H'_{mn} + a_m^{(1)} H'_{mm} - W_1 a_m^{(1)}$$

If we use (3.11-7) for $a_n^{(1)}$ and (3.11-8) for W_1, the last two terms cancel each other with the result

$$W_2 = \sum_{n \neq m} \frac{|H'_{mn}|^2}{E_m - E_n} \tag{3.11-12}$$

Going back to (3.11-11) for the case $k \neq m$, using (3.11-7), (3.11-8), and the result $a_m^{(1)} = 0$ give

$$\underset{k\neq m}{a_k^{(2)}} = \sum_{n\neq m} \frac{H'_{kn}H'_{nm}}{(E_m - E_n)(E_m - E_k)} - \frac{H'_{mm}H'_{km}}{(E_m - E_k)^2}$$

To find $a_m^{(2)}$, we go back to the normalization integral (3.11-9). Adding the second-order correction to ψ gives

$$\int \left[u_m + \sum_n a_n^{(1)}u_n + \sum_n a_n^{(2)}u_n\right]^* \left[u_m + \sum_s a_s^{(1)}u_s + \sum_s a_s^{(2)}u_s\right] dv = 1$$

If we use the result $a_m^{(1)} = 0$, the last equation yields

$$a_m^{(2)} = -\tfrac{1}{2}\sum_n |a_n^{(1)}|^2 = -\tfrac{1}{2}\sum_{n\neq m} \frac{|H'_{nm}|^2}{(E_m - E_n)^2} \tag{3.11-13}$$

The eigenfunction and the energy, to second order, can be written as

$$\psi = u_m + \sum_{k\neq m} \frac{H'_{km}}{E_m - E_k} u_k + \sum_{k\neq m} \left\{\left[\sum_{n\neq m} \frac{H'_{kn}H'_{nm}}{(E_m - E_n)(E_m - E_k)} - \frac{H'_{mm}H'_{km}}{(E_m - E_k)^2}\right] u_k - \frac{|H'_{km}|^2}{2(E_m - E_k)^2} u_m\right\} \tag{3.11-14}$$

$$W = E_m + H'_{mm} + \sum_{n\neq m} \frac{|H'_{mn}|^2}{E_m - E_n} \tag{3.11-15}$$

Notice that the second-order correction tends, according to (3.11-12), to increase the energy separation $|E_m - E_n|$. This fact is often expressed in the physics jargon as "Energy levels repel each other."

3.12 TIME-DEPENDENT PERTURBATION THEORY—RELATION TO LINE BROADENING

Another class of problems that is often treated with the aid of perturbation theory is one in which the Hamiltonian operator varies with time. This happens, for example, when the system interacts with a monochromatic radiation field, so that the Hamiltonian is perturbed harmonically, or when the Hamiltonian is changed by an abrupt step from one value to another. These two situations will be treated in this section and it will be shown that in both cases the perturbation induces transitions between the stationary states of the nonperturbed Hamiltonian.

The perturbation Hamiltonian is taken as $H'(t)$ so that the total Hamiltonian operator can be written as

$$H = H_0 + H'(t) \tag{3.12-1}$$

where H_0 is the nonperturbed Hamiltonian obeying

$$H_0 u_n = E_n u_n \tag{3.12-2}$$

The solution $\psi(t)$ of the Schrödinger equation

$$\frac{\partial \psi}{\partial t} = -\frac{i}{\hbar} H\psi \tag{3.12-3}$$

can be expanded, at any time t, in terms of the complete orthonormal set u_n obeying (3.12-2).

$$\psi(t) = \sum a_n(t) u_n e^{-iE_n t/\hbar} \tag{3.12-4}$$

The coefficients of the expansion are chosen as $a_n(t)e^{-iE_nt/\hbar}$ and have the convenient property that if $H'(t)$ is identically zero, the $a_n(t)$ become constants. Substituting (3.12-4) in (3.12-3), we obtain

$$\sum_n u_n \left[a_n \left(-\frac{iE_n}{\hbar} \right) e^{-iE_nt/\hbar} + \dot{a}_n e^{-iE_nt/\hbar} \right] = -\frac{i}{\hbar} \sum_n a_n (H_0 + H') u_n e^{-iE_nt/\hbar}$$

which after multiplying by u_k^* and integrating becomes

$$\dot{a}_k = -\frac{i}{\hbar} \sum_n a_n H'_{kn} e^{i\omega_{kn}t} \tag{3.12-5}$$

where ω_{kn} is defined by

$$\omega_{kn} = \frac{E_k - E_n}{\hbar}$$

Up to this point, the analysis is exact and solving Eqs. (3.12-5) is fully equivalent to a solution of Schrödinger equation. In a manner similar to that used in Section 3.11, we introduce the "turning on" parameter λ by taking the perturbation as $\lambda H'$, so that the Hamiltonian becomes

$$H = H_0 + \lambda H'$$

The power series expansion for a_n is written as

$$a_n = a_n^{(0)} + \lambda a_n^{(1)} + \lambda^2 a_n^{(2)} + \cdots$$

which when substituted in (3.12-5) becomes

$$\dot{a}_k^{(0)} + \lambda \dot{a}_k^{(1)} + \lambda^2 \dot{a}_k^{(2)} + \cdots = -\frac{1}{\hbar} \sum_n [a_n^{(0)} + \lambda a_n^{(1)} + \lambda^2 a_n^{(2)} + \cdots] \lambda H'_{kn} e^{i\omega_{kn}t}$$

Equating the same powers of λ results in the set of relations

$$\dot{a}_k^{(0)} = 0$$

$$\dot{a}_k^{(1)} = -\frac{i}{\hbar}\sum_n a_n^{(0)} H'_{kn}(t) e^{i\omega_{kn}t}$$

$$\dot{a}_k^{(2)} = -\frac{i}{\hbar}\sum_n a_n^{(1)} H'_{kn}(t) e^{i\omega_{kn}t} \tag{3.12-6}$$

$$\vdots$$

$$\dot{a}_k^{(s)} = -\frac{i}{\hbar}\sum_n a_n^{s-1} H'_{kn}(t) e^{i\omega_{kn}t}$$

The solution of the zero-order equation is $a_k^{(0)}$ = constant. The $a_k^{(0)}$ are thus the initial values for the problem. These are chosen as

$$a_m^{(0)} = 1$$

$$a_n^{(0)} = 0 \qquad n \neq m$$

so that at $t = 0$ the system is known with certainty to be at the state m. For this special case, the second of Eqs. (3.12-6) becomes

$$\dot{a}_k^{(1)} = -\frac{i}{\hbar} H'_{km} e^{i\omega_{km}t} \tag{3.12-7}$$

Since at $t = 0$ the system is at the state m, $|a_k(t)|^2$ is the probability that between $t = 0$ and t the system made a transition to the state k.

Harmonic Perturbation

As a special case, we consider a perturbation that is modulated harmonically according to

$$H'(t) = H'e^{-i\omega t} + (H')^\dagger e^{i\omega t} \qquad t > 0$$

$$H'(t) = 0 \qquad t < 0$$

The breakdown of $H'(t)$ into the two parts is done so as to ensure its Hermiticity. The result of substituting $H'(t)$ into (3.12-7) and performing the integration is

$$\begin{aligned} a_k^{(1)}(t) &= \int_0^t \left(-\frac{i}{\hbar}\right) H'_{km}(t') e^{i\omega_{km}t'}\, dt' \\ &= \hbar^{-1}\left[H'_{km}\frac{e^{i(\omega_{km}-\omega)t} - 1}{\omega_{km} - \omega} + H'^{*}_{mk}\frac{e^{i(\omega_{km}+\omega)t} - 1}{\omega_{km} + \omega}\right] \end{aligned} \tag{3.12-8}$$

We limit ourselves next to a case in which ω is nearly equal to $|\omega_{km}|$. This enables us to discard the second term on the right side of (3.12-8). The transition probability from the state m to k is then

$$|a_k^{(1)}|^2 = \frac{4|H'_{km}|^2}{\hbar^2} \frac{\sin^2[\frac{1}{2}(\omega_{km} - \omega)t]}{(\omega_{km} - \omega)^2} \tag{3.12-9}$$

where the cross terms have been left out since, for the conditions of interest, $\omega_{km} \simeq \pm\omega$, their contribution can be neglected. The first term on the right side of (3.12-8) dominates when $E_k > E_m$ and $E_k - E_m \sim \hbar\omega$, whereas the second term dominates when $E_k < E_m$ and $E_m - E_k \sim \hbar\omega$. The harmonic perturbation can thus cause both upward and downward transitions from state m to states k, separated, in energy, by $\sim\hbar\omega$.

To be specific, let us calculate the transition probability from m to a *group* of states clustered about state k where $E_k > E_m$. Let the density of these final states per unit of ω_{km} be $\rho(\omega_{km})$. Using only the first term of (3.12-8), we obtain

$$\sum_k |a_k^{(1)}|^2 = \frac{4}{\hbar^2} \int_{-\infty}^{+\infty} |H'_{km}|^2 \frac{\sin^2[\frac{1}{2}(\omega_{km} - \omega)t]}{(\omega_{km} - \omega)^2} \rho(\omega_{km})\, d\omega_{km} \tag{3.12-10}$$

If $|H'_{km}|^2$ is not a function of the final state, we may take it outside the integral sign. In addition, we limit the discussion to times t large enough so that the width $\Delta(\omega_{km} - \omega) \sim 2\pi/t$ of the function $\sin^2[\frac{1}{2}(\omega_{km} - \omega)t]/(\omega_{km} - \omega)^2$ is much smaller than the width of $\rho(\omega_{km})$. Under this condition, we may take $\rho(\omega_{km} = \omega)$ outside the integral sign, and using the definite integral

$$\int_{-\infty}^{+\infty} \frac{\sin^2(xt/2)}{x^2}\, dx = \frac{\pi t}{2} \tag{3.12-11}$$

obtain

$$|a_k^{(1)}|^2 = \frac{2\pi}{\hbar^2} |H'_{km}|^2 \rho(\omega_{km} = \omega) t \tag{3.12-12}$$

The transition probability rate from m to the continuum of states near k is thus

$$W_{m\to k} = \frac{2\pi}{\hbar} |H'_{km}|^2 \rho(E_k = E_m + \hbar\omega) \tag{3.12-13}$$

where $\rho(E)\, dE = \rho(\omega)\, d\omega$ is the number of states in the energy range dE centered on E. This result, often referred to as the "golden rule," is of central importance in the study of atomic transitions.

The condition that $2\pi/t$ be small compared to the width of $\rho(\omega_{km})$ is equivalent to treating the function $\sin^2[\frac{1}{2}(\omega_{km} - \omega)t]/(\omega_{km} - \omega)^2$ as a narrow sampling function. We can thus substitute

$$\frac{\sin^2[\frac{1}{2}(\omega_{km} - \omega)t]}{(\omega_{km} - \omega)^2} \to \frac{\pi t}{2} \delta(\omega_{km} - \omega)$$

in Eq. (3.12-9) and obtain

$$|a_k^{(1)}|^2 = \frac{2\pi}{\hbar^2} |H'_{km}|^2 t \, \delta(\omega_{km} - \omega)$$

$$= \frac{2\pi}{\hbar} |H'_{km}|^2 t \, \delta(E_k - E_m - \hbar\omega)$$

or

$$W_{m \to k} = \frac{2\pi}{\hbar} |H'_{km}|^2 \, \delta(E_k - E_m - \hbar\omega) \tag{3.12-14}$$

for the transition rate from m to a *single* state k. Equation (3.12-14) is especially useful as a starting point in analyses where the specific details of the density of final states must be taken into account. This is done by multiplying (3.12-14) by the density of states function and integrating over all energies. For the general case where the density of states function in energy (E) space is taken as $\rho(E)$, the integration yields Eq. (3.12-13).

The consideration of downward (i.e., $E_k < E_m$) in addition to upward transitions involves using both terms of (3.12-8) rather than the first one alone, with the result

$$W_{m \to k} = \frac{2\pi}{\hbar} |H'_{km}|^2 \, \delta(E_k - E_m - \hbar\omega) + \frac{2\pi}{\hbar} |H'_{km}|^2 \, \delta(E_k - E_m + \hbar\omega) \tag{3.12-15}$$

for the case where both upward and downward transitions are considered.

As an example, consider the transition rate $W_{m \to k}$ to a *single* state k where, to be specific, $E_k > E_m$. Let the position of the energy E_k be so smeared that we may describe it only by the probability function $f(E_k)\, dE_k$ of finding it between E_k and $E_k + dE_k$ so that

$$\int_{-\infty}^{+\infty} f(E_k)\, dE_k = 1$$

The transition rate is then

$$\begin{aligned} W_{m \to k} &= \frac{2\pi}{\hbar} |H'_{km}|^2 \int_{-\infty}^{+\infty} \delta(E_k - E_m - \hbar\omega) f(E_k)\, dE_k \\ &= \frac{2\pi}{\hbar} |H'_{km}|^2 f(E_k = E_m + \hbar\omega) \end{aligned} \tag{3.12-16}$$

If we replace $f(E)$ by the corresponding function $g(\nu)$ in the frequency domain, we have $g(\nu_k) = 2\pi\hbar f(E_k)$ so that

$$\int_{-\infty}^{+\infty} g(\nu_k)\, d\nu_k = 1$$

and $g(\nu_k)$ is normalized. Equation (3.12-16) takes the form

$$\begin{aligned} W_{m \to k} &= \frac{1}{\hbar^2} |H'_{km}|^2 g(\nu_k = \nu_m + \nu) \\ &= \frac{1}{\hbar^2} |H'_{km}|^2 g(\nu) \end{aligned} \tag{3.12-17}$$

where $g(\nu)$, the (normalized) natural lineshape function, also describes the transition strength (i.e., rate) as a function of the applied frequency ν. The "smearing out" of the transition $m \rightarrow k$ and the (resultant) need to describe it probabilistically by a lineshape function $g(\nu)$ may be due to a variety of causes. Two typical causes are (1) a transition originating in an atom (or ion) inside a crystal depends on the local surrounding crystalline electric field. This field may shift both E_m and E_k so that $E_k - E_m$ varies from one atomic site to another due to, say, strain induced fluctuations of the local field. (2) A transition involving an energy difference $h\nu = (E_k - E_m)$ undergoes a Doppler shift $\Delta\nu = (v/c)\nu$ in a moving atom where v is the component of velocity along the direction of observation. This causes the radiation emitted or absorbed by an ensemble of atoms (or ions) in a gas (or plasma) to have a linewidth of $\Delta\nu \sim \nu/c\sqrt{kT/m}$ where T is the temperature, c the velocity of light, and m the mass of the atom. If the velocity distribution is Maxwellian, the resultant lineshape function $g(\nu)$ is Gaussian. These two forms of broadening belong to a class referred to as *inhomogeneous broadening.* A second class of broadening is due to the finite lifetimes of states m and k involved in the transition. This gives rise, according to the uncertainty principle, to an uncertainty in the energy separation $\Delta(E_m - E_k) \sim \hbar(\tau_m^{-1} + \tau_k^{-1})$ where τ_m and τ_k are the lifetimes of levels m and k. This form of broadening is called *homogeneous broadening.* The basic difference between homogeneous and inhomogeneous broadening is that in the former the concept of a broadened transition applies to a single atom, whereas inhomogeneous broadening results from a consideration of a large number of atoms, each atom with its own different resonance frequency. This point will be discussed in detail in Chapter 8.

Limits of Validity of the Golden Rule. Two conditions were used in deriving Eqs. (3.12-13) and (3.12-15). The first was that $2\pi/t$ be small compared to the width of $\Delta\nu$ of $\rho(\omega_{km})$. The second condition results from our use of first-order perturbation theory and requires that $|a_k^{(1)}(t)|^2 \ll 1$; otherwise, higher-order terms must be considered. This second condition can be stated, using (3.12-9), as

$$\frac{|H'_{km}|}{\hbar} \ll \frac{1}{t}$$

Its physical significance is that the results of first-order perturbation theory are only valid for times short enough so that the probability for transitions out of the initial state m is very small compared to unity. Combining these two conditions leads to

$$\frac{|H'_{km}|}{\hbar} \ll \frac{1}{t} \ll \Delta\nu \qquad (3.12\text{-}18)$$

as the validity limits for Eqs. (3.12-13) and (3.12-15). Cases in which (3.12-18) does not hold have to be treated separately and several such situations will be dealt with in Chapter 15.

Step Function Perturbation. A second case of interest is one in which the perturbation has the form of a step function applied at $t = 0$, that is,

$$\begin{aligned} H'(t) &= 0, \qquad t < 0 \\ H'(t) &= H', \qquad t \geqslant 0 \end{aligned} \tag{3.12-19}$$

This situation may be regarded as a limiting case of the harmonic perturbation discussed above with $\omega \rightarrow 0$.

Using the second of Eqs. (3.12-6) with $a_n^{(1)}(0) = \delta_{nm}$ and repeating the steps leading to (3.12-14) yield

$$\begin{aligned} W_{m \to k} &= \frac{2\pi}{\hbar} |H'_{km}|^2 \, \delta(E_m - E_k) \\ &= \frac{1}{\hbar^2} |H'_{km}|^2 \, \delta(\nu_m - \nu_k) \end{aligned} \tag{3.12-20}$$

for the transition rate from m to k, the transition thus taking place between initial and final states of equal energy. In practice, this form is often used when the system Hamiltonian contains a term describing the interaction between various constituents of the system. We can use (3.12-20) to calculate the rate at which one constituent gains (or loses) energy while the remaining lose (or gain) an equal amount, due to the interaction Hamiltonian, and the total energy remains a constant.

3.13 DENSITY MATRICES—INTRODUCTION

In treating quantum mechanical systems, it is necessary to deal with two types of uncertainty. The first type of uncertainty is that described in Section 1.2 and is due to the probabilistic interpretation of the wavefunction $\psi(\mathbf{r}, t)$ and is manifested in the uncertainty principle. The second type of uncertainty occurs when one does not have sufficient information to determine the state of a quantum mechanical system. The information available about the system does not make it possible to determine exactly the wavefunction. This second type of uncertainty is handled by using the density matrix (or the "density operator" as it is sometimes called).

3.14 THE DENSITY MATRIX

The density matrix formalism is a method of computing expectation values of operators in cases where the precise wavefunction is unknown. In order to introduce this concept, consider a quantum mechanical system in a state $\psi(\mathbf{r}, t)$. From (1.2-14)

$$\psi(\mathbf{r}, t) = \sum c_n(t) u_n(\mathbf{r}) \tag{3.14-1}$$

where

$$c_n(t) = (u_n(\mathbf{r}), \psi(\mathbf{r}, t)) \tag{3.14-2}$$

and the $u_n(\mathbf{r})$ are an arbitrary complete orthonormal set of functions. Let A be an operator corresponding to some observable of the system. The expectation value of A is

$$\begin{aligned}\langle A\rangle &= (\psi(\mathbf{r}, t), A\psi(\mathbf{r}, t)) \\ &= \sum_{m,n} c_m^*(t)(u_m(\mathbf{r}), Au_n(\mathbf{r}))c_n(t)\end{aligned} \tag{3.14-3}$$

By making use of (3.4-1), one obtains

$$\langle A\rangle = \sum_{m,n} c_m^* A_{mn} c_n \tag{3.14-4}$$

Now suppose that the precise state of the system is unknown. This lack of knowledge is reflected in an uncertainty in the values of the c_n's in the expansion of $\psi(\mathbf{r}, t)$. Assume, however, that we have enough information to calculate an ensemble average for $c_m^* c_n$. The average will be denoted by a bar over the quantity in question. Thus, one can compute an average value of the expectation value of A according to

$$\overline{\langle A\rangle} = \sum_{m,n} \overline{c_m^* c_n}\, A_{mn} \tag{3.14-5}$$

It is convenient to define

$$\rho_{nm} = \overline{c_m^* c_n} \tag{3.14-6}$$

The matrix formed by the values of ρ_{nm} is known as the *density matrix*. Using (3.1-1), we obtain

$$\overline{\langle A\rangle} = \sum_n (\rho A)_{nn} \tag{3.14-7}$$

This computation is indicated by the trace, abbreviated "tr." Thus,

$$\overline{\langle A\rangle} = \mathrm{tr}(\rho A) \tag{3.14-8}$$

and is (see Problem 3.2) independent of the choice of the basis functions $u_n(\mathbf{r})$.

It follows from (3.14-6) that $\rho_{mn} = \rho_{nm}^*$ so that ρ is an Hermitian matrix. Another important result is that $\mathrm{tr}\,\rho = \sum_m \overline{c_m^* c_m} = 1$. This follows directly from the normalization condition of $\psi(\mathbf{r}, t)$ in Eq. (3.14-1).

3.15 THE ENSEMBLE AVERAGE

The type of averaging indicated above by a bar is what is known as the ensemble average. This ensemble averaging process can be interpreted physically in the following manner. One prepares an ensemble of N systems (N large) so that the systems are as nearly identical as allowed by one's incomplete information. The systems are then allowed to evolve in time. Each system is thus characterized by a state function

$$\psi_s(\mathbf{r}, t) = \sum_n c_n^{(s)}(t) u_n(\mathbf{r}) \tag{3.15-1}$$

for $s = 1, 2, \ldots, N$. The ensemble average of $c_m^* c_n$ is computed according to the following formula:

$$\rho_{nm}(t) = \overline{c_m^*(t)c_n(t)} = \frac{1}{N}\sum_{s=1}^{N} c_m^{(s)*}(t)c_n^{(s)}(t) \qquad (3.15\text{-}2)$$

Then, the ensemble average is an average over all N systems.

In this physical interpretation, the density matrix represents certain probabilistic aspects of the ensemble. The diagonal term ρ_{nn} is the probability of finding one of the systems in the ensemble in the state $u_n(\mathbf{r})$. The off-diagonal term $\rho_{nm}(t)$ is equal to the ensemble average of $c_m^*(t)c_n(t)$ that will be shown in Chapter 8 to be related to the radiating dipole of the ensemble.

3.16 TIME EVOLUTION OF THE DENSITY MATRIX

Since the wavefunction of each system in the ensemble satisfies Schrödinger's equation, then

$$H\psi(\mathbf{r}, t) = i\hbar \frac{\partial\psi(\mathbf{r}, t)}{\partial t} \qquad (3.16\text{-}1)$$

Substituting (3.14-1) for $\psi(\mathbf{r}, t)$ yields

$$i\hbar \sum_n \frac{\partial c_n(t)}{\partial t} u_n(\mathbf{r}) = \sum_n c_n(t) H u_n(\mathbf{r}) \qquad (3.16\text{-}2)$$

Taking the inner product of (3.16-2) with $u_m(\mathbf{r})$ and using the orthonormality of the u_n's, one obtains

$$i\hbar \frac{\partial}{\partial t} c_m(t) = \sum_n c_n(t) H_{mn} \qquad (3.16\text{-}3)$$

But from (3.14-6)

$$\frac{\partial \rho_{nm}}{\partial t} = \overline{c_n \frac{\partial c_m^*}{\partial t}} + \overline{c_m^* \frac{\partial c_n}{\partial t}} \qquad (3.16\text{-}4)$$

By making the use of (3.16-2) and the Hermiticity of H, (3.16-4) reduces to

$$\frac{\partial \rho}{\partial t} = \frac{i}{\hbar}[\rho, H] \qquad (3.16\text{-}5)$$

A more detailed discussion of the properties of the density matrix and methods for evaluating it for physical systems can be found in Reference 2–4. It will be applied in Chapters 8 and 15 to describe the interaction of a two-level atomic system with a radiation field. In Appendix 4, we will use it to describe nonlinear optical processes.

3.17 THE TIME EVOLUTION OPERATOR—FEYNMAN DIAGRAMS

Another very important formalism in our tool box of perturbation methods is that of the time evolution operator. It is closely related to the topic of the

Feynman diagrams. We shall make extensive use of it in describing multiphoton (2nd, 3rd, etc.) optical processes and, specifically, to obtain an expression in this chapter for the two photon absorption coefficient and in Chapter 16 for obtaining expressions for nonlinear optical coefficients.

The eigenfunction $\Psi(t)$ of an atom subjected to an electromagnetic field can be obtained formally by solving the Schrödinger equation

$$H\psi = i\hbar \frac{\partial \psi}{\partial t} \tag{3.17-1}$$

or, equivalently, by operating on the eigenfunction at time t_a with the evolution operator $u(t_b, t_a)$ according to

$$\psi(t_b) = u(t_b, t_a)\psi(t_a) \tag{3.17-2}$$

where $u(t_b, t_a)$ satisfies

$$i\hbar \frac{\partial u(t_b, t_a)}{\partial t_b} = Hu(t_b, t_a) \tag{3.17-3}$$

If H does not depend on time, then it follows from (3.17-3) and the Hermiticity of H that

$$\begin{aligned} u(t_b, t_a) &= \exp\left[-i \frac{H}{\hbar}(t_b - t_a)\right] \\ &= \sum_m |m\rangle\langle m| \exp[-i\omega_m(t_b - t_a)] \end{aligned} \tag{3.17-4}$$

where $\omega_m \equiv E_m/\hbar$, $|m\rangle$ is the eigenfunction of H with energy E_m, that is, $H|m\rangle = \hbar\omega_m|m\rangle$ and where we made use of the identity $\sum_m |m\rangle\langle m| = I$.

In the cases of interest to us, we take the Hamiltonian as

$$H(t) = H_0 + V(t) \tag{3.17-5}$$

where H_0 is time-dependent and $V(t)$ represents the time-dependent intereaction of the atom with the optical fields. It follows directly from the definition (3.17-2) that u is a unitary operator. To prove the last statement, we start with

$$\langle\psi(t_b)|\psi(t_b)\rangle = 1 \qquad \text{normalization condition}$$

$$\langle u\psi|u\psi\rangle = \langle\psi|u^\dagger u|\psi\rangle = \langle\psi|\psi\rangle = 1$$

where $u^\dagger$ is the Hermitian adjoint of u

$$\Rightarrow u^\dagger u = I(=\text{identity operator}) \tag{3.17-6}$$

and u is a unitary operator. Another property of u is

$$u(t_b, t_a)u(t_a, t_c) = u(t_b, t_c) \tag{3.17-7}$$

From the last two numbered equations, it follows that

$$u(t_a, t_b) = u^\dagger(t_b, t_a) \tag{3.17-8}$$

It can be shown by simple substitution that the formal solution of (3.17-3) when $H(t)$ is given by (3.17-5) is

$$u(t_b, t_a) = u^{(0)}(t_b, t_a) - \frac{i}{\hbar}\int_{t_a}^{t_b} u^{(0)}(t_b, t)V(t)u(t, t_a)\,dt \tag{3.17-9}$$

with

$$u^{(0)}(t_b, t_a) \equiv \exp\left[-\frac{iH_0}{\hbar}(t_b - t_a)\right] \tag{3.17-9}$$

is the propagator operator at the absence of perturbation.

We can use the last result as a basis for a perturbation expansion. We substitute for $u(t_b, t_a)$

$$u(t_b, t_a) = u^{(0)}(t_b, t_a) - \frac{i}{\hbar}\int_{t_a}^{t_b} u^{(0)}(t_b, t_1)V(t_1)u^{(0)}(t_1, t_a)\,dt_1$$

$$+ \left(-\frac{i}{\hbar}\right)^2 \int_{t_a}^{t_b}\int_{t_a}^{t_1} dt_1\,dt_2 u^{(0)}(t_b, t_1)V(t_1)u^{(0)}(t_1, t_2)V(t_2)u(t_2, t_a) \tag{3.17-10}$$

$$t_b > t_1 > t_2 > t_a$$

Note again the sole appearance of u rather than $u^{(0)}$ at the very end of the integral. Repeating the process, we obtain

$$u(t_b, t_a) = u^{(0)}(t_b, t_a) - \frac{i}{\hbar}\int_{t_a}^{t_b} u^{(0)}(t_b, t_1)V(t_1)u^{(0)}(t_1, t_a)\,dt_1$$

$$+ \left(-\frac{i}{\hbar}\right)^2 \int_{t_a}^{t_b}\int_{t_a}^{t_1} dt_1\,dt_2 u^{(0)}(t_b, t_1)V(t_1)u^{(0)}(t_1, t_2)V(t_2)u^{(0)}(t_2, t_a) \tag{3.17-11}$$

$$+ \left(-\frac{i}{\hbar}\right)^3 \int_{t_a}^{t_b}\int_{t_a}^{t_1}\int_{t_a}^{t_2} dt_1\,dt_2\,dt_3 u^{(0)}(t_b, t_1)V(t_1)u^{(0)}(t_1, t_2)V(t_2)u^{(0)}(t_2, t_3)V(t_3)u(t_3, t_a)$$

We can thus write formally

$$u(t_b, t_a) = u^{(0)}(t_b, t_a) + u^{(1)}(t_b, t_a) + u^{(2)}(t_b, t_a) + \cdots u^{(n)}(t_b, t_a) + \cdots \tag{3.17-12}$$

where

$$u^{(0)}(t_b, t_a) = \exp\left[-i\frac{H_0}{\hbar}(t_b - t_a)\right]$$

$$u^{(1)} = \left(-\frac{i}{\hbar}\right)\int_{t_a}^{t_b} u^{(0)}(t_b, t_1)V(t_1)u^{(0)}(t_1, t_a)\,dt_1$$

$$u^{(2)} = \left(-\frac{i}{\hbar}\right)^2 \int_{t_a}^{t_b}\int_{t_a}^{t_1} dt_1\,dt_2 u^{(0)}(t_b, t_1)V(t_1)$$

$$u^{(0)}(t_1, t_2)V(t_2)u^{(0)}(t_2, t_a) \tag{3.17-13}$$

$$u^{(3)} = \left(-\frac{i}{\hbar}\right)^3 \int_{t_a}^{t_b}\int_{t_a}^{t_1}\int_{t_a}^{t_2} dt_1\,dt_2\,dt_3 u^{(0)}(t_b, t_1)$$

$$\times V(t_1)u^{(0)}(t_1, t_2)V(t_2)$$

$$\times u^{(0)}(t_2, t_3)V(t_3)u^{(0)}(t_3, t_a)$$

where

$$t_b > t_1 > t_2 > t_2 > t_3 \cdots > t_n$$

To obtain $\psi(t)$ to, as an example, third order in $V(t)$, we write directly, putting $t_b = t$

$$\psi_{\text{third}}(t) = \left[u^{(0)}(t, t_a) + u^{(1)}(t, t_a) + u^{(2)}(t, t_a) + u^{(3)}(t, t_a)\right]\psi(t_a) \quad (3.17\text{-}14)$$

$$\equiv \psi^{(0)}(t) + \psi^{(1)}(t) + \psi^{(2)}(t) + \psi^{(3)}(t) \quad (3.17\text{-}15)$$

A short inspection of (3.17-13) should enable us to write $u^{(n)}(t, t_a)$ by inspection. We note that it is composed of a sequence of unperturbed propagators

$$u^{(0)}(t, t_1) \cdots u^{(0)}(t_{n-1}, t_n)\, u^{(0)}(t_n, t_a)$$

punctuated by the action of the perturbation $V(t)$ at times $t_1 \cdots t_n$. That the "flow" of time is from right to left is indicated at the end of (3.17-13).

To calculate the expectation value of physical observables as in (1.1-16), we need to obtain an expression for the wavefunction to any desired order of perturbation. The optical field at the atom site is taken as

$$\mathbf{E}(t) = \tfrac{1}{2}\mathbf{E}_1 e^{i\omega_1 t} + \tfrac{1}{2}\mathbf{E}_2 e^{i\omega_2 t} + \text{c.c.} \quad (3.17\text{-}16)$$

and the interaction Hamiltonian as

$$V(t) = \boldsymbol{\mu}\cdot\left(\frac{\mathbf{E}_1}{2}\, e^{i\omega_1 t} + \frac{\mathbf{E}_2}{2}\, e^{i\omega_2 t} + \text{c.c.}\right). \quad (3.17\text{-}17)$$

where μ is the negative of the dipole moment operator. The perturbation is assumed turned on at $t_0 = -\infty$, at which time the atom is in its ground state n. The eigenfunction at a subsequent time t is

$$\psi(t) = \psi^{(0)}(t) + \psi^{(1)}(t) + \cdots \psi^{(m)}(t) + \cdots$$

where

$$\psi^{(m)}(t) = u^{(m)}(t, t_0)|n\rangle$$

Using (3.17-13), we write

$$\psi^{(0)}(t) = \exp\left[-i\,\frac{H_0}{\hbar}\,(t - t_0)\right]|n\rangle = \exp[-i\omega_n(t - t_0)]\,|n\rangle$$

$$\psi^{(1)}(t) = -\frac{i}{\hbar}\int_{t_0}^{t} \exp\left[-\frac{iH_0}{\hbar}\,(t - t_1)\right] V(t_1)$$

$$\cdot \exp\left[-\frac{iH_0}{\hbar}\,(t_1 - t_0)\right] dt_1\,|n\rangle.$$

which when we use (3.17-4) becomes

$$\psi^{(1)}(t) = -\frac{i}{\hbar}\sum_m \int_{t_0}^{t} |m\rangle\langle m|\exp[-i\omega_m(t - t_1)]V(t_1)$$

$$\cdot \exp[-i\omega_n(t_1 - t_0)]\,|n\rangle\, dt_1 \quad (3.17\text{-}18)$$

Since $V(t_1)$ is, according to (3.17-17), a sum of four terms, the integrand in (3.17-18) is made up of four terms. One such term, for example, resulting from the part of $V(t)$ involving $E_1^* \exp(-i\omega_1 t)$ will yield

$$\psi_{-\omega_1}^{(1)}(t) = \frac{i}{2\hbar} \sum_m \int_{t_0 \to -\infty}^{t} (\mu_1)_{mn} E_1^* \exp i(\omega_{mn} - \omega_1)t_1 \cdot \exp(-i\omega_m t) \exp(i\omega_n t_0)\, dt_1\, |m\rangle \tag{3.17-19}$$

where μ_1 is the projection of $\boldsymbol{\mu}$ along E_1 and $\omega_{mn} \equiv \omega_m - \omega_n \equiv (E_m - E_n)/\hbar$. Carrying out the integration leads to

$$\psi_{-\omega_1}^{(1)}(t) = \frac{1}{2\hbar} \sum_m (\mu_1)_{mn} E_1^* \exp(-i\omega_m t)\, |m\rangle \frac{e^{i(\omega_{mn}-\omega_1)t} - e^{i(\omega_{mn}-\omega_1)t_0}}{(\omega_{mn} - \omega_1)}$$

where the factor $\exp(i\omega_n t_0)$ was dropped since it cancels out (through multiplication by its complex conjugate) in the calculation of any physical observables. The factor $\exp[i(\omega_{mn} - \omega_1)t_0]$ becomes zero for $t_0 \to -\infty$ if we allow

$$\omega_m \to \omega_m - i\gamma \qquad (\gamma > 0)$$

This not only eliminates any initial "memory" effects that, having taken place at $t_0 = -\infty$ should not affect the present, but also introduces in a correct fashion the finite natural linewidth γ of the transition. The result is

$$\psi_{-\omega_1}^{(1)}(t) = \frac{1}{2\hbar} \sum_m (\mu_1)_{mn} E_1^* \frac{\exp[i(-\omega_1 - \omega_n)t}{(\omega_{mn} - \omega_1 - i\delta)} |m\rangle$$

In a similar manner, we can calculate the contributions to $\psi^{(1)}(t)$ due to the terms $E_1 \exp(i\omega_1 t)$, $E_2 \exp(i\omega_2 t)$, and $E_2^* \exp(-i\omega_2 t)$ in $V(t)$ obtaining, respectively,

$$\begin{aligned}
\psi^{(1)}(t) &= \psi_{-\omega_1}^{(1)}(t) + \psi_{\omega_1}^{(1)}(t) + \psi_{-\omega_2}^{(1)}(t) + \psi_{\omega_2}^{(1)}(t) \\
&= \frac{1}{2\hbar} \sum_m \left\{ (\mu_1)_{mn} \frac{E_1^* \exp[i(-\omega_1 - \omega_n)t)]\, |m\rangle}{\omega_{mn} - \omega_1 - i\gamma} \right. \\
&\quad + (\mu_1)_{mn} \frac{E_1 \exp[i(\omega_1 - \omega_n)t]\, |m\rangle}{\omega_{mn} + \omega_1 - i\gamma} \\
&\quad + (\mu_2)_{mn} \frac{E_2^* \exp[i(-\omega_2 - \omega_n)t]\, |m\rangle}{\omega_{mn} - \omega_2 - i\gamma} \\
&\quad \left. + (\mu_2)_{mn} \frac{E_2 \exp[i(\omega_2 - \omega_n)t]\, |m\rangle}{\omega_{mn} + \omega_2 - i\gamma} \right\}
\end{aligned} \tag{3.17-20}$$

Using (3.17-13), we write the second-order wave functions $\psi^{(2)}(t)$ as

$$\begin{aligned}
\psi^{(2)}(t) &= u^{(2)}(t, t_0)|n\rangle = \left(-\frac{i}{\hbar}\right)^2 \int_{t_0}^{t} \int_{t_0}^{t_1} dt_1\, dt_2 \\
&\cdot \exp\left[-i\frac{H_0(t - t_1)}{\hbar}\right] V(t_1) \exp\left[-i\frac{H_0(t_1 - t_2)}{\hbar}\right] V(t_2) \\
&\cdot \exp\left[-i\frac{H_0(t_2 - t_0)}{\hbar}\right] |n\rangle = \left(-\frac{i}{\hbar}\right)^2 \sum_m \sum_s \int_{t_0}^{t} \int_{t_0}^{t_1 > t_2}
\end{aligned}$$

$$\cdot \exp[-i\omega_s(t - t_1)]\, |s\rangle\, \langle s|V(t_1) \qquad (3.17\text{-}21)$$

$$\cdot \exp[-i\omega_m(t_1 - t_2)]\, |m\rangle\, \langle m|V(t_2)$$

$$\cdot \exp[-i\omega_n(t_2 - t_0)]\, |n\rangle\, dt_2\, dt_1$$

Since $V(t)$ contains four frequency terms and appears twice in (3.17-21), the full integration in (3.17-21) will yield 16 terms. A typical term, as an example involving the use of $\mu_1 E_1^* \exp(-i\omega_1 t_2)$ at t_2 and $\mu_2 E_2^* \exp(-i\omega_2 t_1)$ at t_1, is

$$\psi^{(2)}_{-\omega_1,-\omega_2}(t) = \sum_m \sum_s \left(\frac{1}{\hbar}\right)^2 \frac{E_1^* E_2^*}{4} \cdot \frac{(\mu_1)_{mn}(\mu_2)_{sm} \exp i(-\omega_n - \omega_1 - \omega_2)t}{(\omega_{mn} - \omega_1 - i\gamma)(\omega_{sn} - \omega_1 - \omega_2 - i\gamma)} |s\rangle \quad (3.17\text{-}23)$$

A convenient way to represent (3.17-23) is through the use of a Feynman diagram, as shown in Figure 3.1. Time increases from the bottom to the top. Each solid line segment represents an eigenstate. The atom starts at t_0 in state n and "scatters" at t_2 into state m by absorbing a photon at ω_1. This scattering is accounted for by the factor $(\mu_1)_{mn}\, E_1^* e^{-i\omega_1 t}/(\omega_{mn} - \omega_1 - i\gamma)$ in (3.17-22). The next scattering is at t_1 and involves an absorption of a photon at ω_2. A negative frequency denotes absorption and is represented by an arrow terminating at a vertex, whereas an arrow starting at a vertex denotes the emission of a photon. A case where the transition from state n to s involves the absorption of a photon at ω_1 and the emission of a photon at ω_2, as an example, is shown in Figure 3.1*b*.

The corresponding contribution to $\psi^{(2)}(t)$ can be written by inspection

$$\psi^{(2)}_{-\omega_1,\omega_2} = \sum_m \sum_s \left(\frac{1}{\hbar}\right)^2 \frac{E_1^* E_2}{4} \cdot \frac{(\mu_1)_{mn}(\mu_2)_{sm} \exp i(-\omega_n - \omega_1 + \omega_2)t}{(\omega_{mn} - \omega_1 - i\gamma)(\omega_{sn} - \omega_1 + \omega_2 - i\gamma)} |s\rangle \quad (3.17\text{-}24)$$

We note that each scattering, that is, each vertex in the diagram, contributes one factor to the denominator, and the factor is equal to the energy in units of $\hbar$ of the atom and field after the scattering minus the initial ($t = t_0$) energy. The second factor in the denominator in (3.17-24), as an example, is obtained

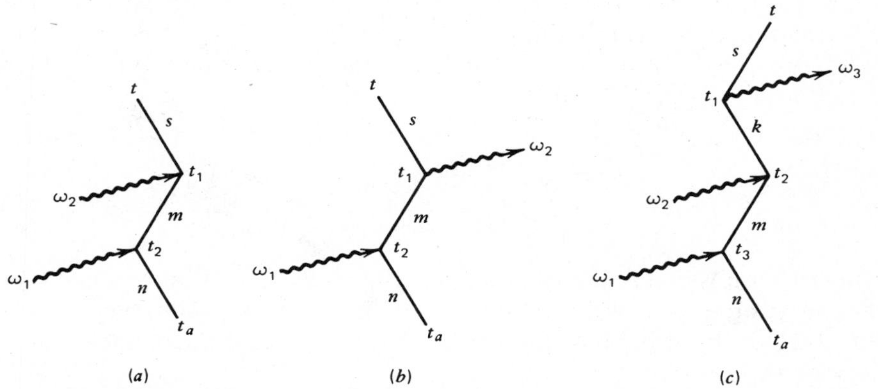

FIGURE 3.1 The Feynman diagrams used to obtain (*a*) $\Psi^{(2)}_{-\omega_1,-\omega_2}(t)$. (*b*) $\Psi^{(2)}_{-\omega_1,+\omega_2}(t)$. (*c*) $\Psi^{(3)}_{-\omega_1,-\omega_2,\omega_3}(t)$

from $\omega_s + \omega_2 - (\omega_1 + \omega_n)$. The remaining 14 wavefunctions are obtained by taking all the possible permutations of $\psi^{(2)}_{\pm\omega_i,\pm\omega_j}$, $i,j = 1, 2$ where our convention is such that $\psi^{(2)}_{+\omega_i,-\omega_j}$, for example, corresponds to emitting an ω_i photon at t_2 and absorbing an ω_j photon at t_1 $(t_1 > t_2)$. We note that $\psi^{(2)}_{-\omega_j,\omega_i}$ is not equal to $\psi^{(2)}_{\omega_i,-\omega_j}$.

Using the diagram technique, we can write the wavefunction to any order of perturbation. As an example, consider the process in which an atom makes a transition from state n to state s while absorbing one photon at ω_1, one photon at ω_2, and emitting a photon at ω_3. The diagram describing this process $\psi^{(3)}_{-\omega_1,-\omega_2,\omega_3}$, is shown in Figure 3.1*c*. We obtain by inspection

$$\psi^{(3)}_{-\omega_1,-\omega_2,\omega_3} = \sum_m \sum_k \sum_s \frac{-E_1^* E_2^* E_3}{8\hbar^3} \cdot \frac{(\mu_1)_{mn}(\mu_2)_{km}(\mu_3)_{sk} \exp i(-\omega_n - \omega_1 - \omega_2 + \omega_3)t}{(\omega_{mn} - \omega_1)(\omega_{kn} - \omega_1 - \omega_2)(\omega_{sn} - \omega_1 - \omega_2 + \omega_3)} |s\rangle$$

The total number of $\psi^{(3)}_{\pm\omega_j,\pm\omega_j,\pm\omega_k}$ combinations is $6^3 = 216$ inside the triple summation. (There are now six terms in $V(t)$, two for each frequency.) The tremendous advantage of the diagrammatic representation is that it enables us to single out and write down very simply the terms that dominate in any given physical situation. This will become clearer in the examples that follow.

Two-Photon Absorption

Here, we will apply our formalism to derive the transition rate for absorption of two photons—one at ω_1 and a second at ω_2—by an atom. By allowing $\omega_1 = \omega_2$, we will obtain the familiar expression for a two-photon absorption coefficient.

We start by writing the second-order wavefunction $\psi^{(2)}(t)$ corresponding to an atom that at $t = 0$ is at state n and that interacts with the radiation field (3.17-17), consisting of fields at ω_1 and ω_2. We will assume that the largest contribution to $\psi^{(2)}$ comes from $\psi^{(2)}_{-\omega_1,-\omega_2}$ and not from $\psi_{-\omega_2,-\omega_1}$ because of the existence of some level in such that $(\omega_{mn} - \omega_1)$ is very small. We thus write

$$\psi^{(2)}_{-\omega_1,-\omega_2}(t) = \sum_l \sum_s \left(\frac{E_1^* E_2^*}{4\hbar^2}\right) \cdot \frac{(\mu_1)_{ln}(\mu_2)_{sl} e^{-i\omega_s t}[\exp i(\omega_{sn} - \omega_1 - \omega_2)t - 1]}{(\omega_{ln} - \omega_1)(\omega_{sn} - \omega_1 - \omega_2)} |s\rangle \quad (3.17\text{-}25)$$

The -1 term is due to the fact that here the problem starts at $t = 0$ and not, as above, at $-\infty$, at which time the atom is in state $|n\rangle$. The integration in (3.17-21) is thus from 0 to t. Since $\omega_{mn} \simeq \omega_1$, the term lm dominates and we rewrite (3.17-25) as

$$\psi^{(2)}_{-\omega_1,-\omega_2}(t) \simeq \frac{E_1^* E_2^*}{4\hbar^2} \sum_s \frac{(\mu_1)_{mn}(\mu_2)_{sm} e^{-i\omega_s t} \exp[i(\omega_{sn} - \omega_1 - \omega_2)t - 1]}{(\omega_{mn} - \omega_1)(\omega_{sn} - \omega_1 - \omega_2)} |s\rangle \quad (3.17\text{-}26)$$

The probability of finding the atom in some state k at time t is $\langle k|\Psi(t)\rangle|^2$ so that the transition probability due to two-photon absorption is

$$\mathcal{P}_k = |\langle k|\psi^{(2)}(t)\rangle|^2 \tag{3.17-27}$$

$$= \frac{|E_1|^2|E_2|^2}{16\hbar^4} \frac{(\mu_1)^2_{mn}(\mu_2)^2_{km} \sin^2\left[\frac{1}{2}(\omega_{kn} - \omega_1 - \omega_2)t\right]}{(\omega_{mn} - \omega_1)^2\left[\frac{1}{2}(\omega_{kn} - \omega_1 - \omega_2)\right]^2} \tag{3.17-28}$$

We note that transitions occur to the state k that conserves energy, that is, to that state where $E_k - E_n = \hbar(\omega_1 + \omega_2)$. If the normalized lineshape function for the transition $n \rightarrow k$ is $g(\omega_{kn})$, then the average value of P_k is obtained by multiplying (3.17-28) by $g(\omega_{kn})$ and integrating from $-\infty$ to ∞. Using

$$\lim_{t\rightarrow\infty} \frac{\sin^2 \frac{xt}{2}}{(x/2)^2} \longrightarrow 2\pi t\, \delta(x)$$

we obtain for the transition rate

$$\begin{aligned} W_{n\rightarrow k} &= \frac{d}{dt}\mathcal{P}_k \\ &= \frac{\pi|E_1|^2|E_2|^2(\mu_1)^2_{mn}(\mu_2)^2_{mk} g(\omega_{kn} = \omega_1 + \omega_2)}{8\hbar^4(\omega_{mn} - \omega_1)^2} \end{aligned} \tag{3.17-29}$$

In the special case of one-frequency two-photon absorption, that is, $\omega_1 = \omega_2$, we obtain the absorption coefficient by equating the change in the intensity I_ω of E_1 to the number of transitions per unit time per unit volume.

$$\frac{dI_\omega}{dz} = 2\hbar\omega\, W_{n\rightarrow k}\,(N_k - N_n) \tag{3.17-30}$$

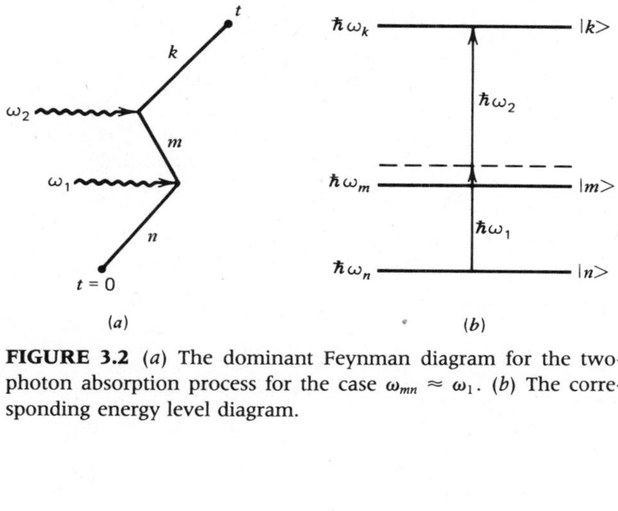

FIGURE 3.2 (*a*) The dominant Feynman diagram for the two-photon absorption process for the case $\omega_{mn} \approx \omega_1$. (*b*) The corresponding energy level diagram.

where I_ω is the intensity at ω, and N_k and N_n are the population densities of level k and n, respectively. Using $I_\omega = (\varepsilon c/n_1)|E_1|^2/2$, we obtain

$$\alpha_{\text{two photon}} = \frac{\pi\omega(N_n - N_k)(\mu_1)^2_{mn}(\mu_1)^2_{km}n_1^2 g(\omega_{kn} = 2\omega)I_\omega}{\hbar^3(\omega_{mn} - \omega)^2\varepsilon^2 c^2}$$

Since $\omega_{mn} \neq \omega_1$, there is no actual transition to and dwelling in the intermediate level $|m\rangle$. To obtain a "real" transition, $n \to m$, $\omega_{mn} - \omega_1$ must, in practice, be smaller than a few times the transition linewidth. We refer to the transition $n \to m$ indicated by Figure 3.2 as "virtual". Also noteworthy is the fact that a transition $n \to k$ takes place even when $\mu_{nk} = 0$ i.e. when the direct one-photon transition may be forbidden.

References

1. E. U. Condon and G. H. Shortley, *The Theory of Atomic Spectra* (New York: Cambridge University Press, 1959), pp. 73–76.
2. C. Kittel, *Elementary Statistical Mechanics* (New York: Wiley, 1958).
3. R. C. Tolman, *Principles of Statistical Mechanics* (London, Oxford University Press, 1938).
4. U. Fano, *Rev. Mod. Phys.* **29,** 74 (1957).

Supplementary References

1. Leighton, R. B., *Principles of Modern Physics* (New York: McGraw-Hill, 1959).
2. Messiah, A., *Quantum Mechanics* (New York: Interscience, 1961).
3. Schiff, L. I., *Quantum Mechanics* (New York: McGraw-Hill, 1959).

Problems

3.1 A function f may be expanded in terms of two arbitrary complete orthonormal sets v_n and u_n in the form

$$f = \sum f_n u_n = \sum f'_n v_n$$

The set v_n can be expanded as

$$v_n(\mathbf{r}) = \sum_k S_{kn} u_k(\mathbf{r})$$

Show that the unitarity of the matrix S can be derived by requiring that $\int f^* f\,dv$ be independent of the set (v_n or u_n) in which it is expanded.

3.2 Show that the trace of a square matrix A

$$\operatorname{tr} A = \sum_k A_{kk}$$

is invariant under matrix transformations, that is, that

$$\operatorname{tr} A = \operatorname{tr}(SAS^{-1})$$

where S is unitary.

3.3 Prove that $(AB)^\dagger = B^\dagger A^\dagger$.

3.4 Prove that a Hermitian matrix remains Hermitian under a unitary transformation. Show that as a consequence, a Hermitian matrix must have real eigenvalues.

3.5 Show that if $A_{kl} = \int v_k^* A v_l \, dv = A_{lk}^*$ where v_n is any orthonormal function set, the operator A is Hermitian.

3.6 Show that the matrix representation of a product of operators is equal to the product of the individual matrices.

3.7 Use the result of Problem 3.6 and the known matrix elements x_{mn} for the harmonic oscillator to evaluate

$$(x^2)_{mn} = \int u_m x^2 u_n \, dv$$

Check the result against the solution as given in Problem 2.3.

3.8 Show that the necessary and sufficient condition that two matrices commute is that the same transformation diagonalize each one of them.

3.9 Prove that $c_{m+1}^- = (c_m^+)^*$ where c_{m+1}^- and c_m^+ are defined by Eq. (3.8-4).

3.10 Prove that $\psi_1^1 = 1/\sqrt{2}\,(\alpha_0^1\beta_1^1 - \alpha_1^1\beta_0^1)$, Eq. (3.10-6), is an eigenfunction of $\mathbf{L}^2$ where $\mathbf{L} = \mathbf{L}_1 + \mathbf{L}_2$. What is the eigenvalue?

3.11 Generate the set of eigenfunctions ψ_m^j resulting from the addition of two angular momenta $j_1 = \frac{1}{2}$ and $j_2 = 1$.

3.12 The potential energy of a harmonic oscillator is perturbed by adding a term bx. (a) What is the correction, to second order, in the energies? (b) Obtain an exact expression for the energies.

3.13 Write by inspection, if you can, the function

$$\psi^{(3)}_{-\omega_1,\omega_2,\omega_3}(t)$$

corresponding to absorption of a photon at ω_1 and a simultaneous emission of photons at ω_2 and ω_3. What is the corresponding Feynman diagram? Assume an electric field

$$E(t) = \frac{1}{2}\left(E_1 e^{i\omega_1 t} + E_2 e^{i\omega_2 t} + E_3 e^{i\omega_3 t} + \text{c.c.}\right)$$

3.14 (a) Calculate the gain coefficient experienced by a field

$$E_2(t) = \frac{1}{2} E_2 e^{i\omega_2 t} + \text{c.c.}$$

in the presence of an intense field

$$E_1(t) = \frac{1}{2} E_2 e^{i\omega_1 t} + \text{c.c.}$$

due to a process in which atoms initially in the ground state in $|n\rangle$ absorb a photon at ω_1 and emit simultaneously a photon at $\omega_2 (\langle \omega_1)$. The atom ends up in a state $|k\rangle$ so that $E_k - E_n = \hbar(\omega_1 - \omega_2)$. (Note the similarity to the two-photon absorption example treated in this chapter.)

(b) Assuming transition linewidths of 10^9 Hz and matrix elements $\mu_{mn} \sim 10^{-30}$ MKS, calculate the gain coefficient experienced by the field E_2 as a function of the intensity I_1.

C H A P T E R 4

Lattice Vibrations and Their Quantization

4.0 INTRODUCTION

The topic of lattice vibrations is treated in this chapter for a number of reasons: (1) The spectrum of lattice vibrations and the formalism for treating it are necessary to explain certain relaxation mechanisms such as the atom-lattice relaxation. (2) The formalism for quantizing the lattice vibration field serves as a model for the quantization of other boson fields such as that of electromagnetic radiation. (3) The formalism will be used in Chapter 18 for treating Brillouin scattering of light by sound and stimulated Brillouin scattering.

4.1 MOTION OF HOMOGENEOUS LINE

Consider the one-dimensional problem of a uniform line of mass density ρ (kg-m^{-1}). Let the displacement of a point x from its equilibrium value be $u(x, t)$. The force at x is taken as a constant times the strain $\partial u/\partial x$

$$F = c\frac{\partial u}{\partial x} \tag{4.1-1}$$

which is equivalent to using Hooke's law for the restoring force of a spring. If we consider an element Δx of the line, the net force acting on it is $F(x + \Delta x) - F(x) = c(\partial^2 u/\partial x^2)\Delta x$. The equation of motion becomes

$$c\frac{\partial^2 u}{\partial x^2} = \rho\frac{\partial^2 u}{\partial t^2} \tag{4.1-2}$$

with a solution, if we assume $e^{i(\omega t + kx)}$ dependence,

$$u = u_0 e^{i(\omega t \pm kx)}$$

$$k = \frac{\omega}{v_s}, \qquad v_s = \sqrt{\frac{c}{\rho}} \tag{4.1-3}$$

According to (4.1-3), the line can support waves with a phase velocity $v_s = \sqrt{c/\rho}$. The dispersion relation is $\omega = kv_s$. Since v_s is a constant, the line is dispersionless.

The general solution of (4.1-2) is a linear superposition of individual solutions that can be written as

$$u(x, t) = \int_{-\infty}^{+\infty} u(\omega) e^{i\omega(t - x/v_s)}\, d\omega \tag{4.1-4}$$

4.2 WAVE MOTION OF A LINE OF SIMILAR ATOMS

As the next improvement in the realism of the model, consider a line of particles of individual masses M separated from each other by a and connected by massless springs with a spring constant β as shown in Figure 4.1. If we let the deviation of the nth atom from its equilibrium position be q_n, the equation of motion for the nth atom becomes

$$\begin{aligned} M\ddot{q}_n &= \beta(q_{n+1} - q_n) - \beta(q_n - q_{n-1}) \\ &= \beta(q_{n+1} + q_{n-1} - 2q_n) \end{aligned} \tag{4.2-1}$$

In direct analogy with the continuous line, we assume the basic harmonic solution to consist of a wave with an effective propagation constant k corresponding to a phase shift (ka) between the sinusoidal motion of two neighboring particles

$$q_{k,n} = \xi_k e^{i(\omega t + kna)} \tag{4.2-2}$$

With this substitution, (4.2-1) becomes

$$-\omega^2 M = \beta(e^{ika} + e^{-ika} - 2) = \beta(e^{ika/2} - e^{-ika/2})^2$$

Since the phase velocity of the wave (4.2-2) is equal to ω/k, we can get both positively and negatively traveling waves by keeping ω positive and letting k assume both positive and negative values. With this convention, the dispersion relation becomes

$$\omega = \left(\frac{4\beta}{M}\right)^{1/2} \left|\sin\frac{ka}{2}\right| = \left(\frac{\beta}{M}\right)^{1/2} \sqrt{2[1 - \cos(ka)]} \tag{4.2-3}$$

A plot of Eq. (4.2-3) is shown in Figure 4.2.

Consider next the solution $q_{k,n}$ for a single wave when k is increased by $l(2\pi/a)$ where l is an integer. From Eq. (4.2-3), we obtain $\omega(k + l2\pi/a) = \omega(k)$. Using this result in (4.2-2) gives

$$q_{(k+l2\pi/a),n} = \xi_k e^{i(\omega t + kna)} e^{i2\pi nl} = \xi_k e^{i(\omega t + kna)} = q_{k,n}$$

so that all the possible waves can be generated by allowing k to roam over a $2\pi/a$ interval. The conventionally chosen interval is $-(\pi/a) \leqslant k \leqslant \pi/a$ and is shown in Figure 4.2. It is called the first Brillouin zone. In a cubic crystal with

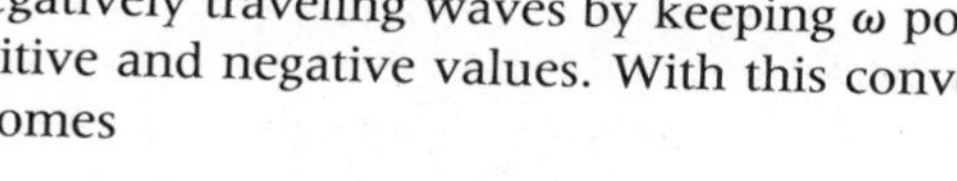

FIGURE 4.1 A line of similar atoms.

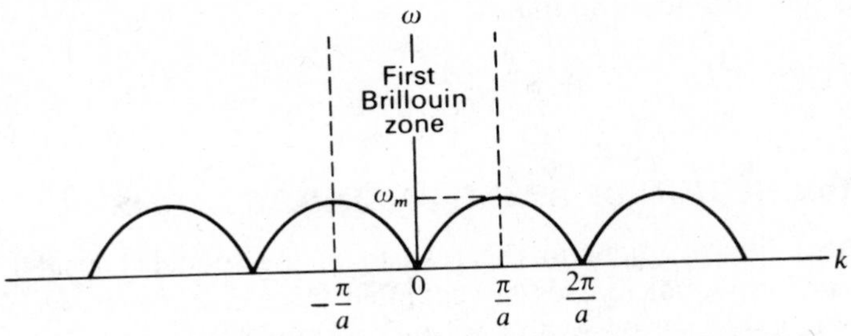

FIGURE 4.2 The dispersion diagram for the discrete line of atoms.

$a = b = c$, for example, the first Brillouin zone will correspond to the region $-(\pi/a) \leqslant k_x, k_y, k_z \leqslant \pi/a$ in **k** space.

The phase velocity, for $ka \ll 1$, is given by

$$v_s = \sqrt{\frac{\beta}{M}}\, a \tag{4.2-4}$$

which is the same as that of the homogeneous line, Section 4.1, if we use the appropriate transformation $\rho \rightarrow M/a$ and $c \rightarrow \beta a$. The condition $ka \ll 1$ means that the acoustic wavelength $\lambda(= 2\pi/k)$ is much larger than the atomic separation a and, consequently, as far as the wave motion is concerned, the line may be considered homogeneous.

The expression for the group velocity is

$$v_g = \frac{d\omega}{dk} = \sqrt{\frac{\beta}{M}}\, a \cos\left(\frac{ka}{2}\right) \tag{4.2-5}$$

so that $v_g = 0$ at the edges of the first Brillouin zone ($ka = \pm\pi$). It is a property of all lossless periodic structures that the frequency cutoff and zero group velocity condition obtain when the phase shift per unit of periodicity (ka in our case) is equal to π radians. This point is discussed extensively by Brillouin.[1] The most profound difference between the homogeneous and discrete lines is that in the latter there exists an upper limit on the wave frequency given, according to (4.2-3), by

$$\omega_m = 2\left(\frac{\beta}{M}\right)^{1/2}$$

This cutoff frequency is equal, according to (4.2-4), to $2v_s/a$ that, if we use the typical values $v_s = 3 \times 10^5$ cm/s (in solids), $a = 3 \times 10^{-8}$ cm, gives

$$f_m = \frac{\omega_m}{2\pi} \approx 3 \times 10^{12} \text{ cps}$$

This corresponds to an electromagnetic (free space) wavelength of 0.1 mm and shows that in typical acoustic experiments, which employ frequencies

[1] L. Brillouin, *Wave Propagation in Period Structures* (New York: Dover, 1953).

through the microwave region, one is still well within the dispersionless, $ka \ll 1$, region.

Mode Enumeration

The individual solution $q_{k,n} = \xi_{k,n} e^{i(\omega t + kna)}$ of (4.2-2) is called an acoustic mode of propagation. The number of independent modes of the discrete line must be equal to the number of particles (atoms) in the line. This implies that only certain values of k are allowed. These can be found by consideration of the boundary conditions. One such condition is that, if the line consists of N atoms, the first and last atom are "clamped" and cannot move. An alternative condition is that the total motion of any atom is reproduced after every N atoms. These two conditions give the same results, so that, for the sake of definiteness, we will use the periodic boundary condition.

The total motion of the nth atom can be described, according to (4.2-2), as

$$q_n(t) = \sum_k q_{k,n} = \sum_k \xi_k e^{i(\omega t + kna)} \qquad (4.2\text{-}6)$$

The periodic boundary condition $q_{n+N} = q_n$ is satisfied if k is restricted to the values

$$k = \frac{2m\pi}{Na} \qquad (4.2\text{-}7)$$

where m is an integer. This can be verified by substitution in (4.2-6). The mode spacing in k space is thus *uniform* and is given by $\Delta k = 2\pi/Na$. Since k is restricted to the first Brillouin zone $-(\pi/a) \leqslant k \leqslant \pi/a$, the total number of modes is $2\pi/(a\,\Delta k) = N$ as required.

4.3 A LINE WITH TWO DIFFERENT ATOMS

Most crystals are made up of a lattice containing a number of different atoms. An ionic crystal like NaCl, for example, can be viewed as a cubic array in which neighboring lattice points are occupied alternately by Na^+ and Cl^- ions. As an approximation to such crystals, we consider a model made of a line in which each atom of mass M has as its neighbors atoms of mass m. Let the M atoms occupy odd lattice points and those of mass m even ones. Using the definitions of Section 4.2, we can write the equations of motion for two neighboring atoms as

$$\begin{aligned} m\ddot{q}_{2n} &= \beta(q_{2n+1} + q_{2n-1} - 2q_{2n}) \\ M\ddot{q}_{2n+1} &= \beta(q_{2n+2} + q_{2n} - 2q_{2n+1}) \end{aligned} \qquad (4.3\text{-}1)$$

The basic solutions are taken as waves of the form

$$\begin{aligned} q_{2n,k} &= \xi_k e^{i(\omega t + 2nka)} \\ q_{2n+1,k} &= \eta_k e^{i[\omega t + (2n+1)ka]} \end{aligned} \qquad (4.3\text{-}2)$$

which substituted into (4.3-1) leads to

$$\begin{aligned} -\omega^2 m\xi_k &= \beta\eta_k(e^{ika} + e^{-ika}) - 2\beta\xi_k \\ -\omega^2 M\eta_k &= \beta\xi_k(e^{ika} + e^{-ika}) - 2\beta\eta_k \end{aligned} \tag{4.3-3}$$

The determinantal equation guaranteeing nontrivial solutions for ξ and η is

$$(2\beta - m\omega^2)(2\beta - M\omega^2) - 4\beta^2\cos^2(ka) = 0 \tag{4.3-4}$$

so that

$$\omega^2 = \beta\left(\frac{1}{m} + \frac{1}{M}\right) \pm \beta\sqrt{\left(\frac{1}{m} + \frac{1}{M}\right)^2 - \frac{4\sin^2(ka)}{Mm}} \tag{4.3-5}$$

An inspection of (4.3-5) reveals that for a given k there are two frequencies. For $ka \ll 1$ we have, to first order in (ka),

$$\omega^2 = \beta\left(\frac{1}{m} + \frac{1}{M}\right) \pm \beta\left(\frac{1}{m} + \frac{1}{M}\right)\left[1 - \frac{2(ka)^2}{Mm(1/m + 1/M)^2}\right]$$

from which we obtain

$$\left.\begin{aligned} \omega_1 &= \sqrt{2\beta\left(\frac{1}{m} + \frac{1}{M}\right)} \\ \omega_2 &= \sqrt{\frac{2\beta}{M + m}}\,ka \end{aligned}\right\} \quad ka \ll 1 \tag{4.3-6}$$

When $ka = \pi/2$, the two frequencies become

$$\begin{aligned} \omega_1 &= \sqrt{\frac{2\beta}{m}} \\ \omega_2 &= \sqrt{\frac{2\beta}{M}} \end{aligned} \tag{4.3-7}$$

The upper branch of the dispersion curve is called the optical branch. Its highest frequency is equal to $[2\beta(1/m + 1/M)]^{1/2}$ and occurs at $ka = 0$, whereas its lowest value is $(2\beta/m)^{1/2}$ when $m < M$, or $(2\beta/M)^{1/2}$ when $M < m$ and occurs at $ka = \pi/2$. The low-frequency branch corresponds to the propagation in a line of similar atoms studied in Section 4.2. It is referred to as the acoustic branch. At low frequencies ($ka \ll 1$), it is dispersionless with a sound velocity, according to (4.3-6), of $[2\beta/(M + m)]^{1/2}a$. The maximum frequency of the acoustic branch occurs at $ka = \pi/2$ and is the lower of $(2\beta/m)^{1/2}$ and $(2\beta/M)^{1/2}$.

The acoustic and optical branches are plotted in Figure 4.3 for the case of $m < M$.

The nature of these modes can be appreciated by considering a number of cases. According to (4.3-2), the relative motion of two neighboring atoms is given by

$$\frac{q_{2n+1}}{q_{2n}} = e^{ika}\frac{\eta_k}{\xi_k}$$

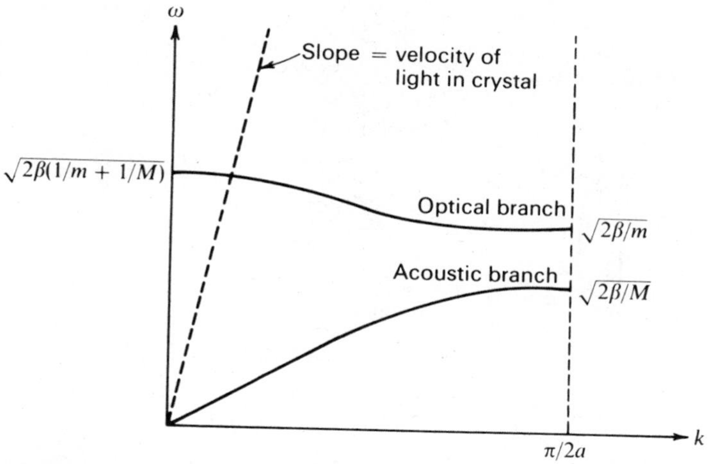

FIGURE 4.3 The dispersion diagram for a one-dimensional crystal containing two different atoms with masses m and $M(m < M)$.

where from (4.3-3)

$$\frac{\eta_k}{\xi_k} = \frac{2\beta - \omega^2 m}{2\beta\cos(ka)}$$

Using these results, we have

$$\left(\frac{q_{2n+1}}{q_{2n}}\right)_{ka \ll 1} = \begin{cases} e^{ika} & \text{for the acoustic branch} \\ -\dfrac{m}{M}e^{ika} & \text{for the optical branch} \end{cases}$$

where the relation $\omega^2 = 2\beta(1/m + 1/M)$ was used for the optical branch. The acoustic branch near $ka \ll 1$ corresponds to equal and, nearly, in-phase excursion of neighboring atoms. In the optical branch, the motions are out of phase with amplitudes that are inversely proportional to the atomic masses. A lattice mode can be excited by an electromagnetic wave of the same frequency and wavelength, that is, of the same phase velocity. Such a point of excitation is shown as the intersection of the ω versus k curve of an electromagnetic wave with that of the sound wave. Since the phase velocity (the slope of the ω versus k curve) of light is approximately 10^5 times that of the velocity of the acoustic mode, the only intersection possible is, as shown in Figure 4.3, with the optical branch at $ka \approx 0$.

In three-dimensional crystals with N atoms, there are $3N$ degrees of freedom. In a crystal with two atoms per unit cell, there are, in general, three acoustic and three optical branches. The three optical, or acoustic, branches involve atomic displacement both along and at right angles to the direction of wave propagation, and in simple cases correspond, respectively, to one longitudinal and two transverse waves. A schematic ω versus k diagram for such a crystal is shown in Figure 4.4.

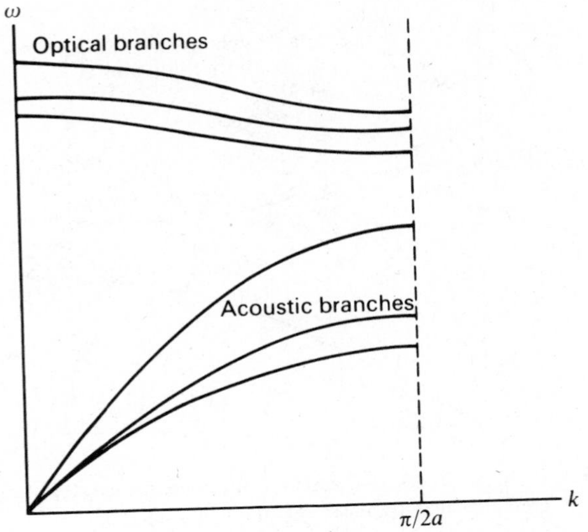

FIGURE 4.4 A schematic plot of ω versus k for a crystal with two atoms per unit cell.

The optical absorption caused by the excitation of the optical modes is studied most often by investigating the radiation reflected from a crystal as a function of wavelength. Figure 4.5 shows the location of three prominent optical modes (near $k = 0$) in quartz.

The acoustic branch does not interact with an electromagnetic radiation since its phase velocity is smaller by $\sim 10^5$ than that of electromagnetic waves. Experiments involving the acoustic mode are done by means of excitation of the surface atoms[2] or of a thin layer[3] (with thickness smaller than the wavelength of sound) where the need for phase velocity synchronism does not arise.

4.4 LATTICE SUMS

In Section 4.2 it was found that the excursion of an arbitrary lattice point n at time t can be expanded [see Eq. (4.2-6)] as

$$q_n(t) = \sum_k \xi_k(t) e^{+ikna}$$

with $k = 2m\pi/Na$ where m takes on the values $m = -(N/2) + 1, -(N/2) + 2, \ldots, (N/2)$. This is a special form of the Fourier series expansion of the function q_n with a periodicity $n = N$. Since the "coordinate" variable n (or rather na) is limited to a set of N points, the summation *contains only* N *terms.*

[2] H. E. Bommel and K. Dransfeld, *Phys. Rev. Letters* **1**, 234–236 (Oct. 1958).
[3] *Proc. IEEE* (Oct. 1965), special issue on ultrasonics; especially articles by M. H. Seavey and by N. F. Foster regarding surface excitation of acoustic phonons.

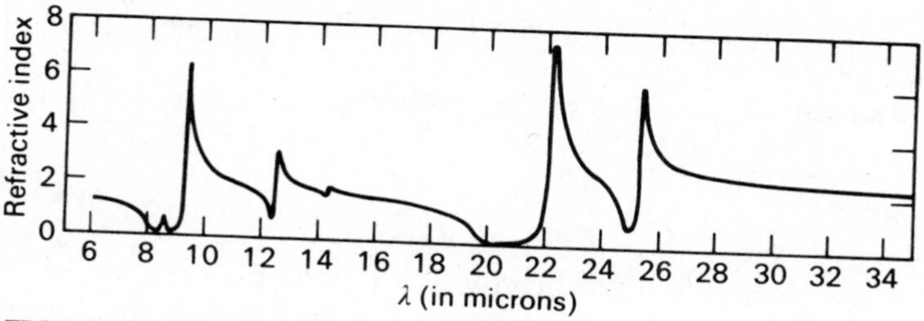

FIGURE 4.5 The refractive index of quartz for the ordinary ray ($E \perp$ to the optic axis) as obtained from the dispersion analysis of the reflectivity. *Source:* W. G. Spitzer and D. A. Kleinman, *Phys. Rev.* **121** 1324 (1961).

Rewriting the summation for q_n in terms of a normalized variable Q_k, we have

$$q_n(t) = \frac{1}{N^{1/2}} \sum_k Q_k(t) e^{+ikna} \tag{4.4-1}$$

The series e^{ikna} is a complete orthonormal series for expanding any arbitrary function of the N lattice points. We may expect a closure property similar to (1.2-11) to exist. The corresponding relation is

$$\sum_k e^{ik(s-n)a} = N\,\delta_{s,n} \tag{4.4-2}$$

Proof:
For $s = n$, the result follows immediately. For $s \neq n$, we define $s - n = l$

$$\sum_k e^{ikal} = \sum_{m=-N/2+1}^{N/2} e^{(i2\pi ma/Na)l}$$

$$= \sum_{m=-N/2+1}^{-1} e^{i2\pi ml/N} + \sum_{m=0}^{N/2} e^{i2\pi ml/N}$$

In the first series change variables to $m' = m + N$

$$\sum_{-N/2+1}^{-1} e^{i2\pi ml/N} = \sum_{m'=N/2+1}^{N-1} e^{i2\pi l(m'-N)/N} = \sum_{m'=N/2+1}^{N-1} e^{i2\pi lm'/N}$$

adding the two series and replacing m' by m give

$$\sum_k e^{ikal} = \sum_{m=0}^{N-1} e^{i2\pi ml/N} = \frac{e^{i2\pi l} - 1}{e^{i2\pi l/N} - 1} = 0$$

This completes the proof.

The inverse Fourier transformation is

$$Q_k = \frac{1}{N^{1/2}} \sum_{s=1}^{N} q_s e^{-iksa} \tag{4.4-3}$$

It can be verified by substituting (4.4-3) for Q_k in (4.4-1) and using the closure property (4.4-2).

Another lattice sum that will prove useful in the next section is

$$\sum_{r=1}^{N} e^{i(k-k')ra} = N\,\delta_{k,k'} \tag{4.4-4}$$

For $k = k'$, the result follows directly. For $k \neq k'$, we have $k - k' = 2\pi/Na\,(m - m') = (2\pi/Na)l$ where $l = m' - m$ is an integer. The summation is identical to that of the last step in the proof of Eq. (4.4-2), thus yielding the desired result.

4.5 QUANTIZATION OF THE ACOUSTIC BRANCH OF LATTICE VIBRATIONS[4,5,6]

The purpose of this section is to show that the *ensemble* of N lattice vibrations in a line of (similar) N atoms is formally equivalent to N independent harmonic oscillators.

The Hamiltonian of the (one-dimensional) crystal is the sum of the kinetic and potential energies of the individual atoms

$$H = \frac{1}{2}\sum_{r=1}^{N}\left[\frac{p_r^2}{m} + \beta(q_{r+1} - q_r)^2\right] \tag{4.5-1}$$

where β is the effective "spring constant." Consider first the kinetic energy term. Expanding p_r in a lattice Fourier expansion as in (4.4-1)

$$p_r = \frac{1}{N^{1/2}}\sum_k P_k e^{-ikra} \qquad \left(P_k = \frac{1}{N^{1/2}}\sum_r p_r e^{+ikra}\right) \tag{4.5-2}$$

gives

$$\sum_{r=1}^{N}\frac{p_r^2}{2m} = \frac{1}{2mN}\sum_{k,k',r} P_k P_{k'} e^{-i(k+k')ra}$$

$$= \frac{1}{2mN}\sum_{k,k'} P_k P_{k'} \sum_r e^{-i(k+k')ra}$$

$$= \frac{1}{2mN}\sum_{k,k'} P_k P_{k'} N\,\delta_{k,-k'}$$

$$= \frac{1}{2m}\sum_k P_k P_{-k}$$

[4] C. Kittel, *Quantum Theory of Solids* (New York: Wiley, 1963).
[5] G. Weinreich, *Solids* (New York: Wiley, 1965).
[6] M. Born and K. Huang, *Dynamical Theory of Crystal Lattices* (Oxford, England: Clarendon Press, 1954).

where, it should be recalled, P_k is an operator, being the transform of the operator p_r. Since p_r and q_r are Hermitian operators, it follows from (4.4-1) that

$$Q_{-k} = Q_k^\dagger \qquad \text{and} \qquad P_{-k} = P_k^\dagger \tag{4.5-3}$$

This makes the sum of the k and $-k$ terms in (4.4-1) an Hermitian operator.

The potential energy is

$$\frac{\beta}{2}\sum_r (q_{r+1} - q_r)^2 = \frac{\beta}{2}\sum_r (q_{r+1}^2 + q_r^2 - 2q_{r+1}q_r)$$

that, when using Eq. (4.4-1), becomes

$$\frac{\beta}{2N}\sum_{r,k,k'} Q_k Q_{k'}[e^{ik(r+1)a}e^{ik'(r+1)a} - e^{ik(r+1)a}e^{ik'ra} - e^{ikra}e^{ik'(r+1)a} + e^{ikra}e^{ik'ra}]$$

$$= \frac{\beta}{2N}\sum_{k,k'}\left\{Q_k Q_{k'}[e^{i(k+k')a} - e^{ika} - e^{ik'a} + 1]\sum_r e^{i(k+k')ra}\right\}$$

If we replace the last summation over r by $N\,\delta_{k,-k'}$ in accordance with (4.4-4), the last expression becomes

$$\beta\sum_k Q_k Q_{-k}[1 - \cos(ka)] \tag{4.5-4}$$

The total Hamiltonian is thus

$$H = \sum_k\left\{\frac{1}{2m}P_k P_{-k} + \beta Q_k Q_{-k}[1 - \cos(ka)]\right\} \tag{4.5-5}$$

Equations (4.4-1) and (4.5-2) can be used to evaluate the commutator of P_k and Q_k

$$\begin{aligned}
[P_{k'}, Q_k] &= \frac{1}{N}\left[\sum_s p_s e^{ik'sa}, \sum_r q_r e^{-ikra}\right] \\
&= \frac{1}{N}\sum_r\sum_s [p_s, q_r]e^{i(k's-kr)a} \\
&= \frac{1}{N}\sum_r\sum_s (-i\hbar\,\delta_{r,s})e^{i(k's-kr)a} \\
&= \frac{(-i\hbar)}{N}\sum_s e^{i(k'-k)sa} = (-i\hbar)\,\delta_{k,k'}
\end{aligned} \tag{4.5-6}$$

where use has been made of (4.4-4) and the commutator relation $[p_s, q_r] = -i\hbar\,\delta_{r,s}$ that applies to the momentum p and displacement q of a particle.

Except for the mixing of the k and $-k$ terms, the Hamiltonian (4.5-5) is in the form of a sum of single harmonic oscillator Hamiltonians. To eliminate the mixing, we introduce the creation and annihilation operators that are defined in correspondence with Eq. (2.2-25) as

$$
\begin{aligned}
a_k^\dagger &= \frac{\alpha\sqrt{g_k}}{\sqrt{2}}\, Q_{-k} - \frac{i}{\sqrt{2g_k}\,\hbar\alpha}\, P_k \\
a_k &= \frac{\alpha\sqrt{g_k}}{\sqrt{2}}\, Q_k + \frac{i}{\sqrt{2g_k}\,\hbar\alpha}\, P_{-k}
\end{aligned}
\tag{4.5-7}
$$

where α is defined, as in Chapter 2, by

$$\alpha^4 = \frac{m\beta}{\hbar^2}$$

and $g_k = \sqrt{2[1 - \cos(ka)]}$.

Equations (4.5-7) are consistent with (4.5-3). Solving for P_k and Q_k gives

$$
\begin{aligned}
P_{-k}^\dagger = P_k &= \frac{i\hbar\alpha\sqrt{g_k}}{\sqrt{2}}\,(a_k^\dagger - a_{-k}) = i\sqrt{\frac{\hbar m\omega_k}{2}}\,(a_k^\dagger - a_k) \\
P_k^\dagger = P_{-k} &= \frac{i\hbar\alpha\sqrt{g_k}}{\sqrt{2}}\,(a_{-k}^\dagger - a_k)
\end{aligned}
\tag{4.5-8}
$$

$$
\begin{aligned}
Q_k^\dagger = Q_{-k} &= \frac{1}{\sqrt{2g_k}\,\alpha}\,(a_k^\dagger + a_{-k}) = \sqrt{\frac{\hbar}{2m\omega_k}}\,(a_k^\dagger + a_k) \\
Q_{-k}^\dagger = Q_k &= \frac{1}{\sqrt{2g_k}\,\alpha}\,(a_{-k}^\dagger + a_k)
\end{aligned}
\tag{4.5-9}
$$

$$\omega_k \equiv \left(\frac{\beta}{m}\right)^{1/2} g_k$$

The commutator of a_k and $a_k^\dagger$ is

$$
\begin{aligned}
[a_k, a_{k'}^\dagger] &= \left[\left(\frac{\alpha\sqrt{g_k}}{\sqrt{2}}\, Q_k + \frac{i}{\sqrt{2g_k}\,\hbar\alpha}\, P_{-k}\right), \left(\frac{\alpha\sqrt{g_{k'}}}{\sqrt{2}}\, Q_{-k'} - \frac{i}{\sqrt{2g_{k'}}\,\hbar\alpha}\, P_{k'}\right)\right] \\
&= -\frac{i}{2\hbar}\,[Q_k, P_{k'}] + \frac{i}{2\hbar}\,[P_{-k}, Q_{-k'}] = \delta_{k,k'}
\end{aligned}
\tag{4.5-10}
$$

where use has been made of Eq. (4.5-6).

Substituting for P_k and Q_k in a single term of H in (4.5-5) results in

$$
\begin{aligned}
H_k &= \frac{1}{2m}\, P_k P_{-k} + \beta Q_k Q_{-k}[1 - \cos(ka)] \\
&= -\frac{\hbar^2\alpha^2 g_k}{4m}\,(a_k^\dagger - a_{-k})(a_{-k}^\dagger - a_k) \\
&\quad + \frac{\beta}{2\alpha^2 g_k}\,(a_k^\dagger + a_{-k})(a_{-k}^\dagger + a_k)[1 - \cos(ka)]
\end{aligned}
$$

Multiplying out the last equation and using the relations $\alpha^2 = \sqrt{m\beta}/\hbar$, $g_k = \sqrt{2[1 - \cos(ka)]}$, $[a_k, a_{k'}^\dagger] = \delta_{k,k'}$, and Eq. (4.2-3)

$$\omega_k = \left(\frac{\beta}{m}\right)^{1/2} \sqrt{2[1 - \cos(ka)]} = \left(\frac{\beta}{m}\right)^{1/2} g_k$$

give, after some algebra,

$$H_k = \frac{\hbar}{2}\,\omega_k(a_k^\dagger a_k + a_{-k}^\dagger a_{-k} + 1) \tag{4.5-11}$$

The total Hamiltonian is

$$\begin{aligned} H = \sum_k H_k &= \sum_k \left[\frac{\hbar\omega_k}{2}\,(a_k^\dagger a_k + \tfrac{1}{2}) + \frac{\hbar\omega_k}{2}\,(a_{-k}^\dagger a_{-k} + \tfrac{1}{2})\right] \\ &= \sum_k \hbar\omega_k(a_k^\dagger a_k + \tfrac{1}{2}) \end{aligned} \tag{4.5-12}$$

since the summation includes, symmetrically, both positive and negative values of k.

Comparing (4.5-12) to the Hamiltonian of the harmonic oscillator, Eq. (2.2-30), and using (4.5-10), we establish that the one-dimensional crystal is equivalent, quantum mechanically, to N independent harmonic oscillators, one harmonic oscillator for each mode k. Each harmonic oscillator is characterized by eigenfunctions u_{n_k} that are functions of Q_k and that obey, in direct analogy with (2.2-27),

$$\begin{aligned} a_k^\dagger u_{n_k} &= \sqrt{n_k + 1}\; u_{n_k+1} \\ a_k u_{n_k} &= \sqrt{n_k}\; u_{n_k-1} \\ a_k^\dagger a_k u_{n_k} &= n_k u_{n_k} \end{aligned} \tag{4.5-13}$$

where n_k, the number of phonons in the k mode, can range from 0 to ∞. The wavefunction of the complete crystal is the product of the single-mode wavefunctions

$$\Phi = u_{n_1}u_{n_2}\cdots u_{n_N} = \prod_{k=1}^{N} u_{n_k} \tag{4.5-14}$$

and, according to (4.5-12) and (4.5-13), corresponds to a total crystal energy (eigenvalue)

$$E = \sum_k \hbar\omega_k(n_k + \tfrac{1}{2})$$

We can obtain the explicit time dependence of the annihilation and creation operators in the Heisenberg representation from

$$\frac{da_j}{dt} = \frac{i}{\hbar}\,[H, a_j] = \frac{i}{\hbar}\left[\sum_k \hbar\omega_k(a_k^\dagger a_k + \tfrac{1}{2}),\, a_j\right]$$

where we used (4.5-12). The only nonvanishing term is $k = j$ so that

$$\frac{da_j}{dt} = i\omega_j[a_j^\dagger a_j,\, a_j] = -i\omega_j a_j$$

and

$$\begin{aligned} a_j(t) &= a_j(0)e^{-i\omega_j t} \\ a_j^\dagger(t) &= a_j^\dagger(0)e^{i\omega_j t} \end{aligned} \tag{4.5-15}$$

The motion of the nth atom can be expressed in terms of the annihilation and creation of the individual modes as

$$\begin{aligned} q_n(t) &= \frac{1}{N^{1/2}} \sum_k Q_k(t)e^{ikna} \\ &= \frac{1}{2N^{1/2}} \sum_k (Q_k e^{ikna} + Q_{-k}e^{-ikna}) \\ &= \frac{1}{2N^{1/2}} \sum_k (Q_k e^{ikna} + Q_k^\dagger e^{-ikna}) \\ &= \frac{1}{N^{1/2}} \sum_k \frac{1}{\sqrt{2g_k}\,\alpha} (a_k^\dagger e^{-ikna} + a_k e^{ikna}) \end{aligned} \tag{4.5-16}$$

where use has been made of (4.5-3), (4.5-8), and (4.5-9). The summation includes both positive and negative values of k. If we substitute for $a_k^\dagger$ and a_k from (4.5-15), we get

$$q_n(t) = \frac{1}{N^{1/2}} \sum_k \frac{1}{\sqrt{2}\, g_k \alpha} [a_k^\dagger(0)e^{i(\omega_k t - kna)} + a_k(0)e^{-i(\omega_k t - kna)}] \tag{4.5-17}$$

In this form, the traveling-wave nature of the individual modes and their explicit time dependence are brought out.

4.6 AVERAGE THERMAL EXCITATION OF LATTICE MODES

According to Section 4.5, a lattice vibrational mode at a (radian) frequency ω has an energy

$$E_n = (n + \tfrac{1}{2})\hbar\omega$$

If the lattice is in thermal equilibrium at a temperature T, the probability of finding a given mode excited to the state n is given by the Boltzmann factor

$$p(n) = \frac{e^{-E_n/kT}}{\sum_{s=0}^{\infty} e^{-E_s/kT}} \tag{4.6-1}$$

where k is the Boltzmann constant. The average excitation energy of the mode is given, consequently, by

$$\overline{E} = \sum_{n=0}^{\infty} E_n p(n) = \frac{\sum_{n=0}^{\infty} (n + \frac{1}{2})\hbar\omega e^{-(n+1/2)\hbar\omega/kT}}{\sum_{s=0}^{\infty} e^{-(s+1/2)\hbar\omega/kT}} = \frac{\hbar\omega}{2} + \frac{\sum_{n=0}^{\infty} n\hbar\omega e^{-n\hbar\omega\beta}}{\sum_{s=0}^{\infty} e^{-s\hbar\omega\beta}} \tag{4.6-2}$$

where $\beta = (kT)^{-1}$.

The denominator of (4.6-2) forms a geometric progression whose sum is

$$\sum_{s=0}^{\infty} e^{-s\hbar\omega\beta} = \frac{1}{1 - e^{-\hbar\omega\beta}}$$

Taking the derivative of the last expression with respect to β gives

$$\sum_{s=0}^{\infty} s\hbar\omega e^{-s\hbar\omega\beta} = \frac{\hbar\omega e^{-\hbar\omega\beta}}{(1 - e^{-\hbar\omega\beta})^2}$$

Substitution in (4.6-2) leads to

$$\overline{E} = \frac{\hbar\omega}{2} + \frac{\hbar\omega}{e^{\hbar\omega/kT} - 1} \tag{4.6-3}$$

The average excitation of a mode is often characterized by the average number of quanta $\overline{n}$

$$\overline{n} = \frac{\overline{E}}{\hbar\omega} = \frac{1}{2} + \frac{1}{e^{\hbar\omega/kT} - 1} \tag{4.6-4}$$

When $kT \gg \hbar\omega$, the average energy per mode is kT, whereas for $kT \ll \hbar\omega$, it is $\hbar\omega/2$. The latter is, of course, just the zero-point energy of the harmonic oscillator.

In Section 4.2 it was shown how the periodic boundary condition $q_{n+N} = q_n$ results in a discrete spectrum for the allowed k values with the difference between adjacent k values being $\Delta k = 2\pi/L$. Extending the periodicity requirement to three dimensions with the basic periodic cell taken as having edges A, B, and C gives

$$\Delta k_x = \frac{2\pi}{A}, \qquad \Delta k_y = \frac{2\pi}{B}, \qquad \Delta k_z = \frac{2\pi}{C} \tag{4.6-5}$$

for the minimum spacings of the three components of the propagation vector **k**. Accordingly, we can associate with each **k** vector a volume

$$(\Delta k)^3 = \frac{(2\pi)^3}{ABC} = \frac{(2\pi)^3}{V} \tag{4.6-6}$$

in **k** space, where V is the volume of the periodic cell in (real) space. The number $N(k)$ of modes between 0 and k is given by the volume of a sphere of radius k divided by the volume per mode, that is,

$$N(k) = \frac{4\pi k^3 V}{3(2\pi)^3}$$

For the case of no acoustic dispersion, we take $k = 2\pi\nu/v_s$ where v_s is the sound velocity. This gives

$$N(\nu) = \frac{4\pi\nu^3 V}{v_s^3} \tag{4.6-7}$$

for the number of modes with frequencies smaller than ν. We have multiplied the result by 3 to account for the three independent acoustic polarizations.

The number of modes per unit volume per unit frequency $p(\nu)$ is

$$p(\nu) = \frac{1}{V}\frac{dN(\nu)}{d\nu} = \frac{12\pi\nu^2}{v_s^3} \tag{4.6-8}$$

Whereas the average vibrational energy at temperature T per unit volume per unit frequency is

$$\rho(\nu) = p(\nu)\overline{E} = \frac{12\pi\nu^2}{v_s^3}\left(\frac{h\nu}{2} + \frac{h\nu}{e^{h\nu/kT} - 1)}\right) \tag{4.6-9}$$

where we used (4.6-3).

In situations where one is concerned with excitation energy, that is, that energy that can be extracted from the oscillators, one neglects the term $h\nu/2$ in (4.6-9) since this zero-field energy cannot be removed from the oscillator.

Problems

4.1 Describe the relative motion of adjacent m and M ions at $ka = \pi/2$ for both the optical and acoustic modes.

4.2 Show that when $m = M$ the dispersion (ω versus k) diagram for the diatomic line is identical to that of the monatomic line.

Hint:
Investigate $d\omega_1/dk_1$ and $d\omega_2/dk_2$ at $k = \pi/2a$ and extend the diagram to $k = \pi/a$.

4.3 A statement was made in Section 4.2 that increasing k by 2A5pl/a, where l is an integer, does not yield new modes (see Figure 4.2). How can one reconcile this statement with the fact that the phase velocity v_p is equal to ω/k and is, consequently, different for modes with different k values (but the same ω)?

Hint:
Consider the physical significance of the "different velocities" in view of the discrete nature of the structure. Can you conceive of an experiment that distinguishes among such different velocities.

4.4 Show, using arguments similar to those used in Section 4.2, that the total number of modes in a "three-dimensional" crystal containing N atoms, two per unit cell, is $3N$.

Hint:
Include both the acoustic and optical branches.

4.5 Assuming that in a line of N similar atoms the initial conditions at $t = 0$ are $q_1(0) = q_2(0) = \Delta$, $\dot{q}_1(0) = \dot{q}_2(0) = 0$,
(a) derive the expression for the deviation $q_n(t)$ of the nth atom.
(b) Assuming $\omega = kv_s$, show that the disturbance propagates as a pulse with a velocity v_s.

C H A P T E R 5

Electromagnetic Fields and Their Quantization

5.0 INTRODUCTION

This chapter provides the main background for the electromagnetic theory used in this book. The following topics are discussed. (1) Propagation of plane waves in homogeneous anisotropic media. This material is used in the treatment of the electrooptic, magnetooptic, and photoelastic manipulation of light. (2) Energy storage and power dissipation. (3) Normal mode expansion in a generalized resonator. This material will be used to derive the laser oscillation condition in Chapter 9 as well as in the treatment of laser noise in Chapter 13. (4) The quantization of the electromagnetic field. This formalism is the background for describing spontaneous emission processes.

5.1 POWER TRANSPORT, STORAGE, AND DISSIPATION IN ELECTROMAGNETIC FIELDS

In this section we derive the formal expressions for the power transport, power dissipation, and energy storage that accompany the propagation of electromagnetic radiation in material media. The starting point is Maxwell's equations (in MKS units)

$$\nabla \times \mathbf{H} = \mathbf{i} + \frac{\partial \mathbf{D}}{\partial t} \tag{5.1-1}$$

$$\nabla \times \mathbf{E} = -\frac{\partial \mathbf{B}}{\partial t} \tag{5.1-2}$$

and the constitutive equations relating the polarization of the medium to the displacement vectors

$$\mathbf{D} = \varepsilon_0 \mathbf{E} + \mathbf{P} \tag{5.1-3}$$

$$\mathbf{B} = \mu_0(\mathbf{H} + \mathbf{M}) \tag{5.1-4}$$

where $\mathbf{i}$ is the current density (amperes per square meter); $\mathbf{E}(\mathbf{r}, t)$ and $\mathbf{H}(\mathbf{r}, t)$ are the electric and magnetic field vectors, respectively; $\mathbf{D}(\mathbf{r}, t)$ and $\mathbf{B}(\mathbf{r}, t)$ are the electric and magnetic displacement vectors; $\mathbf{P}(\mathbf{r}, t)$ and $\mathbf{M}(\mathbf{r}, t)$ are the electric and magnetic polarizations (dipole moment per unit volume) of the medium; and ε_0 and μ_0 are the electric and magnetic permeabilities of vacuum, respectively. For a detailed discussion of Maxwell's equations, the reader is referred to any standard text on electromagnetic theory such as, for example, Reference 1 at the end of the chapter.

Using (5.1-3) and (5.1-4) in (5.1-1) and (5.1-2) leads to

$$\nabla \times \mathbf{H} = \mathbf{i} + \frac{\partial}{\partial t}(\varepsilon_0 \mathbf{E} + \mathbf{P}) \tag{5.1-5}$$

$$\nabla \times \mathbf{E} = -\frac{\partial}{\partial t}\mu_0(\mathbf{H} + \mathbf{M}) \tag{5.1-6}$$

Taking the scalar (dot) product of (5.1-5) and $\mathbf{E}$ gives

$$\mathbf{E} \cdot \nabla \times \mathbf{H} = \mathbf{E} \cdot \mathbf{i} + \frac{\varepsilon_0}{2}\frac{\partial}{\partial t}(\mathbf{E} \cdot \mathbf{E}) + \mathbf{E} \cdot \frac{\partial \mathbf{P}}{\partial t} \tag{5.1-7}$$

where we used the relation

$$\frac{1}{2}\frac{\partial}{\partial t}(\mathbf{E} \cdot \mathbf{E}) = \mathbf{E} \cdot \frac{\partial \mathbf{E}}{\partial t}$$

Next, we take the scalar product of (5.1-6) and $\mathbf{H}$

$$\mathbf{H} \cdot \nabla \times \mathbf{E} = -\frac{\mu_0}{2}\frac{\partial}{\partial t}(\mathbf{H} \cdot \mathbf{H}) - \mu_0 \mathbf{H} \cdot \frac{\partial \mathbf{M}}{\partial t} \tag{5.1-8}$$

Subtracting (5.1-8) from (5.1-7) and using the vector identity

$$\nabla \cdot (\mathbf{A} \times \mathbf{B}) = \mathbf{B} \cdot \nabla \times \mathbf{A} - \mathbf{A} \cdot \nabla \times \mathbf{B} \tag{5.1-9}$$

result in

$$-\nabla \cdot (\mathbf{E} \times \mathbf{H}) = \mathbf{E} \cdot \mathbf{i} + \frac{\partial}{\partial t}\left(\frac{\varepsilon_0}{2}\mathbf{E} \cdot \mathbf{E} + \frac{\mu_0}{2}\mathbf{H} \cdot \mathbf{H}\right) + \mathbf{E} \cdot \frac{\partial \mathbf{P}}{\partial t} + \mu_0 \mathbf{H} \cdot \frac{\partial \mathbf{M}}{\partial t} \tag{5.1-10}$$

We integrate the last equation over an arbitrary volume V and use the Gauss theorem (Reference 1).

$$\int_V (\nabla \cdot \mathbf{A})\, dv = \int_S \mathbf{A} \cdot \mathbf{n}\, da$$

where $\mathbf{A}$ is any vector function, $\mathbf{n}$ is the unit vector normal to the surface enclosing V, and dv and da are the differential volume and surface elements, respectively. The result is

$$\begin{aligned} -\int_V \nabla \cdot (\mathbf{E} \times \mathbf{H})\, dv \\ &= -\int_S (\mathbf{E} \times \mathbf{H}) \cdot \mathbf{n}\, da \\ &= \int_V \left[\mathbf{E} \cdot \mathbf{i} + \frac{\partial}{\partial t}\left(\frac{\varepsilon_0}{2}\mathbf{E} \cdot \mathbf{E}\right) + \frac{\partial}{\partial t}\left(\frac{\mu_0}{2}\mathbf{H} \cdot \mathbf{H}\right) + \mathbf{E} \cdot \frac{\partial \mathbf{P}}{\partial t} + \mu_0 \mathbf{H} \cdot \frac{\partial \mathbf{M}}{\partial t}\right] dv \end{aligned} \tag{5.1-11}$$

According to the conventional interpretation of electromagnetic theory, the left side of (5.1-11), that is,

$$-\int_S (\mathbf{E} \times \mathbf{H}) \cdot \mathbf{n}\, da$$

gives the total power flowing *into* the volume bounded by S. The first term on the right side is the power expended by the field on the moving charges, the sum of the second and third terms corresponds to the rate of increase of the vacuum electromagnetic stored energy $\mathscr{E}_{\text{vac}}$ where

$$\mathscr{E}_{\text{vac}} = \int_V \left(\frac{\varepsilon_0}{2}\mathbf{E}\cdot\mathbf{E} + \frac{\mu_0}{2}\mathbf{H}\cdot\mathbf{H}\right) dv \tag{5.1-12}$$

Of special interest in this book is the next-to-last term

$$\mathbf{E}\cdot\frac{\partial \mathbf{P}}{\partial t}$$

that represents the power per unit volume expended by the field on the electric dipoles. This power goes into an increase in the potential energy stored by the dipoles as well as to supply the dissipation that may accompany the change in **P**. We will return to this subject again in Chapter 8, where we treat the interaction of radiation and atomic systems.

Dipolar Dissipation in Harmonic Fields

According to the discussion in the preceding paragraph, the average power per unit volume expended by the field on the medium electric polarization is

$$\overline{\frac{\text{Power}}{\text{Volume}}} = \overline{\mathbf{E}\cdot\frac{\partial \mathbf{P}}{\partial t}} \tag{5.1-13}$$

where the horizontal bar denotes time averaging. Let us assume for the sake of simplicity that $\mathbf{E}(t)$ and $\mathbf{P}(t)$ are parallel to each other and take their sinusoidally varying magnitudes as

$$E(t) = \text{Re}(Ee^{i\omega t}) \tag{5.1-14}$$

$$P(t) = \text{Re}(Pe^{i\omega t}) \tag{5.1-15}$$

where E and P are the complex amplitudes. The electric susceptibility χ_e is defined by

$$P = \varepsilon_0 \chi_e E \tag{5.1-16}$$

and is thus a complex number. Substituting Eqs. (5.1-14) and (5.1-15) in (5.1-13) and using (5.1-16) give

$$\begin{aligned}\overline{\frac{\text{Power}}{\text{Volume}}} &= \overline{\text{Re}(Ee^{i\omega t})\text{Re}(i\omega Pe^{i\omega t})}\\ &= \tfrac{1}{2}\,\text{Re}(i\omega\varepsilon_0\chi_e EE^*)\\ &= \frac{\omega}{2}\,\varepsilon_0|E|^2\,\text{Re}(i\chi_e)\end{aligned} \tag{5.1-17}$$

Since χ_e is complex, we can write it in terms of its real and imaginary parts as

$$\chi_e = \chi_e' - i\chi_e'' \tag{5.1-18}$$

that, when used in (5.1-17), gives

$$\overline{\frac{\text{Power}}{\text{Volume}}} = \frac{\omega\varepsilon_0\chi_e''}{2}|E|^2 \tag{5.1-19}$$

which is the desired result.

We leave it as an exercise (Problem 5.8) to show that in anisotropic media in which the complex field components are related by

$$P_i = \varepsilon_0 \sum_j \chi_{ij} E_j \tag{5.1-20}$$

the application of (5.1-13) yields

$$\overline{\frac{\text{Power}}{\text{Volume}}} = \frac{\omega}{2}\varepsilon_0 \sum_{i,j} \text{Re}(i\chi_{ij}E_i^* E_j) \tag{5.1-21}$$

The Vector Potential

The electric field $\mathbf{E}(\mathbf{r}, t)$ and magnetic field $\mathbf{H}(\mathbf{r}, t)$ can be derived from a single vector $\mathbf{A}(\mathbf{r}, t)$, the vector potential. The relation $\nabla \cdot \mathbf{B} = 0$ is satisfied by choosing

$$\mathbf{B} = \nabla \times \mathbf{A} \tag{5.1-22}$$

Since $\nabla \cdot \nabla \times$ (any vector) $= 0$, the relation $\nabla \times \mathbf{E} = -\partial\mathbf{B}/\partial t$ is satisfied by letting

$$\mathbf{E} = -\frac{\partial \mathbf{A}}{\partial t} - \nabla \mathbf{V} \tag{5.1-23}$$

since $\nabla \times \nabla \mathbf{V} = 0$ where $\mathbf{V}(\mathbf{r})$ is any scalar function. Our basic defining relations (5.1-22,23) must also satisfy $\nabla \cdot \mathbf{E} = \rho/\varepsilon$ where $\rho(\mathbf{r})$ is the charge density. This leads to

$$-\frac{\partial}{\partial t}(\nabla \cdot \mathbf{A}) - \nabla \cdot \nabla V = \rho/\varepsilon = 0$$

We need to specify $\nabla \cdot \mathbf{A}$, since only $\nabla \times \mathbf{A}$ was specified so far. The choice

$$\nabla \cdot \mathbf{A} = 0 \tag{5.1-24}$$

is called the Coulomb gauge. This ties down the scalar potential $V(\mathbf{r})$ introduced by (5.1-23) that now must satisfy

$$\nabla \cdot \nabla V \equiv \nabla^2 V = 0 \tag{5.1-25}$$

We still need to make sure that $\mathbf{E}$ and $\mathbf{H}$ derived from $\mathbf{A}$ satisfy the second Maxwell equation $\nabla \times \mathbf{H} = \varepsilon\, \partial\mathbf{E}/\partial t$. Using (5.1-22,23), we find that this is the case provided

$$\nabla^2 \mathbf{A} - \mu\varepsilon \frac{\partial^2 \mathbf{A}}{\partial t^2} = 0 \tag{5.1-26}$$

5.2 PROPAGATION OF ELECTROMAGNETIC WAVES IN ANISOTROPIC CRYSTALS

An understanding of wave propagation in anisotropic crystals is a prerequisite to the treatment of a number of important topics. Some of these that are treated in this book are (1) electrooptic, magnetooptic, and acoustooptic modulation and (2) phase matching in nonlinear optical interactions.

In an anisotropic crystal, the polarization induced by an electric field and the field itself are not necessarily parallel. The electric displacement vector **D** and the electric field **E** are consequently related by means of the dielectric tensor ε_{kl} defined by[1]

$$D_k = \varepsilon_{kl} E_l \tag{5.2-1}$$

where the subscripts refer to a Cartesian coordinate ($k, l = x, y, z$), where x, y, z are fixed with respect to the crystal axes and the convention of summation over repeated indices is observed.

Taking the stored electric energy density, as in an isotropic medium, by

$$\omega_e = \tfrac{1}{2}\mathbf{E} \cdot \mathbf{D} = \tfrac{1}{2} E_k \varepsilon_{kl} E_l \tag{5.2-2}$$

we obtain

$$\dot{\omega}_e = \frac{\varepsilon_{kl}}{2} (\dot{E}_k E_l + E_k \dot{E}_l) \tag{5.2-3}$$

According to the derivation of Poynting theorem in Section 5.1, the net power flow into a unit volume is

$$-\nabla \cdot (\mathbf{E} \times \mathbf{H}) = \mathbf{E} \cdot \dot{\mathbf{D}} + \mathbf{H} \cdot \dot{\mathbf{B}}$$

that, if we use (5.2-1) for **D**, can be written as

$$-\nabla \cdot (\mathbf{E} \times \mathbf{H}) = E_k \varepsilon_{kl} \dot{E}_l + \mathbf{H} \cdot \dot{\mathbf{B}} \tag{5.2-4}$$

If the Poynting vector is to correspond to the energy flux in anisotropic media, as it does in the isotropic ones, then the first term on the right side of (5.2-4) must be equal to $\dot{\omega}_e$ and must consequently be the same as $\dot{\omega}_e$ as given by (5.2-3). If we write $E_k \varepsilon_{kl} \dot{E}_l$ as

$$\dot{\omega}_e = \tfrac{1}{2}(\varepsilon_{kl} E_k \dot{E}_l + \varepsilon_{lk} \dot{E}_k E_l)$$

and compare to (5.2-3), it follows that

$$\varepsilon_{kl} = \varepsilon_{lk}$$

and the dielectric tensor ε_{lk} has, in general, only six independent elements.

The electric energy density ω_e can be written, if we use (5.2-1) and (5.2-2), as

$$2\omega_e = \varepsilon_{xx} E_x^2 + \varepsilon_{yy} E_y^2 + \varepsilon_{zz} E_z^2 + 2\varepsilon_{yz} E_y E_z + 2\varepsilon_{xz} E_x E_z + 2\varepsilon_{xy} E_x E_y \tag{5.2-5}$$

[1] Equation (5.2-1) assumes that no dispersion exists so that the instantaneous values of D_k and E_l are related by a single constant. This neglect of dispersion is valid whenever the medium is optically lossless in the spectral region of interest.

A principal axis transformation can be used to diagonalize (5.2-5). In the new coordinate system, ω_e becomes

$$2\omega_e = \varepsilon_x E_x^2 + \varepsilon_y E_y^2 + \varepsilon_z E_z^2 \tag{5.2-6}$$

where the x, y, z symbols now refer to the new axes.

The new coordinate axes are called the *principal dielectric axes*. Since all of our analysis will be carried out in this system, a confusion involving the original coordinates cannot take place and we will retain the x, y, z labeling.

In the principal dielectric coordinate system, the tensor ε_{kl} is diagonal and is given by

$$\begin{vmatrix} D_x \\ D_y \\ D_z \end{vmatrix} = \begin{vmatrix} \varepsilon_x & 0 & 0 \\ 0 & \varepsilon_y & 0 \\ 0 & 0 & \varepsilon_z \end{vmatrix} \begin{vmatrix} E_x \\ E_y \\ E_z \end{vmatrix} \tag{5.2-7}$$

Using (5.2-7) in (5.2-6), we obtain

$$2\omega_e = \frac{D_x^2}{\varepsilon_x} + \frac{D_y^2}{\varepsilon_y} + \frac{D_z^2}{\varepsilon_z} \tag{5.2-8}$$

so that the constant energy (ω_e) surfaces in the space D_x, D_y, D_z are ellipsoids.

Assume next a monochromatic plane wave of radian frequency ω propagating in the crystal with a phase factor

$$\exp\left\{i\omega\left[t - \frac{n}{c}(\mathbf{r}\cdot\mathbf{s})\right]\right\} \tag{5.2-9}$$

This corresponds to a wave vector $\mathbf{k} = (\omega n/c)\mathbf{s}$ where $\mathbf{s}$ is a unit vector normal to the wavefront (plane of constant phase). The phase velocity is then

$$\mathbf{v}_p = \frac{c}{n}\mathbf{s} \tag{5.2-10}$$

so that c is the velocity of light in vacuum and n is the index of refraction. For a monochromatic plane wave, we can formally replace the operator $\boldsymbol{\nabla}$ by $-(i\omega n/c)\mathbf{s}$ so that the Maxwell equations $\boldsymbol{\nabla} \times \mathbf{H} = (\partial\mathbf{D}/\partial t)$ and $\boldsymbol{\nabla} \times \mathbf{E} = -(\partial\mathbf{B}/\partial t)$ become, respectively,

$$\mathbf{D} = -\frac{n}{c}\mathbf{s} \times \mathbf{H} \tag{5.2-11}$$

$$\mathbf{H} = \frac{n}{\mu c}\mathbf{s} \times \mathbf{E} \tag{5.2-12}$$

According to (5.2-11), $\mathbf{D}$ is perpendicular to $\mathbf{H}$ and both $\mathbf{D}$ and $\mathbf{H}$ are perpendicular to $\mathbf{s}$. The direction of energy flow as given by the Poynting vector $\mathbf{E} \times \mathbf{H}$, consequently, is not collinear with the direction of phase propagation $\mathbf{s}$. The vectors $\mathbf{E}$, $\mathbf{D}$, $\mathbf{H}$, $\mathbf{s}$, and $\mathbf{E} \times \mathbf{H}$ are shown in Figure 5.1.

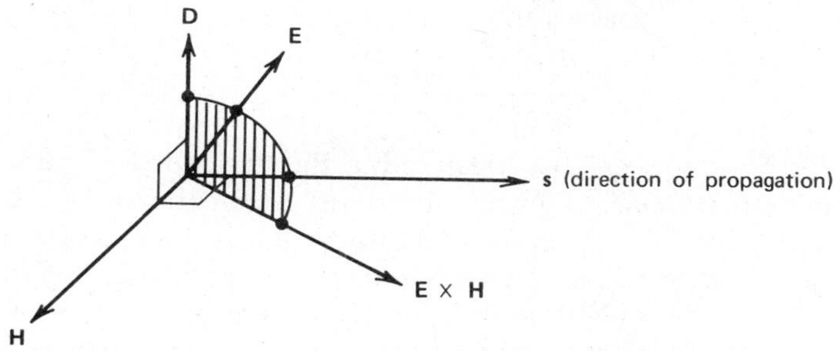

FIGURE 5.1 The relative orientation of **E**, **D**, **H**, **s** and the Poynting vector, **E** × **H**, in an anisotropic crystal. The vectors **D**, **E**, **s**, and **E** × **H** lie in a single plane.

By using (5.2-11), (5.2-12), and the vector identity $\mathbf{A} \times (\mathbf{B} \times \mathbf{C}) = \mathbf{B}(\mathbf{A} \cdot \mathbf{C}) - \mathbf{C}(\mathbf{A} \cdot \mathbf{B})$, we obtain the following expression:

$$\mathbf{D} = -\frac{n^2}{c^2\mu}\mathbf{s} \times \mathbf{s} \times \mathbf{E} = \frac{n^2}{c^2\mu}[\mathbf{E} - \mathbf{s}(\mathbf{s} \cdot \mathbf{E})] = \frac{n^2}{c^2\mu}\mathbf{E}_{\text{transverse}} \tag{5.2-13}$$

and since $\mathbf{s} \cdot \mathbf{D} = 0$ and $n^2/c^2\,\mu = n^2\varepsilon_0$,

$$D^2 = \frac{n^2}{c^2\mu}\mathbf{E} \cdot \mathbf{D} = n^2\varepsilon_0\mathbf{E} \cdot \mathbf{D}$$

By using $D_k = \varepsilon_k E_k$ and $\varepsilon'_k \equiv \varepsilon_k/\varepsilon_0$ in Eq. (5.2-13) and solving for E_k, we obtain

$$E_k = \frac{n^2 s_k(\mathbf{s} \cdot \mathbf{E})}{n^2 - \varepsilon'_k} \qquad k = x, y, z \tag{5.2-14}$$

and after multiplying by s_k and summing over $k = x, y, z$

$$\mathbf{s} \cdot \mathbf{E} = \mathbf{s} \cdot \mathbf{E} \sum_{k=x,y,z} \frac{n^2 s_k^2}{n^2 - \varepsilon'_k}$$

which can also be written as

$$\frac{s_x^2}{n^2 - \varepsilon'_x} + \frac{s_y^2}{n^2 - \varepsilon'_y} + \frac{s_z^2}{n^2 - \varepsilon'_z} = \frac{1}{n^2} \tag{5.2-15}$$

Equation (5.2-15), named after Fresnel, is quadratic in n^2 (see Problem 5.9). The two solutions $\pm n_1$ and $\pm n_2$ (the $\pm$ signs correspond to a mere reversal in the sign of the phase velocity) are the indices of the two independent plane wave propagations that the crystal can support. To complete the solution of the problem, we use the values of n^2, one at a time, in Eqs. (5.2-14). The three equations (resulting from setting k equal to x, y, z) can then be solved for the relative magnitudes of E_x, E_y, and E_z, which are then used with the aid of Eqs. (5.2-11) and (5.2-12) to solve for the allowed directions of **H** and **D**.

From (5.2-14), it follows that

$$\left(\frac{E_k^{1,2}}{E_l^{1,2}}\right) = \frac{s_k[(n^2)^{1,2} - \varepsilon'_l]}{s_l[(n^2)^{1,2} - \varepsilon'_k]}$$

where the superscripts (1, 2) correspond to the two independent solutions. For real values of $(n^2)^{1,2}$, as in nonabsorbing or nonamplifying media, the ratio $E_k^{1,2}/E_l^{1,2}$ is real, so that both solutions are linearly polarized.

To summarize, along an arbitrary direction of propagation **s**, there can exist two independent plane wave, linearly polarized, propagation modes. These propagate with phase velocities $\pm(c/n_1)$ and $\pm(c/n_2)$ where n_1^2 and n_2^2 are the two solutions of Fresnel equation (5.2-15).

In practice, the indices of refraction $(n^2)^{1,2}$ and the direction of **D**, **H**, and **E** are found, most often, not by the procedure outlined above but by the formally equivalent method of the index ellipsoid. This method is discussed in the following section.

5.3 THE INDEX ELLIPSOID

The constant energy density (ω_e) surfaces in **D** space given by (5.2-8) can be written as

$$\frac{D_x^2}{\varepsilon_x'} + \frac{D_y^2}{\varepsilon_y'} + \frac{D_z^2}{\varepsilon_z'} = 2\omega_e\varepsilon_0$$

where ε_x', ε_y', and ε_z' are the (relative) principal dielectric constants. If we replace $\mathbf{D}/\sqrt{2\omega_e\varepsilon_0}$ by **r** and define the principal indices of refraction n_x, n_y, and n_z by $n_k^2 \equiv \varepsilon_k'$ $(k = x, y, z)$, the last equation can be written as

$$\frac{x^2}{n_x^2} + \frac{y^2}{n_y^2} + \frac{z^2}{n_z^2} = 1 \qquad (5.3\text{-}1)$$

This is the equation of a general ellipsoid with major axes parallel to the x, y, and z directions whose respective lengths are $2n_x$, $2n_y$, $2n_z$. The ellipsoid is known as the *index ellipsoid* or, sometimes, as the *optical indicatrix*. The index ellipsoid is used mainly to find the two indices of refraction and the two corresponding directions of **D** associated with the two independent plane waves that can propagate along an *arbitrary* direction **s** in a crystal. This is done by means of the following prescription: Find the intersection ellipse between a plane through the origin that is normal to the direction of propagation **s** and the index ellipsoid (5.3-1). The two axes of the intersection ellipse are equal in length to $2n_1$ and $2n_2$, n_1, and n_2 being the two indices of refraction, that is, the solutions of (5.2-15). These axes are parallel, respectively, to the directions of the $\mathbf{D}_{1,2}$ vectors of the two allowed solutions.

To show that this procedure is formally equivalent to the method of the last section, we follow the treatment of Born and Wolf (Reference 2). The procedure consists of solving for $(r^2)^{1,2}$—the two extrema of the intersection ellipse—and showing that they are equal to $(n^2)^{1,2}$ of (5.2-15). The proof is completed by showing that the radius vectors to these two extrema are parallel to $\mathbf{D}_{1,2}$.

The ellipse is specified by the following two surfaces:

1. The index ellipsoid,

$$\frac{x^2}{\varepsilon_x'} + \frac{y^2}{\varepsilon_y'} + \frac{z^2}{\varepsilon_z'} = 1 \tag{5.3-2}$$

2. The normal to **s**,

$$\mathbf{r} \cdot \mathbf{s} = xs_x + ys_y + zs_z = 0 \tag{5.3-3}$$

The principal semiaxes of the ellipse are given by the extrema of

$$r^2 = x^2 + y^2 + z^2 \tag{5.3-4}$$

subject to conditions (5.3-2) and (5.3-3). The problem of finding extrema subject to auxiliary conditions is handled by means of the Lagrange method of multipliers. We set up a function $F(x, y, z, \lambda_1, \lambda_2)$

$$F = x^2 + y^2 + z^2 + \lambda_1(xs_x + ys_y + zs_z) + \lambda_2\left(\frac{x^2}{\varepsilon_x'} + \frac{y^2}{\varepsilon_y'} + \frac{z^2}{\varepsilon_z'} - 1\right) \tag{5.3-5}$$

where λ_1 and λ_2 are undetermined coefficients. We then solve for the quantities x, y, and z at the extrema of r^2 and the ratio λ_1/λ_2 from the equations

$$\frac{\partial F}{\partial x} = \frac{\partial F}{\partial y} = \frac{\partial F}{\partial z} = 0 \tag{5.3-6}$$

$$\frac{\partial F}{\partial \lambda_1} = \frac{\partial F}{\partial \lambda_2} = 0 \tag{5.3-7}$$

From (5.3-6), we obtain

$$x_k + \frac{\lambda_1}{2} s_k + \frac{\lambda_2 x_k}{\varepsilon_k'} = 0 \qquad k = x, y, z \tag{5.3-8}$$

Equations (5.3-7) merely reproduce the auxiliary conditions (5.3-2) and (5.3-3). Multiplying Eqs. (5.3-8) by x_k, adding them, and using (5.3-2) and (5.3-3) yield

$$r^2 + \lambda_2 = 0 \tag{5.3-9}$$

Multiplying (5.3-8) by s_k, summing over k, using (5.3-2) and (5.3-3), and recalling that $s^2 = 1$ lead to

$$\frac{\lambda_1}{2} + \lambda_2\left(\frac{xs_x}{\varepsilon_x'} + \frac{ys_y}{\varepsilon_y'} + \frac{zs_z}{\varepsilon_z'}\right) = 0 \tag{5.3-10}$$

Using (5.3-9) and (5.3-10) to eliminate λ_1 and λ_2 from (5.3-8) results in the relation

$$x\left(1 - \frac{r^2}{\varepsilon_x'}\right) + s_x r^2\left(\frac{xs_x}{\varepsilon_x'} + \frac{ys_y}{\varepsilon_y'} + \frac{zs_z}{\varepsilon_z'}\right) = 0 \tag{5.3-11}$$

and two similar equations in which x is replaced, respectively, by y and z.

If, as in (5.3-1), we perform the substitution

$$\mathbf{r} \rightarrow \frac{\mathbf{D}}{\sqrt{\mathbf{E} \cdot \mathbf{D} \varepsilon_0}} \tag{5.3-12}$$

to (5.3-11), we have $r^2 \rightarrow D^2/\mathbf{E} \cdot \mathbf{D}\varepsilon_0 = n^2$ and $x/\varepsilon_x \rightarrow E_x/\sqrt{\mathbf{E} \cdot \mathbf{D}\varepsilon_0}$. With these substitutions, (5.3-11) becomes

$$\varepsilon_x E_x = \varepsilon_0 n^2 [E_x - s_x(\mathbf{s} \cdot \mathbf{E})] \tag{5.3-13}$$

which is identical to (5.2-13). Since $r^2 \rightarrow n^2$, it follows that the semiaxes (the extrema of the intersection ellipse) are equal to the values of $n_{1,2}$, that is, to the indices of refraction for the two propagation modes. Second, since the radius vectors $\mathbf{r}$ to the extreme points of the ellipse satisfy (5.3-12), they are parallel to the two allowed $\mathbf{D}$ vectors. This establishes the formal equivalence between the index ellipsoid method and the Maxwell equations solution of electromagnetic propagation in anisotropic crystals.

5.4 PROPAGATION IN UNIAXIAL CRYSTALS

In uniaxial crystals, that is, crystals in which the highest degree of rotational symmetry applies to no more than a single axis, the equation of the index ellipsoid (5.3-1) simplifies to

$$\frac{x^2}{n_o^2} + \frac{y^2}{n_o^2} + \frac{z^2}{n_e^2} = 1 \tag{5.4-1}$$

where the axis of symmetry was chosen, following convention, as the z axis. It is also referred to as the optic axis, n_o is called the ordinary index of refraction, whereas n_e is the extraordinary one. If $n_e < n_o$, we have a negative (optically) uniaxial crystal, whereas in a positive crystal, $n_e > n_o$.

Figure 5.2 shows the index ellipsoid for a positive uniaxial crystal. The direction of propagation is along $\mathbf{s}$. Since the ellipsoid in this case is invariant to a rotation about the z axis, the projection of the $\mathbf{s}$ vector on the x-y plane is chosen without loss of generality to coincide with the y axis.

According to the "prescription" given in Section 5.3, we first find the intersection of the plane through the origin that is normal to $\mathbf{s}$ with the index ellipsoid. The intersection is an ellipse whose plane is cross-hatched in the figure. The length of the semimajor axis OA is equal to the index of refraction $n_e(\theta)$ of the "extraordinary" ray whose electric displacement vector $\mathbf{D}_e(\theta)$ is parallel to $\mathbf{OA}$. The "ordinary" ray is polarized (i.e., has its $\mathbf{D}$ vector) along OB and its index of refraction is equal to n_o.

It is clear from Figure 5.2 that as the angle θ between the optic axis and the direction of propagation $\mathbf{s}$ is changed, the direction of polarization of the ordinary ray remains fixed (along the x axis in the figure) and its index of refraction is always equal to n_o. The direction of $\mathbf{D}_e$, on the other hand, depends, as shown, on θ. The index of refraction varies from $n_e(\theta) = n_o$ for $\theta = 0°$ to $n_e(\theta) = n_e$ for $\theta = 90°$. The index of refraction $n_e(\theta)$ of the extraordi-

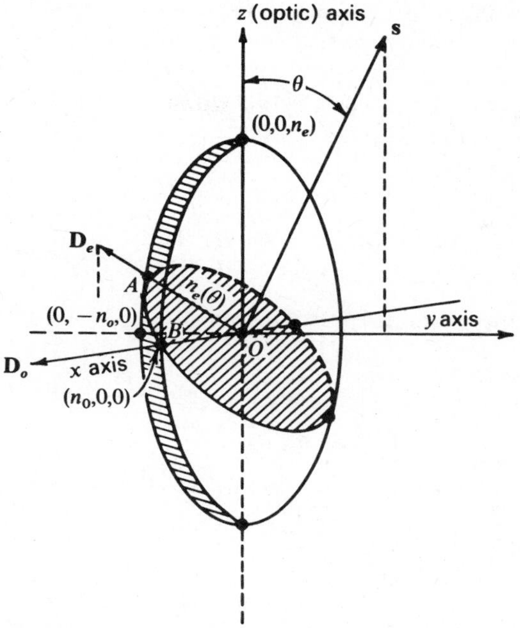

FIGURE 5.2 The construction for finding the indices of refraction and the allowed polarization directions for a given direction of propagation **s**. The figure shown is for a uniaxial crystal with $n_x = n_y = n_o$, $n_z = n_e$.

nary ray is equal to OA, which according to Figure 5.2, is given by

$$\frac{1}{n_e^2(\theta)} = \frac{\cos^2\theta}{n_o^2} + \frac{\sin^2\theta}{n_e^2} \tag{5.4-2}$$

The three-dimensional surfaces giving the indices $n_e(\theta, \phi)$ and $n_o(\theta, \phi)$ as functions of the wave-normal direction (θ, ϕ) are called the normal surfaces. Such surfaces can be constructed from the index ellipsoid by the methods given above. For a uniaxial crystal, the azimuthal angle ϕ is redundant, and the normal surface becomes an ellipsoid of revolution about the z (optic) axis.

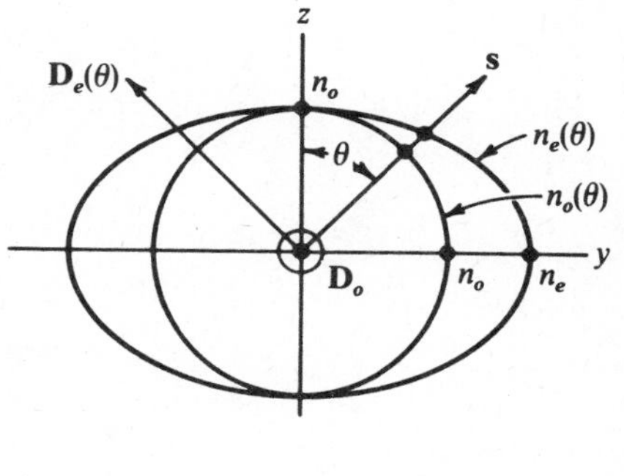

FIGURE 5.3 The intersection of the $s - z$ plane with the normal surfaces for a positive uniaxial crystal ($n_e > n_o$).

The intersection curves of these ellipsoids with the **s**-z plane are shown in Figure 5.3 for a positive uniaxial crystal ($n_e > n_o$). The exterior curve $n_e(\theta)$ is a plot of Eq. (5.4-2). Plots such as Figure 5.3 are very useful since they convey at a glance the two indices. The orientation of the displacement vectors is also shown. $\mathbf{D}_e(\theta)$ is in the **s**-z plane, whereas that of the ordinary ray is at right angles to this plane.

5.5 NORMAL MODE EXPANSION OF THE ELECTROMAGNETIC FIELD IN A RESONATOR

Maxwell's equations and the current continuity equation in the MKS system of units are

$$\nabla \times \mathbf{E} = \frac{-\partial \mathbf{B}}{\partial t}$$

$$\nabla \times \mathbf{H} = \mathbf{i} + \frac{\partial \mathbf{D}}{\partial t}$$

$$\nabla \cdot \mathbf{D} = \rho \tag{5.5-1}$$

$$\nabla \cdot \mathbf{B} = 0$$

$$\nabla \cdot \mathbf{i} = -\frac{\partial \rho}{\partial t}$$

We will limit ourselves, for the moment, to charge-free, isotropic, and homogeneous media so that

$$\mathbf{i} = 0 \qquad \mathbf{B} = \mu\mathbf{H} \qquad \nabla \cdot \mathbf{D} = 0 \qquad \mathbf{D} = \varepsilon\mathbf{E} \tag{5.5-2}$$

where ε is the dielectric constant.

Consider the electric field $\mathbf{E}(\mathbf{r}, t)$ and magnetic field $\mathbf{H}(\mathbf{r}, t)$ inside a volume V bounded by a surface S of perfect conductivity. The tangential component of **E**, $-\mathbf{n} \times \mathbf{n} \times \mathbf{E}$, and the normal component of **H**, $\mathbf{n} \cdot \mathbf{H}$, must both be zero on S (**n** is the unit vector normal to S). We will expand **E** and **H** in terms of two orthogonal sets of vector fields $\mathbf{E}_a$ and $\mathbf{H}_a$, respectively. These sets, which were introduced originally by Slater (Reference 3) obey the relations

$$k_a \mathbf{E}_a = \nabla \times \mathbf{H}_a \tag{5.5-3}$$

$$k_a \mathbf{H}_a = \nabla \times \mathbf{E}_a \tag{5.5-4}$$

where k_a is to be considered, for the moment, a constant. The tangential component of $\mathbf{E}_a$ on S is zero.

$$\mathbf{n} \times \mathbf{E}_a = 0 \qquad \text{on} \qquad S \tag{5.5-5}$$

If we take the curl of both sides of (5.5-3, 4) and use the identity

$$\nabla \times \nabla \times \mathbf{A} = \nabla(\nabla \cdot \mathbf{A}) - \nabla^2 \mathbf{A}$$

they become

$$\nabla^2 \mathbf{E}_a + k_a^2 \mathbf{E}_a = 0$$
$$\nabla^2 \mathbf{H}_a + k_a^2 \mathbf{H}_a = 0 \tag{5.5-6}$$

that is, the familiar wave equation.

It follows from (5.5-3), (5.5-4), and (5.5-5) that the normal component of $\mathbf{H}_a$, $\mathbf{n} \cdot \mathbf{H}_a$ is zero on S. To prove this statement, consider an arbitrary closed contour C on S surrounding a surface S'.

$$\oint_C \mathbf{E}_a \cdot d\mathbf{l} = \oint_C (-\mathbf{n} \times \mathbf{n} \times \mathbf{E}_a) \cdot d\mathbf{l} + \oint_C (\mathbf{n} \cdot \mathbf{E}_a)\mathbf{n} \cdot d\mathbf{l} = 0 \tag{5.5-7}$$

where $\mathbf{E}_a$ is expressed as the vector sum of its tangential $(-\mathbf{n} \times \mathbf{n} \times \mathbf{E}_a)$ and normal $(\mathbf{n} \cdot \mathbf{E}_a)\mathbf{n}$ components. The first term on the right side of (5.5-7) is zero because of (5.5-5) whereas the second one is zero since $\mathbf{n}$ is perpendicular to $d\mathbf{l}$. Using Stokes' theorem on the left side of (5.5-7) gives

$$\oint_C \mathbf{E}_a \cdot d\mathbf{l} = \int_{S'} (\nabla \times \mathbf{E}_a) \cdot \mathbf{n}\, da = k_a \int_{S'} (\mathbf{H}_a \cdot \mathbf{n})\, da = 0$$

and since C is arbitrary, it follows that

$$\mathbf{H}_a \cdot \mathbf{n} = 0 \qquad \text{on} \qquad S \tag{5.5-8}$$

We will next prove that the functions $\mathbf{E}_a$ and $\mathbf{H}_a$ are orthogonal in the sense

$$\int_V \mathbf{E}_a \cdot \mathbf{E}_b\, dv = 0 \qquad a \neq b$$
$$\int_V \mathbf{H}_a \cdot \mathbf{H}_b\, dv = 0 \qquad a \neq b \tag{5.5-9}$$

To prove the first of Eqs. (5.5-9), apply the vector identity $\nabla \cdot (\mathbf{A} \times \mathbf{B}) = \mathbf{B} \cdot \nabla \times \mathbf{A} - \mathbf{A} \cdot \nabla \times \mathbf{B}$ to $(\mathbf{E}_b \times \nabla \times \mathbf{E}_a)$, then to $(\mathbf{E}_a \times \nabla \times \mathbf{E}_b)$ and subtract. The result is

$$\nabla \cdot (\mathbf{E}_b \times \nabla \times \mathbf{E}_a) - \nabla \cdot (\mathbf{E}_a \times \nabla \times \mathbf{E}_b)$$
$$= \nabla \times \mathbf{E}_a \cdot \nabla \times \mathbf{E}_b - \mathbf{E}_b \cdot \nabla \times \nabla \times \mathbf{E}_a - \nabla \times \mathbf{E}_b \cdot \nabla \times \mathbf{E}_a + \mathbf{E}_a \cdot \nabla \times \nabla \times \mathbf{E}_b$$

From (5.5-4), we have $\nabla \times \mathbf{E}_a = k_a \mathbf{H}_a$ and $\nabla \times \nabla \times \mathbf{E}_a = k_a^2 \mathbf{E}_a$, which substituted in the last equation gives

$$k_a \nabla \cdot (\mathbf{E}_b \times \mathbf{H}_a) - k_b \nabla \cdot (\mathbf{E}_a \times \mathbf{H}_b) = (k_b^2 - k_a^2)\mathbf{E}_a \cdot \mathbf{E}_b$$

that, after applying Gauss's theorem (see Section 5.1), becomes

$$\int_S [k_a \mathbf{n} \cdot (\mathbf{E}_b \times \mathbf{H}_a) - k_b \mathbf{n} \cdot (\mathbf{E}_a \times \mathbf{H}_b)]\, da = (k_b^2 - k_a^2) \int_V \mathbf{E}_a \cdot \mathbf{E}_b\, dv$$

The left side of the last equality can be shown, with the aid of the identity $\mathbf{A} \cdot \mathbf{B} \times \mathbf{C} = \mathbf{C} \cdot \mathbf{A} \times \mathbf{B}$ and (5.5-5), to be zero, so that for $k_b \neq k_a$, the first of Eqs. (5.5-9) is proved. If $k_b = k_a$, that is, when $\mathbf{E}_b$ and $\mathbf{E}_a$ are members of a degenerate set, it is possible to construct linear superpositions of the degenerate functions so that orthogonality is preserved. The proof of the orthogonality of the $\mathbf{H}_a$ functions follows along identical lines.

We are free to choose the magnitude of $\mathbf{H}_a$ and $\mathbf{E}_a$ so that they are normalized according to

$$\int_V \mathbf{H}_a \cdot \mathbf{H}_b \, dv = \delta_{a,b}$$
$$\int_V \mathbf{E}_a \cdot \mathbf{E}_b \, dv = \delta_{a,b} \tag{5.5-10}$$

This choice will be used throughout this text.

The total resonator fields of $\mathbf{E}(\mathbf{r}, t)$ and $\mathbf{H}(\mathbf{r}, t)$ can be expanded as

$$\mathbf{E}(\mathbf{r}, t) = -\sum_a \frac{1}{\sqrt{\varepsilon}} p_a(t)\mathbf{E}_a(\mathbf{r})$$
$$\mathbf{H}(\mathbf{r}, t) = \sum_a \frac{1}{\sqrt{\mu}} \omega_a q_a(t)\mathbf{H}_a(\mathbf{r}) \tag{5.5-11}$$

where $\omega_a = k_a/\sqrt{\mu\varepsilon}$. Substituting (5.5-11) in the first of Maxwell equations (5.5-1) and using (5.5-3) result in

$$p_a = \dot{q}_a \tag{5.5-12}$$

and in a similar fashion, from the second of (5.5-1), we have

$$\omega_a^2 q_a = -\dot{p}_a \tag{5.5-13}$$

Eliminating q_a gives

$$\ddot{p}_a + \omega_a^2 p_a = 0 \tag{5.5-14}$$

This identifies $k_a(\mu\varepsilon)^{-1/2} = \omega_a$ as the radian oscillation frequency of the ath mode.

5.6 THE QUANTIZATION OF THE RADIATION FIELD

In this section we will show that the electromagnetic field inside a resonator can be considered, formally, as an ensemble of *independent* harmonic oscillators. The formalism is related closely to that used in Chapter 4, to quantize the spectrum of lattice vibrations. To bring out the similarity (dot) multiply the first of Eqs. (5.5-11) by $\mathbf{E}_b$ and integrate over the resonator volume. The result, after using (5.5-10), is

$$p_b(t) = -\sqrt{\varepsilon} \int_V \mathbf{E}(\mathbf{r}, t) \cdot \mathbf{E}_b(\mathbf{r}) \, dv \tag{5.6-1}$$

This equation is analogous to the second of Eqs. (4.5-2). It follows from (5.6-1) and from the relation

$$q_b(t) = \frac{\sqrt{\mu}}{\omega_b} \int_V \mathbf{H}(\mathbf{r}, t) \cdot \mathbf{H}_b(\mathbf{r}) \, dv \tag{5.6-2}$$

that the state of the classical electromagnetic field can be specified by $\mathbf{H}(\mathbf{r}, t)$ and $\mathbf{E}(\mathbf{r}, t)$ or, alternatively, by the dynamical variables $p_a(t)$ and $q_a(t)$. The total energy (Hamiltonian) is

$$\mathcal{H} = \tfrac{1}{2} \int_V (\mu \mathbf{H} \cdot \mathbf{H} + \varepsilon \mathbf{E} \cdot \mathbf{E}) \, dv \tag{5.6-3}$$

Substituting for **H** and **E** their expansion (5.5-11) gives

$$\mathcal{H} = \sum_a \tfrac{1}{2}(p_a^2 + \omega_a^2 q_a^2) \tag{5.6-4}$$

which has the basic form of a sum of harmonic oscillator Hamiltonians as in (2.2-1). The dynamical variables p_a and q_a constitute canonically conjugate variables. This can be seen by considering Hamilton's equations of motion relating $\dot{p}_a$ to q_a and $\dot{q}_a$ to p_a.

$$\begin{aligned} \dot{p}_a &= -\frac{\partial \mathcal{H}}{\partial q_a} = -\omega_a^2 q_a \\ \dot{q}_a &= \frac{\partial \mathcal{H}}{\partial p_a} = p_a \end{aligned} \tag{5.6-5}$$

These are identical with Eqs. (5.5-12) and (5.5-13) obtained from Maxwell's equations.

The quantization of the electromagnetic radiation is achieved by considering p_a and q_a as *formally equivalent* to the momentum and coordinate operators of a quantum mechanical harmonic oscillator, thus taking the commutator relations connecting the dynamical variables as

$$\begin{aligned} [p_a, p_b] &= [q_a, q_b] = 0 \\ [q_a, p_b] &= i\hbar\, \delta_{a,b} \end{aligned} \tag{5.6-6}$$

In a manner analogous to that used in Chapter 2 [see (2.2-25)], we define the creation operator $a_l^\dagger$ and the annihilation operator a_l by

$$\begin{aligned} a_l^\dagger(t) &= \left(\frac{1}{2\hbar\omega_l}\right)^{1/2} [\omega_l q_l(t) - i p_l(t)] \\ a_l(t) &= \left(\frac{1}{2\hbar\omega_l}\right)^{1/2} [\omega_l q_l(t) + i p_l(t)] \end{aligned} \tag{5.6-7}$$

The commutator relations are found directly from (5.6-6) to be

$$\begin{aligned} [a_l, a_m] &= [a_l^\dagger, a_m^\dagger] = 0 \\ [a_l, a_m^\dagger] &= \delta_{l,m} \end{aligned} \tag{5.6-8}$$

Solving (5.6-7) for p_l and q_l gives

$$\begin{aligned} p_l(t) &= i\left(\frac{\hbar\omega_l}{2}\right)^{1/2} [a_l^\dagger(t) - a_l(t)] \\ q_l(t) &= \left(\frac{\hbar}{2\omega_l}\right)^{1/2} [a_l^\dagger(t) + a_l(t)] \end{aligned} \tag{5.6-9}$$

We can express the Hamiltonian in terms of the operators $a_l^\dagger$ and a_l by substituting (5.6-9) in (5.6-4) and replacing $a_l a_l^\dagger$ by $a_l^\dagger a_l + 1$ [according to (5.6-8)]. The result is

$$\mathcal{H} = \sum_l \hbar\omega_l(a_l^\dagger a_l + \tfrac{1}{2}) \tag{5.6-10}$$

which is the same as Eq. (4.5-12) obtained for the case of the lattice vibrations. The formal analogy between the operators $a_l^\dagger$, a_l and their counterparts in the case of the harmonic oscillators shows that, quantum mechanically, a stationary state of the total radiation field can be characterized by an eigenfunction Φ, which is a product of the eigenfunctions of the individual Hamiltonians $\hbar\omega_l(a_l^\dagger a_l + \frac{1}{2})$

$$\Phi = u_{n_1}u_{n_2}\cdots = \prod_{l=1}^{\infty} u_{n_l} \tag{5.6-11}$$

where

$$\begin{aligned} a_l^\dagger u_{n_l} &= \sqrt{n_l + 1}\; u_{n_l+1} \\ a_l u_{n_l} &= \sqrt{n_l}\; u_{n_l-1} \\ a_l^\dagger a_l u_{n_l} &= n_l u_{n_l} \end{aligned} \tag{5.6-12}$$

The expectation value of the operator $a_l^\dagger a_l$ is

$$\langle\Phi|\, a_l^\dagger a_l \,|\Phi\rangle = \langle n_l|\, a_l^\dagger a_l \,|n_l\rangle = n_l \tag{5.6-13}$$

and is equal to the number of quanta n_l in the lth mode of the resonator.

Plane-Wave Quantization

The discussion just concluded uses a generalized resonator of unspecified shape. We will find it useful to consider the form of the field operators in the case of a plane-wave resonator. Although such a resonator, which requires an infinite cross-sectional area does not exist, most optical resonators that employ curved mirrors as reflectors involve nearly plane-wave propagation.

To be specific, consider the lth mode of a resonator of length L along the axis and mode volume V. Let the electric and magnetic field vectors point along the y and x directions, respectively. Equations (5.5-3, 4, 9) are satisfied by

$$\begin{aligned} \mathbf{E}_l(\mathbf{r}, t) &= -\,i\hat{\mathbf{x}}\left(\frac{\hbar\omega_l}{V\varepsilon}\right)^{1/2} [a_l^\dagger(t) - a_l(t)] \sin k_l z \\ \mathbf{H}_l(\mathbf{r}, t) &= \hat{\mathbf{y}}\left(\frac{\hbar\omega_l}{V\mu}\right)^{1/2} [a_l^\dagger(t) + a_l(t)] \cos k_l z \end{aligned} \tag{5.6-14}$$

where V is the mode volume. In the case of plane wave modes, the total field can be written as a summation over all modes $\mathbf{k}$

$$\begin{aligned} \mathbf{E} &= \sum_{\mathbf{k},\lambda} -\,i\,\hat{\mathbf{e}}_{\mathbf{k},\lambda}\sqrt{\frac{\hbar\omega_k}{2\varepsilon V}}\left(a_{\mathbf{k},\lambda}^\dagger\, e^{-i\mathbf{k}\cdot\mathbf{r}} - a_{\mathbf{k},\lambda}e^{i\mathbf{k}\cdot\mathbf{r}}\right) \\ \mathbf{H} &= \sum_{\mathbf{k},\lambda} -\,i\,\frac{\mathbf{k}}{k}\times\hat{\mathbf{e}}_{\mathbf{k},\lambda}\sqrt{\frac{\hbar\omega_k}{2\mu V}}\left(a_{\mathbf{k},\lambda}^\dagger\, e^{-i\mathbf{k}\cdot\mathbf{r}} + a_{\mathbf{k},\lambda}e^{i\mathbf{k}\cdot\mathbf{r}}\right) \end{aligned} \tag{5.6-15}$$

E and **H** can be derived using (5.1-22, 23) from a quantized vector potential

$$\mathbf{A} = \sum_{\mathbf{k},\lambda} \hat{\mathbf{e}}_{\mathbf{k},\lambda} \sqrt{\frac{\hbar}{2\varepsilon V \omega_k}} \left(a^{\dagger}_{\mathbf{k},\lambda}\, e^{-i\mathbf{k}\cdot\mathbf{r}} + a_{\mathbf{k},\lambda} e^{i\mathbf{k}\cdot\mathbf{r}} \right) \tag{5.6-16}$$

5.7 MODE DENSITY AND BLACKBODY RADIATION

The number of electromagnetic modes with resonant frequencies between ν and $\nu + d\nu$ is, in general, a function of the specific form of the electromagnetic enclosure. If, however, the typical dimensions of the enclosure are large compared to the wavelength of the radiation under consideration, this is no longer true and the mode density can be calculated by using an arbitrarily shaped enclosure. Limiting ourselves to this case, we pick as the resonator a box whose sides are equal to L. The propagation characteristics of a mode are then described by $\exp(i\mathbf{k}\cdot\mathbf{r})$. Using the boundary condition that the field be periodic in L imposes the following condition on the Cartesian components of **k**:

$$k_x = \frac{2\pi l}{L} \qquad k_y = \frac{2\pi m}{L} \qquad k_z = \frac{2\pi n}{L} \tag{5.7-1}$$

where l, m, and n are integers.

From Maxwell's equations, we have

$$\left(\frac{\varepsilon}{\varepsilon_0}\right)\frac{\omega^2}{c^2} = k^2 = (k_x^2 + k_y^2 + k_z^2)$$

From (5.7-1), it follows that with each mode we can associate a volume $dk_x\, dk_y\, dk_z = (2\pi/L)^3$ in **k** space. The number of modes N_k whose wave vectors have magnitudes between 0 and k is derived simply by dividing the total volume in **k** space $(\frac{4}{3})\pi k^3$ by the volume per mode $(2\pi/L)^3$ and by multiplying the result by 2, since for each wave vector **k** there are two possible directions of polarization, each of which corresponds to an independent solution of Maxwell equations. The result is

$$N_k = \frac{k^3 L^3}{3\pi^2}$$

or, if we use $k = 2\pi\nu n/c$, where $n = (\varepsilon/\varepsilon_0)^{1/2}$ is the index of refraction[2]

$$\frac{N_\nu}{V} = \frac{8\pi\nu^3 n^3}{3c^3} \tag{5.7-2}$$

for the total number of modes per unit volume between $\nu = 0$ and ν. The mode density per unit frequency $p(\nu)$ is given by

$$p(\nu) = \frac{1}{V}\frac{dN_\nu}{d\nu} = \frac{8\pi\nu^2 n^3}{c^3} \tag{5.7-3}$$

[2] Not to be confused with the integer n in (5.7-1).

If the enclosure is in thermal equilibrium at a temperature T, the average energy per mode $\overline{E}$ is derived by a calculation identical to that employed in the case of the lattice modes, which led to (4.6-3). We can thus write directly

$$\overline{E} = \frac{h\nu}{2} + \frac{h\nu}{e^{\hbar\nu/kT} - 1} \tag{5.7-4}$$

The thermal radiation (blackbody) energy density per unit frequency width is thus

$$\rho(\nu) = p(\nu)\overline{E} = \frac{8\pi h n^3 \nu^3}{c^3}\left(\frac{1}{2} + \frac{1}{e^{h\nu/kT} - 1}\right)$$

If we integrate $\rho(\nu)$ over all frequencies, we get, because of the presence of the $\frac{1}{2}$ inside the brackets, an infinite result. This situation is not acceptable and does not agree with experiments. This "paradox" may be resolved if we recognize that the term $\frac{1}{2}$ represents the zero-field energy $h\nu/2$ of the radiation oscillators. This is the lowest energy that an oscillator may possess and is thus not available for energy exchange or, equivalently, measurement. From the thermodynamic point of view, therefore, we can write

$$\rho(\nu)\, d\nu = \frac{8\pi h n^3 \nu^3\, d\nu}{c^3}\left(\frac{1}{e^{h\nu/kT} - 1}\right) \tag{5.7-5}$$

for the blackbody radiation density.

5.8 THE COHERENT STATE

The electric field of a single field mode is characterized according to (5.5-11) by the operator $p(t)$ and a field Hamiltonian

$$H = \tfrac{1}{2}(p^2 + \omega^2 q^2) \tag{5.8-1}$$

Since p and q are canonically conjugate "momenta," they are related as in (1.1-25) by

$$q \rightarrow i\hbar \frac{\partial}{\partial p} \tag{5.8-2}$$

so that the Hamiltonian in the p representation becomes

$$H = \tfrac{1}{2}\left(p^2 - \omega^2\hbar^2 \frac{\partial^2}{\partial p^2}\right) \tag{5.8-3}$$

If we define

$$\xi \equiv \alpha p \equiv \frac{p}{\sqrt{\hbar\omega}}, \qquad \Lambda \equiv \frac{2E_n}{\hbar\omega} \tag{5.8-4}$$

the Schrödinger equation $Hu_n = E_n u_n$ assumes the form

$$\frac{\partial^2 u_n}{\partial \xi^2} + [\Lambda_n - \xi^2]u_n = 0 \tag{5.8-5}$$

This is the harmonic oscillator equation (2.2-2) so that the solution u_n can be taken directly from (2.2-14) and (2.2-17) as

$$|n\rangle = u_n(\alpha p) = \left(\frac{\alpha}{2^n n! \sqrt{\pi}}\right)^{1/2} e^{-\alpha^2 p^2/2} H_n(\alpha p) \tag{5.8-6}$$

$$E_n = \hbar\omega(n + \tfrac{1}{2})$$

where $|n\rangle \equiv u_n$ is the same function as that which appears in Eqs. (5.6-11,12) and H_n is the Hermite polynomial. In (5.8-6) it shows the explicit dependence on p.

The most general solution $\psi(p, t)$ of the time-dependent Schrödinger equation

$$H\psi(p, t) = i\hbar \frac{\partial \psi}{\partial t} \tag{5.8-7}$$

is a linear superposition of all $|n\rangle$ with their time exponentials restored

$$\psi(p, t) = \sum_{n=0}^{\infty} c_n e^{-i\omega(n+1/2)t} u_n(\alpha p) \tag{5.8-8}$$

where

$$\int_{-\infty}^{\infty} |\psi|^2 \, dp = \sum_{n=0}^{\infty} |c_n|^2 = 1 \tag{5.8-9}$$

The wave function $\psi(p, t)$ is used, as in the "atomic" quantum mechanics, to obtain any expectation value of the modal field at ω. As an example, using (5.5-11), we obtain

$$\langle \mathbf{E}(\mathbf{r}, t)\rangle = -\frac{\mathbf{E}(\mathbf{r})}{\sqrt{\varepsilon}} \langle \psi(p, t)|p|\psi(p, t)\rangle \tag{5.8-10}$$

If the field is made up of more than one, noninteracting mode, then the above formalism applies separately to each one of these modes since the total Hamiltonian is the sum of one-mode Hamiltonians.

A particular choice of the coefficients c_n plays an extremely important role in quantum optics. The coefficients are given by

$$c_n = \left(\frac{e^{-\lambda}\lambda^n}{n!}\right)^{1/2} e^{-in\phi} \tag{5.8-11}$$

The magnitude squared of c_n

$$|c_n|^2 = \frac{e^{-\lambda}\lambda^n}{n!}$$

can be recognized as the Poisson distribution. The resulting modal wave function

$$\psi_c(t) \equiv |\lambda\rangle = \sum_{n=0}^{\infty} \left(\frac{e^{-\lambda}\lambda^n}{n!}\right)^{1/2} e^{-in\phi} e^{-i\omega(n+1/2)t} |n\rangle \tag{5.8-12}$$

is called the coherent state (References 4 and 5). We dropped the implied p dependence and used $\psi_c(t)$ instead of $\psi_c(p, t)$. Its main claim to fame is that it is the minimum uncertainty packet in the sense of (1.2-19), and it describes the output field of a single mode laser.

We can verify easily from (5.8-11) that

$$\sum_{n=0}^{\infty} |c_n|^2 = 1 \tag{5.8-13}$$

and that the average number of quanta in the coherent state $|\lambda\rangle$ is

$$\langle n \rangle = \sum_{n=0}^{\infty} n|c_n|^2 = \lambda \tag{5.8-14}$$

The average energy is thus $\hbar\omega(\lambda + \frac{1}{2})$.

An important relationship that will be needed in subsequent development is

$$\begin{aligned} \psi_c(0) &= e^{-\lambda/2} e^{\Omega a^\dagger}|0\rangle \\ \Omega &= \lambda^{1/2} e^{-i\phi} \end{aligned} \tag{5.8-15}$$

where $|0\rangle$ is the vacuum state $n = 0$.

The proof of (5.8-15) proceeds as follows. Let

$$\psi(0) = e^{-\lambda/2} e^{\Omega a^\dagger}|0\rangle = e^{-\lambda/2} \sum_{n=0}^{\infty} \frac{(\Omega a^\dagger)^n}{n!} |0\rangle$$

but

$$(a^\dagger)^n|0\rangle = \sqrt{1}\,\sqrt{2} \ldots \sqrt{n}\,|n\rangle = \sqrt{n!}\,|n\rangle$$

so that

$$\psi(0) = e^{-\lambda/2} \sum_{n=0}^{\infty} \frac{\lambda^{n/2} \sqrt{n!}\, e^{-in\phi}}{n!} |n\rangle$$

$$= \sum_{n=0}^{\infty} \left(\frac{e^{-\lambda}\lambda^n}{n!}\right)^{1/2} e^{-in\phi}|n\rangle = \psi_c(0)$$

as stated. It follows immediately from (5.8-15) or (5.8-12) that the vacuum state $|0\rangle$ *is a coherent state* corresponding to $\lambda = 0$.

Another important property whose proof is left as an exercise is that $\psi_c(t)$ is an eigenfunction of the annihilation operator a

$$a|\lambda\rangle = a\psi_c(t) = \lambda^{1/2} e^{-i(\omega t + \phi)} \psi_c(t) \tag{5.8-16}$$

It is instructive to find the classical electric and magnetic fields that correspond to the coherent state $\psi_c(t)$. Consider a plane-wave-like mode with nonvanishing components E_y and H_x. We take the modal function $\mathbf{E}_a(\mathbf{r})$ of (5.5-10) to conform with (5.5-9)

$$\begin{aligned} \mathbf{E}_a(\mathbf{r}) &\rightarrow \hat{\mathbf{e}}_y \sqrt{\frac{2}{V}} \sin kz \\ k &= \frac{\omega n}{c} \end{aligned} \tag{5.8-17}$$

where V is the volume of the mode. The field operator is thus given by a single term (mode) of (5.5-11)

$$E_y = -\sqrt{\frac{2}{V\varepsilon}}\, p \sin kz$$

which corresponds to a classical field

$$\langle E_y\rangle = -\sqrt{\frac{2}{V\varepsilon}} \sin kz\, \langle \psi_c(p,t)|p|\psi_c(p,t)\rangle \tag{5.8-18}$$

Using the first of Eqs. (5.6-9)

$$\langle\lambda|p|\lambda\rangle = i\left(\frac{\hbar\omega}{2}\right)^{1/2} \langle\lambda|a^\dagger - a|\lambda\rangle$$

$$= i\left(\frac{\hbar\omega}{2}\right)^{1/2} \sum_n\sum_m \left(\frac{e^{-\lambda}\lambda^n}{n!}\right)^{1/2}\left(\frac{e^{-\lambda}\lambda^m}{m!}\right)^{1/2} e^{i(n-m)\phi} e^{i\omega t(n-m)} \langle n|a^\dagger - a|m\rangle$$

Making use of (5.6-12) and (5.8-13) leads after some algebra to

$$\langle\lambda|p|\lambda\rangle = -(2\hbar\omega\lambda)^{1/2}\sin(\omega t + \phi) \tag{5.8-19}$$

and from (5.8-18)

$$\mathbf{E}_{\text{classical}} = \hat{\mathbf{e}}_y \sqrt{\frac{4\hbar\omega\lambda}{V\varepsilon}} \sin kz \sin(\omega t + \phi) \tag{5.8-20}$$

Starting with the second of (5.5-10), we can follow a similar path and arrive at

$$\mathbf{H}_{\text{classical}} = -\hat{\mathbf{e}}_x \sqrt{\frac{4\hbar\omega\lambda}{V\mu}} \cos kz \cos(\omega t + \phi) \tag{5.8-21}$$

(Reminder: λ is not the wavelength; it is the average number of quanta in the mode.)

Uncertainties

Using the tools developed above, we can show straightforwardly that using the coherent state $\psi_c(p, t)$

$$\begin{aligned}\langle p^2\rangle &= \langle p\rangle^2 + \frac{1}{2\alpha^2}, \qquad \alpha \equiv (\hbar\omega)^{-1/2}\\ \langle q^2\rangle &= \langle q\rangle^2 + \tfrac{1}{2}\hbar^2\alpha^2\end{aligned} \tag{5.8-22}$$

Taking the mean square deviations of p and q

$$\begin{aligned}(\Delta p)^2 &\equiv \langle (p - \langle p\rangle)^2\rangle = \langle p^2\rangle - \langle p\rangle^2\\ (\Delta q)^2 &\equiv \langle (q - \langle q\rangle)^2\rangle = \langle q^2\rangle - \langle q\rangle^2\end{aligned} \tag{5.8-23}$$

we obtain from (5.8-22)

$$\Delta p = 2^{-1/2}\alpha^{-1}, \qquad \Delta q = 2^{-1/2}\hbar\alpha, \qquad \alpha = (\hbar\omega)^{-1/2}$$
$$\Delta p \Delta q = \frac{\hbar}{2} \tag{5.8-24}$$

which according to (1.2-19) is a minimum uncertainty product.

Since p and q correspond to electric and magnetic fields, it follows that any measurement of the, say, electric field will be subject to an uncertainty

$$\Delta E \sim \sqrt{\frac{2}{V\varepsilon}}\,(\Delta p) = \sqrt{\frac{2}{V\varepsilon}}\,\frac{1}{\sqrt{2}\,\alpha} = \sqrt{\frac{\hbar\omega}{V\varepsilon}}$$

This corresponds to a residual noise energy of

$$\Delta\mathscr{E} \sim \tfrac{1}{2}\varepsilon(\Delta E)^2 V = \frac{\hbar\omega}{2}$$

Ways of "beating" this limit will be considered in the discussion in Chapter 17 of squeezed states.

References

1. See, for example, S. Ramo, J. R. Whinnery, and T. Van Duzer, *Fields and Waves in Communication Electronics* (New York: Wiley, 1965).
2. Born, M. and E. Wolf, *Principles of Optics* (New York: Macmillan, 1964).
3. Slater, J. C., *Microwave Electronics* (Princeton, N.J.: Van Nostrand, 1950).
4. Louisell, W. H., A. Yariv, and A. E. Siegman, "Quantum Fluctuations and Noise in Parametric Processes," *Phys. Rev.* **124,** 1646 (1961).
5. Glauber, R. J., "Coherent and Incoherent States of Radiation Field," *Phys. Rev.* **131,** 2766 (1963).
6. Louisell, W. H., *Radiation and Noise in Quantum Electronics* (New York: McGraw-Hill, 1964).

Problems

5.1 What is the explicit time dependence of $a_1^\dagger(t)$ and $a_1(t)$ when considered as operators in the Heisenberg representation?

5.2 Show that the energy per mode is time invariant, that is, that $\langle a_1^\dagger(t)a_1(t)\rangle =$ constant.

5.3 Consider the subspace of two resonator modes "1" and "2." Let the total Hamiltonian be

$$\mathscr{H} = \hbar\omega_1(a_1^\dagger a_1 + \tfrac{1}{2}) + \hbar\omega_2(a_2^\dagger a_2 + \tfrac{1}{2}) - \hbar K[a_1^\dagger a_2^\dagger e^{-i(\omega t+\phi)} + a_1 a_2 e^{i(\omega t+\phi)}]$$

where K is a real constant and $\omega = \omega_1 + \omega_2$. Solve for $a_1^\dagger(t)$, $a_1(t)$, $a_2^\dagger(t)$, $a_2(t)$ in the Heisenberg representation.

Hint:
At some point, it may be convenient to introduce the variables $A_j^\dagger$, A_j defined by

$$a_j^\dagger(t) = A_j^\dagger(t)e^{i\omega_j t}$$
$$a_j(t) = A_j(t)e^{-i\omega_j t}$$

Answer:

$$a_1(t) = e^{-i\omega_1 t}(a_{10} \cosh Kt + ie^{-i\phi}a_{20}^\dagger \sinh Kt)$$

$$a_2(t) = e^{-i\omega_2 t}(a_{20} \cosh Kt + ie^{-i\phi}a_{10}^\dagger \sinh Kt)$$

5.4 Show that the equations of motion (5.6-5) are the same if we consider p and q as the quantum mechanical "momentum" and "coordinate" operators and use $\mathcal{H}$ as given by (5.6-4).

5.5 Repeat the derivation of (5.7-3) using standing waves instead of traveling waves.

5.6 Supply the missing steps in the derivation of (5.6-15).

5.7 Quantize the simple electrical LC circuit consisting of an inductance L and a capacitance C in parallel with it. Specifically,
(a) What are the canonically conjugate momenta p and q for this problem?
(b) What is the form of the Hamiltonian when expressed in terms of p and q and, alternatively, in terms of the operators $a^\dagger$ and a?

Clue

Somewhere in the solution, one should obtain

$$a^\dagger = \sqrt{\frac{C}{2\hbar\omega}}\left[V(t) - i\sqrt{\frac{L}{C}}\,I(t)\right]$$

$$a = \sqrt{\frac{C}{2\hbar\omega}}\left[V(t) + i\sqrt{\frac{L}{C}}\,I(t)\right]$$

where $I(t)$ and $V(t)$ are the instantaneous voltage and current in the circuit, and $\omega^2 \equiv (LC)^{-1}$.

5.8 Derive (5.1-21).

5.9 Show that (5.2-15) is a quadratic equation in n^2.

5.10 Find the direction of the power flow $\mathbf{E} \times \mathbf{H}$ of a wave propagating along a direction $\mathbf{s}$ in a uniaxial crystal.
(a) For an ordinary ray.
(b) For an extraordinary ray.
(c) Is the direction of the power flow parallel to $\mathbf{k}$?

5.11 Are the directions of the power flow $\mathbf{E} \times \mathbf{H}$ and the group velocity

$$\mathbf{v}_g(\theta) = \nabla_{\mathbf{k}}\omega(\mathbf{k})$$

identical?

5.12 Prove (5.5-13).

5.13 Show that $\psi_c(p, o)$ is an eigenfunction of the annihilation operator a. What is the eigenvalue?

5.14 Prove (5.8-16).

5.15 Prove (5.8-21) and (5.8-22).

CHAPTER 6

The Propagation of Optical Beams in Homogeneous and Lenslike Media

6.0 INTRODUCTION

We first take up the subject of optical ray propagation through a variety of optical media. These include homogeneous and isotropic materials, thin lenses, dielectric interfaces, curved mirrors, and media with quadratic index or gain variation. Since a ray is, by definition, normal to the optical wavefront, an understanding of the ray behavior makes it possible to trace the evolution of complex optical waves passing through various optical elements. We find that the transit of a ray (or its reflection) through these elements can be described by simple 2×2 matrices. Furthermore, these matrices will describe the evolution of Gaussian beams such as those that are characteristic of the output of lasers, and that exist inside spherical mirror optical resonators. The second half of this chapter is devoted to a formal treatment of the Gaussian beam. An understanding of the behavior of Gaussian beams and the closely related subject of optical resonators is probably the single most important prerequisite to working in quantum electronics.

6.1 THE LENS WAVEGUIDE

Consider a paraxial ray[1] passing through a thin lens of focal length f as shown in Figure 6.1. Taking the cylindrical axis of symmetry as z, denoting the ray distance from the axis by r and its slope dr/dz as r', we can relate the output ray $(r_{\text{out}}, r'_{\text{out}})$ to the input ray $(r_{\text{in}}, r'_{\text{in}})$ by means of

$$\begin{aligned} r_{\text{out}} &= r_{\text{in}} \\ r'_{\text{out}} &= r'_{\text{in}} - \frac{r_{\text{out}}}{f} \end{aligned} \tag{6.1-1}$$

where the first of Eqs. (6.1-1) follows from the definition of a thin lens and the second can be derived from a consideration of the behavior of the undeflected central ray with a slope equal to r'_{in}, as shown in Figure 6.1.

[1] By paraxial ray, we mean a ray whose angular deviation from the cylindrical (z) axis is small enough that the sine and tangent of the angle can be approximated by the angle itself.

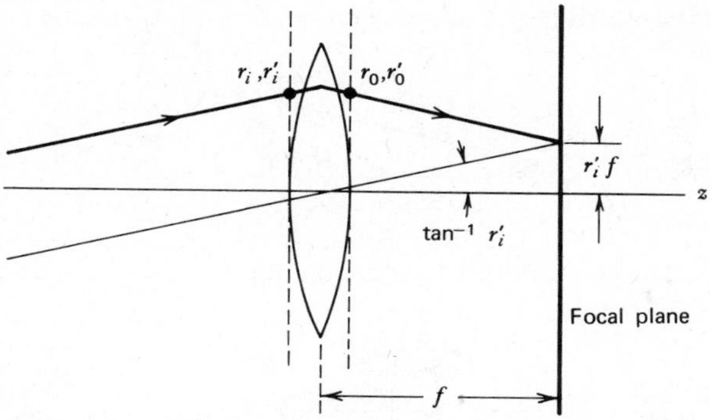

FIGURE 6.1 Passage of a ray through a thin lens.

Representing a ray at any position z as a column matrix

$$\tilde{r}(z) = \begin{vmatrix} r(z) \\ r'(z) \end{vmatrix}$$

we can rewrite (6.1-1) using the rules for matrix multiplication (see references 1–3) as

$$\begin{vmatrix} r_{\text{out}} \\ r'_{\text{out}} \end{vmatrix} = \begin{vmatrix} 1 & 0 \\ -1/f & 1 \end{vmatrix} \begin{vmatrix} r_{\text{in}} \\ r'_{\text{in}} \end{vmatrix} \tag{6.1-2}$$

where $f > 0$ for a converging lens and is negative for a diverging one.

The ray matrices for a number of other optical elements are shown in Table 6.1.

Consider as an example the propagation of a ray through a straight section of a homogeneous medium of length d followed by a thin lens of focal length f. This corresponds to propagation between planes n and $n + 1$ in Figure 6.2. Since the effect of the straight section is merely that of increasing r by dr', using (6.1-2) we can relate the output (at $n + 1$) and input (at n) rays by

$$\begin{vmatrix} r_{\text{out}} \\ r'_{\text{out}} \end{vmatrix} = \begin{vmatrix} 1 & d \\ -1/f & (1 - d/f) \end{vmatrix} \begin{vmatrix} r_{\text{in}} \\ r'_{\text{in}} \end{vmatrix} \tag{6.1-3}$$

The matrix corresponds to the product of the thin lens matrix times the straight section matrix as given in Table 6.1.

We are now in a position to consider the propagation of a ray through a biperiodic lens system made up of lenses of focal lengths f_1 and f_2 separated by d as shown in Figure 6.2. This will be shown in the next chapter to be formally equivalent to the problem of Gaussian-beam propagation inside an optical resonator with mirrors of radii of curvature $R_1 = 2f_1$ and $R_2 = 2f_2$ that are separated by d.

The section between the planes $n - 1$ and $n + 1$ can be considered the basic unit cell of the periodic lens sequence. If we limit ourselves, at the

TABLE 6.1. Ray Matrices for Some Common Optical Elements and Media

(1) Homogeneous Medium: Length d	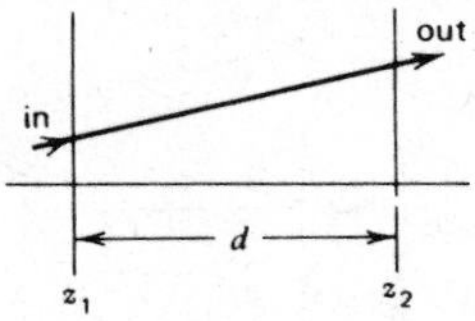	$\begin{bmatrix} 1 & d \\ 0 & 1 \end{bmatrix}$
(2) Thin Lens: Focal length f ($f>0$, converging; $f<0$, diverging)	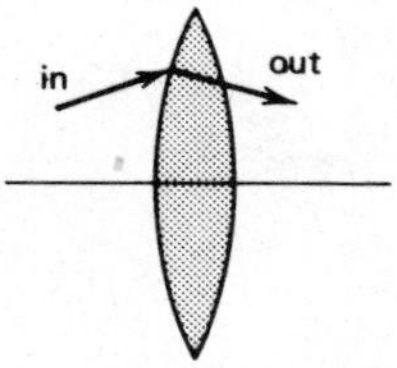	$\begin{bmatrix} 1 & 0 \\ \frac{-1}{f} & 1 \end{bmatrix}$
(3) Dielectric Interface: Refractive indices n_1, n_2	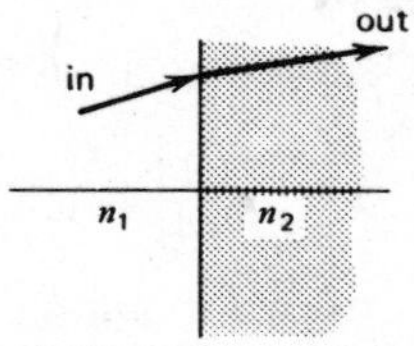	$\begin{bmatrix} 1 & 0 \\ 0 & \frac{n_1}{n_2} \end{bmatrix}$
(4) Spherical Dielectric Interface: Radius R	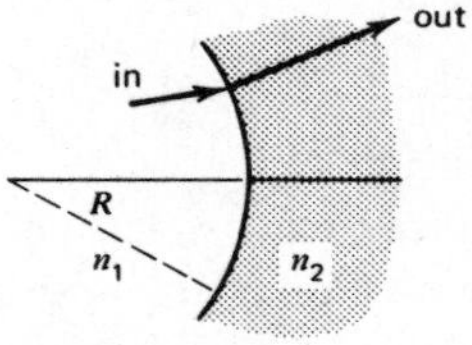	$\begin{bmatrix} 1 & 0 \\ \frac{n_2-n_1}{n_2}\frac{1}{R} & \frac{n_1}{n_2} \end{bmatrix}$
(5) Spherical Mirror: Radius of curvature R	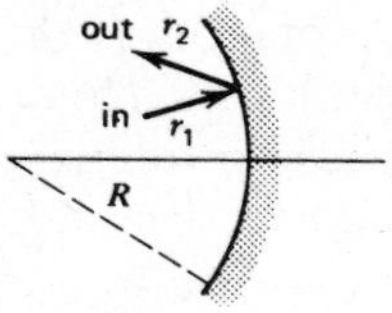	$\begin{bmatrix} 1 & 0 \\ \frac{-2}{R} & 1 \end{bmatrix}$
(6) A medium with a quadratic index profile	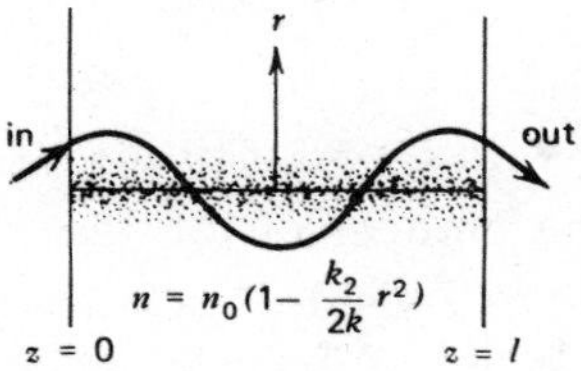	$\begin{bmatrix} \cos\left(\sqrt{\frac{k_2}{k}}\,l\right) & \sqrt{\frac{k}{k_2}}\sin\left(\sqrt{\frac{k_2}{k}}\,l\right) \\ -\sqrt{\frac{k_2}{k}}\sin\left(\sqrt{\frac{k_2}{k}}\,l\right) & \cos\left(\sqrt{\frac{k_2}{k}}\,l\right) \end{bmatrix}$

moment, to planes $n-1, n+1, n+3, \ldots,$ and denote them as planes s, $s+1, s+2, \ldots$ so that $\Delta s = 2\Delta n$, from (6.1-3) we have

$$\begin{vmatrix} r_{s+1} \\ r'_{s+1} \end{vmatrix} = \begin{vmatrix} 1 & d \\ -\frac{1}{f_2} & \left(1-\frac{d}{f_2}\right) \end{vmatrix} \begin{vmatrix} 1 & d \\ -\frac{1}{f_1} & \left(1-\frac{d}{f_1}\right) \end{vmatrix} \begin{vmatrix} r_s \\ r'_s \end{vmatrix} \tag{6.1-4}$$

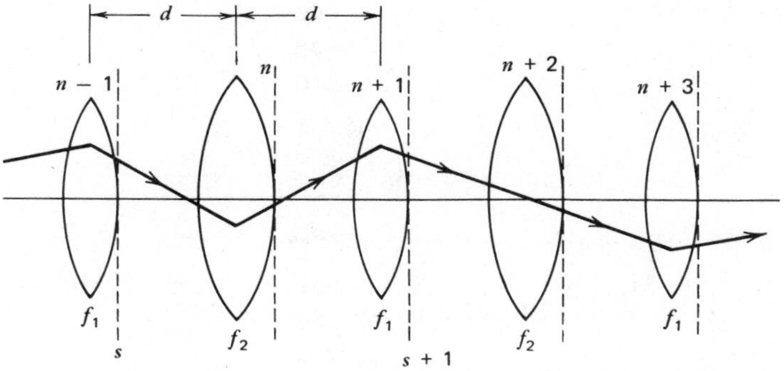

FIGURE 6.2 Propagation of an optical ray through a biperiodic lens sequence.

or, in equation form,

$$\begin{aligned} r_{s+1} &= Ar_s + Br'_s \\ r'_{s+1} &= Cr_s + Dr'_s \end{aligned} \tag{6.1-5}$$

where A, B, C, and D are the elements of the matrix resulting from multiplying the two square matrices in (6.1-4) and are given by

$$\begin{aligned} A &= 1 - \frac{d}{f_2} \\ B &= d\left(2 - \frac{d}{f_2}\right) \\ C &= -\left[\frac{1}{f_1} + \frac{1}{f_2}\left(1 - \frac{d}{f_1}\right)\right] \\ D &= -\left[\frac{d}{f_1} - \left(1 - \frac{d}{f_1}\right)\left(1 - \frac{d}{f_2}\right)\right] \end{aligned} \tag{6.1-6}$$

From the first of (6.1-5), we get

$$r'_s = \frac{1}{B}(r_{s+1} - Ar_s) \tag{6.1-7}$$

and thus,

$$r'_{s+1} = \frac{1}{B}(r_{s+2} - Ar_{s+1}) \tag{6.1-8}$$

Using the second of (6.1-5) in (6.1-8) and substituting for r'_s from (6.1-7) gives

$$r_{s+2} - (A + D)r_{s+1} + (AD - BC)r_s = 0 \tag{6.1-9}$$

for the difference equation governing the evolution through the lens waveguide. Using (6.1-6), we can show that $AD - BC = 1$. We can consequently rewrite (6.1-9) as

$$r_{s+2} - 2br_{s+1} + r_s = 0 \tag{6.1-10}$$

where

$$b = \tfrac{1}{2}(A + D) = \left(1 - \frac{d}{f_2} - \frac{d}{f_1} + \frac{d^2}{2f_1f_2}\right) \tag{6.1-11}$$

Equation (6.1-10) is the equivalent, in terms of difference equations, of the differential equation $r'' + Ar = 0$, which has solutions $r_{\pm}(z) = \rho_{\pm}\exp[\pm i\sqrt{A}\,z]$. We are thus led to try a solution in the form of

$$r_s = \rho e^{is\theta} \tag{6.1-12}$$

that, when substituted in (6.1-10), leads to

$$e^{2i\theta} - 2be^{i\theta} + 1 = 0$$

and therefore,

$$e^{i\theta} = b \pm i\sqrt{1 - b^2} \tag{6.1-13}$$

so that for $b^2 \leqslant 1$, $\cos\theta = b = \tfrac{1}{2}(A + D)$.

The general solution can be taken as a linear superposition of $\exp(is\theta)$ and $\exp(-is\theta)$ solutions or, since r_s is real, as

$$r_s = r_{\max}\sin(s\theta + \delta) \tag{6.1-14}$$

where $r_{\max} = r_0/\sin\delta$ and $\tan\delta = \dfrac{r_0\theta}{dr_0'}$.

The condition for a stable—that is, confined—ray is that θ be a real number since in this case the ray radius r_s oscillates as a function of the cell number s between $r_{\max}$ and $-r_{\max}$. According to (6.1-13), the necessary and sufficient condition for θ to be real is that

$$|b| \leqslant 1 \tag{6.1-15}$$

In terms of the system parameters, we can use (6.1-11) to reexpress (6.1-15) as

$$-1 \leqslant 1 - \frac{d}{f_1} - \frac{d}{f_2} + \frac{d^2}{2f_1f_2} \leqslant 1$$

or

$$0 \leqslant \left(1 - \frac{d}{2f_1}\right)\left(1 - \frac{d}{2f_2}\right) \leqslant 1 \tag{6.1-16}$$

If, on the other hand, the confinement condition $|b| \leqslant 1$ is violated, we can obtain a solution in the form of

$$r_s = ce^{(\alpha^+)s} + de^{(\alpha^-)s} \tag{6.1-17}$$

where $e^{\alpha\pm} = b \pm \sqrt{b^2 - 1}$, and since the magnitude of either $\exp(\alpha^+)$ or $\exp(\alpha^-)$ exceeds unity, the beam radius will increase as a function of (distance) s.

6.2 THE IDENTICAL-LENS WAVEGUIDE

The simplest case of a lens waveguide is one in which $f_1 = f_2 = f$; that is, all the lenses are identical.

The analysis of this situation is considerably simpler than that used for a biperiodic lens sequence. The reason is that the periodic unit cell (the smallest part of the sequence that can, on translation, recreate the whole sequence) contains a single lens only. The (A, B, C, D) matrix for the unit cell is given by the square matrix in (6.1-3). If we follow exactly the steps leading to (6.1-11) through (6.1-14), the confinement condition becomes

$$0 \leqslant d \leqslant 4f \tag{6.2-1}$$

and the beam radius at the nth lens is given by

$$r_n = r_{max} \sin(n\theta + \delta)$$

$$\cos\theta = \tfrac{1}{2}(A + D) = \left(1 - \frac{d}{2f}\right) \tag{6.2-2}$$

Because of the algebraic simplicity of this problem, we can express r_{max} and δ in (6.2-2) in terms of the initial conditions r_0 and r_0', obtaining

$$(r_{max})^2 = \frac{4f}{4f - d}(r_0^2 + dr_0r_0' + dfr_0'^2) \tag{6.2-3}$$

$$\tan\delta = \sqrt{\frac{4f}{d} - 1}\bigg/\left(1 + 2f\frac{r_0'}{r_0}\right) \tag{6.2-4}$$

The derivation of the last two equations is left as an exercise.

The stability criteria can be demonstrated experimentally by tracing the behavior of a laser beam that is injected at an angle to the axis as it propagates down a sequence of lenses spaced uniformly. One can easily notice the rapid "escape" of the beam once condition (6.2-1) is violated.

6.3 THE PROPAGATION OF RAYS BETWEEN MIRRORS

Another important application of the formalism just developed concerns the bouncing of a ray between two curved mirrors. Since the reflection at a mirror with a radius of curvature R is equivalent, except for the folding of the path, to passage through a lens with a focal length $f = R/2$, we can use the formalism of the preceding section to describe the propagation of a ray between two curved reflectors with radii of curvature R_1 and R_2, which are separated by d. Let us consider the simple case of a ray that is injected into a symmetric two-mirror system as shown in Figure 6.3a. Since the x and y coordinates of the ray are independent variables, we can take them according to (6.1-12) in the form of (see Reference 6)

$$\begin{aligned} x_n &= x_{max} \sin(n\theta + \delta_x) \\ y_n &= y_{max} \sin(n\theta + \delta_y) \end{aligned} \tag{6.3-1}$$

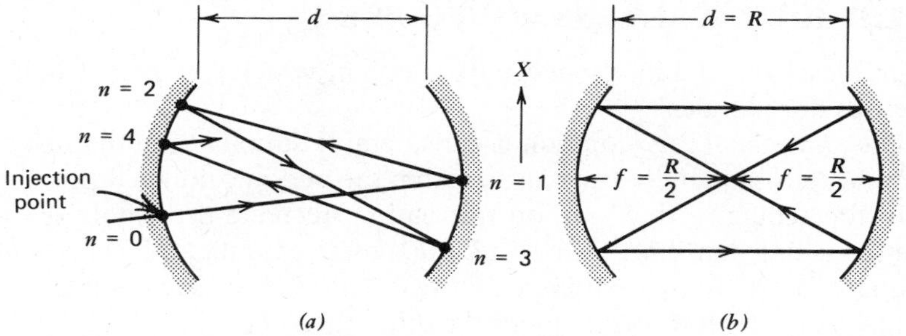

FIGURE 6.3 (*a*) Path of a ray injected in the plane of the figure into the space between two mirrors. (*b*) Reentrant ray in a symmetric confocal ($d = R$) mirror configuration repeating its pattern after two round trips.

where n refers to the ray parameter immediately following the nth reflection. According to (6.3-1), the locus of the points x_n, y_n on a given mirror lies on an ellipse.

Reentrant Rays

If θ in (6.3-1) satisfies the condition

$$2\nu\theta = 2l\pi \tag{6.3-2}$$

where ν and l are any two integers, a ray will return to its starting point following ν round trips and will thus continuously retrace the same pattern on the mirrors. If we consider as an example the simple case of $l = 1$, $\nu = 2$, so that $\theta = \pi/2$, from (6.2-2) we obtain $d = 2f = R$; that is, if the mirrors are separated by a distance equal to their radius of curvature R, the trapped ray will retrace its pattern after two round trips ($\nu = 2$). This situation ($R = d$) is referred to as symmetric confocal, since the two mirrors have a common focal point $f = R/2$. It will be discussed in detail in the next chapter. The ray pattern corresponding to $\nu = 2$ is illustrated in Figure 6.3*b*.

6.4 RAYS IN LENSLIKE MEDIA

The basic physical property of lenses that is responsible for their focusing action is the fact that the optical path across them $\int n(r, z)\, dz$ (where n is the index of refraction of the medium) is a quadratic function of the distance r from the z axis. Using ray optics, we account for this fact by a change in the ray's slope as in (6.1-1). This same property can be represented by relating the complex field amplitude of the incident optical field $E_R(x, y)$ immediately to the right of an ideal thin lens to that immediately to the left $E_L(x, y)$ by (see Reference 7)

$$E_R(x, y) = E_L(x, y) \exp\left(+ik\frac{x^2 + y^2}{2f}\right) \tag{6.4-1}$$

where f is the focal length and $k = 2\pi n/\lambda$ (λ = vacuum wavelength).

The effect of the lens, therefore, is to cause a phase shift $k(x^2 + y^2)/2f$, which increases quadratically with the distance from the axis. We consider next the closely related case of a medium whose index of refraction n varies according to[2]

$$n(x, y) = n_0\left[1 - \frac{k_2}{2k}(x^2 + y^2)\right] \tag{6.4-2}$$

where k_2 is a constant. Since the phase delay of a wave propagating through a section dz of a medium with an index of refraction n is $2\pi n\, dz/\lambda$, it follows directly that a thin slab of the medium described by (6.4-2) will act as a thin lens, introducing [as in (6.4-1)] a phase shift that is proportional to $(x^2 + y^2)$. The behavior of a ray in this case is described by the differential equation that applies to ray propagation in an optically inhomogeneous medium (see Reference 8)

$$\frac{d}{ds}\left(n\frac{d\mathbf{r}}{ds}\right) = \nabla n \tag{6.4-3}$$

where s is the distance along the ray measured from some fixed position on it and $\mathbf{r}$ is the position vector of the point at s. For paraxial rays, we may replace d/ds by d/dz and, using (6.4-2), obtain approximately

$$\frac{d^2r}{dz^2} + \left(\frac{k_2}{k}\right)r = 0 \tag{6.4-4}$$

If at the input plane $z = 0$ the ray has a radius r_0 and slope r_0', we can write the solution of (6.4-4) directly as

$$\begin{aligned} r(z) &= \cos\left(\sqrt{\frac{k_2}{k}}\,z\right)r_0 + \sqrt{\frac{k}{k_2}}\sin\left(\sqrt{\frac{k_2}{k}}\,z\right)r_0' \\ r'(z) &= -\sqrt{\frac{k_2}{k}}\sin\left(\sqrt{\frac{k_2}{k}}\,z\right)r_0 + \cos\left(\sqrt{\frac{k_2}{k}}\,z\right)r_0' \end{aligned} \tag{6.4-5}$$

so that the lenslike medium of length l is described by a ray matrix

$$\begin{aligned} A &= \cos(\sqrt{k_2/k}\,l) \qquad B = \sqrt{k/k_2}\sin(\sqrt{k_2/k}\,l) \\ C &= -\sqrt{k_2/k}\sin(\sqrt{k_2/k}\,l) \qquad D = A \end{aligned} \tag{6.4-6}$$

That is, the ray oscillates back and forth across the axis, as shown in Figure 6.4. A section of the quadratic index medium acts as a lens. This can be proved by showing, using (6.4-5), that a family of parallel rays entering at $z = 0$ at different radii will converge upon emerging at $z = l$ to a common focus at a distance

$$h = \frac{1}{n_0}\sqrt{\frac{k}{k_2}}\cot\left(\sqrt{\frac{k_2}{k}}\,l\right) \tag{6.4-7}$$

[2] Equation (6.4-2) can be viewed as consisting of the first two terms in the Taylor-series expansion of $n(x, y)$ for the radial symmetric case.

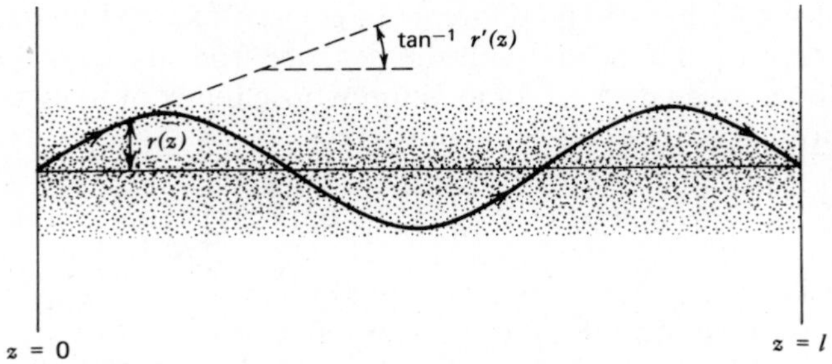

FIGURE 6.4 Path of a ray in a medium with a quadratic index variation.

from the exit plane. The factor n_0 accounts for the refraction at the boundary, if we assume the medium at $z > l$ possesses an index $n = 1$ and a small angle of incidence. The derivation of (6.4-7) is left as an exercise.

Equations (6.4-5) apply to a focusing medium with $k_2 > 0$. In a medium where $k_2 < 0$—that is, where the index increases with the distance from the axis—the solutions for $r(z)$ and $r'(z)$ become

$$
\begin{aligned}
r(z) &= \cosh\left(\sqrt{\frac{k_2}{k}}\,z\right)r_0 + \sqrt{\frac{k}{k_2}}\sinh\left(\sqrt{\frac{k_2}{k}}\,z\right)r_0' \\
r'(z) &= \sqrt{\frac{k_2}{k}}\sinh\left(\sqrt{\frac{k_2}{k}}\,z\right)r_0 + \cosh\left(\sqrt{\frac{k_2}{k}}\,z\right)r_0'
\end{aligned}
\qquad (6.4\text{-}8)
$$

so that $r(z)$ increases with distance and eventually escapes. A section of such a medium acts as a negative lens. Physical situations giving rise to quadratic index variation include the following:

1. Propagation of laser beams with Gaussian intensity profile in a slightly absorbing medium. The absorption heating gives rise, because of the dependence of n on temperature T, to an index profile (Reference 10). If $dn/dT < 0$, as is the case for most materials, the index is smallest on the axis where the absorption heating is highest. This corresponds to a $k_2 < 0$ in (6.4-2), and the beam spreads with distance z. If $dn/dT > 0$, as in certain lead glasses (Reference 10), the beams are focused.
2. The absorption of pump light in solid laser rods, such as ruby, gives rise to an $n(r)$ that decreases with r (for $dn/dT < 0$) and hence causes pumped laser rods to act as lenses.
3. Dielectric waveguides made by sandwiching a layer of index n_1 between two layers with index $n_2 > n_1$. This situation will be discussed further in Chapter 22.
4. Optical fibers produced by cladding a thin optical fiber (whose radius is comparable to λ) of an index n_1 with a sheath of index $n_2 < n_1$. Such fibers are used as light pipes.
5. Optical waveguides consisting of glasslike rods or fibers with large radii compared to λ, whose index decreases with increasing r (References 11,

12). Such waveguides can be used for the simultaneous transmission of a number of laser beams, which are injected into the waveguide at different angles. It follows from (6.4-5) that the beams will emerge, each along a unique direction, and consequently can be easily separated. Furthermore, in view of its previously discussed lens properties, the waveguide can be used to transmit optical image information in much the same way as images are transmitted by a multielement lens systems to the image plane of a camera. The properties of such waveguides will be considered further in Section 6.10.

6.5 THE WAVE EQUATION IN QUADRATIC INDEX MEDIA

The most widely encountered optical beam is one where the intensity distribution at planes normal to the propagation direction is Gaussian. To derive its characteristics, we start with the Maxwell equations in an isotropic charge-free medium.

$$\begin{aligned} \nabla \times \mathbf{H} &= \varepsilon \frac{\partial \mathbf{E}}{\partial t} \\ \nabla \times \mathbf{E} &= -\mu \frac{\partial \mathbf{H}}{\partial t} \\ \nabla \cdot (\varepsilon \mathbf{E}) &= 0 \end{aligned} \tag{6.5-1}$$

Taking the curl of the second of (6.5-1) and substituting the first results in

$$\nabla^2 \mathbf{E} - \mu\varepsilon \frac{\partial^2 \mathbf{E}}{\partial t^2} = -\nabla\left(\frac{1}{\varepsilon} \mathbf{E} \cdot \nabla\varepsilon\right) \tag{6.5-2}$$

where we used $\nabla \times \nabla \times \mathbf{E} \equiv \nabla(\nabla \cdot \mathbf{E}) - \nabla^2 \mathbf{E}$. If we assume the field quantities to vary as $\mathbf{E}(x, y, z, t) = \mathrm{Re}[\mathbf{E}(x, y, z)e^{i\omega t}]$ and neglect the right side of (6.5-2)[3]

$$\nabla^2 \mathbf{E} + k^2(\mathbf{r})\mathbf{E} = 0 \tag{6.5-3}$$

where

$$k^2(\mathbf{r}) = \omega^2 \mu\varepsilon(r)[1 - i\sigma(r)/\omega\varepsilon(r)] \tag{6.5-4}$$

where we allowed for the possible dependence of ε on position $\mathbf{r}$. We have also taken k as a complex number to allow for the possibility of losses ($\sigma > 0$) or gain ($\sigma < 0$) in the medium.[4]

We limit our derivation to the case in which $k^2(\mathbf{r})$ is given by

$$k^2(\mathbf{r}) = k^2 - kk_2 r^2 \tag{6.5-5}$$

[3] This neglect is justified if the fractional change of ε in one optical wavelength is small.
[4] If k is complex (e.g., $k_r + ik_i$), then a traveling electromagnetic planewave has the form of $\exp[i(\omega t - kz)] = \exp[k_i z + i(\omega t - k_r z)]$.

where, according to (6.5-4),

$$k^2 = k^2(0) = \omega^2\mu\varepsilon(0)\left[1 - i\,\frac{\sigma(0)}{\omega\varepsilon(0)}\right].$$

and k_2 is some constant. Furthermore, we assume a solution whose transverse dependence is on $r = \sqrt{x^2 + y^2}$ only so that in (6.5-3) we can replace ∇^2 by

$$\nabla^2 = \nabla_t^2 + \frac{\partial^2}{\partial z^2} = \frac{\partial^2}{\partial r^2} + \frac{1}{r}\frac{\partial}{\partial r} + \frac{\partial^2}{\partial z^2} \tag{6.5-6}$$

The kind of propagation we are considering is that of a nearly plane wave in which the flow of energy is predominantly along a single (e.g., z) direction so that we may limit our derivation to a single transverse field component E. Taking E as

$$E = \psi(x, y, z)e^{-ikz} \tag{6.5-7}$$

we obtain from (6.5-3) and (6.5-5) in a few simple steps

$$\nabla_t^2\psi - 2ik\psi' - kk_2r^2\psi = 0 \tag{6.5-8}$$

where $\psi' = \partial\psi/\partial z$ and where we assume that the longitudinal variation is slow enough that $k\psi' \gg \psi'' \ll k^2\psi$.

Next, we take ψ in the form of

$$\psi = \exp\{-i[P(z) + \tfrac{1}{2}Q(z)r^2]\} \tag{6.5-9}$$

that, when substituted into (6.5-8) and after using (6.5-6), gives

$$-Q^2r^2 - 2iQ - kr^2Q' - 2kP' - kk_2r^2 = 0 \tag{6.5-10}$$

If (6.5-10) is to hold for all r, the coefficients of the different powers of r must each be equal to zero. This leads to (Reference 7)

$$\begin{aligned} Q^2 + kQ' + kk_2 &= 0 \\ P' &= -\frac{iQ}{k} \end{aligned} \tag{6.5-11}$$

The wave equation (6.5-3) is thus reduced to Eqs. (6.5-11).

6.6 THE GAUSSIAN BEAM IN A HOMOGENEOUS MEDIUM

If the medium is homogeneous, we can, according to (6.5-5), put $k_2 = 0$, and (6.5-11) becomes

$$Q^2 + kQ' = 0 \tag{6.6-1}$$

Introducing the function $s(z)$ by the relation

$$Q = k\,\frac{s'}{s} \tag{6.6-2}$$

we obtain directly from (6.6-1)

$$s'' = 0$$

so that

$$s' = a \qquad s = az + b$$

or, using (6.6-2),

$$Q(z) = k\frac{a}{az + b} \tag{6.6-3}$$

where a and b are arbitrary constants. We will find it more convenient to deal with a parameter q, where

$$q(z) = \frac{k}{Q(z)} = \frac{2\pi n}{\lambda Q(z)} \tag{6.6-4}$$

so that we may rewrite (6.6-3) in the form

$$q = z + q_0 \tag{6.6-5}$$

From (6.5-11) and (6.6-4), we have

$$P' = -\frac{i}{q} = -\frac{i}{z + q_0}$$

so that

$$P(z) = -i\ln\left(1 + \frac{z}{q_0}\right) \tag{6.6-6}$$

where the arbitrary constant of integration is chosen as zero.[5]

Combining (6.6-5) and (6.6-6) in (6.5-9), we obtain

$$\psi = \exp\left\{-i\left[-i\ln\left(1 + \frac{z}{q_0}\right) + \frac{k}{2(q_0 + z)}r^2\right]\right\} \tag{6.6-7}$$

We take the arbitrary constant of integration q_0 to be purely imaginary and reexpress it in terms of a new constant ω_0 as

$$q_0 = i\frac{\pi\omega_0^2 n}{\lambda} \qquad \lambda = \frac{2\pi n}{k} \tag{6.6-8}$$

The choice of an imaginary q_0 will be found to lead to physically meaningful waves whose energy density is confined near the z axis. With this last substitution let us consider, one at a time, the two factors in (6.6-7). The first one becomes

$$\exp\left[-\ln\left(1 - i\frac{\lambda z}{\pi\omega_0^2 n}\right)\right] = \frac{1}{\sqrt{1 + \dfrac{\lambda^2 z^2}{\pi^2\omega_0^4 n^2}}}\exp\left[i\tan^{-1}\left(\frac{\lambda z}{\pi\omega_0^2 n}\right)\right] \tag{6.6-9}$$

[5] The constant of integration will merely modify the phase of the field solution (6.5-7). Since the time origin is arbitrary, the phase can be taken as zero.

where we used $\ln(a + ib) = \ln\sqrt{a^2 + b^2} + i\tan^{-1}(b/a)$. Substituting (6.6-8) in the second term of (6.6-7) and separating the exponent into its real and imaginary parts, we obtain

$$\exp\left[\frac{-ikr^2}{2(q_0 + z)}\right] = \exp\left\{\frac{-r^2}{\omega_0^2\left[1 + \left(\dfrac{\lambda z}{\pi\omega_0^2 n}\right)^2\right]} - \frac{ikr^2}{2z\left[1 + \left(\dfrac{\pi\omega_0^2 n}{\lambda z}\right)^2\right]}\right\} \tag{6.6-10}$$

If we define the following parameters:

$$\omega^2(z) = \omega_0^2\left[1 + \left(\frac{\lambda z}{\pi\omega_0^2 n}\right)^2\right] = \omega_0^2\left(1 + \frac{z^2}{z_0^2}\right) \tag{6.6-11}$$

$$R = z\left[1 + \left(\frac{\pi\omega_0^2 n}{\lambda z}\right)^2\right] = z\left(1 + \frac{z_0^2}{z^2}\right) \tag{6.6-12}$$

$$\eta(z) = \tan^{-1}\left(\frac{\lambda z}{\pi\omega_0^2 n}\right) = \tan^{-1}\left(\frac{z}{z_0}\right) \tag{6.6-13}$$

$$z_0 \equiv \frac{\pi\omega_0^2 n}{\lambda}$$

we can combine (6.6-9) and (6.6-10) in (6.6-7) and, recalling that $E(x, y, z) = \psi(x, y, z)\exp(-ikz)$, obtain

$$\begin{aligned} E(x, y, z) &= E_0\frac{\omega_0}{\omega(z)}\exp\left\{-i[kz - \eta(z)] - i\frac{kr^2}{2q(z)}\right\} \\ &= E_0\frac{\omega_0}{\omega(z)}\exp\left\{-i[kz - \eta(z)] - r^2\left[\frac{1}{\omega^2(z)} + \frac{ik}{2R(z)}\right]\right\} \end{aligned} \tag{6.6-14}$$

so that if we use (6.5-9) and (6.6-4)

$$\frac{1}{q(z)} = \frac{1}{R(z)} - i\frac{\lambda}{\pi n\omega^2(z)} \tag{6.6-14a}$$

This is our basic result. We refer to it as the fundamental Gaussian-beam solution since we have excluded the more complicated solutions of (6.5-3) (i.e., those with azimuthal variation) by limiting ourselves to transverse dependence involving $r = (x^2 + y^2)^{1/2}$ only. These higher-order modes will be discussed separately.

From (6.6-14), the parameter $\omega(z)$, which evolves according to (6.6-11), is the distance r at which the field amplitude is down by a factor $1/e$ compared to its value on the axis. We will consequently refer to it as the beam "spot size." The parameter ω_0 is the minimum spot size. It is the beam spot size at the plane $z = 0$. The parameter R in (6.6-14) is the radius of curvature of the very nearly spherical wavefronts[6] at z. We can verify this statement by deriving the radius of curvature of the constant phase surfaces (wavefronts) or,

[6] Actually, it follows from (6.6-14) that, with the exception of the immediate vicinity of the plane $z = 0$, the wavefronts are parabolic since they are defined by $k[z + (r^2/2R)] = \text{const}$. For $r^2 \ll z^2$, the distinction between parabolic and spherical surfaces is not important.

more simply, by considering the form of a spherical wave emitted by a point radiator placed at $z = 0$. It is given by

$$\begin{aligned} E &\propto \frac{1}{R} e^{-ikR} = \frac{1}{R} \exp(-ik\sqrt{x^2 + y^2 + z^2}) \\ &\simeq \frac{1}{R} \exp\left(-ikz - ik\frac{x^2 + y^2}{2R}\right) \qquad x^2 + y^2 \ll z^2 \end{aligned} \tag{6.6-15}$$

since z is equal to R, the radius of curvature of the spherical wave. Comparing (6.6-15) with (6.6-14), we identify R as the radius of curvature of the Gaussian beam. The convention regarding the sign of R is the same as that adopted in Table 6.1; that is, $R(z)$ is negative if the center of curvature occurs at $z' > z$ and vice versa.

The form of the fundamental Gaussian beam is, according to (6.6-14), uniquely determined once its minimum spot size ω_0 and its location—that is, the plane $z = 0$—are specified. Its spot size ω and radius of curvature R at any plane z are then found from (6.6-11) and (6.6-12). Some of these characteristics are displayed in Figure 6.5. The hyperbolas shown in this figure correspond to the ray direction and are intersections of planes that include the z axis and the hyperboloids

$$x^2 + y^2 = \text{const. } \omega^2(z) \tag{6.6-16}$$

They correspond to the direction of energy propagation. The spherical surfaces shown have radii of curvature given by (6.6-12). For large z, the hyperboloids $x^2 + y^2 = \omega^2$ are asymptotic to the cone

$$r = \sqrt{x^2 + y^2} = \frac{\lambda}{\pi\omega_0 n} z \tag{6.6-17}$$

whose half-apex angle, which we take as a measure of the angular beam spread, is

$$\theta_{\text{beam}} = \tan^{-1}\left(\frac{\lambda}{\pi\omega_0 n}\right) \simeq \frac{\lambda}{\pi\omega_0 n} \tag{6.6-18}$$

This last result is a rigorous manifestation of wave diffraction according to which a wave that is confined in the transverse direction to an aperture

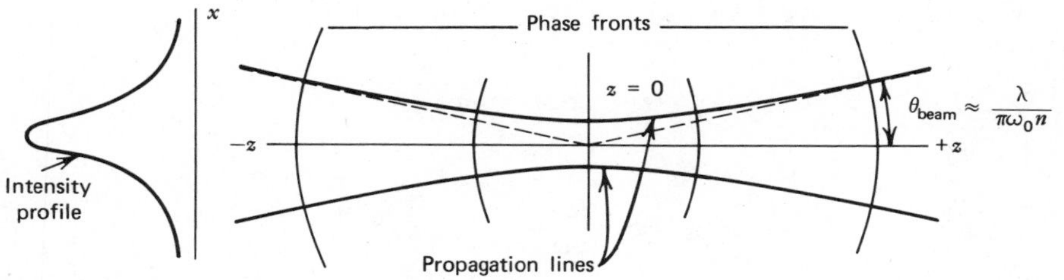

FIGURE 6.5 Propagating Gaussian beam.

of radius ω_0 will spread (diffract) in the far field ($z \gg \pi\omega_0^2 n/\lambda$) according to (6.6-18).

6.7 THE FUNDAMENTAL GAUSSIAN BEAM IN A LENSLIKE MEDIUM—THE *ABCD* LAW

We now return to the general case of a lenslike medium so that $k_2 \neq 0$. The P and Q functions of (6.5-9) obey, according to (6.5-11),

$$\begin{aligned} Q^2 + kQ' + kk_2 &= 0 \\ P' &= -iQ/k \end{aligned} \tag{6.7-1}$$

Using the change of variables

$$Q = \frac{ks'}{s} \tag{6.7-2}$$

we obtain from (6.7-1)

$$s'' + s\frac{k_2}{k} = 0$$

so that

$$\begin{aligned} s(z) &= a \sin \sqrt{\frac{k_2}{k}}\, z + b \cos \sqrt{\frac{k_2}{k}}\, z \\ s'(z) &= a\sqrt{\frac{k_2}{k}} \cos \sqrt{\frac{k_2}{k}}\, z - b\sqrt{\frac{k_2}{k}} \sin \sqrt{\frac{k_2}{k}}\, z \end{aligned} \tag{6.7-3}$$

where a and b are arbitrary constants.

Using (6.7-3) in (6.7-2) and expressing the result in terms of an input value $q_0 = k/Q(0)$ gives the following result for the complex beam radius $q(z)$:

$$q(z) \equiv \frac{k}{Q(z)} = \frac{\cos\left(\sqrt{\frac{k_2}{k}}\, z\right) q_0 + \sqrt{\frac{k}{k_2}} \sin\left(\sqrt{\frac{k_2}{k}}\, z\right)}{-\sin\left(\sqrt{\frac{k_2}{k}}\, z\right)\sqrt{\frac{k_2}{k}}\, q_0 + \cos\left(\sqrt{\frac{k_2}{k}}\, z\right)} \tag{6.7-4}$$

The physical significance of $q(z)$ in this case can be extracted from (6.5-9). We expand the part of $\psi(r, z)$ that involves r. The result is

$$\psi \propto e^{-iQ(z)r^2/2} = e^{-ikr^2/2q(z)}$$

If we express the real and imaginary parts of $q(z)$ by

$$\frac{1}{q(z)} = \frac{1}{R(z)} - i\frac{\lambda}{\pi n \omega^2(z)} \tag{6.7-5}$$

we obtain

$$\psi \propto \exp\left[\frac{-r^2}{\omega^2(z)} - i\frac{kr^2}{2R(z)}\right]$$

so that $\omega(z)$ is the beam spot size and R its radius of curvature, as in the case of a homogeneous medium, which is described by (6.6-14). For the special case of a homogeneous medium ($k_2 = 0$), (6.7-4) reduces to (6.6-5).

The Transformation of the Gaussian Beam (The ABCD Law)

We have derived in Section 6.7 the transformation law (6.7-4) of a Gaussian beam propagating through a generalized lenslike medium that is characterized by k_2. We note first by comparing (6.7-4) to (6.4-5, 6) that the transformation can be described by

$$q_2 = \frac{Aq_1 + B}{Cq_1 + D} \tag{6.7-6}$$

where A, B, C, D are the elements of the ray matrix characterizing the *same* medium. It follows immediately that the propagation through, or reflection from, any of the elements shown in Table 6.1 also obeys (6.7-6) since these elements can all be viewed as special cases of a lenslike medium. For future reference, we note that by applying (6.7-6) to a thin lens of focal length f, we obtain

$$\frac{1}{q_2} = \frac{1}{q_1} - \frac{1}{f} \tag{6.7-7}$$

so that using (6.7-5) yields

$$\begin{aligned} \omega_2 &= \omega_1 \\ \frac{1}{R_2} &= \frac{1}{R_1} - \frac{1}{f} \end{aligned} \tag{6.7-8}$$

These results apply, as well, to reflection from a mirror with a radius of curvature R if we replace f by $R/2$.

Consider next the propagation of a Gaussian beam through two lenslike media that are adjacent to each other. The ray matrix describing the first one is (A_1, B_1, C_1, D_1), whereas that of the second one is (A_2, B_2, C_2, D_2). Taking the input beam parameter as q_1 and the output beam parameter as q_3, we have from (6.7-6)

$$q_2 = \frac{A_1q_1 + B_1}{C_1q_1 + D_1}$$

for the beam parameter at the output of medium 1 and

$$q_3 = \frac{A_2q_2 + B_2}{C_2q_2 + D_2}$$

and after combining the last two equations, we obtain

$$q_3 = \frac{A_Tq_1 + B_T}{C_Tq_1 + D_T} \tag{6.7-9}$$

where (A_T, B_T, C_T, D_T) are the elements of the ray matrix relating the output plane (3) to the input one (1), that is,

$$\begin{vmatrix} A_T & B_T \\ C_T & D_T \end{vmatrix} = \begin{vmatrix} A_2 & B_2 \\ C_2 & D_2 \end{vmatrix} \begin{vmatrix} A_1 & B_1 \\ C_1 & D_1 \end{vmatrix} \tag{6.7-10}$$

It follows by induction that (6.7-9) applies to the propagation of a Gaussian beam through any arbitrary number (e.g., n) of lenslike media and elements. The matrix (A_T, B_T, C_T, D_T) is the product of the n matrices characterizing the individual members of the chain.

The great power of the *ABCD* law is that it enables us to trace the Gaussian beam parameter $q(z)$ through a complicated sequence of lenslike elements. The beam radius $R(z)$ and spot size $\omega(z)$ at any plane z can be recovered through the use of (6.7-5). The application of this method will be made clear by the following example.

Example: Gaussian Beam Focusing. As an example of the application of the *ABCD* law, we consider the case of a Gaussian beam that is incident at its waist on a thin lens of focal length f as shown in Figure 6.6. We will find the location of the waist of the output beam and the beam radius at that point.

At the input plane (1) $\omega = \omega_{01}$, $R_1 = \infty$ so that

$$\frac{1}{q_1} = \frac{1}{R_1} - i\frac{\lambda}{\pi\omega_{01}^2 n} = -i\frac{\lambda}{\pi\omega_{01}^2 n}$$

Using (6.7-8) leads to

$$\frac{1}{q_2} = \frac{1}{q_1} - \frac{1}{f} = -\frac{1}{f} - i\frac{\lambda}{\pi\omega_{01}^2 n}$$

$$q_2 = \frac{1}{-\dfrac{1}{f} - i\dfrac{\lambda}{\pi\omega_{01}^2 n}} = \frac{-a + ib}{a^2 + b^2}$$

$$a \equiv \frac{1}{f} \qquad b \equiv \frac{\lambda}{\pi\omega_{01}^2 n}$$

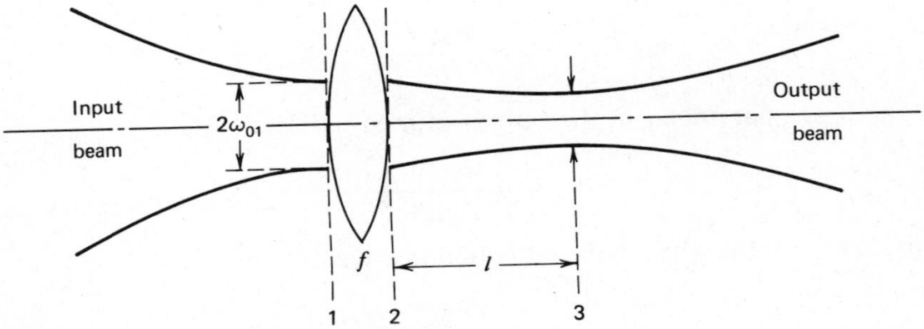

FIGURE 6.6 Focusing of a Gaussian beam.

At plane (3) we obtain, using (6.6-5),

$$q_3 = q_2 + l = \frac{-a}{a^2 + b^2} + l + \frac{ib}{a^2 + b^2}$$

$$\frac{1}{q_3} = \frac{1}{R_3} - i\frac{\lambda}{\pi\omega_3^2 n}$$

$$= \frac{\left(\frac{-a}{a^2 + b^2} + l\right) - i\frac{b}{a^2 + b^2}}{\left(\frac{-a}{a^2 + b^2} + l\right)^2 + \left(\frac{b}{a^2 + b^2}\right)^2}$$

Since plane (3) is, according to the statement of the problem, to correspond to the output beam waist, $R_3 = \infty$. Using this fact in the last equation leads to

$$l = \frac{a}{a^2 + b^2} = \frac{f}{1 + \left(\frac{f}{\pi\omega_{01}^2 n/\lambda}\right)^2} = \frac{f}{1 + \left(\frac{f}{z_{01}}\right)^2} \tag{6.7-11}$$

as the location of the new waist, and to

$$\frac{\omega_3}{\omega_{01}} = \frac{\frac{f\lambda}{\pi\omega_{01}^2 n}}{\sqrt{1 + \left(\frac{f\lambda}{\pi\omega_{01}^2 n}\right)^2}} = \frac{\frac{f}{z_{01}}}{\sqrt{1 + \left(\frac{f}{z_{01}}\right)^2}} \tag{6.7-12}$$

for the output beam waist. The confocal beam parameter

$$z_{01} \equiv \frac{\pi\omega_{01}^2 n}{\lambda}$$

is, according to (6.6-11), the distance from the waist in which the beam spot size increases by $\sqrt{2}$ and is a convenient measure of the convergence of the input beam.

6.8 A GAUSSIAN BEAM IN A LENS WAVEGUIDE

As another example of the application of the *ABCD* law, we consider the propagation of a Gaussian beam through a sequence of thin lenses, as shown in Figure 6.2. The matrix, relating a ray in plane $s + 1$ to the plane $s = 1$, is

$$\begin{vmatrix} A_T & B_T \\ C_T & D_T \end{vmatrix} = \begin{vmatrix} A & B \\ C & D \end{vmatrix}^s \tag{6.8-1}$$

where (A, B, C, D) is the matrix for propagation through a single unit cell $(\Delta s = 1)$ and is given by (6.1-6). We can use a well-known formula for the

s'th power of a matrix with a unity determinant (unimodular) to obtain

$$
\begin{aligned}
A_T &= \frac{A \sin(s\theta) - \sin[(s-1)\theta]}{\sin \theta} \\
B_T &= \frac{B \sin(s\theta)}{\sin \theta} \\
C_T &= \frac{C \sin(s\theta)}{\sin \theta} \\
D_T &= \frac{D \sin(s\theta) - \sin[(s-1)\theta]}{\sin \theta}
\end{aligned}
\tag{6.8-2}
$$

where

$$
\cos \theta = \tfrac{1}{2}(A + D) = \left(1 - \frac{d}{f_2} - \frac{d}{f_1} + \frac{d^2}{2f_1 f_2}\right) \tag{6.8-3}
$$

and then use (6.8-2) in (6.7-6) with the result

$$
q_{s+1} = \frac{\{A \sin(s\theta) - \sin[(s-1)\theta]\}q_1 + B \sin(s\theta)}{C \sin(s\theta) q_1 + D \sin(s\theta) - \sin[(s-1)\theta]} \tag{6.8-4}
$$

The condition for the confinement of the Gaussian beam by the lens sequence is, from (6.8-4), that θ be real; otherwise, the sine functions will yield growing exponentials. From (6.8-3), this condition becomes $|\cos \theta| \leqslant 1$, or

$$
0 \leqslant \left(1 - \frac{d}{2f_1}\right)\left(1 - \frac{d}{2f_2}\right) \leqslant 1 \tag{6.8-5}
$$

that is, the same as condition (6.1-16) for stable-ray propagation.

6.9 HIGH-ORDER GAUSSIAN BEAM MODES IN A HOMOGENEOUS MEDIUM

The Gaussian mode treated up to this point has a field variation that depends only on axial distance z and distance r from the axis. If we do not impose the condition $\partial/\partial\phi = 0$ [where ϕ is the azimuthal angle in a cylindrical coordinate system (r, ϕ, z)] and take $k_2 = 0$, the wave equation (6.5-3) has solutions in the form of (Supplementary Reference 1 and Reference 13).

$$
\begin{aligned}
E_{l,m}(x, y, z) &= E_0 \frac{\omega_0}{\omega(z)} H_l\left[\sqrt{2}\frac{x}{\omega(z)}\right] H_m\left[\sqrt{2}\frac{y}{\omega(z)}\right] \\
&\quad \times \exp\left[-ik\frac{x^2 + y^2}{2q(z)} - ikz + i(m + n + 1)\eta\right] \\
&= E_0 \frac{\omega_0}{\omega(z)} H_l\left[\sqrt{2}\frac{x}{\omega(z)}\right] H_m\left(\sqrt{2}\frac{y}{\omega(z)}\right] \\
&\quad \times \exp\left[-\frac{x^2 + y^2}{\omega^2(z)} - \frac{ik(x^2 + y^2)}{2R(z)} - ikz + i(l + m + 1)\eta\right]
\end{aligned}
\tag{6.9-1}
$$

where H_l is the Hermite polynomial of order l, and $\omega(z)$, $R(z)$, $q(z)$, and η are given by (6.6-11) through (6.6-13).

We note for future reference that the phase shift on the axis is

$$\theta = kz - (l + m + 1)\tan^{-1}(z/z_0)$$
$$z_0 = \pi\omega_0^2 n/\lambda \tag{6.9-2}$$

The transverse variation of the electric field along x (or y) is seen to be of the form

$$E_l\left(\frac{\sqrt{2}\,x}{\omega}\right) \propto H_n(\xi)e^{-\xi^2/2} \tag{6.9-3}$$

with $\xi = \sqrt{2}\,x/\omega$. According to (6.9-3), the solution $E_l(\sqrt{2}\,x/\omega)$ is the same as that of the harmonic oscillator wavefunction $u_l(\xi)$ obtained in Chapter 2. We can, thus, use the field $u_l(\xi)$ and intensity $|u_l(\xi)|^2$ plots of Figure 2.1 to describe the distribution of the optical Gaussian fields. Photographs of actual field patterns are shown in Figure 6.7. Note that the first four pictures correspond to the intensity $|u_l(\xi)|^2$ plots ($l = 0, 1, 2, 3$) of Figure 2.1.

6.10 HIGH-ORDER GAUSSIAN BEAM MODES IN QUADRATIC INDEX MEDIA

In Section 6.7 we treated the propagation of a circularly symmetric Gaussian beam in lenslike media. Here, we extend the treatment to higher-order modes and limit our attention to steady-state [i.e., $q(z)$ = const.] solution in media whose index of refraction can be described by

$$n^2(\mathbf{r}) = n^2\left(1 - \frac{n_2}{n}r^2\right) \tag{6.10-1}$$

that is consistent with (6.5-5) if we put $k_2 = 2\pi n_2/\lambda$.

The vector-wave equation (6.5-3) becomes

$$\nabla^2\mathbf{E} + k^2\left(1 - \frac{n_2}{n}r^2\right)\mathbf{E} = 0$$

where $k = 2\pi n/\lambda$ is the propagation constant in a homogeneous medium of index n.

If we assume a scalar field in the form $E(\mathbf{r}) = \psi(\mathbf{r})\exp(-i\beta z)$, the last equation becomes

$$\frac{\partial^2\psi}{\partial x^2} + \frac{\partial^2\psi}{\partial y^2} + \left[k^2\left(1 - \frac{n_2}{n}r^2\right) - \beta^2\right]\psi = 0 \tag{6.10-2}$$

This equation is easily separated by taking $\psi = f(x)g(y)$. Each of the two resulting differential equations then has a form identical to the harmonic oscillator differential equation (2.2-2). In order to use directly the results of Section 2.2, we perform the linear change of variables

$$\xi \to \frac{\sqrt{2}\,x}{\omega} \qquad \omega = \left(\frac{\lambda}{\pi}\right)^{1/2}\left(\frac{1}{nn_2}\right)^{1/4} \tag{6.10-3}$$

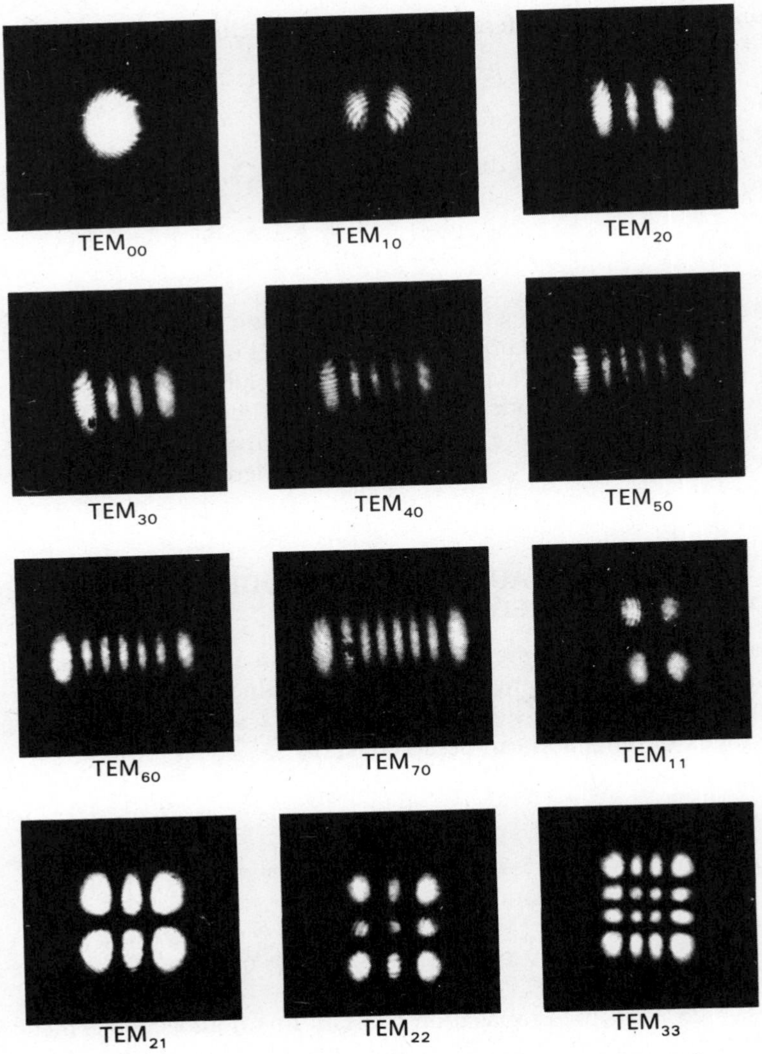

FIGURE 6.7 Some low-order optical-beam modes. *Source:* Reference 9.

and obtain directly from (2.2-14)

$$\psi_{l,m}(x, y) = E_{l,m}(\mathbf{r})e^{i\beta_{l,m}z} = E_0 H_l\left(\sqrt{2}\,\frac{x}{\omega}\right) H_m\left(\sqrt{2}\,\frac{y}{\omega}\right) \exp\left(-\frac{x^2 + y^2}{\omega^2}\right) \quad (6.10\text{-}4)$$

where H_l is the Hermite polynomial of order l. The eigenvalue $\beta_{l,m}$ is obtained from (2.2-8a) and (6.10-3)

$$\beta_{l,m} = k\left[1 - \frac{2}{k}\sqrt{\frac{n_2}{n}}\,(l + m + 1)\right]^{1/2} \quad (6.10\text{-}5)$$

Group Velocity Dispersion of Quadratic Index Media

Glass fibers with an index variation in the form of (6.10-1) are important potential candidates for optical communication purposes because of their mode dispersion properties (References 11 and 12). Consider a short pulse of light entering such a fiber and exciting simultaneously a large number of modes $E_{l,m}$. The propagation velocity of the pulse in each one of these modes is given by the group velocity

$$(v_g)_{l,m} = \frac{d\omega}{d\beta_{l,m}} \tag{6.10-6}$$

For the pulse width not to increase as it propagates through the guiding medium, it is necessary that $(v_g)_{l,m}$ be independent of m and l.

In the case considered here and for fibers of small index variation so that

$$\frac{1}{k}\sqrt{\frac{n_2}{n}}\,(l + m + 1) \ll 1 \tag{6.10-7}$$

we can approximate (6.10-5) as

$$\beta_{l,m} \cong k - \sqrt{\frac{n_2}{n}}\,(l + m + 1) - \frac{n_2}{2kn}\,(l + m + 1)^2 \tag{6.10-8}$$

so that according to (6.10-6),

$$(v_g)_{l,m} = \frac{c/n}{\left[1 + \dfrac{(n_2/n)}{2k^2}\,(l + m + 1)^2\right]} \tag{6.10-9}$$

We thus conclude that subject to condition (6.10-7), the effect of mode dispersion on pulse broadening is of second order in $[(n_2/n)/2k^2]^{1/2}(l + m + 1)$.

One important consequence of the group velocity dependence on ω (group velocity dispersion) is the spreading with distance of optical pulses. Consider a pulse of duration τ. Its spectral width is thus $\Delta\omega \sim (1/\pi\tau)$. If the pulse is used to excite the $(l,\ m)$ mode of the waveguide, its width will increase in a distance L by

$$\Delta\tau \approx \frac{L}{v_g^2}\frac{dv_g}{d\omega}\left(\frac{1}{\pi\tau}\right) \tag{6.10-10}$$

where from (6.10-7) and (6.10-9),

$$\frac{dv_g}{d\omega} \approx \frac{(n_2/n)}{k^3}\,(l + m + 1)^2 \tag{6.10-11}$$

This spreading limits the pulse repetition rate and, thus, the information transmission rate, which can be employed in a channel of length L to a value where the pulse spread $\Delta\tau$ does not exceed the separation between pulses.

6.11 PROPAGATION IN MEDIA WITH A QUADRATIC GAIN PROFILE

In many laser media, the gain is a strong function of position. This variation can be due to a variety of causes including (1) the radial distribution of

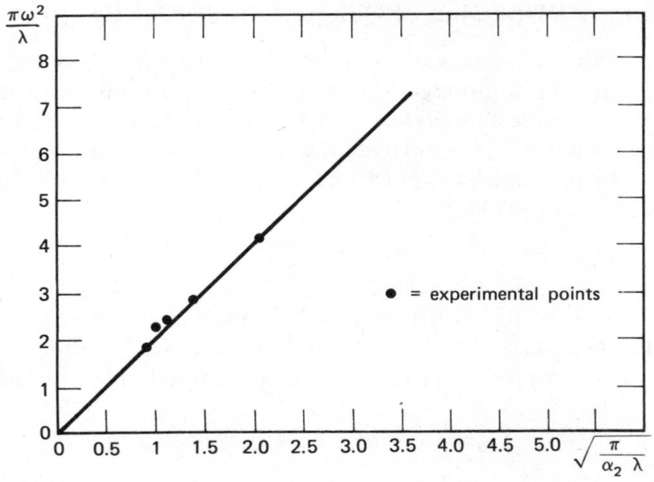

FIGURE 6.8 Theoretical curve showing the dependence of beam radius on quadratic gain constant α_2. Experimental points were obtained in a xenon 3.39 μm laser in which α_2 was varied by controlling the unsaturated laser gain. *Source:* Reference 15.

energetic electrons in the plasma region of gas lasers (References 15 and 16), (2) the variation of pumping intensity in solid-state lasers, and (3) the dependence of the degree of gain saturation on the radial position in the beam.

We can account for an optical medium with quadratic gain (or loss) variation by taking the complex propagation constant $k(r)$ in (6.5-5) as

$$k(r) = k \pm i(\alpha_0 - \tfrac{1}{2}\alpha_2 r^2) \tag{6.11-1}$$

where the plus (minus) sign applies to the case of gain (loss). Assuming $k_2 r^2 \ll k$ in (6.5-5), we have $k_2 = i\alpha_2$. Using this value in (6.5-11) to obtain the steady-state[7] ($Q' = 0$) solution of the complex beam radius yields

$$\frac{1}{q} = -i\sqrt{\frac{k_2}{k}} = -i\sqrt{\frac{i\alpha_2}{k}} \tag{6.11-2}$$

The steady-state beam radius and spot size are obtained from (6.7-5) and (6.11-2)

$$\begin{aligned} \omega^2 &= 2\sqrt{\frac{\lambda}{\pi n \alpha_2}} \\ R &= 2\sqrt{\frac{\pi n}{\lambda \alpha_2}} \end{aligned} \tag{6.11-3}$$

We thus find that the steady-state solution corresponds to phase fronts with a constant spot size but with a finite radius of curvature.

[7] "Steady-state" here refers not to the intensity, which according to (6.11-1) is growing or decaying, but to the beam radius of curvature and spot size.

The general (nonsteady state) behavior of the Gaussian beam in a quadratic gain medium is described by (6.7-4) where $k_2 = i\alpha_2$.

Experimental data showing a decrease of the beam spot size with increasing gain parameter α_2 in agreement with (6.11-3) are shown in Figure 6.8.

6.12 ELLIPTIC GAUSSIAN BEAMS

All the beam solutions considered up to this point have one feature in common. The exponential part of the field drops off as in (6.9-1), according to

$$E_{m,n} \propto \exp\left[-\frac{x^2+y^2}{\omega^2(z)}\right] \tag{6.12-1}$$

We will refer to such beams as circular Gaussian beams.

The wave equation (6.5-3) also allows solutions in which the variation in the x and y directions is characterized by

$$E_{m,n} \propto \exp\left[-\frac{x^2}{\omega_x^2(z)} - \frac{y^2}{\omega_y^2(z)}\right] \tag{6.12-2}$$

with $\omega_x \neq \omega_y$. Such beams, which we name elliptic Gaussian, result when a circular Gaussian beam passes through a cylindrical lens or, as a second example, when a laser beam emerges from an astigmatic resonator, that is, one whose mirrors possess different radii of curvature in the z-y and z-x planes.

We will not repeat the whole derivation for this case but will indicate the main steps.

Instead of (6.5-9), we assume a solution

$$\psi = \exp\left\{-i\left[P(z) + \frac{Q_x(z)x^2}{2} + \frac{Q_y(z)y^2}{2}\right]\right\} \tag{6.12-3}$$

that results, in a manner similar to (6.5-11), in[8]

$$Q_x^2 + k\frac{dQ_x}{dz} + kk_{2x} = 0$$
$$Q_y^2 + k\frac{dQ_y}{dz} + kk_{2y} = 0 \tag{6.12-4}$$

and

$$\frac{dP}{dz} = -i\left(\frac{Q_x+Q_y}{2k}\right) \tag{6.12-5}$$

Defining

$$q_{\substack{x\\y}}(z) = \frac{k}{Q_{\substack{x\\y}}(z)} \tag{6.12-6}$$

[8] The parameters k_{2x} and k_{2y} are defined by

$$k^2(x, y) = k^2 - kk_{2x}x^2 - kk_{2y}y^2$$

which is a generalization of (6.5-5).

we obtain in the case of a homogeneous ($k_{2x} = k_{2y} = 0$) beam as in (6.6-5)

$$q_x(z) = z + C_x \tag{6.12-7}$$

where C_x is an arbitrary constant of integration. We find it useful to write C_x as

$$C_x = -z_x + q_{0x}$$

where z_x is real and q_{0x} is imaginary. The physical significance of these two constants will become clear in what follows. A similar result with $x \to y$ is obtained for $q_y(z)$. Using the solutions of $q_x(z)$ and $q_y(z)$ in (6.12-5) gives

$$P = -\frac{i}{2}\left[\ln\left(1 + \frac{z - z_x}{q_{0x}}\right) + \ln\left(1 + \frac{z - z_y}{q_{0y}}\right)\right]$$

Proceeding straightforwardly as in the derivation connecting (6.6-6) through (6.6-14) results in

$$\begin{aligned} E(x, y, z) &= E_0 \frac{\sqrt{\omega_{0x}\omega_{0y}}}{\sqrt{\omega_x(z)\omega_y(z)}} \exp\left\{-i[kz - \eta(z)] - \frac{ikx^2}{2q_x(z)} - \frac{iky^2}{2q_y(z)}\right\} \\ &= E_0 \frac{\sqrt{\omega_{0x}\omega_{0y}}}{\sqrt{\omega_x(z)\omega_y(z)}} \exp\left\{-i[kz - \eta(z)] - x^2\left[\frac{1}{\omega_x^2(z)} + \frac{ik}{2R_x(z)}\right]\right. \\ &\qquad \left. - y^2\left[\frac{1}{\omega_y^2(z)} + \frac{ik}{2R_y(z)}\right]\right\} \end{aligned} \tag{6.12-8}$$

where

$$\begin{aligned} q_{0x} &= i\frac{\pi\omega_{0x}^2 n}{\lambda} \\ \omega_x^2(z) &= \omega_{0x}^2\left\{1 + \left[\frac{\lambda(z - z_x)}{\pi\omega_{0x}^2 n}\right]^2\right\} \\ R_x(z) &= (z - z_x)\left\{1 + \left[\frac{\pi\omega_{0x}^2 n}{\lambda(z - z_x)}\right]^2\right\} \end{aligned} \tag{6.12-9}$$

with similar expression in which $x \to y$ for q_{0y}, ω_y, R_y.

The angle $\eta(z)$ in (6.12-8) is now given by

$$\eta(z) = \tfrac{1}{2}\tan^{-1}\left[\frac{\lambda(z - z_x)}{\pi\omega_{0x}^2 n}\right] + \tfrac{1}{2}\tan^{-1}\left[\frac{\lambda(z - z_y)}{\pi\omega_{0y}^2 n}\right] \tag{6.12-10}$$

It follows that all the results derived for the case of circular Gaussian beams apply, separately, to the x-z and to the y-z behavior of the elliptic Gaussian beam. For the purpose of analysis, the elliptic beam can be considered two independent "beams." The position of the waist, $z = 0$, is not necessarily the same for these two beams. It occurs at $z = z_x$ for the x-z beam and at $z = z_y$ for the y-z beam in the example of Figure 6.9 where z_x and z_y are arbitrary.

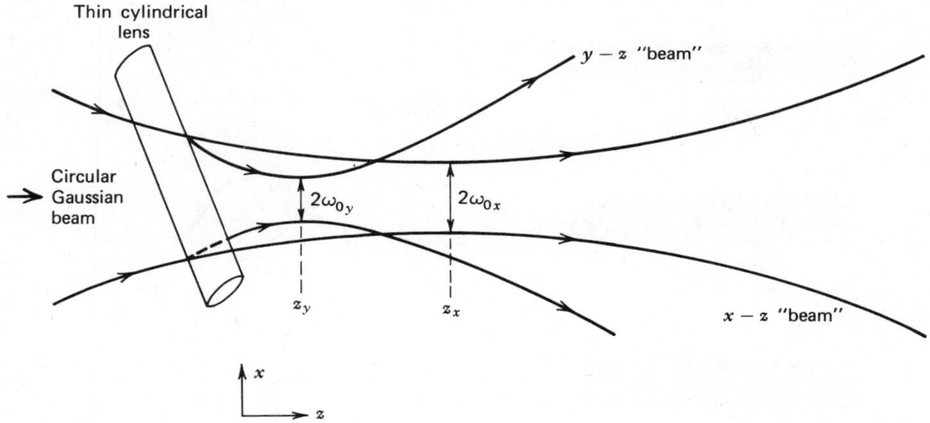

FIGURE 6.9 An illustration of an elliptic beam produced by cylindrical focusing of a circular Gaussian beam.

It also follows from the similarity between (6.12-4) and (6.5-11) that the *ABCD* transformation law (6.7-9) can be applied separately to $q_x(z)$ and $q_y(z)$ that, according to (6.12-8), are given by

$$\frac{1}{q_x(z)} = \frac{1}{R_x(z)} - i\frac{\lambda}{\pi n \omega_x^2(z)}$$
$$\frac{1}{q_y(z)} = \frac{1}{R_y(z)} - i\frac{\lambda}{\pi n \omega_y^2(z)} \qquad (6.12\text{-}11)$$

Elliptic Gaussian Beams in a Quadratic Lenslike Medium

Here, we consider the *steady-state* elliptic beam propagating in a medium whose index of refraction is given by

$$n^2(\mathbf{r}) = n^2\left(1 - \frac{n_{2x}}{n}x^2 - \frac{n_{2y}}{n}y^2\right) \qquad (6.12\text{-}12)$$

The derivation proceeds along the same lines as in Section 6.10, resulting in

$$E_{l,m}(\mathbf{r}) = E_0 e^{-i\beta_{l,m}z} H_l\left(\sqrt{2}\frac{x}{\omega_x}\right) H_m\left(\sqrt{2}\frac{y}{\omega_y}\right)\exp\left(-\frac{x^2}{\omega_x^2} - \frac{y^2}{\omega_y^2}\right) \qquad (6.12\text{-}13)$$

where

$$\omega_x = \left(\frac{\lambda}{\pi}\right)^{1/2}\left(\frac{1}{nn_{2x}}\right)^{1/4}$$
$$\omega_y = \left(\frac{\lambda}{\pi}\right)^{1/2}\left(\frac{1}{nn_{2y}}\right)^{1/4} \qquad (6.12\text{-}14)$$

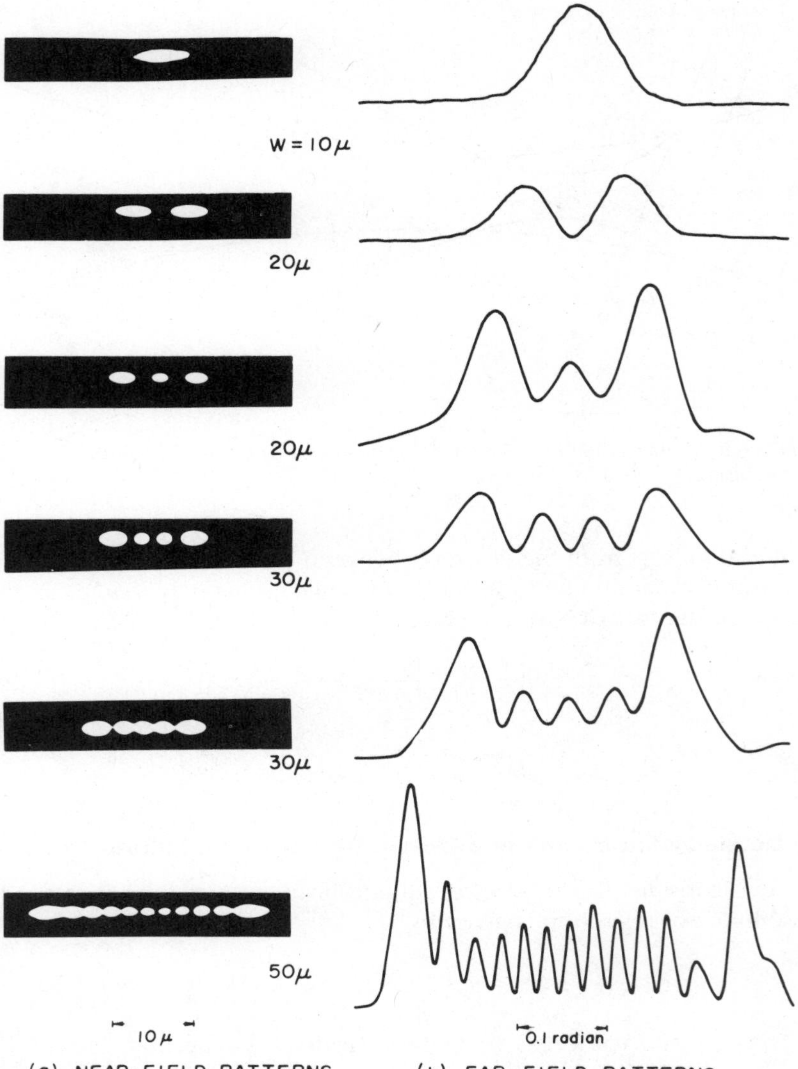

FIGURE 6.10 (*a*) Near-field and (*b*) far-field intensity distributions of the output of stripe contact GaAs-GaA1As lasers. *Source:* Reference 20.

$$\beta_{l,m} = k\left\{1 - \frac{2}{k}\left[\sqrt{\frac{n_{2x}}{n}}\,(l + \tfrac{1}{2}) + \sqrt{\frac{n_{2y}}{n}}\,(m + \tfrac{1}{2})\right]\right\}^{1/2} \qquad (6.12\text{-}15)$$

Elliptic Gaussian beams have been observed experimentally in the output of stripe-geometry gallium arsenide junction lasers (References 18, 19, and 20). Near and far field experimental intensity distributions corresponding to some $(0, m)$ modes are shown in Figure 6.10.

References

1. Pierce, J. R., *Theory and Design of Electron Beams,* 2d ed. (New York: Van Nostrand, 1954), Chap. 11.
2. Ramo, S., J. R. Whinnery, and T. Van Duzer, *Fields and Waves in Communication Electronics* (New York: Wiley, 1965), p. 576.
3. Yariv, A., *Introduction to Optical Electronics* (New York: Holt, Rinehart and Winston, 1971).
4. Siegman, A. E., *An Introduction to Lasers and Masers* (New York: McGraw-Hill, 1968).
5. Kogelnik, H. and T. Li, "Laser Beams and Resonators," *Proc. IEEE* **54** (1966), p. 1312.
6. Herriot, D., H. Kogelnik, and R. Kompfner, "Off-Axis Paths in Spherical Mirror Interferometers," *Appl. Opt.* **3** (1964), p. 523.
7. Kogelnik, H., "On the Propagation of Gaussian Beams of Light Through Lenslike Media Including Those with a Loss and Gain Variation," *Appl. Opt.* **4** (1965), p. 1562.
8. Born, M. and E. Wolf, *Principles of Optics,* 3rd ed. (New York: Pergamon, 1965).
9. Kogelnik, H. and W. Rigrod, *Proc. IRE* **50** (1962), p. 230.
10. Dabby, F. W. and J. R. Whinnery, "Thermal Self-Focusing of Laser Beams in Lead Glasses," *Appl. Phys. Lett.* **13** (1968), p. 284.
11. Kawakami, S. and J. Nishizawa, "An Optical Waveguide with the Optimum Distribution of the Refractive Index with Reference to Waveform Distortion, *IEEE Trans. Microwave Theory and Technique* **MTT-16,** 10 (1968), p. 814.
12. Marcuse, D., "The Impulse Response of an Optical Fiber with Parabolic Index Profile," *Bell System. Tech. Jour.* **52** (1973), p. 1169.
13. Casperson, L., "Modes and Spectra of High Gain Lasers," Ph.D. Thesis, California Institute of Technology, Pasadena, Calif. (1971).
14. Tien, P. K., J. P. Gordon, and J. R. Whinnery, "Focusing of a Light Beam of Gaussian Field Distribution in Continuous and Periodic Lenslike Media," *Proc. IEEE* **53** (1965), p. 129.
15. Casperson, L. and A. Yariv, "The Gaussian Mode in Optical Resonators with a Radial Gain Profile," *Appl. Phys. Lett.* **12** (1968), p. 355.
16. Bennett, W. R., "Inversion Mechanisms in Gas Lasers," *Appl. Opt., Suppl. 2, Chemical Lasers* **3** (1965).
17. Casperson, L., "Gaussian Light Beams in Inhomogeneous Media," *Appl. Opt.* **12** (1973), p. 2434.
18. Zachos, T. H., "Gaussian Beams from GaAs Junction Lasers," *Appl. Phys. Lett.* **12** (1969), p. 318.
19. Zachos, T. H. and J. E. Ripper, "Resonant Modes of GaAs Junction Lasers," *IEEE J. Quan. Elect.* **QE-5** (1969), p. 29.
20. H. Yonezu et al., "A GaAs—$Al_xGa_{1-x}As$ Double Heterostructure Planar Stripe Laser," *Jap. Journ. of Appl. Phys.* **12** (1973), p. 1585.

Supplementary References

1. Marcuse, D., *Light Transmission Optics* (New York: Van Nostrand, 1973).
2. Arnaud, J. A., "Hamiltonian Theory of Beam Mode Propagation," *Progress in Optics,* E. Wolf, ed., (Amsterdam: North Holland, 1973.)

Problems

6.1 Derive Eqs. (6.2-1) through (6.2-4).

6.2 Show that the eigenvalues λ of the equation

$$\begin{vmatrix} A & B \\ C & D \end{vmatrix} \begin{vmatrix} r_s \\ r'_s \end{vmatrix} = \lambda \begin{vmatrix} r_s \\ r'_s \end{vmatrix}$$

are $\lambda = e^{\pm i\theta}$ with $\exp(\pm i\theta)$ given by (6.1-13). Note that, according to (6.1-5), the foregoing matrix equation can also be written as

$$\begin{vmatrix} r_{s+1} \\ r'_{s+1} \end{vmatrix} = \lambda \begin{vmatrix} r_s \\ r'_s \end{vmatrix}$$

6.3 Make a plausibility argument to justify (6.4-1) by showing that it holds for a plane-wave incident on a lens.

6.4 Derive Eq. (6.4-7).

6.5 Show that a lenslike medium occupying the region $0 \leqslant z \leqslant l$ will image a point on the axis at $z < 0$ onto a single point. (If the image point occurs at $z < l$, the image is virtual.)

6.6 Derive the ray matrices of Table 6.1.

6.7 Solve the problem leading up to Eqs. (6.7-11) and (6.7-12) for the case where the lens is placed in an arbitrary position relative to the input beam (i.e., not at its waist).

6.8 (a) Assume a Gaussian beam incident normally on a solid prism with an index of refraction n as shown.
What is the far-field diffraction angle of the output beam?

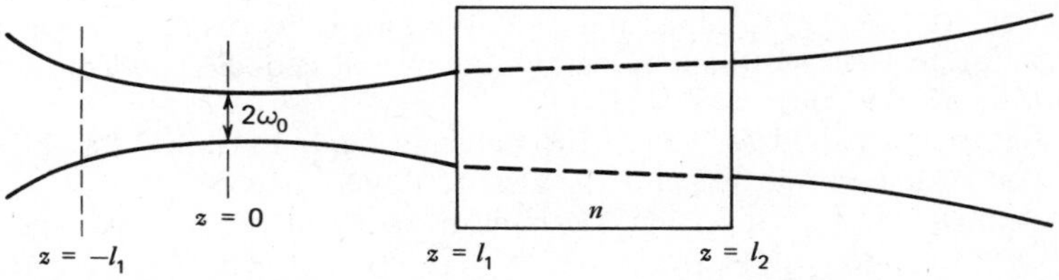

(b) Assume that the prism is moved to the left until its input face is at $z = -l_1$. What is the new beam waist and what is its location? (Assume that the crystal is long enough so that the beam waist is inside the crystal.)

6.9 A Gaussian beam with a wavelength λ is incident on a lens placed at $z = l$ as shown.

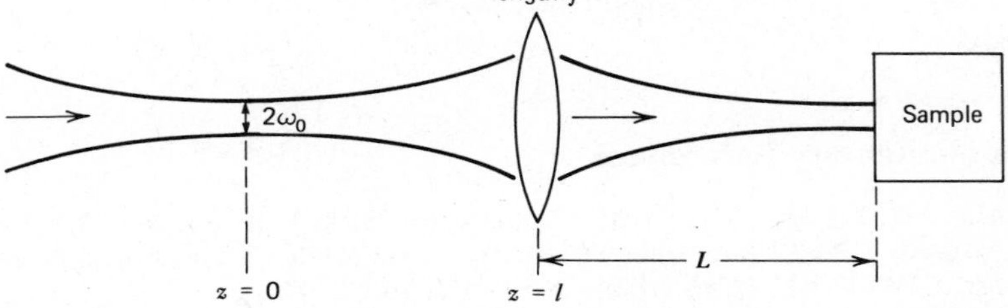

Calculate the lens focal length f so that the output beam has a waist at the front surface of the sample crystal. Show that (given l and L) up to two solutions may exist. Sketch the beam behavior for each of these solutions.

6.10 Complete all the missing steps in the derivation of Section 6.12.

6.11 Prove Eq. (6.10-10).

Hint:
Consider the optical pulse field as the product of a carrier and envelope functions

$$E(z, t) = E_0 e^{i(\omega_0 t - k_0 z)} \int_{-\infty}^{\infty} G(\Delta\omega) e^{i(\Delta\omega t - \Delta k z)} d(\Delta\omega)$$

where $\Delta\omega \equiv \omega - \omega_0$, $\Delta k = k(\omega) - k_0$.

6.12 Find the beam spot size and the maximum number of pulses per second that can be carried by an optical beam ($\lambda = 1\ \mu$m) propagating in a quadratic index glass fiber 1000 m long with $n = 1.5$, $n_2 = 5 \times 10^2$ cm^{-2}.
(a) In the case of a single mode excitation $l = m = 0$.
(b) $l = m = 5$.

6.13 Derive Eqs. (6.10-4) and (6.10-5).

C H A P T E R 7

Optical Resonators

7.0 INTRODUCTION

Optical resonators, like their lower frequency (e.g., microwave) counterparts, are needed for two related main purposes: (1) to build up large field intensities at specified (resonance) frequencies with moderate power inputs; (2) to act as spatial and frequency filters responding selectively to fields with prescribed spatial variation and frequency. The ability of a resonator to perform these two tasks is measured by a universal figure of merit, the quality factor Q. This point will be taken up at the end of the chapter. Our first task is to determine the field distribution inside some common configurations of optical resonators.

7.1 SPHERICAL MIRROR RESONATORS

At microwave frequencies, as an example, it is possible to restrict the number of resonances within a given, reasonably narrow, frequency interval to one or at most a few. This is done by choosing cavity dimensions comparable to a wavelength. At optical frequencies where $\lambda \sim 10^{-4}$ cm, this is not usually feasible. As a result, an enclosed optical resonator will, according to (5.7-3), possess

$$N \cong \frac{8\pi\nu^2 n^3}{c^3}\, d\nu \tag{7.1-1}$$

modes within a frequency interval $d\nu$ per unit resonator volume. For the case of $V = 1$ cm^3, $\nu = 3 \times 10^{14}$ Hz, and $d\nu = 3 \times 10^{10}$ Hz, as an example, Eq. (7.1-1) yields $N \sim 2 \times 10^9$ modes. If the resonator were closed, as in microwave frequencies, all these modes would have comparable values of Q. This situation is to be avoided in the case of lasers since it will cause the atoms to emit power (thus causing oscillation) into a large number of modes, which may differ in their frequencies as well as their spatial characteristics.

This objection is overcome to a large extent by the use of open resonators, which consist of a pair of opposing flat or curved reflectors. In such resonators, the energy of the vast majority of the modes does not travel at right angles to the mirrors and will thus be lost in essentially a single traversal. These modes will consequently possess a very low Q. If the mirrors are curved, the few surviving modes will, as shown below, have their energy localized near the axis; thus, the diffraction losses caused by the open sides can be made small compared with other loss mechanisms such as mirror transmission.

The earliest proposal for using open resonators seems to have been that of Dicke (Reference 1). Schawlow and Townes suggested in their original laser article (Reference 2) that such a resonator will discriminate heavily against modes whose energy propagates along directions other than that normal to the reflectors.

Figure 7.1 illustrates several mirror configurations, which are discussed in more detail below. In order for such a resonator to be able to support low loss (i.e., high Q field modes), it must satisfy two criteria (Reference 3). First, there must be a family of rays that, on suffering sequential specular reflections from the two reflectors, do not miss either reflector before making a reasonable number (e.g., 20–100) of traversals. Second, the dimensions of the reflectors must satisfy the relation

$$\frac{a_1 a_2}{\lambda l} \gtrsim 1$$

where a_1 and a_2 are half the widths of the two reflectors, respectively, in any arbitrary direction perpendicular to the resonator axis, and l is the distance between the reflectors. The first of these criteria, which follows from consideration of geometrical optics, is valid because the reflecting areas we are concerned with are large compared to a wavelength and have radii of curvature

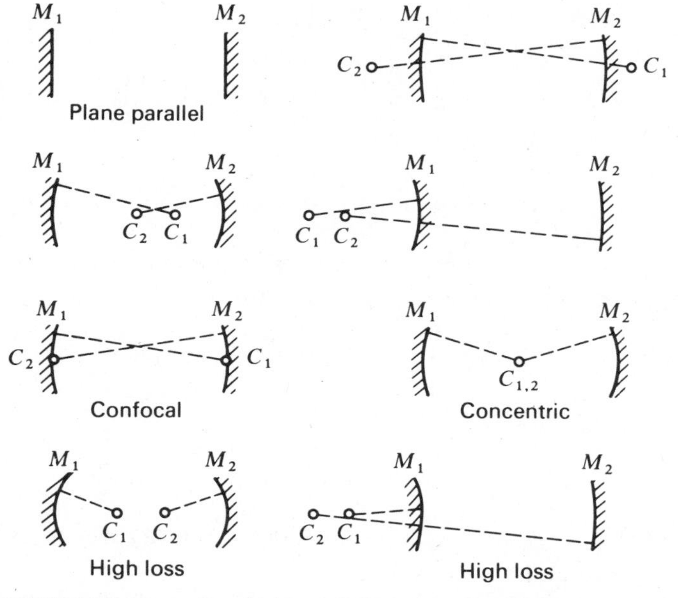

FIGURE 7.1 Examples of mirror configurations for optical masers. All except the bottom two exhibit low-loss resonant modes. *Source:* Reference 3.

that are also large compared to a wavelength. The second criterion follows from considerations of physical optics. We may think of it as requiring that the half-angle subtended by one reflector at the second (i.e., a_1/l) be somewhat greater than the half-angle of the far-field diffraction pattern of a nearly plane wave originating at and restricted to the dimension of the second (i.e., $\lambda/2a_2$).

The problem of finding the transverse modes has been attacked in a number of ways. One is to seek simple solutions to Maxwell's equations that take the form of narrow beams and then to make the reflection surfaces intersect the beam along phase fronts (i.e., everywhere perpendicular to the local direction of propagation), thus ensuring the reflection of the wave *back exactly on itself*. The other method (Reference 4) is to use the scalar formulation of Huygen's principle to compute the field at one mirror caused by the illumination of the other. The return field configuration is similarly calculated and is then required to match, within a constant, the initial field configuration. Solutions of the resulting integral equation yield the modes and the diffraction losses (References 5, 6). This method is adaptable to numerical machine calculations for situations in which analytical methods are not available. These two methods yield similar results. A third important approach is outlined by Problem 7.6.

We will use the first of these methods that readily lends itself to analysis.

We start with a description of the propagating beam modes in a homogeneous medium of index n. The pertinent relations taken from Section 6.9 are

$$E_{l,m}(\mathbf{r}) = E_0 \frac{\omega_0}{\omega(z)} H_l\left[\sqrt{2}\frac{x}{\omega(z)}\right] H_m\left[\sqrt{2}\frac{y}{\omega(z)}\right] \times \exp\left[-\frac{x^2+y^2}{\omega^2(z)} - ik\frac{x^2+y^2}{2R(z)} - ikz + i(l+m+1)\eta\right] \quad (7.1\text{-}2)$$

where the spot size $\omega(z)$ is

$$\omega(z) = \omega_0\left[1 + \left(\frac{z}{z_0}\right)^2\right]^{1/2} \qquad z_0 = \frac{\pi\omega_0^2 n}{\lambda} \quad (7.1\text{-}3)$$

and where ω_0, the minimum spot size, is a parameter characterizing the beam. The radius of curvature of the wavefronts is

$$R(z) = z\left[1 + \left(\frac{\pi\omega_0^2 n}{\lambda z}\right)^2\right] = \frac{1}{z}(z^2 + z_0^2) \quad (7.1\text{-}4)$$

and the phase factor η is

$$\eta = \tan^{-1}\left(\frac{\lambda z}{\pi\omega_0^2 n}\right) = \tan^{-1}\left(\frac{z}{z_0}\right)$$

The sign of $R(z)$ is taken as positive when the center of curvature is to the left of the wavefront and vice versa.

Given a beam of the type described by the last group of equations, we can form an optical resonator merely by inserting at points z_1 and z_2 two reflectors with radii of curvature that match those of the propagating beam spherical phase fronts at these points. Since the surfaces are normal to the direction of energy propagation as shown in Figure 6.5, the reflected beam retraces itself; thus, if the phase shift between the mirrors is some multiple of π radians, a *self-reproducing stable field* configuration results.

Alternatively, given two mirrors, with spherical radii of curvature R_1 and R_2 and some distance of separation l, we can, under certain conditions to be derived later, adjust the position $z = 0$ and the parameter ω_0 so that the mirrors coincide with two spherical wavefronts of the propagating beam defined by the position of the waist ($z = 0$) and ω_0. If, in addition, the mirrors can be made large enough to intercept the majority (e.g., 99%) of the incident beam energy in the fundamental ($l = m = 0$) transverse mode, we may expect this mode to have a larger Q than higher-order transverse modes, which, according to Figure 2.1, have fields extending farther from the axis and consequently lose a larger fraction of their energy by "spilling" over the mirror edges (diffraction losses).

Optical Resonator Algebra

As mentioned in the preceding paragraphs, we can form an optical resonator by using two reflectors, one at z_1 and the other at z_2, chosen so that their radii of curvature are the same as those of the beam wavefronts at the two locations. The propagating beam mode (7.1-2) is then reflected back and forth between the reflectors without a change in its transverse profile. The requisite radii of curvature are thus given by Eq. (7.1-4) as

$$R_1 = z_1 + \frac{z_0^2}{z_1}$$

$$R_2 = z_2 + \frac{z_0^2}{z_2}$$

from which we get

$$\begin{aligned} z_1 &= \frac{R_1}{2} \pm \tfrac{1}{2}\sqrt{R_1^2 - 4z_0^2} \\ z_2 &= \frac{R_2}{2} \pm \tfrac{1}{2}\sqrt{R_2^2 - 4z_0^2} \end{aligned} \qquad (7.1\text{-}5)$$

For a given arbitrary minimum spot size $\omega_0 = (\lambda z_0/\pi n)^{1/2}$, we can use (7.1-5) to find the positions z_1 and z_2 at which to place mirrors with curvatures R_1 and R_2, respectively. In practice, we most often start with given mirror curvatures R_1 and R_2 and a mirror separation l. The problem is then to find the minimum spot size ω_0, its location with respect to the reflectors, and the mirror spot sizes ω_1 and ω_2. Taking the mirror spacing as $l = z_2 - z_1$, we solve (7.1-5) for z_0^2, obtaining

$$z_0^2 = \frac{l(-R_1 - l)(R_2 - l)(R_2 - R_1 - l)}{(R_2 - R_1 - 2l)^2} \qquad (7.1\text{-}6)$$

where z_2 is to the right of z_1 (so that $l = z_2 - z_1 > 0$) and the mirror curvature is taken as positive when the center of curvature is to the left of the mirror.

The minimum spot size is $\omega_0 = (\lambda z_0/\pi n)^{1/2}$ and its position is next determined from (7.1-5). The mirror spot sizes $\omega(z_1)$ and $\omega(z_2)$ are then calculated by the use of (7.1-3).

The Symmetrical Mirror Resonator

The special case of a resonator with symmetrically (about $z = 0$) placed mirrors merits a few comments. The planar phase front at which the minimum spot size occurs is, by symmetry, at $z = 0$. Putting $R_2 = -R_1 = R$ in (7.1-6) gives

$$z_0^2 = \frac{(2R - l)l}{4} \tag{7.1-7}$$

and

$$\omega_0 = \left(\frac{\lambda z_0}{\pi n}\right)^{1/2} = \left(\frac{\lambda}{\pi n}\right)^{1/2} \left(\frac{l}{2}\right)^{1/4} \left(R - \frac{l}{2}\right)^{1/4} \tag{7.1-8}$$

that, when substituted in (7.1-3) with $z = l/2$, yields the following expression for the spot size at the mirrors:

$$\omega_{1,2} = \left(\frac{\lambda l}{2\pi n}\right)^{1/2} \left[\frac{2R^2}{l(R - l/2)}\right]^{1/4} \tag{7.1-9}$$

A comparison with (7.1-8) shows that, for $R \gg l$, $\omega_{1,2} \simeq \omega_0$ and the beam spread inside the resonator is small.

The value of R (for a given l), for which the mirror spot size is a minimum, is readily found from (7.1-9) to be $R = l$. When this condition is fulfilled, we have what is called a symmetrical *confocal resonator*, since the two foci, occurring at a distance of $R/2$ from the mirrors, coincide. From (7.1-7) and the relation $\omega_0 = (\lambda z_0/\pi n)^{1/2}$, we obtain

$$(\omega_0)_{\text{conf}} = \left(\frac{\lambda l}{2\pi n}\right)^{1/2} \tag{7.1-10}$$

whereas from (7.1-9), we get

$$(\omega_{1,2})_{\text{conf}} = \sqrt{2}\,(\omega_0)_{\text{conf}} \tag{7.1-11}$$

so the beam spot size increases by $\sqrt{2}$ between the center and the mirrors.

Numerical Example: Design of a Symmetrical Resonator. Consider the problem of designing a symmetrical resonator for $\lambda = 10^{-4}$ cm with a mirror separation $l = 2$m. If we were to choose the confocal geometry with $R = l = 2$m, the minimum spot size at the resonator center, taking $n = 1$, would be from (7.1-10)

$$(\omega_0)_{\text{conf}} = \left(\frac{\lambda l}{2\pi}\right)^{1/2} = 0.056 \text{ cm}$$

whereas, if we use (7.1-11), the spot size at the mirrors would have the value

$$(\omega_{1,2})_{\text{conf}} = \omega_0\sqrt{2} \simeq 0.08 \text{ cm}$$

Assume next that a mirror spot size $\omega_{1,2} = 0.3$ cm is desired. Using this value in (7.1-9) and assuming $R \gg l$, we get

$$\frac{\omega_{1,2}}{(\lambda l/2\pi)^{1/2}} = \frac{0.3}{0.056} = \left(\frac{2R}{l}\right)^{1/4}$$

Hence,

$$R \simeq 400l \simeq 800 \text{ m}$$

so that the assumption $R \gg l$ is valid. The minimum beam spot size ω_0 is found, through using (7.1-8), to be

$$\omega_0 = 0.9994\omega_{1,2} \simeq 0.3 \text{ cm}$$

Thus, to increase the mirror spot size from its minimum (confocal) value of 0.08 cm to 0.3 cm, we must use exceedingly plane mirrors ($R = 800$ m). This also shows that even small mirror curvatures (i.e., large R) give rise to "narrow" beams.

The numerical example we have worked out applies equally well to the case in which a plane mirror is placed at $z = 0$. The beam pattern is equal to that existing in the corresponding half of the symmetric resonator in the example, so the spot size on the planar reflector is ω_0.

7.2 MODE STABILITY (CONFINEMENT) CRITERIA AND THE SELF-CONSISTENT RESONATOR SOLUTIONS

The ability of an optical resonator to support low (diffraction) loss[1] modes depends on the mirror separation l and their radii of curvature R_1 and R_2. To illustrate this point, consider first the symmetric resonator with $R_2 = R_1 = R$.

The ratio of the mirror spot size at a given l/R to its minimum confocal ($l/R = 1$) value, given by the ratio of (7.1-9) to (7.1-11), is

$$\frac{\omega_{1,2}}{(\omega_{1,2})_{\text{conf}}} = \left\{\frac{1}{(l/R)[2 - (l/R)]}\right\}^{1/4} \tag{7.2-1}$$

This ratio is plotted in Figure 7.2. For $l/R = 0$ (plane-parallel mirrors) and $l/R = 2$ (two concentric mirrors), the spot size becomes infinite. It is clear that the diffraction losses for these cases are very high, since most of the beam energy "spills over" the reflector edges.

According to Table 6.1, the reflection of a Gaussian beam from a mirror with a radius of curvature R is formally equivalent to its transmission through a lens with a focal length $f = R/2$. The problem of the existence of confined optical modes in a resonator is thus formally the same as that of the existence

[1] By diffraction loss we refer to the fact that due to the beam spread [see Eq. (7.1-3)] a fraction of the Gaussian beam energy "misses" the mirror and is not reflected and is thus lost.

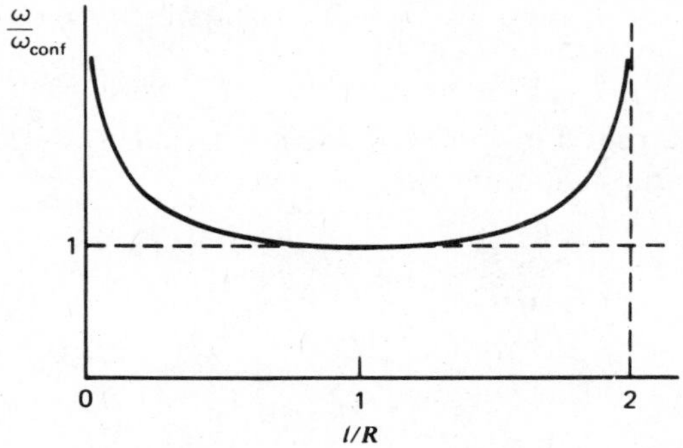

FIGURE 7.2 Ratio of beam spot size at the mirrors of a symmetrical resonator to its confocal ($l/R = 1$) value.

of confined solutions for the propagation of a Gaussian beam in a biperiodic lens sequence as shown in Figure 7.3. This problem was considered in Section 6.8 and led to the confinement condition (6.8-5).

If, in (6.8-5), we replace f_1 by $R_1/2$ and f_2 by $R_2/2$,[2] we obtain the confinement condition for optical resonators

$$0 \leqslant \left(1 - \frac{l}{R_1}\right)\left(1 - \frac{l}{R_2}\right) \leqslant 1 \tag{7.2-2}$$

A convenient representation of the confinement condition (7.2-2) is by means of the diagram (Reference 6) shown in Figure 7.4. From this diagram, for example, it can be seen that the symmetric concentric ($R_1 = R_2 = l/2$), confocal ($R_1 = R_2 = l$), and the plane-parallel ($R_1 = R_2 = \infty$) resonators are all on the verge of nonconfinement and thus may become extremely lossy by small deviations of the parameters.

Modes in a Generalized Resonator—The Self-Consistent Method

Up to this point, we treated resonators consisting of two opposing spherical mirrors. We may, sometimes, wish to consider the properties of more complex resonators made up of an arbitrary number of lenslike elements such as those shown in Table 6.1. A simple case of such a resonator may involve placing a lens between two spherical reflectors or constructing an off-axis three-reflector resonator. Yet another case is that of a traveling wave resonator in which the beam propagates in one direction only.

In each of these cases, we need to find if low loss (i.e., confined) modes

[2] This causes the sign convention of R_1 and R_2 to be different from that used in the preceding sections. The sign of R is the same as that of the focal length of the equivalent lens. This makes R_1 (or R_2) positive when the center of curvature of mirror 1 (or 2) is in the direction of mirror 2 (or 1), and negative otherwise.

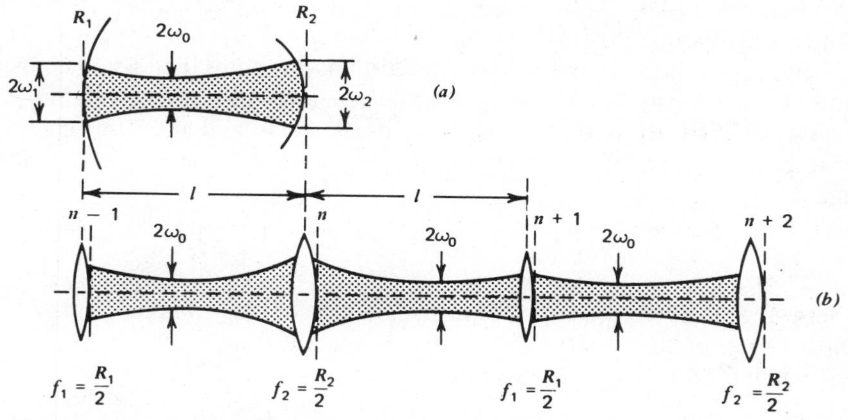

FIGURE 7.3 (*a*) Asymmetric resonator ($R_1 \neq R_2$) with mirror curvatures R_1 and R_2. (*b*) Biperiodic lens system (lens waveguide) equivalent to the resonator shown in (*a*).

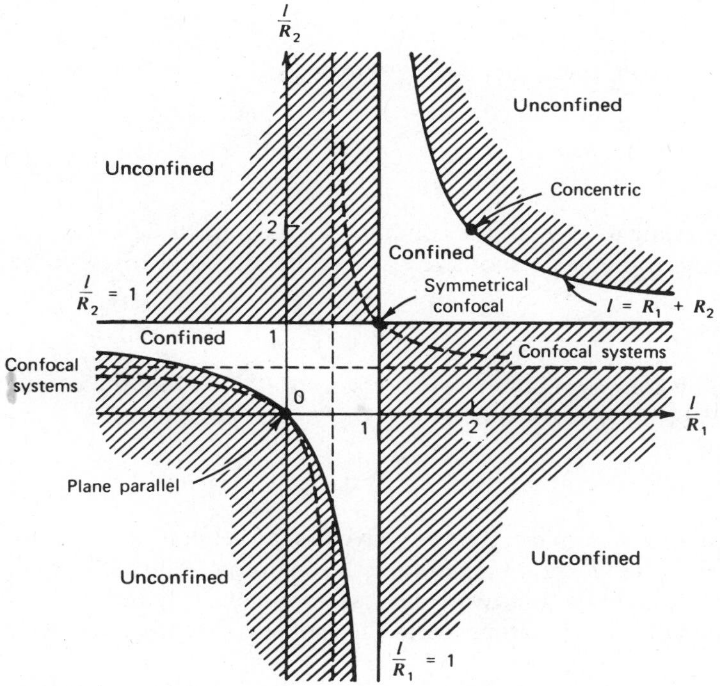

FIGURE 7.4 Confinement diagram for optical resonators. Shaded (high-loss) areas are those in which the confinement condition $0 \leq (1 - l/R_1)(1 - l/R_2) \leq 1$ is violated, and the clear (low-loss) areas are those in which it is fulfilled. The sign convention of R_1 and R_2 is discussed in footnote 2. *Source:* Reference 6.

exist in the complex resonator, and if so to solve for the spot size $\omega(z)$ and the radius of curvature $R(z)$ everywhere.

We apply the self-consistency condition and require that a stable eigenmode of the resonator is one that *reproduces itself after one round trip*. We choose an *arbitrary* reference plane in the resonator, denote the complex beam parameter at this plane as q, and, using the *ABCD* law (6.7-6), require that

$$q = \frac{Aq + B}{Cq + D} \tag{7.2-3}$$

where A, B, C, D are the "ray" matrix elements for one complete round trip, starting and ending at the chosen reference plane.

Solving (7.2-3) for $1/q$ gives

$$\frac{1}{q} = \frac{(D - A) \pm \sqrt{(D - A)^2 + 4BC}}{2B} \tag{7.2-4}$$

When the individual elements in the resonator are unimodular, that is, $A_iD_i - B_iC_i = 1$ (see Table 6.1), it follows that the matrix A, B, C, D, which is the product of individual matrices, satisfies

$$AD - BC = 1$$

and (7.2-4) can, consequently, be written as

$$\frac{1}{q^{(\pm)}} = \frac{D - A}{2B} \pm i \frac{\sqrt{1 - \left(\frac{D + A}{2}\right)^2}}{B} \tag{7.2-5}$$

According to (7.1-2), the condition for a confined Gaussian beam is that the square of the beam spot size ω^2 be real and positive. Recalling that

$$\frac{1}{q} = \frac{1}{R} - i\frac{\lambda}{\pi\omega^2 n}$$

we find by comparing the last expression to (7.2-5) that the condition for a confined beam is satisfied by either $1/q^{(+)}$ or $1/q^{(-)}$, provided

$$\left|\frac{D + A}{2}\right| \leqslant 1 \tag{7.2-6}$$

Equation (7.2-6) can thus be viewed as the generalization of the confinement condition (7.2-2) to an arbitrary resonator. When applied to the case of a resonator composed of two spherical resonators, it reduces to (7.2-2).

The radius of curvature R and the spot size ω at the reference plane are then

$$\begin{aligned} R &= \frac{2B}{D - A} \\ \omega &= \left(\frac{\lambda}{\pi n}\right)^{1/2} \frac{|B|^{1/2}}{\left[1 - \left(\frac{D + A}{2}\right)^2\right]^{1/4}} \end{aligned} \tag{7.2-7}$$

and, by application of the *ABCD* law (6.7-6), can be used to obtain ω and R at any other plane.

For a discussion of the stability of the steady state beam solution (7.2-5), see Problem (7.10) and Reference 15.

7.3 THE RESONANCE FREQUENCIES

Up to this point, we considered only the dependence of the spatial mode characteristics on the resonator mirrors (their radii of curvature and separation). Another important consideration is that of determining the resonance frequencies of a given spatial mode.

The frequencies are determined by the requirement that the complete round trip phase delay be some multiple of 2π. This requirement is the equivalent of the requirement in microwave waveguide resonators that the resonator length be equal to an integer number of half guide wavelengths (Reference 7). It makes it possible for a stable standing wave pattern to establish itself along the axis with a transverse field distribution equal to that of the propagating mode.

If we consider a spherical mirror resonator with mirrors at z_2 and z_1, the resonance condition for the m, n mode can be written as[3]

$$\theta_{m,n}(z_2) - \theta_{m,n}(z_1) = q\pi \tag{7.3-1}$$

where q is some integer and $\theta_{m,n}(z)$, the phase, is given according to (6.9-2) by

$$\theta_{m,n}(z) = kz - (m + n + 1)\tan^{-1}\frac{z}{z_0} \tag{7.3-2}$$

($z_0 = \pi\omega_0^2 n_0/\lambda$). In this section we will use n_0 to denote the index of refraction to avoid confusion with the integer n.

The resonance condition (7.3-1) is thus

$$k_q l - (m + n + 1)\left(\tan^{-1}\frac{z_2}{z_0} - \tan^{-1}\frac{z_1}{z_0}\right) = q\pi \tag{7.3-3}$$

where $l = z_2 - z_1$ is the resonator length. It follows that

$$k_{q+1} - k_q = \frac{\pi}{l}$$

or using $k = 2\pi\nu n_0/c$, we obtain

$$\nu_{q+1} - \nu_q = \frac{c}{2n_0 l} \tag{7.3-4}$$

for the intermode frequency spacing.

Let us consider, next, the effect of varying the transverse mode indices m and n in a mode with a fixed q. We notice from (7.3-3) that the resonant frequencies depend on the sum $(m + n)$ and not on m and n separately, so for

[3] In obtaining (7.3-1), we did not allow for the possibility of phase shift upon reflection. This correction does not affect any of the results of this section.

a given q, all the modes with the same value of $m + n$ are degenerate (i.e., they have the same resonance frequencies). Considering (7.3-3) at two different values of $m + n$ gives

$$k_1 l - (m + n + 1)_1 \left(\tan^{-1}\frac{z_2}{z_0} - \tan^{-1}\frac{z_1}{z_0}\right) = q\pi$$

$$k_2 l - (m + n + 1)_2 \left(\tan^{-1}\frac{z_2}{z_0} - \tan^{-1}\frac{z_1}{z_0}\right) = q\pi$$

and, by subtraction,

$$(k_1 - k_2)l = [(m + n + 1)_1 - (m + n + 1)_2] \left(\tan^{-1}\frac{z_2}{z_0} - \tan^{-1}\frac{z_1}{z_0}\right) \quad (7.3\text{-}5)$$

and using $k_1 - k_2 = (\omega_1 - \omega_2)n_0/c = 2\pi(\Delta\nu)n_0/c$ yields

$$\Delta\nu = \frac{c}{2\pi n_0 l}\,\Delta(m + n)\left(\tan^{-1}\frac{z_2}{z_0} - \tan^{-1}\frac{z_1}{z_0}\right) \quad (7.3\text{-}6)$$

for the change $\Delta\nu$ in the resonance frequency caused by a change $\Delta(m + m)$ in the sum $(m + n)$. As an example, in the case of a confocal resonator ($R = l$), we have, according to (7.1-7), $z_2 = -z_1 = z_0$; therefore, $\tan^{-1}(z_2/z_0) = -\tan^{-1}(z_1/z_0) = \pi/4$, and (7.3-6) becomes

$$\Delta\nu_{\text{conf}} = \tfrac{1}{2}[\Delta(m + n)]\,\frac{c}{2n_0 l} \quad (7.3\text{-}7)$$

Comparing (7.3-7) to (7.3-4), we find that in the confocal resonator the resonance frequencies of the transverse modes, resulting from changing m and n, either coincide or fall halfway between those that result from a change of the longitudinal mode index q. This situation is depicted in Figure 7.5.

To see what happens to the transverse resonance frequencies (i.e., those due to a variation of m and n) in a nonconfocal resonator, we may consider the nearly planar resonator in which $|z_1|$ and z_2 are small compared to z_0 (i.e., $l \ll R_1$ and R_2). In this case, Eq. (7.3-6) becomes

$$\Delta\nu \simeq \frac{c}{2\pi n_0 z_0}\,\Delta(m + n) \quad (7.3\text{-}8)$$

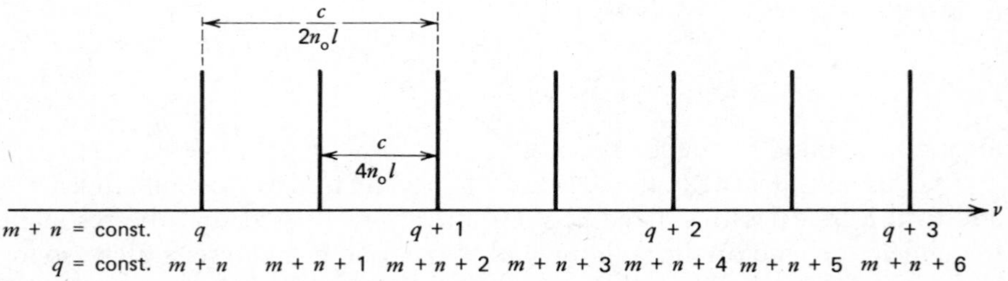

FIGURE 7.5 Position of resonance frequencies of a confocal ($l = R$) optical resonator as a function of the mode indices m, n, and q.

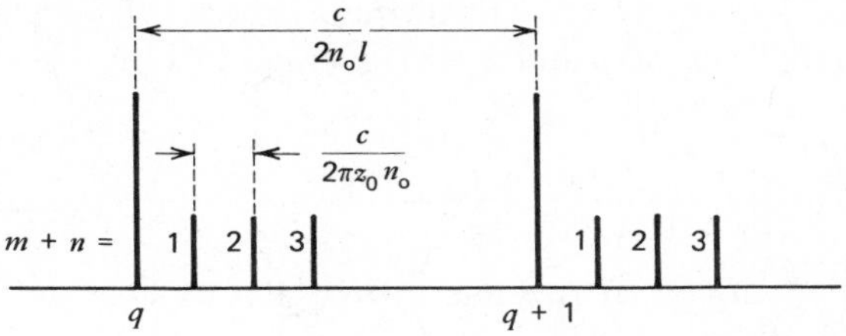

FIGURE 7.6 Resonant frequencies of a near-planar ($R \gg l$) optical resonator as a function of the mode indices m, n, and q.

where the n inside the parentheses is an integer, not to be confused with the index of refraction n_0 appearing in the denominator. The mode grouping for this case is illustrated in Figure 7.6.

The situation depicted in Figure 7.6 is highly objectionable if the resonator is to be used as a scanning interferometer. The reason is that in reconstructing the spectral profile of the unknown signal, an ambiguity is caused by the simultaneous transmission at more than one frequency. This ambiguity is resolved by using a confocal etalon with a mode spacing as shown in Figure 7.5 and by choosing l to be small enough that the intermode spacing $c/4n_0 l$ exceeds the width of the spectral region scanned.

7.4 LOSSES IN OPTICAL RESONATORS

An understanding of the mechanisms by which electromagnetic energy is dissipated in optical resonators and the ability to control them are of major importance in understanding and operating a variety of optical devices. For historical reasons as well as for reasons of convenience, these losses are often characterized by a number of different parameters. This book uses, in different places, the concepts of loss per pass, photon lifetime, and quality factor Q to describe losses in resonators. Let us see how these quantities are related to each other.

The decay lifetime (photon lifetime) t_c of a cavity mode is defined by means of the equation

$$\frac{d\mathscr{E}}{dt} = -\frac{\mathscr{E}}{t_c} \tag{7.4-1}$$

where $\mathscr{E}$ is the energy stored in the mode. If the fractional (intensity) loss per pass is L and the length of the resonator is l, then the fractional loss per unit time is cL/nl; therefore;

$$\frac{d\mathscr{E}}{dt} = -\frac{cL}{nl}\mathscr{E}$$

and, from (7.4-1),

$$t_c = \frac{nl}{cL} \tag{7.4-2}$$

For the case of a resonator with mirrors' reflectivities R_1 and R_2 and an average distributed loss constant α, the average loss per pass is $L = \alpha l - \ln \sqrt{R_1 R_2}$, for $L \ll 1$, so that

$$t_c = \frac{n}{c(\alpha - 1/l \ln \sqrt{R_1 R_2})_{R_1 R_2 \to 1}} \approx \frac{nl}{c[\alpha l + (1 - \sqrt{R_1 R_2})]} \tag{7.4-3}$$

The quality factor of the resonator $\alpha l \ll 1$ is defined universally as

$$Q = \frac{\omega \mathscr{E}}{P} = -\frac{\omega \mathscr{E}}{d\mathscr{E}/dt} \tag{7.4-4}$$

where $\mathscr{E}$ is the stored energy and $P = -d\mathscr{E}/dt$ is the power dissipated. By comparing (7.4-4) and (7.4-1), we obtain

$$Q = \omega t_c \tag{7.4-5}$$

The Q factor is related to the full width $\Delta\nu_{1/2}$ (at the half-power points) of the resonator's Lorentzian response curve as (Reference 8)

$$\Delta\nu_{1/2} = \frac{\nu}{Q} = \frac{1}{2\pi t_c} \tag{7.4-6}$$

so that, according to (7.4-3),

$$\Delta\nu_{1/2} = \frac{c\left(\alpha - \frac{1}{l} \ln \sqrt{R_1 R_2}\right)}{2\pi n} \tag{7.4-7}$$

The most common loss mechanisms in optical resonators are the following:

1. *Loss Resulting from Nonperfect Reflection.* Reflection loss is unavoidable since without some transmission no power output is possible. In addition, no mirror is ideal; and even when mirrors are made to yield the highest possible reflectivities, some residual absorption and scattering reduce the reflectivity to somewhat less than 100%.
2. *Absorption and Scattering in the Laser Medium.* Transitions from some of the atomic levels, which are populated in the process of pumping, to higher-lying levels constitute a loss mechanism in optical resonators when they are used as laser oscillators. Scattering from inhomogeneities and imperfections is especially serious in solid-state laser media.
3. *Diffraction losses.* From Eq. (7.1-2) or from Figure 2.1, we find that the energy of propagating-beam modes extends to considerable distances from the axis. When a resonator is formed by "trapping" a propagating beam between two reflectors, it is clear that for finite-dimension reflectors some of the beam energy will not be intercepted by the mirrors and will therefore be lost. For a given set of mirrors, this loss will be greater the higher the transverse mode indices m, n since, in this case, the energy extends farther. This fact is used to prevent the oscillation of higher-order modes by inserting apertures into the laser resonator whose opening is large enough to allow most of the fundamental (0, 0, q) mode energy through but small enough to increase substantially the losses of the higher-order

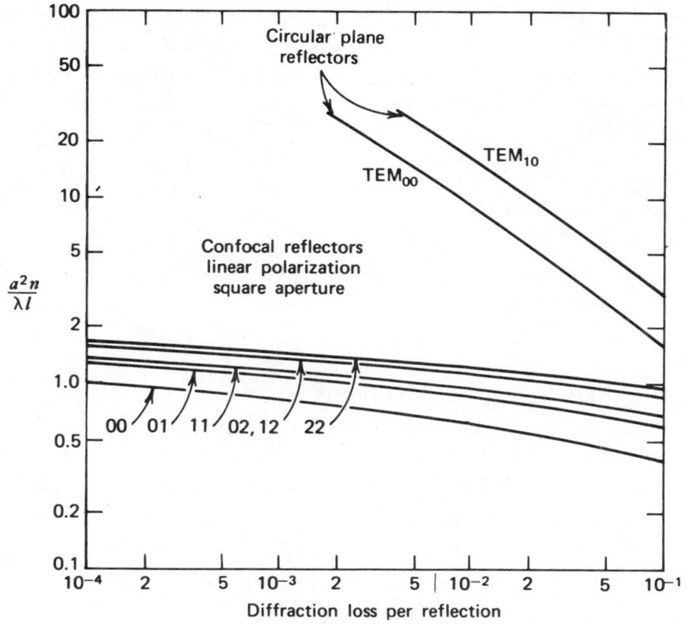

FIGURE 7.7 Diffraction losses for a plane-parallel and several low-order confocal resonators; a is the mirror radius and l is their spacing. The pairs of numbers under the arrows refer to the transverse-mode indices m, n. *Source:* Reference 5.

modes. Figure 7.7 shows the diffraction losses of a number of low-order confocal resonators. Of special interest is the dramatic decrease of the diffraction losses that results from the use of spherical reflectors instead of the plane-parallel ones.

7.5 UNSTABLE OPTICAL RESONATORS

As the resonator parameters approach the shaded regions of Figure 7.4, the beam spot size at the mirrors increases, reaching infinity as the boundary is crossed. When this happens, the finite dimensions of the reflectors must be included in the analysis.

There exist a number of important laser applications where the large mode volume and diffraction losses attendant upon operation in the nonconfining region are acceptable or even desirable. Some of the reasons are the following:

1. Operation in the confining regime has been shown (see example of Section 7.1) to lead to narrow Gaussian beams. This situation is not compatible with the need for high power output that requires large lasing volumes.
2. The losses in "unstable" resonators are dominated by diffraction (i.e., beam power "missing" the reflectors) and are thus desirable in situations

where the high gain prescribes large output coupling ratios (see Chapter 9).

3. The nature of the coupling results in an output beam with a large aperture that is consequently well collimated without the use of telescoping optics. This point will be made clear by the following discussion.

The more sophisticated theoretical analyses of this problem (References 9 through 14) make use of Huygen's integral method to derive the diffraction losses and field distribution of the modes of the unstable resonator. In the following brief treatment, we will use a geometrical output analysis advanced by Siegman (Reference 9) that emphasizes the essential physical characteristics of the resonator and that yields results in fair agreement with experiments.

Referring to Figure 7.8, we assume that the right-going wave leaving mirror M_1 is a spherical wave originating in a virtual center P_1 that is not, in general, the center of curvature of mirror M_1. This wave is incident on M_2 from which a fraction of the original intensity is reflected as a uniform spherical wave coming from a virtual center P_2. For self-consistency this wave, then, is reflected from M_1 as if it originated at P_1. This self-consistency condition is satisfied if the virtual image of P_1 upon reflection from M_2 is at P_2 and vice versa.

If we apply the imaging formulas of geometrical optics to the configuration of Figure 7.8, the self-consistency condition becomes

$$\frac{1}{r_1} - \frac{1}{r_2 + 1} = -\frac{2l}{R_1} = 2(g_1 - 1)$$
$$\frac{1}{r_2} - \frac{1}{r_1 + 1} = -\frac{2l}{R_2} = 2(g_2 - 1) \qquad (7.5\text{-}1)$$

where $g_i \equiv 1 - l/R_i$ $(i = 1, 2)$, and the sign of R is as discussed in footnote 2, so that in the example of Figure 7.8, R_1 and R_2 are negative.

Solving (7.5-1) for r_1 and r_2 gives

$$r_1 = \frac{\pm\sqrt{g_1 g_2(g_1 g_2 - 1)} - g_1 g_2 + g_2}{2g_1 g_2 - g_1 - g_2}$$
$$r_2 = \frac{\pm\sqrt{g_1 g_2(g_1 g_2 - 1)} - g_1 g_2 + g_1}{2g_1 g_2 - g_1 - g_2} \qquad (7.5\text{-}2)$$

The expressions (7.5-2) for r_1 and r_2 can be used to calculate, in the geometrical optics approximations, the loss per round trip of the unstable resonator. To demonstrate this, we return to Figure 7.8 and consider first for simplicity the case of a strip geometry where the mirrors are infinitely long in the y direction and have curvature only along their width as shown.

The self-consistent wave, immediately following reflection from mirror 1, is taken to have a total energy of unity. Upon reflection from M_2, the total

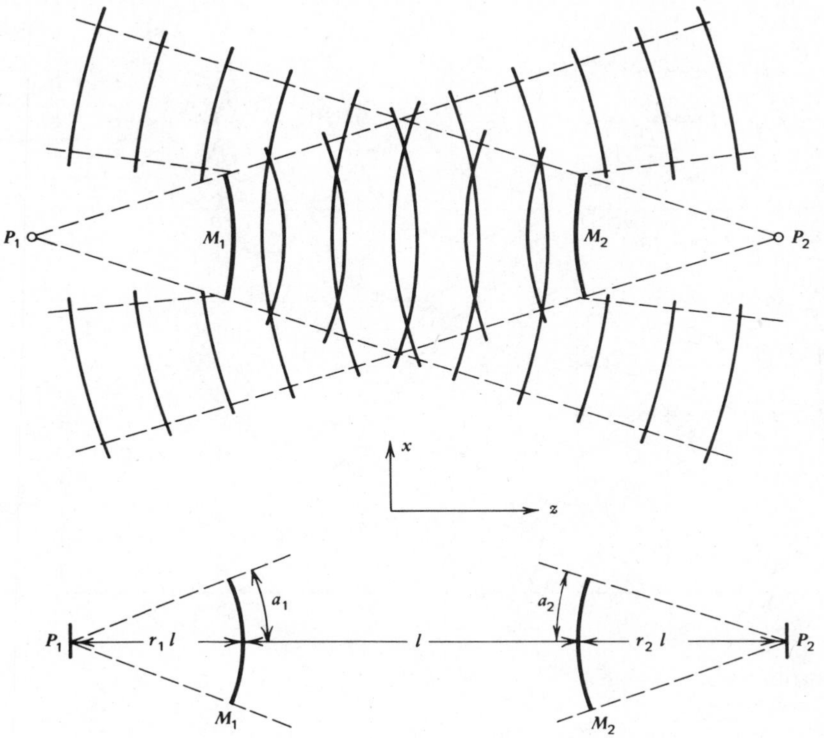

FIGURE 7.8 Spherical-wave picture of the mode in an unstable resonator. Points P_1 and P_2 are the virtual centers of the spherical waves. Each wave diverges so that a sizable fraction of its energy spills past the opposite mirror. *Source:* Reference 9.

energy is reduced to

$$(\Gamma_1)_{\text{strip}} = \frac{r_2 a_2}{(r_1 + 1)a_1} \tag{7.5-3}$$

To arrive at the last result, we took into account the two dimensional beam spread due to virtual emanation from P_1 between M_1 and M_2. The total transmission factor per round trip due to both mirrors is thus

$$\Gamma_{\text{strip}} = (\Gamma_1 \Gamma_2)_{\text{strip}} = \frac{r_1 r_2}{(r_1 + 1)(r_2 + 1)} \tag{7.5-4}$$

and for spherical mirrors (with curvature in both planes)

$$\Gamma_{1,2} = (\Gamma_{1,2})^2_{\text{strip}}$$

and

$$\Gamma = \Gamma_1 \Gamma_2 = \frac{r_1^2 r_2^2}{(r_1 + 1)^2 (r_2 + 1)^2} \tag{7.5-5}$$

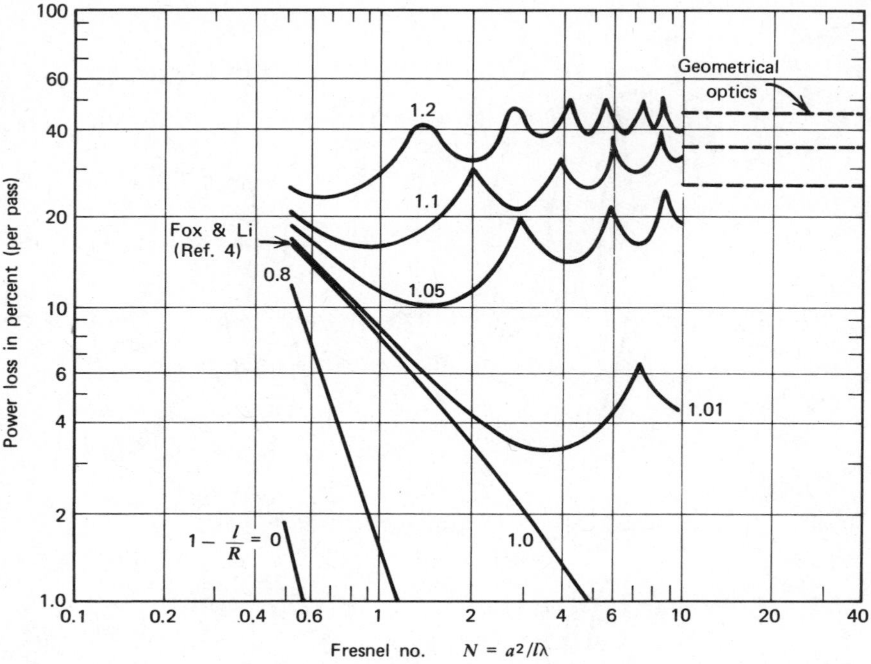

FIGURE 7.9 Loss per bounce versus Fresnel number for stable and unstable resonators. *Source:* Reference 9.

The average fractional power loss *per pass* may be taken as

$$\bar{\delta} = 1 - \Gamma^{1/2} = [1 - (\Gamma_1\Gamma_2)^{1/2}] \tag{7.5-6}$$

The unstable resonator loss is thus independent of the mirror dimensions, depending only on their radii of curvature and separation.

A plot of the losses of some symmetric unstable resonators that is obtained from Eqs. (7.5-2) and (7.5-5) is shown in Figure 7.9.

References

1. Dicke, R. H., "Molecular Amplification and Generation Systems and Methods," U.S. Patent No. 2,851,652 (Sept. 9, 1958).
2. Schawlow, A. L. and C. H. Townes, "Infrared and Optical Masers," *Phys. Rev.* **112,** 1940 (1958).
3. Yariv, A. and J. P. Gordon, "The Laser," *Proc. IEEE* **51,** 4 (1963).
4. Fox, A. G. and T. Li, "Resonant Modes in a Maser Interferometer," *Bell Sys. Tech. J.* **40,** 453 (1961).
5. Boyd, G. D. and J. P. Gordon, "Confocal Multimode Resonator for Millimeter through Optical Wavelength Masers," *Bell Sys. Tech. J.* **40,** 489 (1961).
6. Boyd, G. D. and H. Kogelnik, "Generalized Confocal Resonator Theory," *Bell Sys. Tech. J.* **41,** 1347 (1962).

7. See, for example, S. Ramo, J. R. Whinnery, and T. Van Duzer, *Fields and Waves in Communication Electronics* (New York: Wiley, 1965).
8. Born, M. and E. Wolf, *Principles of Optics* (New York: Macmillan, 1964). Also, A. Yariv, *Introduction to Optical Electronics* (New York: Holt, Rinehart and Winston, 1971).
9. Siegman, A. E., "Unstable Optical Resonators for Laser Applications," *Proc. IEEE* **53,** 277 (1965).
10. Siegman, A. E. and H. Y. Miller, "Unstable Optical Resonator Loss Calculations Using the Prony Method," *Appl. Opt.* **9,** 2729 (1970).
11. Reilly, J. P., "Single-Mode Operation of a High-Power Pulsed N_2/CO_2 Laser," *IEEE J. Quant. Elect.* **QE-8,** 136 (1972).
12. Anan'ev, Y. A., "Unstable Resonators and Their Applications," *Sov. J. Quant. Elect.* **1,** 565 (1972).
13. Freiburg, R. J., P. P. Chenausky, and C. J. Buczek, "Unidirectional Unstable Ring Lasers," *Appl. Opt.* **12,** 1140 (1973).
14. Chodzko, R. A., H. Mirels, F. S. Roehrs, and R. J. Pedersen, "Application of a Single Frequency Unstable Cavity to a CW HF Laser," *IEEE J. Quant. Electr.* **QE-9,** 523 (1973).
15. Casperson, L. W., "Mode Stability of Lasers and Periodic Optical Systems" *IEEE J. Quant. Elect.* **QE-10,** 629 (1974).

Problems

7.1 Compare typical Q values that are available from optical resonators ($R \sim 0.99$) with those of microwave cavities.

7.2 Design a resonator with $R_1 = 20$ cm, $R_2 = -32$ cm, $l = 16$ cm, $\lambda = 10^{-4}$ cm. Determine
(a) the minimum spot size ω_0.
(b) Its location.
(c) The spot size ω_1 and ω_2 at the mirrors.
(d) the ratios of ω_0, ω_1, and ω_2 to their respective confocal ($R_1 = -R_2 = l$) values.

7.3 Consider a confocal resonator with $l = 16$ cm, $\lambda = 10^{-4}$ cm, and reflectivities $R_1 = R_2 = 0.995$. Using Figure 7.7, choose the mirror's aperture for which the total losses of the first high-order mode (TE_{01}) exceed 3%. For this choice of aperture, what is the loss of the fundamental mode? How can the choice of apertures be used to quench the oscillation of high-order transverse modes?

7.4 Show why the confinement diagram, Figure 7.4, is a graphic representation of the condition $0 \leqslant [1 - (l/R_1)][1 - (l/R_2)] \leqslant 1$. Locate the eight resonators of Figure 7.1 on this diagram.

7.5 According to Figure 7.4 [or Eq. (7.2-2), we can get stable modes with $|R_2| = R_1$, $R_2 < 0$, that is, an alternating sequence of equally converging and diverging lenses. Explain on physical grounds why this leads to net focusing.

Hint:
Consider the distance from the axis at which a ray traverses both types of lenses.

7.6 Use the *ABCD* law to derive the mode-characteristics (minimum spot size ω_0 and the mirror spot sizes $\omega_{1,2}$) of a symmetric resonator with a mirror separation l and a radius of curvature R.

Hint:
Show that the radius of curvature (of the self-consistent beam solution) of the phase front at the mirror positions is equal to that of the mirrors.

7.7 Show that an optical resonator formed by "replacing" *any* two phase fronts of a propagating Gaussian beam by reflectors [i.e., placing reflectors at z_1 and z_2 with radii of curvature equal to $R(z_1)$ and $R(z_2)$, respectively] is stable.

7.8 Obtain the mode confinement condition of an optical resonator formed by two identical mirrors with radii of curvature R and a separation l and a thin lens f at its center.

7.9 Show that the self-consistent beam parameter q (7.2-5) leads to beam radii of curvature at the mirrors' positions that are identical to those of the respective mirrors, that is, $R(z_2) = R_2$, $R(z_1) = R_1$.

7.10 Show that after one round trip, a perturbation $\Delta(1/q)$ of the complex beam parameter (from its steady-state value (7.2-5) becomes $\delta(1/q) = e^{\mp i2\theta}\ \Delta(1/q)$ where $\cos\theta = \frac{1}{2}(A + D) \cdot \Delta(1/q)$ is thus neutrally stable, $|\delta(1/q| = |\Delta(1/q)|$, in confined beams satisfying (7.2-6).

CHAPTER 8

Interaction of Radiation and Atomic Systems

8.0 INTRODUCTION

In this chapter we consider the laws governing the interaction between atomic systems and electromagnetic radiation and ponder some of their consequences. We will make heavy use of the density matrix formalism introduced in Chapter 3. Some of the main topics considered include atomic susceptibilities, induced transitions, spontaneous transitions, amplification by an inverted atomic population, broadening mechanisms, and gain saturation.

We will keep the discussion general and not limit it to a specific atomic system so that the results may be widely applied. Some of the energy levels and transitions involved in individual laser materials will be considered in Chapter 10.

8.1 DENSITY MATRIX DERIVATION OF THE ATOMIC SUSCEPTIBILITY

In this section we will apply the density matrix formalism developed in Sections 3.14 and 3.15 to derive an expression for the susceptibility of an ensemble of atoms (or spins, ions, etc.) interacting with a time-harmonic electromagnetic field. The assumption is made that *only two levels,* with energies E_1 and E_2, are involved in the interaction as shown in Figure 8.1. This assumption is justified when the angular frequency ω of the field satisfies $\omega \sim (E_2 - E_1)/\hbar$. As a result, the density matrix (3.14-6) is reduced to a 2×2 matrix with elements $\rho_{11}, \rho_{12}, \rho_{21}, \rho_{22}$.

We assume that the interaction Hamiltonian $\mathcal{H}'(t)$ is of the dipole type and can be written as

$$\mathcal{H}' = -\mu E(t) \tag{8.1-1}$$

where μ is the component of the dipole operator along the direction of field $E(t)$. In our initial analysis, field $E(t)$ will be considered a classical variable. The diagonal matrix elements of $\mathcal{H}'$ are taken as zero

$$\mu_{11} = \mu_{22} = 0 \tag{8.1-2}$$

as appropriate to transitions between states of definite parity. The phases of the eigenfunctions $|2\rangle$ and $|1\rangle$ are taken, without loss of generality, such that

$$\mu_{21} = \mu_{12} \equiv \mu \tag{8.1-3}$$

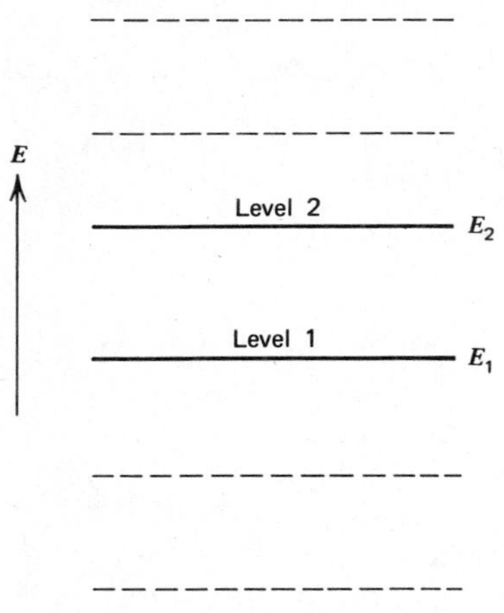

FIGURE 8.1 A two-level atomic system interacting with a radiation field whose frequency ω is approximately equal to $(E_2 - E_1)/h$. Other nonresonant levels (shown by broken lines) are assumed to play no role in the interaction except in determining the equilibrium populations N_{20} and N_{10}.

The total Hamiltonian of the two-level system is

$$\mathcal{H} = \mathcal{H}_0 + \mathcal{H}' \tag{8.1-4}$$

where $\mathcal{H}_0$ is the Hamiltonian of the system in the absence of any field.

Our task consists of solving for the ensemble average $\langle \mu \rangle$ of the dipole moment of the atom that is induced by field $E(t)$. The value of $\langle \mu \rangle$ is given, according to (3.14-8), by

$$\langle \mu \rangle = \text{tr}(\rho\mu) = \rho_{12}\mu_{21} + \rho_{21}\mu_{12} + \rho_{11}\mu_{11} + \rho_{22}\mu_{22}$$

and using (8.1-2) and (8.1-3), we obtain

$$\langle \mu \rangle = \mu(\rho_{12} + \rho_{21}) \tag{8.1-5}$$

We represent the density matrix operator in terms of the eigenfunctions ψ_n of the unperturbed Hamiltonian $\mathcal{H}_0$ so that $\mathcal{H}_0\psi_n = E_n\psi_n$.

Using (3.16-5), we obtain

$$\begin{aligned}\frac{d\rho_{21}}{dt} &= -\frac{i}{\hbar}[(\mathcal{H}_0 + \mathcal{H}'), \rho]_{21} = -\frac{i}{\hbar}(\mathcal{H}'_{21}\rho_{11} + E_2\rho_{21} - E_1\rho_{21} - \rho_{22}\mathcal{H}'_{21})\\ &= -\frac{i}{\hbar}[\mathcal{H}'_{21}(\rho_{11} - \rho_{22}) + (E_2 - E_1)\rho_{21}]\end{aligned}$$

and after using (8.1-1) and defining the resonance frequency $\omega_0 = (E_2 - E_1)/\hbar$,

$$\frac{d\rho_{21}}{dt} = -i\omega_0\rho_{21} + i\frac{\mu}{\hbar}E(t)(\rho_{11} - \rho_{22}) \tag{8.1-6}$$

In a similar manner, we obtain

$$\frac{d\rho_{22}}{dt} = -i\frac{\mu}{\hbar}E(t)(\rho_{21} - \rho^*_{21})$$

and

$$\frac{d}{dt}(\rho_{11} - \rho_{22}) = 2i\frac{\mu}{\hbar}E(t)(\rho_{21} - \rho_{21}^*) \tag{8.1-7}$$

where the last equation follows from the normalization condition $\rho_{11} + \rho_{22} = 1$.

The Inclusion of Collision Terms

If we pause to retrace the steps leading to Eqs. (8.1-5), (8.1-6), and (8.1-7), we recognize that the density matrix method of obtaining $\langle\mu\rangle$ is formally equivalent to the conventional procedure whereby $\langle\mu(t)\rangle = \int \psi^*\mu\psi\, dv$ [see Eq. (1.1-16)]. Indeed, there is no particular merit to the use of the density matrix unless we take advantage of the fact that, according to Eq. (3.15-2), it is defined as an ensemble average. First, consider Eq. (8.1-6). When perturbation field $E(t)$ is turned off, we would expect from (3.15-2) that ρ_{21} would decrease and eventually approach zero as the relative phase coherence among the N eigenfunctions in the ensemble is lost via "collisions." These collisions are characterized by the fact that they conserve the average energy (or level occupation) but cause a loss of (ensemble) information involving the phase ϕ_n in the wavefunction

$$\psi_n(\mathbf{r}, t) = u_n(\mathbf{r}) \exp[-i(E_n t/\hbar + \phi_n)]$$

Such collisions were considered first in magnetic resonance (Reference 1) and are referred to as the "spin-spin" relaxation time T_2 (Reference 2). In our case, they will be considered in Chapter 10 in connection with pressure broadening in molecular lasers in which this type of collision determines the absorption linewidth.

We will incorporate the loss of phase coherence into the density matrix formalism by modifying (8.1-6) to

$$\frac{d\rho_{21}}{dt} = -i\omega_0\rho_{21} + i\frac{\mu}{\hbar}(\rho_{11} - \rho_{22})E(t) - \frac{\rho_{21}}{T_2} \tag{8.1-8}$$

If we use (3.14-6), it follows that ρ_{ii} is the probability of finding an atom in the ith state. If N is the density of atoms, $N(\rho_{11} - \rho_{22}) \equiv \Delta N$ becomes the (average) density of the population difference between the two levels. Let the equilibrium $[E(t) = 0]$ value of $\rho_{11} - \rho_{22}$ be denoted by $(\rho_{11} - \rho_{22})_0$ and let us assume that when $E(t)$ is turned off, the population difference ΔN relaxes toward its equilibrium value $N(\rho_{11} - \rho_{22})_0$[1] with a time constant τ.[2] We may consequently rewrite (8.1-7) as

$$\frac{d}{dt}(\rho_{11} - \rho_{22}) = \frac{2i\mu E(t)}{\hbar}(\rho_{21} - \rho_{21}^*) - \frac{(\rho_{11} - \rho_{22}) - (\rho_{11} - \rho_{22})_0}{\tau} \tag{8.1-9}$$

[1] This is not necessarily the thermal equilibrium value since some "pump" mechanism may be present that causes ΔN at equilibrium to have some fixed value that is different from its thermal equilibrium value.

[2] Since an inelastic collision also causes a loss of phase coherence, in cases where $\tau < T_2$, we use τ instead of T_2. This point is discussed in Section 8.6.

Next, we consider the special case when the local perturbing field $E(t)$ is time-harmonic so that

$$E(t) = E_0 \cos \omega t = \frac{E_0}{2}(e^{i\omega t} + e^{-i\omega t}) \tag{8.1-10}$$

In addition, we see from (8.1-8) that the nondriven [i.e., $E(t) = 0$] behavior of ρ_{21} is $\rho_{21} = \rho_{21}(0)e^{[-i\omega_0-(1/T_2)]t}$ so that, for $\omega \approx \omega_0$, it is useful to define new "slowly" varying variables σ_{21} and σ_{12} through the relations

$$\begin{aligned} \rho_{21}(t) &= \sigma_{21}(t)e^{-i\omega t} \\ \rho_{12}(t) &= \sigma_{12}(t)e^{i\omega t} = \rho_{21}^* \; (\Rightarrow \sigma_{12} = \sigma_{21}^*) \end{aligned} \tag{8.1-11}$$

Using (8.1-10) and (8.1-11), we rewrite Eqs. (8.1-8) and (8.1-9) as

$$\frac{d\sigma_{21}}{dt} = i(\omega - \omega_0)\sigma_{21} + \frac{i\mu E_0}{2\hbar}(\rho_{11} - \rho_{22}) - \frac{\sigma_{21}}{T_2} \tag{8.1-12}$$

$$\frac{d}{dt}(\rho_{11} - \rho_{22}) = \frac{i\mu E_0}{\hbar}(\sigma_{21} - \sigma_{21}^*) - \frac{(\rho_{11} - \rho_{22}) - (\rho_{11} - \rho_{22})_0}{\tau} \tag{8.1-13}$$

In deriving (8.1-12), we kept only terms with $\exp(-i\omega t)$ time dependence, whereas in (8.1-13) we kept only the terms with no exponential time dependence, thus ignoring factors with time dependence $\exp(2i\omega t)$ and $\exp(-2i\omega t)$. This neglect of the nonsynchronous terms is physically justified since their contribution averages out to zero in times that are short compared to those of interest (but long compared to $2\pi/\omega$).

Equations (8.1-12) and (8.1-13) are formally analogous to the Bloch equations (Reference 1) of magnetic resonance for the magnetization.

If we use (8.1-11) in Eq. (8.1-5), we obtain

$$\langle \mu \rangle = \mu(\sigma_{12}e^{i\omega t} + \sigma_{21}e^{-i\omega t})$$

and since $\sigma_{21} = \sigma_{12}^*$

$$\langle \mu(t) \rangle = 2\mu[\text{Re}\,\sigma_{21}(t)\cos \omega t + \text{Im}\,\sigma_{21}(t)\sin \omega t] \tag{8.1-14}$$

Steady-State Solutions

To obtain the steady-state solutions of the density matrix, we set the left side of (8.1-12) and (8.1-13) equal to zero. By obvious manipulations involving the addition and subtraction of (8.1-12) and its complex conjugate and then using (8.1-13), we obtain

$$\begin{aligned}\text{Im}\,\sigma_{21} &= \frac{\Omega T_2(\rho_{11}-\rho_{22})_0}{1+(\omega-\omega_0)^2T_2^2+4\Omega^2T_2\tau}\\ \text{Re}\,\sigma_{21} &= \frac{(\omega_0-\omega)T_2^2\Omega(\rho_{11}-\rho_{22})_0}{1+(\omega-\omega_0)^2T_2^2+4\Omega^2T_2\tau}\\ (\rho_{11}-\rho_{22}) &= (\rho_{11}-\rho_{22})_0\frac{1+(\omega-\omega_0)^2T_2^2}{1+(\omega-\omega_0)^2T_2^2+4\Omega^2T_2\tau}\end{aligned} \tag{8.1-15}$$

where the "precession" frequency Ω is defined by $\Omega \equiv \mu E_0/2\hbar$.

The macroscopic (oscillating) polarization is $P = N\langle\mu\rangle$ so that, according to (8.1-14),

$$P = \frac{\mu^2\,\Delta N_0 T_2}{\hbar}E_0\left[\frac{\sin\omega t+(\omega_0-\omega)T_2\cos\omega t}{1+(\omega-\omega_0)^2T_2^2+4\Omega^2T_2\tau}\right] \tag{8.1-16}$$

whereas the population difference (per unit volume) is

$$\Delta N = \Delta N_0\frac{1+(\omega-\omega_0)^2T_2^2}{1+(\omega-\omega_0)^2T_2^2+4\Omega^2T_2\tau} \tag{8.1-17}$$

where $\Delta N_0 \equiv N(\rho_{11}-\rho_{22})_0$ is the population difference at zero field.

If, as in (5.1-18), we define the atomic susceptibility by $\chi = \chi' - i\chi''$, then

$$\begin{aligned}P(t) &= \text{Re}(\varepsilon_0\chi E_0 e^{i\omega t})\\ &= E_0(\varepsilon_0\chi'\cos\omega t+\varepsilon_0\chi''\sin\omega t)\end{aligned} \tag{8.1-18}$$

and from (8.1-16),

$$\begin{aligned}\chi''(\omega) &= \frac{\mu^2T_2\,\Delta N_0}{\varepsilon_0\hbar}\,\frac{1}{1+(\omega-\omega_0)^2T_2^2+4\Omega^2T_2\tau} = \frac{\mu^2}{2\varepsilon_0\hbar}\,\Delta N g(\nu)\\ \chi'(\omega) &= \frac{\mu^2T_2\,\Delta N_0}{\varepsilon_0\hbar}\,\frac{(\omega_0-\omega)T_2}{1+(\omega-\omega_0)^2T_2^2+4\Omega^2T_2\tau} = \frac{\mu^2(\omega_0-\omega)T_2}{2\varepsilon_0\hbar}\,\Delta N g(\nu)\end{aligned} \tag{8.1-19}$$

where the normalized lineshape function $g(\nu)$ is

$$g(\nu) = \frac{2T_2}{1+4\pi^2(\nu-\nu_0)^2T_2^2} = \frac{(\Delta\nu/2\pi)}{(\nu-\nu_0)^2+\left(\dfrac{\Delta\nu}{2}\right)^2} \tag{8.1-20}$$

with a full width at half maximum $\Delta\nu = (\pi T_2)^{-1}$.

We note that χ'', which according to (5.1-19) is proportional to the absorption, and χ' are in the form of

$$\begin{aligned}\chi''(\nu) &\propto \Delta N g(\nu)\\ \chi'(\nu) &\propto \Delta N(\nu_0-\nu)g(\nu)\end{aligned} \tag{8.1-21}$$

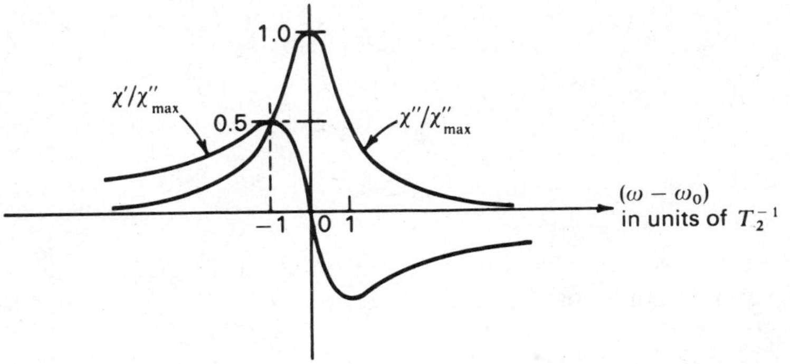

FIGURE 8.2 A plot of the real (χ') and imaginary (χ'') parts of the susceptibility for negligible saturation ($\mu^2 E_0^2 T_2 \tau/\hbar^2 \ll 1$).

where ΔN is the inversion in the presence of the optical field as given by (8.1-17). We will refer to $g(\nu)$ as given by (8.1-20) as the *normalized Lorentzian lineshape function.* The normalization constant was chosen so that

$$\int_{-\infty}^{\infty} g(\nu)\, d\nu = 1 \tag{8.1-22}$$

The derivation leading to (8.1-19) shows that the Lorentzian lineshape is characteristic of collision (τ, T_2) dominated transitions. A plot of the Lorentzian absorption (χ'') and dispersion (χ') in the limit $4\Omega^2 T_2 \tau \ll 1$ is shown in Figure 8.2.

Saturation

One consequence of (8.1-17) and (8.1-19) is that the population difference ΔN as well as χ' and χ'' decrease with increasing field intensity. This phenomenon, which is called saturation, becomes noticeable when $4\Omega^2 T_2 \tau > 1 + (\omega - \omega_0)^2 T_2^2$, or, if we use $\Omega \equiv \mu E_0/2\hbar$, when

$$\frac{\mu^2 E_0^2 T_2 \tau}{\hbar^2} > 1 + (\omega - \omega_0)^2 T_2^2 \tag{8.1-23}$$

Another consequence of saturation is a broadening of the Lorentzian lineshape function from a zero field value of $\Delta\nu = (\pi T_2)^{-1}$ to

$$\Delta\nu_{sat} = \Delta\nu \sqrt{1 + \frac{\mu^2 E_0^2 T_2 \tau}{\hbar^2}} \tag{8.1-24}$$

We will return to this topic in Section 8.7 in connection with gain saturation.

The Kramers–Kronig Relations

According to a fundamental theorem of the theory of complex variables, the real and imaginary parts of a complex function $f(z)$ that has no poles in the lower (or upper) z plane are related by the Hilbert transformation (Reference 3). When applied to the complex susceptibility function $\chi(\omega) = \chi'(\omega) - i\chi''(\omega)$, these transformations for the case of $\chi(\infty) = 0$ are

$$\chi'(\omega) = \frac{1}{\pi} \text{P.V.} \int_{-\infty}^{+\infty} \frac{\chi''(\omega')}{\omega' - \omega} d\omega'$$

$$\chi''(\omega) = -\frac{1}{\pi} \text{P.V.} \int_{-\infty}^{+\infty} \frac{\chi'(\omega')}{\omega' - \omega} d\omega' \tag{8.1-25}$$

where "P.V." stands for the Cauchy principal value of the integral that follows. The equations in (8.1-25) are derived in Appendix 1. In the present context, they are known as the Kramers–Kronig relations (Reference 4).

We next inquire whether the solutions for $\chi'(\omega)$ and $\chi''(\omega)$, (8.1-19), satisfy (8.1-25). To do this, we must find out if $\chi(\omega)$ is analytic in the lower half of the ω plane. From (8.1-19), we have

$$\begin{aligned}\chi(\omega) &= \chi'(\omega) - i\chi''(\omega)\\ &= -\frac{\mu^2\,\Delta N_0}{\varepsilon_0\hbar}\,\frac{\omega - [\omega_0 - (i/T_2)]}{\{\omega - [\omega_0 - (i/T_2)(1+s^2)^{1/2}]\}\{\omega - [\omega_0 + (i/T_2)(1+s^2)^{1/2}]\}}\end{aligned} \tag{8.1-26}$$

where $s^2 \equiv \mu^2 E_0^2 T_2 \tau/\hbar^2$. In the absence of saturation, $s = 0$, $\chi(\omega)$ has a single pole at $\omega = \omega_0 + i/T_2$. For this case, $\chi'(\omega)$ and $\chi''(\omega)$ obey (8.1-25). The actual demonstration is left as an exercise. For $s \neq 0$, $\chi(\omega)$ has poles at $\omega = \omega_0 \pm (i/T_2)(1 + s^2)^{1/2}$ and the Kramers–Kronig relations do not apply. The significance of the presence or absence of poles from the point of view of the transient behavior of the system is discussed in Appendix 1.

Connection with Magnetic Resonance. The treatment leading to (8.1-19) is formally identical to the Bloch equations solutions of magnetic resonance (Reference 1). In the simple case of $m_J = \pm\frac{1}{2}$ transitions, we associate $|2\rangle$ with $|m_J = \frac{1}{2}\rangle$ and $|1\rangle$ with $|m_J = -\frac{1}{2}\rangle$. To complete the analogy, we use

$$E_0 \to H_1$$

$$\boldsymbol{\mu} \to \gamma\mathbf{J} = \gamma(\mathbf{L} + \mathbf{S})$$

$$\mu \to \langle -\tfrac{1}{2}|\,\gamma J_x\,|\tfrac{1}{2}\rangle = \frac{\gamma\hbar}{2}$$

$$-\Delta N_0 \to \frac{2M_0}{\gamma\hbar}, \qquad \gamma = \frac{g_J\beta}{\hbar}$$

$$\tau \to T_1, \qquad \mathbf{P}(t) \to \mathbf{M}(t)$$

where H_1 is the amplitude of the linearly polarized *rf* magnetic field, β the Bohr magneton, $\mathbf{J}$ the total angular momentum operator, M_0 the equilibrium magnetization, T_1 the "spin-lattice" relaxation time, γ the magnetogyric ratio, and g_J the Landé g factor.

With these substitutions and by defining $M_x(t) = \text{Re}[\chi(\omega) H_1 e^{i\omega t}]$, (8.1-16) and (8.1-17) yield

$$M_z = M_0 \frac{1 + (\omega - \omega_0)^2 T_2^2}{1 + (\omega - \omega_0)^2 T_2^2 + \frac{1}{4}\gamma^2 H_1^2 T_1 T_2}$$

$$\chi'(\omega) = \frac{(\frac{1}{2})\,|\gamma|\,(\omega_0 - \omega) T_2^2 M_0}{1 + (\omega - \omega_0)^2 T_2^2 + \frac{1}{4}\gamma^2 H_1^2 T_1 T_2} \tag{8.1-27}$$

$$\chi''(\omega) = \frac{(\frac{1}{2})\,|\gamma|\,T_2 M_0}{1 + (\omega - \omega_0)^2 T_2^2 + \frac{1}{4}\gamma^2 H_1^2 T_1 T_2}$$

that are the same as the conventional solutions of Bloch equations (Reference 2).

8.2 THE SIGNIFICANCE OF $\chi(\nu)$

According to (5.1-3), the electric displacement vector is defined by

$$\mathbf{D} = \varepsilon_0 \mathbf{E} + \mathbf{P} + \mathbf{P}_{\text{transition}} = \varepsilon \mathbf{E} + \varepsilon_0 \chi \mathbf{E}$$

where the complex notation is used and the polarization is separated into a resonant component $\mathbf{P}_{\text{transition}}$ due to the specific atomic transition and a nonresonant component $\mathbf{P}$ that accounts for all the other contributions to the polarization. We can rewrite the last equation as

$$\mathbf{D} = \varepsilon\left[1 + \frac{\varepsilon_0}{\varepsilon}\chi(\omega)\right]\mathbf{E} = \varepsilon'(\omega)\mathbf{E} \tag{8.2-1}$$

so that the complex dielectric constant becomes

$$\varepsilon'(\omega) = \varepsilon\left[1 + \frac{\varepsilon_0}{\varepsilon}\chi(\omega)\right] \tag{8.2-2}$$

We have thus accounted for the effect of the atomic transition by modifying ε according to (8.2-2). Having derived $\chi(\omega)$, using detailed atomic information as in Section 8.1, we can now ignore its physical origin and proceed to treat the wave propagation in the medium with ε' given by (8.2-2), using Maxwell's equations.

As an example of this point of view, we consider the propagation of a plane electromagnetic wave in a medium with a dielectric constant $\varepsilon'(\omega)$. According to (5.2-9), the wave has the form of

$$E(z, t) = \text{Re}[E e^{i(\omega t - k' z)}]$$

where

$$k' = \omega\sqrt{\mu\varepsilon'} \simeq k\left(1 + \frac{\varepsilon_0}{2\varepsilon}\chi\right), \qquad |\chi| \ll 1 \tag{8.2-3}$$

and $k = \omega\sqrt{\mu\varepsilon}$.

Expressing $\chi(\nu)$ in terms of its real and imaginary components, $\chi = \chi' -$

$i\chi''$, (8.2-3) becomes

$$k' \simeq k\left[1 + \frac{\chi'(\nu)}{2n^2}\right] - i\frac{k\chi''(\nu)}{2n^2} \tag{8.2-4}$$

where $n = (\varepsilon/\varepsilon_0)^{1/2}$ is the index of refraction in the medium far from resonance. Substituting (8.2-4) back into (8.2-3), we find that in the presence of the atomic transition the wave propagates according to

$$E(z, t) = \text{Re}[Ee^{i\omega t - i(k+\Delta k)z + (\gamma/2)z}] \tag{8.2-5}$$

The result of the atomic polarization is thus to change the phase delay per unit length from k to $k + \Delta k$, where

$$\Delta k = \frac{k\chi'(\nu)}{2n^2} \tag{8.2-6}$$

as well as to cause the amplitude to vary exponentially with distance according to $e^{(\gamma/2)z}$, where

$$\gamma(\nu) = -\frac{k\chi''(\nu)}{n^2} \tag{8.2-7}$$

It is quite instructive to rederive (8.2-7) using a different approach. According to (5.1-19), the average power absorbed per unit volume from an electromagnetic field with a y component is

$$\overline{\frac{\text{Power}}{\text{Volume}}} = \overline{E_y(t)\frac{dP_y(t)}{dt}} = \tfrac{1}{2}\,\text{Re}[E(i\omega P)^*] \tag{8.2-8}$$

where E and P are the complex electric field and polarization amplitudes in the y direction, respectively, and horizontal bars denote time averaging. Using $P = \varepsilon_0\chi E$ in (8.2-8), we obtain

$$\overline{\frac{\text{Power}}{\text{Volume}}} = \frac{\omega\varepsilon_0}{2}\chi''\,|E|^2 \tag{8.2-9}$$

The absorption of energy at a rate given by (8.2-9) must lead to a change of the wave intensity I, according to

$$I(z) = I_0 e^{\gamma(\nu)z} \tag{8.2-10}$$

where

$$\gamma(\nu) = I^{-1}\frac{dI}{dz} \tag{8.2-11}$$

Conservation of energy thus requires that

$$\frac{dI}{dz} = -(\text{power absorbed per unit volume}) = -\frac{\omega\varepsilon_0}{2}\chi''\,|E|^2$$

Using the last result in (8.2-11) as well as the relations

$$I = \frac{c\varepsilon}{2n}|E|^2, \qquad \frac{\varepsilon}{\varepsilon_0} = n^2$$

where c is the velocity of light in vacuum gives

$$\gamma(\nu) = -\frac{k\chi''(\nu)}{n^2}$$

in agreement with (8.2-7).

8.3 SPONTANEOUS AND INDUCED TRANSITIONS

In the formalism of Section 8.1 the electromagnetic field was taken as a classical variable. This is sufficient for treating the coherent interaction of atoms with strong optical fields. There are, however, several aspects of this problem that require the quantization of the electromagnetic field as well as that of the atomic variables. The most notable of these is the phenomenon of spontaneous emission. This term is used to indicate the transition from an excited atomic state, say level 2 in Figure 8.1, to a lower level (e.g., 1) in the *absence* of any externally applied inducing field. This transition is accompanied by the emission of a photon of energy $E_2 - E_1$.

We start by considering an atom excited initially to level 2 that is placed inside a large optical enclosure. We will calculate first the rate for the process indicated in Figure 8.3 in which the atom undergoes a transition from 2 to 1 due to its interaction with a single radiation mode, for example, l (of the enclosure). The mode l, simultaneously, makes a transition from state $|n_l\rangle$ to $|n_l + 1\rangle$.

The interaction Hamiltonian is (see Appendix 5)

$$\mathcal{H}' = -e\mathbf{E}_l(\mathbf{r}, t) \cdot \mathbf{r} \tag{8.3-1}$$

Let the mode l correspond to a plane-wave propagating along $\mathbf{k}$ with a polarization λ. If we use (5.6-15), the interaction Hamiltonian is

$$\mathcal{H}' = ie\sqrt{\frac{\hbar\omega_{\mathbf{k}}}{2V\varepsilon}}\,[a^{\dagger}_{\mathbf{k},\lambda}e^{-i\mathbf{k}\cdot\mathbf{r}} - a_{\mathbf{k},\lambda}e^{i\mathbf{k}\cdot\mathbf{r}}]\hat{\mathbf{e}}_{\mathbf{k},\lambda} \cdot \mathbf{r} \tag{8.3-2}$$

Consider the transition depicted in Figure 8.3. The initial state is $|2, n_k\rangle$ in which the atom is in level 2 and the mode $(\mathbf{k}, \lambda)$ has $n_{\mathbf{k}}$ quanta. The final state finds the atom in level 1, whereas the mode has gained a quantum and is in the state $(n_{\mathbf{k}} + 1)$. The transition rate of the system from the initial state $|2, n_{\mathbf{k}}\rangle$ to the final $|1, n_{\mathbf{k}} + 1\rangle$ state is obtained from (3.12-14) as

$$W = \frac{2\pi}{\hbar}|\mathcal{H}'_{fi}|^2\delta(E_{\text{initial}} - E_{\text{final}}) \tag{8.3-3}$$

$$= \frac{2\pi e^2}{\hbar}\left(\frac{\hbar\omega_{\mathbf{k}}}{2V\varepsilon}\right)|\langle 1, n_{\mathbf{k}} + 1|a^{\dagger}_{\mathbf{k},\lambda}\hat{\mathbf{e}}_{\mathbf{k},\lambda} \cdot \mathbf{r}|2, n_{\mathbf{k}}\rangle|^2\,\delta(E_2 - E_1 - \hbar\omega_{\mathbf{k}})$$

$$= \frac{\pi e^2\omega_{\mathbf{k}}}{V\varepsilon}|\langle 1|\hat{\mathbf{e}}_{\mathbf{k},\lambda} \cdot \mathbf{r}|2\rangle|^2(n_{\mathbf{k}} + 1)\,\delta(E_2 - E_1 - \hbar\omega_{\mathbf{k}}) \tag{8.3-4}$$

where we used

$$\langle n_{\mathbf{k}} + 1|a^{\dagger}_{\mathbf{k},\lambda}|n_{\mathbf{k}}\rangle = \sqrt{n_{\mathbf{k}} + 1}$$

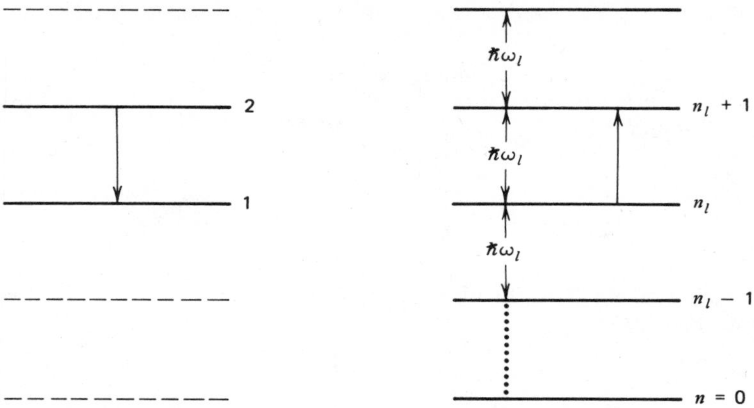

FIGURE 8.3 The atomic levels (left) and those of the radiation mode (right) involved in an emission process.

The part of the induced rate W that is proportional to $n_{\mathbf{k}}/V$, that is, the field intensity (or photon density), is called the induced rate, whereas the field-independent term is called the spontaneous rate (into one mode)

$$W_{\text{induced/mode}} = \frac{\pi e^2 \omega_{\mathbf{k}}}{V\varepsilon} |\hat{\mathbf{e}}_{\mathbf{k},\lambda} \cdot \mathbf{r}_{12}|^2 \, \delta(E_2 - E_1 - \hbar\omega_{\mathbf{k}}) \, n_{\mathbf{k}} \tag{8.3-5}$$

$$W_{\text{spont/mode}} = \frac{\pi e^2 \omega_{\mathbf{k}}}{V\varepsilon} |\hat{\mathbf{e}}_{\mathbf{k},\lambda} \cdot \mathbf{r}_{12}|^2 \, \delta(E_2 - E_1 - \hbar\omega_{\mathbf{k}}) \tag{8.3-6}$$

$$W_{\text{mode}} = W_{\text{spont/mode}} + W_{\text{induced/mode}}$$

where $r_{12} \equiv \langle 1|\mathbf{r}|2\rangle$.

We note that the ratio of the induced to the spontaneous transition rates into a single mode is equal to the number of quanta $n_{\mathbf{k}}$ in that mode—a result of central importance.

We repeat the above development for the transition rate from an initial $|1, n_{\mathbf{k}} + 1\rangle$ to $|2, n_{\mathbf{k}}\rangle$, that is, a process in which the field loses a quantum while the atom is elevated to a higher energy state. We find that in this case $W_{\text{spont}} = 0$, whereas W_{induced} is again given by (8.3-5). In words, *an atomic system does not spontaneously make an upward (in energy) transition. The induced transition rate between a given pair of eigenstates is the same for the upward and the downward directions.*

We have in (8.3-6) above an expression for the spontaneous rate for a downward electron transition due to its interaction with a *single* radiation mode. To determine the total spontaneous rate of the electron, we need merely sum the result over all modes (this assumes that interactions with different modes are not correlated).

We recall (see Section 5.7) that the number of electromagnetic modes with $\mathbf{k}$ vectors within a differential volume $d^3\mathbf{k}$ in $\mathbf{k}$ space is

$$p(\mathbf{k})\,d^3\mathbf{k} = d^3\mathbf{k}/(8\pi^3 V) = \frac{\mathbf{k}^2\,dk\,\sin\theta\,d\theta\,d\Phi\,V}{8\pi^3} \tag{8.3-7}$$

Summing (8.3-6) over all modes

$$W_{\text{spont}} = \sum_{\mathbf{k},\lambda} \frac{\pi e^2 \omega_{\mathbf{k}}}{V\varepsilon} |\hat{\mathbf{e}}_{\mathbf{k},\lambda} \cdot \mathbf{r}_{12}|^2\, \delta(E_2 - E_1 - \hbar\omega_{\mathbf{k}}) \tag{8.3-8}$$

For a sufficiently dense set of modes, we can replace the summation by an integral using the recipe

$$\sum_{\mathbf{k}} F(\mathbf{k}) = \iiint F(\mathbf{k})\, p(\mathbf{k})\, d^3\mathbf{k}$$

where $p(\mathbf{k})$ is the mode density (in $\mathbf{k}$ space) and $F(\mathbf{k})$ is any well behaved function of $\mathbf{k}$. Using (8.3-7,8), we obtain

$$W_{\text{spont}} = \frac{e^2}{8\pi^2\varepsilon\hbar} \sum_{\lambda=1,2} \int_0^\infty \int_0^\pi \int_0^{2\pi} \omega_{\mathbf{k}} |\hat{\mathbf{e}}_{\mathbf{k},\lambda} \cdot \mathbf{r}_{12}|^2\, \delta(\omega_0 - \omega_{\mathbf{k}}) k^2\, dk\, \sin\theta\, d\theta\, d\Phi$$

where we used $\delta(E) = \delta(\omega_0)\hbar^{-1}$ and $\hbar\omega_0 \equiv E_2 - E_1$. Using $\omega_{\mathbf{k}} = kc/n$, we first integrate over k

$$W_{\text{spont}} = \frac{e^2 \omega_0^3 n^3}{8\pi^2 \varepsilon c^3 \hbar} \sum_{\lambda=1,2} \int_0^\pi \int_0^{2\pi} |\hat{\mathbf{e}}_{\mathbf{k},\lambda} \cdot \mathbf{r}_{12}|^2\, \sin\theta\, d\theta\, d\Phi \tag{8.3-8a}$$

To carry out the remaining integration, we use a spherical coordinate system in which $\mathbf{r}_{12}$ is parallel to the k_z axis as shown in Figure 8.4. We choose the polarization direction $\hat{\mathbf{e}}_{\mathbf{k},1}$ to lie in the plane of $\mathbf{r}_{12}$ and $\mathbf{k}$. Since $\hat{\mathbf{e}}_{\mathbf{k},2}$ is perpendicular to both $\mathbf{k}$ and $\hat{\mathbf{e}}_{\mathbf{k},1}$, it is also perpendicular to $\mathbf{r}_{12}$ so

$$|\hat{\mathbf{e}}_{\mathbf{k},1} \cdot \mathbf{r}_{12}|^2 = |\mathbf{r}_{12}|^2 \sin^2\theta$$

$$|\mathbf{r}_{12}|^2 \equiv \mathbf{r}_{12} \cdot \mathbf{r}_{12}^*$$

$$\hat{\mathbf{e}}_{\mathbf{k},2} \cdot \mathbf{r}_{12} = 0$$

and

$$W_{\text{spont}} = \frac{e^2 \omega_0^3 n^3 |\mathbf{r}_{12}|^2}{8c^3\pi^2\hbar\varepsilon} \int_0^\pi \int_0^{2\pi} \sin\theta\, d\theta\, d\Phi$$

$$= \frac{e^2\omega_0^3 n^3}{3\pi c^3 \hbar\varepsilon} |\mathbf{r}_{12}|^2 = \frac{1}{t_{\text{spont}}} \tag{8.3-9}$$

$$|\mathbf{r}_{12}|^2 = |x_{12}|^2 + |y_{12}|^2 + |z_{12}|^2$$

The commonly used spontaneous emission lifetime is $t_{\text{spont}} = W_{\text{spont}}^{-1}$.

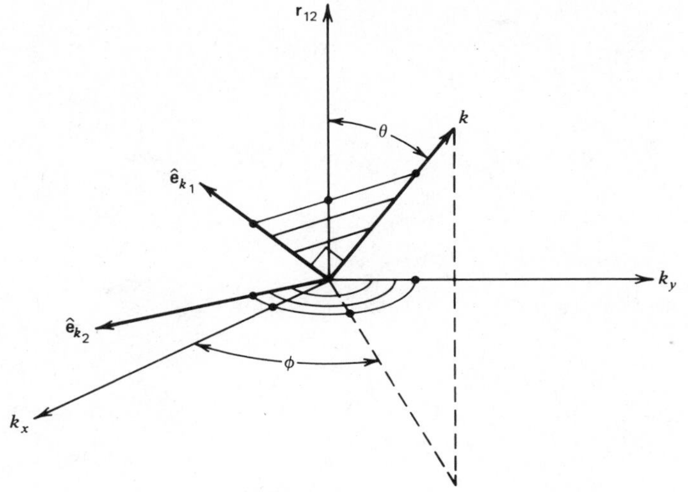

FIGURE 8.4 The coordinate system used to evaluate the integral in Eq. (8.3-8a).

Example. We will calculate here the spontaneous emission lifetime for a $|2p\rangle \to |1s\rangle$ transition in a hydrogenic atom—that is, in an atom with a nuclear charge $+Ze$ and a single electron.

The $|2p\rangle$ threefold degenerate spatial eigenfunctions are taken as (Reference 11)

$$u_{210} = R_{21}(r)\left(\frac{3}{4\pi}\right)^{1/2} \cos\theta = \frac{1}{\sqrt{32\pi}}\left(\frac{Z}{a_0}\right)^{5/2} z$$

$$\begin{aligned} u_{211} &= R_{21}(r)\left(\frac{3}{8\pi}\right)^{1/2} \cos\theta\, e^{i\Phi} = \frac{1}{\sqrt{64\pi}}\left(\frac{Z}{a_0}\right)^{5/2} e^{-Zr/2a_0}(x+iy) \\ u_{21-1} &= R_{21}(r)\left(\frac{3}{8\pi}\right)^{1/2} \cos\theta\, e^{-i\Phi} = \frac{1}{\sqrt{64\pi}}\left(\frac{Z}{a_0}\right)^{5/2} e^{-Zr/2a_0}(x-iy) \end{aligned} \tag{8.3-10}$$

$$R_{21}(r)\left(\frac{Z}{2a_0}\right)^{3/2} = \frac{Zr}{a_0\sqrt{3}}\, e^{-Zr/2a_0}$$

The ground state $|1\rangle$ is

$$u_{100} = \left(\frac{1}{\pi}\right)^{1/2} = \left(\frac{Z}{a_0}\right)^{3/2} e^{-Zr/a_0}$$

where $a_0 = 0.5292 \times 10^{-10}$ m is the Bohr radius of the electron.

We shall find it more convenient to use the following linear combination of the degenerate $|2p\rangle$ states:

$$
\begin{aligned}
|x\rangle &\equiv \frac{1}{\sqrt{2}}(u_{211} + u_{21-1}) \frac{1}{\sqrt{32\pi}}\left(\frac{Z}{a_0}\right)^{5/2} e^{-Zr/2a_0}\, x \\
|y\rangle &\equiv \frac{1}{\sqrt{2}}(u_{211} - u_{21-1}) \frac{i}{\sqrt{32\pi}}\left(\frac{Z}{a_0}\right)^{5/2} e^{-Zr/2a_0}\, y \qquad (8.3.11) \\
|z\rangle &= u_{210} = \frac{1}{\sqrt{32\pi}}\left(\frac{Z}{a_0}\right)^{5/2} e^{-Zr/2a_0}\, z
\end{aligned}
$$

Referring to the ground state u_{100} as $|1\rangle$, we have

$$
\langle 1|x|x\rangle = \frac{1}{\sqrt{32\pi^2}} \frac{1}{(a_0')^4} \int_0^\infty r^4 e^{-3r/2a_0'}\, dr \int_0^\pi \sin^3\theta\, d\theta \int_0^{2\pi} \cos^2\Phi\, d\Phi
$$

where $a_0' \equiv a_0/Z$. Using

$$
\int_0^\infty e^{-3r/2a_0'}\, dr = \frac{\Gamma(5)}{(3/2a_0')^5}
$$

leads to

$$
\langle 1|x|x\rangle = \langle 1|y|y\rangle = \langle 1|z|z\rangle = 0.7449\, \frac{a_0}{Z}\, y
$$

All the other matrix elements connecting the upper and lower states are zero as determined by symmetry consideration, so we can write

$$
\begin{aligned}
|x_{12}|^2 &= |\langle 1|x|x\rangle|^2 = 1.5539 \times 10^{-21}/Z^2 \\
|y_{12}|^2 &= |\langle 1|y|y\rangle|^2 = 1.5539 \times 10^{-21}/Z^2 \\
|z_{12}|^2 &= |\langle 1|z|z\rangle|^2 = 1.5539 \times 10^{-21}/Z^2
\end{aligned}
$$

To finish the calculation, we recall that for a hydrogenic transition

$$
\begin{aligned}
\hbar\omega_0(n = 2 \rightarrow n = 1) &= \left(\frac{3}{4}\right) \frac{\mu Z^2 e^4}{32\pi^2 \varepsilon_0^2 \hbar^2} \\
a_0 &= \frac{4\pi\varepsilon_0 \hbar^2}{\mu e^2} \\
\mu^{-1} &= m_e^{-1} + m_n^{-1}
\end{aligned}
$$

which when used in (8.3-9) gives

$$
t_{\text{spont}|2p\rangle \rightarrow |1s\rangle} = \frac{1}{W_{\text{spont}}} = \frac{1.595 \times 10^{-9}}{Z^4}\ \text{sec.} \qquad (8.3\text{-}12)
$$

The Induced Transition Rate due to a Monochromatic Field

The induced rate for a $2 \rightarrow 1$ transition is given by (8.3-5) for the case of an atom interacting with a *single* mode. This result can be used to obtain the

transition rate due to a traveling monochromatic wave at frequency ν. We start by recalling that expression (8.3-5) applies to the case where the transition energy is exactly $E_2 - E_1$. In general, $E_2 - E_1$ is not known precisely, and the probability that $E_2 - E_1$ occurs in interval $E \to E + dE$ is given by $g(E)\,dE = (1/h)g(\nu)\,dE$ where $g(\nu)$ is the normalized lineshape function for the $2 \to 1$ transition. We thus multiply (8.3-5) by $h^{-1}g[\nu = (E_2 - E_1)/h]\,dE$ and integrate over all energies. We allow for a degeneracy g_1 of level 1 and get

$$(W_{21})_i = \frac{\pi e^2 y_{12}^2 \omega_l n_l g_1}{hV\varepsilon} g(\nu_l) \qquad (8.3\text{-}13)$$

where we replaced $\sin^2 k_l z$ by a spatial average of 1/2 as appropriate for a traveling wave with the *same energy density as that of the* l*th mode.*

Finally, we relate the mode excitation number n_l in (8.3-13) to the wave intensity I_{ν_l} by

$$I_{\nu_l} = \frac{c n_l h \nu_l}{nV}$$

and eliminate y_{12}^2 in (8.3-13) through the use of (8.3-9). This leads to, in the case $x_{12}^2 = y_{12}^2 = z_{12}^2$

$$(W_{21})_i = \frac{\eta \lambda^2 I_\nu}{8\pi h\nu n^2 t_{\text{spont}}} g(\nu) \qquad (8.3\text{-}14)$$

$$\eta = \frac{3y_{12}^2}{x_{12}^2 + y_{12}^2 + z_{12}^2}$$

which is the key result of this section. It points out the proportionality between the induced transition rate and the intensity I_ν of the inducing field as well as the functional dependence on $g(\nu)$, the lineshape function. In the case of a collision broadened transition such as that leading to (8.1-20), only frequencies within $\sim\Delta\nu$ of line center are thus effective in inducing a transition.

If the interacting atom is initially in the lower level 1, the relevant matrix element, which would appear in (8.3-4), is

$$\langle 2, n_l - 1|ya_l|1, n_l\rangle = \langle 2|y|1\rangle\sqrt{n_l}$$

and thus if $n_l = 0$, the transition rate is zero. It follows that no spontaneous transitions exist from a low level to a (energetically) higher level. In addition, the $2 \to 1$ and $1 \to 2$ *induced* transition rates are the same. We can thus use (8.3-9) to obtain

$$(W_{12})_i = (W_{21})_i\left(\frac{g_2}{g_1}\right) = \left(\frac{g_2}{g_1}\right)\frac{\eta \lambda^2 I_\nu}{8\pi h\nu n^2 t_{\text{spont}}} g(\nu) \qquad (8.3\text{-}15)$$

The factor g_2/g_1 accounts for the degeneracy g_2 of level 2, since (8.3-14) was derived by assuming in (8.3-6) a degeneracy $g_1 = 1$ for level 1.

8.4 THE GAIN COEFFICIENT

Consider the passage of a monochromatic wave of frequency ν through an assembly of atoms. The atom denisty is N_2(atoms/m^3) in level 2 and N_1(atoms/m^3) in level 1.

The excess of induced $2 \to 1$ over $1 \to 2$ transitions per unit volume per unit time gives rise to an induced power

$$\frac{\text{Power}}{\text{Volume}} = [N_2(W_{21})_i - N_1(W_{12})_i]h\nu \tag{8.4-1}$$

where spontaneous transitions are ignored. Using (8.3-14) and (8.3-15) leads to

$$\frac{\text{Power}}{\text{Volume}} = \left(N_2 - \frac{g_2}{g_1}N_1\right)\frac{\lambda^2 g(\nu)I_\nu}{8\pi n^2 t_{\text{spont}}}\eta \tag{8.4-2}$$

If we assume that this power is added to the inducing wave, it follows that the latter grows according to

$$\frac{dI_\nu(z)}{dz} = (\text{power/volume}) = \gamma(\nu)I_\nu(z) \tag{8.4-3}$$

where, using (8.3-9),

$$\gamma = \frac{2\pi^2 e^2 y_{12}^2 n \left(N_2 - N_1\dfrac{g_2}{g_1}\right)g(\nu)}{h\lambda\varepsilon} = \frac{\left(N_2 - N_1\dfrac{g_2}{g_1}\right)\lambda^2\eta}{8\pi n^2 t_{\text{spont}}}g(\nu) \tag{8.4-4}$$

If N_2 and N_1 are independent of z, a situation that prevails when the pumping is uniform and saturation effects are negligible, the wave intensity grows exponentially according to

$$I_\nu(z) = I_\nu(0)e^{\gamma(\nu)z} \tag{8.4-5}$$

The problem of exponential amplification has also been treated in Section 8.2. It can be readily shown, using (8.1-17), (8.1-19), (8.1-20), and the equivalence

$$\Delta N \to N_1\frac{g_2}{g_1} - N_2$$

that $\gamma(\nu)$ as given by (8.2-7) is the same as that of (8.4-4).

Example: Gain in a Ruby Crystal. Consider an Al_2O_3 crystal with 0.5 wt.% of Cr_2O_3 added. This is "pink" ruby. The crystal contains about 2.4×10^{19} chromium atoms per cubic centimeter. This crystal is a common laser material and will be described in some detail in Chapter 10.

We assume that some pump agency causes an inversion

$$N_2 - N_1\frac{g_2}{g_1} = 5 \times 10^{17}\ \text{cm}^{-3}$$

Using the following data:

$$t_{\text{spont}} = 3 \times 10^{-3} \text{ sec}$$

$$\lambda = 0.6943 \ \mu\text{m}$$

$$n = 1.77 , \eta \approx 1$$

$$\Delta\nu \equiv 1/g(\nu_0) = \sim 2 \times 10^{11} \text{ Hz at } 300°\text{K}$$

in (8.4-4) leads to a gain constant

$$\gamma(\nu_0) \sim 5 \times 10^{-2} \text{ cm}^{-1}$$

8.5 THE EINSTEIN TREATMENT OF INDUCED AND SPONTANEOUS TRANSITIONS

The results of Section 8.3 can also be derived using classical arguments as was done originally by Einstein (Reference 5).

Consider the interaction of an assembly of identical atoms with a radiator field whose energy density is distributed uniformly in frequency in the vicinity of the transition frequency. Let the energy density per unit frequency be $\rho(\nu)$. We assume that the induced transition rates per atom from $2 \rightarrow 1$ and $1 \rightarrow 2$ are both proportional to $\rho(\nu)$ and take them as

$$\begin{aligned} (W'_{21})_i &= B_{21}\rho(\nu) \\ (W'_{12})_i &= B_{12}\rho(\nu) \end{aligned} \tag{8.5-1}$$

where B_{21} and B_{12} are constants to be determined. The total downward $(2 \rightarrow 1)$ transition rate is the sum of the induced and spontaneous contributions

$$W'_{21} = B_{21}\rho(\nu) + A \tag{8.5-2}$$

The spontaneous rate A was discussed in Section 8.3. The total upward $(1 \rightarrow 2)$ transition rate is

$$W'_{12} = (W'_{12})_i = B_{12}\rho(\nu) \tag{8.5-3}$$

Our first task is to obtain an expression for B_{12} and B_{21}. Since the magnitude of the coefficients B_{21} and B_{12} depends on the atoms and not on the radiation field, we consider, without loss of generality, the case where the atoms are in thermal equilibrium with a blackbody (thermal) radiation field at temperature T. In this case, the radiation density is given by (5.7-5) as

$$\rho(\nu) = \frac{8\pi n^3 h\nu^3}{c^3}\left(\frac{1}{e^{h\nu/kT} - 1}\right) \tag{8.5-4}$$

Since at thermal equilibrium the average populations of levels 2 and 1 are constant with time, it follows that the number of $2 \rightarrow 1$ transitions in a given time interval is equal to the number of $1 \rightarrow 2$ transitions; that is,

$$N_2 W'_{21} = N_1 W'_{12} \tag{8.5-5}$$

where N_1 and N_2 are the population densities of level 1 and 2, respectively. Using (8.5-2) and (8.5-3) in (8.5-5), we obtain

$$N_2[B_{21}\rho(\nu) + A] = N_1 B_{12}\rho(\nu)$$

and, substituting for $\rho(\nu)$ from (8.5-4),

$$N_2\left[B_{21}\frac{8\pi n^3 h\nu^3}{c^3(e^{h\nu/kT}-1)} + A\right] = N_1\left[B_{12}\frac{8\pi n^3 h\nu^3}{c^3(e^{h\nu/kT}-1)}\right] \qquad (8.5\text{-}6)$$

Since the atoms are in thermal equilibrium, the ratio N_2/N_1 is given by the Boltzmann factor

$$\frac{N_2}{N_1} = \frac{g_2}{g_1}e^{-h\nu/kT} \qquad (8.5\text{-}7)$$

Equating (N_2/N_1) as given by (8.5-6) to (8.5-7) gives

$$\frac{8n^3\pi h\nu^3}{c^3(e^{h\nu/kT}-1)} = \frac{A(g_2/g_1)}{B_{12}e^{h\nu/kT} - B_{21}(g_2/g_1)} \qquad (8.5\text{-}8)$$

The last equality can be satisfied only when

$$B_{12} = B_{21}\frac{g_2}{g_1} \qquad (8.5\text{-}9)$$

and, simultaneously,

$$\frac{A}{B_{21}} = \frac{8\pi n^3 h\nu^3}{c^3} \qquad (8.5\text{-}10)$$

We can, using (8.5-10), rewrite the induced transition rate (8.5-1) as

$$(W'_{21})_i = \frac{Ac^3}{8\pi n^3 h\nu^3}\rho(\nu) = \frac{c^3}{8\pi n^3 h\nu^3 t_{\text{spont}}}\rho(\nu) \qquad (8.5\text{-}11)$$

where $t_{\text{spont}} = 1/A$.

Equation (8.5-11) gives the transition rate per atom due to a field with a uniform (white) spectrum with energy density per unit frequency $\rho(\nu)$. In quantum electronics our main concern is in the transition rates that are induced by a monochromatic (i.e., single-frequency) field of frequency ν. Let us denote this transition rate as $(W_{21})_i$. We have established in Section 8.3 that the strength of interaction of a monochromatic field of frequency ν with an atomic transition is proportional to the lineshape function $g(\nu)$, so $(W_{21})_i \propto g(\nu)$. Furthermore, we would expect $(W_{21})_i$ to go over into $(W'_{21})_i$ as given by (8.5-11) if the spectral width of the radiation field is gradually increased from zero to a point at which it becomes large compared to the transition linewidth. These two requirements are satisfied if we take $(W_{21})_i$ as

$$(W_{21})_i = \frac{c^3\rho_\nu}{8\pi n^3 h\nu^3 t_{\text{spont}}} g(\nu) \qquad (8.5\text{-}12)$$

where ρ_ν is the energy density of the electromagnetic field inducing the transitions. To show that $(W_{21})_i$ as given by (8.5-12) indeed goes over smoothly into (8.5-11) as the spectrum of the field broadens, we may consider the broad spectrum field as made up of a large number of closely spaced monochromatic components at ν_k with random phases and then by adding the individual transition rates obtained from (8.5-12) obtain

$$(W'_{21})_i = \sum_{\nu_k} (W_{21})_i(\nu_k) = \frac{c^3}{8\pi n^3 h t_{\text{spont}}} \sum_k \frac{\rho_{\nu_k}}{\nu_k^3} g(\nu_k) \tag{8.5-13}$$

where ρ_{ν_k} is the energy density of the field component oscillating at ν_k. We can replace the summation of (8.5-13) by an integral if we replace ρ_{ν_k} by $\rho(\nu)\,d\nu$ where $\rho(\nu)$ is the energy density per unit frequency; thus, (8.5-13) becomes

$$(W'_{21})_i = \frac{c^3}{8\pi n^3 h t_{\text{spont}}} \int_{-\infty}^{+\infty} \frac{\rho(\nu) g(\nu)\, d\nu}{\nu^3} \tag{8.5-14}$$

In situations where $\rho(\nu)$ is sufficiently broad compared with $g(\nu)$, and thus the variation of $\rho(\nu)/\nu^3$ over the region of interest [where $g(\nu)$ is appreciable] can be neglected, we can pull $\rho(\nu)/\nu^3$ outside the integral sign, obtaining

$$(W'_{21})_i = \frac{c^3}{8\pi n^3 h \nu^3 t_{\text{spont}}} \rho(\nu)$$

where we used the normalization condition

$$\int_{-\infty}^{\infty} g(\nu)\, d\nu = 1$$

This agrees with (8.5-11).

Returning to our central result, (8.5-12), we can rewrite it in terms of the intensity $I_\nu = c\rho_\nu/n$ (W/m^2) of the optical wave as

$$(W_{21})_i = \frac{Ac^2 I_\nu}{8\pi n^2 h \nu^3} g(\nu) = \frac{\lambda^2 I_\nu}{8\pi n^2 h \nu t_{\text{spont}}} g(\nu) \tag{8.5-15}$$

where c is the velocity of propagation of light in vacuum. This is the same result as that given by (8.3-9).

8.6 HOMOGENEOUS AND INHOMOGENEOUS BROADENING

The term broadening is used to denote the finite spectral width of the response of atomic systems to electromagnetic fields. The broadening may manifest itself, as an example, in a plot of the absorption as a function of frequency or in the frequency dependence of the gain of a laser medium. Such plots are included in Chapter 10.

We distinguish between two main classes of broadening mechanisms.

Homogeneous Broadening (Reference 6)

In this case, the atoms are indistinguishable and have the same transition energy $E_2 - E_1$. The broadening is due to one or a combination of the following factors: (1) inelastic collisions with phonons or with other atoms (or molecules); (2) transitions to other levels—these may be spontaneous radiative transitions or nonradiative; (3) elastic phase-destroying collisions; and (4) broadening due to the interaction with an electromagnetic field (power broadening).

The case of homogeneous broadening is the one considered in Section 8.1 where it was found to give rise to a Lorentzian response curve of the form

$$\chi''(\omega) \propto \frac{1}{1 + (\omega - \omega_0)^2 T_2^2 + \dfrac{\mu^2 E_0^2}{\hbar^2} T_2 \tau} \tag{8.6-1}$$

with a power-dependent width

$$\Delta\nu_{\text{sat}} = \Delta\nu \sqrt{1 + \frac{\mu^2 E_0^2 T_2 \tau}{\hbar^2}} \tag{8.6-2}$$

where $\Delta\nu = (\pi T_2)^{-1}$. T_2 is the time constant characterizing the loss of atomic coherence as in (8.1-8) and is thus given by

$$\frac{1}{T_2} = \sum_i \frac{1}{\tau_i} \tag{8.6-3}$$

where the summation is over all the processes (collisions, transitions) that interrupt the coherent field-atom interaction.

Inhomogeneous Broadening (Reference 6)

In this case, the atoms are distinguishable, and the broadening reflects a spread in the individual resonant (transition) energies of the atoms. Two main examples of this type of broadening are that of impurity ions in a host crystal and of molecules in low-pressure gases.

In the first case, the energy levels, hence the transition frequencies, depend on the immediate crystalline surrounding of each atom. The ever-present random strain, as well as other types of crystal imperfections, cause the crystal surroundings to vary from one ion to the next, thus effecting a spread in the transition frequencies.

In the second example, the transition frequency ν of a gaseous atom (or molecule) is Doppler-shifted due to the finite velocity of the atom according to

$$\nu = \nu_0 + \frac{v_x}{c}\nu_0 \tag{8.6-4}$$

where v_x is the component of the velocity along the direction connecting the observer with the moving atom, c is the velocity of light, and ν_0 is the frequency corresponding to a stationary atom. The Maxwell velocity distribution function of a gas with atomic mass M that is at equilibrium at temperature T is

$$f(v_x, v_y, v_z) = \left(\frac{M}{2\pi kT}\right)^{3/2} \exp\left[-\frac{M}{2kT}(v_x^2 + v_y^2 + v_z^2)\right] \tag{8.6-5}$$

$f(v_x, v_y, v_z)\, dv_x\, dv_y\, dv_z$ corresponds to the fraction of all the atoms whose x component of velocity is contained in the interval v_x to $v_x + dv_x$, while, simultaneously, their y and z components lie between v_y and $v_y + dv_y$, v_z and $v_z + dv_z$, respectively. Alternatively, we may view $f(v_x, v_y, v_z)\, dv_x\, dv_y\, dv_z$ as the *a priori* probability that the velocity vector $\mathbf{v}$ of any given atom terminates within the differential volume $dv_x\, dv_y\, dv_z$ centered on $\mathbf{v}$ in velocity space so that

$$\iiint_{-\infty}^{\infty} f(v_x, v_y, v_z)\, dv_x\, dv_y\, dv_z = 1 \tag{8.6-6}$$

According to (8.6-4), the probability $g(\nu)\, d\nu$ that the transition frequency is between ν and $\nu + d\nu$ is equal to the probability that v_x will be found between $v_x = (\nu - \nu_0)(c/\nu_0)$ and $(\nu + d\nu - \nu_0)(c/\nu_0)$ irrespective of the values of v_y and v_z [since if $v_x = (\nu - \nu_0)(c/\nu_0)$, the Doppler-shifted frequency will be equal to ν regardless of v_y and v_z]. This probability is thus obtained by substituting $v_x = (\nu - \nu_0)c/\nu_0$ in $f(v_x, v_y, v_z)\, dv_x\, dv_y\, dv_z$, and then integrating over all values of v_y and v_z. The result is

$$g(\nu)\, d\nu = \left(\frac{M}{2\pi kT}\right)^{3/2} \int_{-\infty}^{\infty}\int_{-\infty}^{\infty} e^{-(M/2kT)(v_y^2+v_z^2)}\, dv_y\, dv_z\, e^{-(M/2kT)(c^2/\nu_0^2)(\nu-\nu_0)^2} \left(\frac{c}{\nu_0}\right) d\nu \tag{8.6-7}$$

Using the definite integral

$$\int_{-\infty}^{\infty} e^{-(M/2kT)v_z^2}\, dv_z = \left(\frac{2\pi kT}{M}\right)^{1/2}$$

we obtain, from (8.6-7),

$$g(\nu) = \frac{c}{\nu_0}\left(\frac{M}{2\pi kT}\right)^{1/2} e^{-(M/2kT)(c^2/\nu_0^2)(\nu - \nu_0)^2} \tag{8.6-8}$$

for the *normalized Doppler-broadened lineshape*. The functional dependence of $g(\nu)$ in (8.6-8) is referred to as Gaussian. The width of $g(\nu)$ in this case is taken as the frequency separation between the points where $g(\nu)$ is down to half its peak value. It is obtained from (8.6-8) as

$$\Delta\nu_D = 2\nu_0\sqrt{\frac{2kT}{Mc^2}\ln 2} \tag{8.6-9}$$

where the subscript D stands for "Doppler." We can reexpress $g(\nu)$ in terms of $\Delta\nu_D$ as

$$g(\nu) = \frac{2(\ln 2)^{1/2}}{\pi^{1/2}\,\Delta\nu_D} e^{-[4(\ln 2)(\nu - \nu_0)^2/\Delta\nu_D^2]} \tag{8.6-10}$$

8.7 GAIN SATURATION IN SYSTEMS WITH HOMOGENEOUS AND INHOMOGENEOUS BROADENING

The most important difference between atomic systems with homogeneous and inhomogeneous broadening manifests itself in their power saturation. Specifically, when such systems are used as laser media, their gain decreases with increasing field intensity. The amount of decrease and its spectral dependence are different in these two cases and will be considered below.

Homogeneous Broadening

Gain saturation here is due to the decrease of the population inversion with field intensity. The gain is given by (8.4-4) as

$$\gamma(\nu) = \Delta N \frac{\lambda^2 \eta}{8\pi n^2 t_{\text{spont}}} g(\nu) \tag{8.7-1}$$

where $g(\nu)$ is the normalized lineshape function

$$g(\nu) = \frac{2T_2}{1 + 4\pi^2(\nu - \nu_0)^2 T_2^2} \tag{8.7-2}$$

as derived in Section 8.1. The population inversion density $\Delta N \equiv N_2 - N_1(g_2/g_1)$ is given by (8.1-17) as

$$\Delta N = \Delta N_0 \frac{1 + (\omega - \omega_0)^2 T_2^2}{1 + (\omega - \omega_0)^2 T_2^2 + 4\Omega^2 T_2 \tau} \tag{8.7-3}$$

where $\Omega^2 = (\mu E_0/2\hbar)^2 g_1$. We substitute (8.7-2) and (8.7-3) in (8.7-1). In the resulting expression, we eliminate the matrix element $\mu = e y_{12}$ through the use of (8.3-9). The result is

$$\begin{aligned} \gamma(\nu) &= -\frac{\Delta N_0 \lambda^2 g(\nu) \eta}{8\pi n^2 t_{\text{spont}}} \left[\frac{1}{1 + \dfrac{I_\nu}{I_s(\nu)}} \right] \\ &\equiv \frac{\gamma_0(\nu)}{1 + \dfrac{I_\nu}{I_s(\nu)}} \end{aligned} \tag{8.7-4}$$

where $\gamma_0(\nu)$ is the unsaturated ($E_0 = 0$) gain, I_ν is the intensity (W/m^2) given by

$$I_\nu = \frac{c n \varepsilon_0 E_0^2}{2}$$

and $I_s(\nu)$, the intensity at which the gain at ν is reduced by a factor of 1/2 (compared to the zero intensity case), is called the "saturation intensity" and is given by

$$I_s(\nu) = \frac{c n \varepsilon_0 \hbar^2}{\mu^2 \tau g(\nu) \eta} = \frac{4\pi n^2 h\nu}{(\tau/t_{\text{spont}}) \lambda^2 g(\nu) \eta} \tag{8.7-5}$$

The degeneracy factor g_1 does not appear here since according to (8.3-9), it is included in t_{spont}. The inversion lifetime τ as defined by (8.1-9) is equal in most cases to the actual lifetime (i.e., not necessarily the radiative lifetime) of the upper laser level.

Returning to (8.7-5), we note that $I_s(\nu)$, the saturation intensity, is inversely proportional to $g(\nu)$ so that saturation becomes increasingly difficult off line-center.

Example: Gain Saturation in a Ruby Laser. We use the ruby example of Section 8.4. We take $\tau = t_{\text{spont}}$, $1/g(\nu_0) \equiv \Delta\nu \sim 2 \times 10^{11}$ Hz at 300°K. Using these data in (8.7-5) gives

$$I_s(\nu_0) \sim 467 \text{ W/cm}^2$$

Inhomogeneous Broadening

In the first part of this section, we considered the reduction in optical gain—that is, saturation—due to the optical field in a homogeneous laser medium. In what follows we treat the problem of gain saturation in inhomogeneous systems.

According to the above discussion, in an inhomogeneous atomic system the individual atoms are distinguishable, with each atom having a unique transition frequency $(E_2 - E_1)/h$. We can thus imagine the inhomogeneous medium as made up of classes of atoms each designated by a center frequency ν_ξ. Furthermore, we define a function $p(\nu_\xi)$ so that the a priori probability that an atom has its center frequency between ν_ξ and $\nu_\xi + d\nu_\xi$ is $p(\nu_\xi)\, d\nu_\xi$. Then

$$\int_{-\infty}^{\infty} p(\nu_\xi)\, d\nu_\xi = 1 \qquad (8.7\text{-}6)$$

since any atom has a unit probability of having its ν_ξ between $-\infty$ and ∞.

The atoms within a given ν_ξ are considered homogeneously broadened, having a lineshape function $g^\xi(\nu)$ that is normalized so that

$$\int_{-\infty}^{\infty} g^\xi(\nu)\, d\nu = 1 \qquad (8.7\text{-}7)$$

We can define the transition lineshape $g(\nu)$ by taking $g(\nu)\, d\nu$ to represent the a priori probability that a spontaneous emission will result in a photon whose frequency is between ν and $\nu + d\nu$. Using this definition, we obtain

$$g(\nu)\, d\nu = \left[\int_{-\infty}^{\infty} p(\nu_\xi) g^\xi(\nu)\, d\nu_\xi\right] d\nu \qquad (8.7\text{-}8)$$

which is a statement of the fact that the probability of emitting a photon of frequency between ν and $\nu + d\nu$ is equal to the probability $g^\xi(\nu)\, d\nu$ of this occurrence, given that the atom belongs to class ξ, summed up over all the classes.

If the total unsaturated inversion is ΔN_0 (atom/m³), then the inversion due to atoms in $d\nu_\xi$ is $\Delta N_0 p(\nu_\xi)\, d\nu_\xi$ and the contribution of that class *alone* to the exponential gain constant at ν is given by (8.7-4) as

$$\gamma_\xi(\nu) = -\frac{\Delta N_0 \lambda^2 \eta}{8\pi n^2 t_{\text{spont}}} \left[\frac{p(\nu_\xi)\, d\nu_\xi}{\dfrac{1}{g^\xi(\nu)} + \dfrac{I_\nu \phi \lambda^2 \eta}{4\pi n^2 h\nu}} \right] \tag{8.7-9}$$

where $\phi \equiv \tau/t_{\text{spont}}$. It follows from the definition of the gain constant (8.4-3) that the contribution of the various classes ν_ξ to $\gamma(\nu)$ are *additive* so that

$$\gamma(\nu) = -\frac{\Delta N_0 \lambda^2 \eta}{8\pi n^2 t_{\text{spont}}} \int_{-\infty}^{\infty} \frac{p(\nu_\xi)\, d\nu_\xi}{[1/g^\xi(\nu)] + (\phi\lambda^2 \eta I_\nu / 4\pi n^2 h\nu)} \tag{8.7-10}$$

This is our basic result.

As a first check on (8.7-10), we shall consider the case in which $I_\nu \ll 4\pi n^2 h\nu/\phi\lambda^2 g^\xi(\nu)$ and therefore, the effects of saturation can be ignored. Using (8.7-8) in (8.7-10) yields

$$\gamma(\nu) = -\frac{\Delta N_0 \lambda^2 \eta}{8\pi n^2 t_{\text{spont}}} g(\nu)$$

which is the same as (8.7-4) with $I_\nu = 0$. This shows that in the absence of saturation the expressions for the gain of a homogeneous and an inhomogeneous atomic system are identical.

Our main interest in this treatment is in deriving the saturated gain constant for an inhomogeneously broadened atomic transition. If we assume that in each class ξ all the atoms are identical (homogeneous broadening), we can use (8.1-20) for the lineshape function $g^\xi(\nu)$,

$$g^\xi(\nu) = \frac{\Delta\nu}{2\pi[(\Delta\nu/2)^2 + (\nu - \nu_\xi)^2]} \tag{8.7-11}$$

where $\Delta\nu$ is called the homogeneous linewidth of the inhomogeneous line. Atoms with transition frequencies that are clustered within $\Delta\nu$ of each other can be considered indistinguishable. The term "homogeneous packet" is often used to describe them. Using (8.7-11) in (8.7-10) leads to

$$\gamma(\nu) = -\frac{\Delta N_0 \lambda^2\, \Delta\nu\, \eta}{16\pi^2 n^2 t_{\text{spont}}} \int_{-\infty}^{\infty} \frac{p(\nu_\xi)\, d\nu_\xi}{(\nu - \nu_\xi)^2 + (\Delta\nu/2)^2 + (\phi\lambda^2 I_\nu\, \Delta\nu / 8\pi^2 n^2 h\nu)} \tag{8.7-12}$$

In the extreme inhomogeneous case, the width of $p(\nu_\xi)$ is by definition very much larger than the remainder of the integrand in (8.7-12), and thus, it is essentially a constant over the region in which the integrand peaks. In this case, we can pull $p(\nu_\xi)_{\nu_\xi=\nu} = p(\nu)$ outside the integral sign in (8.7-12), obtaining

$$\gamma(\nu) = -\frac{\Delta N_0 \lambda^2\, \Delta\nu\, \eta}{16\pi^2 n^2 t_{\text{spont}}} p(\nu) \int_{-\infty}^{\infty} \frac{d\nu_\xi}{(\nu - \nu_\xi)^2 + (\Delta\nu/2)^2 + (\phi\lambda^2\, \eta\, \Delta\nu I_\nu / 8\pi^2 n^2 h\nu)} \tag{8.7-13}$$

Using the definite integral

$$\int_{-\infty}^{\infty} \frac{dx}{x^2 + a^2} = \frac{\pi}{a}$$

to evaluate (8.7-13), we obtain

$$\gamma(\nu) = -\frac{\Delta N_0 \lambda^2 p(\nu)}{8\pi n^2 t_{\text{spont}}} \frac{1}{\sqrt{1 + (\phi\lambda^2 \eta I_\nu / 2\pi^2 n^2 h\nu\, \Delta_\nu)}} \tag{8.7-14}$$

$$= \gamma_0(\nu) \frac{1}{\sqrt{1 + (I_\nu / I_s)}} \tag{8.7-15}$$

and

$$I_s = \frac{2\pi^2 n^2 h\nu\, \Delta\nu}{\phi\lambda^2\eta} \tag{8.7-16}$$

is the saturation intensity of the inhomogeneous line. A comparison of (8.7-15) and (8.7-16) to (8.7-4) and (8.7-5) reveals two essential differences between the saturation behavior of homogeneous and inhomogeneous systems.

1. The inhomogeneous system saturates more "slowly" as indicated by the square root in (8.7-15). This can be explained by the fact that although the inversion per class (packet) ν_ξ decreases as in (8.7-4), this is partly compensated for by the fact that more classes are brought into the interaction as I_ν increases in accordance with (8.6-2). If we multiply the form of (8.6-2) by (8.7-4), the result is the inverse square law dependence of (8.7-15).
2. The saturation intensity in the inhomogeneous case does not depend on the position in the lineshape. That is, I_s in (8.7-16) does not depend on $g(\nu)$ as does the saturation intensity of the homogeneous case (8.7-5).

"Hole" Burning

To further appreciate the difference between the saturation behavior of homogeneous and inhomogeneous media, consider the following case. A strong field at ν is applied to the medium and, simultaneously, a very weak probing signal at ν' is used to measure the gain $\gamma(\nu')$. Our task is to determine the form of $\gamma(\nu')$ for both homogeneous and inhomogeneous media.

Consider the homogeneous case first. The gain at ν' is given according to (8.7-1) by

$$\gamma(\nu') = \Delta N \frac{\lambda^2 \eta}{8\pi n^2 t_{\text{spont}}} g(\nu')$$

where ΔN is the inversion *in the presence of the strong field at* ν. ΔN is given by (8.7-3). That, when used in the last equation, leads to

$$\gamma(\nu') = \gamma_0(\nu') \left[\frac{1 + 4\pi^2(\nu - \nu_0)^2 T_2^2}{1 + 4\pi^2(\nu - \nu_0)^2 T_2^2 + \dfrac{\mu^2 E_0^2}{\hbar^2} T_2 \tau g_1}\right] \tag{8.7-17}$$

where E_0 is the amplitude of the strong field at ν and $\gamma_0(\nu')$ is the nonsaturated ($E_0 = 0$) gain function

$$\gamma_0(\nu') = \Delta N_0 \frac{\lambda^2 \eta}{8\pi n^2 t_{\text{spont}}} g(\nu')$$

The important conclusion is that $\gamma(\nu')$ has the *same frequency dependence* as $\gamma_0(\nu')$ but is *reduced* in magnitude by the factor inside the square brackets of (8.7-17).

In the case of an inhomogeneously broadened gain medium, the situa-

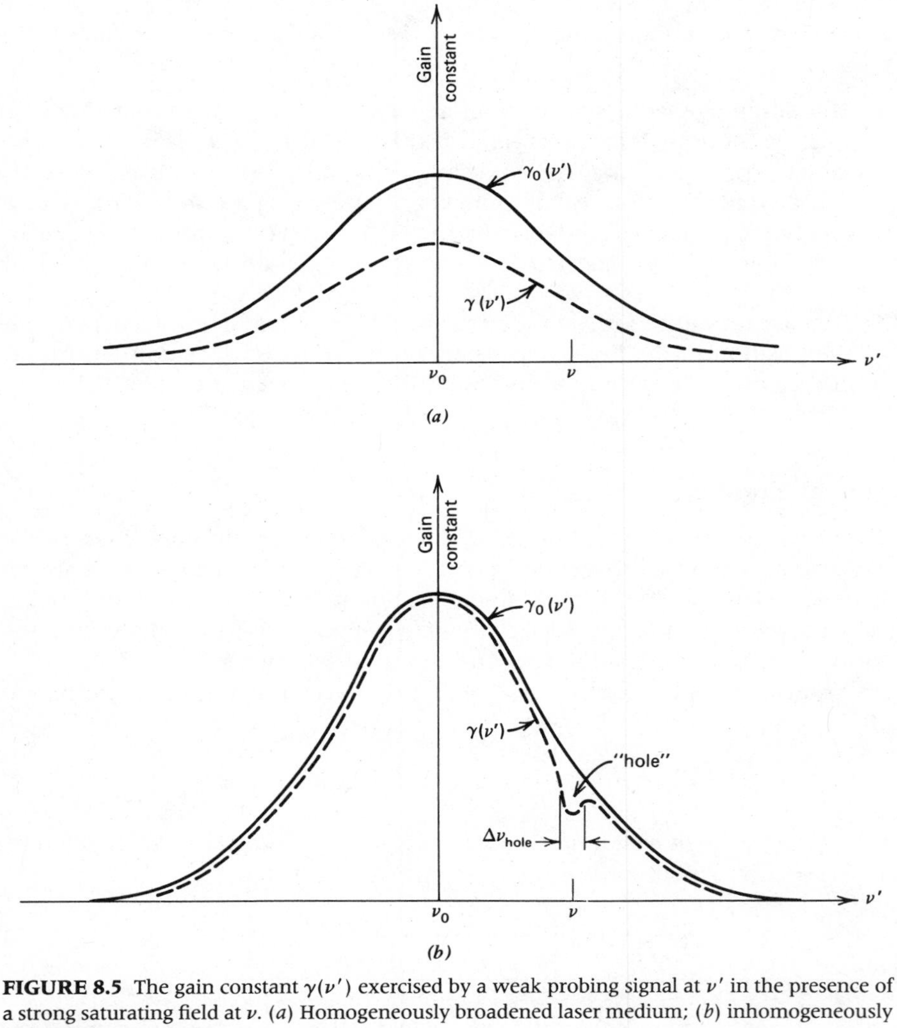

FIGURE 8.5 The gain constant $\gamma(\nu')$ exercised by a weak probing signal at ν' in the presence of a strong saturating field at ν. (*a*) Homogeneously broadened laser medium; (*b*) inhomogeneously broadened laser medium.

tion is more complicated. We start by using (8.7-11) to rewrite (8.7-10) as

$$\gamma(\nu) = -\frac{\Delta N_0 \lambda^2 \eta}{8\pi n^2 t_{\text{spont}}} \int_{-\infty}^{\infty} d\nu_\xi p(\nu_\xi) g^\xi(\nu) \left[\frac{\left(\frac{\Delta\nu}{2}\right)^2 + (\nu - \nu_\xi)^2}{\left(\frac{\Delta\nu}{2}\right)^2 + (\nu - \nu_\xi)^2 + \frac{\phi\lambda^2 \Delta\nu I_\nu}{8\pi^2 n^2 h\nu}} \right]$$

The integrand is clearly proportional to the contribution to the gain at ν due to the atomic packet centered on ν_ξ. It follows directly that the quantity inside the square bracket represents the factor by which this contribution is reduced due to the saturating field at ν. The gain exercised by the weak probing signal at ν' is thus the unsaturated gain $\gamma_0(\nu')$ multiplied by this local reduction factor, that is,

$$\gamma(\nu') = \gamma_0(\nu') \left[\frac{\left(\frac{\Delta\nu}{2}\right)^2 + (\nu - \nu')^2}{\left(\frac{\Delta\nu}{2}\right)^2 + (\nu - \nu')^2 + \frac{\phi\lambda^2 \Delta\nu I_\nu}{8\pi^2 n^2 h\nu}} \right] \tag{8.7-18}$$

The main features of (8.7-18) is that $\gamma(\nu')$ is essentially identical to $\gamma_0(\nu')$ except for frequencies ν' in the vicinity of the saturating frequency ν. Here, the gain is depressed over a frequency interval approximately equal to

$$\Delta\nu_{\text{hole}} = \Delta\nu \sqrt{1 + \frac{I_\nu}{I_s}} \tag{8.7-19}$$

and the gain at $\nu' = \nu$ is reduced by a factor $(1 + I_\nu/I_s)^{-1}$. I_s is the saturation intensity defined by (8.7-16). This depressed region is usually referred to as a "hole" and the phenomenon just described as "hole burning."

The gain profile $\gamma(\nu')$ of the probing signal is sketched in Figure 8.5 for both cases of broadening.

A different approach to the problem of gain and gain saturation based on rate equations is given in Reference 7. More recent application (Reference 10) of the techniques of nonlinear optics to the problem of the gain experienced by a weak probe beam at ν' due to a strong pump beam at ν shows that the signal may exercise a small hole even in the case of a homogeneously broadened transition. The effect is subtle and the theoretical treatment relies on the method developed in Chapter 18 for treating the field-induced nonlinear polarization in atoms.

References

1. Bloch, F., W. W. Hansen, and M. Packard, "Nuclear Induction," *Phys. Rev.* **70,** 960 (1946)
2. Bloembergen, N., *Nuclear Magnetic Relaxation* (New York: Benjamin, 1901), p. 43.
3. See, for example, P. M. Morse and H. Feshback, *Methods of Theoretical Physics* (New York: McGraw-Hill, 1953), p. 372.
4. Kronig, R. L., *J. Opt. Soc. Am.* **12,** 547 (1926); Kramers, H. A., *Atti. Congr. Intern. Fis.* **2,** 545 (1927).

5. Einstein, A, "Die Quanten Theorie der Strahlung," *Phys. Leit.* **18,** 121 (1917).
6. Portis, A. M., "Inhomogeneous Line-Broadening in F Centers," *Phys. Rev.* **91,** 1071 (1953).
7. Yariv, A., *Introduction to Optical Electronics* (New York: Holt, Rinehart and Winston, 1971).
8. Yariv, A., *Quantum Electronics,* 1st ed. (New York: Wiley, 1967).
9. R. W. Ditchburn, *Light* (New York: Wiley-Interscience, 1962).
10. Boyd, R. W., S. Mukamel, "Origin of Spectral Holes in Pump-Probe Studies of Homogeneously Broadened Lines," *Phys. Rev. A* **29,** 1973 (1984).
11. Yariv, A., *Theory and Applications of Quantum Mechanics* (New York: Wiley, 1982).

Problems

8.1 Show that relation (8.5-10)

$$\frac{A}{B_{21}} = \frac{8\pi h\nu^3 n^3}{c^3}$$

is consistent with Eq. (8.3-4) according to which

$$\frac{W_{\text{induced}} \text{ per mode}}{W_{\text{spont}} \text{ per mode}} = n$$

where n = number of quanta in the mode.

8.2 Determine the peak absorption coefficient $\alpha(\nu_0)$ due to a transition at $\nu_0 = 3 \times 10^{14}$ Hz where $N_2 \simeq 0$, $N_1 = 10^{18}$ cm^{-3}, the full width of the Gaussian absorption curve is 400 cm^{-1}, and $t_{\text{spont}} = 10^{-4}$ sec. Defining the optical density as

$$\log_{10} \frac{\text{I}_{\text{in}}}{I_{\text{out}}}$$

where I denotes intensity, what is the optical density at ν_0 for a 1-cm path length of material? At what temperature will the rate for the transition induced by blackbody radiation equal the spontaneous emission rate?

8.3 Calculate the classical lifetime t_{class} = energy/(radiated power) of an electron oscillating so that $r = r_0 \cos(2\pi\nu t)$ where r is the electron position.

8.4 (a) Acquaint yourself with the concept of the oscillator strength of a transitions (References 8 and 9).
(b) What is the oscillator strength of the transition described in 8.2?
(c) Show that the oscillator strength f_{21} for a transition $1 \leftrightarrow 2$ at frequency ν is equal to $t_{\text{classical}}/3t_{\text{spont}}$.

8.5 Derive (8.1-15)

8.6 Show that $\chi'(\omega)$ and $\chi''(\omega)$ in (8.1-19) obey the Kramers–Kronig relations in the limit of negligible saturation ($\Omega = 0$).

C H A P T E R 9

Laser Oscillation

9.0 INTRODUCTION

Proposals for using stimulated emission from a system of inverted population for microwave amplification were made, independently, by Weber (Reference 1), Gordon, Zeiger, and Townes (Reference 2), and Basov and Prokhorov (Reference 3). The first operation of such an amplifier was by the Columbia University group of Gordon, Zeiger, and Townes. This group is responsible for the name "maser," an acronym for "microwave amplification by the stimulated emission of radiation." The first maser utilized a microwave transition in the ammonia (NH_3) molecule. The feasibility of maser action at optical and near optical frequencies was considered in a paper by Schawlow and Townes (Reference 4) in 1958. In 1960, less than two years later, Maiman (Reference 5) succeeded in operating a pulsed ruby laser (acronym for "light amplification by stimulated emission of radiation"). The first continuous wave (cw) laser was a He-Ne gas laser announced in 1961 (Reference 6). Laser action in semiconductors was demonstrated in 1962 and described in Chapter 10.

In the last chapter we found that a medium with an inverted atomic population is capable of amplifying radiation at frequencies near that of the atomic transition. In this chapter we consider what happens if such an amplifying medium is placed within an optical resonator. The ever-present zero point fluctuation fields of the resonator modes will now be amplified. The few modes whose losses are sufficiently low (i.e., those with a high Q) may experience net amplification. These modes will grow in intensity until their gain saturates at a level equal to the loss and steady-state oscillation prevails.

In this chapter we will derive the condition for laser oscillation, determine the frequencies at which such oscillation can take place, and consider the problem of power extraction from the laser.

9.1 THE LASER OSCILLATION CONDITION[1]

In this section we will derive the laser oscillation condition. Specifically, we will determine the density of inverted population of the laser medium at which a laser will start oscillating and the oscillation frequency.

To gain a better insight into the nature of oscillation, we will derive the laser condition from two, seemingly different, points of view.

[1] As a simple introduction to this subject, the reader is advised to read the treatment in the author's book, *Introduction to Optical Electronics* (New York: Holt, Rinehart and Winston), p. 148.

In the first of these derivations, which will be considered in this section, we will "unleash" a Gaussian propagating beam mode inside an optical resonator made up of lenslike media and elements (including the gain medium), then trace its internal propagation using the *ABCD* law. The laser oscillation condition will emerge from the requirement that the beam reproduce itself in *shape, amplitude,* and *phase,* after each round trip.

The second point of view is developed in Section 9.2.

We find it convenient to use the function

$$q(z)e^{-i\theta(z)} \tag{9.1-1}$$

to characterize the beam at z. The parameter $q(z)$ appearing in (9.1-1) is the complex beam radius defined as in (6.7-5) by

$$\frac{1}{q(z)} = \frac{1}{R(z)} - i\frac{\lambda}{\pi\omega^2(z)n} \tag{9.1-2}$$

and the factor $\exp[-i\theta(z)]$ is the complex amplitude of the wave at z. θ is complex so that if we take $\theta = \theta_r + i\theta_i$, the beam power at z relative to its value at $z = 0$ is given by $\exp[2\theta_i(z)]$, and its phase by $-\theta_r(z)$.

The passage of a Gaussian beam through some lenslike element, labeled by s, previously described by (6.7-6) as

$$q_2 = \frac{A_s q_1 + B_s}{C_s q_1 + D_s} \tag{9.1-3}$$

is now modified to

$$q_2 e^{-i\theta_2} = \frac{A_s q_1 + B_s}{C_s q_1 + D_s} e^{-i(\theta_s + \theta_1)} \tag{9.1-4}$$

where $\exp(-i\theta_s)$ is the complex amplitude transmission factor of the sth element. A lenslike element is now characterized by its A, B, C, D matrix as well as by its transmission factor $\exp(-i\theta)$.

The θ of an homogeneous medium extending between z_1 and z_2 is given by

$$\theta_{\text{hom}} = k'(z_2 - z_1) - \left[(l + m + 1)\left(\tan^{-1}\frac{z_2}{z_0} - \tan^{-1}\frac{z_1}{z_0}\right)\right] \tag{9.1-5}$$

where k' is the complex propagation constant given by (8.2-4).

The spherical mirror, as an example, with an A, B, C, D matrix as in Table 6.1, has a transmission factor $re^{-i\theta_m}$ where $|r|^2$ is the fraction of the incident power reflected by the mirror (reflectivity) and θ_m the phase shift upon reflection.

The propagation of a Gaussian beam through a sequence of N lenslike elements previously given by (6.7-9) will now be taken in the form

$$q_{\text{out}} e^{-i\theta_{\text{out}}} = \frac{Aq_{\text{in}} + B}{Cq_{\text{in}} + D} e^{-i\theta_{\text{in}}} \prod_{s=1}^{N} e^{-i\theta_s} \tag{9.1-6}$$

where A, B, C, D are the elements of the product of the N individual A_s, B_s, C_s, D_s matrices.

We can now apply this formalism to deriving the condition of laser oscillation. Consider the resonator shown in Figure 9.1 that consists of two mirrors with (amplitude) reflectances $r_1 \exp(-i\theta_{m1})$ and $r_2 \exp(-i\theta_{m2})$, respectively. The resonator is filled with an amplifying medium whose complex propagation constant is k'.

Following the evolution of the Gaussian *laser* beam through one round trip, we obtain

$$q_5 e^{-i\theta_5} = \frac{Aq_1 + B}{Cq_1 + D} e^{-i(\theta_1 + \theta)} \tag{9.1-7}$$

where using (9.1-5), (9.1-6) and taking $l = m = 0$ yield

$$e^{-i\theta} = e^{-i2[k'l - \tan^{-1}(z_2/z_0) + \tan^{-1}(z_1/z_0)]} r_1 r_2 e^{-i(\theta_{m1} + \theta_{m2})} \tag{9.1-8}$$

For self-reproducing oscillation, we require that the beam shape as well as its *complex* amplitude return to their original values after one round trip. This happens when

$$\begin{aligned} q_5 &= q_1 \\ \theta_5 &= \theta_1 + 2m\pi \end{aligned} \tag{9.1-9}$$

where m is an integer. The first of conditions (9.1-9) is satisfied if

$$q_1 = \frac{Aq_1 + B}{Cq_1 + D}$$

and was used in Section 7.2 to obtain q_1. The second condition is satisfied when

$$e^{-i\theta} = e^{-i2m\pi}$$

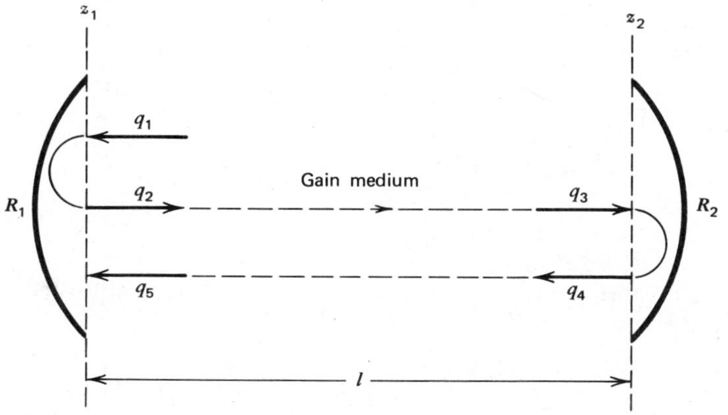

FIGURE 9.1 The propagation of a Gaussian beam with a complex radius q through one round trip inside an optical resonator. The mirror reflectivities, R_1 and R_2, are related to the field reflectances, r_1 and r_2, through $R_1 \equiv r_1^2$, $R_2 \equiv r_2^2$.

which if we use (9.1-8) can be written as

$$e^{-i2[k'l-\tan^{-1}(z_2/z_0)+\tan^{-1}(z_1/z_0)]} r_1 r_2 e^{-i(\theta_{m1}+\theta_{m2})} = e^{-i2m\pi} \tag{9.1-10}$$

This is the laser oscillation condition. It is an intuitively obvious statement of the requirement that in steady-state oscillation, the beam amplitude and phase return to their initial values after each round trip inside the resonator. It will be used, in what follows, to determine the threshold inversion density as well as the oscillation frequency.

The Threshold Inversion

The threshold gain condition is obtained by equating the magnitudes of both sides of (9.1-10)

$$\left|e^{-i\theta}\right| = 1 \tag{9.1-11}$$

It ensures that after a complete round trip the beam amplitude returns to its original value. If we use (9.1-10), it can be written as

$$\left|e^{-i2k'l} r_1 r_2\right| = 1 \tag{9.1-12}$$

The *complex* propagation constant k' is, from (8.2-4),

$$k' = k\left[1 + \frac{\chi'(\nu)}{2n^2}\right] - ik\frac{\chi''(\nu)}{2n^2} - i\frac{\alpha}{2} \tag{9.1-13}$$

where α is the distributed absorption coefficient of the medium due to all the loss mechanisms except the resonant laser transition. The laser transition contributes the two terms involving $\chi'(\nu)$ and $\chi''(\nu)$.

From (9.1-13), the exponential gain constant of the medium is given by

$$\gamma = -k\frac{\chi''(\nu)}{n^2} \tag{9.1-14}$$

so that the oscillation condition (9.1-12) becomes

$$e^{(\gamma_t-\alpha)l} r_1 r_2 = 1 \tag{9.1-14a}$$

or

$$\gamma_t = \alpha - \frac{1}{l}\ln r_1 r_2 \tag{9.1-15}$$

where the subscript t indicates threshold.

We can use (8.4-4) to convert (9.1-15) into a condition for the threshold inversion density at line center

$$\begin{aligned} \Delta N_t \equiv \left(N_2 - N_1\frac{g_2}{g_1}\right)_t &= \frac{8\pi n^2 t_{\text{spont}}}{\eta g(\nu_0)\lambda^2}\left(\alpha - \frac{1}{l}\ln r_1 r_2\right) \\ &= \frac{8\pi n^2 t_{\text{spont}}\Delta\nu}{\eta\lambda^2}\left(\alpha - \frac{1}{l}\ln r_1 r_2\right) \end{aligned} \tag{9.1-16}$$

where $\Delta\nu \equiv 1/g(\nu_0)$ is the gain linewidth. For an alternative expression for ΔN_t, see footnote 2.

Example: Inversion in a He-Ne Laser. To get an order of magnitude estimate of the threshold inversion, consider the case of a He-Ne laser oscillating at $\lambda = 6.328 \times 10^{-5}$ cm. If we use $\Delta\nu \sim [g(\nu_0)]^{-1} \sim 1.5 \times 10^9$ Hz, $\alpha = 0$, $r_1 r_2 \sim 0.98$, $\eta \approx 1$, $t_{\text{spont}} \sim 10^{-7}$ sec, $l = 10$ cm in (9.1-16) gives

$$\left(N_2 - N_1 \frac{g_2}{g_1}\right)_t \sim 1.9 \times 10^9 \text{ cm}^{-3}$$

In solid-state lasers, on the other hand, the combination of a broader transition and, typically, longer spontaneous lifetimes results in much larger threshold inversions. The example of Section 10.2 shows that a typical inversion for the ruby ($Cr^{3+} : Al_2O_3$) laser is $\sim 10^{17}$ cm^{-3}.

The Oscillation Frequency

The oscillation frequency is obtained by equating the phases on both sides of (9.1-10). This ensures that the round-trip phase delay is some multiple (m) of 2π.

$$\frac{\omega n}{c} l \left[1 + \frac{\chi'(\nu)}{2n^2}\right] - \tan^{-1}\frac{z_2}{z_0} + \tan^{-1}\frac{z_1}{z_0} + \frac{\theta_{m1} + \theta_{m2}}{2} = m\pi \qquad (9.1\text{-}17)$$

The mth resonant frequency of the "cold" ($\chi = 0$) resonator is obtained from (9.1-17) by putting $\chi' = 0$ and is

$$\nu_m = \frac{mc}{2nl} + \frac{c}{2\pi n l}\left(\tan^{-1}\frac{z_2}{z_0} - \tan^{-1}\frac{z_1}{z_0} - \frac{\theta_{m1} + \theta_{m2}}{2}\right) \qquad (9.1\text{-}18)$$

We use it to rewrite (9.1-17) as

$$\nu\left[1 + \frac{\chi'(\nu)}{2n^2}\right] = \nu_m \qquad (9.1\text{-}19)$$

From (8.1-19), we have

$$\chi'(\nu) = \frac{2(\nu_0 - \nu)}{\Delta\nu}\chi''(\nu)$$

and from (9.1-14),

$$\gamma(\nu) = -k\frac{\chi''(\nu)}{n^2}$$

[2] If we use the expression (see Section 7.4) $t_c^{-1} = (c/n)[\alpha - (1/l)\ln r_1 r_2]$ for the cavity loss rate, Eq. (9.1-16) for ΔN_t can be recast in an often encountered form

$$\Delta N_t = \frac{8\pi n^3 \nu^2 t_{\text{spont}}}{\eta c^3 t_c g(\nu_0)} = \frac{8\pi n^3 \nu^2 t_{\text{spont}} \Delta\nu}{\eta c^3 t_c}$$

With these relations, (9.1-19) becomes

$$\nu\left[1 - \frac{(\nu_0 - \nu)}{\Delta\nu}\frac{\gamma_t(\nu)}{k}\right] = \nu_m \tag{9.1-20}$$

We anticipate that ν will turn out to be much closer to ν_m than to ν_0 and, consequently, replace $\gamma_t(\nu)$ in (9.1-20) by $\gamma_t(\nu_m)$. This results in

$$\nu \approx \nu_m - (\nu - \nu_0)\frac{c\gamma_t(\nu_m)}{2\pi n\Delta\nu}$$

At threshold, the gain $\gamma_t(\nu_m)$ is given by (9.1-15) so that

$$\nu \approx \nu_m - (\nu_m - \nu_0)\frac{c\left[\alpha - \frac{1}{l}\ln(r_1 r_2)\right]}{2\pi n\Delta\nu}$$

which if we use (7.4-7) becomes

$$\nu \approx \nu_m - (\nu_m - \nu_0)\left(\frac{\Delta\nu_{1/2}}{\Delta\nu}\right) \tag{9.1-21}$$

where $r_1 r_2 = \sqrt{R_1 R_2}$ and $\Delta\nu_{1/2}$ is the full width of the passive optical resonator.

If the atomic resonance frequency ν_0 does not coincide with the passive resonance frequency ν_m, the laser frequency will, according to (9.1-21), be shifted away from ν_m *toward* ν_0. This phenomenon is called "frequency pulling." Since typically $\Delta\nu_{1/2} \ll \Delta\nu$, the laser tends to oscillate near ν_m.

Example: Frequency Pulling in a He-Ne Laser. In a typical He-Ne 0.6328 μm laser, as an example, we have $l = 30$ cm, $R = 0.99$, $\Delta\nu \simeq 1.5 \times$

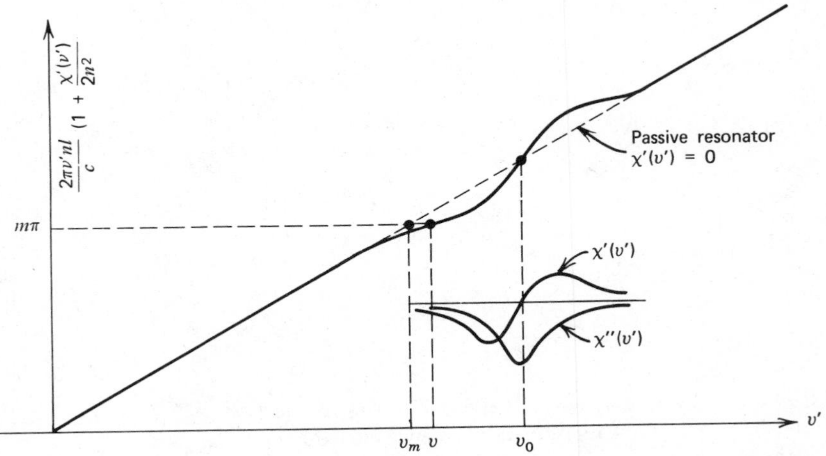

FIGURE 9.2 A graphical illustration of the laser frequency condition [Eq. 9.1-17) showing how the atomic dispersion $\chi'(\nu)$ "pulls" the laser oscillation frequency ν from the passive resonator value ν_m toward that of the atomic resonance at ν_0.

10^9 Hz, $\alpha \simeq 0$. Using (7.4-7) gives $\Delta\nu_{1/2} \simeq 1.6 \times 10^6$ Hz so that $\Delta\nu_{1/2}/\Delta\nu \simeq 10^{-3}$. If ν_m and ν_0 differ by, say 10^8 Hz, for example, the oscillation frequency will be pulled by $10^8 \times 10^{-3} = 10^5$ Hz away from ν_m. This is a small, but for many applications, noticeable effect.

A graphical solution of the oscillation condition (9.1-17) is shown in Figure 9.2. In this case, we assumed that in (9.1-17) $z_2/z_0 \approx 0$, $z_1/z_0 \approx 0$ and $\theta_{m1} + \theta_{m2} \approx 0$. The inclusion of these neglected terms would cause a slight shift of the passive oscillation frequency ν_m without, to first order, changing any of the results.

9.2 LASER OSCILLATION—GENERAL TREATMENT

The derivation of the oscillation condition leading to (9.1-10) is limited to lasers employing Gaussian beams and, consequently, is of great utility in the many practical systems that fit this category.

In this section we will present an alternative derivation. It is cast in a general form so as to bring out the basic and common features of laser oscillation that are independent of geometry.

We consider a model of a generalized resonator that contains an inverted laser medium. We will assume that some mode (e.g., l) of the resonator is excited and is oscillating. This mode with an electric field $\mathbf{E}_l(\mathbf{r}, t)$ induces a coherent polarization field $\mathbf{P}_l(\mathbf{r}, t) = \varepsilon_0 \chi \mathbf{E}_l(\mathbf{r}, t)$, where the susceptibility χ is given by (8.1-19). The circle is completed by requiring that $\mathbf{P}_l(\mathbf{r}, t)$ acting as a driving source gives rise to an oscillation field $\mathbf{E}_l(\mathbf{r}, t)$. This approach, modeled after Lamb's self-consistent analysis of an inhomogeneous laser (Reference 7), is illustrated by Figure 9.3.

We start with the Slater mode expansion of a resonator field as in Section 5.5.

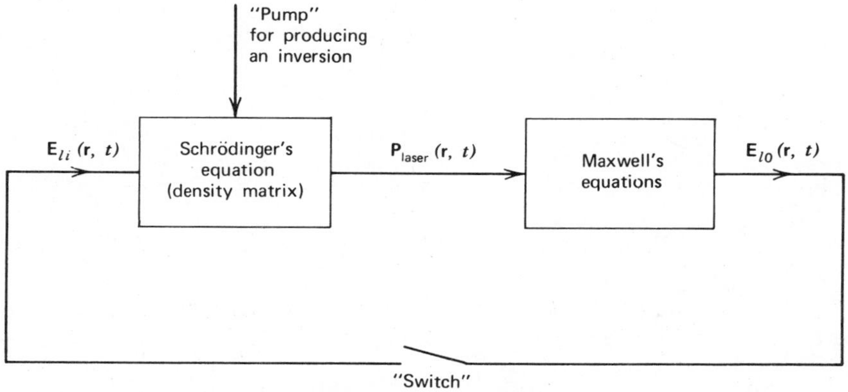

FIGURE 9.3 A schematic flow diagram of the self-consistent laser analysis. Self-consistency is obtained by closing the "switch," thus making $\mathbf{E}_{l0}(\mathbf{r}, t) = \mathbf{E}_{li}(\mathbf{r}, t)$.

$$\mathbf{E}(\mathbf{r}, t) = \sum_a -\frac{1}{\sqrt{\varepsilon}} p_a(t)\mathbf{E}_a(\mathbf{r})$$
$$\mathbf{H}(\mathbf{r}, t) = \sum_a \frac{1}{\sqrt{\mu}} \omega_a q_a(t)\mathbf{H}_a(\mathbf{r}) \tag{9.2-1}$$

where

$$\nabla \times \mathbf{H}_a = k_a \mathbf{E}_a$$
$$\nabla \times \mathbf{E}_a = k_a \mathbf{H}_a \tag{9.2-2}$$

The resonator contains a distributed polarization field $\mathbf{P}_{\text{laser}}(\mathbf{r}, t)$ due to the laser medium so that Maxwell's equations can be written as

$$\nabla \times \mathbf{H} = \mathbf{i} + \frac{\partial}{\partial t}(\varepsilon_0 \mathbf{E} + \mathbf{P}_{\text{nonresonant}} + \mathbf{P}_{\text{laser}})$$
$$= \sigma \mathbf{E} + \varepsilon \frac{\partial \mathbf{E}}{\partial t} + \frac{\partial}{\partial t}\mathbf{P}_{\text{laser}} \tag{9.2-3}$$
$$\nabla \times \mathbf{E} = -\mu \frac{\partial \mathbf{H}}{\partial t}$$

σ is the effective conductivity that is introduced to account for the losses, and ε is the dielectric constant. If we use (9.2-1) and (9.2-2), (9.2-3) becomes

$$\sum_a \frac{1}{\sqrt{\mu}} \omega_a q_a k_a \mathbf{E}_a = -\sigma \sum_a \frac{1}{\sqrt{\varepsilon}} p_a \mathbf{E}_a - \sum_a \sqrt{\varepsilon}\, \dot{p}_a \mathbf{E}_a + \frac{\partial}{\partial t}\mathbf{P}_{\text{laser}}(\mathbf{r}, t) \tag{9.2-4}$$

Dot multiplying (9.2-4) by $\mathbf{E}_l$, integrating over the resonator volume, and using the orthonormality condition (5.5-10) lead to

$$\omega_l^2 q_l + \frac{\sigma}{\varepsilon} p_l + \dot{p}_l - \frac{1}{\sqrt{\varepsilon}} \frac{\partial}{\partial t} \int_V \mathbf{P}_{\text{laser}} \cdot \mathbf{E}_l \, dv = 0 \tag{9.2-5}$$

$$\omega_l^2 \dot{q}_l + \frac{\sigma}{\varepsilon} \dot{p}_l + \ddot{p}_l - \frac{1}{\sqrt{\varepsilon}} \frac{\partial^2}{\partial t^2} \int_V \mathbf{P}_{\text{laser}} \cdot \mathbf{E}_l \, dv = 0 \tag{9.2-6}$$

From (9.2-1), the second equation in (9.2-2), and (9.2-4), we obtain

$$p_a = \dot{q}_a$$

so that (9.2-6) becomes

$$\omega_l^2 p_l + \ddot{p}_l + \frac{\sigma}{\varepsilon} \dot{p}_l = \frac{1}{\sqrt{\varepsilon}} \frac{\partial^2}{\partial t^2} \int_V \mathbf{P}_{\text{laser}}(\mathbf{r}, t) \cdot \mathbf{E}_l(\mathbf{r}) \, dv \tag{9.2-7}$$

This is an equation of a classical harmonic oscillator with the right side representing the driving term. If the driving is zero, the solution is

$$p_l(t) = p_l(0) e^{-i\omega_l[1-(1/8)(\sigma^2/\omega_l^2\varepsilon^2)]t} e^{-(\sigma/2\varepsilon)t} \tag{9.2-8}$$

from which we identify ω_l as the resonant frequency in the lossless limit and ε/σ as the decay time constant t_c of the lth mode energy in the passive resonator. It follows from (7.4-5) that

$$t_c = \frac{\varepsilon}{\sigma} = \frac{Q}{\omega_l} \tag{9.2-9}$$

In a high Q resonator, $p_l(t)$ cannot vary rapidly so that it is reasonable to take it as

$$p_l(t) = p_{l0}(t)e^{i\omega t} \tag{9.2-10}$$

where $p_{l0}(t)$ is a "slowly" varying amplitude so that

$$\ddot{p}_{l0} \ll \omega \dot{p}_{l0} \tag{9.2-11}$$

and ω, the laser oscillation frequency, is to be determined.

Substituting (9.2-10) in (9.2-7) and using (9.2-11) to justify the discard of the term $\ddot{p}_{l0}$ lead to an adiabatic form of the oscillation equation

$$\left\{\left[(\omega_l^2 - \omega^2) + i\frac{\sigma\omega}{\varepsilon}\right]p_{l0}(t) + \left(2i\omega + \frac{\sigma}{\varepsilon}\right)\dot{p}_{l0}\right\}e^{i\omega t} = \frac{1}{\sqrt{\varepsilon}}\frac{\partial^2}{\partial t^2}\int_V (\mathbf{P}_{\text{laser}} \cdot \mathbf{E}_l)\, dv \tag{9.2-12}$$

To derive the oscillation threshold condition, we may assume that only one mode (e.g., l) is sufficiently near threshold to be appreciably excited. We thus have from (9.2-1)

$$\mathbf{E}(\mathbf{r}, t) = -\frac{1}{\sqrt{\varepsilon}}p_l(t)\mathbf{E}_l(\mathbf{r})$$

so that according to (9.2-10) and (8.1-18),

$$\mathbf{P}_{\text{laser}}(\mathbf{r}, t) = -\frac{\varepsilon_0}{\sqrt{\varepsilon}}\chi(\omega)p_{l0}e^{i\omega t}\mathbf{E}_l(\mathbf{r}) \tag{9.2-13}$$

Equation (9.2-12) expresses $p_{l0}(t)$ (i.e., the field) as a function of the polarization $\mathbf{P}_{\text{laser}}(\mathbf{r}, t)$. In (9.2-13) the reverse is true. The laser oscillation condition is obtained by making (9.2-12) and (9.2-13) self-consistent.

We limit ourselves to the steady state, $\dot{p}_{l0} = 0$, and substitute (9.2-13) in (9.2-12) with $\partial^2/\partial t^2 \to -\omega^2$. The result is

$$(\omega_l^2 - \omega^2) + i\frac{\sigma\omega}{\varepsilon} = \frac{\omega^2\varepsilon_0 f}{\varepsilon}(\chi' - i\chi'') \tag{9.2-14}$$

where we used $\chi = \chi' - i\chi''$ and introduced the filling factor $f \leq 1$ by

$$f = \int_{V_{\text{laser medium}}} \mathbf{E}_l \cdot \mathbf{E}_l\, dv$$

so that $f = 1$ in the case of a completely filled resonator (and uniform inversion). Equation (9.2-14) is the laser oscillation condition.

It will be left as an exercise to demonstrate that (9.2-14) is fully equivalent to the oscillation condition (9.1-10).

9.3 POWER OUTPUT FROM LASERS

In the last two sections we considered the threshold oscillation condition. We obtained, in (9.1-16), an expression for the minimum inversion necessary to start oscillation. In this section we treat the problem of converting pump

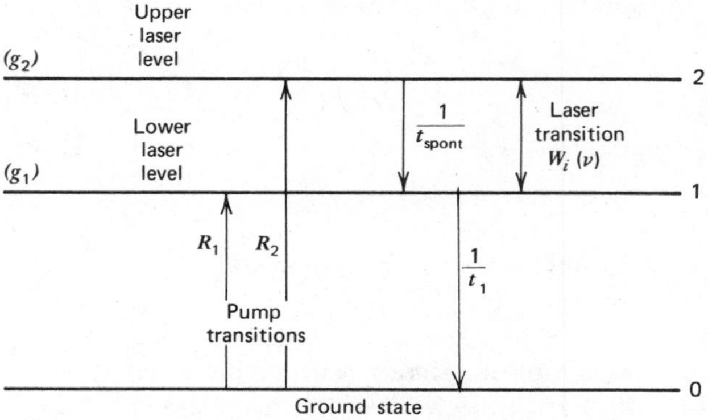

FIGURE 9.4 Energy levels and transition and relaxation rates of a four-level laser system.

power into coherent laser output above *threshold.* Specifically, we derive expressions relating the laser power output to atomic, optical, and pumping parameters.

The Rate Equations

We start with the basic four-level[3] atomic system shown in Figure 9.4. The laser transition is $2 \rightarrow 1$. Level 0 is the ground state. The actual lifetimes of level 2 and 1 are t_2 and t_1, respectively. The lifetime t_2 of the upper laser level may be due to spontaneous radiative transitions to level 1 whose rate is denoted as t_{spont}^{-1}, nonradiative transition to level 1, as well as to radiative and nonradiative transitions to other levels. This fact can be accounted for by taking

$$\frac{1}{t_2} = \frac{1}{t_{21}} + \text{(transition rates to other levels)}$$

where $\qquad$ (9.3-1)

$$\frac{1}{t_{21}} = \frac{1}{t_{spont}} + \left(\frac{1}{t_{21}}\right)_{\text{nonradiative}}$$

The density of atoms in levels 1 and 2 is taken as N_1 and N_2, respectively, and the level degeneracies are g_1 and g_2. The pumping rates into the two levels (atoms/second-volume) are R_1 and R_2. Pumping at a rate R_1 into the lower laser level is not desirable since it causes a reduction of the optical gain. In many situations, especially those involving pumping by discharge electrons or chemical reactions, some degree of lower level pumping is inevitable and must be included in the analysis.

[3] The designation "four-level laser" is due to the fact that in many real lasers the upper level excitations R_2 proceeds through some intermediate state that is not shown in Figure 9.4.

The induced transition rates between levels 1 and 2 are taken, following (8.3-14, 15), as

$$W_{2\to1} \equiv W_i(\nu) = \frac{\lambda^2 g(\nu)}{8\pi n^2 h\nu t_{\text{spont}}} I_\nu \tag{9.3-2}$$

$$W_{1\to2} = \frac{g_2}{g_1} W_i(\nu) \tag{9.3-3}$$

The equations describing the change in level populations due to the combined effect of pumping, spontaneous and induced radiative transitions, and relaxation processes in a homogeneously broadened medium are

$$\begin{aligned}\frac{dN_2}{dt} &= R_2 - \frac{N_2}{t_2} - \left(N_2 - \frac{g_2}{g_1} N_1\right) W_i(\nu) \\ \frac{dN_1}{dt} &= R_1 - \frac{N_1}{t_1} + \frac{N_2}{t_{21}} + \left(N_2 - \frac{g_2}{g_1} N_1\right) W_i(\nu)\end{aligned} \tag{9.3-4}$$

Equations (9.3-4) can be solved for the equilibrium ($d/dt = 0$) inversion, yielding

$$\Delta N \equiv N_2 - \frac{g_2}{g_1} N_1 = \frac{R_2 t_2 - (R_1 + \delta R_2) t_1 \dfrac{g_2}{g_1}}{1 + \left[t_2 + (1 - \delta) t_1 \dfrac{g_2}{g_1}\right] W_i(\nu)} \tag{9.3-5}$$

where

$$\delta \equiv \frac{t_2}{t_{21}} \tag{9.3-6}$$

The equilibrium inversion at the absence of an optical field is obtained from (9.3-5) by setting $W_i(\nu) = 0$ and is

$$\Delta N_0 \equiv \left(N_2 - \frac{g_2}{g_1} N_1\right)_0 = R_2 t_2 - (R_1 + \delta R_2) t_1 \frac{g_2}{g_1} \tag{9.3-7}$$

Using the last expression, we rewrite (9.3-5) as

$$\Delta N = \frac{\Delta N_0}{1 + \phi t_{21} W_i(\nu)} \tag{9.3-8}$$

$$\phi = \delta\left[1 + (1 - \delta) \frac{t_1 g_2}{t_2 g_1}\right] \tag{9.3-9}$$

The complicated but *unavoidable* dependence of the inversion on relaxation rates and degeneracies has been lumped into ϕ, which is a constant in a given atomic system. Before proceeding, let us consider an idealized simple case where $t_2 = t_{21} (\delta = 1)$ and $R_1 = 0$. In this case,

$$\Delta N_0 \equiv \left(N_2 - \frac{g_2}{g_1} N_1\right)_0 = R_2\left(t_2 - t_1 \frac{g_2}{g_1}\right)$$

so that gain ($\Delta N_0 > 0$) obtains when

$$t_2 > t_1 \frac{g_2}{g_1} \tag{9.3-10}$$

Note that when $g_1 > g_2$, it is possible to have gain ($\Delta N_0 > 0$) even when $t_2 < t_1$ (so that $N_2 < N_1$), provided (9.3-10) is satisfied. This is simply a reflection of the fact that, as stated by (9.3-2) and (9.3-3), when $g_1 > g_2$, the induced $2 \to 1$ rate is larger than the $1 \to 2$ rate. This gives rise to a net emission of power even though $N_2 < N_1$.

In the large majority of the practical laser systems, the condition $t_1 g_2/t_2 g_1 \ll 1$ is satisfied. When this is true, $\phi \approx \delta \equiv t_2/t_{21}$ and (9.3-8) becomes

$$\Delta N = \frac{\Delta N_0}{1 + W_i(\nu) t_2} \tag{9.3-11}$$

This is the starting point for the analysis of laser power output that follows.

Power and Optimum Coupling

Consider the gain prevailing inside a laser resonator in the presence of the laser field. Using (9.3-11) in (8.4-4) leads to

$$\gamma = \frac{\gamma_0}{1 + W_i(\nu) t_2} \tag{9.3-12}$$

where

$$\gamma_0 = \Delta N_0 \frac{\lambda^2 \eta}{8\pi n^2 t_{\text{spont}}} g(\nu)$$

Inside the laser oscillator at steady state, the average gain constant cannot exceed the threshold value

$$\gamma_t = \alpha - \frac{1}{l} \ln r_1 r_2 \tag{9.3-12a}$$

since when $\gamma > \gamma_t$, the gain per pass exceeds the total losses (per pass) so that the field intensity increases with time. For $\gamma < \gamma_t$, the reverse is true so that at steady state, $\gamma = \gamma_t$. Putting $\gamma = \gamma_t$ in (9.3-12) and solving for $W_i(\nu)$ *above threshold* give

$$W_i(\nu) = \frac{1}{t_2}\left(\frac{\gamma_0 l}{\alpha l - \ln r_1 r_2} - 1\right) \tag{9.3-13}$$

The total power emitted by the atomic population due to stimulated emission is $P_e = \Delta N h\nu V_m W_i$ where V_m is the mode volume. The inversion density ΔN above threshold is clamped at its threshold value (9.1-16):

$$\Delta N_t = \frac{8\pi n^2 t_{\text{spont}}}{g(\nu_0)\lambda^2 \eta}\left(\alpha - \frac{1}{l} \ln r_1 r_2\right) = \frac{8\pi n^3 \nu^2 t_{\text{spont}}}{c^3 t_c g(\nu_0)} \tag{9.3-14}$$

since $\Delta N \propto \gamma$ and γ above threshold is clamped at a value of γ_t. The stimulated power emitted by the atoms above threshold is thus

$$P_e = \Delta N_t h\nu V_m W_i$$
$$= \frac{8\pi n^2 hc(V_m/l)(\alpha l - \ln r_1 r_2)}{g(\nu_0)\lambda^3 \dfrac{t_2}{t_{\text{spont}}}\eta}\left(\frac{\gamma_0 l}{\alpha l - \ln r_1 r_2} - 1\right) \tag{9.3-15}$$

We pause for a moment to consider the implication of some of the factors in (9.3-15). Take, as an example, a low gain laser with equal reflectivity mirrors $R_1 = R_2 \equiv R \approx 1$. Here, $-\ln r_1 r_2 = -\ln \sqrt{R_1 R_2} \approx T$ where $T = 1 - R$ is the mirror transmittance. In this case, $-\ln r_1 r_2$ is the fraction of the internal power coupled as output. In the same limit, αl is the fractional loss in intensity per pass and $\gamma_0 l$ is the fractional gain per pass. We simplify the notation by adopting

$$\alpha l \rightarrow L_i = \text{internal loss factor}$$
$$-\ln r_1 r_2 \rightarrow T = \text{useful coupling factor}$$
$$\gamma_0 l \rightarrow g_0 = \text{unsaturated gain factor per pass}$$

The *useful* power output from the laser is thus

$$P_0 = P_e \frac{T}{L_i + T} = \frac{8\pi n^2 hcA}{\eta g(\nu_0)\lambda^3 \dfrac{t_2}{t_{\text{spont}}}}\left(\frac{g_0}{L_i + T} - 1\right)T \tag{9.3-16}$$

where $A = V_m/l$ is the cross-sectional area of the laser mode.

Maximizing P_0 with respect to coupling T yields

$$T_{\text{opt}} = -L_i + \sqrt{g_0 L_i} \tag{9.3-17}$$

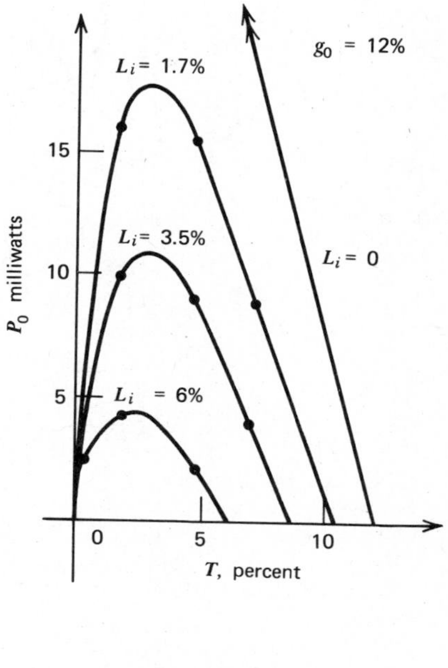

FIGURE 9.5 Useful power output (P_0) versus mirror transmission T for various values of internal loss L_i in an He-Ne 6328 Å laser. *Source:* Reference 8.

for the coupling factor resulting in maximum power output. The value of the optimum power is obtained by substituting (9.3-17) in (9.3-16)

$$(P_0)_{\text{opt}} = \frac{8\pi n^2 hcA}{\eta g(\nu_0)\lambda^3(t_2/t_{\text{spont}})}(\sqrt{g_0} - \sqrt{L_i})^2 = 2I_s A(\sqrt{g_0} - \sqrt{L_i})^2 \quad (9.3\text{-}18)$$

where I_s is the saturation intensity (8.7-5).

Theoretical plots of (9.3-16) with the internal loss factor L_i as a parameter are shown in Figure 9.5. Also shown are some experimental data points of a He-Ne 6328 Å laser (Reference 8). Note that the value of g_0 is given by the intercept of the $L_i = 0$ curve and is equal to 12%. The existence of an optimum coupling resulting in a maximum power output for each L_i is evident.

It is instructive to consider what happens to the energy $\mathscr{E}$ stored in the laser resonator as coupling T is varied. This energy is proportional to P_0/T. A plot of P_0 (taken from Figure 9.5) and $\mathscr{E} \propto P_0/T$ as a function of coupling T is shown in Figure 9.6. As we may expect, $\mathscr{E}$ is a monotonically decreasing function of coupling T.

The Effect of Spontaneous Emission

The derivation of the laser output power just concluded considered only the stimulated power. According to this analysis, which leads to (9.3-16), the power output at threshold ($g_0 = L_i + T$) is zero. A careful measurement,

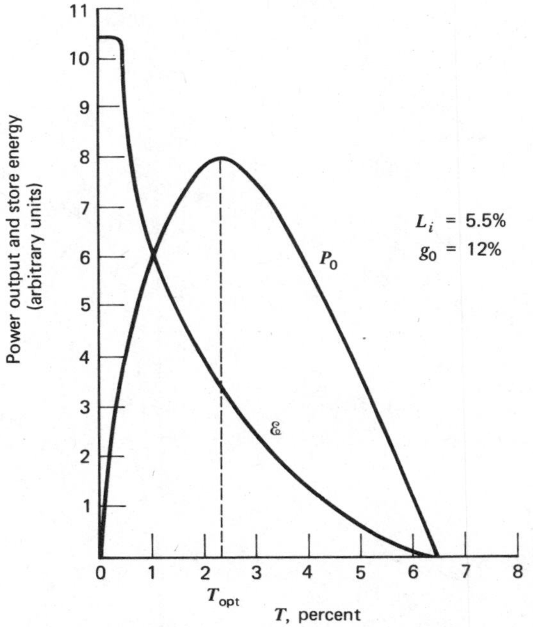

FIGURE 9.6 Power output P_0 and stored energy $\mathscr{E}$ plotted against mirror transmission T.

however, will show that laser power is present at any pumping level and that the distinction between "below threshold" and "above threshold" is not as sharp as (9.3-16) may lead us to conclude. This discrepancy is due to our neglect of the spontaneous emission power. The role of spontaneous and emitted power cannot be easily separated in this discussion, as will be evident from a study of Chapter 13. At the risk of oversimplification, we present an argument that predicts the proper dependence of the average laser power. The more subtle aspects of spectral behavior are left for the discussion of Chapter 13.

At any pumping level, the total power emitted by the atoms includes, in addition to the stimulated emission power given by (9.3-16), a contribution due to spontaneous emission.

$$P_{\text{spont}} = \kappa N_2 h\nu / t_{\text{spont}} \tag{9.3-19}$$

where N_2 is the density of atoms in the upper level and κ is some constant. For the purpose of this discussion, we will assume that $N_2 \propto \Delta N$.[4] It follows that, in a given laser,

$$P_{\text{spont}} = K \Delta N \tag{9.3-20}$$

where K is a new constant.

The total power (in some mode) is thus the sum of the stimulated emission and the spontaneous powers

$$\begin{aligned} P_e &= \Delta N h\nu V_m W_i + K\Delta N \\ &= \frac{\Delta N_0}{1 + W_i t_2}(h\nu V_m W_i + K) \end{aligned} \tag{9.3-21}$$

where we used (9.3-11) for ΔN. The constant K can be determined by noting that according to (8.3-4),

$$\frac{\text{Induced emission rate/mode}}{\text{Spontaneous emission rate/mode}} = \frac{h\nu V_m W_i}{K} = n_m$$

where n_m is the number of quanta in the mode under consideration (m). The result is

$$K = \frac{hc^3\eta}{8\pi n^3 \nu \Delta\nu t_{\text{spont}}} \tag{9.3-22}$$

where $\Delta\nu \equiv g(\nu_0)^{-1}$. The derivation of (9.3-22) is left as an exercise.

We will consider two main regimes.

1. *Below Threshold.* Here, we use the definition of "threshold" (9.1-15)

$$\gamma_t l = \alpha l - \ln r_1 r_2$$

so that below threshold it follows, from (9.3-13), that $W_i = 0$. Using this

[4] This, according to (9.3-5), is true when $t_1 \ll t_2$.

fact in (9.3-21) leads to

$$(P_e)_{\substack{\text{below threshold} \\ \text{per mode}}} = \Delta N_0 K = \frac{\Delta N_0 V_m h\nu}{t_{\text{spont}} p} \eta \tag{9.3-23}$$

where

$$p = \frac{8\pi\nu^2 \Delta\nu n^3 V_m}{c^3} \tag{9.3-24}$$

We thus find that the total spontaneous emission power below threshold $\Delta N_0 V_m h\nu / t_{\text{spont}}$ is divided, more or less equally, among the p enclosure modes that are, according to (5.7-3), on "speaking terms" with the atomic transition (i.e., modes with resonant frequencies within $\Delta\nu$ of ν_0).

As a further simplification, we use the relation in footnote 2 for the threshold inversion to rewrite (9.3-23) as

$$(P_e)_{\substack{\text{below threshold} \\ (\Delta N_0 < \Delta N_t)\ \text{per mode}}} = \frac{\Delta N_0}{\Delta N_t} \frac{h\nu}{t_c} = \frac{g_0}{(L_i + T)} \frac{h\nu}{t_c} \tag{9.3-25}$$

where t_c is the decay lifetime of the passive resonator mode. This equation shows that at threshold ($\Delta N_0 = \Delta N_t$), the (spontaneous) mode power is equivalent to a stored energy of $h\nu$ (*one quantum*) per mode.

2. *Above Threshold.* Here, $h\nu V_m W_i \gg K$ (in 9.3-21) and $\Delta N = \Delta N_t$. The power in this case is the sum of the spontaneous (9.3-25 with $\Delta N_0 = \Delta N_t$) and the stimulated power (9.3-15)

$$(P_e)_{\substack{\text{above threshold} \\ (\Delta N_0 > \Delta N_t)}} = \frac{\Delta N_t h\nu V_m}{t_2} \left(\frac{\Delta N_0}{\Delta N_t} - 1 \right) + \frac{h\nu}{t_c} = \frac{\Delta N_t h\nu V_m}{t_2} \left(\frac{g_0}{L_i + T} - 1 \right) + \frac{h\nu}{t_c} \tag{9.3-26}$$

Neglecting the small spontaneous term $h\nu/t_c$, we find that both below threshold (9.3-25) and above it (9.3-26), the output power varies linearly with the pumping rate ΔN_0.

The ratio of the slopes in both regions, however, is given by

$$\frac{[dP_e/d(\Delta N_0)]_{\text{above threshold}}}{[dP_e/d(\Delta N_0)]_{\text{below threshold}}} = \frac{V_m t_c \Delta N_t}{t_2} = \frac{8\pi\nu^2 n^3 \Delta\nu V_m}{c^3 (t_2/t_{\text{spont}})} = \frac{p}{(t_2/t_{\text{spont}})\eta} \tag{9.3-27}$$

where we use the expression in footnote 2 for ΔN_t and (9.3-24) for p.

The interpretation of (9.3-27) is straightforward and is very revealing. Consider, for simplicity, a laser where $\eta = 1$ and $t_2 = t_{\text{spont}}$. Both below threshold and above it, the total pumping power is converted into radiation. Below threshold, this radiation is divided among the p (with $p \gtrsim 10^8$) modes on "speaking terms" with the transition. The incremental pumping power above threshold, however, is channeled into *one* mode (here we assume a homogeneously broadened ideal laser). This leads directly to (9.3-27).

Using a set of typical laser values such as $\nu = 3 \times 10^{14}$, $n = 1$, $\Delta\nu = 5 \times 10^9$, $V_m = 1\ \text{cm}^3$, $t_2 = t_{\text{spont}}$ gives $p \sim 4 \times 10^8$. It is because of this large difference in slopes that most experimental plots fail to show the "below threshold" region. A theoretical plot based on (9.3-25) and (9.3-26) is shown in Figure 9.7. An experimental power versus pump plot is shown in Figure 9.8.

In addition to the dramatic increase in the slope of the power-pumping curve, there are very important differences in the spectrum of the emitted power above and below threshold. This problem will be discussed in detail in Chapter 13. We will anticipate some of the results by stating that below threshold, the noise spectrum is that of broad white noise source passed by a Lorentzian filter. The bandwidth of the laser emission is

$$(\Delta\nu_l)_{\substack{\text{below}\\ \text{threshold}}} = \frac{\pi h\nu(\Delta\nu_{1/2})^2}{P_e} \frac{N_2}{\left(N_2 - N_1 \dfrac{g_2}{g_1}\right)} \tag{9.3-28}$$

where $\Delta\nu_{1/2}$ is the width of the passive resonator response curve as given by (7.4-6, 7).

Above threshold, the intensity fluctuations are essentially damped out by gain saturation and the noise is that due to phase fluctuations in a nonlinear (saturated) oscillator. The spectral width of the laser field is now given by

$$(\Delta\nu_l)_{\substack{\text{above}\\ \text{threshold}}} = \frac{\pi h\nu(\Delta\nu_{1/2})^2}{P_e} \frac{N_{2t}}{\left(N_2 - N_1 \dfrac{g_2}{g_1}\right)_t} \tag{9.3-29}$$

The huge increase in P_e attendant on operating only slightly above threshold manifests itself according to (9.3-28, 29) as a dramatic spectral

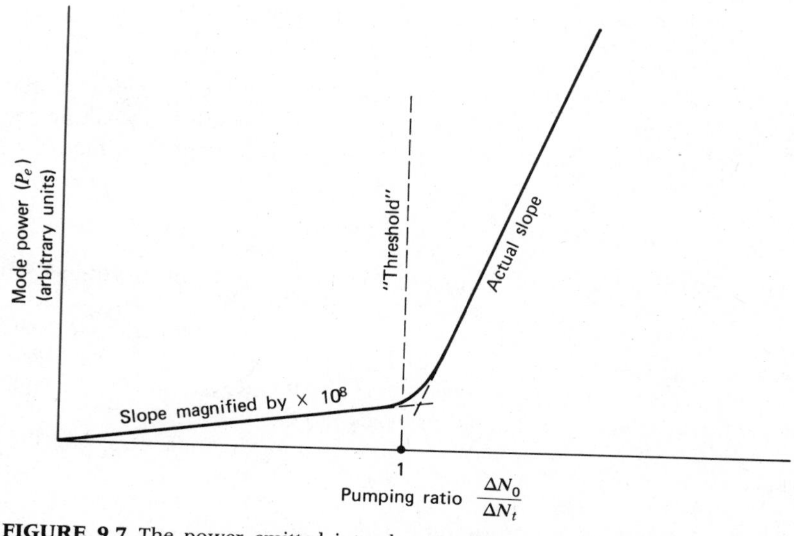

FIGURE 9.7 The power emitted into the (single) laser mode by the atoms as a function of the pumping ratio (normalized to threshold). In a real laser, the slope above threshold is many orders of magnitude larger than below threshold, so that in the present plot, the "below threshold" curve will not be distinguishable from the abscissa.

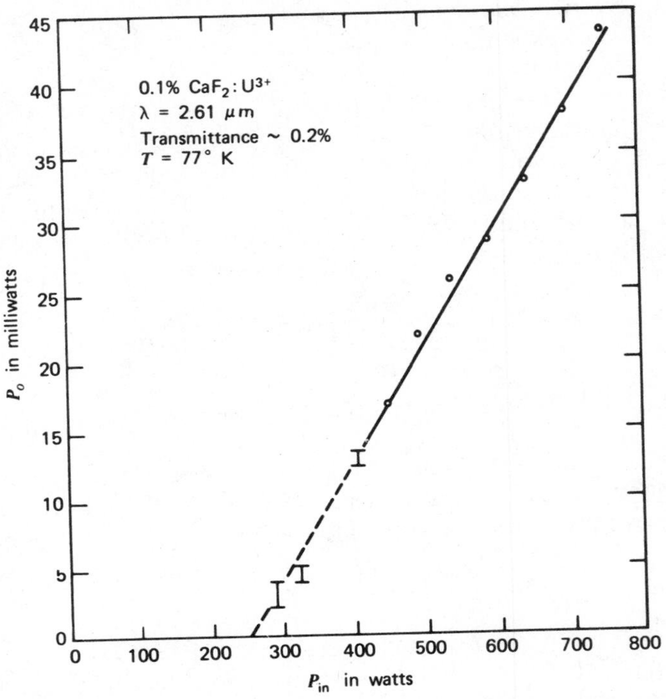

FIGURE 9.8 A plot of the output power versus the electrical input to a xenon lamp in a cw 0.1% $CaF_2 : U^{3+}$ laser. The mirror transmittance at 2.61 μ is 0.2% and corresponds to severe undercoupling.

narrowing. It is easy to convince ourselves that for the example used above, operating at 10% above threshold ($\Delta N_0 = 1.1 \Delta N_t$) should lead to a narrowing by a factor of, approximately, 10^8. This situation is illustrated in Figure 13.5.

References

1. Weber, J., "Amplification of Microwave Radiation by Substances Not in Thermal Equilibrium," *IRE Trans. Prof. Group on Electron Devices* **3,** 1 (1953).
2. Gordon, J. P., H. J. Zeiger, and C. H. Townes, "Molecular Microwave Oscillator and New Hyperfine Structure in the Microwave Spectrum of NH_3," *Phys. Rev.* **95,** 282 (1954); also, "The Maser—New Type of Microwave Amplifier, Frequency Standard, and Spectrometer," ibid. **99,** 1264 (1955).
3. Basov, N. G. and A. M. Prokhorov, "Application of Molecular Beams to the Radiospectroscopic Study of the Rotation Spectrum of Molecules," *J. Expt. Theoret. Phys. (USSR)* **27,** 431 (1954); also, "On the Possible Methods of Producing Active Molecules for a Molecular Generator," ibid. **28,** 249 (1955).
4. Schawlow, A. L. and C. H. Townes, "Infrared and Optical Masers," *Phys. Rev.* **112,** 1940 (1958).
5. Maiman, T. H., "Stimulated Optical Radiation in Ruby," *Nature* **187,** 493 (1960).
6. Javan, A., W. B. Bennett Jr., and D. R. Herriott, "Population Inversion and Contin-

uous Optical Maser Oscillation in Gas Discharge Containing a He-Ne Mixture," *Phys. Rev. Letters* **6,** 106 (1961).
7. Lamb, W. E., Jr., "Theory of an Optical Maser, *Phys. Rev.* **134,** (6A), A1429 (1964).
8. Laures, P., "Variation of the 6328 Å Gas Laser Output Power with Mirror Transmission," *Phys. Letters* **10,** 61 (1964).

Problems

9.1 Derive Eq. (9.3-22).

9.2 Discuss the effect on the resonant frequency of a laser oscillator due to the transverse Gaussian confinement of the beam modes. What is the change in the resonant frequency between the confocal mode and one with $z_0 \gg l$ in a resonator of length l?

9.3 Show that if a Fabry–Perot resonator is filled with an atomic medium with a susceptibility $\chi(\omega)$, the intermode frequency spacing is given by

$$\omega_m - \omega_{m-1} = \frac{\pi c}{nl\left[1 + \dfrac{\omega}{2n^2}\dfrac{\partial \chi'(\omega)}{\partial \omega}\right]_{\omega=\omega_m}}$$

9.4 Consider the effect of atomic dispersion on the group velocity of an optical pulse propagating in an atomic medium with a center (pulse) frequency equal to the atomic resonance ω_0
(a) For an amplifying medium.
(b) For an absorbing medium.
Express the group velocity as a function of the peak gain for the case of a Lorentzian line. Ignore hole burning and assume that the spectrum of the pulse is narrow compared to $\Delta\nu$.

9.5 Show that (9.2-14) is equivalent to (9.1-12).

9.6 Derive (9.3-5).

CHAPTER 10

Some Specific Laser Systems

10.0 INTRODUCTION

In the preceding chapters we studied some of the general properties of atomic systems interacting with electromagnetic modes. In this chapter we will see how some specific atomic systems are used to produce laser oscillation.

A number of pumping schemes including optical, electrical discharge, and current injection are considered in detail.

10.1 PUMPING AND LASER EFFICIENCY

Figure 10.1 shows the pumping-oscillation cycle of some (hypothetical) representative laser. The pumping agent elevates the atoms into some excited state 3 from which they relax into the upper laser level 2. The stimulated laser transition takes place between levels 2 and 1 and results in the emission of a photon of frequency ν_{21}.

It is evident from this figure that the minimum energy input per output photon is $h\nu_{30}$, so the power efficiency of the laser cannot exceed

$$\eta_{\text{atomic}} = \frac{\nu_{21}}{\nu_{30}} \tag{10.1-1}$$

to which quantity we will refer as the "atomic quantum efficiency." The overall laser efficiency depends on the fraction of the total pump power that is effective in transferring atoms into level 3 and on the pumping quantum efficiency defined as the fraction of the atoms that, once in 3, make a transition to 2. The product of the last two factors, which constitutes an upper limit on the efficiency of optically pumped lasers, ranges from about 1% for solid-state lasers such as Nd^{3+} : YAG to about 30% in the CO_2 laser and to near unity in the GaAs junction laser. We will discuss these factors when we get down to some specific laser systems. We may note, however, that according to (10.1-1), in an efficient laser system ν_{21} and ν_{30} must be of the same order of magnitude, so the laser transition should involve low-lying levels.

10.2 THE RUBY LASER

The first material in which laser action was demonstrated (Reference 1) and still one of the most useful laser materials is ruby, whose output is at $\lambda = 0.6943\ \mu m$. The active laser particles are Cr^{3+} ions present as impurities in

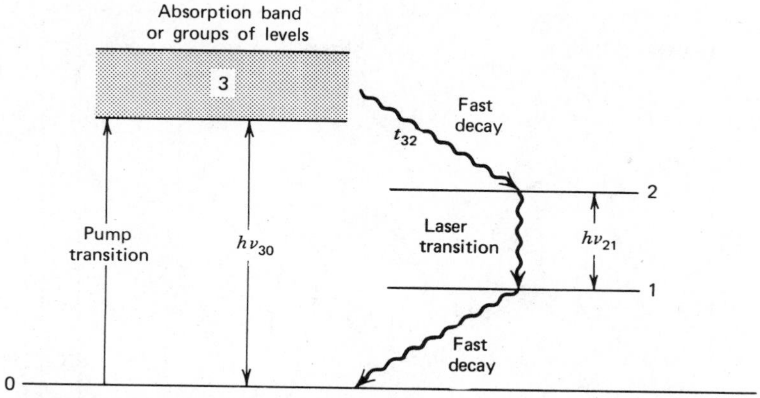

FIGURE 10.1 Pumping-oscillation cycle of a typical laser.

Al_2O_3 crystal. Typical Cr^{3+} concentrations are ~0.05% by weight. The pertinent energy level diagram is shown in Figure 10.2.

The pumping of ruby is performed usually by subjecting it to the light of intense flashlamps (quite similar to the types used in flash photography). A portion of this light that corresponds in frequency to the two absorption bands $^4F_2$ and $^4F_1$ is absorbed, thereby causing Cr^{3+} ions to be transferred into these levels. The ions proceed to decay, within an average time of $\omega_{32}^{-1} \simeq 5 \times 10^{-8}$ seconds (Reference 2), into the upper laser level ^2E. The level ^2E is composed of two levels $2\bar{A}$ and $\bar{E}$ separated by 29 cm^{-1}.[1] The lower of these two, $\bar{E}$, is the upper laser level. The lower laser level is the ground state. The lifetime of atoms in the upper laser level $\bar{E}$ is $t_2 \simeq 3 \times 10^{-3}$ sec. Each decay very nearly results in the (spontaneous) emission of a photon, so $t_2 \simeq t_{spont}$.

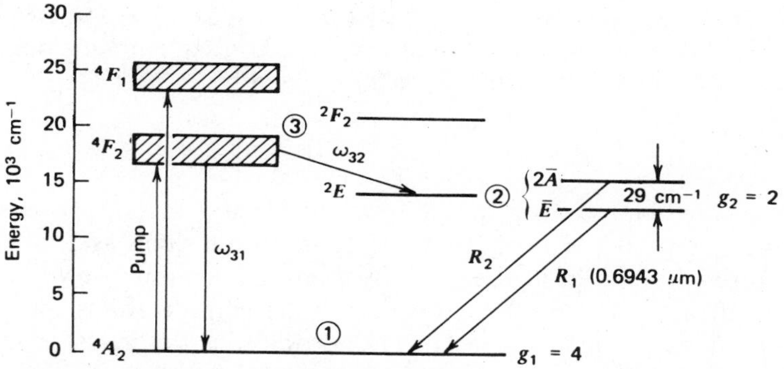

FIGURE 10.2 Energy levels pertinent to the operation of a ruby laser. *Source:* Reference 2.

[1] The unit 1 cm^{-1} (one wavenumber) is the frequency corresponding to $\lambda = 1$ cm, so 1 cm^{-1} is equivalent to $\nu = 3 \times 10^{10}$ Hz. It is also used as a measure of energy where 1 cm^{-1} corresponds to the energy $h\nu$ of a photon with $\nu = 3 \times 10^{10}$ Hz.

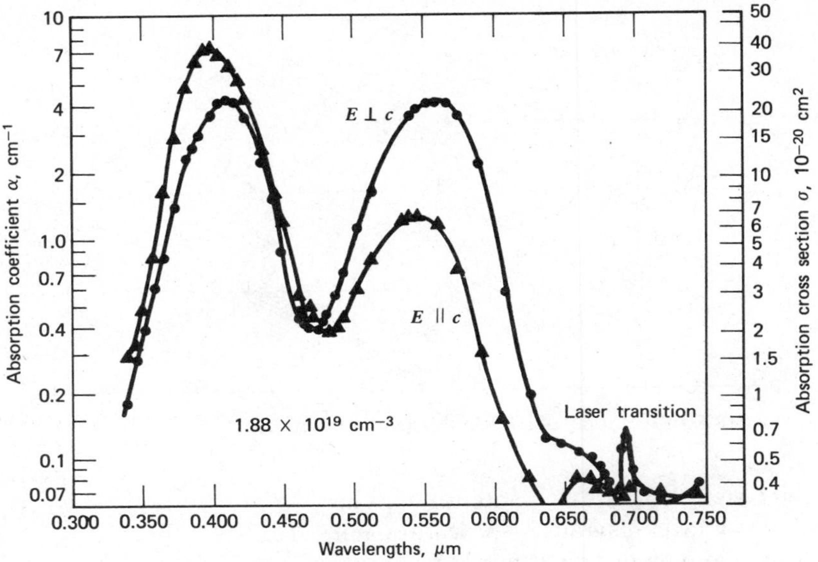

FIGURE 10.3 Absorption coefficient and absorption cross section as functions of wavelength for $E \parallel c$ and $E \perp c$. The 300°K data were derived from transmittance measurements on pink ruby with an average Cr ion concentration of 1.88×10^{19} cm^{-3}. *Source:* Reference 3.

An absorption spectrum of a typical ruby with two orientations of the optical field relative to the c (optic) axis is shown in Figure 10.3. The two main peaks correspond to absorption in the useful $^4F_1$ and $^4F_2$ bands, which are responsible for the characteristic (ruby) color.

The ordinate is labeled in terms of the absorption coefficient and in terms of the transition cross section σ that may be defined as the absorption coefficient per unit inversion per unit volume and has consequently the dimension of area. According to this definition, $\alpha(\nu)$ is given by

$$\alpha(\nu) = \left(N_1 \frac{g_2}{g_1} - N_2\right) \sigma(\nu) \tag{10.2-1}$$

A more detailed plot of the absorption near the laser emission wavelength is shown in Figure 10.4. The width $\Delta\nu$ of the laser transition as a function of temperature is shown in Figure 10.5. At room temperature, $\Delta\nu = 11$ cm^{-1}.

We can use ruby to illustrate some of the considerations involved in optical pumping of solid-state lasers. Figure 10.6 shows a typical setup of an optically pumped laser, such as ruby. The helical flashlamp surrounds the ruby rod. The flash excitation is provided by the discharge of a capacitor bank across the lamp.

The typical flash output consists of a pulse of light of duration $t_{\text{flash}} \simeq 5 \times 10^{-4}$ sec. Let us, for the sake of simplicity, assume that the flash pulse is rectangular in time and of duration t_{flash} and that it results in an optical flux at

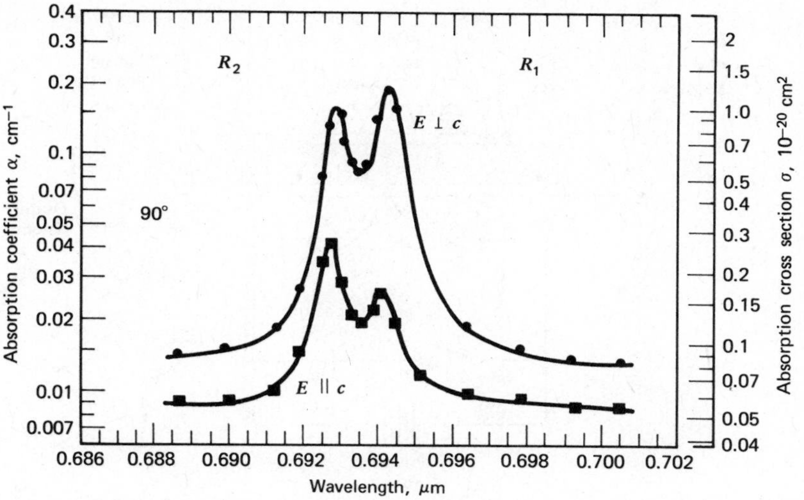

FIGURE 10.4 Absorption coefficient and absorption cross section as functions of wavelength for $E \parallel c$ and $E \perp c$. Sample was a pink ruby laser rod having a 90° c axis orientation with respect to the rod axis and a Cr^{3+} concentration of 1.58×10^{19} cm^{-3}. *Source:* Reference 3.

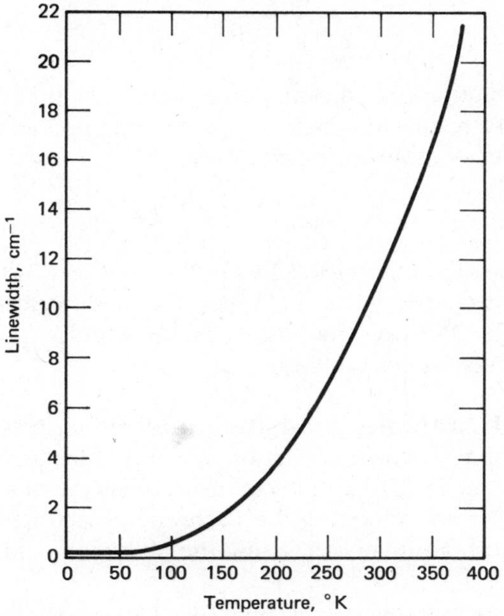

FIGURE 10.5 Linewidth of the R_1 line of ruby as a function of temperature. *Source:* Reference 4.

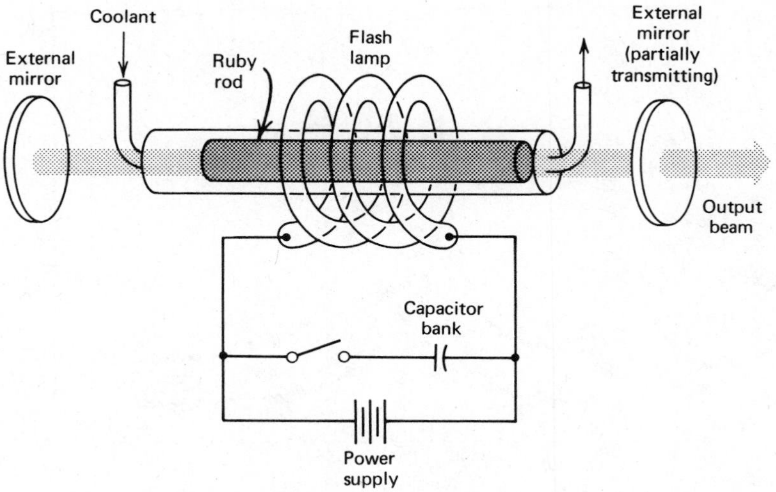

FIGURE 10.6 Typical setup of a pulsed ruby laser using flashlamp pumping and external mirrors.

the crystal surface having $s(\nu)$ watts per unit area per unit frequency at the frequency ν. If the absorption coefficient of the crystal is $\alpha(\nu)$, then the amount of energy absorbed by the crystal per unit volume is[2]

$$t_{\text{flash}} \int_0^\infty s(\nu)\alpha(\nu)\, d\nu$$

If the absorption quantum efficiency (the probability that the absorption of a pump photon at ν results in transferring one atom into the upper laser level) is $\eta(\nu)$, the number of atoms pumped into level 2 per unit volume is

$$N_2 = t_{\text{flash}} \int_0^\infty \frac{s(\nu)\alpha(\nu)\eta(\nu)}{h\nu}\, d\nu \qquad (10.2\text{-}2)$$

Since the lifetime of atoms in level 2 ($t_2 \simeq 3 \times 10^{-3}$ sec) is considerably longer than the flash duration ($\sim 5 \times 10^{-4}$ sec), we may neglect the spontaneous decay out of level 2 during the time of the flash pulse, so N_2 represents the population of level 2 after the flash.

Numerical Example: Flash Pumping of a Pulsed Ruby Laser. Consider the problem of flash excitation of a pink ruby laser with a chromium ion density of $N_0 = 2 \times 10^{19}$ atoms/cm³ using a flash with a duration of $t_{\text{flash}} = 5 \times 10^{-4}$ sec. Before estimating the flash energy at threshold, we need to estimate the threshold inversion. Using the following data, we obtain

[2] We assume that the total absorption in passing the crystal is small, so $s(\nu)$ is taken to be independent of the distance through the crystal.

$$\Delta\nu \equiv \frac{1}{g(\nu_0)} = 12\ \text{cm}^{-1}, \qquad n = 1.77 \quad \text{(From Figure 10.5 at 300°K)}$$

$$t_2 \approx t_{\text{spont}} \sim 3 \times 10^{-3}\ \text{sec}$$

$$\nu = 14{,}422\ \text{cm}^{-1}, \qquad \eta_m \sim 1$$

$$g_2 = g(\bar{E}) = 2$$

$$g_1 = g(^4A_2) = 4$$

$$t_c = 10^{-8}\ \text{sec} \quad \text{(consistent with } l = 10\ \text{cm and a loss of 0.04 per pass)}$$

From footnote 2 on page 187, we get

$$\Delta N_t = \left(N_2 - N_1 \frac{g_2}{g_1}\right)_t = 10^{17}\ \text{cm}^{-3}$$

Using $N(2\bar{A})/N(\bar{E}) = e^{-\Delta E/kT} = 0.87$ where $\Delta E = 29\ \text{cm}^{-1}$ and $N(^4A_2) + N(\bar{E}) + N(2A) = 2 \times 10^{19}\ \text{cm}^{-3}$, yields

$$N(2\bar{A}) = 0.454 \times 10^{19}\ \text{cm}^{-3}$$

$$N(\bar{E}) = 0.522 \times 10^{19}\ \text{cm}^{-3}$$

$$N(^4A_2) = 1.025 \times 10^{19}\ \text{cm}^{-3}$$

If the useful absorption is limited to relatively narrow spectral regions, we may approximate (10.2-2) by

$$N_2 = \frac{t_{\text{flash}}\, \overline{s(\nu)}\, \overline{\alpha(\nu)}\, \overline{\eta(\nu)}\, \overline{\Delta\nu}}{h\bar{\nu}} \tag{10.2-3}$$

where the bars represent average values over the useful absorption region whose width is $\overline{\Delta\nu}$.

From Figure 10.3, we deduce an average absorption coefficient of $\overline{\alpha(\nu)} \simeq 2\ \text{cm}^{-1}$ over the two central peaks. Using $\bar{\nu} \simeq 5 \times 10^{14}$ Hz and $N_2 = N(2\bar{A}) + N(\bar{E}) \simeq 10^{19}\ \text{cm}^{-3}$, (10.2-3) yields

$$\bar{s}\ \overline{\Delta\nu}\, t_{\text{flash}} \simeq 1.5\ \text{J/cm}^2$$

for the pump energy in the useful absorption region that must fall on each square centimeter of crystal surface in order to obtain threshold inversion. To calculate the total lamp energy that is incident on the crystal, we need to know the spectral characteristics of the lamp output. Typical data of this sort are shown in Figure 10.7. The mercury-discharge lamp is seen to contain considerable output in the useful absorption regions (near 4000 and 5500 Å) of ruby. If we estimate the usefully absorbed fraction of the lamp output at 10%, the fraction of the lamp light actually incident on the crystal as 20%,

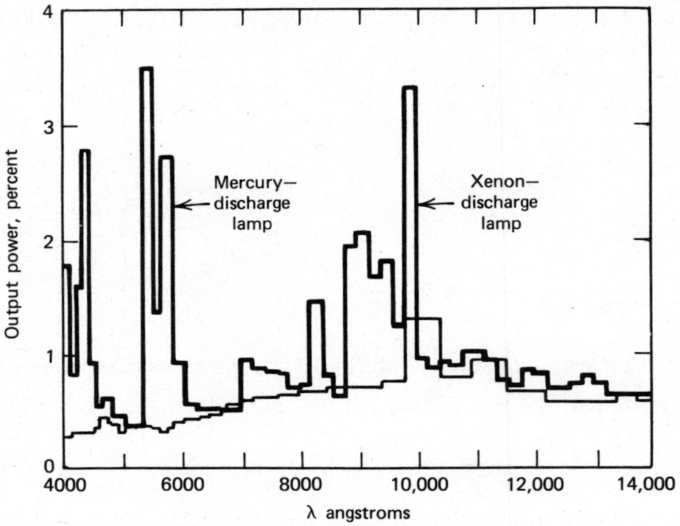

FIGURE 10.7 Spectral output characteristics of two commercial high-pressure lamps. Output is plotted as a fraction of electrical input to lamp over certain wavelength intervals (mostly 200 Å) between 0.4 and 1.4 μm. *Source:* Reference 5.

and the conversion of electrical-to-optical energy as 50%, we find the threshold electric energy input to the flashlamp per square centimeter of laser surface is

$$\frac{1.5}{0.1 \times 0.2 \times 0.5} = 150 \text{ J/cm}^2$$

10.3 THE Nd^{3+} : YAG LASER

One of the most important laser systems is that using trivalent neodymium ions (Nd^{3+}) that are present as impurities in yttrium aluminum garnet (YAG = $Y_3Al_5O_{12}$); see References 6 and 7. The laser emission occurs at λ = 1.0641 μm at room temperature. The relevant energy levels are shown in Figure 10.8. The lower laser level is at $E_1 \simeq 2111$ cm^{-1} from the ground state so that at room temperature its population is down by a factor of $\exp(-E_2/kT) \simeq e^{-10}$ from that of the ground state and can be neglected. We thus have $N_{2t} \approx \Delta N_t$. Lasers with this property are often called "four-level" lasers.

The spontaneous emission spectrum of the laser transition is shown in Figure 10.9. The width of the gain linewidth at room temperature is $\Delta\nu \simeq 6$ cm^{-1}. The spontaneous lifetime for the laser transition has been measured (Reference 7) as $t_{\text{spont}} = 5.5 \times 10^{-4}$ sec. The room-temperature cross section at the center of the laser transition is $\sigma = 9 \times 10^{-19}$ cm^2. If we compare this number to $\sigma = 1.22 \times 10^{-20}$ cm^2 in ruby (see Figure 10.4), we expect that at a

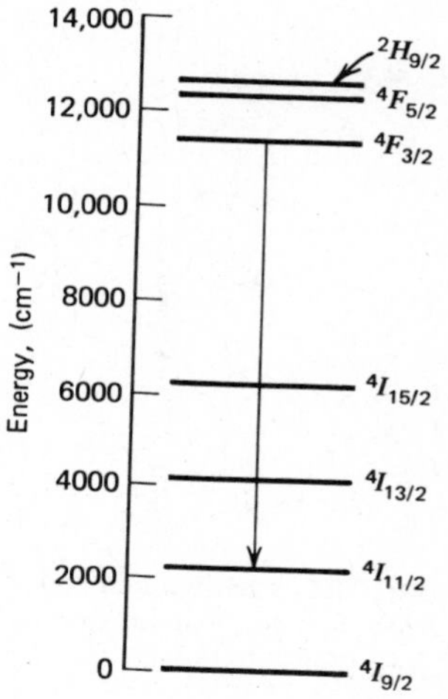

FIGURE 10.8 Energy-level diagram of Nd^{3+} in YAG. *Source:* Reference 6.

given inversion the optical gain constant γ in Nd^{3+} : YAG is approximately 75 times that of ruby. This causes the oscillation threshold to be very low and explains the easy continuous (CW) operation of this laser compared to ruby.

The absorption responsible for populating the upper level takes place in a number of bands between 13,000 and 25,000 cm^{-1}.

Numerical Example: Threshold of an Nd^{3+} : YAG Laser. (a) *Pulsed Threshold.* First, we estimate the energy needed to excite a typical Nd^{3+} : YAG laser on a pulse basis so that we can compare it with that of ruby. We use the following data:

$$l = 20 \text{ cm} \quad \text{(length optical resonator)}$$

$$L = 4\% \ (= \text{loss per pass}), \qquad t_c \simeq \frac{l}{Lc} = 1.6 \times 10^{-8} \text{ sec}$$

$$n = 1.5$$

$$\lambda = 1.064 \ \mu\text{m}$$

$$t_{\text{spont}} = 5.5 \times 10^{-4} \text{ sec}$$

$$\Delta\nu = 6 \text{ cm}^{-1}, \qquad \eta \approx 1$$

Using the foregoing data in (9.1-16) gives

$$N_{2t} \approx \Delta N_t = \frac{8\pi n^2 t_{\text{spont}} \Delta\nu}{\eta \ c t_c \lambda^2} \simeq 1.03 \times 10^{15} \text{ cm}^{-3}$$

Estimating that 5% of the exciting light energy falls within the useful absorption bands, that 5% of this light is actually absorbed by the crystal, that the

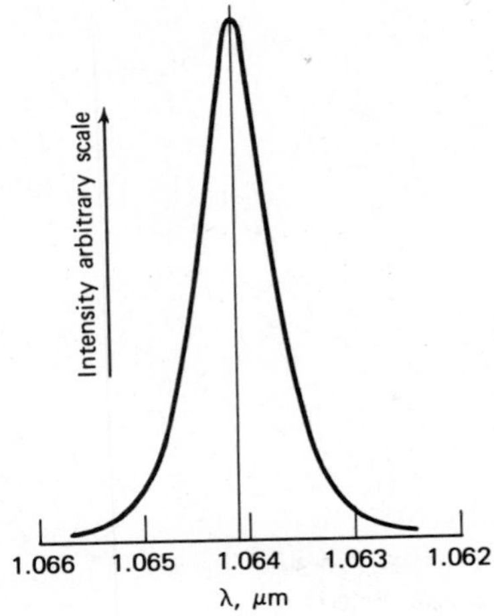

FIGURE 10.9 Spontaneous-emission spectrum of Nd^{3+} in YAG near the laser transition at $\lambda = 1.064$ μm. *Source:* Reference 7.

average ratio of laser frequency to the pump frequency is 0.5, and that the lamp efficiency (optical output/electrical input) is 0.5, we obtain

$$\mathscr{E}_{\text{lamp}} = \frac{N_{2t} h\nu_{\text{laser}}}{5 \times 10^{-2} \times 5 \times 10^{-2} \times 0.5 \times 0.5} \simeq 0.31 \text{ J/cm}^3$$

for the energy input to the lamp at threshold.

It is interesting to compare this last number to the figure of 300 J/cm² of surface area obtained in the ruby example of Section 10.2. For reasonable dimension crystals (e.g., length = 5 cm, $r = 2$ mm), we obtain $\mathscr{E}_{\text{lamp}} = 0.2$ J. We expect the ruby threshold to exceed that of Nd^{3+} : YAG by three orders of magnitude, which is indeed the case.

(b) *Continuous Operation*. The minimum power needed to maintain N_{2t} atoms (per unit volume) in level 2 is just prior to attaining threshold

$$P_{\min} = \frac{N_{2t} h\nu}{t_2}$$

which for $t_2 \approx t_{\text{spont}}$, as is the case here, gives

$$P_{\min} \approx \frac{N_t h\nu}{t_{\text{spont}}} \simeq 0.35 \text{ W/cm}^3$$

Taking the crystal diameter as 0.25 cm and its length as 3 cm and using the same efficiency factors assumed in the first part of this example, we can estimate the power input to the lamp at threshold as

$$P_{(\text{to lamp})} = \frac{0.35 \times (\pi/4) \times (0.25)^2 \times 3}{5 \times 10^{-2} \times 5 \times 10^{-2} \times 0.5 \times 0.5} \simeq 82 \text{ W}$$

which is in reasonable agreement with experimental values (Reference 6).

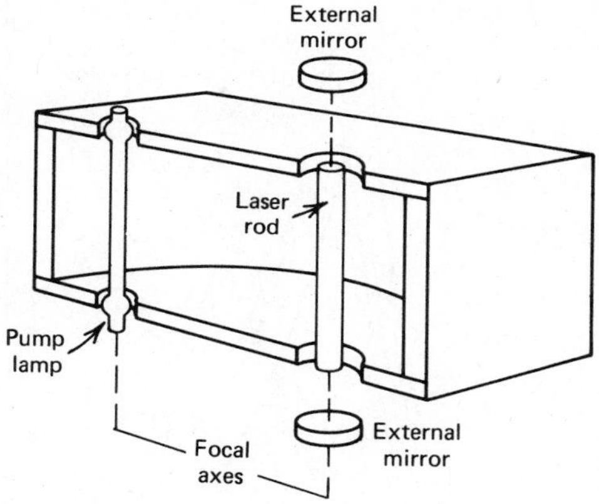

FIGURE 10.10 Typical continuous solid-state laser setup employing an elliptic cylinder housing for concentrating lamplight onto laser.

A typical arrangement used in continuous solid-state lasers is shown in Figure 10.10. The highly polished elliptic cylinder is used to concentrate the light from the lamp, which is placed along one focal axis, onto the laser rod, which occupies the other axis. This configuration guarantees that most of the light emitted by the lamp passes through the laser rod. The reflecting mirrors are placed outside the cylinder.

10.4 THE NEODYMIUM-GLASS LASER

One of the most useful laser systems is that which results when the Nd^{3+} ion is present as an impurity atom in glass (Reference 8).

The energy levels involved in the laser transition in a typical glass are shown in Figure 10.11. The laser emission wavelength is at $\lambda = 1.059\ \mu m$,

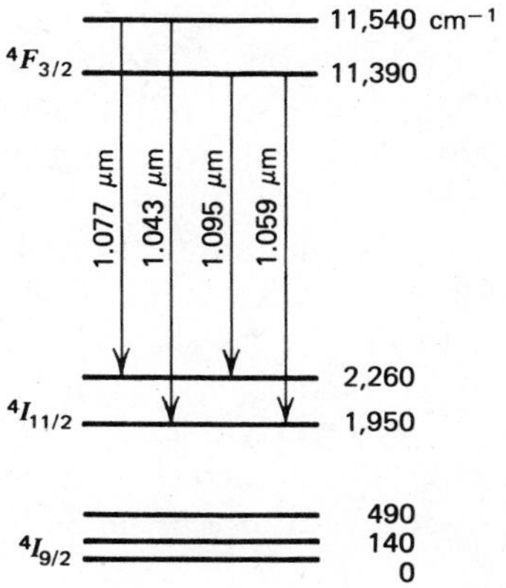

FIGURE 10.11 Energy-level diagram for the ground state and the states involved in laser emission near 1.059 μm for Nd^{3+} in a rubidium potassium barium silicate glass. *Source:* Reference 8.

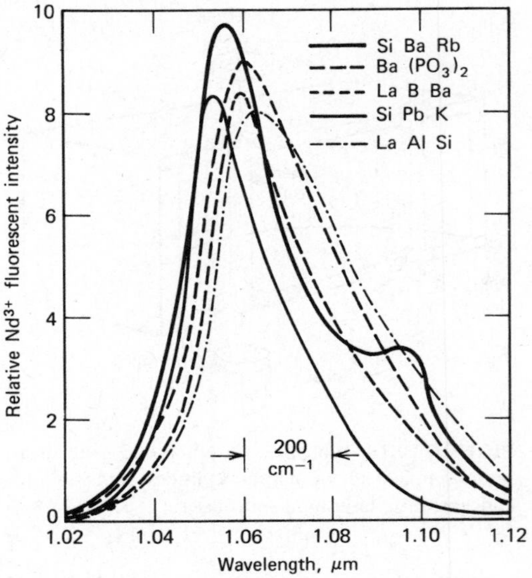

FIGURE 10.12 Fluorescent emission of the 1.06 μm line of Nd^{3+} at 300°K in various glass bases. *Source:* Reference 8.

and the lower level is approximately 1950 cm^{-1} above the ground state. As in the case of Nd^{3+} : YAG described in Section 10.3, we have here a four-level laser since the thermal population of the lower laser level is negligible. The fluorescent emission near $\lambda = 1.06\ \mu$m is shown in Figure 10.12. The fluorescent linewidth can be measured off directly and ranges, for the glasses shown, around 300 cm^{-1}. This width is approximately a factor of 50 larger than that of Nd^{3+} in YAG. This is due to the amorphous structure of glass, which causes different Nd^{3+} ions to "see" slightly different surroundings. This makes their energy splittings vary slightly. Different ions consequently radiate at slightly different frequencies, causing a broadening of the spontaneous emission spectrum. The absorption bands responsible for pumping the laser level are shown in Figure 10.13. The probability that the absorption of a photon in any of these bands will result in pumping an atom to the upper laser level (i.e., the absorption quantum efficiency) has been estimated (Reference 8) at about 0.4.

The lifetime t_2 of the upper laser level depends on the host glass and on the Nd^{3+} concentration. This variation in two glass series is shown in Figure 10.14.

Numerical Example: Threshold for CW and Pulsed Operation of Nd^{3+} Glass Lasers. Let us estimate first the threshold for continuous (CW) laser action in a Nd^{3+} glass laser using the following data:

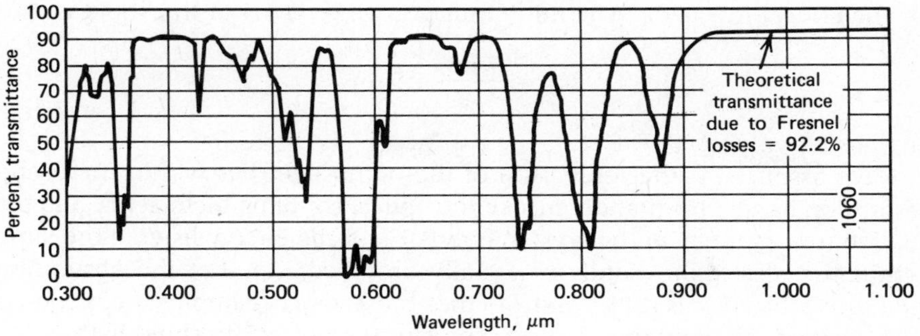

FIGURE 10.13 Nd^{3+} absorption spectrum for a sample of glass 6.4 mm thick with the composition 66 wt. % SiO_2, 5 wt. % Nd_2O_3, 16 wt. % Na_2O, 5 wt. % BaO, 2 wt. % Al_2O_3, and 1 wt. % Sb_2O_3. *Source:* Reference 8.

$$\Delta\nu = 200 \text{ cm}^{-1} \quad \text{(see Figure 10.12)}$$

$$n = 1.5$$

$$t_{\text{spont}} \simeq t_2 = 3 \times 10^{-4} \text{ sec}$$

$$\left.\begin{aligned} l &= \text{length of resonator} = 20 \text{ cm} \\ L &= \text{loss per pass} = 2\% \end{aligned}\right\} \quad t_c \simeq \frac{l}{Lc} = 3.3 \times 10^{-8} \text{ sec}$$

Using (9.1-16), we obtain

$$N_{2t} \approx \Delta N_t = \frac{8\pi\, t_{\text{spont}} n^2\, \Delta\nu}{c t_c \lambda^2} = 9 \times 10^{15} \text{ atoms/cm}^3$$

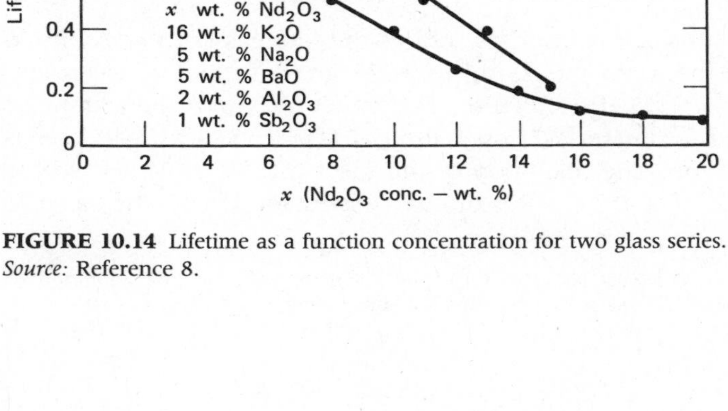

FIGURE 10.14 Lifetime as a function concentration for two glass series. *Source:* Reference 8.

for the critical inversion. The minimum pumping power at threshold is thus

$$P_{\min} \approx \frac{N_t h\nu V}{t_{\text{spont}}} = 5.6 \text{ W}$$

in a crystal volume $V = 1 \text{ cm}^3$.

We assume (a) that only 10% of the pump light lies within the useful absorption bands; (b) that because of the optical coupling inefficiency and the relative transparency of the crystal, only 10% of the energy leaving the lamp within the absorption bands is actually absorbed; (c) that the absorption quantum efficiency is 40%; and (d) that the average pumping frequency is twice that of the emitted radiation. The lamp ouput at threshold is thus

$$\frac{2 \times 5.6}{0.1 \times 0.1 \times 0.4} = 2800 \text{ W}$$

If the efficiency of the lamp in converting electrical to optical energy is about 50%, we find that continuous operation of the laser requires about 5 kW of power. This number is to be contrasted with a threshold of approximately 100 W for the Nd : YAG laser, which helps explain why Nd : glass lasers are not operated continuously.

If we consider the pulsed operation of a Nd : glass laser by flash excitation, we have to estimate the minimum energy needed to pump the laser at threshold. Let us assume here that the losses (attributable mostly to the mirror transmittance) are $L = 20\%$.[3] A recalculation of N_{2t} gives

$$N_{2t} = 9 \times 10^{16} \text{ atoms/cm}^3$$

The minimum energy needed to pump N_t atoms into level 2 is then

$$\frac{\mathscr{E}_{\min}}{V} = N_{2t}(h\nu) = 1.7 \times 10^{-2} \text{ J/cm}^3$$

Assuming a crystal volume $V = 10 \text{ cm}^3$ and the same efficiency factors used in the CW example above, we find that the input energy to the flashlamp at threshold $\simeq 1.7 \times 2 \times 10^{-2} \times 10/(0.1 \times 0.1 \times 0.4) = 85$ J. Typical Nd^{3+} : glass lasers with characteristics similar to those used in this example are found to require an input of about 150–300 J at threshold.

10.5 THE He-Ne LASER

The first continuous laser, as well as the first gas laser, was one in which a transition between the $2S$ and the $2p$ levels in atomic Ne resulted in the emission of 1.15 μm radiation (Reference 9). Since then, transitions in Ne were used to obtain laser oscillation at $\lambda = 0.6328$ μm (Reference 10) and $\lambda = 3.39$ μm. The operation of this laser can be explained with the aid of Figure 10.15. A DC (or RF) discharge is established in the gas mixture con-

[3] Because of the higher pumping rate available with flash pumping, optimum coupling (see Section 9.3) calls for larger mirror transmittances compared to the CW case.

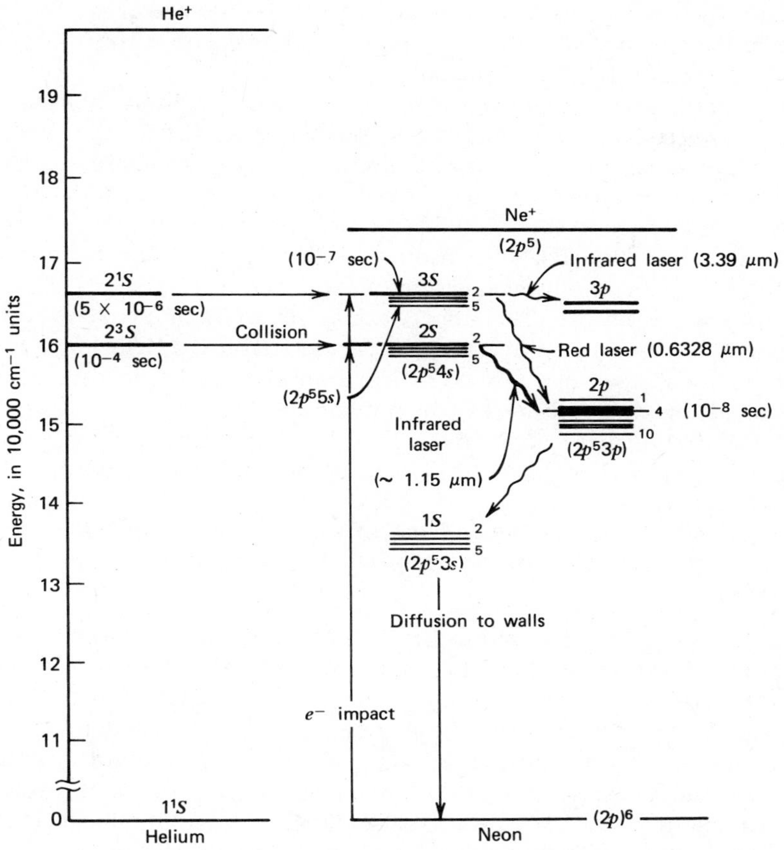

FIGURE 10.15 He-Ne energy levels. The dominant excitation paths for the red and infrared laser transitions are shown. *Source:* Reference 11.

taining, typically, 1.0 mmHg of He and 0.1 mm of Ne. The energetic electrons in the discharge excite helium atoms into a variety of excited states. In the normal cascade of these excited atoms down to the ground state, many collect in the long-lived metastable states 2^3S and 2^1S whose lifetimes are 10^{-4} sec and 5×10^{-6} sec, respectively. Since these long-lived (metastable) levels nearly coincide in energy with the $2S$ and $3S$ levels of Ne, they can excite Ne atoms into these two excited states. This excitation takes place when an excited He atom collides with an Ne atom in the ground state and exchanges energy with it. The small difference in energy ($\sim$400 cm^{-1} in the case of the $2S$ level) is taken up by the kinetic energy of the atoms after the collision. This is the main pumping mechanism in the He-Ne system.

1. *0.6328 μm Oscillation.* The upper level is one of the Ne $3S$ levels, whereas the terminal level belongs to the $2p$ group. The terminal ($2p$) level decays radiatively with a time constant of about 10^{-8} sec into the long-lived $1S$

state. This time is much shorter than the 10^{-7} sec lifetime of the upper laser level 3S. The condition $t_1 < t_2$ for population inversion in the 3S-2p transition (see Section 9.3) is thus fulfilled.

Another important point involves the level 1S. Because of its long life, it tends to collect atoms reaching it by radiative decay from the lower laser level 2p. Atoms in 1S collide with discharge electrons and are excited back into the lower laser level 2p. This reduces the inversion. Atoms in the 1S states relax back to the ground state mostly in collisions with the wall of the discharge tube. For this reason, the gain in the 0.6328 μm transition is found to increase with decreasing tube diameter.

2. *1.15 μm Oscillation.* The upper laser level 2S is pumped by resonant (energy-conserving) collisions with the metastable 2^3S He level. It uses the same lower level as the 0.6328 μm transition and, consequently, also depends on wall collisions to depopulate the 1S Ne level.
3. *3.39 μm Oscillation.* This involves a 3S-3p transition and thus uses the same upper level as the 0.6328 μm oscillation. It is remarkable for the fact that it provides a small-signal[4] optical gain of about 50 dB/m. This large gain reflects partly the dependence of γ on λ^2 [see Eq. (8.4-4)] as well as the short lifetime of the 3p level, which allows the buildup of a large inversion.

Because of the high gain in this transition, oscillation would normally occur at 3.39 μm rather than at 0.6328 μm. The reason is that the threshold condition will be reached first at 3.39 μm and, once that happens, the gain "clamping" will prevent any further buildup of the population of 3S. The 0.6328 μm laser overcomes this problem by introducing into the optical path elements, such as glass or quartz Brewster windows, that absorb strongly at 3.39 μm but not at 0.6328 μm. This raises the threshold pumping level for the 3.39 μm oscillation above that of the 0.6328 μm oscillation.

A typical gas-laser setup is illustrated by Figure 10.16. The gas envelope windows are tilted at Brewster's angle θ_B, so radiation with the electric field vector in the plane of the paper suffers no reflection losses at the windows. This causes the output radiation to be polarized in the sense shown, since the orthogonal polarization (the **E** vector out of the plane of the paper) undergoes reflection losses at the windows and, consequently, has a higher threshold.

10.6 THE CARBON DIOXIDE LASER

The lasers described so far in this chapter depend on electronic transitions between states in which the electronic orbitals (i.e., charge distributions around the atomic nucleus) are different. An example, consider the red (0.6328 μm) transition in Ne shown in Figure 10.15. It involves levels $2p^55s$ and $2p^53p$ so that in making a transition from the upper to the lower laser

[4] This is not the actual gain that exists inside the laser resonator, but the one-pass gain exercised by a very small input wave propagating through the discharge. In the laser, the gain per pass is reduced by saturation until it equals the loss per pass.

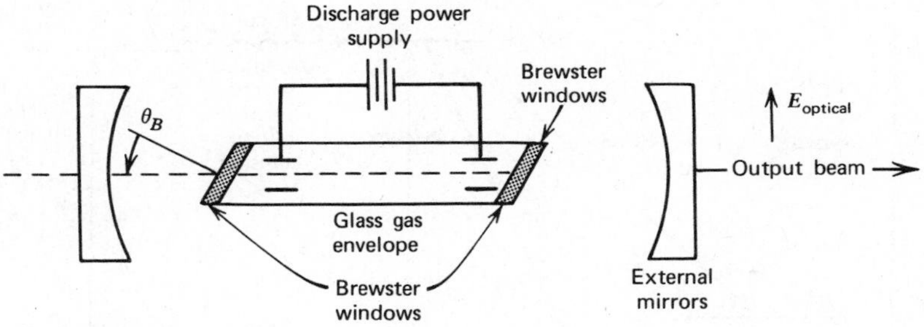

FIGURE 10.16 Typical gas laser.

level, one of the six outer electrons changes from a hydrogenlike state 5*s* (i.e., $n = 5$, $l = 0$) to one in which $n = 3$ and $l = 1$.

The CO_2 laser (Reference 12) is representative of the so-called molecular lasers in which the energy levels of concern involve the internal vibration of the molecules—that is, the relative motion of the constituent atoms. The atomic electrons remain in their lowest energetic states and their degree of excitation is not affected.

As an illustration, consider the simple case of the nitrogen molecule. The molecular vibration involves the relative motion of the two atoms with respect to each other. This vibration takes place at a characteristic frequency of $\nu_0 = 2326\ \text{cm}^{-1}$, which depends on the molecular mass as well as the elastic restoring force between the atoms (Reference 13). The quantum mechanical features of this system closely resemble those of the simple harmonic oscillator treated in Section 2.2. The degrees of vibrational excitation are discrete (i.e., quantized) and the energy of the molecule can take on the values $h\nu_0(v + \frac{1}{2})$, where $v = 0, 1, 2, 3, \ldots$. The energy-level diagram of N_2 (in its lowest electronic state) would then ideally consist of an equally spaced set of levels with a spacing of $h\nu_0$. The ground state ($v = 0$) and the first excited state ($v = 1$) are shown on the right side of Figure 10.17.

The CO_2 molecule presents a more complicated case. Since it consists of three atoms, it can execute three basic internal vibrations, the so-called normal modes of vibration. These are shown in Figure 10.18. In (*a*) the molecule is at rest. In (*b*) the atoms vibrate along the internuclear axis in a symmetric manner. In (*c*) the molecules vibrate symmetrically along an axis perpendicular to the internuclear axis—the bending mode. In (*d*) the atoms vibrate asymmetrically along the internuclear axis. This mode is referred to as the asymmetric stretching mode. In the first approximation, one can assume that the three normal modes are independent of each other, so the state of the CO_2 molecule can be described by a set of three integers (v_1, v_2, v_3) that correspond, respectively, to the degree of excitation of the three modes described. The total vibrational energy of the molecule is thus

$$E(v_1, v_2, v_3) = h\nu_1(v_1 + \tfrac{1}{2}) + h\nu_2(v_2 + \tfrac{1}{2}) + h\nu_3(v_3 + \tfrac{1}{2}) \qquad (10.6\text{-}1)$$

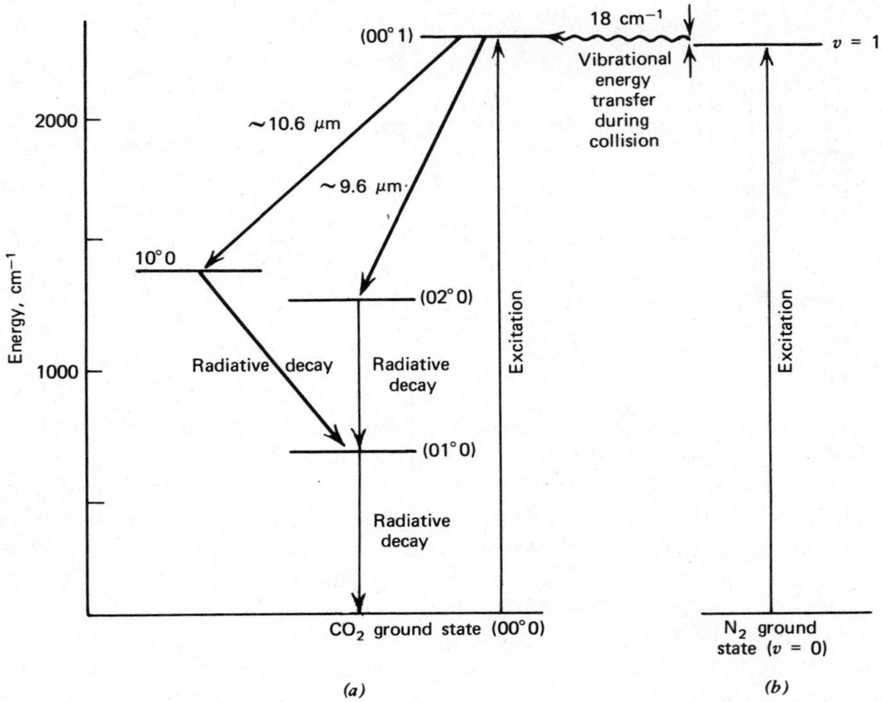

FIGURE 10.17 (*a*) Some of the low-lying vibrational levels of the carbon dioxide (CO_2) molecule, including the upper and lower levels for the 10.6 and 9.6μm laser transitions. (*b*) Ground state ($\nu = 0$) and first excited state ($\nu = 1$) of the nitrogen molecule, which plays an important role in the selective excitation of the (00°1)CO_2 level.

where ν_1, ν_2, and ν_3 are the frequencies of the symmetric stretch and bending and asymmetric stretch modes, respectively.

Some of the low vibrational levels of CO_2 are shown in Figure 10.17. The upper laser level (00°1) is thus one in which only the asymmetric stretch mode, Figure 10.18*d*, is excited and contains a single quantum ($h\nu_3$) of energy.

The laser transition at 10.6 μm takes place between the (00°1) and (10°0) levels of CO_2. The radiative lifetime of the upper laser level is $t_{\text{spont}} \sim 3$ sec. Its actual lifetime t_2 is determined by molecular collisions and ranges from $t_2 \sim 10^{-5}$ sec at a few torrs of pressure to $t_2 \sim 10^{-7}$ sec at atmospheric pressure. The exact value depends on the partial pressure of the gases involved and can be calculated from published collision cross-section data (Reference 15).

The laser transition has a low-pressure (≲5 torr) Doppler linewidth at $T = 300°K$ of $\Delta\nu_D \sim 60$ MHz. Pressure broadening sets in at ≳5 torr. This point is discussed in detail below. The excitation is provided usually in a plasma discharge that, in addition to CO_2, typically contains N_2 and He. The CO_2 laser possesses a high overall working efficiency of about 30%. This efficiency results primarily from three factors: (a) the laser levels are all near the ground state, and the atomic quantum efficiency ν_{21}/ν_{30}, which was discussed in Section 10.1, is about 45%; (b) a large fraction of the CO_2 molecules excited by electron impact cascade down the energy ladder from their original level of excitation and tend to collect in the long-lived (00°1) level; (c) a very large fraction of the N_2 molecules that are excited by the discharge tend to collect in the $v = 1$ level. Collisions with ground state CO_2 molecules result in transferring their excitation to the latter, thereby exciting them to the (00°1) state as shown in Figure 10.17. The slight deficiency in energy (about 18 cm^{-1}) is made up by a decrease of the total kinetic energy of the molecules following the collision. This collision can be represented by

$$(v = 1) + (00°0) + \text{K.E.} = (v = 0) + (00°1) \tag{10.6-2}$$

and has a sufficiently high cross section[5] that at the pressures and temperatures involved in the operation of a CO_2 laser, most of the N_2 molecules in the $v = 1$ lose their excitation energy by this process.

Carbon dioxide lasers are not only efficient but can emit large amounts of power. Laboratory-size lasers with discharge envelopes of a few feet in length can yield an output of a few kilowatts. This is due not only to the very *selective* excitation of the low-lying upper laser level but also to the fact that once a molecule is stimulated to emit a photon, it returns quickly to the ground state where it can be used again. This is accomplished mostly through collisions with other molecules—such as that of He, which is added to the gas mixture.

Inversion in Vibrational-Rotational Transitions

The CO_2 laser system involves transitions between states characterized not only by the vibrational quantum number v as discussed above, but also by the molecular rotational quantum numbers J, m. An eigenstate of the molecule, thus, has to be specified by v_1, v_2, v_3, J, m where v_1, v_2, v_3 are the vibrational quantum numbers corresponding to the three degrees of freedom of Figure 10.18. The rotational energies of a given vibrational state i (specified by v_1, v_2, v_3) relative to the $J = 0$ level is (Reference 13)

$$\frac{E_{i,J}}{hc} = B_i J(J+1) - DJ^2(J+1)^2 \tag{10.6-3}$$

where B_i is a constant of the ith vibrational state and $D \ll B_i$ is a spectroscopic constant of the molecule. A more detailed discussion of this topic is contained in Appendix 3.

Some vibrational rotational levels of the CO_2 laser transition near $\lambda = 10.6\ \mu m$ are shown in Figure 10.19.

Using (8.4-4) and (8.6-10), we can write the low-pressure peak gain due

[5] The cross section σ was defined in Section 10.2. In the present context, it follows directly from that definition that the number of collisions of the type described by (10.6-2) per unit volume per unit time is equal to $N(v = 1)N(00°0)\sigma\mathbf{v}$ where $N(v = 1)$ and $N(00°0)$ are the densities of molecules in the states $v = 1$ of N_2 and (00°0) of CO_2, respectively, and $\mathbf{v}$ is the (mean) relative velocity of the colliding molecules.

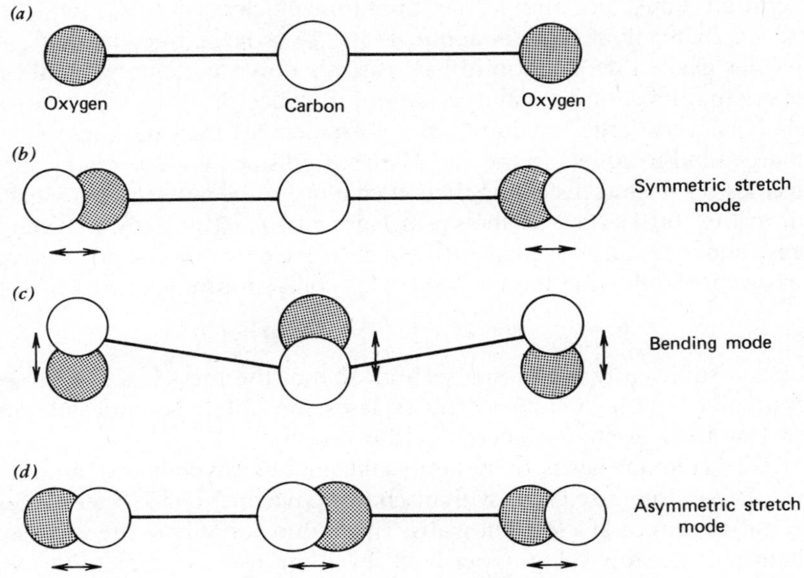

FIGURE 10.18 (*a*) Unexcited CO_2 molecule, (*b*), (*c*), and (*d*) The three normal modes of vibration of the CO_2 molecule. *Source:* Reference 14.

to some 2, $J \rightarrow 1, J \pm 1$ transition as

$$\gamma_{2,J\rightarrow 1,J\pm 1} = \left(N_{2,J} - N_{1,J\pm 1}\frac{g_J}{g_{J\pm 1}}\right)\frac{\lambda^2}{4\pi(t_{\text{spont}})_{2,J\rightarrow 1,J\pm 1}}\left(\frac{\ln 2}{\pi}\right)^{1/2}\left(\frac{1}{\Delta\nu_D}\right) \quad (10.6\text{-}4)$$

Here, we denote the upper vibrational state by $i = 2$ and the lower one by $i = 1$. λ is the transition wavelength and $(t_{\text{spont}})_{2,J\rightarrow 1,J\pm 1}$ is the spontaneous lifetime for the specific $(2, J) \rightarrow (1, J \pm 1)$ transition. The degeneracy factor g_J is equal to (Reference 13)

$$g_J = 2J + 1$$

The rotational population density $N_{i,J}$ $(i = 1, 2)$ is

$$N_{i,J} = \frac{N_i g_J e^{-E_{i,J}/kT_{\text{rot}}}}{\sum_J g_J e^{-E_{i,J}/kT_{\text{rot}}}} \quad (10.6\text{-}5)$$

so that N_i is the total population density of the ith vibrational state.

The time constant for achieving rotational thermal equilibrium, by molecule-molecule collisions, is usually extremely short so that the rotational level occupation may be characterized by a Boltzman distribution with a temperature $T_{\text{rot}} > 0$ even under dynamic excitation conditions. The vibrational level population, however, may be inverted ($N_2 > N_1$) so that $T_{\text{vib}} < 0$. The translational kinetic energy of the molecules may be characterized by another temperature T_{trans} that, under excitation conditions, is not necessarily the same as

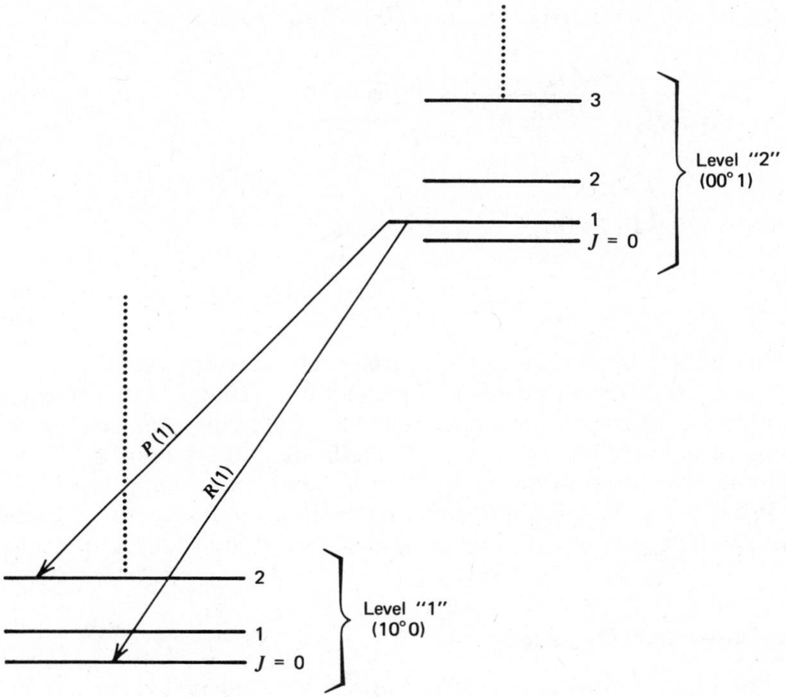

FIGURE 10.19 The vibrational-rotational level scheme of a CO_2 laser near 10.6 μ. Transitions from $J \rightarrow J + 1$ are named $P(J)$ those from $J \rightarrow (J - 1)$ are $R(J)$. Only some of the low J (rotational) levels are shown.

T_{rot}, so that the rotational occupation is given by

$$\frac{N_{i,J}}{N_{i,J'}} = e^{-(E_{i,J}-E_{i,J'})/kT_{rot}}\left(\frac{g_J}{g_{J'}}\right) \tag{10.6-6}$$

For large T_{rot} and small B_i, the denominator of (10.6-5) is closely approximated by integration [using (10.6-3)]

$$\sum_J g_J e^{-E_{i,J}/kT_{rot}} \approx \frac{kT_{rot}}{hcB_i} \tag{10.6-7}$$

The spontaneous transition rate is given by (see Appendix 3)

$$\left(\frac{1}{t_{spont}}\right)_{2,J\rightarrow 1,J\pm 1} = \frac{G_{12}}{\lambda^3}\frac{(J + \frac{1}{2} \pm \frac{1}{2})}{g_J} \tag{10.6-8}$$

where G_{12} is a constant that depends only on the vibrational quantum numbers. The upper sign is to be taken with $P(J \rightarrow J + 1)$ transitions, whereas the lower one goes with $R(J \rightarrow J - 1)$ transitions.

Using (10.6-5), (10.6-6), and (10.6-8) in (10.6-4) gives

$$\gamma_{2,J\to 1,J\pm 1} \frac{G_{12}hc}{8\pi kT_{rot}\left(\dfrac{2\pi kT_{trans}}{M}\right)^{1/2}} (J + \tfrac{1}{2} \pm \tfrac{1}{2})(N_2 B_2 e^{-E_{2,J}/kT_{rot}} - N_1 B_1 e^{-E_{1,J\pm 1}/kT_{rot}}) \tag{10.6-9}$$

where we used (8.6-9)

$$\Delta\nu_D = \frac{2}{\lambda}\left(\frac{2kT_{trans}}{M}\right)^{1/2} (\ln 2)^{1/2}$$

in relating the Doppler width to the translational temperature.

Figure 10.20 shows normalized gain plots of (10.6-9) as a function of the upper level J with N_2/N_1 as a parameter. The following is noted. When $N_2 > N_1$ (complete inversion), gain is available from all the P branch ($J \to J + 1$) transitions. The R transitions give rise to gain only up to some J_{max} (Reference 16). When $N_2 < N_1$ (partial inversion), amplification results in P transitions above some J_{min}, whereas no gain is available at any of the R transitions.

High-Pressure CO_2 Lasers

CO_2 lasers are also operated in the pressure broadened regime, that is, at pressures where the transition linewidth

$$\Delta\nu = \Delta\nu_D + \sum_i \frac{1}{\pi\tau_i}$$

is much larger than the Doppler linewidth, $\Delta\nu_D$. Here, τ_i is the mean collision time of a CO_2 molecule due to the ith molecular species. For a large range of pressures (Reference 15), the collision cross section is a constant so that τ_i^{-1} and $\Delta\nu$ are proportional to the pressure as shown in Figure 10.21*a*.

Consider now the problem of maintaining the laser oscillation in a high-pressure discharge. First, to achieve a given gain, we need, according to (8.7-1), to increase the inversion density by an amount proportional to the pressure P in order to compensate for the increase of $\Delta\nu$. Second, since the lifetime t_2 varies as P^{-1}, the pumping power per molecule increases as P. The result is that the pumping power, for a given gain, increases as P^2. It follows that the output power, along with the excitation power, increases with P^2. This conclusion follows formally, from (9.3-16), for the power output

$$P_0 = \frac{8\pi hcn^2\,\Delta\nu A}{(t_2/t_{spont})\lambda^3}\, T\left(\frac{g_0}{L_i + T} - 1\right)$$

since $\Delta\nu \propto P$ and $t_2 \propto 1/P$. We also used $1/g(\nu_0) \equiv \Delta\nu$.

The increase of power with pressure is seen in Figure 10.21*b*. The roll-off near $P = 150$ torr reflects the reduction in gain at the higher pressures. A

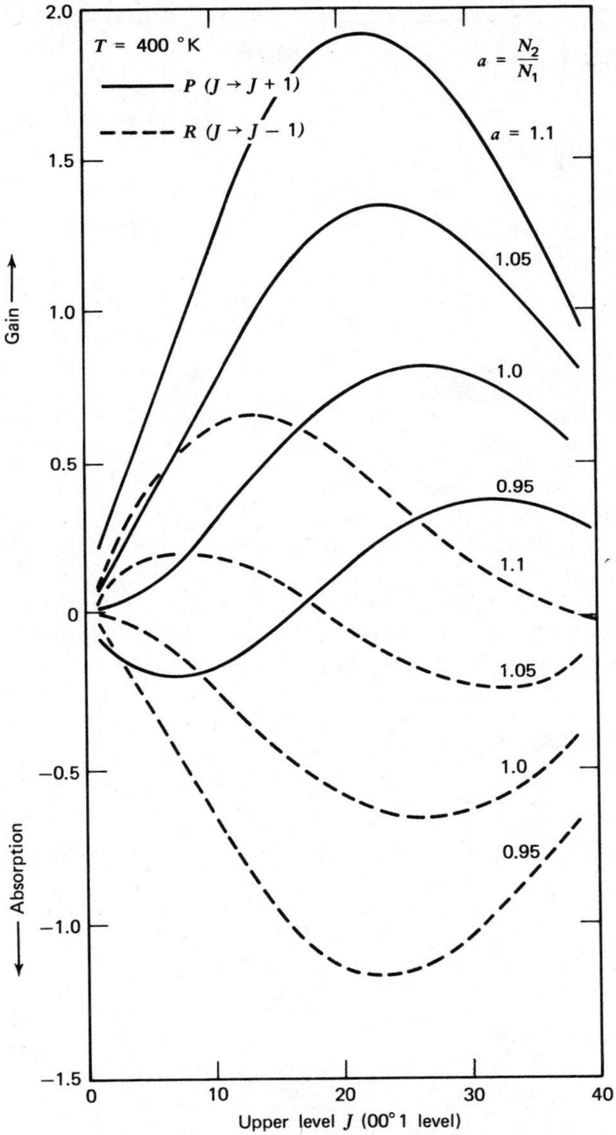

FIGURE 10.20 Normalized gain as a function of upper-level J number for the P and R branches for $N_2/N_1 = 0.95$, 1, 1.05, 1.1 and $T_{\text{rot}} = T_{\text{vibrational}} = 400°\text{K}$. *Source:* Patel, Reference 12.

more fundamental measure of the pressure effects is the variation of the saturation intensity [see (8.7-5)]

$$I_s = \frac{4\pi n^2 \,\Delta\nu h\nu}{(t_2/t_{\text{spont}})\lambda^2}$$

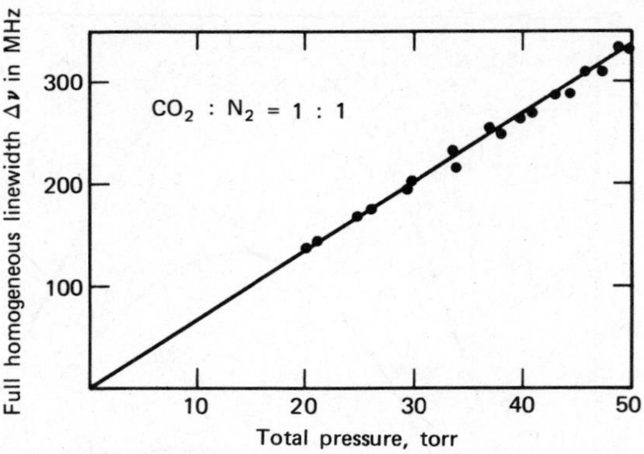

FIGURE 10.21*a* The 10.6 μm transition linewidth versus pressure for a gas mixture with equal partial pressures of CO_2 and N_2.

that, for the reasons given above, should increase as P^2. Experimental data of I_s versus P are shown in Figure 10.22.

High-energy pulsed operation of CO_2 lasers (Reference 18) at atmospheric pressure has been responsible for large and simple lasers suitable for many industrial uses.

10.7 ORGANIC-DYE LASERS

Many organic dyes (i.e., organic compounds that absorb strongly in certain visible-wavelength regions) also exhibit efficient luminescence, which often spans a large wavelength region in the visible portion of the spectrum. This

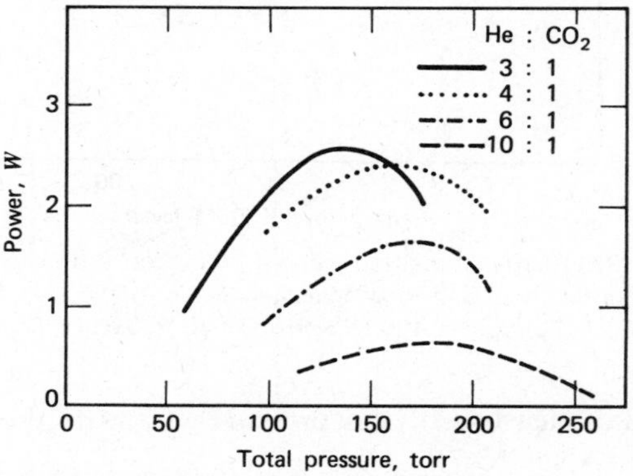

FIGURE 10.21*b* Output power versus total pressure under optimum pumping for He : CO_2 mixtures. *Source:* Reference 17.

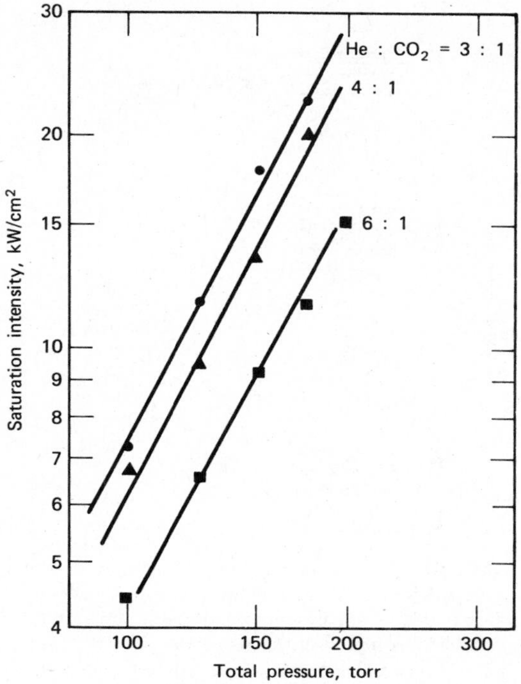

FIGURE 10.22 Measured saturation intensity versus pressure in a CO_2 laser. *Source:* Reference 17.

last property makes it possible to obtain an appreciable tuning range from dye lasers (References 19, 20, and 21).

A schematic representation of an organic dye molecule (such as rhodamine 6G, e.g.) is shown in Figure 10.23.

State S_0 is the ground state. S_1, S_2, T_1, and T_2 are excited electronic states—that is, states in which one ground-state electron is elevated to an excited orbit. Typical energy separation, such as $S_0 - S_1$, is about 20,000 cm^{-1}. In a singlet (S) state, the magnetic spin of the excited electron is antiparallel to the spin of the remaining molecule. In a triplet (T) state, the spins are parallel. Singlet → triplet or triplet → singlet transitions thus involve a spin flip and are far less likely than transitions between two singlet or between two triplet states.

Transitions between two singlet states or between two triplet states, which are spin-allowed (i.e., they do not involve a spin flip), give rise to intense absorption and fluorescence. The characteristic color of organic dyes is due to the $S_0 \rightarrow S_1$ absorption.

The singlet and triplet states are split further into vibrational levels shown as heavy horizontal lines in Figure 10.23. These correspond to the quantized vibrational states of the organic molecule, as discussed in detail in Section

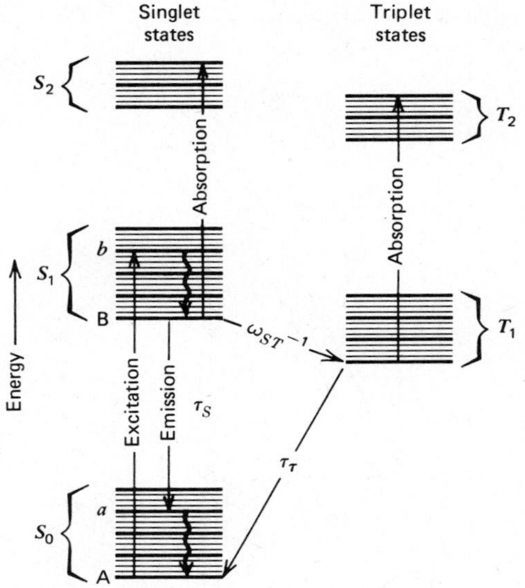

FIGURE 10.23 Schematic representation of the energy levels of an organic dye molecule. The heavy horizontal lines represent vibrational states, and the lighter lines represent the rotational fine structure. Excitation and laser emission are represented by the transitions $A \rightarrow b$ and $B \rightarrow a$, respectively.

10.6. Typical energy separation between two adjacent vibrational levels within a given singlet or triplet state is about 1500 cm^{-1}. The fine splitting shown corresponds to rotational levels whose spacing is about 15 cm^{-1}.

In the process of pumping the laser, the molecule is first excited, by absorbing a pump photon, into a rotational-vibrational state b within S_1. This is followed by a very fast decay to the bottom of the S_1 state with the excess energy taken up by the vibrational and rotational energy of the molecules. Most of the excited molecules will then decay spontaneously to state a, emitting a photon of energy $\nu = (E_B - E_a)/h$. The lifetime for this process is τ_S.

There is, however, a small probability, approximately $\omega_{ST}\tau_S$, that an excited molecule will decay, instead, to the triplet state T_1, where ω_{ST} is the rate per molecule for undergoing an $S_1 \rightarrow T_1$ transition. Since this is a spin-forbidden transition, its rate is usually much smaller than the spontaneous decay rate τ_S^{-1}, so that $\omega_{ST}\tau_S \ll 1$. The lifetime τ_T for decay of T_1 to the ground state is relatively long (since this too is a spin-forbidden transition) and may vary from 10^{-7} to 10^{-3} sec, depending on the experimental conditions (Reference 22). Owing to its relatively long lifetime, the triplet state T_1 acts as a trap for excited molecules. The absorption of molecules due to a $T_1 \rightarrow T_2$ transition is spin-allowed and is therefore very strong. If the wavelength region of this absorption coincides with that of the laser emission [at $\nu \simeq (E_B - E_a)/h$], an accumulation of molecules in T_1 increases the laser losses and at some critical

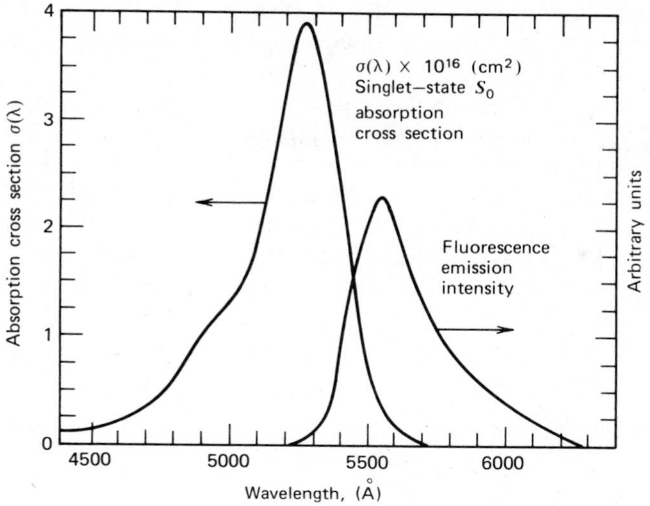

FIGURE 10.24 Singlet-state absorption and fluorescence spectra of rhodamine 6G obtained from measurements with a 10^{-4} molar ethanol solution of the dye. *Source:* Reference 22.

value quenches the laser oscillation. For this reason, many organic-dye lasers operate only on a pulsed basis. In these cases, fast-rise-time pump pulses—often derived from another laser (Reference 20)—cause a buildup of the S_1 population with oscillation taking place until an appreciable buildup of the T_1 population occurs.

Another basic property of molecules is that the peak of the absorption spectrum usually occurs at shorter wavelengths than the peak of the corresponding emission spectrum. This is illustrated in Figure 10.24 that shows the absorption and emission spectra of rhodamine 6G that when dissolved in H_2O is used as a CW laser medium (Reference 23). Laser oscillation occurring near the peak of the emission curve is thus absorbed weakly.

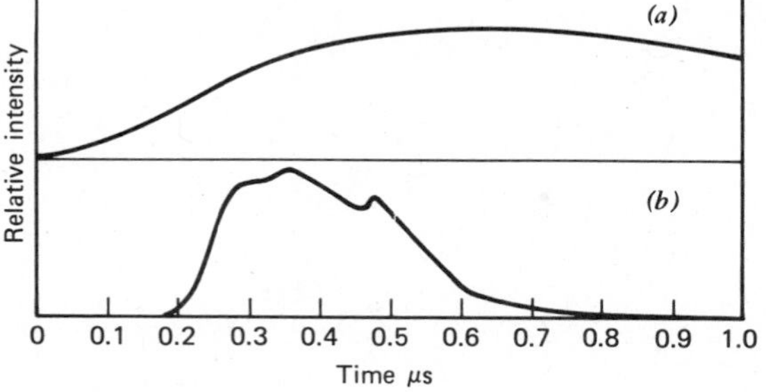

FIGURE 10.25 (*a*) Pumping pulse produced by a linear xenon flashlamp in a low-inductance circuit. (*b*) Laser pulse from a 10^{-3}-molar solution of rhodamine 6G in methanol. *Source:* Reference 21.

Typical excitation and oscillation waveforms of a dye laser are shown in Figure 10.25. The possibility of quenching the laser action by triplet state absorption is evident.

A list of some common laser dyes is given in Table 10.1. The broad fluorescence spectrum of the organic dyes suggests a broad tunability range for lasers using them as the active material. The spectrum in Figure 10.21, as an example, corresponds to a width of $\Delta\nu \simeq 1000\ \text{cm}^{-1}$. One elegant solution for realizing this tuning range (Reference 24) consists of replacing one of the laser mirrors with a diffraction grating. A diffraction grating has the property that (for a given order) an incident beam will be reflected back *exactly* along

TABLE 10.1. Molecular Structure, Laser Wavelength, and Solvents for Some Laser Dyes

Dye	*Structure*	*Solvent*	*Wavelength Emitted (nm)*
Acridine red	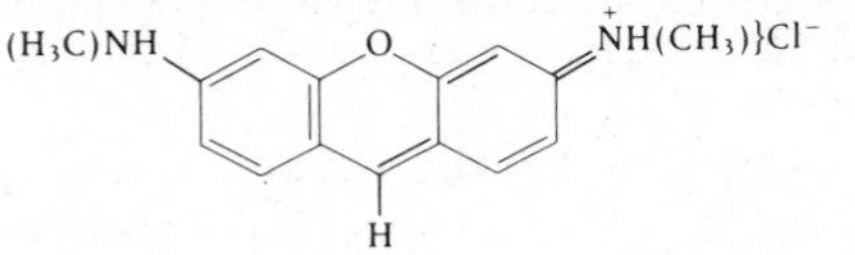	EtOH	Red 600–630
Puronin B	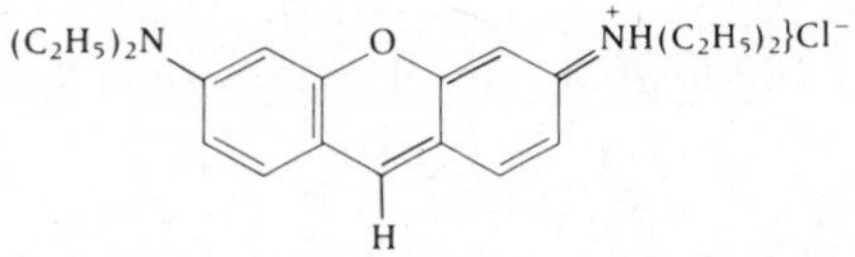	MeOH H_2O	Yellow
Rhodamine 6G	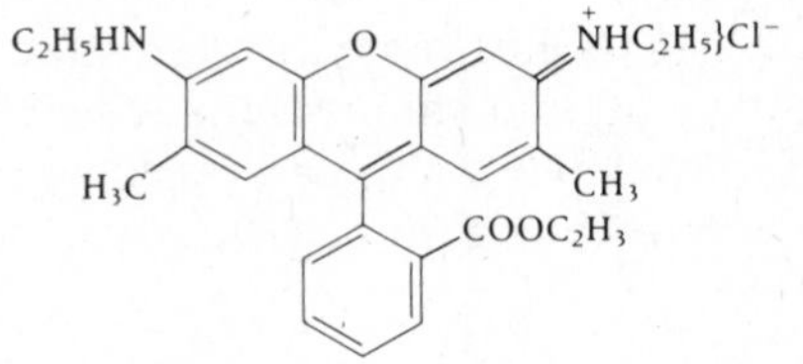	EtOH MeOH H_2O DMSO Polymethyl-methacrylate	Yellow 570–610
Rhodamine B	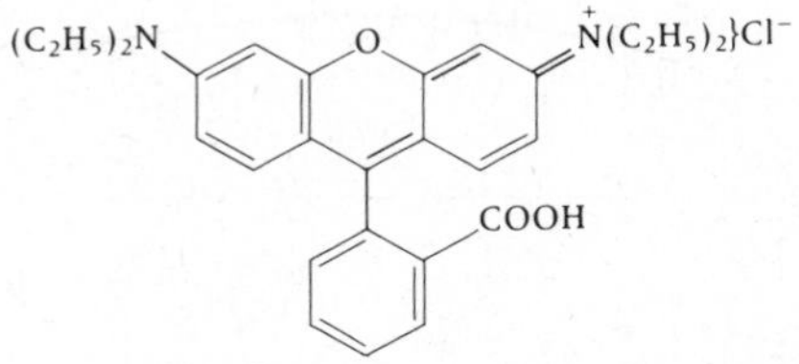	EtOH MeOH Polymethyl-methacrylate	Red 605–635
Na-fluorescein	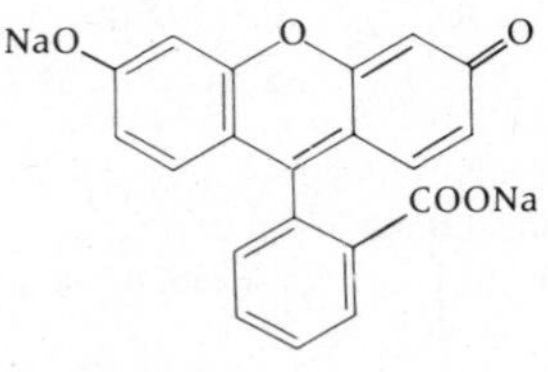	EtOH H_2O	Green 530–560

TABLE 10.1. (*continued*)

Dye	*Structure*	*Solvent*	*Wavelength Emitted (nm)*
2,7-Dichloro-fluorescein	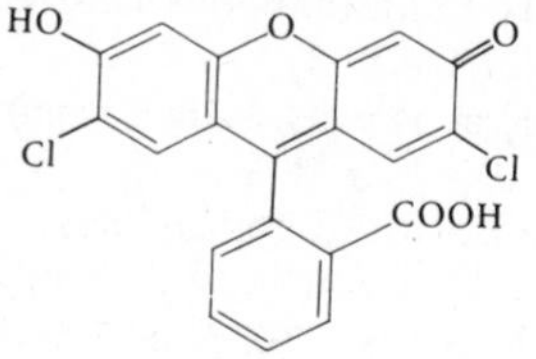	EtOH	Green 530–560
7-Hydroxycoumarin	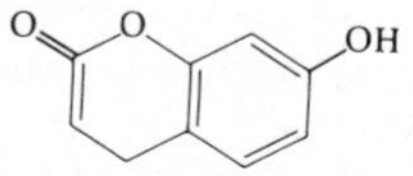	H_2O (pH ~ 9)	Blue 450–470

Source: Reference 22.

the direction of incidence, provided

$$2d \cos \theta = m\lambda \qquad m = 1, 2, \ldots \tag{10.7-1}$$

where d is the ruling distance, θ the angle between the incident ray direction and its projection on the grating surface, λ the optical wavelength, and m the order of diffraction. This type of operation of a grating is usually referred to as the Littrow arrangement. When a grating is used as one of the laser mirrors, it is clear that the oscillation wavelength will be that which satisfies (10.7-1).

This tuning can also be achieved by inserting a wavelength selective element (prism, filter, etc.) in the optical path (Reference 25).

References

1. Maiman, T. H., "Stimulated Optical Radiation in Ruby Masers," *Nature* **187,** 493 (1960).
2. Maiman, T. H., "Optical and Microwave-Optical Experiments in Ruby," *Phys. Rev. Letters* **4,** 564 (1960).
3. Cronemeyer, D. C., "Optical Absorption Characteristics of Pink Ruby," *J. Opt. Soc. Amer.* **56,** 1703 (1966).
4. Schawlow, A. L., "Fine Structure and Properties of Chromium Fluorescence," in *Advances in Quantum Electronics,* J. R. Singer, ed. (New York: Columbia University Press, 1961), p. 53.
5. Yariv, A., "Energy and Power Considerations in Injection and Optically Pumped Lasers," *Proc. IEEE* **51,** 1723 (1963).
6. Geusic, J. E., H. M. Marcos, and L. G. Van Uitert, "Laser Oscillations in Nd-Doped Yttrium Aluminum, Yttrium Gallium, and Gadolinium Garnets," *Appl. Phys. Letters* **4,** 182 (1964).
7. Kushida, T., H. M. Marcos, and J. E. Geusic, "Laser Transition Cross Section and Fluorescence Branching Ratio for Nd^{3+} in Yttrium Aluminum Garnet," *Phys. Rev.* **167,** 1289 (1968).

8. Snitzer, E. and C. G. Young, "Glass Lasers," in *Lasers*, Vol. 2, A. K. Levine, ed. (New York: Marcel Dekker, 1968), p. 191.
9. Javan, A., W. R. Bennett, Jr., and D. R. Herriott, "Population Inversion and Continuous Optical Maser Oscillation in a Gas Discharge Containing a He–Ne Mixture," *Phys. Rev. Letters* **6,** 106 (1961).
10. White, A. D. and J. D. Rigden, "Simultaneous Gas Maser Action in the Visible and Infrared," *Proc. IRE* **50,** 2366 (1962).
11. Bennett, W. R., "Gaseous Optical Masers," *Appl. Optics*, Suppl. 1, *Optical Masers*, p. 24 (1962).
12. Patel, C. K. N., "Interpretation of CO_2 Optical Maser Experiments," *Phys. Rev. Letters* **12,** 588 (1964); also, "Continuous-Wave Laser Action on Vibrational Rotational Transitions of CO_2," *Phys. Rev.* **136,** A1187 (1964). Also, F. LeGay and N. LeGay-Sommaire, *Compt. Rendu.* **260,** 3339 (1964).
13. Herzberg, G. H., *Spectra of Diatomic Molecules* (Princeton, N.J.: Van Nostrand, 1963).
14. Patel, C. K. N., "High Power CO_2 Lasers," *Sci. Amer.* **219,** 22 (Aug. 1968).
15. Taylor, R. L. and S. Bitterman, "Survey of Vibrational and Relaxation Data for Processes Important in the CO_2–N_2 Laser System," *Rev. Mod. Phys.* **41,** 26 (1969).
16. Basov, N. G., V. I. Igoshin, E. P. Markin, and A. N. Orevskii, "Dynamics of Chemical Lasers," *Sov. J. Quant. Elect.* **1,** 119 (1971).
17. Abrams, R. L. and W. B. Bridges, "Characteristics of Sealed-Off Waveguide CO_2 Lasers," *IEEE J. Quant. Elect.* **QE-9,** 940 (1973).
18. Beaulieu, J. A., "High Peak Power Gas Lasers," *Proc. IEEE* **59,** 667 (1971).
19. Stockman, D. L., W. R. Mallory, and K. F. Tittel, "Stimulated Emission in Aromatic Organic Compounds," *Proc. IEEE* **52,** 318 (1964).
20. Sorokin, P. P. and J. R. Lankard, "Stimulated Emission Observed from an Organic Dye, Chloroaluminum Phtalocyanine," *IBM J. Res. Develop.* **10,** 162 (1966).
21. Schafer, F. P., W. Schmidt, and J. Volze, "Organic Dye Solution Laser," *Appl. Phys. Letters* **9,** 306 (1966).
22. Snavely, B. B., "Flashlamp-Excited Dye Lasers," *Proc. IEEE* **57,** 1374 (1969).
23. Peterson, O. G., S. A. Tuccio, and B. B. Snavely, "CW Operation of an Organic Dye Laser," *Appl. Phys. Letters* **17,** 266 (1970).
24. Soffer, B. H. and B. B. McFarland, "Continuously Tunable, Narrow Band Organic Dye Lasers," *Appl. Phys. Letters* **10,** 266 (1967).
25. Dienes, A., E. P. Ippen, and C. V. Shank, "High-Efficiency Tunable CW Dye Laser," *IEEE J. Quant. Elect.* **QE-8,** 388 (1972).

Problems

10.1 Consider the problem of hole burning in a He–Ne laser for the 1.5 μm transition. We have

$$t_1 \equiv 10^{-8} \text{ sec} \qquad t_2 = 10^{-7} \text{ sec}$$

where 1 and 2 refer to the $2p_4$ and $2S_2$ levels, respectively.

(a) Calculate the Doppler width $\Delta\nu_D$ (take $T = 300°$K).

(b) Calculate the homogeneous line width $\Delta\nu_h$. What is $\Delta\nu_D/\Delta\nu_h$?

(c) Calculate the power flux (inside the optical resonator) at which $\Delta\nu_{\text{hole}} = 2\ \Delta\nu_h$ ($\Delta\nu_h$ = homogeneous linewidth).

(d) What is the corresponding power output (assume some reasonable mirror transmittance and mode diameter)?

(e) At what power level (internal) does the whole line become power broadened?

(f) How long can the resonator be before the "holes" of adjacent longitudinal modes overlap (1) for near zero power, (2) for $P_{\text{internal}} = 10\ \text{W/cm}^2$?

10.2 Using data from this chapter and from Reference 15, estimate the saturation intensity of a CO_2 laser using the following partial pressures: CO_2 10 torr, He 50 torr, N_2 20 torr.

10.3 Discuss qualitatively the process of amplification in a one-pass laser amplifier (i.e., no feedback) when the input intensity is far larger than the saturation intensity of the laser medium. How does the signal power grow with distance in the amplifier?

C H A P T E R 11

Semiconductor Diode Lasers

11.0 INTRODUCTION

Stimulated emission from GaAs semiconductor diode lasers was observed simultaneously by three groups (References 1–3) following theoretical analyses and proposals by Basov et al. and Dumke (Reference 5).

In the intervening years, the semiconductor diode laser (SDL) has come to dominate the laser field in its technological importance and has become the key element in an increasing number of applications, most notably in optical fiber communication and optical data storage. This "success" is due to the fact that semiconductor lasers are simply pumped by passing a current through them at voltage (and current) levels that are compatible with those of integrated circuits, and because they can be modulated directly at frequencies in excess of 20 GHz (Reference 6). SDL's can be mass-produced by the same photolithographic techniques as electronic circuits and can be integrated monolithically with the latter. This combination has given birth to the field of integrated optoelectronic circuits (Reference 7).

11.1 SOME SEMICONDUCTOR BACKGROUND

In this section, we will state some of the more elementary results of semiconductor theory. A detailed treatment of this topic can be found in numerous books dealing with the wave mechanics of solids (References 8 and 9).

The wave function of an electron in a given band (e.g., the valence band) can be written as

$$\Psi_v(\mathbf{r}) = u_{v\mathbf{k}}(\mathbf{r})e^{i\mathbf{k}\cdot\mathbf{r}} \tag{11.1-1}$$

where $u_{vk}(\mathbf{r})$ has the periodicity of the crystalline lattice. The "propagation" constants $\mathbf{k}$ are quantized in a manner similar to that of (4.2-7) so that we have

$$k_i = \frac{2\pi s}{L_i} \tag{11.1-2}$$

where $i = x, y, z$; s is an integer and L_i is the length of the crystal in the i direction. The volume, in $\mathbf{k}$ space, per electronic state is thus $8\pi^3/V$ where $V = L_xL_yL_z$. The number of electron states per band with a value of k between k and $k + dk$ is thus given by the volume of a spherical shell of radius k and

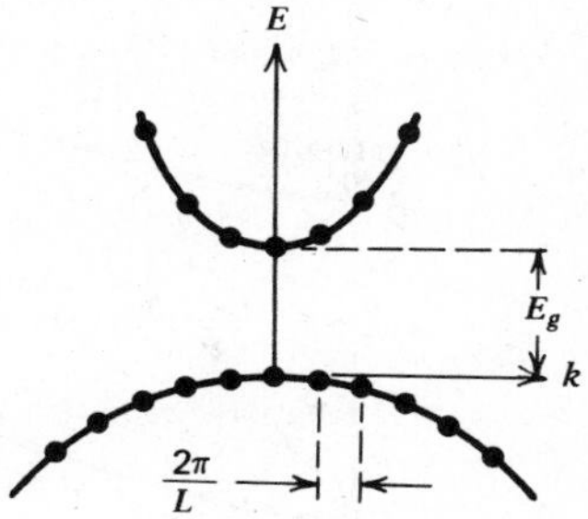

FIGURE 11.1*a* A typical energy band structure for a direct semiconductor with $m_c < m_v$. The uniformly spaced dots correspond to electron states.

thickness dk divided by the volume per state, that is,

$$\rho(k)dk = \frac{k^2 V}{\pi^2}\, dk \qquad (11.1\text{-}3)$$

where a factor of 2 was added to account for the two spin states associated with each k eigenvalue.

The energy above the band edge associated with a given state k (e.g., in the conduction band) is

$$E(k) = \frac{\hbar^2 k^2}{2m_c} \qquad (11.1\text{-}4)$$

in the parabolic band approximation and is thus a function of k rather than $\mathbf{k}$. m_c is the effective mass for electrons in the conduction band. $E(k)$ is measured from the bottom of the band.

Figure 11.1*a* shows a typical energy band structure E versus k for a direct semiconductor, that is, for a semiconductor in which the conduction band minimum and the valence band maximum occur at the same point in $\mathbf{k}$ (k_x, k_y, k_z) space. The direction of $\mathbf{k}$ must be specified in general except in the parabolic band approximation. The black dots correspond to allowed energy states and are spaced uniformly in k. According to (11.1-4), the situation depicted by Figure 11.1*a* corresponds to $m_v > m_c$.

From the expression for the density of states in $\mathbf{k}$ space (11.1-3) and from (11.1-4), we obtain readily the expression for the density of states per unit energy interval

$$\begin{aligned}\rho_v(E) &= \frac{1}{V}\rho_v(k)\frac{dk}{dE} = \frac{1}{2\pi^2}\left(\frac{2m_v}{\hbar^2}\right)^{3/2} E^{1/2} \\ \rho_c(E) &= \frac{1}{V}\rho_c(k)\frac{dk}{dE} = \frac{1}{2\pi^2}\left(\frac{2m_c}{\hbar^2}\right)^{3/2} E^{1/2}\end{aligned} \qquad (11.1\text{-}5)$$

where the subscripts c and v refer to the conduction and valence bands, respectively, and where the energy E is measured from the band edge extre-

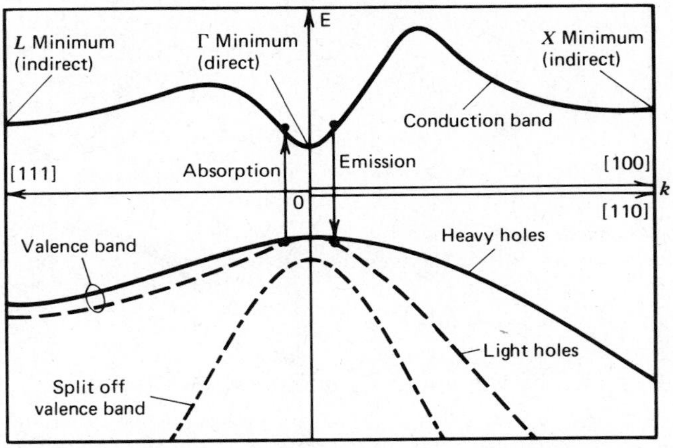

FIGURE 11.1*b* Plot of energy versus wave-vector for electrons in condition and valence bands of direct-gap semiconductor, for example, GaAs (diagrammatic). *Source:* Reference 10.

mum. The actual energy diagram of GaAs near $k = 0$ is shown in Figure 11.1*b*.

The Fermi–Dirac Distribution Law

The probability that an electron state at energy E is occupied by an electron is given by the Fermi–Dirac law

$$f(E) = \frac{1}{e^{(E-E_F)/kT} + 1} \tag{11.1-6}$$

where E_F is the Fermi energy and T the temperature. In thermal equilibrium, a single Fermi energy applies to both the valence and conduction bands. Under conditions in which the thermal equilibrium is disturbed, such as in a *p-n* junction with a current flow or a bulk semiconductor, in which a large population of conduction electrons and holes is created by photoexcitation, separate Fermi levels E_{Fc} and E_{Fv}, called quasi-Fermi levels, are used for each of the bands. The concept of quasi-Fermi levels in excited systems is valid whenever the carrier scattering time within a band is much shorter than the equilibrium time between bands. This is usually true at the large carrier densities used in *p-n* junction lasers.

A good estimate of the Fermi level E_{Fc} can be obtained by deriving its value at $T = 0$. In this case, it follows from (11.1-6) that all the energy levels up to E_{Fc} are occupied and those above it are empty. In this case, we obtain from (11.1-5)

$$N = \int_0^{E_{Fc}} \rho_c(E)\, dE = \frac{1}{2\pi^2}\left(\frac{2m_c}{\hbar^2}\right)^{3/2} \int_0^{E_{Fc}} E^{1/2}\, dE$$

where N is the density of electrons (of all energies) in the conduction band.

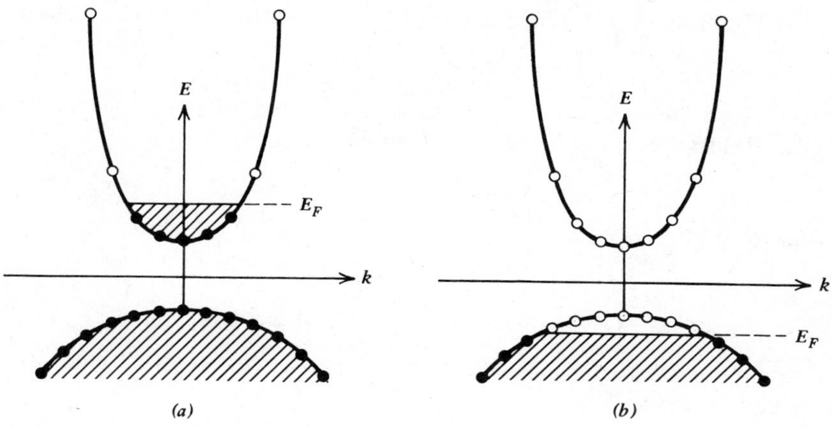

FIGURE 11.2 (*a*) Energy band of degenerate *n* type of semiconductor at 0°K. (*b*) A degenerate *p* type of semiconductor at 0°K. The cross hatching represents regions in which all the electron states are filled. Empty circles indicate unoccupied states.

Carrying out the integration gives

$$E_{Fc} = \frac{\hbar^2}{2m_c}[3\pi^2 N]^{2/3} \tag{11.1-6a}$$

In very highly doped semiconductors, the Fermi level is forced into either (a) the conduction band for donor impurity doping or (b) into the valence band for acceptor impurity doping. The semiconductor is said to be degenerate. This situation is demonstrated by Figure 11.2. According to (11.1-6) at 0 K, all the states below E_F are filled, whereas those above it are unoccupied as shown in the figure. In this respect, the degenerate semiconductor behaves like a metal where the conductivity does not disappear at low temperatures. The unoccupied states in the valence band are referred to as "holes," and they are treated like the electrons except that their charge, corresponding to an electron deficiency, is positive and their energy is measured downward.

The Reciprocal Lattice Expansion

In a crystal with primitive translation vectors **a**, **b**, **c**, which threesome determines the primitive unit cell, we can define three reciprocal lattice vectors

$$\mathbf{A} \equiv \frac{2\pi\mathbf{b}\times\mathbf{c}}{\mathbf{a}\cdot\mathbf{b}\times\mathbf{c}}, \qquad \mathbf{B} \equiv \frac{2\pi\mathbf{c}\times\mathbf{a}}{\mathbf{a}\cdot\mathbf{b}\times\mathbf{c}}, \qquad \mathbf{C} \equiv \frac{2\pi\mathbf{a}\times\mathbf{b}}{\mathbf{a}\cdot\mathbf{b}\times\mathbf{c}} \tag{11.1-7}$$

$$\mathbf{a}\cdot\mathbf{b}\times\mathbf{c} = \mathbf{b}\cdot\mathbf{c}\times\mathbf{a} = \mathbf{c}\cdot\mathbf{a}\times\mathbf{b}$$

so that

$$\begin{aligned} \mathbf{a}\cdot\mathbf{B} &= \mathbf{a}\cdot\mathbf{C} = 0, & \mathbf{a}\cdot\mathbf{A} &= 2\pi \\ \mathbf{b}\cdot\mathbf{A} &= \mathbf{b}\cdot\mathbf{C} = 0, & \mathbf{b}\cdot\mathbf{B} &= 2\pi \\ \mathbf{c}\cdot\mathbf{A} &= \mathbf{c}\cdot\mathbf{B} = 0, & \mathbf{c}\cdot\mathbf{C} &= 2\pi \end{aligned} \tag{11.1-8}$$

The set of points in the real lattice that can be reached by

$$\boldsymbol{\rho}(h, k, l) = h\mathbf{a} + k\mathbf{b} + l\mathbf{c} \tag{11.1-9}$$

where h, k, l are any integers are all equivalent. Similarly, a vector in the reciprocal lattice can be defined by any triplet of integers, say m, n, p

$$\mathbf{G}(m, n, p) = m\mathbf{A} + n\mathbf{B} + p\mathbf{C} \tag{11.1-10}$$

so that from (11.1-8)

$$\mathbf{G}(m, n, p) \cdot \boldsymbol{\rho}(h, k, l) = (mh + nk + pl)2\pi$$

$$\exp[i\mathbf{G}(m, n, p) \cdot \boldsymbol{\rho}(h, k, l)] = 1 \tag{11.1-11}$$

Any function that possesses the same periodicity as that of the crystal lattice can be expanded as

$$f(\mathbf{r}) = \sum_{\mathbf{G}} a_{\mathbf{G}} e^{i\mathbf{G}\cdot\mathbf{r}} \tag{11.1-12}$$

where $\mathbf{G} = \mathbf{G}(m, n, p)$. This can be verified by expanding

$$f[\mathbf{r} + \boldsymbol{\rho}(h, k, l)] = \sum_{\mathbf{G}} a_{\mathbf{G}} e^{i\mathbf{G}(\mathbf{r}+\boldsymbol{\rho})} = f(\mathbf{r}) \tag{11.1-13}$$

where the last equality follows from (11.1-11)

11.2 OPTICALLY INDUCED BAND-TO-BAND TRANSITIONS IN SEMICONDUCTORS

The interaction Hamiltonian of an electron with an electromagnetic field will be taken as

$$H' = -\frac{e}{m}\mathbf{A} \cdot \mathbf{p} \tag{11.2-1a}$$

where $\mathbf{A}(\mathbf{r}, t)$ is the vector potential of the optical field as discussed in Section (5.1) and $\mathbf{p}$ is the electron momentum operator.

It can be shown, the proof being assigned as a problem, that in a periodic crystal we can also express H' by

$$H' = -e\mathbf{E} \cdot \mathbf{r} \tag{11.2-1b}$$

where $\mathbf{E}$ is the electric field.

We assume an "x" polarized optical field

$$\mathbf{A}(\mathbf{r}, t) = \hat{\mathbf{x}}\frac{A_0}{2} e^{i(\omega t - \mathbf{k}_{\text{opt}}\cdot\mathbf{r})} \tag{11.2-2}$$

so that

$$H' = i\hbar\frac{eA_0}{2m} e^{-i(\omega t - \mathbf{k}_{\text{opt}}\cdot\mathbf{r})}\frac{\partial}{\partial x} + \text{Hermitian Adjoint}$$

The induced transition rate of an electron from the conduction to the valence band, or in reverse, will thus involve the matrix element

$$H'_{vc} \propto \int_{\text{crystal}} u^*_{v\mathbf{k}} e^{-i(\mathbf{k}+\mathbf{k}_{\text{opt}})\cdot\mathbf{r}} \frac{\partial}{\partial x}\left(u_{c\mathbf{k}}\, e^{i\mathbf{k}'\cdot\mathbf{r}}\right) dr \tag{11.2-2}$$

Since $u_{v\mathbf{k}}$ and $u_{c\mathbf{k}'}$ are periodic, we can expand them according to (11.1-12)

$$u_{v\mathbf{k}} = \sum_{\mathbf{G}} a_{v,\mathbf{k},\mathbf{G}} e^{i\mathbf{G}\cdot\mathbf{r}}$$

$$u_{c\mathbf{k}'} = \sum_{\mathbf{G}'} a_{c,\mathbf{k}',\mathbf{G}'} e^{i\mathbf{G}'\cdot\mathbf{r}}$$

so that

$$H'_{vc} \propto \sum_{\mathbf{G}} \sum_{\mathbf{G}'} a^*_{v,\mathbf{k},\mathbf{G}} a_{c,\mathbf{k}',\mathbf{G}'} (G'_x + k'_x) \int_{\text{crystal}} \exp\Big[i(-\mathbf{k} - \mathbf{k}_{\text{opt}} + \mathbf{k}' - \mathbf{G} + \mathbf{G}')\cdot\mathbf{r}\Big]\, dr \tag{11.2-3}$$

We can express the position $\mathbf{r}$ with the unit cell

$$\mathbf{r} = \boldsymbol{\rho}(h, k, l) + \delta\mathbf{r}$$

so that $\delta\mathbf{r}$ is the position relative to the "address" $\boldsymbol{\rho}(h, k, l)$ of the unit cell h, k, l. The integral in (11.2-3) can then be expressed using (11.1-11) as

$$\sum_{h,k,l} \int_{\text{unit cell}(h,k,l)} \exp\Big\{i[-\mathbf{k} - \mathbf{k}_{\text{opt}} + \mathbf{k}' - \mathbf{G} + \mathbf{G}']\cdot[\boldsymbol{\rho}(h, k, l) + \delta\mathbf{r}]\Big\}\, dr$$

$$= \int_{\text{any unit cell}} \exp[i(-\mathbf{k} - \mathbf{k}_{\text{opt}} + \mathbf{k}' - \mathbf{G} + \mathbf{G}')\cdot\delta\mathbf{r}]\, dr$$

$$\sum_{h,k,l} \exp[i(-\mathbf{k} - \mathbf{k}_{\text{opt}} + \mathbf{k}')\cdot\boldsymbol{\rho}(h, k, l)]$$

The last summation, in general, amounts to adding up complex phasors with arbitrary phases, a procedure leading to a "small" number unless

$$\mathbf{k}' = \mathbf{k} + \mathbf{k}_{\text{opt}} \tag{11.2-4}$$

in which case, the summation is equal to the number of unit cells in the crystal.

Since, in general, $k_{\text{opt}} \sim 10^5\ \text{cm}^{-1}$ while $k, k' \sim 10^6$–$10^7\ \text{cm}^{-1}$, except at the very bottom of the bands, we neglect k_{opt} in (11.2-4) and write

$$\mathbf{k}' = \mathbf{k} \tag{11.2-5}$$

as a necessary condition for optical transitions in a semiconductor.

Optical Gain and Loss in Semiconductor Media

We are now in a position to treat the problem of absorption or gain in a semiconductor. Consider the transition between the states (a) and (b) shown in Figure 11.3. The transition conserves "momentum" and is represented in Figure 11.3 by a vertical arrow. The sample is assumed to be in thermal

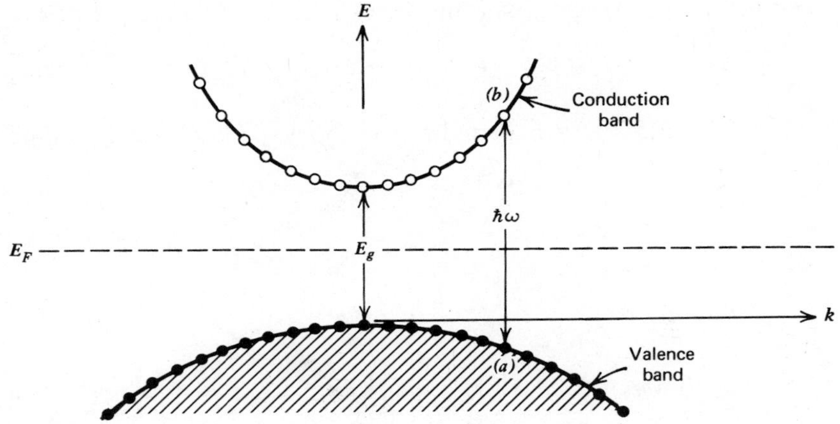

FIGURE 11.3 The absorption of a photon in a semiconductor due to a transition of an electron from an occupied state (a) in the valence band to an empty state (b) in the conduction band.

equilibrium, and the Fermi level is assumed to lie sufficiently far from the band edges so that all the levels in the valence band are full and those in the conduction band are empty. We find it most convenient to start with Eq. (3.12-14) for the transition rate.

$$W_{ab} = \frac{2\pi}{\hbar} |H'_{ab}|^2 \delta(E_b - E_a - \hbar\omega) \tag{11.2-6}$$

for the probability rate for a transition from state a to b. According to (11.1-4) and (11.2-5) and Figure 11.2, we have

$$E_b - E_a = \frac{\hbar^2 k^2}{2}\left(\frac{1}{m_c} + \frac{1}{m_v}\right) + E_g$$

for the transition indicated by an arrow in the figure. The probability rate for this *single* transition can thus be written as

$$W(k) = \frac{2\pi}{\hbar} |H'_{vc}(k)|^2 \delta\left(\frac{\hbar^2 k^2}{2m_r} + E_g - \hbar\omega\right) \tag{11.2-7}$$

where $m_r = m_v m_c/(m_v + m_c)$ is the reduced effective mass. The total number N of transitions per second in a crystal of volume V is given by multiplying (11.2.7) by $\rho(k)$, the number of states per unit k in (V), and then integrating over all values of k. If we use (11.1-3), the result is

$$N = \frac{2V}{\pi\hbar} \int_0^{+\infty} |H'_{vc}(k)|^2 \delta\left(\frac{\hbar^2 k^2}{2m_r} + E_g - \hbar\omega\right) k^2\, dk \tag{11.2-8}$$

Introducing the variable X

$$X \equiv \frac{\hbar^2 k^2}{2m_r} + E_g - \hbar\omega$$

the last integral becomes

$$\begin{aligned} N &= \frac{2V}{\pi\hbar}\int |H'_{vc}(k)|^2 \frac{m_r}{\hbar^2}\,\delta(X)\,\sqrt{\frac{2m_r}{\hbar^2}(X+\hbar\omega-E_g)}\,dX \\ &= \frac{V}{\pi}|H'_{vc}(k)|^2\frac{(2m_r)^{3/2}}{\hbar^4}(\hbar\omega-E_g)^{1/2} \end{aligned} \tag{11.2-9}$$

where $\hbar^2k^2/2m_r + E_g = \hbar\omega$. The absorption coefficient $\alpha(\omega)$ is given by

$$\alpha(\omega) = \frac{\text{power absorbed per unit volume}}{\text{power crossing a unit area}}$$

$$= \frac{N\hbar\omega/V}{\varepsilon_0 n E_0^2 c/2}$$

where n is the index of refraction, c the velocity of light in vacuum, and E_0 the field amplitude. Using (11.2-9), we obtain

$$\alpha_0(\omega) = \frac{\omega e^2 x_{vc}^2 (2m_r)^{3/2}}{2\pi\varepsilon_0 n c\hbar^3}(\hbar\omega - E_g)^{1/2} \tag{11.2-10}$$

where we replaced, in conformity with (11.2-1b), $H'_{vc}(k)$ by $eE_0x_{vc}/2$ with $x_{vc} \equiv \langle u_{v\mathbf{k}'}|x|u_{c\mathbf{k}}\rangle$. In practice, the numerical coefficients are lumped together and $\alpha_0(\omega)$ is expressed as

$$\alpha_0(\omega) = K(\hbar\omega - E_g)^{1/2} \tag{11.2-11}$$

where K can be determined from absorption data. In gallium arsenide (GaAs), for example, the following data apply:

using

$$\hbar\omega = 1.5eV$$

$$m_v = .46\ m_{\text{electron}}$$

$$m_c = .067\ m_{\text{electron}}$$

$$x_{vc} = 3.2\ \text{Å}$$

$$n = 3.64$$

We get

$$K = 11700\ \text{cm}^{-1}(eV)^{-1/2}$$

so that at a frequency whose photon energy, as an example, exceeds the gap energy E_g by 0.01 eV, the absorption coefficient is $\alpha_0(\omega) = 1.17 \times 10^4 \times 10^{-1} = 1170\ \text{cm}^{-1}$.

Now assume that by some means we can prepare a semiconducting crystal in which the states up to some level in the conduction band are all occupied and those above a certain level in the valence band are empty. This situation is illustrated in Figure 11.4 where at 0 K all the conduction states up to the quasi-Fermi level E_{Fc} are occupied, whereas all the valence band states down to the quasi-Fermi level E_{Fv} are empty. The calculation of the absorption coefficient at ω is *identical* to that leading to (11.2-10) except that, here,

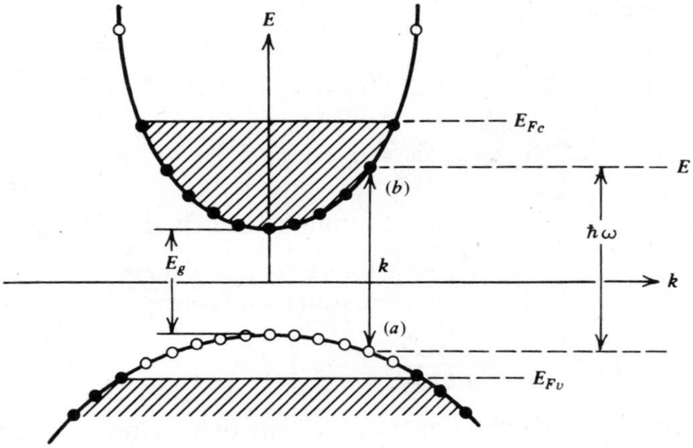

FIGURE 11.4 A semiconductor with degenerate electron and hole populations. At a frequency $\hbar\omega$, as shown, the sign of the absorption is reversed with respect to that of Figure 11.3 and amplification results. Full dots denote states occupied by electrons, whereas empty circles denote vacant states.

the upper states are full, whereas the lower ones are empty, so that the sign of the absorption coefficient is reversed, that is, the radiation is amplified rather than absorbed. We thus have for the semiconductor depicted by Figure 11.4

$$\hbar\omega < E_g \qquad \alpha(\omega) = 0$$

$$E_g < \hbar\omega < E_{Fc} - E_{Fv} \qquad \alpha(\omega) = -\alpha_0(\omega) = -K(\hbar\omega - E_g)^{1/2}(\text{amplification})$$

$$E_{Fc} - E_{Fv} < \hbar\omega \qquad \alpha(\omega) = \alpha_0(\omega) = K(\hbar\omega - E_g)^{1/2}(\text{absorption})$$

This is the argument used by Dumke (Reference 5) in predicting and estimating the possibility of laser action in semiconductors. This argument shows that in the case of the numerical example considered above, in which $\hbar\omega - E_g = 0.01\ eV$ in GaAs, if, by some manner, we invert the populations as in Figure 11.4, the exponential loss coefficient, that is, α_0 in $I = I_0 e^{-\alpha_0 z}$, is $-1170\ \text{cm}^{-1}$ (provided $\hbar\omega < E_{Fc} - E_{Fv}$) corresponding to amplification.

We will next extend our analysis to the general case of gain or loss in a semiconductor medium at any temperature. We will use a slightly different point of view that is more general and elegant than that leading to (11.2-10).

We start with the expression (8.1-19) for the complex susceptibility of an atomic, or electronic, medium

$$\chi(\omega) = \frac{\mu^2 T_2(N_1 - N_2)[(\omega_0 - \omega)T_2 - i]}{\varepsilon_0 \hbar[1 + (\omega - \omega_0)^2 T_2^2]} \tag{11.2-13}$$

N_1 and N_2 are the population densities of the lower and upper transition levels, respectively. μ is the matrix element for, say, x polarized field

$$\mu \equiv \mu_{12} = -\frac{\hbar e}{m\omega}\left\langle u_{v\mathbf{k}}\left|\frac{\partial}{\partial x}\right|u_{c\mathbf{k}}\right\rangle = e\langle u_{v\mathbf{k}}|x|u_{c\mathbf{k}}\rangle \tag{11.2-14}$$

T_2 is the coherence time of ρ_{12} and ω_0 is the transition frequency. To obtain the expression for $\chi(\omega)$ in the case of a semiconductor, we need to replace in (11.2-13) the inversion density $(N_2 - N_1)$ by the appropriate inversion density in a semiconductor. If we consider first only the subset of electrons with k values within dk, their contribution to the inversion density is

$$d(N_2 - N_1) \rightarrow \frac{\rho_c(k)\,dk}{V}\{\overbrace{f_c(E_b)[1 - f_v(E_a)]}^{\propto N_2} - \overbrace{f_v(E_a)[1 - f_c(E_b)]}^{\propto N_1}\} \tag{11.2-15}$$

$\rho_c(k)\,dk$ is the number of electrons within dk, whereas the Fermi factors simply insure that only occupied states are counted as the initial states of a transition and only empty states are counted for the final states. This is necessary since the electrons obey Fermi–Dirac statistics. If we substitute (11.2-15) for $N_1 - N_2$ and integrate over all k, the total susceptibility becomes

$$\chi(\omega) = \frac{\mu^2 T_2}{\varepsilon_0 \hbar}\int \frac{dk\rho_c(k)}{V}[f_v(\hbar\omega_0) - f_c(\hbar\omega_0)]\frac{[\omega_0(k) - \omega]T_2 - i}{1 + (\omega - \omega_0)^2 T_2^2} \tag{11.2-16}$$

The Fermi functions are taken at the electron and hole energies that are separated by $\hbar\omega_0$ according to

$$\hbar\omega_0 = E_g + \frac{\hbar^2 k^2}{2}(m_c^{-1} + m_v^{-1}) \equiv E_g + E_c(k) + E_v(k) \tag{11.2-17}$$

From the second of (11.1-5) and replacing E by $\hbar\omega_c = \hbar^2k^2/2m_c$ so that $\hbar\omega_c$ is the energy of an electron relative to the bottom of the conduction band, we obtain

$$\frac{1}{V}\rho_c(k)\,dk = \rho(\omega_c)\,d\omega_c \tag{11.2-18}$$

$$\rho(\omega_c) = \frac{1}{2\pi^2}\left(\frac{2m_c}{\hbar}\right)^{3/2}\omega_c^{1/2} \tag{11.2-19}$$

From (11.2-17), it follows that $d\omega_c = (m_r/m_c)d\omega_0$ where $m_r^{-1} = m_c^{-1} + m_v^{-1}$. We thus have

$$\rho(\omega_c)\,d\omega_c = \frac{1}{2\pi^2}\left(\frac{2m_r}{\hbar}\right)^{3/2}(\omega_0 - E_g/\hbar)^{1/2}\,d\omega_0 \equiv \rho_j(\omega_0)\,d\omega_0 \tag{11.2-20}$$

$\rho_j(\omega_0)$ is the joint density of states and is a convenient concept to employ when our independent variable is not the electron energies but $\hbar\omega_0$, the electron-hole transition energy. Using (11.2-18,20) in (12.-15) leads to the basic result

$$\chi(\omega) = \frac{1}{\varepsilon_0\hbar}\int_{E_g/\hbar}^{\infty}\mu^2(\omega_0)T_2\rho_j(\omega_0)[f_v(\omega_0) - f_c(\omega_0)]\frac{(\omega_0 - \omega)T_2 - i}{1 + (\omega - \omega_0)^2T_2^2}\,d\omega_0 \tag{11.2-21}$$

From (8.2-7) and (11.2-21), we obtain the gain constant

$$\gamma(\omega) = -\frac{k\chi''(\omega)}{n^2}$$
$$= \frac{k\mu^2\pi}{n^2\varepsilon_0\hbar}\int_{E_g/\hbar}^{\infty} \rho_j(\omega_0)[f_c(\omega_0) - f_v(\omega_0)]\frac{T_2\,d\omega_0}{\pi[1 + (\omega - \omega_0)^2T_2^2]} \tag{11.2-22}$$
$$\chi(\omega) = \chi'(\omega) - i\chi''(\omega)$$

where we took $k = \omega n/c$ and we assumed that over the significant range $\Delta\omega_0 \sim T_2^{-1}$ of the integrand, μ and T_2 are constant. It is instructive to show that for $T_2 \to \infty$, in which case

$$\frac{T_2}{\pi[1 + (\omega - \omega_0)^2T_2^2]} \to \delta(\omega - \omega_0)$$

the gain $\gamma(\omega)$ becomes, after integration,

$$\gamma(N, \omega) = \frac{\mu^2}{\lambda_0\varepsilon_0 n\hbar}\left(\frac{2m_r}{\hbar}\right)^{3/2}[f_c(\omega) - f_v(\omega)](\omega - E_g/\hbar)^{1/2} \tag{11.2-23}$$

$$= \alpha_0(\omega)[f_c(\omega) - f_v(\omega)] \tag{11.2-24}$$

where $\alpha_0(\omega)$ is the loss of the unpumped medium as given by (11.2-10). In the case of a fully inverted system, $f_c(\omega) = 1$, $f_v(\omega) = 0$, $\gamma(N, \omega)$ as given by (11.2-23) is equal to the loss $\alpha_0(\omega)$ of (11.2-10), as it should. The dependence of the gain γ on the carrier density N is contained in the factor $[f_c(\omega) - f_v(\omega)]$.

We note that Eq. (11.2-21) also contains an expression for $\chi'(\omega)$, the real part of the medium susceptibility. We will make use of it in other parts of this book.

It follows from (11.2-23) that a necessary condition for net gain, $\gamma > 0$, is that

$$f_c(\omega) - f_v(\omega) > 0 \tag{11.2-25}$$

If we use the definition (11.1-6), the inversion condition (11.2-24) becomes

$$E_{Fc} - E_{Fv} > \hbar\omega \tag{11.2-26}$$

At thermal equilibrium, $E_{Fc} = E_{Fv}$ and the medium is either transparent or absorbing at all ω.

A calculation of the gain $\gamma(\omega)$ requires, according to (11.2-23), a knowledge of the values of the quasi-Fermi energies E_{Fc} and E_{Fv} which are needed to specify $f_c(\omega)$ and $f_v(\omega)$. E_{Fc}, as an example, is determined from (11.6-6*a*), once the total density N of electrons in the conduction band is specified, from the condition

$$N = \int_0^{\infty} \rho_c(E)f_c(E, E_{Fc})\,dE$$

where $\rho_c(E)\,dE = V^{-1}\rho_c(k)\,dk$ (see 11.2-18) is the volume density of conduction band electrons with energies within dE.

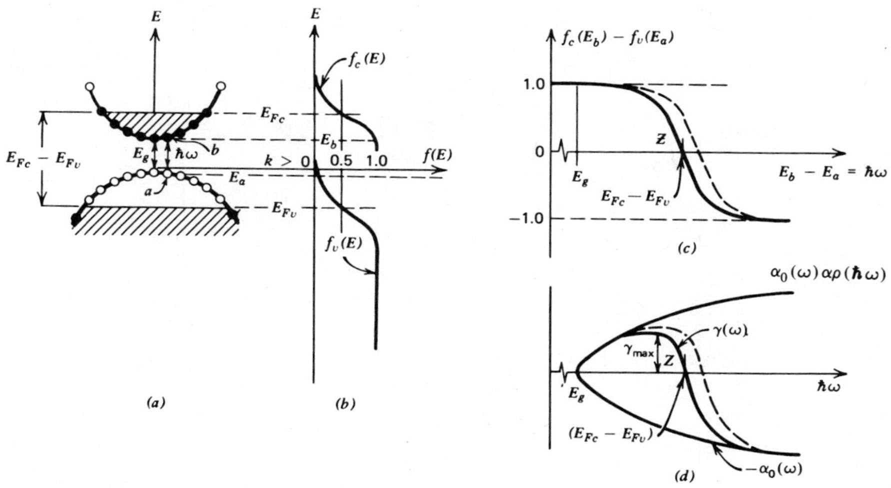

FIGURE 11.5 A graphical step-by-step construction of the gain profile $\gamma(\omega)$, according to (11.2-22), with $T_2 = \infty$. The dashed curves in (*c*) and (*d*) correspond to a higher excitation level. This causes the point *Z*, where $\hbar\omega = E_{Fc} - E_{Fv}$ to move to the right, resulting in an increase of γ_{max} and of the frequency at which it occurs.

A graphic illustration of a gain calculation based on (11.2-23) is shown in Figure 11.5.

Numerical calculation of the gain $\gamma(\omega)$ based on (11.2-22) is shown in Figure 11.6. A calculation of the peak gain vs. the carrier density is shown in Figure 11.7. The electron and hole densities are taken as equal, as dictated by charge neutrality under high injection conditions. We note that there exists a minimum density to achieve gain ($\gamma > 0$), $N_{tr} \sim 1.5 \times 10^{18}$ cm^{-3}. This occurs for radiation whose frequency ω satisfies (11.2-26) with an equal sign.

11.3 DIODE LASERS

The high carrier densities needed to obtain gain in semiconductors can be achieved with moderate currents densities near the junction region of a *p-n* diode. The most important class of semiconductor diode lasers are based on III-V semiconductors. One system is based on GaAs and $Ga_{1-x}Al_xAs$. The active region in this case is GaAs or $Ga_{1-x}Al_xAs$. The subscript *x* indicates the fraction of the Ga atoms in GaAs that are replaced by Al. The resulting lasers emit (depending on the active region molar fraction *x* and its doping) at 0.75 μm $< \lambda <$ 0.88 μm. This spectral region is convenient for the short-haul (<2 km) optical communication in silica fibers.

A second system has $Ga_{1-x}In_xAs_{1-y}P_y$ as its active region. The lasers emit in the 1.1 μm $< \lambda <$ 1.6 μm spectral region depending on *x* and *y*. The region near 1.55 μm is especially favorable, since optical fibers with losses as small as

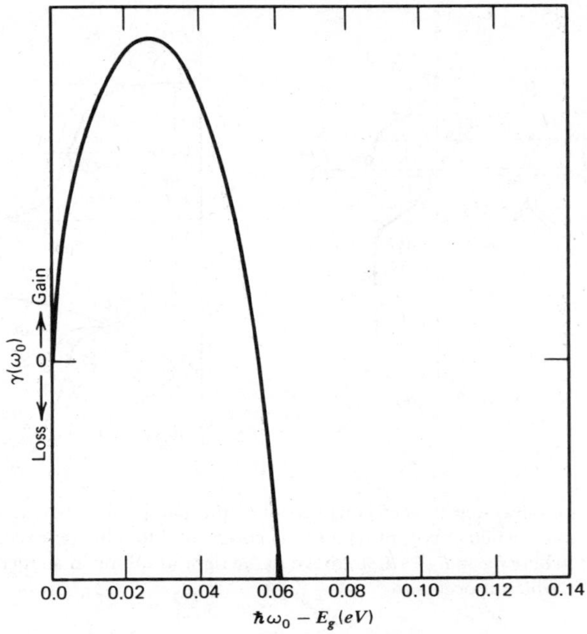

FIGURE 11.6*a* A typical plot of gain $\gamma(\omega_0)$ as a function of frequency for a fixed pumping level N. *Source:* Reference 11.

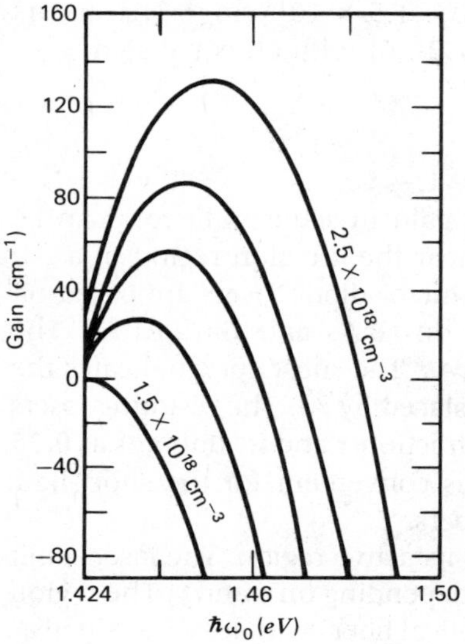

FIGURE 11.6*b* A plot based on (11.2-22) of the photon energy dependence of the optical gain (or loss = negative gain) of GaAs with the injected carrier density as a parameter. *Source:* Reference 11.

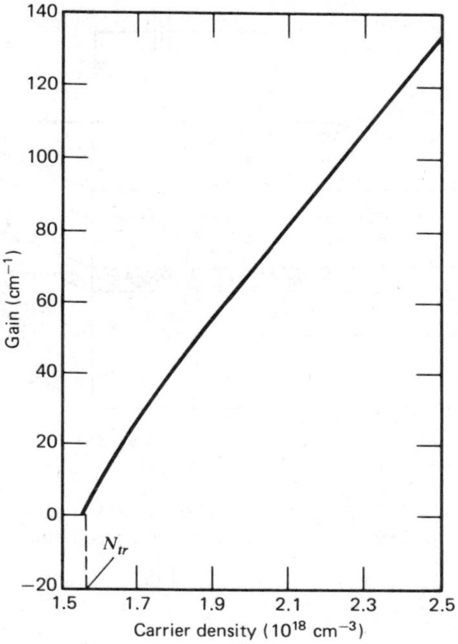

FIGURE 11.7 A plot of the peak gain γ_{max} of Figure 11.6b as a function of the injected carrier density at $T = 300°K$.

0.15 dB/km at this wavelength are available, making it extremely desirable for long-distance high data rate optical communication.

We will start with a description of GaAs/$Ga_{1-x}Al_xAs$ lasers. A generic laser of this type, depicted in Figure 11.8, has a thin (0.1–0.2 μm) region of GaAs that is sandwiched between two regions of $Ga_{1-x}Al_xAs$ of opposite doping, thus forming a double heterojunction. The introduction of this layered structure (References 12–14) is probably the most important single technical improvement on the original invention. The rationale for this structure is explained with the help of Figure 11.9 that shows in (a) the conduction and valence band edges in a typical heterojunction diode at full forward bias. A key element in this structure is the formation of a potential well for electrons of height ΔE_c that coincides spatially with a well for holes of height ΔE_v. The energy gap discontinuity is due to its dependence in $Ga_{1-x}Al_xAs$ on the Al molar fraction x. Under forward bias with $eV_a \sim E_g$, the large densities of injected electrons (from the n side) and holes (from the p side) in the well, cause the inversion condition (11.2-25) in that region to be satisfied so that radiation at ω satisfying (11.2-26) is amplified in the well. This GaAs inner layer where stimulated emission takes place is called the active region. For maximum gain, it is necessary to confine the light as tightly as possible to the active region since light traveling outside this region does not give rise to stimulated emission, thus not contributing to the gain. This confinement is

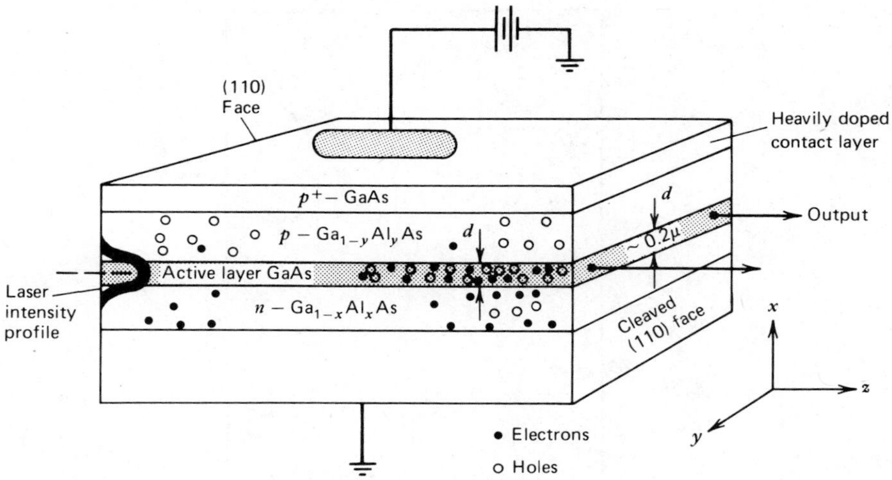

FIGURE 11.8 A typical double heterostructure GaAs-GaAlAs regions, respectively. Frequencies near $\nu = E_g/\hbar$ are amplified by stimulating electron-hole recombination.

brought about by a dielectric waveguiding effect due to the fact that a lowering of the energy gap of a semiconductor causes an increase in the index of refraction so that the $Ga_{1-x}Al_xAs/GaAs/Ga_{1-x}Al_xAs$ sandwich acts as a dielectric waveguide with the modal energy concentrated in the active region. The index distribution and the modal profile of a typical heterojunction laser is shown in Figure 11.9*b*. The index of refraction dependence on x is $\Delta n \approx -0.7x$.

The dependence of the band-gap energy of $Ga_{1-x}Al_xAs$ on the Al mole fraction x is shown in Figure 11.10. This dependence can be approximated for $x < .42$ by

$$E_g = (1.424 + 1.266x)eV$$

The total gap change $\Delta E_g \sim 1.266\, x(eV)$ is "divided" between the conduction and valence band according to

$$\Delta E_c \simeq 0.67 \Delta E_g$$

$$\Delta E_v \simeq 0.33 \Delta E_g$$

When the optical mode is effectively confined to the active region, the gain scales as d^{-1}, where d is the height of the active region. This is due to the fact that as the optical intensity, for a fixed total power, in the guided mode, goes as d^{-1}, so the stimulated emission rate hence the gain scale similarly. To quantify these arguments, we start with the basic definition of the net modal gain at an injected carrier density N in the active volume.[1]

[1] The small discontinuities of the index of refraction at the interfaces are ignored in (11.3-1).

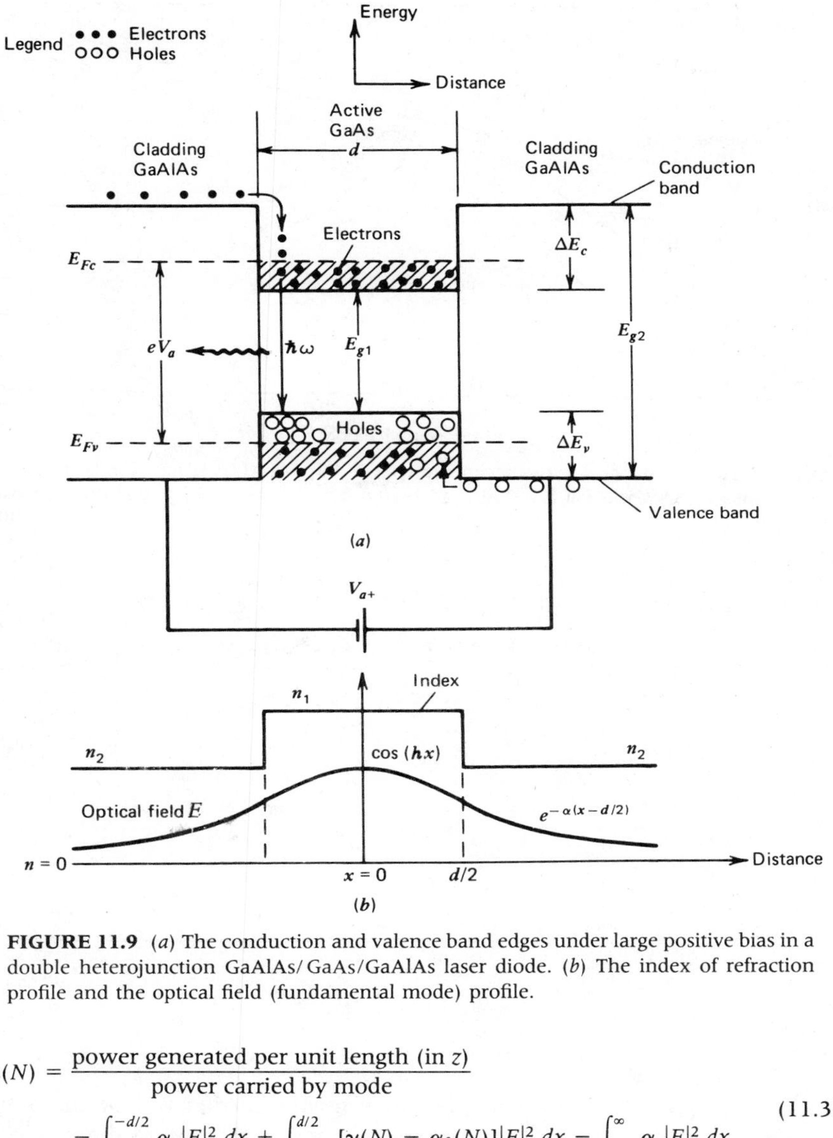

FIGURE 11.9 (*a*) The conduction and valence band edges under large positive bias in a double heterojunction GaAlAs/GaAs/GaAlAs laser diode. (*b*) The index of refraction profile and the optical field (fundamental mode) profile.

$$
\begin{aligned}
g_m(N) &= \frac{\text{power generated per unit length (in } z)}{\text{power carried by mode}} \\
&= \frac{-\int_{-\infty}^{-d/2} \alpha_n |E|^2\,dx + \int_{-d/2}^{d/2} [\gamma(N) - \alpha_{fc}(N)]|E|^2\,dx - \int_{d/2}^{\infty} \alpha_p |E|^2\,dx}{\int_{-\infty}^{\infty} |E|^2\,dx}
\end{aligned}
\tag{11.3-1}
$$

where γ is the bulk gain constant, that is, the gain experienced by a plane wave in a medium whose inversion density is equal to that of the active medium and is given by (11.2-23). α_n is the loss constant of the unpumped n-$Ga_{1-x}Al_xAs$ and is due mostly to free electron absorption $\alpha_{fc}(N)$, the (bulk) loss constant due to the density N of free carriers in the active region. α_p is the loss (by free holes) in the bounding p-$Ga_{1-y}Al_yAs$ region. We note that as $d \to \infty$, $g_m \to \gamma$.

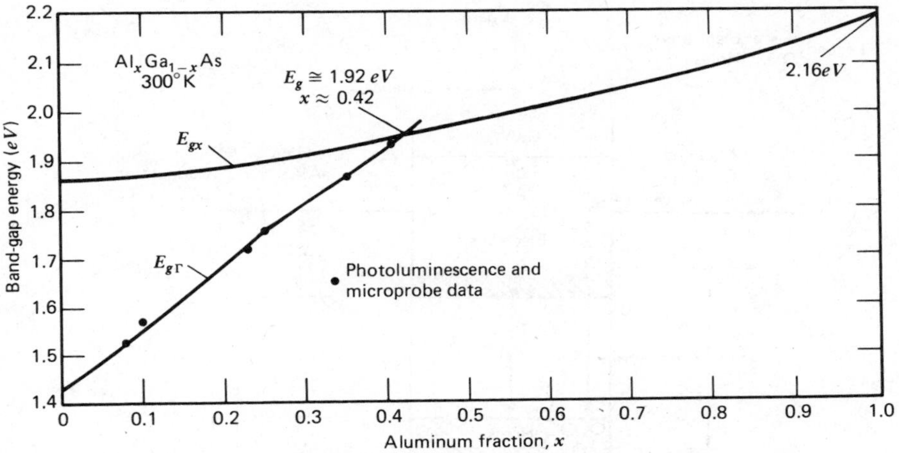

FIGURE 11.10 The magnitude of the energy gap in $Ga_{1-x}Al_xAs$ as a function of the molar fraction x. For $x > 0.37$, the gap is indirect. $\Delta E_c \approx 0.6\ \Delta E_g$, $\Delta E_v \simeq 0.4\ E_g$. *Source:* Reference 15.

It is convenient to rewrite (11.3-1) as

$$g_m = \gamma\Gamma_a - \alpha_{fc}\Gamma_a - \alpha_n\Gamma_n - \alpha_p\Gamma_p \tag{11.3-2}$$

$$\Gamma_a \equiv \frac{\int_{-d/2}^{d/2} |E|^2\, dx}{\int_{-\infty}^{\infty} |E|^2\, dx} \tag{11.3-3}$$

$$\Gamma_n \equiv \frac{\int_{-\infty}^{-d/2} |E|^2\, dx}{\int_{-\infty}^{\infty} |E|^2\, dx} \tag{11.3-4}$$

$$\Gamma_p \equiv \frac{\int_{d/2}^{\infty} |E|^2\, dx}{\int_{-\infty}^{\infty} |E|^2\, dx} \tag{11.3-5}$$

$$\Gamma_a + \Gamma_n + \Gamma_p = 1$$

Γ_a is very nearly the fraction of the mode power carried within the active GaAs layer, whereas Γ_n and Γ_p are, respectively, the fraction of the power in the n and p regions. As long as $\Gamma_a \sim 1$, that is, most of the mode energy is in the active region, the gain g is inversely proportional to the active region thickness d. As d decreases, an increasing fraction of the mode intensity is carried within the lossy n and p bounding regions as can be seen from the modal waveguide solution plotted in Figure 11.11. The resulting decrease of the confinement factor Γ_a eventually dominates over the d^{-1} dependence and the gain begins to decrease with the further decrease of d. A plot of the threshold current dependence on d is given in Figure 11.12.

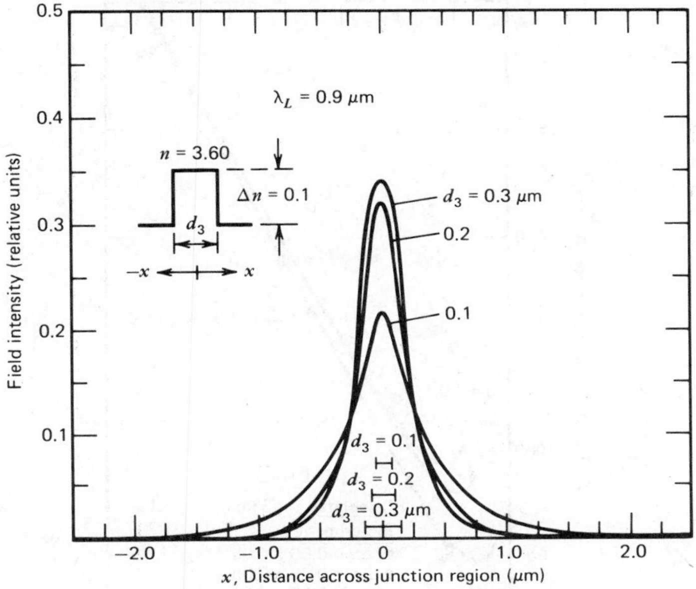

FIGURE 11.11 Calculated near-field intensity distribution of the step discontinuity waveguide for various values of the guiding layer thickness. *Source:* Reference 15.

Numerical Example: Threshold Current Density in Double Heterostructure Lasers. Consider the case of a GaAS/GaAlAs laser of the type illustrated in Figure 11.8. We will use the following parameters: $\tau \approx 4 \times 10^{-9}$ s, $L = 500\ \mu$m. The threshold condition, gain = loss, becomes according to (11.3-2) (we will denote threshold by a $_t$)

$$(\gamma_t - \alpha_{fc}(N_t))\Gamma_a = \alpha_n\Gamma_n + \alpha_p\Gamma_p + \frac{1}{L}\ln\left(\frac{1}{R}\right) + \alpha_s \qquad (11.3\text{-}6)$$

where the term α_s accounts for scattering losses (mostly at heterojunction interfacial imperfections). The largest loss term in lasers with uncoated faces is usually $L^{-1} \ln 1/R$. In our case, taking $R = 0.31$ as due to Fresnel reflectivity at a GaAs ($n = 3.5$) air interface, we obtain

$$-\frac{1}{L}\ln R = 23.4 \text{ cm}^{-1}$$

The rest of the loss terms are assumed to add up to ~ 10 cm^{-1}, so that taking $\Gamma_a \sim 1$ the total gain needed is $\gamma_t \sim 33.4$ cm^{-1}. This requires, according to Figure 11.7, an injected carrier density of $N \sim 1.7 \times 10^{18}$ cm^{-3}.[2] Under steady-state conditions, the rate at which carriers are injected into the active

[2] Here, we assumed for the sake of simplicity that at threshold $\alpha_{fc}(N_t) \ll \gamma$, which is true in most cases except for high reflectivity, $R \approx 1$, very low-loss lasers.

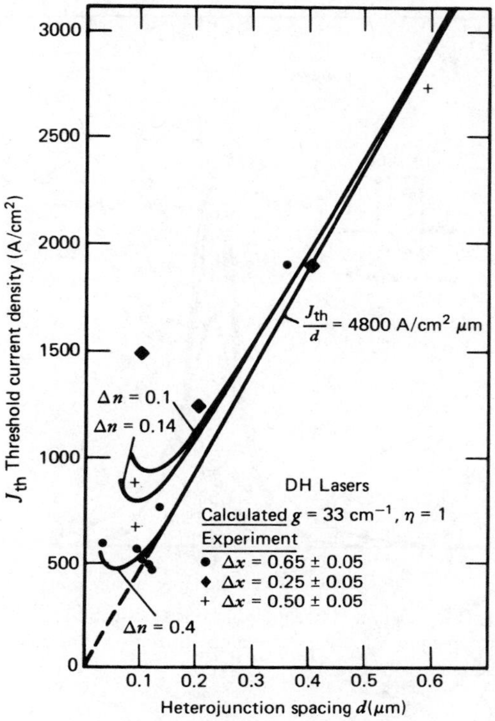

FIGURE 11.12 The dependence of the threshold current density of a GaAlAs/GaAs/GaAlAs on the active region (GaAs) thickness. *Source:* Reference 15.

region must equal the electron-hole recombination rate

$$\frac{J_t}{e} = \frac{Nd}{\tau}$$

Using the above data, we obtain

$$\frac{J_t}{d} = \frac{eN}{\tau} \sim 6.8 \times 10^3 \text{ A/(cm}^2 \times \mu\text{m)}$$

This value of J_t/d is in reasonable agreement with experimental values. If we use this value to estimate the lowest threshold current density that from Figure 11.12 occurs when $d \sim 0.08$ μm, we obtain

$$J_{\min} = 0.68 \times 10^4 \times 0.08 = 544 \text{ A/cm}^2$$

again, close to the range of observed values.

The feasibility of epitaxial growth of $Ga_{1-x}Al_xAs$ on top of GaAs (and vice versa), which is the main reason for the success of double heterostructure lasers, is due to the fact that their lattice constants are the same, to within a fraction of a percent, over the range $0 \leq x \leq 1$. This can be seen from the plot of Figure 11.13, which shows the lattice constant corresponding to various

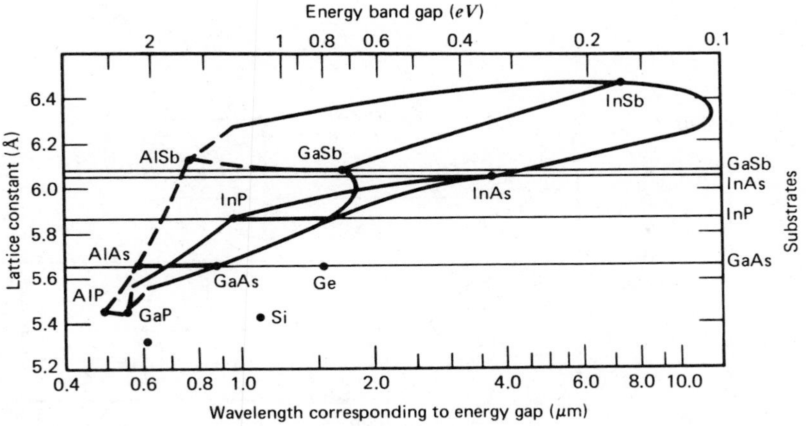

FIGURE 11.13 III-V compounds: Lattice constants versus energy band gaps and corresponding wavelengths. The solid lines correspond to direct gap materials and the dashed lines to indirect gap materials. The binary compounds on the right can be used as substrates to grow the lattice-matched ternary or quaternary compounds indicated by solid dark lines. *Source:* Reference 15.

compositions of III-V semiconductors as a function of the band-gap energy. We note that the line connecting the AlAs ($x = 1$) and the GaAs ($x = 0$) is nearly horizontal, which corresponds to a (very nearly) single value of the lattice constant over this compositional range.

11.4 GaInAsP LASERS

Figure 11.13 shows another possible lattice matched scheme involving GaInAsP and InP with the quaternary GaInAsP possessing an energy gap or a corresponding wavelength in the ~1.0–1.6 μm range. This region spans the choice transmission regions of silica fibers in which the minimum losses (~0.15 db/km) occur near 1.3 μm and the minimum group velocity dispersion near 1.55 μm. This has led to intense development of double heterostructure lasers in which the active region is $Ga_{1-x}In_xAs_{1-y}P_y$ and the cladding layers are *n* and *p* InP, as illustrated in Figure 11.14 (Reference 16).

11.5 SOME REAL LASERS

To further lower the threshold current, but not the current density, it is necessary to also confine the optical mode and the injected electron and holes in the transverse (y direction in Figure 11.8). One popular laser incorporating transverse confinement is the buried heterostructure laser that is illustrated in Figure 11.15. The three layers shown in the middle are grown by molecular beam epitaxy, liquid-phase epitaxy, or metal organic chemical vapor deposition atop the GaAs substrate. A thin rectangular mesa containing the three basic layers is then etched and a "burying" layer of the lower index

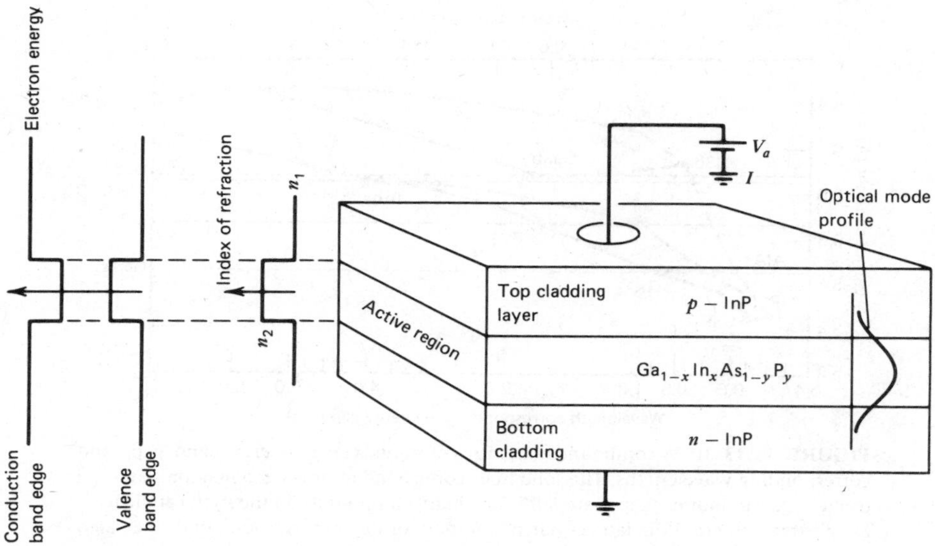

FIGURE 11.14 The basic layered structure of a quaternary GaInAsP/InP laser.

$Ga_{1-y}Al_yAs$ is grown usually by liquid-phase epitaxy on the sides of the mesa. The result is a rectangular 2×0.1 μm GaAs active region prism embedded in a lower index $Ga_{1-y}Al_yAs$ material. Other embedding techniques are also used.

The most important feature of the buried heterostructure laser is that the active GaAs region is surrounded on *all* sides by the lower index GaAlAs, so

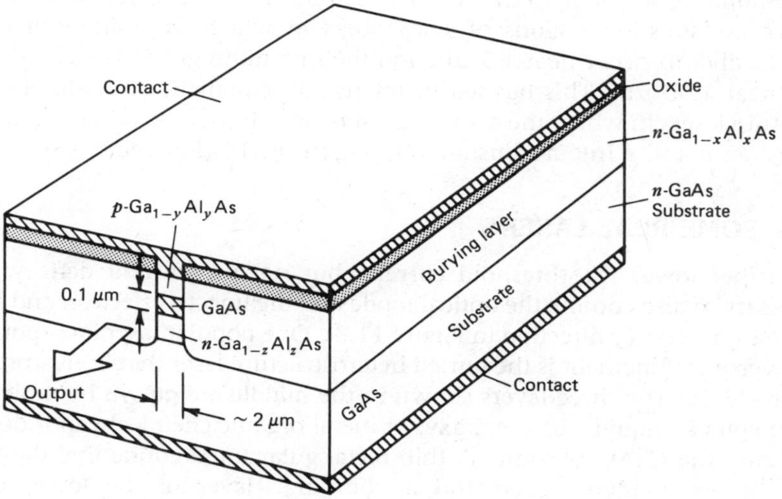

FIGURE 11.15 A buried heterostructure laser. *Source:* Reference 17.

that electromagnetically the structure is that of a rectangular dielectric waveguide. The transverse dimensions of the active region and the index discontinuities (i.e., the molar fractions x, y, and z) are chosen so that only the lowest-order transverse mode can propagate in the waveguide. Another important feature of this laser is the confinement of the injected carriers to the active region by the energy band discontinuity at a GaAs/GaAlAs interface as discussed in the last section. These act as potential barriers, inhibiting carrier escape out of the active region. GaAs semiconductor lasers utilizing this structure or similar buried waveguides have been fabricated with threshold currents of less than 1 milliampere; more typical lasers have thresholds of ~20 milliamperes.

Buried or embedded lasers are often fabricated by a single-step liquid epitaxial growth of the layers on a grooved substrate. Figure 11.17 shows a scanning electron micrograph of an $InP/Ga_{1-y}In_yAs_{1-x}P_xInP$ laser fabricated in this manner. This type of laser is quite popular and is known as a "buried crescent."

Power Output of Injection Lasers

The considerations of saturation and power output in an injection laser are basically the same as that of conventional lasers. As the injection current is increased above the threshold value, the laser oscillation intensity builds up. The resulting stimulated emission shortens the lifetime of the inverted carriers to the point where the magnitude of the inversion is clamped at its threshold value. Taking the probability that an injected carrier recombines radiatively within the active region as η_i, we can write the following expression for the power emitted by stimulated emission:

$$P_e = \frac{(I - I_t)\eta_i}{e} \hbar\omega \tag{11.5-1}$$

Part of this power is dissipated inside the laser resonator, and the rest is coupled out through the end reflectors. These two powers are, according to (11.3-6), proportional to the effective internal loss $\alpha \equiv \alpha_n\Gamma_n + \alpha_p\Gamma_p + \alpha_s + \alpha_{fc}(N_t)\Gamma_a$ and to $\frac{1}{L}\ln(1/R)$, respectively. We can thus write the output power as

$$P_0 = \frac{(I - I_t)\eta_i\hbar\omega}{e} \frac{(1/L)\ln(1/R)}{\alpha + (1/L)\ln(1/R)} \tag{11.5-2}$$

The external differential quantum efficiency η_{ex} is defined as the ratio of the increase in photon output rate that results from an increase in the injection rate (carriers per second) to the increase in the injection rate

$$\eta_{ex} = \frac{d(P_0/\hbar\omega)}{d(I/e)} \tag{11.5-3}$$

Using (11.5-2), we obtain

$$\eta_{ex}^{-1} = \eta_i^{-1}\left(\frac{\alpha L}{\ln(1/R)} + 1\right) \tag{11.5-4}$$

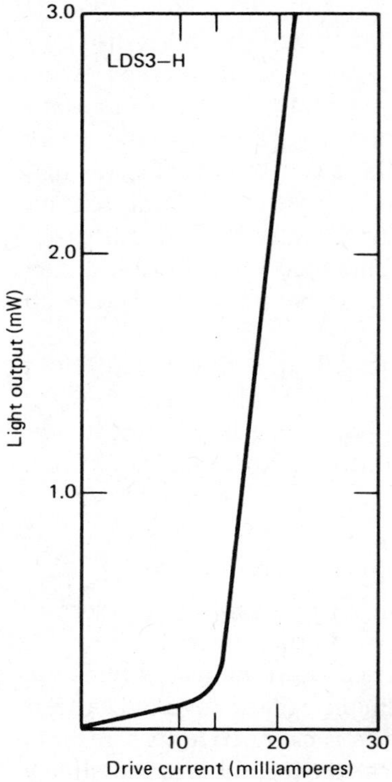

FIGURE 11.16*a* Power versus current plot of a low-threshold (~14 milliamperes) commercial DH GaAs/GaAlAs laser. *Source:* Courtesy of ORTEL Corp., Alhambra, Calif., Product Data Sheet.

External efficiencies ≃80% have been reported, indicating η_i close to unity. The semiconductor laser is thus unique in its near-perfect conversion of pump energy to radiation.

Since the incremental efficiency of converting electrons into useful output photons is η_{ex}, the main remaining loss mechanism degrading the conversion of electrical to optical power is the discrepancy between the energy eV_{appl} supplied to each injected carrier and the photon energy $\hbar\omega$. This discrepancy is due mostly to the series resistance of the laser diode, including the contact. The efficiency of the laser in converting electrical power input to optical power is thus

$$\eta = \frac{P_0}{VI} = \eta_i \frac{I - I_t}{I} \frac{h\nu}{eV_{\text{appl}}} \frac{\ln(1/R)}{\alpha L + \ln(1/R} \tag{11.5-5}$$

In practice, $eV_{\text{appl}} \sim 1.4E_g$ and $\hbar\omega \simeq E_g$. Values of $\eta \sim 70\%$ at 300°K have been achieved.

We conclude this section by showing in Figures 11.16*a* and 11.16*b* typical plots of the power output versus current and the farfield of commercial low-threshold GaAs semiconductor lasers, respectively. A scanning electron micrograph of the cross section of a GaInAsP/InP laser is shown in Figure

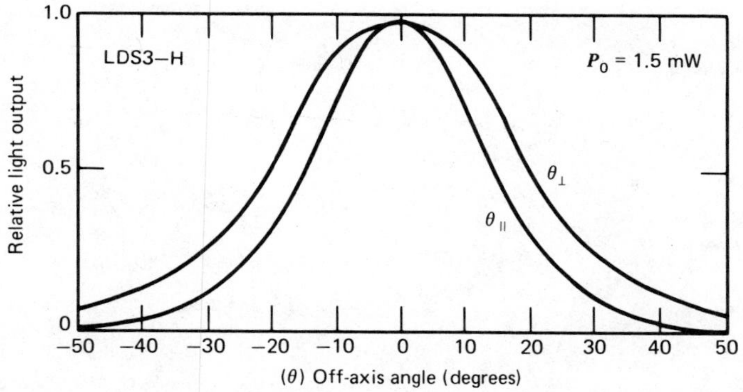

FIGURE 11.16*b* Farfield angular intensity distribution of a low-threshold commercial DH GaAs/GaAlAs laser. *Source:* Courtesy of ORTEL Corp., Alhambra, Calif., Product Data Sheet.

11.17. The narrow waveguiding GaInAsP channel is grown by liquid-phase epitaxy on the grooved InP substrate.[3]

11.6 DIRECT-CURRENT MODULATION OF SEMICONDUCTOR LASERS

Since the main application of semiconductor lasers is as a source for optical communication systems, the problem of the high-speed modulation of their output by the high data rate information is one of great technological importance.

A unique feature of semiconductor lasers is that, unlike other lasers that are modulated externally, the semiconductor laser can be modulated directly by modulating the excitation current. This is especially important in view of the possibility of the monolithic integration of the laser and the modulation electronic circuit [Supplementary Reference 6]. The following treatment matches closely that of Reference 18.

If we denote the photon density inside the active region of a semiconductor laser by P and the injected electron (and hole) density by N, we can write

$$\frac{dN}{dt} = \frac{I}{eV} - \frac{N}{\tau} - A(N - N_{tr})P$$
$$\frac{dP}{dt} = A(N - N_{tr})P\Gamma_a - \frac{P}{\tau_p} \qquad (11.6\text{-}1)$$

where I is the total current, V the volume of the active region, τ the spontaneous recombination lifetime, τ_p the photon lifetime as limited by absorption in the bounding media, scattering, and coupling through the output mirrors and free carrier absorption in the active region.

[3] Courtesy H. Blauvelt, ORTEL Corp., Alhambra, Calif.

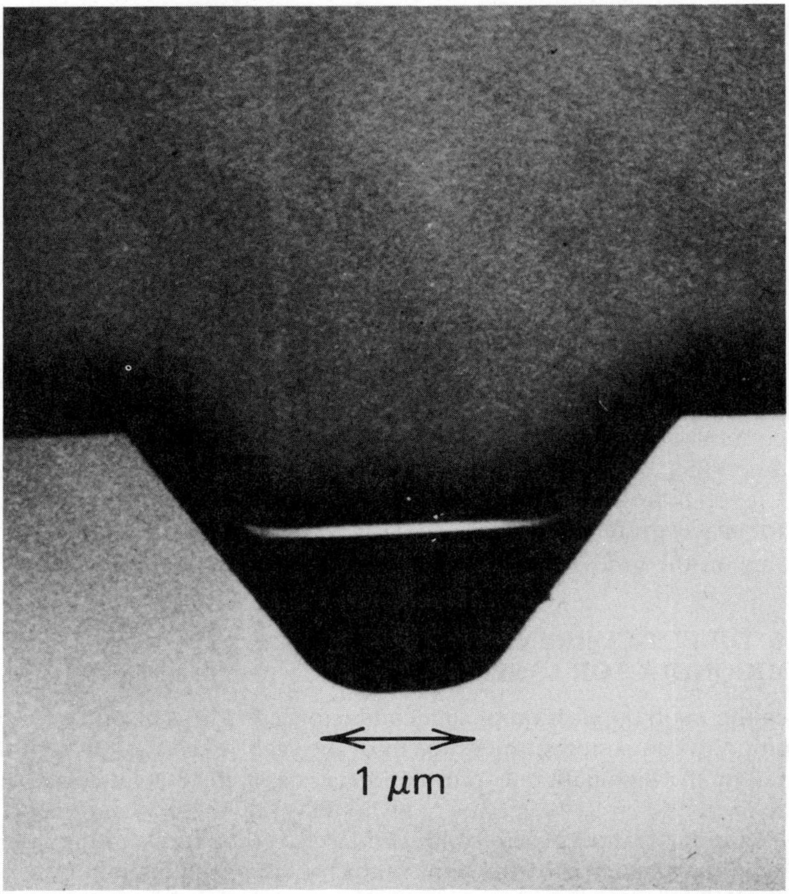

FIGURE 11.17 A buried crescent InP/$Ga_{1-y}In_yAs_{1-x}P_x$/InP laser grown by a single-step liquid-phase eitaxy on an InP grooved substrate. *Source:* Courtesy of T.R. Chen, California Institute of Technology, Pasadena, Calif.

The term $A(N - N_{tr})P$ gives the net rate per unit volume of induced transitions. N_{tr} is the inversion density needed to achieve transparency as illustrated in Figure 11.7. Since A is a temporal rate constant, it is related to the slope B of Figure 11.7 by $A = Bc/n$. Γ_a is the filling factor defined by (11.3-3), and its presence here is merely a matter of bookkeeping to ensure that the total number, rather than the density of electrons undergoing stimulated transitions, is equal to the number of photons emitted. The contribution of spontaneous emission to the photon density is neglected since only a very small fraction ($\lesssim 10^{-4}$) of the spontaneously emitted power enters the lasing mode.

By setting the left side of (11.6-1) equal to zero, we obtain the steady-state solutions N_0 and P_0

$$0 = \frac{I_0}{eV} - \frac{N_0}{\tau} - A(N_0 - N_{tr})P_0$$
$$A(N_0 - N_{tr}) = (\Gamma_a \tau_p)^{-1} \tag{11.6-2}$$

We consider the case where the current is made up of dc and ac components

$$I = I_0 + i_1 e^{i\Omega t} \tag{11.6-3}$$

and define the small-signal modulation n_1 and p_1 by

$$N = N_0 + n_1 e^{i\Omega t} \qquad P = P_0 + p_1 e^{i\Omega t} \tag{11.6-4}$$

where N_0 and P_0 are the solutions of (11.6-2).

Using (11.6-3,4) and the result $A(N_0 - N_{tr}) = (\tau_p \Gamma_a)^{-1}$ from (11.6-2) in (11.6-1) lead to the small-signal algebraic equations

$$-i\Omega n_1 = -\frac{i_1}{eV} + \left(\frac{1}{\tau} + AP_0\right) n_1 + \frac{1}{\tau_p \Gamma_a} p_1$$
$$i\Omega p_1 = AP_0 \Gamma_a n_1 \tag{11.6-5}$$

Our principal interest is in the modulation response $p_1(\Omega)/i_1(\Omega)$ so that from (11.6-5), we obtain

$$p_1(\Omega) = \frac{-(i_1/eV)AP_0\Gamma_a}{\Omega^2 - i\Omega/\tau - i\Omega AP_0 - AP_0/\tau_p} \tag{11.6-6}$$

The main feature of (11.6-6) is a flat (i.e., desirable) response at low frequencies, followed by a peak at

$$\Omega_R = \sqrt{\frac{AP_0}{\tau_p} - \frac{1}{2}\left(\frac{1}{\tau} + AP_0\right)^2} \tag{11.6-7}$$

followed by a steep decline. These features are observed in the experimental data as shown in Figure 11.18. In a typical semiconductor laser with $L = 300$ μm, we have $\tau_p \simeq n/(c[\alpha - (1/L) \ln R])^{-1} \sim 10^{-12}s$, $\tau \sim 4 \times 10^{-9}s$, and $AP_0 \sim 10^9 s^{-1}$ so that to a very good accuracy

$$\Omega_R = \sqrt{\frac{AP_0}{\tau_p}} \tag{11.6-8}$$

The predicted square root dependence in (11.6-8) of Ω_R on the photon density, hence on the power output, is illustrated in Figure 11.18. The last result is extremely useful, since it suggests that to increase ω_R and thus increase the useful linear region of the modulation response $p_1(\Omega)/i_1(\Omega)$, we need to increase the optical gain coefficient A, decrease the photon lifetime τ_p, and operate the laser at as high an internal photon density P_0 as possible. A detailed discussion of the optimum strategy for maximizing Ω_R is given in Reference 18.

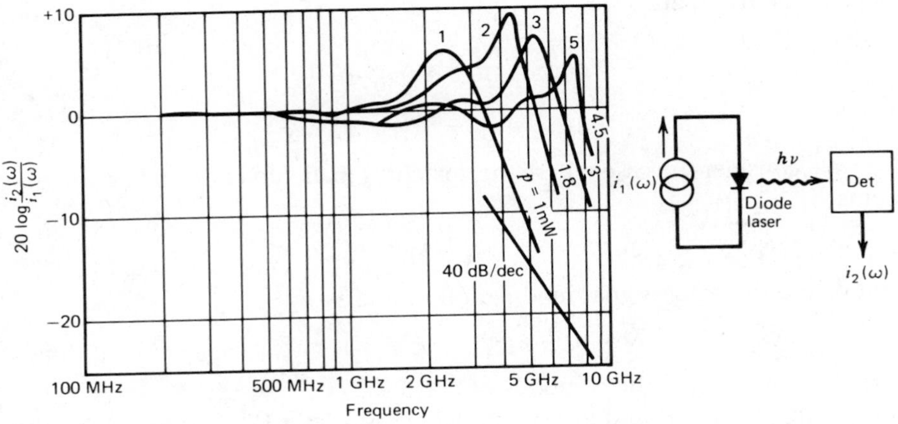

FIGURE 11.18 Modulated light output power versus current modulation frequency of a laser of length = 120 μm. The inset demonstrates the method of measurement. *Source:* Reference 18.

Numerical Example: Modulation Bandwidth in GaAs/GaAlAs Lasers. Here, we will estimate, using (11.6-8), the uppermost useful modulation frequency ω_R of a typical GaAs/GaAlAs laser. We shall assume a typical laser emitting 5×10^{-3} W from a single face with an active area cross section of 3 μm $\times$ 0.1 μm and a facet reflectivity of $R = 0.31$. Solving for P_0 from the relationship

$$\frac{(1 - R)P_0 c\hbar\omega}{n} = \frac{\text{power}}{\text{area}}$$

we obtain $P_0 = 1.21 \times 10^{15}$ photons/cm^3 for the photon density in the laser cavity. The constant A has a typical value of 2×10^{-6} cm^3/s. (This can be checked against the relationship $A = Bc/n$, where B is the spatial gain parameter and is equal to the slope of Figure 11.7.) The photon lifetime τ_p is obtained from

$$\tau_p = \frac{n}{c}\left(\alpha - \frac{1}{L}\ln R\right)^{-1}$$

which for $L = 120$ μm, $\alpha = 10$ cm^{-1}, and $R = 0.31$ yields $\tau_p \sim 1.08 \times 10^{-12}$s. Combining these results gives

$$\nu_R \equiv \frac{\omega_R}{2\pi} = \frac{1}{2\pi}\sqrt{\frac{AP_0}{\tau_p}} = \frac{1}{2\pi}\sqrt{\frac{2 \times 10^{-6} \times 1.2 \times 10^{15}}{1.08 \times 10^{-12}}}$$

$$= 7.53 \times 10^9 \text{ Hz}$$

This value is in the range of the experimental data shown in Figure 11.18, which was obtained on a laser with characteristics similar to that used in our example.

High-speed modulation of GaInAsP lasers has also been achieved (References 19, 20). This is particularly important for long-haul high data rate optical communication via fibers that utilize these lasers as the light source.

Frequency Chirping in Current Modulated Semiconductor Lasers

The current modulation of semiconductor laser causes an attendant and unavoidable modulation of the electron and hole density. This modulation is represented by the term $n_1(\Omega)$ in

$$N = N_0 + n_1(\Omega)e^{i\Omega t}$$

in (11.6-4). A carrier density modulation translates to a modulation of the Fermi levels E_{Fc} and E_{Fv} and hence, according to (11.2-16), to a modulation of $\chi(\omega) = \chi'(\omega) - i\chi''(\omega)$, the optical susceptibility of the gain medium. Since the resonant frequency of the laser oscillator depends on $\chi'(\omega)$ [see Eq. (9.1-17)], the current modulation leads directly to a frequency modulation of the laser. This modulation and the resulting spectral broadening can become a limiting factor in the data rate of long distance optical fiber communication. This issue is discussed in Problem 11.5 and in Supplementary References 6–8.

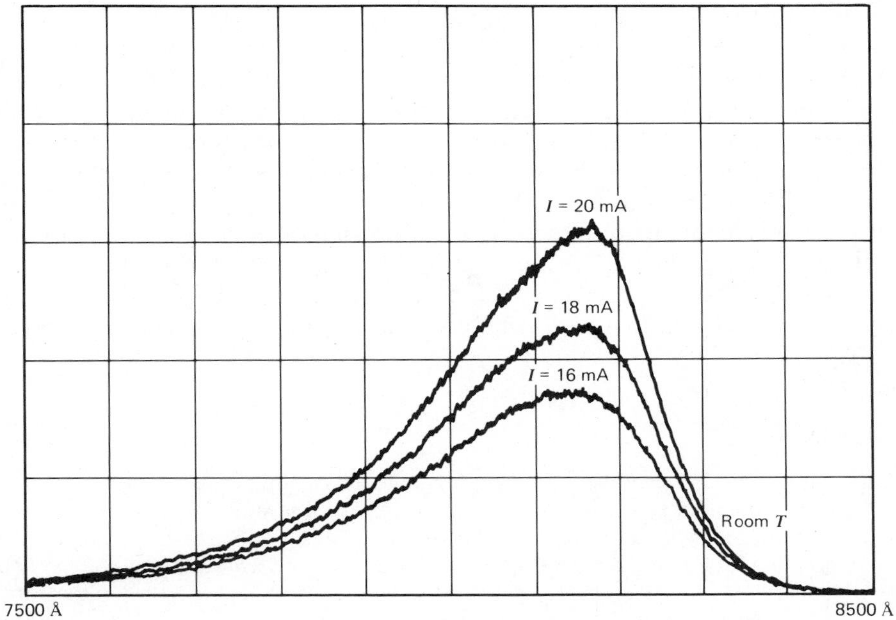

FIGURE 11.19*a* The spontaneous recombination spectrum of a GaAs/GaAlAs laser below threshold.

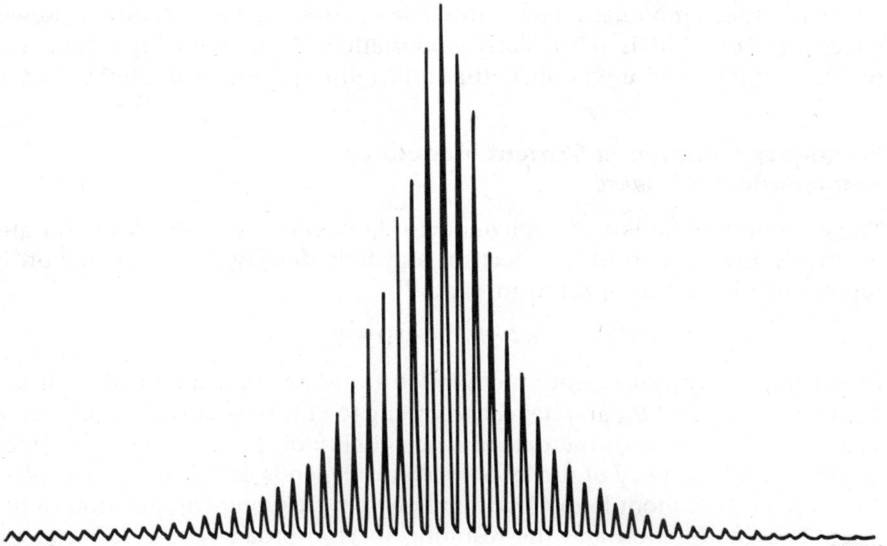

FIGURE 11.19*b* The oscillating mode spectrum of the laser above threshold.

The Recombination (Spontaneous) Emission Spectrum of Semiconductor Lasers

The electrons in semiconductor lasers obey Fermi–Dirac statistics so that no more than one electron can occupy a given eigenstate. It follows that as the injection level (pumping) increases, additional states are occupied. This leads to very wide recombination radiation spectra and, consequently, to the possibility of multimode oscillation. This is illustrated by Figure 11.19 that shows in (*a*) the recombination spectrum below threshold and in (*b*) the laser spectrum that is made up of numerous longitudinal modes. In lasers designed for optical communication, it is necessary to limit the number of modes, often to one mode. This is achieved by incorporating a grating into the laser in what have come to be known as distributed feedback lasers. These are described in Chapter 20.

References

1. Hall, R. N., G. E. Fenner, J. D. Kingsley, T. J. Soltys, and R. O. Carlson, "Coherent Light Emission from GaAs Junctions," *Phys. Rev. Letters* **9,** 366 (1962).
2. Nathan, M. I., W. P. Dumke, G. Burns, F. H. Dills, and G. Lasher, "Stimulated Emission of Radiation from GaAs *p-n* Junctions," *Appl. Phys. Letters* **1,** 62 (1962).
3. Quist, T. M., R. J. Keyes, W. E. Krag, B. Lax, A. L. McWhorter, R. H. Rediker, and H. J. Zeiger, "Semiconductor Maser of GaAs," *Appl. Phys. Letters* **1,** 91 (1962).
4. Basov, N. G., B. M. Vul, and Y. M. Popov, "Quantum Mechanical Semiconductor Generators of Electromagnetic Oscillations," *Zh. Eksp. Theo. Fiz.* **37,** 587 (1959), *Soviet Phys. JETP* **10,** 416 (1959).

5. Dumke, W. P., "Interband Transitions and Maser Action," *Phys. Rev.* **127,** 1559 (1962).
6. Lau, K. Y. and A. Yariv, "Ultra-High-Speed Semiconductor Lasers," *IEEE J. of Quant. Elect.* **QE-21,** 125 (1985).
7. Yariv, A., "The Beginning of Integrated Optoelectronic Circuits," *IEEE Trans. Elect. Devices* **ED-31,** 1656 (1984).
8. See, for example, C. Kittel, *Introduction to Solid State Physics,* 3rd ed. (New York: Wiley, 1967).
9. Smith, R. A., *Wave Mechanics of Crystalline Solids* (London: Chapman and Hall, 1961).
10. Thomson, G. H. B., *Physics of Semiconductor Lasers* (New York: Wiley, 1981).
11. Vahala, K., Calif. Inst. of Technology, Pasadena, Calif., private communication.
12. Alferov, Z. I., V. M. Andreev, D. Z. Garbuzov, Y. V. Zhilayev, E. L. Morozov, E. L. Portnoi, and V. G. Triofim, *Sov. Phys.—Semicond.* **4,** 1573 (1971).
13. Hayashi, J., M. B. Panish, and P. W. Foy, "A Low-Threshold Room-Temperature Injection Laser," *IEEE J. Quant. Elect.* **QE-5,** 211 (1969).
14. Kressel, H. and H. Nelson, "Close Confinement Gallium Arsenide *p-n* Junction Laser with Reduced Optical Loss at Room Temperature," *RCA Rev.* **30,** 106 (1969).
15. Kressel, H. and J. K. Butler, *Semiconductor Lasers and Heterojunction on LED's* (New York: Academic Press, 1977).
16. Hsieh, J. J., J. A. Rossi, and J. P. Donnelly, "Room Temperature CW Operation of GaInAsP/InP Double Heterostructure Diode Lasers Emitting at 1.1 μm," *Appl. Phys. Letters* **28,** 709 (1976).
17. Tsukada, T., "GaAs-$Ga_{1-x}Al_xAs$ Buried Heterostructure Injection Lasers," *J. Appl. Phys.* **45,** 4899 (1974).
18. Lau, K. T., N. Bar-Chaim, I. Ury, and A. Yariv, "Direct Amplitude Modulation of Semiconductor GaAs Lasers up to X-Band Frequencies," *Appl. Phys. Letters* **43,** 11 (1983).
19. Bowers, E. J., B. R. Hemenway, A. H. Gamuck, and D. P. Wilt, "High Speed InGaAsP Constricted Mesa Lasers," *IEEE J. Quant. Elect.* **QE-22,** 833 (1986).
20. Su, C. B. and V. A. Lanzisera, "Ultra High Speed Modulation of 1.3 μm InGaAsP Diode Lasers," *IEEE J. Quant. Elect.* **QE-22,** 1568 (1968).

Supplementary References

1. Kressel, H. and J. K. Butler, *Semiconductor Lasers and Heterojunction LED* (New York: Academic Press, 1977).
2. "Semiconductor Lasers," W. T. Tsang, ed. in Vol. 22 of *Semiconductors and Metals, Lightwave Communication Technology* (New York: Academic Press, 1985).
3. Casey, H. C. and M. B. Banish, *Heterostructure Lasers* (New York: Academic Press, 1978).
4. Agrawal, G. P. and N. K. Dutta, *Long Wavelength Semiconductor Lasers* (New York: Van Nostrand Reinhold Company, 1986).
5. Special Issue on Semiconductor Lasers, P. D. Dapkus, ed., *J. Quant. Elect.* **QE-23** (June 1987).
6. Harder, C., K. Vahala, and A. Yariv, "Measurement of the Linewidth Enhancement Factor α of Semiconductor Lasers," *Appl. Phys. Letters* **42,** 328 (1983).
7. Koch, T. L. and R. A. Linke, "Effect of Nonlinear Gain Reduction on Semiconductor Laser Wavelength Chirping," *Appl. Phys. Letters* **48,** 613 (1986).

8. Yamamoto, Y., T. Mukai, and S. Saito, "Quantum Phase Noise and Linewidth of a Semiconductor Laser," *Elect. Letters* **17,** 327 (1981): also, Okoshi, T., K. Kikuchi, and A. Nakayama, "Novel Method for High Resolution Measurement of Laser Output Spectrum," *Elect. Letters* **6,** 630 (1980).

Problems

11.1 Show that in a (periodic) crystal the interaction Hamiltonian as introduced in chapter 11.2

$$H' = -\frac{e}{m}\mathbf{A}\cdot\mathbf{p}$$

can also be expressed as

$$H' = -e\,\mathbf{E}\cdot\mathbf{r}$$

where $\mathbf{E} = -(\partial\mathbf{A})/(\partial t)$ is the electric field.

11.2 Evaluate and plot:
(a) The gain $\gamma(\omega)$ of an inverted GaAs crystal under the following conditions:

$$N_{\text{elec}} = N_{\text{hole}} = 3 \times 10^{18}\ \text{cm}^{-3}$$

$$m_c = 0.07\, m_{\text{electron}}$$

$$m_h = 0.4\, m_{\text{electron}}$$

$$T = 0\ \text{K}$$

$$E_g = 1.45\ eV$$

$$T_2 = \infty$$

(b) Comment qualitatively on the changes in $\gamma(\omega)$ as the temperature is raised.
(c) What is the effect of a finite T_2 on $\gamma(\omega)$?

11.3 Plot the modulation response function $p_1(\omega)/i_1(\omega)$ of (11.6-6) for power outputs of 1, 5, 10, and 30 mW using the numerical data and diode parameters given in Section 11.6.

11.4 Consider the effect on the modulation response $p_1(\omega)/i_1(\omega)$ of the inclusion of a nonlinear gain term bP in the rate equation (11.6-1)

$$\frac{dN}{dt} = \frac{I}{eV} - \frac{N}{\tau} - A(1 - bP)(N - N_{tr})P$$

$$\frac{dP}{dt} = A(1 - bP)(N - N_{tr})P\Gamma_a - \frac{P}{\tau_p}$$

where $bP \ll 1$. Show that the main effect is a damping of the resonance peak at Ω_R.

11.5 Solve Eqs. (11.6-1):
(a) For the amplitude $n_1(\Omega)$ of the density fluctuation as defined by (11.6-4).

Answer:

$$n_1(\Omega) = -i\frac{i_1}{eV}\frac{\Omega}{\Omega^2 - \dfrac{AP_0}{\tau_p} - i\Omega\left(\dfrac{1}{\tau} + AP_0\right)}$$

(b) Show that if we define the complex index of refraction of the active (gain) layer as

$$n = n' + in''$$

then the gain coefficient can be written as

$$A = \frac{4\pi\nu}{n'}\frac{\partial n''}{\partial N}$$

where N is the carrier density

(c) Show that the result of carrier density modulation with an amplitude $n_1(\Omega)$ is to cause a frequency modulation with a maximum excursion

$$|\Delta\nu|_{\Omega \ll \Omega_R} = \frac{i_1(\Omega)}{eV}\left(\frac{\Gamma_a A\alpha}{4\pi}\right)\frac{\Omega}{\Omega_R^2}$$

$$= \left(\frac{i_1(\Omega)}{eV}\right)\left(\frac{\alpha}{4\pi}\right)\frac{\Omega\tau_p}{P_0}$$

$$\alpha = \frac{\partial n'/\partial N}{\partial n''/\partial N} \qquad \Gamma_a = \text{confinement factor as defined by (11.3-3)}$$

[This is the same α parameter as that defined by (21.6-1,2) and illustrated by Figure 21.8.]

In obtaining $|\Delta\nu|$, assume that the oscillation frequency adjusts instantaneously to the passive resonant frequency so that $\Delta\nu/\nu = -(\Delta n'/n')\Gamma_a$.

11.6 In the threshold condition (11.3-6), assume that near threshold

$$\gamma \cong B(N - N_{tr})$$

$$\alpha_{fc} = aN$$

Estimate B from Figure 11.7 and take $a = 2.5 \times 10^{-18}$ cm^2.

(a) What is the correction to the threshold inversion due to the inclusion of the free carrier loss α_{fc} in the losses?

(b) What is the correction to the differential quantum efficiency due to the inclusion of α_{fc} in the losses?

11.7 Assume that we can make a semiconductor laser with α_s, α_n, and α_p, but not α_{fc}, all zero, and that we can control the facet reflectivity R and the laser length L at will. What strategy will you pursue if your aim is to reduce the *total* threshold current to a minimum?

CHAPTER 12

Quantum Well Lasers

12.0 INTRODUCTION

A thin region of GaAs bounded on either side by $Ga_{1-x}Al_xAs$, as in Figure 11.9, acts as a trap for electrons and holes. If the thickness of the active GaAs region is reduced to the range of, say, $d < 200$ Å, the confined electrons and holes display quantum effects. Their energies and wavefunctions are determined mostly by the confinement distance d and the resulting lasers are called quantum well (QW) lasers.

The major benefit brought about by this significant (factor >10) thinning of the active region is that the total number of carriers needed to achieve transparency density in the active region of the QW laser is reduced relative to that in conventional lasers roughly by the ratio of the active region thicknesses. The (modal) free carrier absorption coefficient which is proportional to the total number of carriers per unit area is also reduced by the same factor. The result is a greatly reduced (factor >10) threshold current.

There are other important consequences of the quantum confinement of the carriers in a QW laser so that a reexamination of the relevant physics is warranted.

12.1 THE QUANTUM MECHANICS

Consider the GaAlAs/GaAs/GaAlAs "sandwich" depicted in Figure 12.1. The conduction band edge is lower by $\Delta E_c \approx 0.67\ \Delta E_g$ in the GaAs inner region compared to the two sides where

$$\Delta E_g \equiv (E_g)_{\mathrm{Ga}_{1-x}\mathrm{Al}_x\mathrm{As}} - (E_g)_{\mathrm{GaAs}} = x \cdot 1.27 \text{ eV}$$

There exists, consequently, a discontinuity in the valence band edge, which is

$$\Delta E_v \approx 0.33\ \Delta E_g$$

The depressed potential region, the quantum well, acts as a trap for electrons and holes.

We will consider the electrons (and holes) as free carriers with effective masses equal to those in the bulk material. We will limit the discussion to electrons since the results apply directly to holes. The Hamiltonian is taken as

$$H_c(\mathbf{r}) = V_c(z) - \frac{\hbar^2}{2m_c}\frac{\partial^2}{\partial z^2} - \frac{\hbar^2}{2m_c}\frac{\partial^2}{\partial r_\perp^2} = V_c(z) - \frac{\hbar^2}{2m_c}\frac{\partial^2}{\partial z^2} + H(\mathbf{r}_\perp) \qquad (12.1\text{-}1)$$

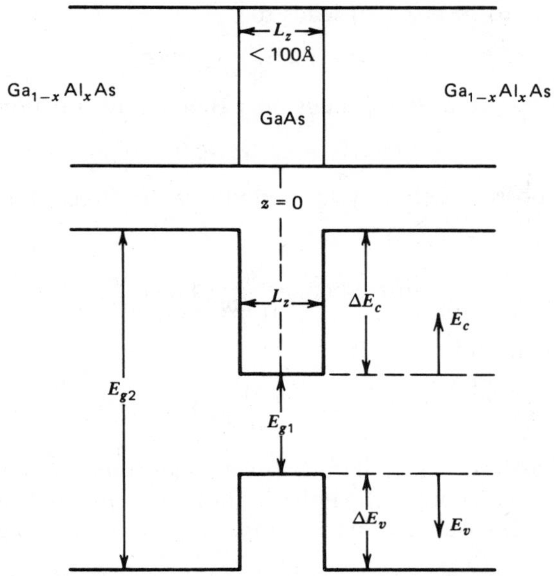

FIGURE 12.1 The layered structure and the band edges of a GaAlAs/GaAs/GaAlAs quantum well laser.

where the subscript c denotes the conduction band (i.e., electrons), z is the direction normal to the interfaces and $\mathbf{r}_\perp$ is the position vector in the plane of the interfaces so that $\mathbf{r} = \hat{\mathbf{z}}z + \mathbf{r}_\perp$. $V_c(z)$ is the conduction band edge, that is, the potential energy curve, as shown in Figure 12.1. We need to solve the Schrödinger equation

$$H_c(\mathbf{r})u(\mathbf{r}) = Eu(\mathbf{r}) \tag{12.1-2}$$

We will separate $u(\mathbf{r})$ according to

$$u(\mathbf{r}) = \Psi_{\mathbf{k}}(\mathbf{r}_\perp)v(z) \tag{12.1-3}$$

where

$$\left[V_c(z) - \frac{\hbar^2}{2m_c}\frac{\partial^2}{\partial z^2}\right] v(z) = E_z v(z) \tag{12.1-4}$$

For sufficiently large ΔE_c, we can to first approximation take the height of $V_c(z)$ in the well region as infinite. In this case,

$$v(z)\begin{cases} \cos \ell \dfrac{\pi}{L_z} z & \ell = 1, 3, 5, \ldots \\ \\ \sin \ell \dfrac{\pi}{L_z} z & \ell = 2, 4, 6, \ldots \end{cases} \tag{12.1-5}$$

$$E_z = \ell^2 \frac{\hbar^2\pi^2}{2m_c L_z^2} = \ell^2 E_{1c} \equiv E_{\ell c} \qquad \ell = 1, 2, 3, \ldots \tag{12.1-6}$$

Using (12.1-3, 4, 6) in (12.1-2) leads to

$$H(\mathbf{r}_\perp)\Psi(\mathbf{r}_\perp) = (E - E_z)\Psi(\mathbf{r}_\perp) \tag{12.1-7}$$

We can take $\Psi(\mathbf{r}_\perp)$ as a two-dimensional Bloch wavefunction

$$\Psi(\mathbf{r}_\perp) = u_{\mathbf{k}_\perp}(\mathbf{r}_\perp)e^{i\mathbf{k}_\perp \mathbf{r}_\perp} \tag{12.1-8}$$

where $u_{\mathbf{k}_\perp}(\mathbf{r}_\perp)$ possesses the crystal periodicity. In direct analogy with bulk semiconductors

$$H(\mathbf{r}_\perp)\Psi(\mathbf{r}_\perp) = \frac{\hbar^2 k_\perp^2}{2m_c}\Psi(\mathbf{r}_\perp) \tag{12.1-9}$$

and from (12.1-6) and (12.1-7),

$$E_c(\mathbf{k}_\perp, \ell) = \frac{\hbar^2 k_\perp^2}{2m_c} + \ell^2 \frac{\hbar^2 \pi^2}{2m_c L_z^2} = \frac{\hbar^2 k_\perp^2}{2m_c} + E_{\ell c} \qquad \ell = 1, 2, 3, \ldots \tag{12.1-10a}$$

$u_{\mathbf{k}_\perp}(\mathbf{r}_\perp)$ is periodic in the lattice (two-dimensional) periodicity. Similar results with $m_c \rightarrow m_v$ apply to the holes in the valence band. We recall that the hole energy E_v is measured downward in our electronic energy diagrams so that

$$E_v(\mathbf{k}_\perp, l) = \frac{\hbar^2 k_\perp^2}{2m_v} + l^2 \frac{\hbar^2 \pi^2}{2m_v L_z^2} = \frac{\hbar^2 k_\perp^2}{2m_v} + E_{lv} \qquad l = 1, 2, 3, \ldots \tag{12.1-10b}$$

The complete wave functions are then

$$\psi_c(\mathbf{r}) = \sqrt{\frac{2}{L_z}}\,\Psi_{\mathbf{k}_{\perp c}}(\mathbf{r}_\perp) CS\left(\ell \frac{\pi}{L_z} z\right) \tag{12.1-11}$$

for electrons and

$$\psi_v(\mathbf{r}) = \sqrt{\frac{2}{L_z}}\,\Psi_{\mathbf{k}_{\perp v}}(\mathbf{r}_\perp) CS\left(\ell \frac{\pi}{L_z} z\right) \tag{12.1-11}$$

for holes. We defined $CS(x) \equiv \cos(x)$ or $\sin(x)$ following (12.1-5). The normalization $\langle \Psi_c(\mathbf{r}) | \Psi_c(\mathbf{r}) \rangle = 1$ follows, provided we choose $\langle \Psi \mathbf{k}_{\perp c} | \Psi \mathbf{k}_{\perp c} \rangle = 1$.

Selection Rules

Assume that an optical field with polarization in the plain of the active region is propagating through the quantum well. In this case, the transition rate between an occupied state in the conduction band and an empty (hole) state in the valence band is proportional, according to (11.2-2), to the squared magnitude of

$$H'_{vc} \propto \int \Psi^*_{\mathbf{k}_{\perp v}}(\mathbf{r}_\perp) CS\left(\ell \frac{\pi}{L_z} z\right)\left(-i\hbar \frac{\partial}{\partial x}\right) \Psi'_{\mathbf{k}_{\perp c}}(\mathbf{r}_\perp) CS\left(\ell' \frac{\pi}{L_z} z\right) dV \tag{12.1-13}$$

The transverse part integral is similar to that of the three-dimensional case considered in Section 11.2 so that, as in (11.2-5), it leads to

$$\mathbf{k}_\perp = \mathbf{k}'_\perp \tag{12.1-14}$$

whereas the z integration yields in this approximation simply

$$\ell = \ell' \tag{12.1-15}$$

Transitions thus take place between electron and hole states with the same transverse momentum and the same quantized well state.

The Density of States

The consideration applying here are similar to those of Section 11.1. Since the electron is "free" in the x and y directions, we apply two-dimensional quantization by assuming the electrons are confined to a rectangle $L_x L_y$. This leads, as in Section 11.2, to

$$k_x = h\frac{\pi}{L_x}, \qquad h = 1, 2, \ldots, \qquad k_y = m\frac{\pi}{L_y}, \qquad m = 1, 2, \ldots$$

The area in $\mathbf{k}_\perp$ space per one eigenstate is thus $A_k = \pi^2/L_x L_y \equiv \pi^2/A_\perp$. We will drop the sub $\perp$ notation from now on so that $k \equiv k_\perp$. The number of states with transverse momenta less than $\hbar k$ is obtained by dividing the area $\pi k^2/4$ by A_k (the factor $\frac{1}{4}$ is due to the fact that $\mathbf{k}_\perp$ and $-\mathbf{k}_\perp$ describe the same state). The result is

$$N(k) = \frac{k^2 A_\perp}{2\pi}$$

where a factor of two for the two spin orientations of each electron was included.

The number of states between k and $k + dk$ is

$$\rho(k)\,dk = \frac{dN(k)}{dk}\,dk = A_\perp \frac{k}{\pi}\,dk$$

so that those with total energies between E and $E + dE$ number

$$\frac{dN(E)}{dE}\,dE = \frac{dN(k)}{dk}\frac{dk}{dE}\,dE \tag{12.1-16}$$

The number of states per unit energy per unit area is thus

$$\frac{1}{A_\perp}\frac{dN(E)}{dE} = \frac{k}{\pi}\frac{dk}{dE}$$

from (12.1-10a) with $\mathbf{k}_\perp \rightarrow \mathbf{k}$, $\ell = 1$

$$k = \sqrt{\frac{2m_c}{\hbar^2}}\,(E - E_{1c})^{1/2}$$

so that the two-dimensional density of states (per unit energy and unit area) is

$$\rho_{QW}(E) \equiv \frac{1}{A_\perp}\frac{dN(E)}{dE} = \frac{m_c}{\pi\hbar^2} \tag{12.1-17}$$

In the reasoning leading to (12.1-17), we considered only one transverse quantum state $\ell = 1$. But once $E > E_{2c}$, as an example, an electron of a given total energy E can be found in either $\ell = 1$ or $\ell = 2$ state so that the density of states doubles. The total density of states thus increases by $m_c/\pi\hbar^2$ at each of the energies $E_{\ell c} = \ell^2 E_{1c}$ of (12.1-6). We can express it mathematically as

$$\rho_{QW}(E) = \sum_{n=1}^{\infty} \frac{m_c}{\pi\hbar^2} H(E - E_{nc}) \qquad (12.1\text{-}18)$$

where $H(x)$, the Heaviside function which is equal to unity when $x > 0$ and is zero when $x < 0$.

The first two steps of the staircase density of states are shown in Figure 12.2. In the figure, we plotted the volumetric density of states of the quantum well medium ρ_{QW}/L_z so that we can compare it to the bulk density of states in a conventional semiconductor medium. It is a straightforward exercise to show that in this case, the QW volumetric density of states equals the bulk value $\rho_{3D}(E)$ at each of the steps as shown in the figure.

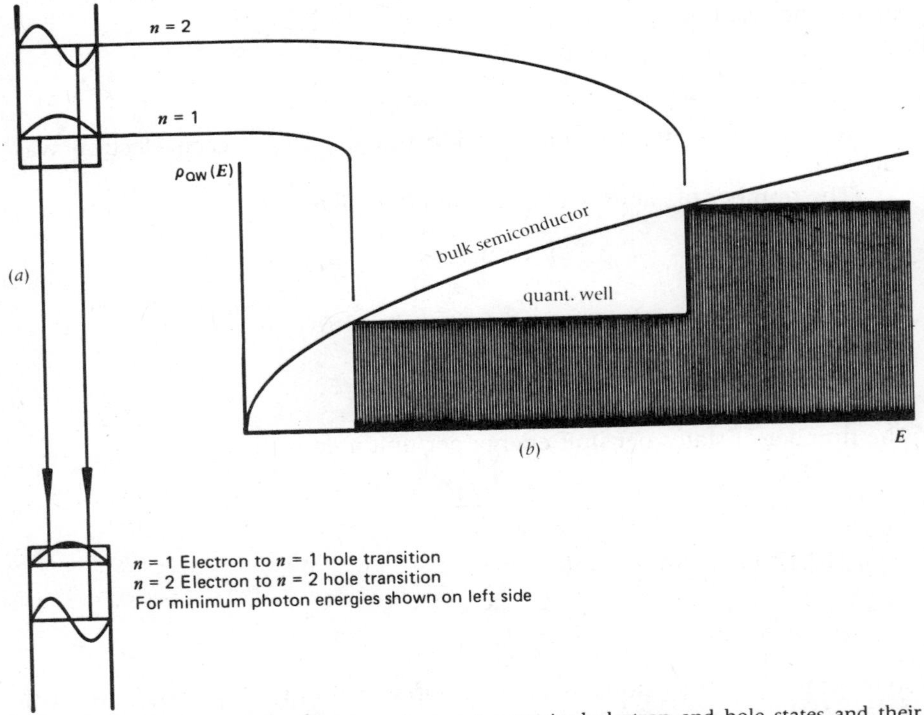

FIGURE 12.2 (*a*) The first two $n = 1$, $n = 2$ quantized electron and hole states and their eigenfunctions in an infinite potential well. (*b*) A plot of the volumetric density of states $(1/AL_z)/[dN(E)/dE]$ (i.e., the number of states per unit area per unit energy divided by the thickness of the action region) of a quantum well and of a bulk semiconductor. *Source:* Courtesy of M. Mittelstein, California Institute of Technology, Pasadena, Calif.

12.2 GAIN IN QUANTUM WELL LASERS

The expression for the complex susceptibility of a quantum well medium can be obtained in a manner similar to that employed in Section 11.2 for a bulk semiconductor. We start with Eqs. (8.1-19), but replace the population inversion density with the density of electrons with energies between E_c and $E_c + dE_c$ in the conduction band minus that of the valence band. So following (11.2-21)

$$d(N_2 - N_1) \rightarrow \frac{1}{L_z} \rho_{\text{QW}}(E_c)[f_c(E_c) - f_v(E_c - \hbar\omega_0)]\,dE_c$$

where the L_z^{-1} factor converts the area density to volumetric density. We thus write as in (11.2-21)

$$\chi(\omega) = \frac{\mu^2 T_2}{\varepsilon_0 \hbar} \int \frac{dE_c}{L_z} \rho_{\text{QW}}(E_c)[f_v(E_c - \hbar\omega_0) - f_c(E_c)] \frac{[\omega_0 - \omega]T_2 - i}{1 + (\omega - \omega_0)^2 T_2^2} \tag{12.2-1}$$

As in the case of the bulk semiconductor in Section 11.2, we will perform the integration not over E_c, the electron energy, but rather over the transition frequency ω_0. First, we define $\hbar\omega_c$ as the electron energy as measured from the bottom of the conduction band and similarly for the holes. Using Eq. (12.1-10)

$$E_c = \frac{\hbar^2}{2m_c}\left(k^2 + \ell^2 \frac{\pi^2}{L_z^2}\right), \qquad E_v = \frac{\hbar^2}{2m_v}\left(k^2 + \ell^2 \frac{\pi^2}{L_z^2}\right) \tag{12.1-10a}$$

$$\hbar\omega_0 \text{ (Energy difference between electron and hole states)} = E_g + E_c + E_v$$

and defining the reduced mass m_r by $m_r^{-1} = m_c^{-1} + m_v^{-1}$, we have

$$E_c = (\hbar\omega_0 - E_g)\frac{m_r}{m_c}, \qquad E_v = (\hbar\omega_0 - E_g)\frac{m_r}{m_v} \tag{12.2-2}$$

and from (12.1-18),

$$\rho_{\text{QW}}(E_c)\,dE_c = \sum_{n=1}^{\infty} \frac{m_c}{\pi\hbar^2} H\left[(\hbar\omega_0 - E_g)\frac{m_r}{m_c} - E_{nc}\right] d(\hbar\omega_0)\frac{m_r}{m_c}$$

which when substituted into (12.2-1) results in

$$\begin{aligned}\chi(\omega) &= \chi_r(\omega) + i\chi_i(\omega)\\ &= \frac{\mu^2 T_2 m_r}{\pi\varepsilon_0\hbar^2 L_z}\int d\omega_0 \sum_{n=1}^{\infty} H[\hbar\omega_0 - (E_g + E_{nc} + E_{nv})][f_v(E_c - \hbar\omega_0) - f_c(E_c)]\\ &\times \frac{\omega_0 T_2 - i}{1 + (\omega - \omega_0)^2 T_2^2}\end{aligned}$$

where we used (12.1-6) and

$$(\hbar\omega_0 - E_g)\frac{m_r}{m_c} - E_{nc} = \frac{m_r}{m_c}[\hbar\omega_0 - E_g - (E_{nc} + E_{nv})]$$

$$H\left[(\hbar\omega_0 - E_g)\frac{m_r}{m_c} - E_{nc}\right] = H[\hbar\omega_0 - (E_g + E_{nc} + E_{nv})]$$

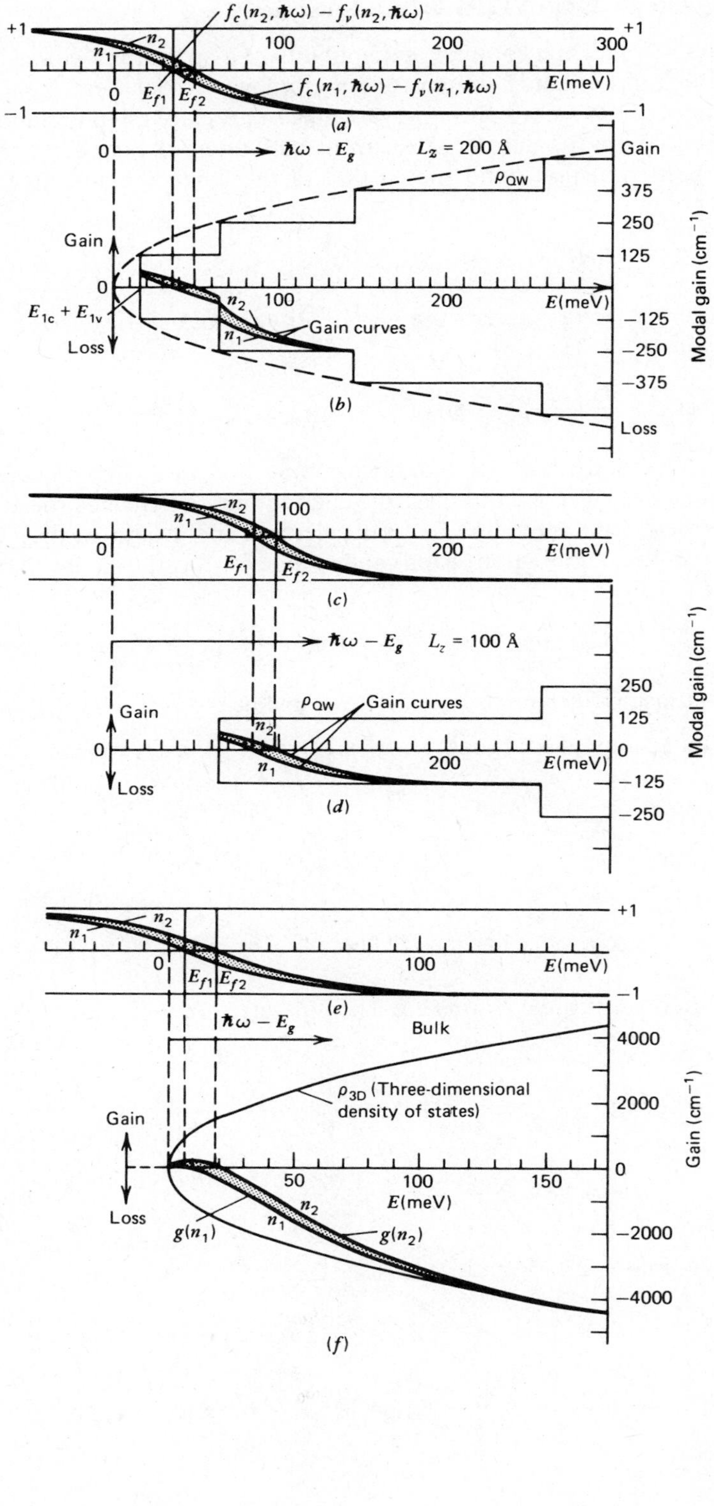

$f_c(n_2, \hbar\omega) - f_v(n_2, \hbar\omega)$
$f_c(n_1, \hbar\omega) - f_v(n_1, \hbar\omega)$
+1
−1
E_{f1} E_{f2}
E(meV)
(a)
$\hbar\omega - E_g$
L_z = 200 Å
ρ_{QW}
Gain
Loss
$E_{1c} + E_{1v}$
Gain curves
(b)
Modal gain (cm^{-1})
(c)
L_z = 100 Å
(d)
(e)
Bulk
ρ_{3D} (Three-dimensional density of states)
$g(n_1)$
$g(n_2)$
Gain (cm^{-1})
(f)

The exponential bulk gain constant (i.e., of a wave completely confined to the quantum well medium) is quantum well medium) is according to (8.2-7)

$$\gamma(\omega) = -k\frac{\chi_i(\omega)}{n^2} = -\frac{2\pi}{\lambda_0 n}\chi_i(\omega)$$

$$= \frac{2\pi\mu^2 m_r}{\lambda_0\varepsilon_0 n\hbar^2 L_z}\int d\omega_0 \sum_{n=1}^{\infty} H[\hbar\omega_0 - (E_g + E_{nc} + E_{nv})] \qquad (12.2\text{-}3)$$

$$\times [f_c(\hbar\omega_0) - f_v(\hbar\omega_0)]\left\{\frac{T_2}{\pi[1 + (\omega - \omega_0)^2 T_2^2]}\right\}$$

where f_c and f_v are evaluated, respectively, at the electron and hole transition energies that are separated by $\hbar\omega_0$.

Expression (12.2-3) is the basic tool for the gain calculations. In the limit of $T_2 \to \infty$, the last factor on the right side approaches $\delta(\omega - \omega_0)$ and the result is

$$\gamma(\omega) = \frac{2\pi\mu^2 m_r}{\lambda_0\varepsilon_0 n\hbar^2 L_z}\Big[f_c(\hbar\omega) - f_v(\hbar\omega)\Big]\sum_{n=1}^{\infty} H[\hbar\omega - (E_g + E_{nc} + E_{nv})\Big] \qquad (12.2\text{-}4)$$

so that $\gamma(\omega)$ is proportional to the product of the inversion factor $[f_c(\hbar\omega) - f_v(\hbar\omega)]$ and the (joint) density of states function

$$\sum_{n=1}^{\infty} H(\omega, n)$$

the latter function is identical to the density of state function ρ_{QW} of (12.1-18) and Figure 12.2, except that its argument is not the electron energy but rather, the photon energy. This causes the steps to occur at photon energies $\hbar\omega = E_g + E_{nc} + E_{nv}$.

A graphical illustration of Eq. (12.2-4) is shown in Figure 12.3. Figure 12.3*b* is a plot of the reduced density of states vs. the photon energy. In (*a*) is plotted the Fermi inversion factor $f_c(\hbar\omega) - f_v(\hbar\omega)$. (*b*) also shows the product of the reduced density of states and the Fermi inversion factor that, according to (12.2-4), is the gain.

12.3 SOME NUMERICAL CONSIDERATIONS

The gain $\gamma(\omega)$ given by (12.2-3) is the bulk gain, that is, the *gain experienced by a wave confined entirely to the active region.* As in the case of the conventional double heterostructure laser (DH laser), the modal gain, that is, the gain

---

FIGURE 12.3 (opposite) A graphical illustration using a realistic computer calculation of the gain expression (12.2-4). (*a*) The Fermi inversion $f_c(n, \hbar\omega) - f_v(n, \hbar\omega)$ at two carriers densities ($n_2 > n_1$) for quantum well thickness $Lz = 200$Å. (*b*) The gain vs. $\hbar\omega$ at n_1 and n_2. (*c*) The same as (*a*) except that $L_z = 100$Å. (*d*) The same as (*b*) except that $d = 100$Å. (*e*) The same as (*a*) except in a bulk semiconductor. (*f*) The same as (*b*) except in a bulk semiconductor L_z. The energy E_f in (*a*) and (*c*) corresponds to the photon energy $\hbar\omega$ for which $f_c - f_v = 0$ which is the transparency condition. (Courtesy of M. Mittelstein) experienced by a mode propagating along the junction plane, say along y, is obtained as in (11.3-3) by multiplying the bulk gain by the confinement factor Γ_a

$$\gamma_{\text{mode}}(\omega) = \gamma(\omega)\Gamma_a$$

$$\Gamma_a = \frac{\int_{-L_z/2}^{L_z/2} |E|^2 \, dz}{\int_{-\infty}^{\infty} |E|^2 \, dz} \approx \frac{L_z}{W_{\text{mode}}}$$

(note that the direction ⊥ to the junction plane is z) where W_{mode} is the width of the EM mode. If we limit ourselves to injection levels small enough so that only the first energy level $n = 1$ contributes as in cases 1, 2, of Figure 12.3 (this is the usual operating condition), it follows from (12.2-4) where we take only the first term

$$\gamma_{\text{mode}}(\omega) \simeq \frac{2\pi\mu^2 m_r}{\lambda_0 \varepsilon_0 n \hbar^2 W_{\text{mode}}} [f_c(\hbar\omega) - f_v(\hbar\omega)]$$

The factor

$$\gamma_0 \equiv \frac{2\pi\mu^2 m_r}{\lambda_0 \varepsilon_0 n \hbar^2 W_{\text{mode}}} \tag{12.3-1}$$

is thus the maximum modal gain available from a quantized level ($n = 1$). For practical purposes, it is the maximum gain available from a single-well QW laser.

A knowledge of γ_0 is important in designing QW lasers. The only unknown parameter in (12.3-1) is the matrix element μ. We notice from (11.2-

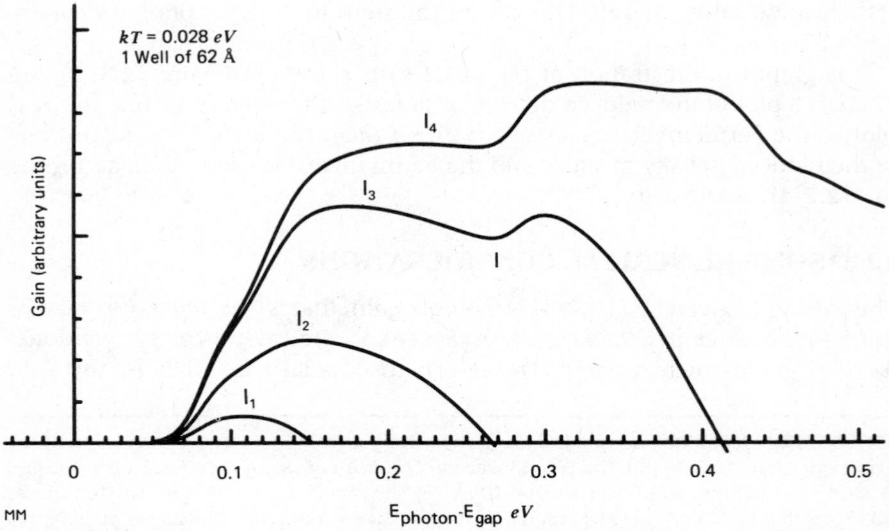

FIGURE 12.4 Gain vs. photon energy of a quantum well laser with (increasing) pumping current as a parameter. *Source:* Courtesy of M. Mittelstein, California Institute of Technology, Pasadena, Calif.

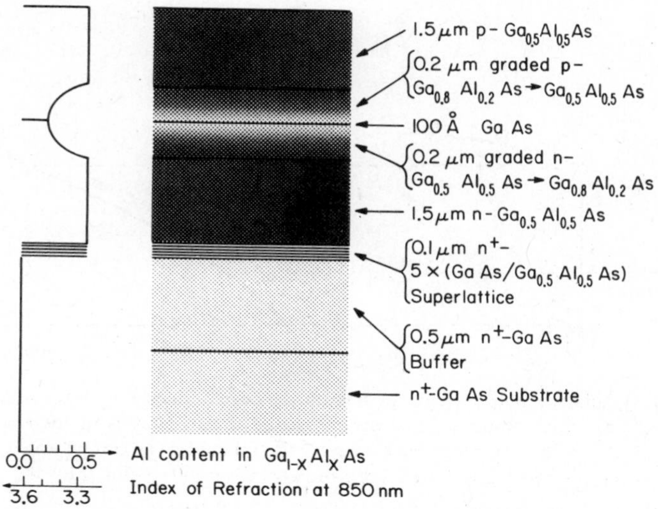

FIGURE 12.5 The actual layered structure grown by molecular beam epitaxy used to fabricate QW lasers. *Source:* Courtesy M. Mittelstein, L. Anders and P. Derry, California Institute of Technology., Pasadena, Calif.

23) that μ^2 figures also in the expression for the gain $\gamma(\omega)$ or loss $[\alpha(\omega) = -\gamma(\omega)]$ of a semiconductor medium. By fitting (11.2-23) to measured loss data $\alpha(\omega)$ vs. ω of a bulk semiconductor, we obtain

$$\mu = 5.2 \times 10^{-29} (\text{MKS})$$

and using the following data: $m_v = .47\, m_e$, $m_c = 0.067\, m_e$, $\lambda_0 = 0.85\ \mu\text{m}$, $n = 3.5$, $W_{\text{mode}} = 2500$ Å leads to

$$\gamma_0 \approx 125 \text{ cm}^{-1}$$

Figure 12.4 shows a calculated modal gain vs. ω curve. The rounding of the sharp corners is due to the inclusion of finite T_2 (a value of $T_2 = 0.1$ ps was used). The pump current increases as $1 \rightarrow 2 \rightarrow 3 \rightarrow 4$.

The actual layered structure of a QW laser grown by molecular beam epitaxy is shown in Figure 12.5. The 100 Å GaAs active region corresponds to the thin spike at the bottom of the parabolic region. The graded aluminum region gives rise to a dielectric waveguide that "traps" the optical mode. Such a structure, (Reference 1, 5), has become a work horse in QW laser applications.

Ultra-Low-Threshold QW Lasers

In a conventional semiconductor laser, most of the threshold current is needed to achieve transparency and only a minority (20–30%) goes toward overcoming the internal losses and the loss due to output coupling. Since the transparency density of the QW lasers (with thickness $L_z \sim 60$ Å) and DH

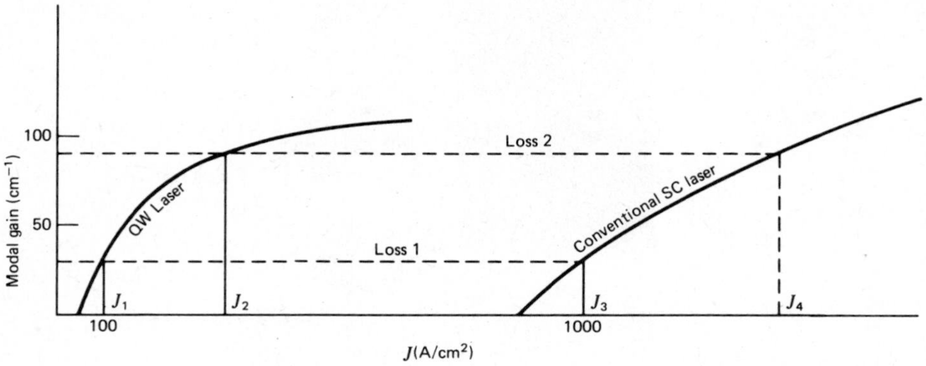

FIGURE 12.6 Qualitative plots of modal gain vs. pump current density for a DH laser and a QW well laser. The large ratio of threshold currents is due to large ratios of their active region thicknesses. The horizontal lines represent two loss values that must be compensated by the gain. The requisite current densities are J_1 and J_2 for the QW laser and J_3 and J_4 for the DH laser.

lasers are similar, it follows that the current density needed to reach transparency in the QW laser is smaller than that in a DH laser by roughly the ratio of the volumes of their active regions (a ratio that is typically ~ 20). These ideas are illustrated graphically in Figure 12.6 that also shows why an increased reflectivity can reduce the threshold current density of a QW laser by large factors, from J_2 to J_1, but only by a small amount, from J_4 to J_3, in a conventional DH laser. The effect of increased reflectivity on the QW laser threshold is illustrated in Figure 12.7.

Multiquantum Well Lasers

Since a QW laser with one well operating in the $n = 1$ transition is limited according to (12.3-1) to a gain $\gamma_0 \approx 100\ \text{cm}^{-1}$, we may employ a number of quantum wells, grown on top of each other in order to increase the maximum gain. At low currents, however, the multiquantum well laser yields lower (modal) gains, which is due to the smallness of the inversion factor $(f_c - f_v)$ in (12.2-3) at small carrier densities. This point is illustrated in Figure 12.8.

If we compare the gain of a QW laser with a single quantum well ($N = 1$) to that of one with N wells, then at a *given* current injection, the ratio of the modal gains is

$$\frac{g_N^{\text{mode}}}{g_{N=1}^{\text{mode}}} = \frac{\left[f_c\left(\frac{n}{N}, \omega\right) - f_v\left(\frac{n}{N}, \omega\right)\right] N}{f_c(n, \omega) - f_n(n, \omega)} \qquad (12.3\text{-}2)$$

where n is the carrier density in the single well laser and we assumed that the confinement factor per well is the same and, somewhat simplistically, a density-independent carrier lifetime τ so that the carrier density in the wells is given by

$$n = \frac{J\tau}{eL_zN}$$

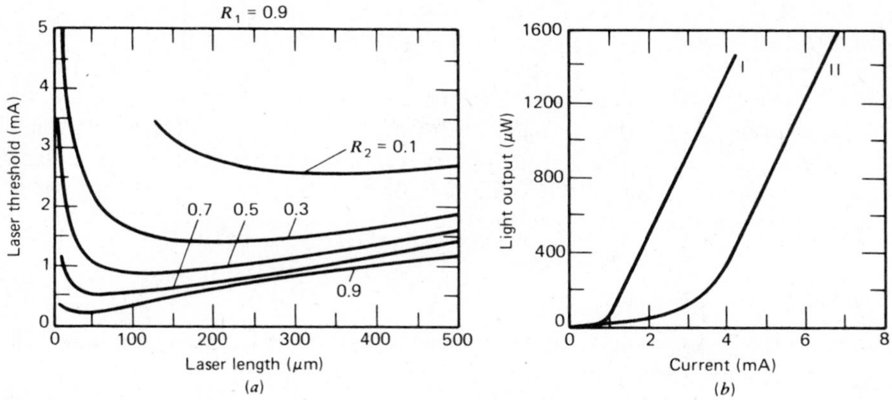

FIGURE 12.7 (*a*) Calculated threshold current for a buried single QW laser with various cavity lengths and facet reflectivities. (*b*) Light output vs. current curve for a 250-μm long buried single QW laser: I, High-reflectivity dielectric coated cleaved facets; II, uncoated cleaved facets. Both curves are for the same laser. *Source:* Reference 2.

and at a given current scales as N^{-1}. At sufficiently high-injection currents, the Fermi levels cross over the $n = 1$ step in both the conduction and valence bands so that $f_c - f_v > 0$ both in the single and multiquantum well case. In this case, $g_N = Ng_{N=1}$. At low currents, however, the situation is reversed as shown in the theoretical plot of Figure 12.8. The optimal number of wells for

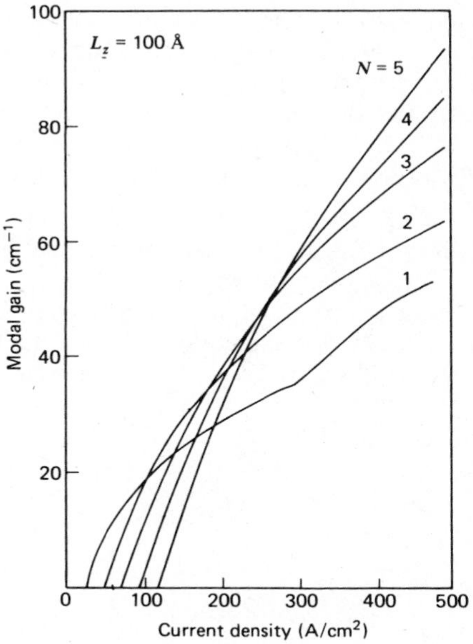

FIGURE 12.8 The modal gain versus current density for a laser with 1, 2, 3, 4, or 5 quantum wells. *Source:* Reference 3.

the lowest threshold then depends on the required modal gain that, in turn, is equal to the losses.

The introduction of a chapter on QW lasers is due to the belief on the author's part that they will in the near future come to dominate the semiconductor laser field. In addition to the low-threshold currents, QW lasers display improved modulation characteristics and a reduced spectral linewidth that is due to a smaller α parameter (Reference 3). A comprehensive recent review of their properties is given in Reference 4.

References

1. W. T. Tsang, "Extremely Low Threshold (AlGa)as Modified Multiquantum Well Heterostructure Laser Grown by MBE," *Appl. Phys. Lett.* **39,** 786 (1981).
2. Derry, P. L., A. Yariv, K. Lau, N. Bar Chaim, K. Lee, and J. Rosenberg "Ultra Low Threshold Graded Index Single Quantum Well DH Lasers with High Reflectivity Coatings," *Appl. Phys. Letters* **50,** 1773 (1987).
3. Arakawa, Y. and A. Yariv, "Theory of Gain, Modulation Response and Spectral Linewidth in AlGaAs Quantum Well Lasers," *IEEE J. Quant. Elect.* **QE-21,** 1666 (1985).
4. Okamoto, H., "Semiconductor Quantum Well Lasers for Optoelectronic Applications," *Jap. J. of Applied Phys.* **26,** 315 (1987).
5. Special Issue on Semiconductor Quantum Wells and Superlattices: Physics and Applications" *IEEE J. Quant. Elect.* **QE-22** (Sept. 1986).

Problems

12.1 (a) Calculate the energy levels, eigenfunctions, and selection rules of a "quantum wire" semiconductor laser in which an active region of a GaAs in the form of a square cylinder (cross section $L \times L$) is embedded in a higher-energy gap GaAlAs. Assume infinite confining potential walls at the interfaces.

(b) What is the density of states (carriers per energy interval per unit length of the quantum wire).

12.2 (a) Obtain an expression for the bulk gain (i.e., gain of wave entirely confined to wire material) of a quantum wire laser.

(b) Assuming the same matrix elements as in the example of the quantum well laser and $L_x = L_y = 70$ Å, estimate the total modal gain available from a quantum wire laser if we assume a confinement factor $\Gamma \approx 10^{-3}$.

CHAPTER 13

The Free-Electron Laser

13.0 INTRODUCTION

The free-electron laser (FEL) is a device that converts the kinetic energy of free (unbound) electrons to electromagnetic radiation. It is closely related to a general class of devices called traveling wave tube amplifiers and oscillators (Reference 1) and in its present embodiment, which employs a spatially alternating static magnetic field, it was proposed in 1951 by Motz and collaborators (References 2 and 3). A microwave version, the "Ubitron," was operated in 1960 by Phillips (Reference 4).

The revival of interest in devices capable of operating in the visible and ultraviolet regions of the spectrum followed an analysis by Madey (References 5 and 6) and a demonstration at 10.6 and 3.4 μm by Elias et al. (Reference 7) and Deacon et al. (Reference 8).

13.1 THE KINEMATICS OF FREE-ELECTRON-PHOTON INTERACTION

The radiation of electromagnetic energy in free-electron lasers (FEL) is due to the acceleration of electrons. The most familiar form of such radiation is the synchrotron radiation of electrons that are confined to a circular orbit by a magnetic field. Such radiation covers a wide spectrum and is not suitable for laser oscillation.

In the FEL, the electrons are forced to undulate in the transverse direction (x or y) while moving with near relativistic velocities along the main (z) axis. The resulting radiation field unlike synchrotron radiation can have most of its energy concentrated in a narrow band of frequencies. These frequencies are essentially those at which the slower ($v_z < c$) electron *falls back by one optical wavelength while undergoing one cycle of its transverse undulatory motion.* The radiation emitted by such an electron at each point along its trajectory is then in step with that emitted at earlier times so that a cumulative buildup of the field takes place. Another, and equivalent, point of view is to consider the exchange of power $E_x(\mathbf{r}, t)v_x(\mathbf{r}, t)$ between the moving electron and the traveling electromagnetic wave with an electric field $E_x(\mathbf{r}, t)$. The synchronism condition derived above insures that $E_x v_x$ does not change sign since a reversal of the sign v_x occurs at the same time as the reversal of the sign of E_x. The power flow $v_x E_x$ from the electron beam to the electromagnetic wave is thus continuous. (This power flow can be negative, that is, the electron gains

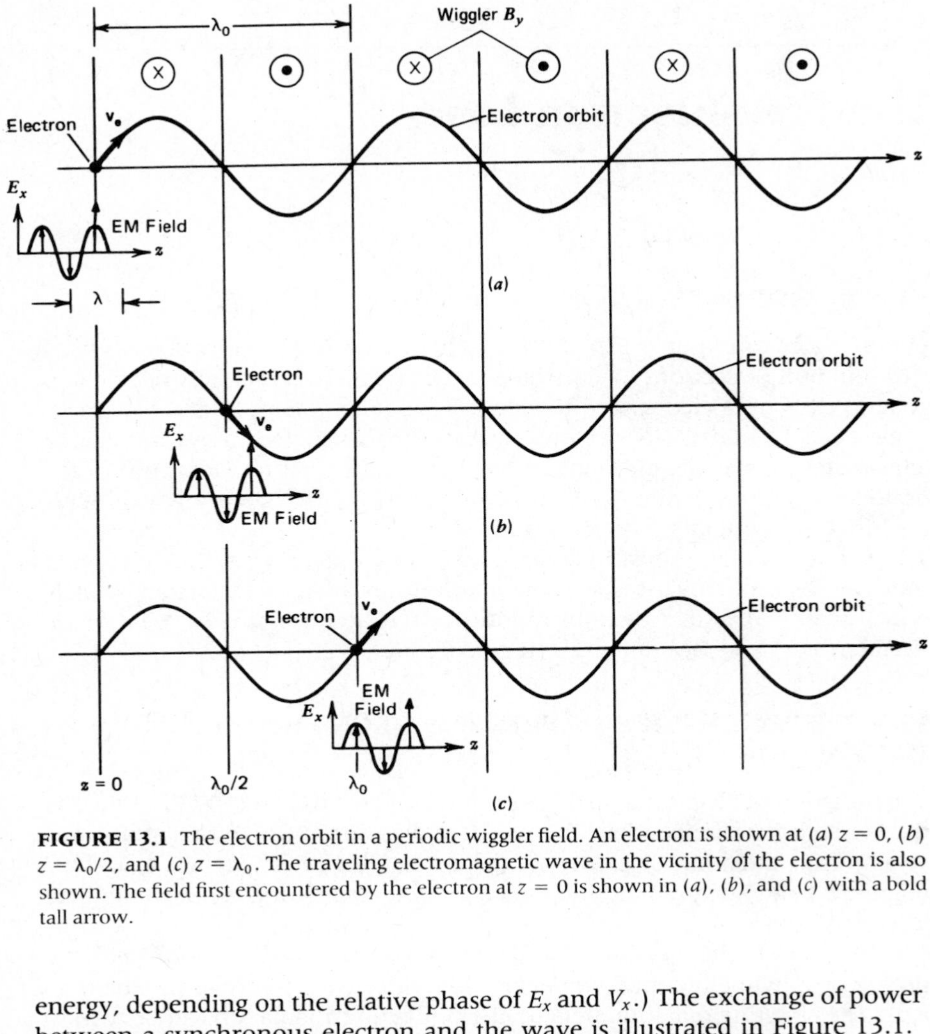

FIGURE 13.1 The electron orbit in a periodic wiggler field. An electron is shown at (*a*) $z = 0$, (*b*) $z = \lambda_0/2$, and (*c*) $z = \lambda_0$. The traveling electromagnetic wave in the vicinity of the electron is also shown. The field first encountered by the electron at $z = 0$ is shown in (*a*), (*b*), and (*c*) with a bold tall arrow.

energy, depending on the relative phase of E_x and V_x.) The exchange of power between a synchronous electron and the wave is illustrated in Figure 13.1.

We will return soon to the point of view described above; in the meantime, however, we will take a small pedagogic detour and describe the interaction between the electrons and the *EM* wave using the concept of transitions between occupied and empty electron states. This point of view, though more in keeping with laser theory, has not proven as useful as the (more) classical interaction picture and will not be pursued beyond this section. As a matter of fact, the reader can opt to omit it altogether.

Consider a free electron with an energy E_1 emitting or absorbing a photon (energy E_{ph}) from a field $\mathbf{E}_0 \cos(\omega t - \mathbf{k} \cdot \mathbf{r})$ and ending up with an energy E_2. The conservation of total energy and momentum can be stated as

$$E_1 - E_2 = (\pm)E_{ph} \tag{13.1-1}$$

$$\mathbf{p}_1 - \mathbf{p}_2 = (\pm)\mathbf{p}_{ph} = \pm\hbar\mathbf{k} \tag{13.1-2}$$

$$E_{ph} = cp_{ph} \tag{13.1-3}$$

$$E_{1,2} = \sqrt{m^2c^4 + c^2p_{1,2}^2} \tag{13.1-4}$$

where $\mathbf{p}$ is the momentum vector, m the mass of the electron, and c the velocity of light in vacuum. $\mathbf{p}_{ph} = \hbar\mathbf{k}$ is the photon momentum and $E_{ph} = \hbar\omega$ is the photon energy. The upper sign refers to an emission of a photon, the lower one to absorption. The set of equations (13.1-2,3) cannot be satisfied simultaneously; to illustrate the reason for this failure, consider the one-dimensional case in which, from (13.1-4), we obtain

$$\frac{dE_e}{dp_e} = \frac{c^2p_e}{\sqrt{m^2c^4 + c^2p_e^2}} < c \tag{13.1-5}$$

while from (13.1-3),

$$\frac{dE_{ph}}{dp_{ph}} = c \tag{13.1-6}$$

so that referring to Figure 13.2, we see that the change in energy ΔE_e of an electron that makes a transition from a momentum state p_1 to a state p_2 is always smaller than the energy $c(p_1 - p_2)$ of the photon with momentum $(p_1 - p_2)$. This argument remains true also in the case of three dimensions.

There are two obvious solutions to the problem of conserving simultaneously energy and momentum. The first is to carry out the interaction inside a medium in which the velocity of light is reduced to c/n ($n > 1$). In this case,

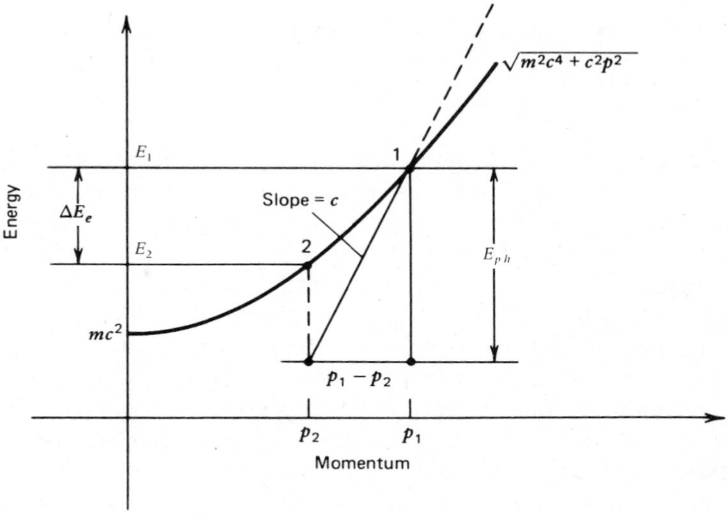

FIGURE 13.2 The dispersion curve for electrons and (straight line) for photons illustrating why the momentum and energy cannot both be conserved simultaneously in a radiative transition involving free electrons.

we replace c in (13.1-3) and (13.1-6), but not elsewhere, by c/n. This situation is called Çerenkov radiation and is the origin of the blue radiation observed in reactor pools where energetic electrons moving through water ($n = 1.33$) emit light. A second, less obvious solution is to mediate the interaction between the electron and the light beam (photons) by some spatially periodic "perturbation" that, in a manner equivalent to "Umklapp" processes in crystals (Reference 1), can soak up or deliver momentum in multiples of $2\pi/L$ where L is the period. The periodic perturbation can involve primarily the photons or the electrons or both. To see how it works, consider, for example, the case used in commercial microwave traveling wave tubes where the electromagnetic field propagates in a periodic structure, thus assuming the Floquet (Bloch) form (Reference 1)

$$E(z, t) = F(z) \exp[i(\omega t - \beta z)]$$

where

$$F(z + L) = F(z)$$

Expanding $F(z)$ in a Fourier series yields

$$F(z) = \sum_m a_m \exp\left(-im \frac{2\pi}{L} z\right)$$

so that

$$E(z, t) = \sum_{m=-\infty}^{+\infty} a_m \exp\left\{i\left[\omega t - \left(k + m\frac{2\pi}{L}\right) z\right]\right\} \tag{13.1-7}$$

The conservation of momentum condition (13.1-2) is replaced in this case by

$$\mathbf{p}_1 - \mathbf{p}_2 = \pm\hbar\left(\mathbf{k} + \hat{\mathbf{e}}m\frac{2\pi}{L}\right), \qquad m = 0, \pm 1, \pm 2, \ldots \tag{13.1-8}$$

Since the wave contains components with propagation constants, $\mathbf{k}_m = \mathbf{k} + \hat{\mathbf{e}}m2\pi/L$ where $\hat{\mathbf{e}}$ is a unit vector along the direction of spatial modulation z in our one-dimensional example. It is clear that our ability to choose m and the sign of m can be used to satisfy (13.1-8).

In the case of a free-electron laser, it is the electron motion that is modulated periodically by employing a spatially periodic magnetic field. This causes the electron wavefunction to assume a Floquet form and we are led again to Equation (13.1-8). We can, of course, "modulate" spatially the *EM* field by using a spatially periodic waveguide. This is done, for example, in traveling wave tubes and in linear charged particle accelerators. We will now return to the classical point of view. Consider an electron with velocity $\mathbf{v}$ interacting with a traveling electromagnetic field whose electric and magnetic fields are, respectively, $\mathbf{E}(z, t)$ and $\mathbf{B}(z, t)$. We will need the relations

$$\mathbf{v} = \boldsymbol{\beta}c, \gamma^{-2} \equiv 1 - \boldsymbol{\beta}\cdot\boldsymbol{\beta}, E_e = \sqrt{p^2c^2 + m^2c^4} = \gamma mc^2, \mathbf{p} = \gamma m\mathbf{v} \tag{13.1-9}$$

and the Lorentz equations

$$\frac{d}{dt}(\gamma\boldsymbol{\beta}) = -\frac{e}{mc}(\mathbf{E} + \boldsymbol{\beta} \times \mathbf{B}) \tag{13.1-10}$$

$$\frac{d\gamma}{dt} = -\frac{e}{mc}\boldsymbol{\beta} \cdot \mathbf{E} \tag{13.1-11}$$

In a single exception to this book, we employed the Gaussian (CGS) system of units to conform with the majority of the research literature in this field. In MKS units, we need modify only (13.1-10) to read

$$\frac{d}{dt}(\gamma\boldsymbol{\beta}) = -\frac{e}{mc}(\mathbf{E} + c\boldsymbol{\beta} \times \mathbf{B})$$

whereas (13.1-11) remains unaltered.

The Synchronism Condition

From (13.1-11), it follows that for an exchange of energy between the electron (whose energy is γmc^2) and a field $\mathbf{E}$ to take place, the electron velocity $\mathbf{v}$ must have a nonvanishing component along $\mathbf{E}$. In the case of a plane electromagnetic wave propagating along z, $E_z = 0$ and, say, $E_x \neq 0$ so that there must exist a transverse component v_x of the electron velocity. This is illustrated by Figure 13.1 that shows the electron launched with a finite transverse velocity. Since $v_z < c$, the electron will slip back relative to the wave and the sign of $\boldsymbol{\beta} \cdot \mathbf{E}$ will change. This will keep happening every $\lambda/(c - v_z)$ sec so that the net exchange of energy between the electron and the beam will average out to near zero. A solution to this problem is to force the electron to change its transverse velocity every $\lambda/(c - v_z)$ sec, that is, every $(\lambda v_z)/(c - v_z)$ cm so that it is always moving in the same sense relative to the transverse field. This can be achieved by employing a steady (dc) spatially alternating transverse magnetic field with a period $\lambda_0 = \lambda v_z/(c - v_z)$.

The resonance condition of an electron moving through a spatially periodic (period λ_0) transverse magnetic field—wiggler—(B_y) in the presence of a plane *EM* wave (wavelength λ) is illustrated in Figure 13.1. At (*a*) the electron bold velocity vector encounters the *EM* field at $z = 0$ and has a transverse velocity parallel to the field direction $(v_x \| E_x)$ so that $v_x E_x > 0$. The same electron is shown at two additional positions. The portion of the optical (electric) field first encountered by the electron at $z = 0$ is shown in a bold arrow at each of the three positions. At $z = \lambda_0/2$, plot (*b*), V_x is negative, but the faster *EM* field has moved ahead of the electron by $\lambda/2$ so that $E_x < 0$ and $v_x E_x > 0$. At $z = \lambda_0$, plot (*c*), $v_x > 0$ but the *EM* field has gained by another $\lambda/2$ relative to the electron so that $E_x > 0$. It follows that $v_x E_x > 0$ everywhere, and the electron is continuously "braked" down and loses energy to the optical field. The resonance condition (13.1-2) that is illustrated in the figure is thus one in which while traversing λ_0 (magnetic period), the electron falls back by one optical wavelength λ relative to the optical wave so that the sign of $\boldsymbol{\beta} \cdot \mathbf{E}$ remains a constant and, according to (13.1-11), the electron either loses (or gains) energy continuously throughout the interaction path. This resonance was shown above to be expressible as

$$\lambda_0 = \lambda \frac{v_z}{c - v_z} = \lambda \frac{\beta_z}{1 - \beta_z} \tag{13.1-12}$$

where $\lambda = 2\pi/k$ is the optical wavelength and $\beta_z \equiv v_z/c$. For highly relativistic electrons, $\beta_z \approx 1$ and (13.1-12) can be written using the first two relations in (13.1-9) as

$$\lambda = \frac{\lambda_0}{2}\left(\frac{1}{\gamma^2} + \beta_\perp^2\right) \tag{13.1-13}$$

In what follows, we will show that in a periodic magnetic field

$$B_y = B_0 \sin k_0 z \qquad \left(k_0 \equiv \frac{2\pi}{\lambda_0}\right) \tag{13.1-14}$$

$$\beta_\perp = \beta_x = (-eB_0\lambda_0/2\pi\gamma mc^2) \cos k_0 z$$

so that

$$\lambda = \frac{\lambda_0}{2}\left(\frac{1}{\gamma^2} + \overline{\beta_\perp^2}\right) = \frac{\lambda_0}{2\gamma^2}\left[1 + \tfrac{1}{2}\left(\frac{eB_0\lambda_0}{2\pi mc^2}\right)^2\right] = \frac{\lambda_0}{2\gamma^2}\left(1 + \tfrac{1}{2}a_w^2\right) \tag{13.1-15}$$

where

$$a_w \equiv eB_0\lambda_0/2\pi mc^2$$

where the spatial averaging, denoted by a bar on top of $\beta_\perp^2$ was employed since $\beta_\perp^2$ is not a constant of the motion. In the case of a circularly polarized magnetic field, however,

$$B_y = B_0 \sin k_0 z \tag{13.1-16}$$

and

$$B_x = B_0 \cos k_0 z$$

$\beta_\perp^2 = \beta_x^2 + \beta_y^2$ is a constant. The synchronism condition (13.1-15) becomes

$$\lambda = \frac{\lambda_0}{2\gamma^2}(1 + a_w^2) \tag{13.1-17}$$

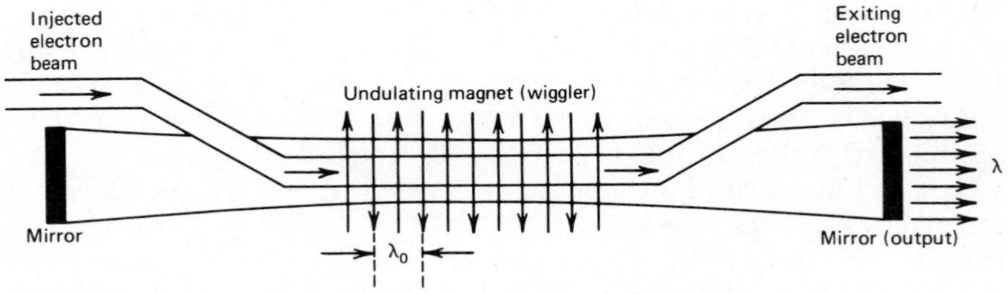

FIGURE 13.3 An experimental arrangement of a typical free-electron laser.

The mechanism described above gives rise to an amplification of an electromagnetic wave so that in the presence of feedback provided by an optical resonator, it may lead to laser oscillation. A typical wiggler arrangement is illustrated in Figure 13.3.

Equation (13.1-15) is the tuning condition that relates the wavelength λ of the amplified radiation to the electron energy $\gamma(mc^2)$, to the magnetic field period λ_0 and its strength B_0.

13.2 THEORY OF OPTICAL GAIN IN FREE-ELECTRON LASERS

In this section we will consider the problem of amplifying an optical field that travels along the axis of a device of the type shown in Figure 13.3. The wave encounters an electron beam in an interaction region consisting of a spatially periodic transverse magnetic field that is popularly referred to as the "wiggler." A typical wiggler magnet design is depicted in Figure 13.4. The magnetic field along the axis ($x = y = 0$) is taken as

$$B_y = B_0 \sin k_0 z, \qquad B_x = B_z = 0 \tag{13.2-1}$$

We will first solve for the motion of an electron in the wiggler field while neglecting the optical field. Since $\mathbf{E} = 0$, it follows from (13.1-11) that $\gamma =$ constant and (13.1-10) becomes

$$\gamma \frac{d\boldsymbol{\beta}}{dt} = -\frac{e}{mc} \boldsymbol{\beta} \times \mathbf{B}$$

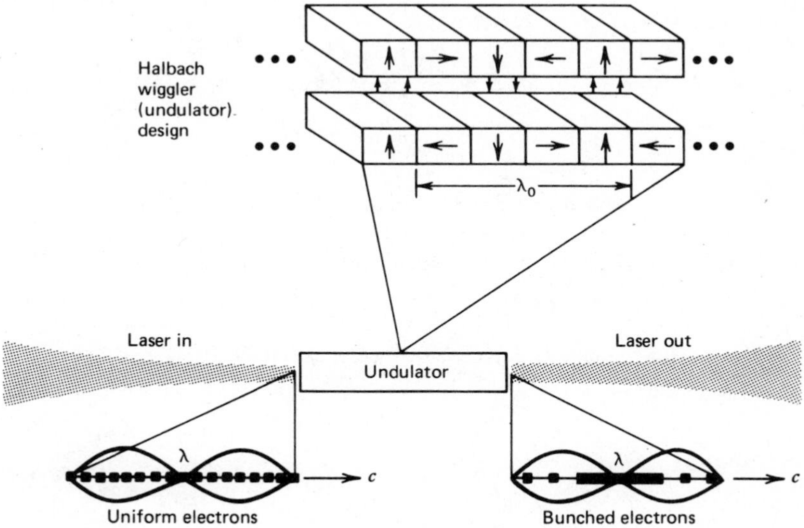

FIGURE 13.4 A practical design (the Halbach magnet) for constructing the wiggler is shown. Eight permanent magnets are used in one wiggler period. *Source:* Reference 9.

that with the aid of (13.2-1) becomes

$$\frac{d\beta_x}{dt} = \frac{e}{\gamma mc}\beta_z B_0 \sin k_0 z \tag{13.2-2}$$

In the case of highly relativistic electron beam $\gamma \gg 1$, $\beta_z \approx 1$. [If we were to employ a helical magnetic field ($B_0 \cos k_0 z$, $B_0 \sin k_0 z$, 0), then β_z will be rigorously a constant equal to $(1 - \gamma^{-2} - \beta_\perp^2)^{1/2}$]. We thus put $z = v_0 t$, $\beta_z = v_0/c$ and rewrite (13.2-2) as

$$\frac{d\beta_x}{dt} = \frac{e\beta_z B_0}{\gamma mc} \sin k_0(v_0 t)$$

so that

$$\beta_x = -\frac{e\beta_z B_0}{\gamma mck_0 v_0} \cos k_0(v_0 t) = -\frac{eB_0\lambda_0}{2\pi\gamma mc^2} \cos(\omega_0 t) \tag{13.2-3}$$

where in the last equality we used $k_0 = 2\pi/\lambda_0$, $v_0 = \beta_z c$ and the definition

$$\omega_0 \equiv k_0 v_0 \tag{13.2-4}$$

so that $\omega_0/2\pi$ is the temporal frequency of the wiggler magnetic field as seen in the frame of the moving electron.

The optical electric field is taken as

$$\mathbf{E}_x(z, t) = \hat{\mathbf{x}}E \cos(kz - \omega t + \Phi) \tag{13.2-5}$$

that when substituted in (13.1-11) gives

$$\frac{d\gamma}{dt} = -\frac{e}{mc}\beta_x(t)E_x[z(t),t] = \left(\frac{e}{mc}\right)\frac{eB_0\lambda_0 E}{2\pi\gamma mc^2} \cos \omega_0 t \cos\,[kz(t) - \omega t + \Phi] \tag{13.2-6}$$

which if we use $z(t) = v_0 t$ leads to

$$\frac{d\gamma}{dt} = \frac{ea_w E}{2\gamma mc}\cos[(\omega_0 + kv_0 - \omega)t + \Phi] + \frac{ea_w E}{2\gamma mc}\cos[(\omega_0 - kv_0 + \omega)t - \Phi] \tag{13.2-7}$$

$$a_w \equiv \frac{eB_0\lambda_0}{2\pi mc^2} = \text{wiggler parameter} \tag{13.2-8}$$

is the so-called wiggler parameter. Recalling that $k = \omega/c$, we rewrite (13.2-7) as

$$\frac{d\gamma}{dt} = \frac{ea_w E}{2\gamma mc}\Big(\cos\{[\omega_0 - \omega(1 - \beta_z)]t + \Phi\} + \cos\{[\omega_0 + \omega(1 - \beta_z)]t - \Phi\}\Big) \tag{13.2-9}$$

The frequency

$$\Omega \equiv \omega_0 - \omega(1 - \beta_z) \tag{13.2-10}$$

is the frequency at which the product $\beta_x E_x$ of a moving electron changes sign. The condition $\Omega = 0$ thus ensures that $\boldsymbol{\beta} \cdot \mathbf{E}$ does not change sign, which, according to (13.1-11), signifies continuous power gain (or loss) by the electron. It is thus of no surprise that the condition $\Omega = 0$ is equivalent to the synchronism condition (13.1-12) derived using kinematic considerations. The proof of this statement is left as an exercise.

Since we are still dealing with a single electron, we can without loss of generality take $t = 0$ as the arrival time of the electron at the beginning of the interaction region at $z = 0$. The phase Φ in (13.2-5) thus determines the initial position of the electron relative to the optical wave. We will thus use Φ *to label the electrons* so that a $\Phi = 0$ electron encounters the peak field at $z = 0$. In case of a uniform electron beam, the individual electrons are distributed, initially, uniformly over $0 \leq \Phi \leq 2\pi$.

We now return to the energy relation (13.2-9). Near synchronism, $\Omega \cong 0$ and $\omega_0 \approx \omega(1 - \beta_z)$, the frequency of the second cosine term is $\approx 2\omega_0$ so that the contribution of that term averages out to zero in a few periods $(2\pi/\omega_0)$, or equivalently in a distance that includes a few wiggler periods. We are thus left with

$$\frac{d\gamma}{dt} = \frac{ea_w E}{2\gamma mc} \cos(\Omega t + \Phi) \tag{13.2-11}$$

and using $z = v_0 t$, we obtain

$$\Delta\gamma(z, \Phi) = \frac{ea_w E}{2\gamma mc\Omega} \sin(\Omega t + \Phi) \Big|_0^{z/v_0} = \frac{ea_w E}{2\gamma mc\Omega} \left[\sin\left(\frac{\Omega z}{v_0} + \Phi\right) - \sin\Phi\right] \tag{13.2-12}$$

If we now average $\Delta\gamma(z)$ over all the phases,

$$\langle \Delta\gamma(z) \rangle_\Phi \equiv \frac{1}{2\pi} \int_0^{2\pi} \Delta\gamma(z, \Phi)\, d\Phi = 0 \tag{13.2-13}$$

Thus, to first order there is no net transfer of power between the electron beam and the *EM* wave. The second-order correction will consist of accounting for the fact that as an individual electron (with a phase Φ) gains or loses in energy, its position relative to the unperturbed $z = v_0 t$ position is advanced or retarded. This will be accomplished by replacing $z(t) = v_0 t$ in (13.2-5) with

$$z(t) = \int_0^t v_z(t')dt' = \int_0^t [v_0 + c\Delta\beta_z(t')]\, dt' \tag{13.2-14}$$

where $\Delta\beta_z(t')$ is the change calculated to first order. We use

$$\beta_z^2 = 1 - \beta_\perp^2 - \gamma^{-2}$$

since from (13.2-3),

$$\beta_\perp^2 = \beta_x^2 = \left(\frac{eB_0\lambda_0}{2\pi\gamma mc^2}\right)^2 \cos^2 k_0 z$$

we will average $\beta_\perp^2$ over a few wiggler periods (we recall that for a helical magnetic field, β_x^2 is a constant and the averaging is not needed), thus replacing $\cos^2 k_0 z$ by $\frac{1}{2}$.

$$\overline{\beta_\perp^2} = \frac{1}{2\gamma^2}\left(\frac{eB_0\lambda_0}{2\pi mc^2}\right)^2 = \frac{a_w^2}{2\gamma^2}, \qquad \overline{\beta_z^2} = 1 - \gamma^{-2}\left(1 + \frac{a_w^2}{2}\right)$$

$$a_w \equiv \frac{eB_0\lambda_0}{2\pi mc^2}$$

so that

$$\Delta\beta_z = \frac{(1 + \frac{1}{2}a_w^2)}{\gamma^3\beta_z}\,\Delta\gamma \approx \frac{(1 + a_w^2/2)}{\gamma^3}\,\Delta\gamma = D[\sin(\Omega t + \Phi) - \sin\Phi] \quad (13.2\text{-}15)$$

where we used (13.2-12), $\beta_z \approx 1$, and defined

$$D \equiv \frac{ea_wE(1 + a_w^2/2)}{2\gamma^4 mc\Omega} \quad (13.2\text{-}16)$$

Using (13.2-15) in (13.2-14), we obtain

$$z(t) = v_0 t + cD\int_0^t [\sin(\Omega t' + \Phi) - \sin\Phi]\,dt'$$

$$= v_0 t - cD\left[\frac{\cos(\Omega t + \Phi) - \cos\Phi}{\Omega} + t\sin\Phi\right]$$

The field "seen" by an electron is obtained by using $z(t)$ in (13.2-5)

$$E_x(t) = E\cos\left\{kv_0 t - \frac{kcD}{\Omega}[\cos(\Omega t + \Phi) - \cos\Phi + \Omega t\sin\Phi] - \omega t + \Phi\right\} \quad (13.2\text{-}17)$$

Using the corrected field expression in (13.2-6) and discarding, as before, the nonsynchronous term lead to

$$\frac{d\gamma}{dt} = \frac{ea_wE}{2\gamma mc}\cos\left\{(\Omega t + \Phi) - \frac{\omega D}{\Omega}[\cos(\Omega t + \Phi) - \cos\Phi + \Omega t\sin\Phi]\right\} \quad (13.2\text{-}18)$$

A comparison to (13.2-11) reveals that the first-order correction consists of a phase slippage

$$\Delta\Phi = -\frac{\omega D}{\Omega}[\cos(\Omega t + \Phi) - \cos\Phi + \Omega t\sin\Phi]$$

Since $D \propto E$, $\Delta\Phi$ can be made arbitrarily small so that for $\Delta\Phi \ll \pi$, we expand (13.2-18)

$$\frac{d\gamma}{dt} = \frac{eEa_w}{2\gamma mc}\left\{\cos(\Omega t + \Phi) + \frac{\omega D}{\Omega}\sin(\Omega t + \Phi)[\cos(\Omega t + \Phi) - \cos\Phi + \Omega t\sin\Phi]\right\} \quad (13.2\text{-}19)$$

and after averaging over all phases Φ (we can do it by inspection keeping only $\sin^2\Phi$ and $\cos^2\Phi$ terms)

$$\left\langle\frac{d\gamma}{dt}\right\rangle_\Phi = b\left[-\frac{\sin\Omega t}{2} + \frac{\Omega t\cos\Omega t}{2}\right] \quad (13.2\text{-}20)$$

$$b \equiv \frac{ea_wE\omega D}{2\gamma mc\Omega} \quad (13.2\text{-}21)$$

Expression (13.2-20) is integrated over the electron transit time through the wiggler $\tau = L/v_0$ to obtain the average change in γ per electron

$$\langle \Delta\gamma \rangle_\Phi = \int_0^{\tau = L/v_0} \left\langle \frac{d\gamma}{dt} \right\rangle_\Phi dt = -\frac{b}{2\Omega}(2 - 2\cos\Omega\tau - \Omega\tau\sin\Omega\tau) \quad (13.2\text{-}22)$$

Using (13.2-8,16,21), the tuning relation $\lambda = \lambda_0(1 + \frac{1}{2}a_w^2)/2\gamma^2$, and the expression for the *EM* power

$$P = c\frac{E^2}{8\pi}\text{ Area}$$

we obtain for the change of the *EM* power in one transit

$$\Delta P = -\left(\frac{I}{e}\right) mc^2 \langle \Delta\gamma \rangle_\Phi$$

where I is the average beam current. Using (13.2-22) leads finally to

$$\frac{\Delta P}{P} = I\frac{e^3 B_0^2 \lambda_0}{\gamma^3 m^3 \text{ Area } c^4}\tau^3 \left[\frac{2 - 2\cos\Omega\tau - \Omega\tau\sin\Omega\tau}{(\Omega\tau)^3}\right] \quad (13.2\text{-}23)$$

which is our main result.

The function

$$g(\Omega\tau) = \frac{2 - 2\cos\Omega\tau - \Omega\tau\sin\Omega\tau}{\Omega^3\tau^3} \quad (13.2\text{-}24)$$

is plotted in Figure 13.5. It is antisymmetric in $\Omega\tau$, and has peaks of 0.135 at

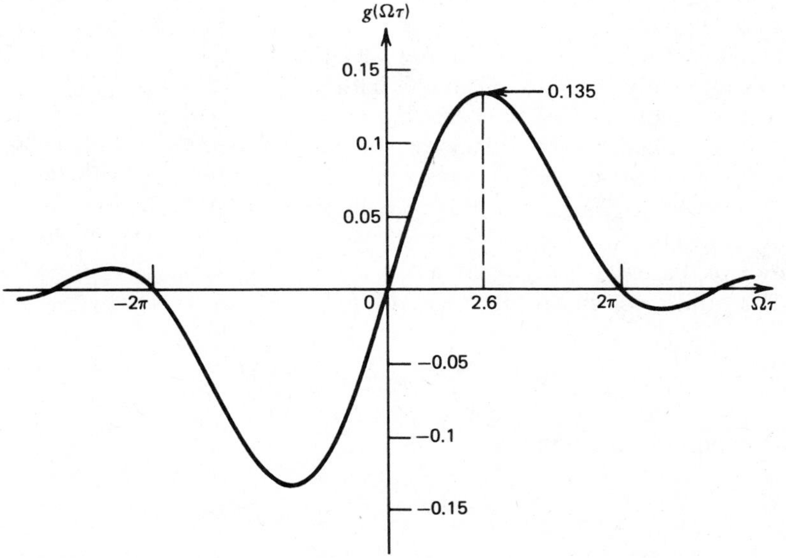

FIGURE 13.5 The normalized gain function $g(\Omega\tau)$ of Eq. (13.2-24) as a function of the phase slippage in the per transit in the ponder motive potential ($\Omega\tau$).

$\Omega\tau = 2.6$. The main gain region thus obtains for $\Omega\tau > 0$ that if we use our definition

$$\Omega \equiv [\omega_0 - \omega(1 - \beta_z)]$$

is seen to correspond to electrons traveling at velocities exceeding the synchronous ($\Omega = 0$) value. It should also be noted that since Ω depends on the optical frequency ω, the function $g(\Omega\tau)$ describes the frequency dependence of the gain.

A basic feature that distinguishes the FEL from atomic lasers is that increasing the interaction length L, that is, $\tau = L/v_0$ in (13.2-24), does not necessarily increase the gain and might even cause it to disappear altogether or become negative. This reflects the interference nature of the interaction as described in Section 13.1. As L increases, the frequency for maximum gain approaches the resonant value $\omega_r = \omega_0/(1 - \beta_z)$ (i.e., $\Omega = 0$) and the frequency width of the gain curve decreases. If, with slight arbitrariness, we take the width of the gain function $g(\Omega\tau)$ as $\Delta(\Omega\tau) = \pi$, then for $v_z \approx c$

$$(\Delta\omega)_{\text{gain}} \approx \frac{\pi c}{L(1 - \beta_z)}$$

and using the tuning relation (13.1-12) with $\beta_z \approx 1$, we obtain

$$\frac{(\Delta\omega)_{\text{gain}}}{\omega} = \frac{\lambda_0}{2L} = \frac{1}{2N} \qquad (13.2\text{-}25)$$

where N is the number of wiggler periods.

Another interesting feature is the dependence of the gain on the third power of the interaction length L that is contained in the factor $\tau^3 = (L/v_0)^3$. The bookkeeping involved in the steps leading from (13.2-22) to (13.2-23) assumes implicitly that the *EM* mode diameter exceeds or is equal to that of the electron beam since all the beam electrons were included. If we shrink the *EM* mode diameter beyond this point, the gain $\Delta P/P$ becomes independent of the mode diameter since the increase in the optical intensity is offset by the loss of contributing electrons. To properly account for this effect, we should take Area in (13.2-23) as the larger of the cross-sectional area of the optical mode or of that of the electron beam.

The tuning condition of Eq. (13.1-15)

$$\lambda = \frac{\lambda_0}{2\gamma^2}\left[1 + \frac{1}{2}\left(\frac{eB_0\lambda_0}{2\pi mc^2}\right)^2\right]$$

can be expressed in convenient units as

$$\lambda = \frac{\lambda_0}{2\gamma^2}\,[1 + 4.4 \times 10^{-3}\lambda_0^2(\text{cm})B_0^2(\text{kG})] \qquad (13.2\text{-}26)$$

In a typical configuration with a wiggler period of $\lambda_0 = 5$ cm and $B_0 = 3$ kG,

we obtain

Electron energy (MeV)	10	100	1000
$\lambda(\mu m)$	129.9	1.299	0.01299

a wavelength region extending from the far infrared to the x-ray region.

13.3 THE PONDERMOTIVE POTENTIAL

A useful point of view that is used extensively in accelerator theory is as follows:

We start with the energy equation (13.2-7). We discard the (nonsynchronous) second-order terms and restore the original spatial dependence $z = v_0 t$.

$$\frac{d\gamma}{dt} = \left(\frac{e}{mc}\right)\frac{eB_0\lambda_0 E}{4\pi\gamma mc^2}\cos[(k_0 + k)z - \omega t + \Phi] \tag{13.3-1}$$

If we compare this result to the energy equation (13.1-11),

$$\frac{d\gamma}{dt} = -\frac{e}{mc}\beta_z E_z \tag{13.3-2}$$

which describes the interaction of an electron with an axial field E_z. We find that, in our case, the interaction with the magnetic field and the transverse optical field is equivalent to one with an axial electric field

$$(E_z)_{\text{eff}} = \frac{eB_0\lambda_0 E}{4\pi\beta_z\gamma mc^2}\cos[(k_0 + k)z - \omega t + \Phi] \tag{13.3-3}$$

The corresponding electron potential energy

$$V = e\int_0^z (E_z)_{\text{eff}}\, dz' = \frac{-e^2B_0\lambda_0 E}{4\pi(k + k_0)\beta_z\gamma mc^2}\sin[(k_0 + k)z - \omega t + \Phi] \tag{13.3-4}$$

is called the "pondermotive potential" wave. It is the potential seen by the moving electron. The amplitude of this potential is proportional to the product of the optical field E and B_0—the wiggler magnetic field. The resonance condition (13.1-12) or, equivalently, (13.1-17) is equivalent to stating that the axial velocity v_0 of the electron is equal to the phase velocity of the pondermotive potential

$$v_0 = \frac{\omega}{k_0 + k} \tag{13.3-5}$$

The energy exchange between an electron and the EM wave can be treated very conveniently using the pondermotive potential approach. This can also be used to extend the analysis of the interaction beyond the second-order perturbation employed above. We start with the energy equation (13.2-7), keeping only the first near-synchronous term on the right side

$$\frac{d\gamma}{dz} = \frac{ea_w E}{2\gamma mc^2}\cos[(k_0 + k)z - \omega t + \Phi] \tag{13.3-6}$$

where we used $d/dz \approx (1/c)[d/(dt)]$ for a moving electron.

Defining the electron "phase" Ψ

$$\Psi = (k_0 + k)z - \omega t + \pi/2 + \Phi \tag{13.3-7}$$

we write (13.3-6) as

$$\frac{d\gamma}{dz} = \frac{ea_w E}{2\gamma mc^2}\sin\Psi \tag{13.3-8}$$

Ψ is thus the *phase of the electron relative to the pondermotive potential.* At synchronism, Ψ = constant. Since the resonance condition (13.1-13) can be written as

$$k_0 = \frac{k}{2\gamma^2}\left(1 + \tfrac{1}{2}a_w^2\right)$$

where $k_0 = 2\pi\lambda_0^{-1}$, $k = 2\pi\lambda^{-1}$, a deviation from the synchronism thus leads to a change of Ψ at a spatial rate

$$\frac{d\Psi}{dz} = k_0 - \frac{k}{2\gamma^2}\left(1 + \tfrac{1}{2}a_w^2\right) \tag{13.3-9}$$

If we define the resonant value of γ as that for which $(d\Psi)/(dz) = 0$ and let $\gamma = \gamma_r + \Delta\gamma$ then from (13.3-9)

$$\frac{d\Psi}{dz} = 2k_0\frac{\Delta\gamma}{\gamma_r}$$

$$\frac{d^2\Psi}{dz^2} = 2k_0\frac{d}{dz}\left(\frac{\Delta\gamma}{\gamma_r}\right) = \frac{2k_0}{\gamma_r}\frac{d\gamma}{dz} \tag{13.3-10}$$

which using (13.3-8) yield

$$\frac{d^2\Psi}{dz^2} = +k_\Psi^2\sin\Psi \tag{13.3-11}$$

$$k_\Psi^2 = \frac{2e^2E_0B_0}{(\gamma_r mc^2)^2} \tag{13.3-12}$$

$$\gamma_r^2 = k\left(1 + \tfrac{1}{2}a_w^2\right)/2k_0 \tag{13.3-13}$$

The pendulum equation (13.3-11) describes the evolution of the phase Ψ of a single electron. It can be solved formally given the boundary conditions

$$\Psi(0) = \phi + \frac{\pi}{2}$$

and the condition

$$\Psi'(0) \equiv \left.\frac{d\Psi}{dz}\right|_{z=0} = 2k_0 \frac{\gamma_{\text{init}} - \gamma_r}{\gamma_r}$$

which follows from (13.3-10). Before considering the more general case, we take up the case of electrons injected near $\Psi = n\pi + \delta$ with n an integer and $\delta \ll \pi$, that is, near an extremum of the pondermotive potential, with $\Delta\gamma \propto \Psi'(0) = 0$. Equation (13.3-11) becomes

$$\frac{d^2\delta}{dz^2} = k_\Psi^2 \delta \qquad n = 0, 2, 4, \ldots \tag{13.3-14}$$

and

$$\frac{d^2\delta}{dz^2} = -k_\Psi^2 \delta \qquad n = 1, 3, 5, \ldots \tag{13.3-15}$$

Equation (13.3-15) corresponds to the electrons injected near the pondermotive potential minimum. These electrons execute a simple harmonic motion about the minimum with a spatial period $l_B = 2\pi k_\Psi^{-1}$. Electrons injected near the peaks ($n = 1, 3, \ldots$) are unstable as evidenced from the positive sign of the right side of (13.3-14).

The formal solution of the pendulum equation involves elliptic integral functions. The equation, however, lends itself readily to numerical integration.

A convenient representation of the results is in the phase diagram $\Psi'(z)[\propto\Delta\gamma(z)]$ vs. Ψ. An example of such a diagram is shown in Figure 13.6. The following two different cases are illustrated: (*a*) closed orbits corresponding to electrons oscillating about $\gamma = \gamma_r$ ($\Psi' = 0$) or physically to electrons executing small near-harmonic motion about $\Psi = 0, 2\pi, 4\pi, \ldots$. (*b*) Electrons that are not tied down to a given potential well with a mean energy that is not equal to γ_r. The orbit dividing the two regions is called the separatrix. Its height is proportional to the height of the pondermotive potential and thus to the product of E and B_0. A large pondermotive potential thus makes possible the trapping, in closed orbits, of a large range of electron energies.

In Figure 13.6 we show the result of a solution of the pendulum equation (13.3-11) in the case of an electron beam. The electrons that initially are spread uniformly over $0 < \Psi < 2\pi$ enter with a velocity that exceeds the resonance value, that is, they travel somewhat faster than the pondermotive potential, so that $\Delta\gamma_{\text{initial}} > 0$. The figure shows the electrons' phases and energies initially as well as at $z = l_B/2$ and $z = l_B$. We see that on the average, the electrons fall back with distance and are bunched, whereas the average energy $\gamma < \gamma_{\text{initial}}$ so that energy is lost to the *EM* field. The figure illustrates that to extract a large fraction of the electrons' energy, it is necessary to have a large separatrix, that is, deep pondermotive potential wells. This is the reason why FEL amplifiers rather than oscillators have become the focus of recent work. In these amplifiers, the optical beam is fed into the interaction region with a large initial amplitude.

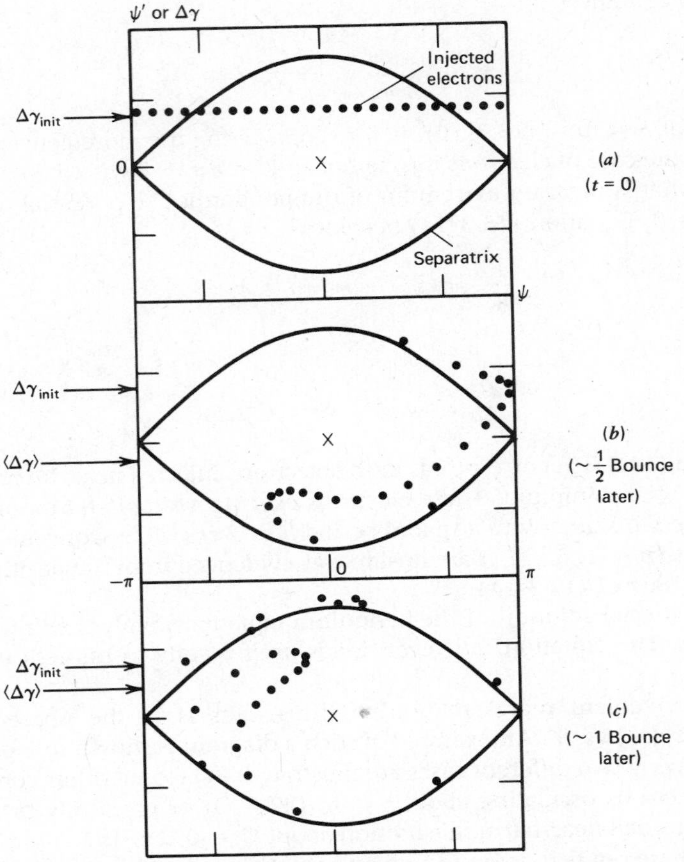

FIGURE 13.6 Injection and bunching of electrons in a traveling ponder-motive potential. *Source:* Reference 10, courtesy of Addison-Wesley, Reading, Mass.

Spontaneous Emission and Gain in the Free-Electron Laser

The transverse periodic acceleration of the electron that is undulating in the wiggler magnetic field leads to spontaneous radiation. The spectrum of this radiation obeys the familiar grating relations (References 11 and 12)

$$|F(\Omega)|^2 \propto \frac{\sin^2(N\Omega t_0/2)}{(\Omega t_0/2)^2} \tag{13.3-16}$$

where N is the number of grating periods, $t_0 \equiv \lambda_0/v_0$ is the electron transit time per period and Ω is given, as in 13.2-10, by

$$\Omega = \omega_0 - \omega(1 - \beta_z)$$

If we multiply the right side of (13.3-16) by (N^2/N^2), the result is

$$|F(\Omega)|^2 \propto \frac{\sin^2(\Omega\tau/2)}{(\Omega\tau/2)^2} \equiv f_{\text{spont}}(\Omega\tau)$$

From (12.2-24), it follows that

$$g(\Omega\tau) = -\frac{1}{2}\frac{d}{d(\Omega\tau)} f_{\text{spont}}(\Omega\tau) \tag{13.3-17}$$

The gain lineshape function is thus proportional to the derivative of the spontaneous emission spectrum—a result first derived by Madey (Reference 5). The connection to spontaneous emission is surprising since we did not make use of any spontaneous radiation physics. A quantum mechanical derivation of the gain expression, otherwise an extremely tedious affair, leads however to (13.3-17) in an intuitive manner. The gain is obtained as the difference between the stimulated emission rate by electrons of radiation and that for stimulated absorption. Because of the dispersion relation for electrons and the need to conserve total energy and momentum (see Section 13.1), the lineshape functions for these two processes are similar but peak at slightly different frequencies, as illustrated by Figures 13.7 and 13.8. We thus have

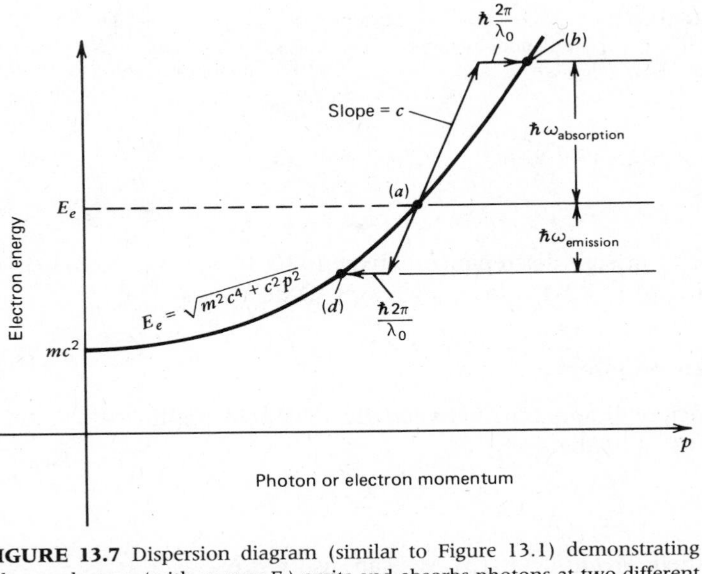

FIGURE 13.7 Dispersion diagram (similar to Figure 13.1) demonstrating why an electron (with energy E_e) emits and absorbs photons at two different frequencies. The horizontal segments of length $\hbar 2\pi/\lambda_0$ represent the momentum supplied (in absorption) or taken up (in emission) by the periodic structure (wiggler) of period λ_0. Since the final electron states (*b*) or (*d*) must fall on the electron dispersion curve E_e vs. p and the wiggler "momentum" $2\pi\hbar\lambda_0^{-1}$ is fixed, the frequencies of the photons absorbed or emitted by the electrons at (*a*) are uniquely determined as shown.

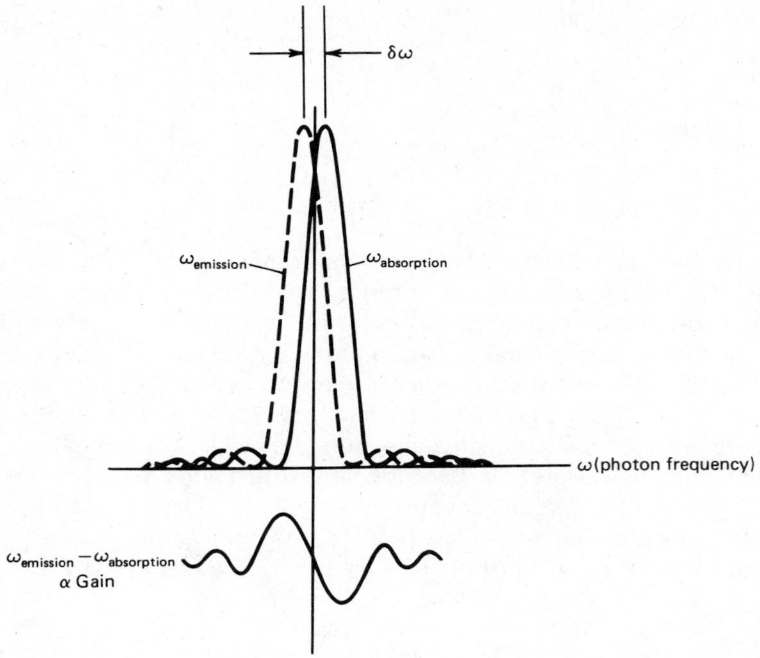

FIGURE 13.8 A construction explaining the origin of the derivative relation (13.3-7) between spontaneous emission and stimulated gain in free-electron Lasers. The shift $\delta\omega = \omega_{\text{absorption}} - \omega_{\text{emission}}$ is determined from Figure 13.7.

$$\text{gain}(\omega) \propto W_{em}(\omega) - W_{ab}(\omega) = W_{em}(\omega) - W_{em}(\omega - \delta\omega) = \frac{\partial W_{em}}{\partial \omega}\,\delta\omega \tag{13.3-18}$$

$$\delta\omega \equiv \omega_{ab} - \omega_{em}$$

The apparent sign discrepancy between (13.3-18) and (13.3-17) is because, according to (13.2-10), as ω increases, Ω decreases.

Tapered Wigglers

The efficient interaction between the *EM* field requires that the condition $\Omega \approx 0$ or equivalently

$$k_0 \approx \frac{k}{2\gamma^2}(1 + a_w^2/2)$$

be satisfied. From (13.1-15), we can also write it as

$$\lambda = \frac{\lambda_0}{2\gamma^2}\left[1 + \frac{1}{2}\left(\frac{eB_0\lambda_0}{2\pi mc^2}\right)^2\right] \tag{13.3-19}$$

As the electrons lose, on the average, energy to the field γ decreases and the resonance condition (13.3-19) is violated. Once $\langle\Delta\gamma\rangle/\gamma_r \gtrsim (2N)^{-1}$, the energy

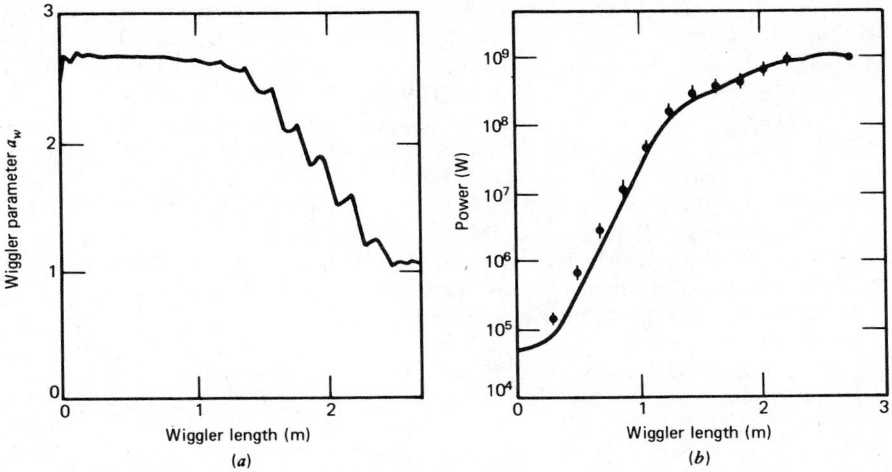

FIGURE 13.9 (*a*) The profile of the wiggler parameter a_w used in a tapered wiggler experiment at $f = 3.46 \times 10^{10}$ Hz. (*b*) The buildup of the power along the wiggler, solid curve-theoretical, points-experimental. *Source:* Reference 15.

exchange is reduced, according to (13.2-25), to near zero. From (13.3-19), it follows that the resonance can be maintained by tapering either the magnet period λ_0 or B_0 or both in such a way that

$$\lambda = \frac{\lambda_0(z)}{2\gamma^2(z)} \left\{1 + \frac{1}{2}\left[\frac{eB_0(z)\lambda_0(z)}{2\pi mc^2}\right]^2\right\}$$

A survey of recent FEL progress and experiments is given in Reference 13. Among the noteworthy results described is the Stanford FEL utilizing a 66 MeV electron beam with a peak current of 2.5 A. A peak power output of ~1.2 MW was $\sim 10^{10}$ times the peak power of the spontaneous wiggler radiation. Extraction efficiencies at 10.6 μm of ~1–2% were reported recently (Reference 14) by tapering the wiggler field downward along the interaction path. Figure 13.9*a* shows the taper profile of a_w along the wiggler, whereas in (*b*), the amplified power level of an injected signal at $f = 3.46 \times 10^{10}$ Hz is shown.

An excellent monograph by T. C. Marshall (see Reference 15) summarizes much of the theoretical and experimental progress up to 1985. A good review of the field up to 1985 is given in Reference 16.

References

1. See, for example, J. R. Pierce, "Traveling Wave Tubes," (New York: Van Nostrand, 1950).
2. Motz, H., "Applications of the Radiation from Fast Electron Beams," *J. Appl. Phys.* **22,** 527 (1951).

3. Motz, H., W. Thon, and R. N. Whitehurst, "Experiments on Radiation by Fast Electron Beams," *J. Appl. Phys.* **24,** 826 (1953).
4. Phillips, R. M., "The Ubitron, a High-Power Traveling-Wave Tube Based on a Periodic Beam Interaction in Unloaded Waveguide," *IRE Trans. Electron. Dev.* **7,** 231 (1960).
5. Madey, J. M. J., "Stimulated Emission of Bremstrahlung in a Periodic Magnetic Field," *J. Appl. Phys.* **42,** 1906 (1971).
6. Madey, J. M. J., "Stimulated Emission of Radiation in Periodically Deflected Electron Beams," U.S. Patent 3,822,410 (1974).
7. Elias, L. R., W. M. Fairbank, J. M. J. Madey, H. A. Schwettman, and T. I. Smith, "Observation of Stimulated Emission of Radiation by Relativistic Electrons in a Spatially Periodic Transverse Magnetic Field," *Phys. Rev. Letters* **36,** 717 (1976).
8. Deacon, D. A. G., L. R. Elias, J. M. J. Madey, G. J. Ramian, H. A. Schwettman, and T. I. Smith, "First Operation of a Free-Electron Laser," *Phys. Rev. Letters* **38,** 892 (1977).
9. Colson, W. B. and A. M. Sessler, "Free Electron Lasers," *Ann. Rev. Nucl. Sci.* **35,** 25 (1985).
10. Morton, P. L., "Free Electron Lasers," *Phys. of Quant. Elect.* **8,** 132 (1982).
11. Colson, W. B., "One Body Analysis of Free Electron Lasers," in *Physics of Quantum Electronics,* Vol. 5 (Reading, Mass.: Addison Wesley, 1978).
12. The astute student may notice the similarity between (13.3-16) and the mode locked pulse shapes in (14.13-4). The similarity is due to formal analogy between one magnet wiggle and one mode of a multiequispaced mode laser. A similar result is obtained for the radiation pattern of light diffracted from a grating and many phenomena involving waves in periodic environments.
13. Goss Levi, B., "Free Electron Lasers Take Small Steps toward Distant Goal," *Phys. Today* **17** (1987).
14. Orzechowsky, T. J., "High Efficiency Extraction of Microwave Radiation from a Tapered Wiggler Free Electron Laser," *Phys. Rev. Letters* **57,** 2172 (1986).
15. Marshall, T. C., *Free Electron Lasers* (New York: MacMillan, 1985).
16. Numerous articles on Free Electron Lasers can be found in special issues of the *J. Quant. Elect.* **QE-17** (Aug. 1981) and **QE-21** (July 1985).

Problems

13.1 (a) Derive the power gain of a longitudinal free-electron laser in which an electromagnetic beam

$$E_z(z, t) = E_0 \cos(\omega t - \beta z)$$

interacts with an electron beam travelling with a velocity v_0 along the z axis (i.e., no wiggling).

13.2 (a) Could you simulate the *EM* wave of Problem 13.1 by using a succession of microwave resonators as shown in the figure below. The $\Phi_1, \ldots, \Phi_N$ are adjustable phase delays, ω the microwave frequency, $v_0 = \beta_z c$ the electron average velocity.

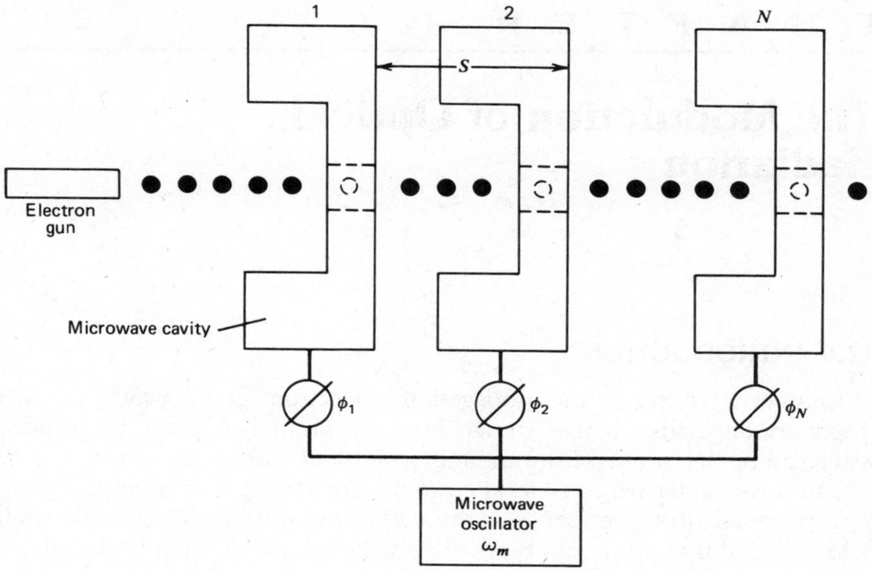

(b) Discuss the resonance condition for this configuration.

13.3 Show that the condition $\Omega = 0$ where $\Omega \equiv \omega_0 - \omega(1 - \beta_z)$ is equivalent to the synchronism condition (13.1-12).

CHAPTER 14

The Modulation of Optical Radiation

14.0 INTRODUCTION

In Chapter 5 we treated the propagation of electromagnetic waves in anisotropic crystal media. It was shown how the properties of the propagating wave can be determined from the index ellipsoid surface.

In this chapter we consider the problem of propagation of optical radiation in crystals in the presence of an applied electric field or acoustic strain field. We find that, in certain types of crystals, it is possible to effect a change in the index of refraction that is proportional to the field. These are referred to, respectively, as the electrooptic and photoelastic effects. They afford a convenient and widely used means of controlling the intensity or phase of the propagating radiation. This modulation is used in an ever-expanding number of applications including the impression of information onto optical beams, Q-switching of lasers (Chapter 20) for generation of giant optical pulses, mode locking, and optical beam deflection. Some of these applications will be discussed further in this chapter. An extended treatment of all the material in this chapter can be found in Reference 25.

14.1 THE ELECTROOPTIC EFFECT

In Chapter 5 we found that, given a direction in a crystal, in general two possible linearly polarized modes exist—the so-called rays of propagation. Each mode possesses a unique direction of polarization (i.e., direction of **D**) and a corresponding index of refraction (i.e., a velocity of propagation). The mutually orthogonal polarization directions and the indices of the two rays are found most easily by using the index ellipsoid

$$\frac{x^2}{n_x^2} + \frac{y^2}{n_y^2} + \frac{z^2}{n_z^2} = 1 \qquad (14.1\text{-}1)$$

where the directions x, y, and z are the principal dielectric axes—that is, the directions in the crystal along which **D** and **E** are parallel. The existence of an "ordinary" and an "extraordinary" ray with different indices of refraction is called birefringence.

The linear electrooptic effect is the change in the indices of the ordinary and extraordinary rays that is caused by and is proportional to an applied electric field. This effect exists only in crystals that do not possess inversion symmetry.[1] This statement can be justified as follows: Assume that in a crystal possessing an inversion symmetry, the application of an electric field E along some direction causes a change $\Delta n_1 = sE$ in the index, where s is a constant characterizing the linear electrooptic effect. If the direction of the field is reversed, the change in the index is given by $\Delta n_2 = s(-E)$, but because of the inversion symmetry, the two directions are physically equivalent, so $\Delta n_1 = \Delta n_2$. This requires that $s = -s$, which is possible only for $s = 0$, so no linear electrooptic effect can exist. The division of all crystal classes into those that do and those that do not possess an inversion symmetry is an elementary consideration in crystallography and this information is widely tabulated (Reference 1).

Since the propagation characteristics in crystals are fully described by means of the index ellipsoid (14.1-1), the effect of an electric field on the propagation is expressed most conveniently by giving the changes in the constants $1/n_x^2$, $1/n_y^2$, $1/n_z^2$ of the index ellipsoid.

We take the equation of the index ellipsoid in the presence of an electric field as

$$\left(\frac{1}{n^2}\right)_1 x^2 + \left(\frac{1}{n^2}\right)_2 y^2 + \left(\frac{1}{n^2}\right)_3 z^2 + 2\left(\frac{1}{n^2}\right)_4 yz + 2\left(\frac{1}{n^2}\right)_5 xz + 2\left(\frac{1}{n^2}\right)_6 xy = 1 \qquad (14.1\text{-}2)$$

If we choose x, y, and z to be parallel to the principal dielectric axes of the crystal, then with zero applied field, (14.1-2) must reduce to (14.1-1); therefore,

$$\left.\left(\frac{1}{n^2}\right)_1\right|_{E=0} = \frac{1}{n_x^2} \qquad \left.\left(\frac{1}{n^2}\right)_2\right|_{E=0} = \frac{1}{n_y^2}$$

$$\left.\left(\frac{1}{n^2}\right)_3\right|_{E=0} = \frac{1}{n_z^2} \qquad \left.\left(\frac{1}{n^2}\right)_4\right|_{E=0} = \left.\left(\frac{1}{n^2}\right)_5\right|_{E=0} = \left.\left(\frac{1}{n^2}\right)_6\right|_{E=0} = 0$$

The linear change in the coefficients

$$\left(\frac{1}{n^2}\right)_i \qquad i = 1, \ldots, 6$$

due to an arbitrary "low-frequency" electric field $\mathbf{E}(E_x, E_y, E_z)$ is defined by

$$\Delta\left(\frac{1}{n^2}\right)_i = \sum_{j=1}^{3} \mathbf{r}_{ij} E_j \qquad (14.1\text{-}3)$$

where in the summation over j we use the convention $1 = x$, $2 = y$, $3 = z$.

[1] If a crystal contains a regular lattice of points such that inversion (replacing each atom at $\mathbf{r}$ by one at $-\mathbf{r}$, with $\mathbf{r}$ being the position vector relative to the point) about any one of these points leaves the crystal structure invariant, the crystal is said to possess inversion symmetry.

Equation (14.1-3) can be expressed in a matrix form as

$$\begin{vmatrix} \Delta\left(\frac{1}{n^2}\right)_1 \\ \Delta\left(\frac{1}{n^2}\right)_2 \\ \Delta\left(\frac{1}{n^2}\right)_3 \\ \Delta\left(\frac{1}{n^2}\right)_4 \\ \Delta\left(\frac{1}{n^2}\right)_5 \\ \Delta\left(\frac{1}{n^2}\right)_6 \end{vmatrix} = \begin{vmatrix} \mathbf{r}_{11} & \mathbf{r}_{12} & \mathbf{r}_{13} \\ \mathbf{r}_{21} & \mathbf{r}_{22} & \mathbf{r}_{23} \\ \mathbf{r}_{31} & \mathbf{r}_{32} & \mathbf{r}_{33} \\ \mathbf{r}_{41} & \mathbf{r}_{42} & \mathbf{r}_{43} \\ \mathbf{r}_{51} & \mathbf{r}_{52} & \mathbf{r}_{53} \\ \mathbf{r}_{61} & \mathbf{r}_{62} & \mathbf{r}_{63} \end{vmatrix} \begin{vmatrix} E_1 \\ E_2 \\ E_3 \end{vmatrix} \tag{14.1-4}$$

where, using the rules for matrix multiplication, we have, for example,

$$\Delta\left(\frac{1}{n^2}\right)_6 = r_{61}E_1 + r_{62}E_2 + r_{63}E_3$$

The 6×3 matrix with elements r_{ij} is called the electrooptic tensor. We have argued above that in crystals possessing an inversion symmetry (centrosymmetric), $r_{ij} = 0$. The form, but not the magnitude, of the tensor r_{ij} can be derived from symmetry considerations (Reference 1), which dictate which of the 18 r_{ij} coefficients are zero, as well as the relationships that exist between the remaining coefficients. In Table 14.1, we give the form of the electrooptic tensor for all the noncentrosymmetric crystal classes. The electrooptic coefficients of some crystals are given in Table 14.2.

In general, the principal axes of the new ellipsoid (14.1-2) do not coincide with the original ($\mathbf{E} = 0$) principal axes (x, y, z). We thus need to find the direction and magnitude of the new principal axes. The procedure is the familiar one of principal axis transformation of quadratic forms (Reference 1) and consists first of finding the eigenvalues of the matrix

$$\begin{vmatrix} \left(\frac{1}{n^2}\right)_1 & \left(\frac{1}{n^2}\right)_6 & \left(\frac{1}{n^2}\right)_5 \\ \left(\frac{1}{n^2}\right)_6 & \left(\frac{1}{n^2}\right)_2 & \left(\frac{1}{n^2}\right)_4 \\ \left(\frac{1}{n^2}\right)_5 & \left(\frac{1}{n^2}\right)_4 & \left(\frac{1}{n^2}\right)_3 \end{vmatrix} \tag{14.1-5}$$

which is made up of the constants of the index ellipsoid (14.1-2). The three eigenvalues correspond to the new values of $(1/n^2)_{i'i'}$. These are then used to determine the new directions, $x_{i'}$, of the principal axes.

TABLE 14.1. The Form of the Electrooptic Tensor for All Crystal Symmetry Classes

Symbols:

• zero element ●—● equal nonzero elements

● nonzero element ●—○ equal nonzero elements, but opposite in sign

The symbol at the upper left corner of each tensor is the conventional symmetry group designation.

Centrosymmetric—All elements zero.

Triclinic

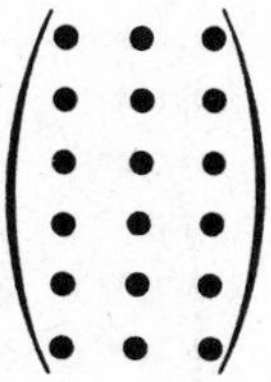

Monoclinic

2 (parallel to x_2)

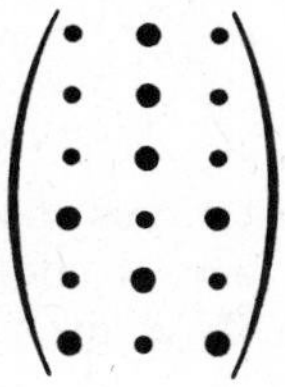

(parallel to x_3)

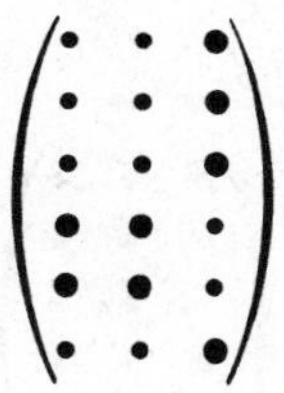

m (perpendicular to x_2)

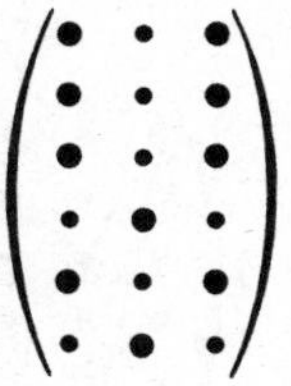

(perpendicular to x_3)

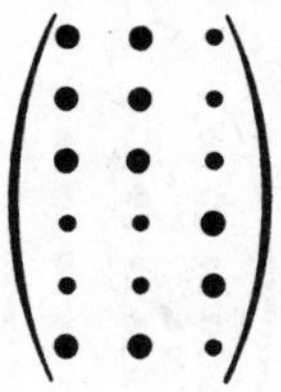

Orthorhombic

222

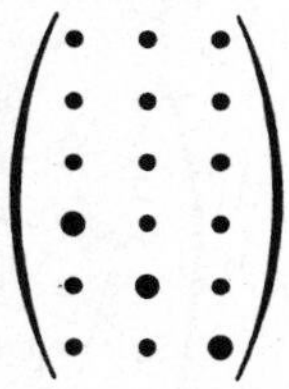

$mm2$

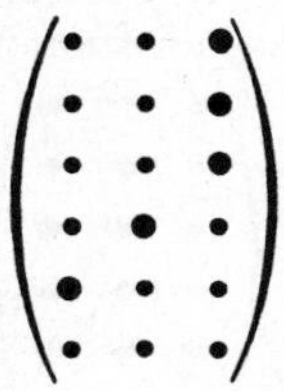

TABLE 14.1. (*Continued*)

Tetragonal

4

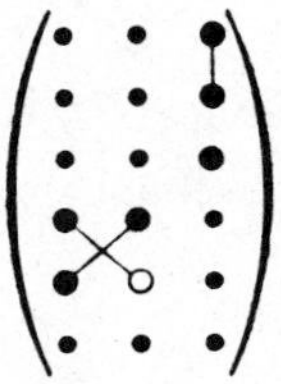

$\bar{4}$

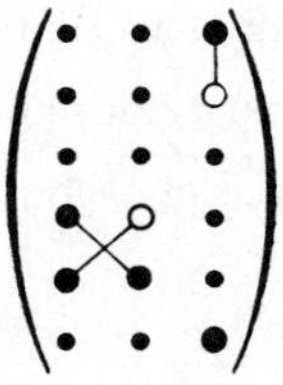

422

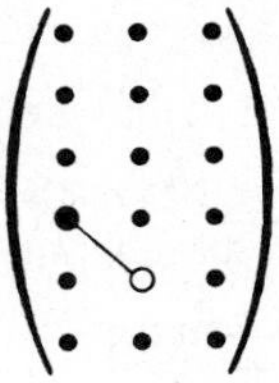

4*mm*

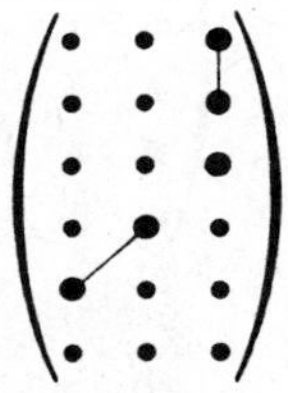

Example:
$BaTiO_3$

$\bar{4}2m$ (2 parallel to x_1)

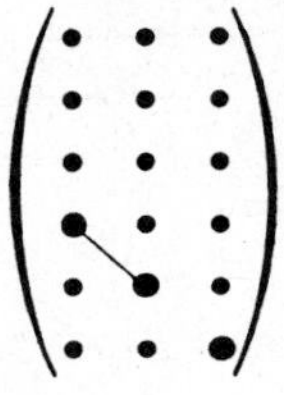

Example:
KH_2PO_4(KDP)

Cubic

$\bar{4}3m$, 23

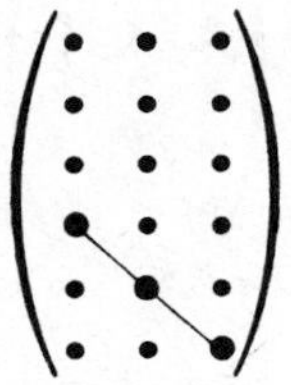

Examples: Crystals of the zinc blende class: GaAs, InAs, CdTe

432

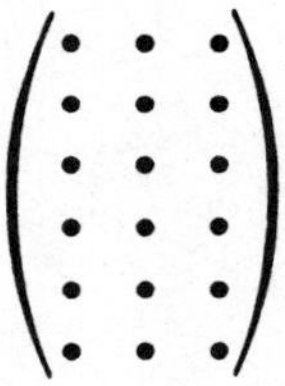

Trigonal

3

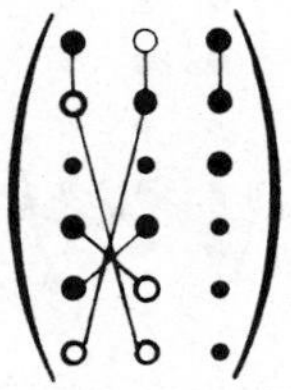

32

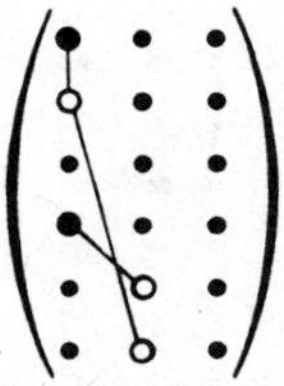

Examples: Te, quartz

3*m* (*m* perpendicular to x_1)
standard orientation

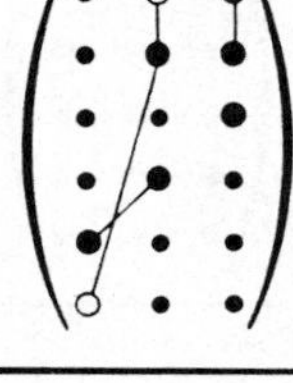

3*m* (*m* perpendicular to x_2)

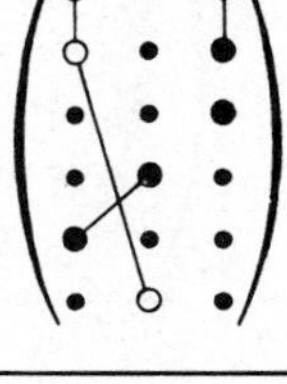

Example:
$LiNbO_3$
$LiTaO_3$

TABLE 14.1. ***(Continued)***

Hexagonal

6

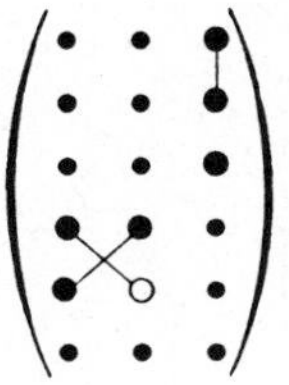

$6mm$

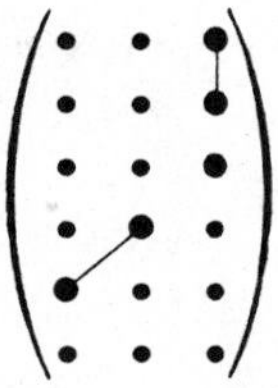

(same as $4mm$)

Example:
CdS

622

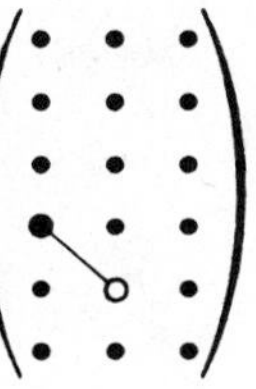

$\bar{6}$

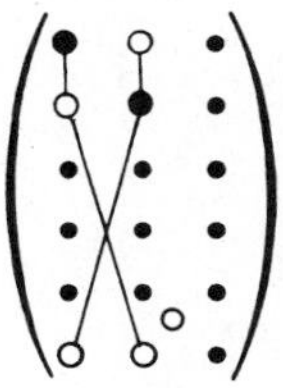

$\bar{6}m2$ (m perpendicular to x_1 standard orientation)

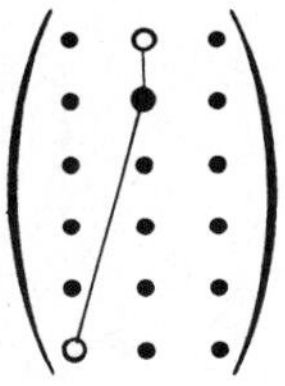

(m perpendicular to x_2)

Example: The Electrooptic Effect in KH_2PO_4. Consider the specific example of a crystal of potassium dihydrogen phosphate (KH_2PO_4), also known as KDP. The crystal has a fourfold axis of symmetry[2] that, by strict convention, is taken as the z (optic) axis, as well as two mutually orthogonal twofold axes of symmetry that lie in the plane normal to z. These are designated as the x and y axes. The symmetry group[3] of this crystal is $\bar{4}2m$. Using

[2] That is, a rotation by $2\pi/4$ about this axis leaves the crystal structure invariant.

[3] The significance of the symmetry group symbols and a listing of most known crystals and their symmetry groups are found in any basic book on crystallography.

TABLE 14.2. Some Electrooptic Materials and Their Properties

Material	*Room Temperature Electrooptic Coefficients in Units of* 10^{-12} *m/V*	*Index of Refraction*[a]	$n_o^3 r$, *in Units of* 10^{-12} *m/V*	$\varepsilon/\varepsilon_0$ *(Room Temperature)*	*Point-Group Symmetry*
KDP (KH_2PO_4)	$r_{41} = 8.6$ $r_{63} = 10.6$	$n_o = 1.51$ $n_e = 1.47$	29 34	$\varepsilon \parallel c = 20$ $\varepsilon \perp c = 45$	$\bar{4}2m$
KD_2PO_4	$r_{63} = 23.6$	~ 1.50	80	$\varepsilon \parallel c \sim 50$ at 24°C	$\bar{4}2m$
ADP ($NH_4H_2PO_4$)	$r_{41} = 28$ $r_{63} = 8.5$	$n_o = 1.52$ $n_e = 1.48$	95 27	$\varepsilon \parallel c = 12$	$\bar{4}2m$
Quartz	$r_{41} = 0.2$ $r_{63} = 0.93$	$n_o = 1.54$ $n_e = 1.55$	0.7 3.4	$\varepsilon \parallel c \sim 4.3$ $\varepsilon \perp c \sim 4.3$	32
CuCl	$r_{41} = 6.1$	$n_o = 1.97$	47	7.5	$\bar{4}3m$

ZnS	$r_{41} = 2.0$	$n_o = 2.37$	27	~ 10	$\bar{4}3m$
GaAs at 10.6 μ	$r_{41} = 1.6$	$n_o = 3.34$	59	11.5	$\bar{4}3m$
ZnTe at 10.6 μ	$r_{41} = 3.9$	$n_o = 2.79$	77	7.3	$\bar{4}3m$
CdTe at 10.6 μ	$r_{41} = 6.8$	$n_o = 2.6$	120		$\bar{4}3m$
$LiNbO_3$	$r_{33} = 30.8$ $r_{13} = 8.6$ $r_{22} = 3.4$ $r_{42} = 28$	$n_o = 2.29$ $n_e = 2.20$	$n_e^3 r_{33} = 328$ $n_o^3 r_{22} = 37$ $\frac{1}{2}(n_e^3 r_{33} - n_o^3 r_{13}) = 112$	$\varepsilon \perp c = 98$ $\varepsilon \parallel c = 50$	$3m$
GaP	$r_{41} = 0.97$	$n_o = 3.31$	$n_o^3 r_{41} = 29$		$\bar{4}3m$
$LiTaO_3$ (30°C)	$r_{33} = 30.3$ $r_{13} = 5.7$	$n_o = 2.175$ $n_e = 2.180$	$n_e^3 r_{33} = 314$	$\varepsilon \parallel c = 43$	$3m$
$BaTiO_3$ (30°C)	$r_{33} = 23$ $r_{13} = 8.0$ $r_{42} = 820$	$n_o = 2.437$ $n_e = 2.365$	$n_e^3 r_{33} = 334$	$\varepsilon \perp c = 4300$ $\varepsilon \parallel c = 106$	$4mm$

[a] Typical value.
Source: References 2 and 3.

Table 14.1, we write the electrooptic tensor in the form of

$$r_{ij} = \begin{vmatrix} 0 & 0 & 0 \\ 0 & 0 & 0 \\ 0 & 0 & 0 \\ r_{41} & 0 & 0 \\ 0 & r_{41} & 0 \\ 0 & 0 & r_{63} \end{vmatrix} \tag{14.1-6}$$

so that the only nonvanishing elements are $r_{41} = r_{52}$ and r_{63}. Using (14.1-2), (14.1-4), and (14.1-6), we obtain the equation of the index ellipsoid in the presence of a field $\mathbf{E}(E_x, E_y, E_z)$ as

$$\frac{x^2}{n_o^2} + \frac{y^2}{n_o^2} + \frac{z^2}{n_e^2} + 2r_{41}E_x yz + 2r_{41}E_y xz + 2r_{63}E_z xy = 1 \tag{14.1-7}$$

where the constants involved in the first three terms do not depend on the field and, since the crystal is uniaxial, are taken as $n_x = n_y = n_o$, $n_z = n_e$. We thus find that the application of an electric field causes the appearance of "mixed" terms in the equation of the index ellipsoid. These are the terms with xy, xz, yz. This means that the major axes of the ellipsoid, with a field applied, are no longer parallel to the x, y, and z crystal axes. It becomes necessary, then, to find the directions of the new axes, and the magnitudes of the respective indices in the presence of $\mathbf{E}$, so that we may determine the effect of the field on the propagation. To be specific, we choose the direction of the applied field parallel to the z axis, so (14.1-7) becomes

$$\frac{x^2 + y^2}{n_o^2} + \frac{z^2}{n_e^2} + 2r_{63}E_z xy = 1 \tag{14.1-8}$$

The problem is one of finding a new coordinate system (x', y', z') in which the equation of the ellipsoid (14.1-8) contains no mixed terms; that is, it is of the form

$$\frac{x'^2}{n_{x'}^2} + \frac{y'^2}{n_{y'}^2} + \frac{z'^2}{n_{z'}^2} = 1 \tag{14.1-9}$$

x', y', and z' are then the directions of the major axes of the ellipsoid in the presence of an external field applied parallel to z. The length of the major axes of the ellipsoid is, according to (14.1-9), $2n_{x'}$, $2n_{y'}$, and $2n_{z'}$ and these will, in general, depend on the applied field.

In the case of (14.1-8), it is clear from inspection that in order to put it in a diagonal form, we need to choose a coordinate system x', y', z', where z' is parallel to z. Because of the symmetry of (14.1-8) in x and y, x' and y' are related to x and y by a 45° rotation as shown in Figure 14.1. The transformation relations from x, y to x', y' are thus

$$x = x' \cos 45^\circ - y' \sin 45^\circ$$

$$y = x' \sin 45^\circ + y' \cos 45^\circ$$

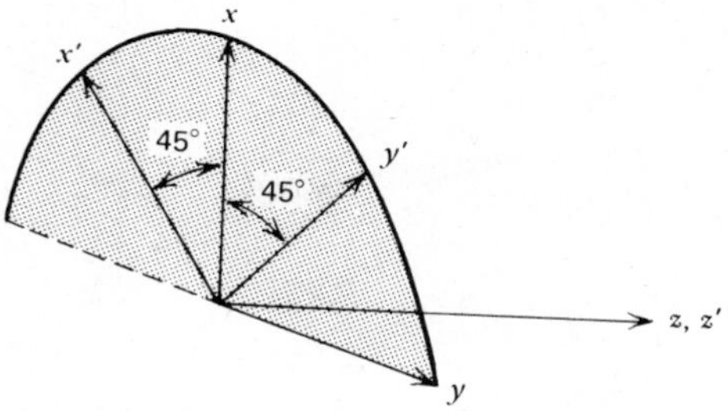

FIGURE 14.1 The x, y, and z axes of $\bar{4}2m$ crystals (such as KH_2PO_4) and the x', y', and z' axes, where z is the fourfold optic axis and x and y are the twofold symmetry axes of crystals with $\bar{4}2m$ symmetry.

that, upon substitution in (14.1-8), yield

$$\left(\frac{1}{n_o^2} + r_{63}E_z\right)x'^2 + \left(\frac{1}{n_o^2} - r_{63}E_z\right)y'^2 + \frac{z^2}{n_e^2} = 1 \qquad (14.1\text{-}10)$$

Equation (14.1-10) shows that x', y', and z are indeed the principal axes of the ellipsoid when a field is applied along the z direction. According to (14.1-10), the length of the x' axis of the ellipsoid is $2n_{x'}$, where

$$\frac{1}{n_{x'}^2} = \frac{1}{n_o^2} + r_{63}E_z$$

that, assuming $r_{63}E_z \ll n_o^{-2}$ and using the differential relation

$$dn = -(n^3/2)d(1/n^2)$$

gives

$$n_{x'} = n_o - \frac{n_o^3}{2} r_{63}E_z \qquad (14.1\text{-}11)$$

and, similarly,

$$n_{y'} = n_o + \frac{n_o^3}{2} r_{63}E_z \qquad (14.1\text{-}12)$$

$$n_z = n_e \qquad (14.1\text{-}13)$$

14.2 ELECTROOPTIC RETARDATION

The index ellipsoid for KDP with **E** applied parallel to z is shown in Figure 14.2. If we consider propagation along the z direction, then, according to the procedure described in Section 5.3, we need to determine the ellipse formed by the intersection of the plane $z = 0$ (in general, the plane that contains the origin and is normal to the propagation direction) and the ellipsoid. The equation of this ellipse is obtained from (14.1-10) by putting $z = 0$ and is

$$\left(\frac{1}{n_o^2} + r_{63}E_z\right)x'^2 + \left(\frac{1}{n_o^2} - r_{63}E_z\right)y'^2 = 1 \qquad (14.2\text{-}1)$$

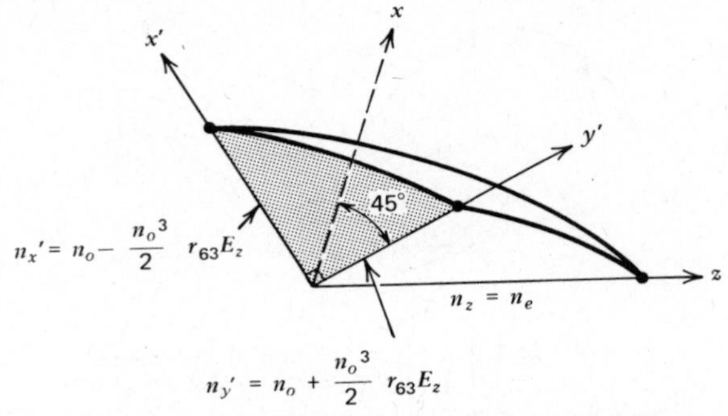

FIGURE 14.2 A section of the index ellipsoid of KDP, showing the principal dielectric axes x', y', and z due to an electric field applied along the z axis. The directions x' and y' are defined by Figure 14.1.

One quadrant of the ellipse is shown (shaded) in Figure 14.2 along with its minor and major axes, which in this case coincide with x' and y', respectively. It follows from Section 5.7 that the two allowed directions of polarization are x' and y' and that the corresponding indices of refraction are $n_{x'}$ and $n_{y'}$, which are given by (14.1-11) and (14.1-12).

We are now in a position to take up the concept of retardation. We consider an optical field that is incident normally on the $x'y'$ plane with its **E** vector along the x direction. We can resolve the optical field at $z = 0$ (input plane) into two mutually orthogonal components polarized along x' and y'. The x' component propagates as

$$E_{x'} = Ae^{i[\omega t - (\omega/c) n_{x'} z]}$$

that, if we use (14.1-11), becomes

$$E_{x'} = Ae^{i\{\omega t - (\omega/c)[n_o - (n_o^3/2) r_{63} E_z]z\}} \tag{14.2-2}$$

whereas the y' component is given by

$$E_{y'} = Ae^{i\{\omega t - (\omega/c)[n_o + (n_o^3/2) r_{63} E_z]z\}} \tag{14.2-3}$$

The phase difference at the output plane $z = l$ between the two components is called the *retardation*. It is given by the difference of the exponents in (14.2-2) and (14.2-3) and is equal to

$$\Gamma = \phi_{x'} - \phi_{y'} = \frac{\omega n_o^3 r_{63} V}{c} \tag{14.2-4}$$

where $V = E_z l$ and $\phi_{x'} = -(\omega n_{x'}/c)l$.

Figure 14.3 shows $E_{x'}(z)$ and $E_{y'}(z)$ at some moment in time. Also shown are the curves traversed by the tip of the optical field vector at various points along the path. At $z = 0$, the retardation is $\Gamma = 0$ and the field is linearly polarized along x. At point e, $\Gamma = \pi/2$; thus, omitting a common phase factor,

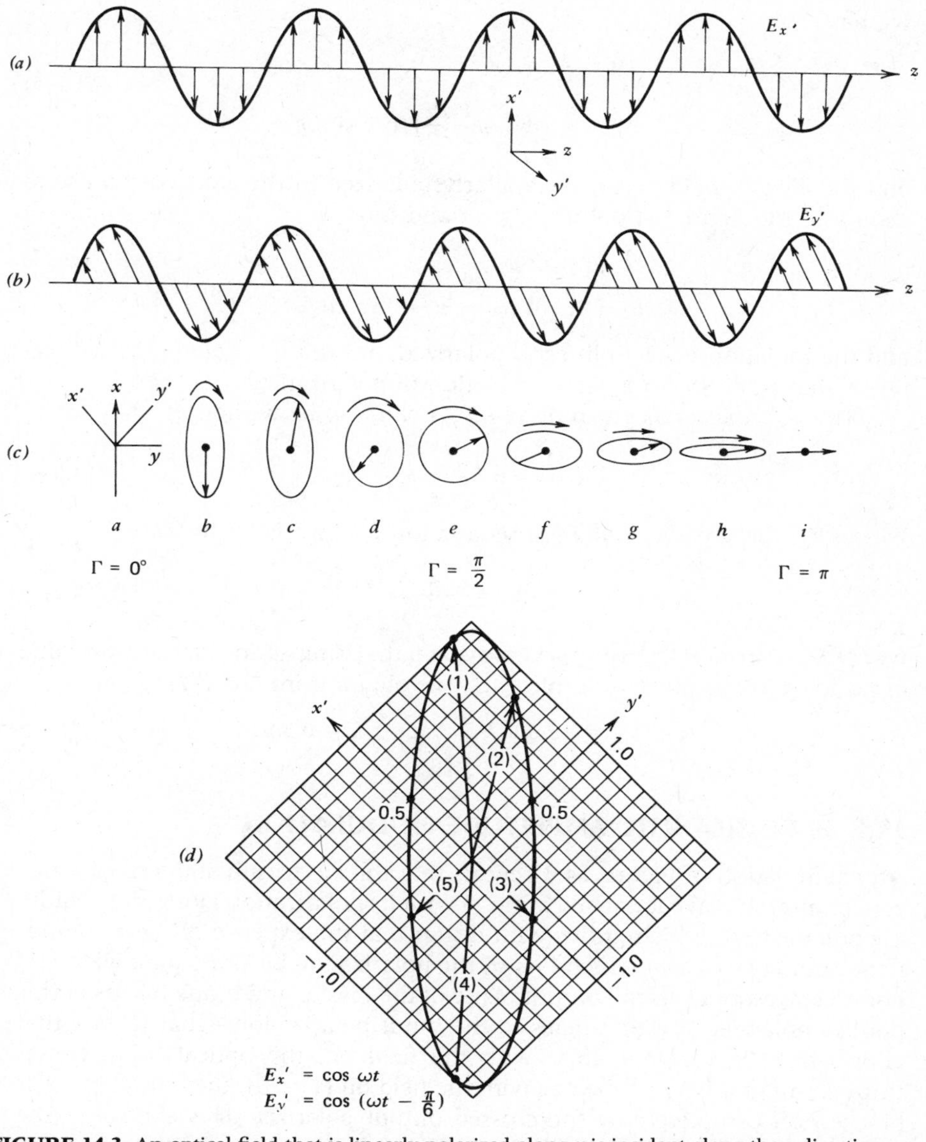

FIGURE 14.3 An optical field that is linearly polarized along x is incident along the z direction on an electrooptic crystal having its electrically induced principal axes along x' and y'. (This is the case in KH_2PO_4 when an electric field is applied along its z axis. (*a*) The component, $E_{x'}$, at some time t as a function of the position, z, along the crystal. (*b*) $E_{y'}$ as a function of z at the same value of t as in (*a*). (*c*) The ellipses in the x'-y' plane traversed by the tip of the optical electric field at various points (*a* through *i*) along the crystal during one optical period. The arrow shows the instantaneous field vector at time t, whereas the curved arrow gives the sense in which the ellipse is traversed. (*d*) A plot of the polarization ellipse due to two orthogonal components with a retardation of $\Gamma = \pi/6$ [i.e., $E_{x'} = \cos \omega t$ and $E_{y'} = \cos(\omega t - \pi/6)$]. Also shown are the instantaneous field vectors at (1) $\omega t = 0°$, (2) $\omega t = 60°$, (3) $\omega t = 120°$, (4) $\omega t = 210°$, and (5) $\omega t = 270°$.

we have

$$E_{x'} = A \cos \omega t$$
$$E_{y'} = A \cos\left(\omega t - \frac{\pi}{2}\right) = A \sin \omega t \tag{14.2-5}$$

and the electric field vector is circularly polarized in the clockwise sense as shown in the figure. At point i, $\Gamma = \pi$ and thus

$$E_{x'} = A \cos \omega t$$
$$E_{y'} = A \cos(\omega t - \pi) = -A \cos \omega t$$

and the radiation is again linearly polarized, but this time along the y direction—that is, at 90° to its input direction of polarization.

The retardation as given by (14.2-4) can also be written as

$$\Gamma = \pi \frac{E_z l}{V_\pi} = \pi \frac{V}{V_\pi} \tag{14.2-6}$$

where V_π, the voltage yielding a retardation $\Gamma = \pi$,[4] is in this case

$$V_\pi = \frac{\lambda}{2n_o^3 r_{63}} \tag{14.2-7}$$

where $\lambda = 2\pi c/\omega$ is the free-space wavelength. Using as an example the value of r_{63} for ADP, as given in Table 14.2, we obtain from (14.2-7)

$$(V_\pi)_{\text{ADP}} = 10{,}000 \text{ V} \qquad \text{at} \qquad \lambda = 0.5\ \mu\text{m}$$

14.3 ELECTROOPTIC AMPLITUDE MODULATION

An examination of Figure 14.3 reveals that the electrically induced birefringence causes a wave launched at $z = 0$ with its polarization along x to acquire a y polarization, which grows with distance at the expense of the x component until at point i, at which $\Gamma = \pi$, the polarization becomes parallel to y. If point i corresponds to the output plane of the crystal and if one inserts at this point a polarizer at right angles to the input polarization—that is, one that allows only E_y to pass—then with the field on, the optical beam passes through unattenuated, whereas with the field off ($\Gamma = 0$), the output beam is blocked off completely by the crossed output polarizer. This control of the optical energy flow serves as the basis of the electrooptic amplitude modulation of light.

A typical arrangement of an electrooptic amplitude modulator is shown in Figure 14.4. It consists of an electrooptic crystal placed between two crossed polarizers, which are at an angle of 45° with respect to the electrically induced birefringent axes x' and y'. To be specific, we show how this arrange-

[4] V_π is referred to as the "half-wave" voltage since, as can be seen in Figure 14.3i, it causes the two waves that are polarized along x' and y' to acquire a relative spatial displacement of half a wavelength.

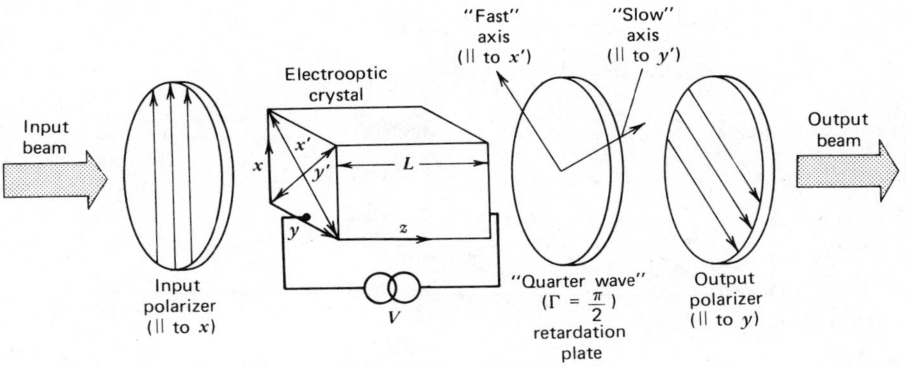

FIGURE 14.4 A typical electrooptic amplitude modulator. The total retardation Γ is the sum of the fixed retardation bias ($\Gamma_B = \pi/2$) introduced by the "quarter-wave" plate and that caused by the electrooptic crystal.

ment is achieved using a KDP crystal. Also included in the optical path is a naturally birefringent crystal that introduces a fixed retardation, so the total retardation Γ is the sum of the retardation due to this crystal and the electrically induced one. The incident field is parallel to x at the input face of the crystal, thus having equal in-phase components along x' and y' that we take as

$$E_{x'} = A \cos \omega t$$

$$E_{y'} = A \cos \omega t$$

or, using the complex amplitude notation, we obtain

$$E_{x'}(0) = A$$

$$E_{y'}(0) = A$$

The incident intensity is thus

$$I_i \propto \mathbf{E} \cdot \mathbf{E}^* = |E_{x'}(0)|^2 + |E_{y'}(0)|^2 = 2A^2 \tag{14.3-1}$$

Upon emerging from the output face $z = l$, the x' and y' components have acquired, according to (14.2-4), a relative phase shift (retardation) of Γ radians, so we may take them as

$$\begin{aligned} E_{x'}(l) &= A \\ E_{y'}(l) &= Ae^{-i\Gamma} \end{aligned} \tag{14.3-2}$$

The total (complex) field emerging from the output polarizer is the sum of the y components of $E_{x'}(l)$ and $E_{y'}(l)$

$$(E_y)_0 = \frac{A}{\sqrt{2}} (e^{-i\Gamma} - 1) \tag{14.3-3}$$

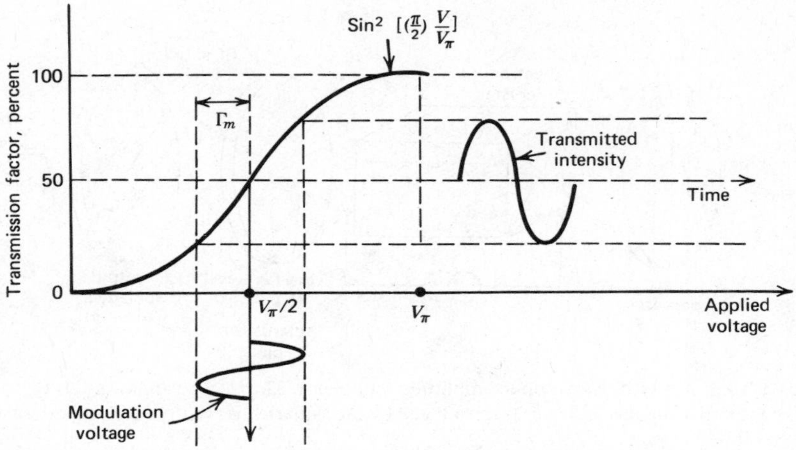

FIGURE 14.5 Transmission factor of a cross-polarized electrooptic modulator as a function of an applied voltage. The modulator is biased to the point $\Gamma = \pi/2$, which results in a 50% intensity transmission. A small applied sinusoidal voltage modulates the transmitted intensity about the bias point.

that corresponds to an output intensity

$$I_0 \propto [(E_y)_0(E_y^*)_0]$$

$$= \frac{A^2}{2}[(e^{-i\Gamma} - 1)(e^{i\Gamma} - 1)] = 2A^2 \sin^2 \frac{\Gamma}{2}$$

where the proportionality constant is the same as in (14.3-1). The ratio of the output intensity to the input is thus

$$\frac{I_0}{I_i} = \sin^2 \frac{\Gamma}{2} = \sin^2 \left[\left(\frac{\pi}{2}\right)\frac{V}{V_\pi}\right] \tag{14.3-4}$$

The second equality in (14.3-4) was obtained using (14.2-6). The transmission factor (I_0/I_i) is plotted in Figure 14.5 against the applied voltage.

The process of amplitude modulation of an optical signal is also illustrated in Figure 14.5. The modulator is usually biased[5] with a fixed retardation $\Gamma_B = \pi/2$ to the 50% transmission point. A small sinusoidal modulation voltage would then cause a nearly sinusoidal modulation of the transmitted intensity as shown.

To treat mathematically the situation depicted by Figure 14.5, we take

$$\Gamma = \frac{\pi}{2} + \Gamma_m \sin \omega_m t \tag{14.3-5}$$

[5] This bias can be achieved by applying a voltage $V = V_\pi/2$ or, more conveniently, by using a naturally birefringent crystal as in Figure 14.4 to introduce a phase difference (retardation) of $\pi/2$ between the x' and y' components.

where the retardation bias is taken as $\pi/2$, and Γ_m is related to the amplitude V_m of the modulation voltage $V_m \sin \omega_m t$ by (14.2-6); thus, $\Gamma_m = \pi(V_m/V_\pi)$.
Using (14.3-4), we obtain

$$\frac{I_0}{I_i} = \sin^2\left(\frac{\pi}{4} + \frac{\Gamma_m}{2} \sin \omega_m t\right) \tag{14.3-6}$$

$$= \tfrac{1}{2}[1 + \sin(\Gamma_m \sin \omega_m t)] \tag{14.3-7}$$

that, for $\Gamma_m \ll 1$, becomes

$$\frac{I_0}{I_i} \simeq \tfrac{1}{2}(1 + \Gamma_m \sin \omega_m t) \tag{14.3-8}$$

so that the intensity modulation is a linear replica of the modulating voltage $V_m \sin \omega_m t$. If the condition $\Gamma_m \ll 1$ is not fulfilled, it follows from Figure 14.5 or from (14.3-7) that the intensity variation is distorted and will contain an appreciable amount of the higher (odd) harmonics. The dependence of the distortion on Γ_m is discussed further in Problem 14.3.

14.4 PHASE MODULATION OF LIGHT

In the preceding section we saw how the modulation of the state of polarization, from linear to elliptic, of an optical beam by means of the electrooptic effect can be converted, using polarizers, to intensity modulation. Here, we consider the situation depicted by Figure 14.6, in which, instead of having equal components along the induced birefringent axes (x' and y' in Figure 14.4), the incident beam is polarized parallel to one of them—x', for example. In this case, the application of the electric field along the z direction does not

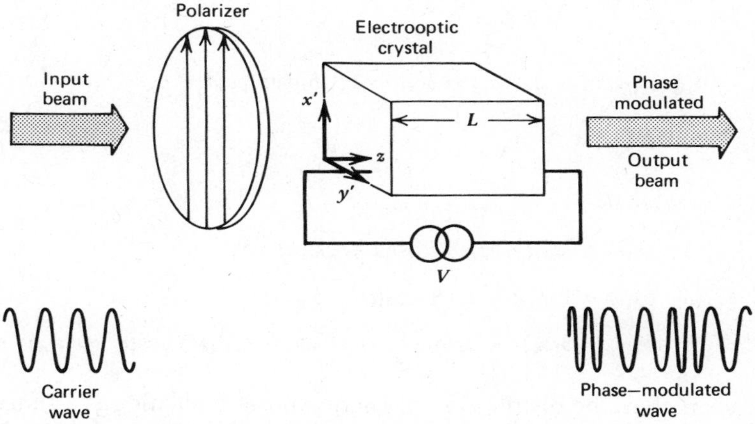

FIGURE 14.6 An electrooptic phase modulator. The crystal orientation and applied directions are appropriate to KDP. The optical polarization is parallel to an electrically induced principal dielectric axis (x').

change the state of polarization but merely changes the output phase by

$$\Delta\phi'_x = -\frac{\omega l}{c}\,\Delta n_{x'}$$

so that, from (14.1-11),

$$\Delta\phi'_x = \frac{\omega n_o^3 r_{63}}{2c} E_z l \tag{14.4-1}$$

If the bias field is sinusoidal and is taken as

$$E_z = E_m \sin \omega_m t \tag{14.4-2}$$

then an incident optical field that, at the input ($z = 0$) face of the crystal varies as $E_{\text{in}} = A \cos \omega t$, will emerge according to (14.2-2) as

$$E_{\text{out}} = A \cos\left[\omega t - \frac{\omega}{c}\left(n_0 - \frac{n_o^3}{2} r_{63} E_m \sin \omega_m t\right) l\right]$$

where l is the length of the crystal. Dropping the constant phase factor, which is of no consequence here, we rewrite the last equation as

$$E_{\text{out}} = A \cos[\omega t + \delta \sin \omega_m t] \tag{14.4-3}$$

where

$$\delta = \frac{\omega n_o^3 r_{63} E_m l}{2c} = \frac{\pi n_o^3 r_{63} E_m l}{\lambda} \tag{14.4-4}$$

is referred to as the phase modulation index. The optical field is thus phase-modulated with a modulation index δ. If we use the Bessel function identities

$$\cos(\delta \sin \omega_m t) = J_0(\delta) + 2J_2(\delta) \cos 2\omega_m t + 2J_4(\delta) \cos 4\omega_m t + \cdots$$

and

$$\sin(\delta \sin \omega_m t) = 2J_1(\delta) \sin \omega_m t + 2J_3(\delta) \sin 3\omega_m t + \cdots$$

we can rewrite (14.4-3) as

$$\begin{aligned} E_{\text{out}} = A[&J_0(\delta) \cos \omega t + J_1(\delta) \cos(\omega + \omega_m)t \\ &-J_1(\delta) \cos(\omega - \omega_m)t + J_2(\delta) \cos(\omega + 2\omega_m)t \\ &+J_2(\delta) \cos(\omega - 2\omega_m)t + J_3(\delta) \cos(\omega + 3\omega_m)t \\ &-J_3(\delta) \cos(\omega - 3\omega_m)t + J_4(\delta) \cos(\omega + 4\omega_m)t + J_4(\delta) \cos(\omega - 4\omega_m)t + \cdots] \end{aligned}$$

which form gives the distribution of energy in the sidebands as a function of the modulation index δ. We note that, for $\delta = 0$, $J_0(0) = 1$ and $J_n(\delta) = 0$, $n \neq 0$. Another point of interest is that the phase modulation index δ as given by (14.4-4) is one-half the retardation Γ as given by (14.2-4).

14.5 TRANSVERSE ELECTROOPTIC MODULATORS

In the examples of electrooptic retardation discussed in the two preceding sections, the electric field was applied along the direction of light propagation. This is the so-called longitudinal mode of modulation. A more desirable mode of operation is the transverse one, in which the field is applied normal to the direction of propagation. The reason is that in this case, the field electrodes do not interfere with the optical beam, and the retardation, being proportional to the product of the field times the crystal length, can be increased by the use of longer crystals. In the longitudinal case the retardation, according to (14.2-4), is proportional to $E_z l = V$ and is independent of the crystal length l. Figures 14.1 and 14.2 suggest how transverse retardation can be obtained using a KDP crystal with the actual arrangement shown in Figure 14.7. The light propagates along y', and its polarization is in the x'-z plane at 45° from the z axis. The retardation, with a field applied along z, is, from (14.1-11) and (14.1-13),

$$\Gamma = \phi_z - \phi'_x$$

$$= \frac{\omega l}{c}\left[(n_o - n_e) - \frac{n_o^3}{2} r_{63}\left(\frac{V}{d}\right)\right] \qquad (14.5\text{-}1)$$

where d is the crystal dimension along the direction of the applied field. We note that Γ contains a term that does not depend on the applied voltage. The resultant degradation of the modulation caused by this term is discussed in Problem 14.7. A detailed example of transverse electrooptic modulation using $\bar{4}3m$, cubic zinc-blende type crystals is considered next.

Example: The Electrooptic Effect in Cubic $\bar{4}3m$ Crystals. As an example of transverse modulation and of the application of the electrooptic effect, we consider the case of crystals of the $\bar{4}3m$ symmetry group. Examples of this group are InAs, CuCl, GaAs, and CdTe. The last two are used for modulation in the infrared, since they remain transparent beyond 10 μm (References 5 and 6). These crystals are cubic and have axes of fourfold

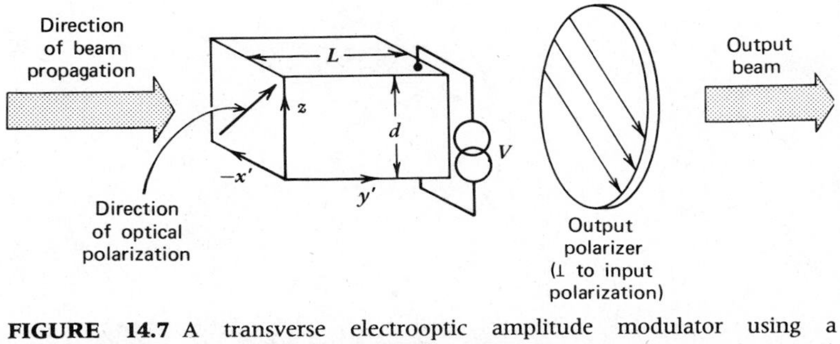

FIGURE 14.7 A transverse electrooptic amplitude modulator using a KH_2PO_4(KDP) crystal in which field is applied normal to the direction of propagation.

symmetry along the cube edges (⟨100⟩ directions) and threefold axes of symmetry along the cube diagonals ⟨111⟩.

To be specific, we apply the field in the ⟨111⟩ direction—that is, along a threefold-symmetry axis. Taking the field magnitude as E, we have

$$\mathbf{E} = \frac{E}{\sqrt{3}}(\mathbf{i} + \mathbf{j} + \mathbf{k}) \tag{14.5-2}$$

where **i**, **j**, and **k** are unit vectors directed along the cube edges x, y, and z, respectively. The three nonvanishing electrooptic tensor elements are, according to Table 14.1 (see $\bar{4}3m$ group), r_{41}, $r_{52} = r_{41}$, and $r_{63} = r_{41}$. Thus, using (14.1-2) through (14.1-4) with

$$\left(\frac{1}{n^2}\right)_1 = \left(\frac{1}{n^2}\right)_2 = \left(\frac{1}{n^2}\right)_3 \equiv \frac{1}{n_o^2}$$

we obtain

$$\frac{x^2 + y^2 + z^2}{n_o^2} + \frac{2r_{41}E}{\sqrt{3}}(xy + yz + xz) = 1 \tag{14.5-3}$$

as the equation of the index ellipsoid. One can proceed formally at this point to derive the new directions x', y', and z' of the principal axes of the ellipsoid. A little thought, however, will show that the ⟨111⟩ direction along which the field is applied will continue to remain a threefold-symmetry axis, whereas the remaining two orthogonal axes can be chosen *anywhere* in the plane normal to ⟨111⟩. Thus, (14.5-3) is an equation of an ellipsoid of revolution about ⟨111⟩. To prove this, we choose ⟨111⟩ as the z' axis, so

$$z' = \frac{1}{\sqrt{3}}x + \frac{1}{\sqrt{3}}y + \frac{1}{\sqrt{3}}z \tag{14.5-4}$$

and take

$$x' = \frac{1}{\sqrt{2}}y - \frac{1}{\sqrt{2}}z$$

and choose y' to be normal to both z' and x' so that

$$y' = -\frac{2}{\sqrt{6}}x + \frac{1}{\sqrt{6}}y + \frac{1}{\sqrt{6}}z \tag{14.5-5}$$

Therefore,

$$\begin{aligned} x &= -\frac{2}{\sqrt{6}}y' + \frac{1}{\sqrt{3}}z' \\ y &= \frac{1}{\sqrt{2}}x' + \frac{1}{\sqrt{6}}y' + \frac{1}{\sqrt{3}}z' \\ z &= -\frac{1}{\sqrt{2}}x' + \frac{1}{\sqrt{6}}y' + \frac{1}{\sqrt{3}}z' \end{aligned} \tag{14.5-6}$$

Substituting (14.5-6) in (14.5-3), we obtain the equation of the index ellip-

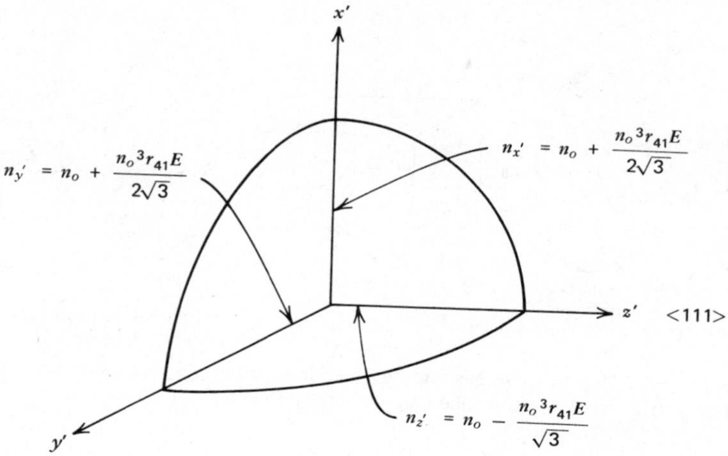

FIGURE 14.8 The intersection of the index ellipsoid of 43m crystals (with E_{dc} parallel to ⟨111⟩) with the planes $x' = 0$, $y' = 0$, $z' = 0$. The principal indices of refraction for this case are $n_{x'}$, $n_{y'}$, and $n_{z'}$.

soid in the x', y', z' coordinate system as

$$(x'^2 + y'^2)\left(\frac{1}{n_o^2} - \frac{r_{41}}{\sqrt{3}}E\right) + \left(\frac{1}{n_o^2} + \frac{2r_{41}}{\sqrt{3}}E\right)z'^2 = 1 \tag{14.5-7}$$

so the principal indices of refraction become

$$\begin{aligned} n_{y'} = n_{x'} &= n_o + \frac{n_o^3 r_{41} E}{2\sqrt{3}} \\ n_{z'} &= n_o - \frac{n_o^3 r_{41} E}{\sqrt{3}} \end{aligned} \tag{14.5-8}$$

It is clear from (14.5-7) that other choices of x' and y', as long as they are normal to z' and to each other, are also acceptable since x' and y' enter (14.5-7) as the combination $x'^2 + y'^2$, which is invariant to rotations about the z' axis. The principal axes of the index ellipsoid (14.5-7) are shown in Figure 14.8.

An amplitude modulator based on the foregoing situation is shown in Figure 14.9. The fractional intensity transmission is given by (14.3-4) as

$$\frac{I_0}{I_i} = \sin^2\frac{\Gamma}{2}$$

where the retardation if we use (14.5-8), is

$$\Gamma = \phi_{z'} - \phi_{y'} = \frac{(\sqrt{3}\,\pi)n_o^3 r_{41}}{\lambda}\left(\frac{Vl}{d}\right) \tag{14.5-9}$$

A graphic summary of the electrooptic properties of $\bar{4}3m$ crystals is shown in Table 14.3.

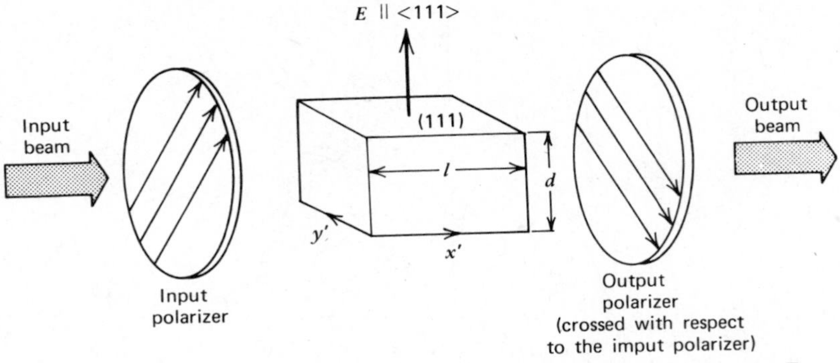

FIGURE 14.9 A transverse electrooptic modulator using a zinc-blende-type ($\bar{4}3m$) crystal with **E** parallel to the cube diagonal ⟨111⟩ direction.

14.6 HIGH-FREQUENCY MODULATION CONSIDERATIONS

In the examples considered in the three preceding sections, we derived expressions for the retardation caused by electric fields of low frequencies. In many practical situations, the modulation signal is often at very high frequencies and, in order to utilize the wide frequency spectrum available with lasers, may occupy a large bandwidth. In this section we consider some of the basic factors limiting the highest usable modulation frequencies in a number of typical experimental situations.

Consider first the situation described by Figure 14.10. The electrooptic crystal is placed between two electrodes with a modulation field containing frequencies near $\omega_0/2\pi$ applied to it. R_s is the internal resistance of the modulation source, and C represents the parallel-plate capacitance due to the electrooptic crystal. If $R_s > (\omega_0 C)^{-1}$, most of the modulation voltage drop is across R_s and is thus wasted, since it does not contribute to the retardation. This can be remedied by resonating the crystal capacitance with an inductance L, where $\omega_0^2 = (LC)^{-1}$, as shown in Figure 14.10. In addition, a shunting resistance R_L is used so that at $\omega = \omega_0$ the impedance of the parallel RLC circuit is R_L, which is chosen to be larger than R_s, so most of the modulation voltage appears across the crystal. The resonant circuit has a finite bandwidth—that is, its impedance is high only over a frequency interval $\Delta\omega/2\pi \simeq 1/(2\pi R_L C)$ (centered on ω_0). Therefore, the maximum modulation bandwidth (the frequency spectrum occupied by the modulation signal) must be less than

$$\frac{\Delta\omega}{2\pi} \simeq \frac{1}{2\pi R_L C} \tag{14.6-1}$$

if the modulation field is to be a faithful replica of the modulation signal.

In practice, the size of the modulation bandwidth $\Delta\omega/2\pi$ is dictated by the specific application. In addition, one requires a certain peak retardation

TABLE 14.3. Electrooptical Properties and Retardation in $\bar{4}3m$ (Zinc Blende Structure) Crystals for Three Directions of Applied Field.

	$E \perp (001)$ plane $E_x = E_y = 0,\ E_z = E$	$E \perp (110)$ plane $E_x = E_y = \frac{E}{\sqrt{2}},\ E_z = 0$	$E \perp (111)$ plane $E_x = E_y = E_z = \frac{E}{\sqrt{3}}$
Index ellipsoid	$\frac{x^2+y^2+z^2}{n_o^2} + 2 r_{41} E xy = 1$	$\frac{x^2+y^2+z^2}{n_o^2} + \sqrt{2}\, r_{41} E(yz + zx) = 1$	$\frac{x^2+y^2+z^2}{n_o^2} + \frac{2}{\sqrt{3}} r_{41} E(yz + zx + xy) = 1$
n_x'	$n_o + \frac{1}{2} n_o^3 r_{41} E$	$n_o + \frac{1}{2} n_o^3 r_{41} E$	$n_o + \frac{1}{2\sqrt{3}} n_o^3 r_{41} E$
n_y'	$n_o - \frac{1}{2} n_o^3 r_{41} E$	$n_o - \frac{1}{2} n_o^3 r_{41} E$	$n_o + \frac{1}{2\sqrt{3}} n_o^3 r_{41} E$
n_z'	n_o	n_o	$n_o - \frac{1}{\sqrt{3}} n_o^3 r_{41} E$
$x'y'z'$ coordinates	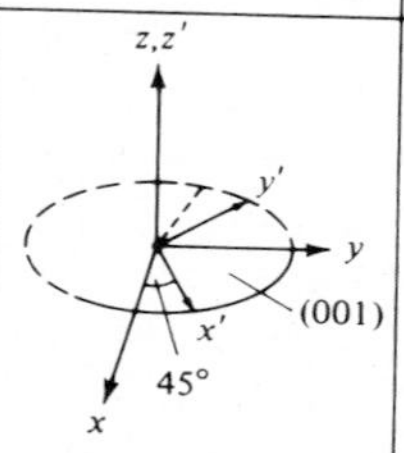	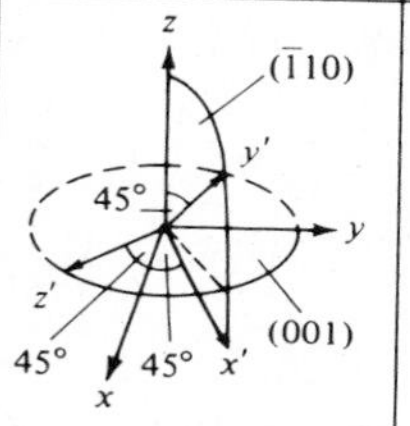	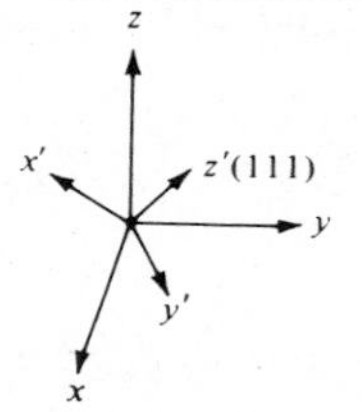
Directions of optical path and axes of crossed polarizer	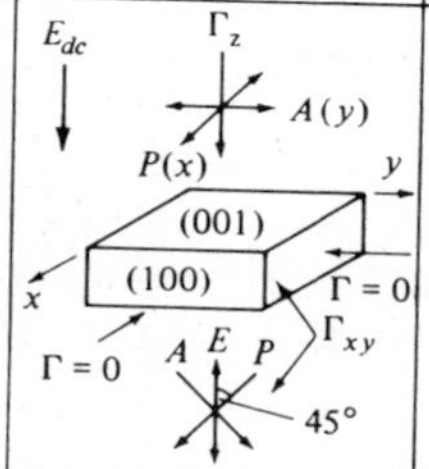	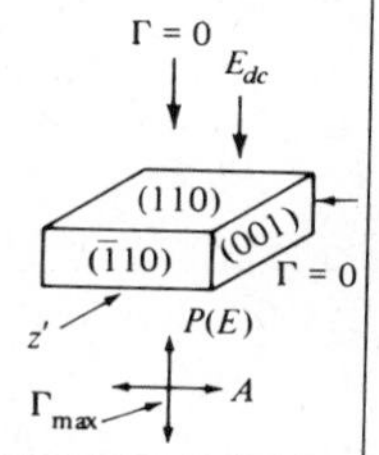	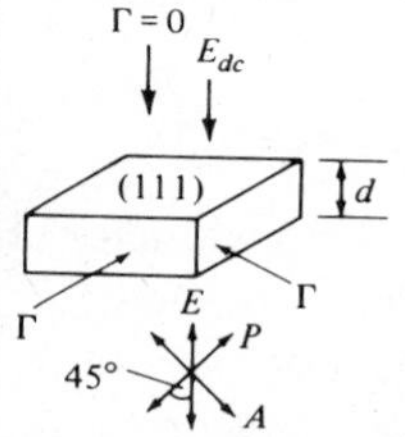
Retardation phase difference $r(V = Ed)$	$\Gamma_z = \frac{2\pi}{\lambda} n_o^3 r_{41} V$ $\Gamma_{xy} = \frac{\pi}{\lambda} \frac{l}{d} n_o^3 r_{41} V$	$\Gamma_{max} = \frac{2\pi}{\lambda} \frac{l}{d} n_o^3 r_{41} V$	$\Gamma = \frac{\sqrt{3}\pi}{\lambda} \frac{l}{d} n_o^3 r_{41} V$

Source: Reference 6.

Γ_m. Using (14.2-4) to relate Γ_m to the peak modulation voltage $V_m = (E_z)_m l$, we can show, with the aid of (14.6-1), that the power $V_m^2/2R_L$ needed in KDP-type crystals to obtain a peak retardation Γ_m is related to the modulation bandwidth $\Delta\nu = \Delta\omega/2\pi$ as

$$P = \frac{\Gamma_m^2 \lambda^2 A \varepsilon\, \Delta\nu}{4\pi l n_o^6 r_{63}^2} \qquad (14.6\text{-}2)$$

where l is the length of the optical path in the crystal, A is the cross-sectional area of the crystal normal to l, and ε is the dielectric constant at the modulation frequency ω_0.

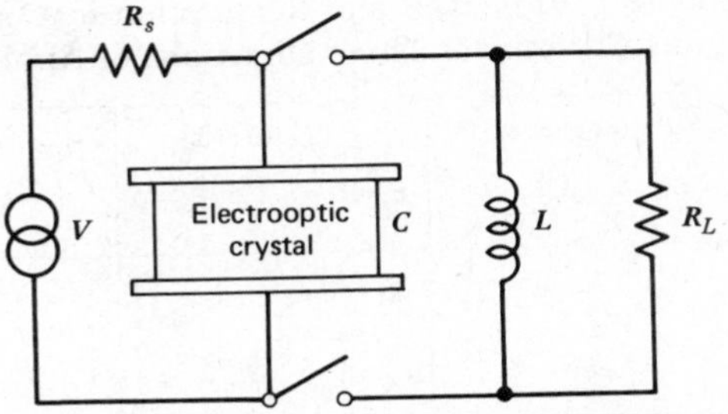

FIGURE 14.10 Equivalent circuit of an electrooptic modulation crystal in a parallel-plate configuration.

Transit-Time Limitations to High-Frequency Electrooptic Modulation

According to (14.2-4), the electrooptic retardation due to a field E can be written as

$$\Gamma = aEl \tag{14.6-3}$$

where $a = \omega n_o^3 r_{63}/c$ and l is the length of the optical path in the crystal. If the field E changes appreciably during the transit time, $\tau_d = nl/c$, of light through the crystal, we must replace (14.6-3) by

$$\Gamma(t) = a\int_0^l E(t')\,dz = \frac{ac}{n}\int_{t-\tau_d}^{t} E(t')\,dt' \tag{14.6-4}$$

where c is the velocity of light in vacuum and $E(t')$ is the instantaneous (low-frequency) electric field. In the second integral, we replace integration over z by integration over time, recognizing that the portion of the wave that reaches the output face $z = l$ at time t entered the crystal at time $t - \tau_d$. We also assumed that at any given moment, the field $E(t)$ has the same value throughout the crystal.

Taking $E(t')$ as a sinusoid

$$E(t') = E_m e^{i\omega_m t'}$$

we obtain from (14.6-4)

$$\begin{aligned}\Gamma(t) &= \frac{ac}{n}E_m\int_{t-\tau_d}^{t} e^{i\omega_m t'}\,dt' \\ &= \Gamma_0\left(\frac{1 - e^{-i\omega_m\tau_d}}{i\omega_m\tau_d}\right)e^{i\omega_m t}\end{aligned} \tag{14.6-5}$$

where $\Gamma_0 = (ac/n)\tau_d E_m = alE_m$ is the peak retardation, which obtains when $\omega_m\tau_d \ll 1$. The factor

$$r = \frac{1 - e^{-i\omega_m\tau_d}}{i\omega_m\tau_d} \tag{14.6-6}$$

gives the decrease in peak retardation resulting from the finite transit time. For $r \simeq 1$ (i.e., no reduction), the condition $\omega_m \tau_d \ll 1$ must be satisfied, so the transit time must be small compared to the shortest modulation period. The factor r is plotted in Figure 14.11.

If, somewhat arbitrarily, we take the highest useful modulation frequency as that for which $\omega_m \tau_d = \pi/2$ (at this point, according to Figure 14.11, $|r| = 0.9$), and we use the relation $\tau_d = ln/c$, we obtain

$$(\nu_m)_{\max} = \frac{c}{4ln} \tag{14.6-7}$$

that, using a KDP crystal ($n \simeq 1.5$) and a length $l = 1$ cm, yields $(\nu_m)_{\max} = 5 \times 10^9$ Hz.

Traveling-Wave Modulators

One method that can, in principle, overcome the transit-time limitation involves applying the modulation signal in the form of a traveling wave (Reference 7), as shown in Figure 14.12. If the optical and modulation field phase velocities are equal to each other, then a portion of an optical wavefront will experience the same instantaneous modulating electric field, which corresponds to the field it encounters at the entrance face, as it propagates through the crystal, and the transit-time problem discussed above is eliminated. This form of modulation can be used mostly in the transverse geometry as discussed in the preceding section, since the *RF* field in most propagating structures is predominantly transverse.

Consider an element of the optical wavefront that *enters* the crystal at $z = 0$ at time t. The position z of this element at some later time t' is

$$z(t') = \frac{c}{n}(t' - t) \tag{14.6-8}$$

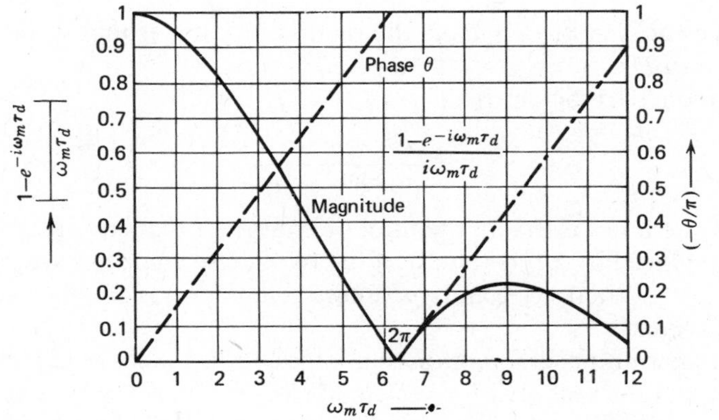

FIGURE 14.11 Phase and magnitude of the transit-time reduction factor $(1 - e^{-i\omega_m \tau_d})/(i\omega_m \tau_d)$.

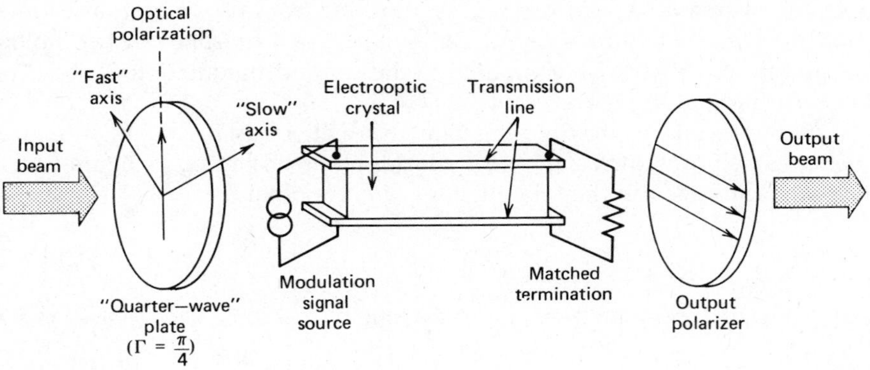

FIGURE 14.12 A traveling-wave electrooptic modulator.

The retardation exercised by this element is given similarly to (14.6-4) by

$$\Gamma(t) = a\frac{c}{n}\int_t^{t+\tau_d} E[t', z(t')]\, dt' \tag{14.6-9}$$

where $E[t', z(t')]$ is the instantaneous modulation field as seen by an observer traveling with the phase front. Taking the traveling modulation field as

$$E(t', z) = E_m e^{i(\omega_m t' - k_m z)}$$

we obtain, using (14.6-8),

$$E[t', z(t')] = E_m e^{i[\omega_m t' - k_m (c/n)(t'-t)]} \tag{14.6-10}$$

Recalling that $k_m = \omega_m / c_m$, where c_m is the phase velocity of the modulation field, we substitute (14.6-10) in (14.6-9) and, carrying out the integration, obtain

$$\Gamma(t) = \Gamma_0 e^{i\omega_m t}\left[\frac{e^{i\omega_m \tau_d(1-c/nc_m)} - 1}{i\omega_m \tau_d(1 - c/nc_m)}\right] \tag{14.6-11}$$

where $\Gamma_0 = alE_m = a(c/n)\tau_d E_m$ is the retardation that would result from a dc field equal to E_m.

The reduction factor

$$r = \frac{e^{i\omega_m \tau_d(1-c/nc_m)} - 1}{i\omega_m \tau_d(1 - c/nc_m)} \tag{14.6-12}$$

is of the same form as that of the lumped constant modulator (14.6-6) except that τ_d is replaced by $\tau_d(1 - c/nc_m)$. If the two phase velocities are made equal so that $c/n = c_m$, then $r = 1$ and maximum retardation is obtained *regardless* of the crystal length.

The maximum useful modulation frequency is taken, as in the treatment leading to (14.6-7), as that for which $\omega_m \tau_d(1 - c/nc_m) = \pi/2$, yielding

$$(\nu_m)_{\max} = \frac{c}{4\,nl(1 - c/nc_m)} \tag{14.6-13}$$

that, upon comparison with (14.6-7), shows an increase in the frequency limit or useful crystal length of $(1 - c/nc_m)^{-1}$. The problem of designing traveling wave electrooptic modulators is considered in References 8–10. A more rigorous treatment of traveling wave modulators is included in Reference 25.

14.7 ELECTROOPTIC BEAM DEFLECTION

The electrooptic effect is also used to deflect light beams (References 11 and 12). The operation of such a beam deflector is shown in Figure 14.13. Imagine an optical wavefront incident on a crystal in which the optical path length depends on the transverse position x. This could be achieved by having the velocity of propagation—that is, the index of refraction n—depend on x, as in Figure 14.13. Taking the index variation to be a linear function of x, the upper ray A "sees" an index $n + \Delta n$ and hence transverses the crystal in a time.

$$T_A = \frac{l}{c}(n + \Delta n)$$

The lower portion of the wavefront (i.e., ray B) sees an index n and has a transit time

$$T_B = \frac{l}{c}n$$

The difference in transit times results in a lag of ray A with respect to B of

$$\Delta y = \frac{c}{n}(T_A - T_B) = l\frac{\Delta n}{n}$$

that corresponds to a deflection of the beam-propagation axis, as measured inside the crystal, at the output face of

$$\theta' = -\frac{\Delta y}{D} = -\frac{l\,\Delta n}{Dn} = -\frac{l}{n}\frac{dn}{dx} \tag{14.7-1}$$

where we replaced $\Delta n/D$ by dn/dx. The external deflection angle θ, measured

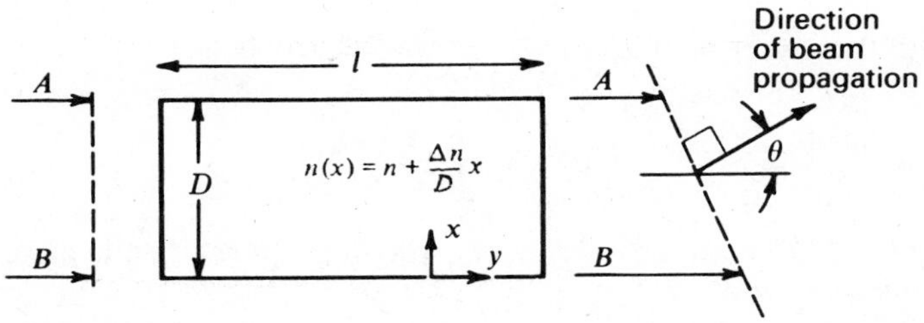

FIGURE 14.13 Schematic diagram of a beam deflector. The index of refraction varies linearly in the x direction as $n(x) = n_0 + ax$. Ray B "gains" on ray A in passing through the crystal axis, thus causing a tilting of the wavefront by θ.

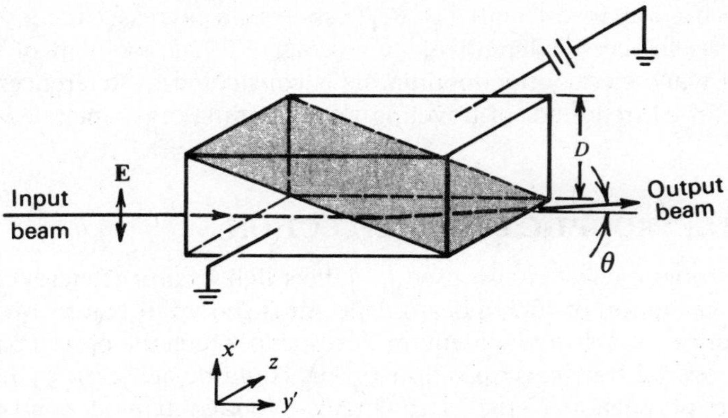

FIGURE 14.14 Double-prism KDP beam deflector. Upper and lower prisms have their z axes reversed with respect to each other. The deflection field is applied parallel to z.

with respect to the horizontal axis, is related to θ' by Snell's law

$$\frac{\sin\theta}{\sin\theta'} = n$$

that if we use (14.7-1) and assume $\sin\theta \simeq \theta \ll 1$, yields

$$\theta = \theta' n = -l\frac{\Delta n}{D} = -l\frac{dn}{dx} \qquad (14.7\text{-}2)$$

A simple realization of such a deflector using a KH_2PO_4(KDP) crystal is shown in Figure 14.14. It consists of two KDP prisms with edges along the x', y', and z directions.[6] The two prisms have their z axes opposite to one another but are otherwise similarly oriented. The electric field is applied parallel to the z direction and the light propagates in the y' direction with its polarization along x'. For this case, the index of refraction seen by ray A, which propagates entirely in the upper prism, is given by (14.1-11) as

$$n_A = n_o - \frac{n_o^3}{2} r_{63} E_z$$

whereas in the lower prism, the sign of the electric field with respect to the z axis is reversed so that

$$n_B = n_o + \frac{n_o^3}{2} r_{63} E_z$$

If we use (14.7-2) with $\Delta n = n_A - n_B$, the deflection angle is given by

$$\theta = \frac{l}{D} n_o^3 r_{63} E_z \qquad (14.7\text{-}3)$$

[6] These are the principal axes of the index ellipsoid when an electric field is applied along the z direction as described in Section 14.1.

According to (6.6-18), every optical beam has a finite, farfield divergence angle that we call θ_{beam}. It is clear that a fundamental figure of merit for the deflector is not the angle of deflection θ that can be enlarged by a lens but the factor N by which θ exceeds θ_{beam}. If one were, as an example, to focus the output beam, then N would correspond to the number of resolvable spots that can be displayed in the focal plane using fields with a magnitude up to E_z.

To obtain an expression for N, we assume that the crystal is placed at the "waist" of a Gaussian (fundamental) beam with a spot size ω_0. According to (6.6-18), the farfield diffraction angle is

$$\theta_{\text{beam}} = \frac{\lambda}{\pi n \omega_0}$$

Such a beam can be passed through a crystal with height $D \geqslant 2\omega_0$ so that, if we take $n = n_o$ and use (14.7-3), the number of resolvable spots is

$$N = \frac{\theta}{\theta_{\text{beam}}} = \frac{\pi l n_o^4 r_{63}}{2\lambda} E_z \tag{14.7-4}$$

It follows directly from (14.7-4), the details being left as a problem, that an electric field that induces a birefringent retardation (in a distance l) $\Delta\Gamma = \pi$ will yield $N \simeq 1$. Therefore, fundamentally, the electrooptic extinction of a beam, which according to (14.3-4) requires $\Gamma = \pi$, is equivalent to a deflection by one spot diameter.

14.8 THE PHOTOELASTIC EFFECT

The linear photoelastic effect involves the first-order changes in the optical properties of insulators due to acoustic strain. In a manner analogous to the electrooptic effect, and specifically to (14.1-3), the effect is characterized by fourth-rank tensor p_{idkl}, the *photoelastic tensor*, via the relation (Reference 1),

$$\Delta\left(\frac{1}{n^2}\right)_{id} = p_{idkl} S_{kl} \tag{14.8-1}$$

where $(1/n^2)_{id}$ is a constant of the index ellipsoid (14.1-2) and S_{kl} is the strain (Reference 13) component

$$S_{kl}(\mathbf{r}) = \frac{1}{2}\left[\frac{\partial u_k(\mathbf{r})}{\partial x_l} + \frac{\partial u_l(\mathbf{r})}{\partial x_k}\right]$$

where $u_k(\mathbf{r})$ is the deviation (from equilibrium) of the point $\mathbf{r}$ in the crystal projected along the direction k (k, l = 1, 2, 3).

The effect of a given strain on the optical propagation can be determined using the same formalism as in Sections 14.1–14.3. We will, however, find it more profitable to adopt another but formally equivalent point of view.

First, we need to relate $\Delta(1/n^2)_{id}$ appearing in (14.8-1) to the dielectric tensor ε_{ij} of (5.2-1). We start with the equation for the constant energy surface (5.2-6) in **D** space

$$2\omega_e = D_i E_i = \varepsilon_{ij} E_i E_j \tag{14.8-2}$$

We need to replace the electric field components E_i in (14.8-2) by D_i. We use the relations

$$\begin{aligned} D_i &= \varepsilon_{ij} E_j \\ E_i &= g_{ij} D_j \end{aligned} \tag{14.8-3}$$

where $g = (\varepsilon)^{-1}$ is the inverse of the matrix ε. Taking advantage of the fact that $\varepsilon_{ij}(i \neq j) \ll \varepsilon_{ii}$, we use the rule for matrix inversion to obtain

$$\begin{aligned} g_{ii} &= (\varepsilon_{ii})^{-1} \\ \underset{i \neq j}{g_{ij}} &\simeq -\frac{\varepsilon_{ji}}{\varepsilon_{ii}\varepsilon_{jj}} = -\frac{\varepsilon_{ij}}{\varepsilon_{ii}\varepsilon_{jj}} \end{aligned} \tag{14.8-4}$$

since $\varepsilon_{ij} = \varepsilon_{ji}$. Using (14.8-4) and defining $\varepsilon'_{ij} \equiv \varepsilon_{ij}/\varepsilon_0$, we can write (14.8-2) as

$$2\omega_e \varepsilon_0 = \frac{D_x^2}{\varepsilon'_{11}} + \frac{D_y^2}{\varepsilon'_{22}} + \frac{D_z^2}{\varepsilon'_{33}} - 2\frac{\varepsilon'_{32}}{\varepsilon'_{33}\varepsilon'_{22}} D_z D_y - 2\frac{\varepsilon'_{31}}{\varepsilon'_{33}\varepsilon'_{11}} D_z D_x - 2\frac{\varepsilon'_{21}}{\varepsilon'_{22}\varepsilon'_{11}} D_x D_y$$

Substituting, as in Section 5.3, $\mathbf{D} = \sqrt{2\omega_e \varepsilon_0}\, \mathbf{r}$ in the last equation results in

$$\frac{x^2}{\varepsilon'_{11}} + \frac{y^2}{\varepsilon'_{22}} + \frac{z^2}{\varepsilon'_{33}} - 2\frac{\varepsilon'_{32}}{\varepsilon'_{33}\varepsilon'_{22}} zy - 2\frac{\varepsilon'_{31}}{\varepsilon'_{33}\varepsilon'_{11}} zx - 2\frac{\varepsilon'_{21}}{\varepsilon'_{22}\varepsilon'_{11}} xy = 1 \tag{14.8-5}$$

This is the equation of the indicatrix. If the initial choice of x, y, z is such that $\varepsilon'_{ij} = 0$ when $i \neq j$, we may view the off-diagonal terms in (14.8-5) as due to the perturbing field (strain, electric field, etc.). Equating it term by term to (14.1-2), we have

$$\begin{aligned} \left(\frac{1}{n^2}\right)_{\substack{ij \\ i \neq j}} &= -\frac{\varepsilon'_{ij}}{\varepsilon'_i \varepsilon'_j} \\ \left(\frac{1}{n^2}\right)_{ii} &= \frac{1}{\varepsilon'_i} \end{aligned} \tag{14.8-6}$$

where $\varepsilon'_i \equiv \varepsilon'_{ii}$. Equations (14.8-6) form a bridge between the description of dielectric phenomena in terms of the index ellipsoid formalism and in terms of the dielectric tensor ε_{ij}.

Returning to the photoelastic effect, we can reexpress (14.8-1) using (14.8-6) as

$$\Delta \varepsilon'_{id} = -\varepsilon'_i \varepsilon'_d p_{idkl} S_{kl} \tag{14.8-7}$$

Equation (14.8-7) is formally equivalent to (14.8-1) as a description of the photoelastic effect. Next, consider a situation where an optical field $\mathbf{E}$ and a strain field S_{kl} exist simultaneously in a crystal. From (14.8-3), we have

$$D_i = \varepsilon_{id} E_d = \varepsilon_0 E_i + P_i$$

Solving for the polarization P_i, we get

$$P_i = \varepsilon_{id} E_d - \varepsilon_0 E_i = (\varepsilon_{id} - \varepsilon_0 \delta_{id}) E_d$$

so that the polarization induced by the strain is

$$\Delta P_i = \Delta\varepsilon_{id} E_d = \varepsilon_0 \Delta\varepsilon'_{id} E_d$$

that, using (14.8-7), becomes

$$\Delta P_i = -\frac{\varepsilon_i \varepsilon_d}{\varepsilon_0} p_{idkl} S_{kl} E_d \tag{14.8-8}$$

This is our key result for this section. This equation shows how a strain S_{kl} conspires with a field component E_d to generate a polarization component ΔP_i along the direction i that is proportional to the product, $E_d S_{kl}$. We can thus view the photoelastic effect as a nonlinear effect since it relates one field component (ΔP_i) to the product of two other fields. We will put (14.8-8) to work in the next section.

14.9 BRAGG DIFFRACTION OF LIGHT BY ACOUSTIC WAVES

The physical problem considered in this section is the following: an input acoustic wave propagates through some optical medium (liquid, crystal). An optical beam is incident on the same medium. Under the proper conditions, to be derived, part of the input optical beam is diffracted into a new direction while simultaneously being shifted in *frequency* by an amount equal to the acoustic frequency (References 14, 16, 17), the shift being positive or negative. The basic interaction thus involves two optical fields at frequencies designated as ω_i and $\omega_d = \omega_i \pm \omega_s$ and a sound field at ω_s.

In deriving the equations that describe this interaction, we will ignore the question of which field is the input and which one is the output. This question is settled easily when the boundary conditions are applied to the general solutions.

Since we have here a case in which power is exchanged between two optical fields of different frequencies, we may properly characterize the effect as nonlinear.

We start with the wave equation for an optical field propagating in a medium supporting simultaneously some acoustic field.

$$\nabla^2 \mathbf{E} = \mu\varepsilon \frac{\partial^2 \mathbf{E}}{\partial t^2} + \mu \frac{\partial^2}{\partial t^2}(\Delta \mathbf{P}) \tag{14.9-1}$$

where $\Delta\mathbf{P}$ is the *change* in polarization caused by the presence of the sound wave as given by (14.8-8). Let the optical field consist of two plane waves: one polarized along i and one along d. The two waves propagate along $\mathbf{k}_i$ and $\mathbf{k}_d$, respectively, where $\mathbf{k}_i$ in general is not parallel to $\mathbf{k}_d$.

$$\begin{aligned} E_i(\mathbf{r}, t) &= \tfrac{1}{2} E_i(r_i) e^{i(\omega_i t - \mathbf{k}_i \cdot \mathbf{r})} + \text{c.c.} \\ E_d(\mathbf{r}, t) &= \tfrac{1}{2} E_d(r_d) e^{i(\omega_d t - \mathbf{k}_d \cdot \mathbf{r})} + \text{c.c.} \end{aligned} \tag{14.9-2}$$

r_i, r_d are distances as measured along $\mathbf{k}_i$ and $\mathbf{k}_d$, respectively. Two differentiations lead to

$$\nabla^2 E_i(\mathbf{r}, t) \simeq -\frac{1}{2}\left(k_i^2 E_i + 2ik_i \frac{dE_i}{dr_i}\right)e^{i(\omega_i t - \mathbf{k}_i\cdot\mathbf{r})} + \text{c.c.} \tag{14.9-3}$$

where $k_i = \omega_i\sqrt{\mu\varepsilon_i}$ and we assume $\nabla^2 E_i \ll k_i\, dE_i/dr_i$.

We now substitute (14.9-3) for the left side of (14.9-1). On the right-hand side we have, from (14.8-8),

$$\Delta P_i(\mathbf{r}, t) = -\tfrac{1}{2}\varepsilon_0\varepsilon_i'\varepsilon_d' p_{idkl} S_{kl}(\mathbf{r}, t) E_d(r_d) e^{i(\omega_d t - \mathbf{k}_d\cdot\mathbf{r})}$$

The input strain field $S_{kl}(\mathbf{r}, t)$ is next taken to be that of an acoustic wave propagating along some arbitrary direction $\mathbf{k}_s$ at a frequency ω_s

$$S_{kl}(\mathbf{r}, t) = \frac{S_{kl}}{2} e^{i(\omega_s t - \mathbf{k}_s\cdot\mathbf{r})} + \text{c.c.} \tag{14.9-4}$$

that, in conjunction with the last equation, gives

$$\Delta P_i(\mathbf{r}, t) = -\tfrac{1}{4}\varepsilon_0\varepsilon_i'\varepsilon_d' p_{idkl} E_d(r_d)[e^{i(\omega_d t - \mathbf{k}_d\cdot\mathbf{r})} + \text{c.c.}] S_{kl}[e^{i(\omega_s t - \mathbf{k}_s\cdot\mathbf{r})} + \text{c.c.}] \tag{14.9-5}$$

In order that ΔP_i on the right side of (14.9-1) act as a synchronous driving term for E_i, it must, like E_i, have a term varying as $\exp[i(\omega_i t - \mathbf{k}_i \cdot \mathbf{r})]$, otherwise, any contributions from ΔP_i are localized (in time and space) and average out to zero over long times and/or distances. The above requirement can be satisfied, according to (14.9-5), if

$$\omega_i = \omega_d \pm \omega_s \tag{14.9-6a}$$

$$\mathbf{k}_i \simeq \mathbf{k}_d \pm \mathbf{k}_s \tag{14.9-6b}$$

We will assume first that the equalities in (14.9-6) involve the plus (+) sign. Using the synchronous term of (14.9-5) in (14.9-1) as well as (14.9-3) and (14.9-6) gives

$$\frac{dE_i}{dr_i} = \frac{i}{4}\,\omega\sqrt{\mu\varepsilon_0\varepsilon_i'}\,\varepsilon_d' p_{idkl} S_{kl} E_d e^{-i(\mathbf{k}_d + \mathbf{k}_s - \mathbf{k}_i)\cdot\mathbf{r}} \tag{14.9-7}$$

Using $\omega_i\sqrt{\mu\varepsilon_0} = 2\pi/\lambda$ and taking $\varepsilon_i' \simeq \varepsilon_d' \equiv n^2$, we can write the last equation as

$$\frac{dE_i}{dr_i} = i\eta_{id} E_d e^{i(\mathbf{k}_i - \mathbf{k}_s - \mathbf{k}_d)\cdot\mathbf{r}}$$

and, similarly, (14.9-8)

$$\frac{dE_d}{dr_d} = i\eta_{di} E_i e^{-i(\mathbf{k}_i - \mathbf{k}_s - \mathbf{k}_d)\cdot\mathbf{r}}$$

$$\eta_{di} = \eta_{id} \simeq \frac{\pi n^3}{2\lambda}\, p_{idkl} S_{kl} \tag{14.9-9}$$

$\lambda_i \simeq \lambda_d \equiv \lambda.$

The reason for the phase-matching condition

$$\mathbf{k}_i = \mathbf{k}_s + \mathbf{k}_d \qquad (14.9\text{-}10)$$

is now clear. If (14.9-10) is not satisfied, the contributions from E_i to E_d (and vice versa) reverse sign according to (14.9-8) with a spatial period l_c where

$$l_c \equiv \frac{\pi}{|\mathbf{k}_i - \mathbf{k}_s - \mathbf{k}_d|}$$

and no cumulative interaction can take place over distances $l > l_c$, l_c is referred to as the "coherence distance." A more appropriate term may be the "beat distance."

The use of the plus (+) sign in (14.9-6) leads to a frequency condition

$$\omega_i = \omega_s + \omega_d \qquad (14.9\text{-}11)$$

The interaction in this case can be viewed in the following terms. A photon with energy $\hbar\omega_i$ and momentum $\hbar\mathbf{k}_i$ is incident on a sound wave of frequency ω_s and wave momentum $\hbar\mathbf{k}_s$. The incident photon is annihilated, giving rise instead to a new photon at ω_d, $\mathbf{k}_d$, and a phonon ω_s, $\mathbf{k}_s$. In this case, (14.9-10) is a statement of total momentum conservation whereas (14.9-11) is that of energy conservation.

This situation is depicted in Figure 14.15 drawn, for simplicity, for the case of $n_i = n_d$. In this case, since $\omega_s \ll \omega_i$, $k_i \simeq k_d = k$. The vector triangle of Figure 14.15*a* yields

$$k_s = 2k \sin \theta$$

that, if we use $k_s = 2\pi/\lambda_s$, $k = 2\pi n/\lambda$, can be written as

$$2\lambda_s \sin \theta = \frac{\lambda}{n} \qquad (14.9\text{-}12)$$

This condition is identical to the first-order Bragg condition (References 13 and 18) for the scattering of x rays in crystals

$$2d \sin \theta = m\frac{\lambda}{n} \qquad m = 1, 2, \ldots$$

for $m = 1$. The periodic nature of the sound wave plays a role similar to that of a regular arrangement of atomic planes that are separated by d. The lack of higher-order diffraction angles ($m = 2, 3, \ldots$) in (14.9-12) is due to the sinusoidal nature of the sound disturbance. The atomic planes are localized, thus giving rise to more orders. (This point is explored further in Problem 14.8.) Another difference is the lack of a frequency shift in x-ray diffraction. This is due to the stationary nature of the atomic planes while the acoustic "lattice" moves with a velocity v_s. As a matter of fact, if we calculate the Doppler shift of a wave ω_i incident at an angle θ satisfying (14.9-12) on a plane moving at v_s we find that $\omega_d = \omega_i + \omega_s$.

We now return to the solution of the coupled equations (14.9-8). We assume that the angle of incidence θ is chosen so that (14.9-10) is satisfied.

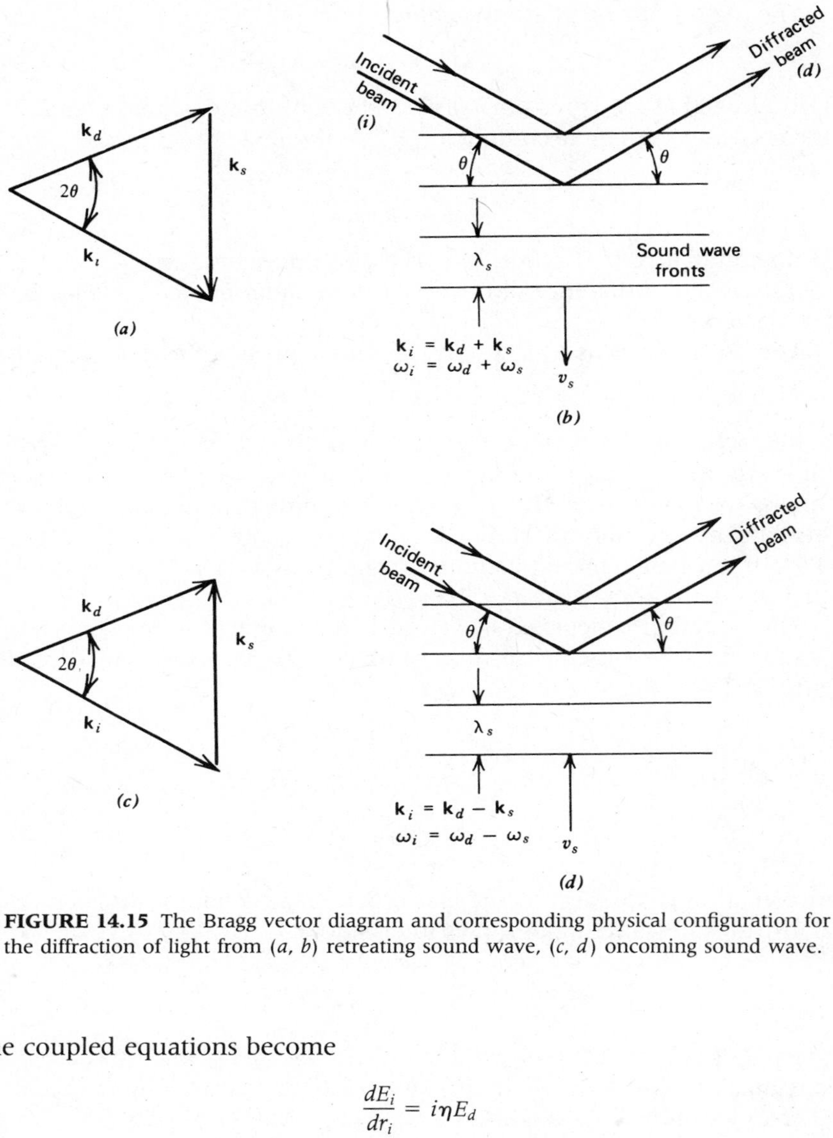

FIGURE 14.15 The Bragg vector diagram and corresponding physical configuration for the diffraction of light from (*a*, *b*) retreating sound wave, (*c*, *d*) oncoming sound wave.

The coupled equations become

$$\frac{dE_i}{dr_i} = i\eta E_d$$

$$\frac{dE_d}{dr_d} = i\eta E_i \tag{14.9-13}$$

where, since $\omega_i \simeq \omega_d$, we took $\eta_{id} = \eta_{di} \equiv \eta$.

Equations (14.9-13) are our main result. An apparent difficulty in solving them is the fact that they involve two different spatial coordinates r_i and r_d measured along the two respective ray directions. This difficulty can be resolved by transforming to a coordinate ζ measured along the bisector of the

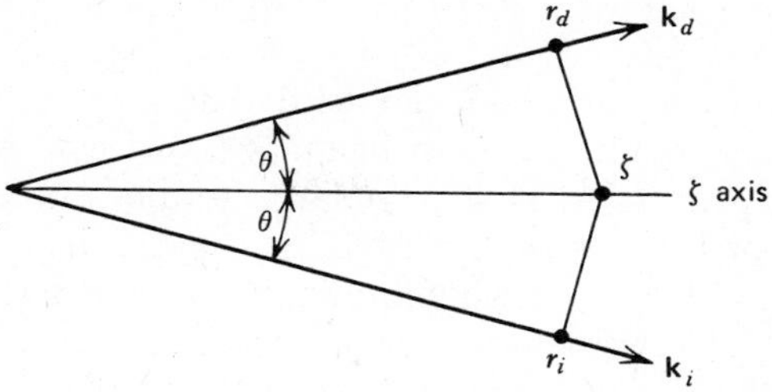

FIGURE 14.16 The directions and angles appearing in the diffraction equations (14.9-15).

angle formed between $\mathbf{k}_i$ and $\mathbf{k}_d$, as shown in Figure 14.16. Defining the values of r_d and r_i, which correspond to a given ζ as the respective projections of ζ along $\mathbf{k}_d$ and $\mathbf{k}_i$, we have

$$r_i = \zeta \cos\theta \qquad r_d = \zeta \cos\theta \tag{14.9-14}$$

so that (14.9-13) become

$$\begin{aligned} \frac{dE_i}{d\zeta} &= \frac{dE_i}{dr_i}\cos\theta = i\eta E_d \cos\theta \\ \frac{dE_d}{d\zeta} &= i\eta E_i \cos\theta \end{aligned} \tag{14.9-15}$$

whose solutions are

$$\begin{aligned} E_i(\zeta) &= E_i(0)\,\cos(\eta\zeta\cos\theta) + iE_d(0)\,\sin(\eta\zeta\cos\theta) \\ E_d(\zeta) &= E_d(0)\,\cos(\eta\zeta\cos\theta) + iE_i(0)\,\sin(\eta\zeta\cos\theta) \end{aligned}$$

Using the correspondence between ζ, r_i, and r_d defined above, we can rewrite the solutions as

$$\begin{aligned} E_i(r_i) &= E_i(0)\,\cos(\eta r_i) + iE_d(0)\,\sin(\eta r_i) \\ E_d(r_d) &= E_d(0)\,\cos(\eta r_d) + iE_i(0)\,\sin(\eta r_d) \end{aligned} \tag{14.9-16}$$

which is the desired result. It is of sufficient generality to describe the interaction between two input fields at ω_i and ω_d with arbitrary phases [$E_i(0)$ and $E_d(0)$ are complex] and arbitrary amplitudes as long as the Bragg condition (14.9-10) and the frequency condition $\omega_i = \omega_s + \omega_d$ are fulfilled. In the special case of a single frequency input at ω_i, $E_d(0) = 0$ and

$$\begin{aligned} E_i(r_i) &= E_i(0)\,\cos(\eta r_i) \\ E_d(r_d) &= iE_i(0)\,\sin(\eta r_d) \end{aligned} \tag{14.9-17}$$

we note that

$$|E_i(r_i)|^2 + |E_d(r_d = r_i)|^2 = |E_i(0)|^2 \tag{14.9-18}$$

so that the total optical power carried by both waves is conserved (Reference 20).[7]

If the interaction distance between the two beams is such that $\eta r_i = \eta r_d = \pi/2$, the total power of the incident beam is transferred into the diffracted beam. Since this process is used in a large number of technological and scientific applications, it may be worthwhile to gain some appreciation for the diffraction efficiencies possible using known acoustic media and conveniently available acoustic power levels.

The fraction of the power of the incident beam transferred in a distance l into the diffracted beam is given, using (14.9-17), by

$$\frac{I_{\text{diffracted}}}{I_{\text{incident}}} = \frac{E^2_{\text{diffracted}}}{E_i^2(0)} = \sin^2(\eta l) = \sin^2\left(\frac{\pi n^3}{2\lambda} pSl\right) \tag{14.9-19}$$

where

$$pS \equiv p_{idkl}S_{kl} \tag{14.9-20}$$

It is useful to express (14.9-19) in a form that is more amenable to practical application. In practice, we often know the acoustic intensity I_{acoustic} in watts per square meter. The acoustic strain amplitude S is related to I_{acoustic} by[8]

$$S = \sqrt{\frac{2I_{\text{acoustic}}}{\rho v_s^3}} \tag{14.9-21}$$

where v_s is the velocity of sound in the medium and ρ is the mass density (kg/m^3). Combining (14.9-20) and (14.9-21), we obtain

$$\frac{I_{\text{diffracted}}}{I_{\text{incident}}} = \sin^2\left(\frac{\pi l}{\sqrt{2}\,\lambda}\sqrt{\frac{n^6p^2}{\rho v_s^3}I_{\text{acoustic}}}\right) \tag{14.9-22}$$

[7] More rigorously, it is the total number of photons at ω_i and ω_d that is conserved and not the total optical power. But since $\omega_i \simeq \omega_d$ (to within a factor of $\omega_s/\omega_i \lesssim 10^{-6}$), we can talk about a conservation of the total optical power. To show that the total optical power is not strictly conserved, we need use the relationship $\eta_{di} = (\omega_d/\omega_i)\eta_{id}$, starting with (14.9-8) instead of taking, as we did, $\eta_{id} = \eta_{di} \equiv \eta$.

[8] The (elastic) potential energy per unit volume due to an instantaneous strain $s(t)$ is $\frac{1}{2}TS^2(t)$ where T is the bulk modulus (elastic stiffness constant). The time averaged energy per unit volume due to the propagation of a sound wave with a strain amplitude S is the sum of the (equal) average potential and kinetic energy densities

$$\mathscr{E}/\text{vol} = 2(\tfrac{1}{2})T\overline{S^2}(t) = \tfrac{1}{2}TS^2$$

since $\overline{S^2}(t) = \frac{1}{2}S^2$, the bar denoting time averaging. Using the relation $I_{\text{acoustic}} = v_s\mathscr{E}/\text{vol}$ and $T/\rho = v_s^2$ where ρ is the mass density and v_s the velocity of sound, we get

$$I_{\text{acoustic}} = \tfrac{1}{2}\rho v_s^3 S^2$$

or

$$S = \sqrt{\frac{2I_{\text{acoustic}}}{\rho v_s^3}}$$

which is the result stated in (14.9-21).

and if we use the following definition for the diffraction figure of merit

$$M \equiv \frac{n^6 p^2}{\rho v_s^3} \tag{14.9-23}$$

(14.9-22) becomes

$$\frac{I_{\text{diffracted}}}{I_{\text{incident}}} = \sin^2\left(\frac{\pi l}{\sqrt{2}\,\lambda}\sqrt{M I_{\text{acoustic}}}\right) \tag{14.9-24}$$

Example: Acoustic Diffraction in Water. If we take water as an example, an optical wavelength of $\lambda = 0.6328\ \mu\text{m}$ and the constants (taken from Table 14.4)

$$n = 1.33$$

$$p = 0.31$$

$$v_s = 1.5 \times 10^3\ \text{m/s}$$

$$\rho = 1000\ \text{kg/m}^3$$

Equation (14.9-24) gives

$$\left(\frac{I_{\text{diffracted}}}{I_{\text{incident}}}\right)_{\substack{H_2O \\ \text{at } \lambda=0.6328\ \mu m}} = \sin^2(1.4 l \sqrt{I_{\text{acoustic}}}) \tag{14.9-25}$$

TABLE 14.4. A List of Some Materials Commonly Used in the Diffraction of Light by Sound and Some of Their Relevant Properties. ρ is the Density, v_s the Velocity of Sound, n the Index of Refraction, p the Photoelastic Constant, and M_ω is the Relative Diffraction Constant Defined Above.

Material	ρ (Mg/m^3)	v_s (km/s)	n	p	M_ω
Water	1.0	1.5	1.33	0.31	1.0
Extra-dense flint glass	6.3	3.1	1.92	0.25	0.12
Fused quartz (SiO_2)	2.2	5.97	1.46	0.20	0.006
Polystyrene	1.06	2.35	1.59	0.31	0.8
KRS-5	7.4	2.11	2.60	0.21	1.6
Lithium niobate ($LiNbO_3$)	4.7	7.40	2.25	0.15	0.012
Lithium fluoride (LiF)	2.6	6.00	1.39	0.13	0.001
Rutile (TiO_2)	4.26	10.30	2.60	0.05	0.001
Sapphire (Al_2O_3)	4.0	11.00	1.76	0.17	0.001
Lead molybdate ($PbMO_4$)	6.95	3.75	2.30	0.28	0.22
Alpha iodic acid (HIO_3)	4.63	2.44	1.90	0.41	0.5
Tellurium dioxide (TeO_2) (slow shear wave)	5.99	0.617	2.35	0.09	5.0

Source: Reference 19.

TABLE 14.5. A List of Materials Commonly Used in Acoustooptic Interactions and Some of Their Relevant Properties. $M = n^6p^2/\rho n_s^3$ is the Figure of Merit, Defined by (14.9-23) and Is Given in MKS Units.

Material	$\lambda(\mu)$	n	$\rho(g/cm^3)$	*Acoustic Wave Polarization and Direction*	$v_s(10^5\ cm/s)$	*Opt. Wave Polarization and Direction*[a]	$M = n^6p^2/\rho v_s^3$ *in Units of* 10^{-15}
Fused quartz	0.63	1.46	2.2	Long.	5.95	$\perp$	1.51
Fused quartz	0.63			Trans.	3.76	$\parallel$ or $\perp$	0.467
GaP	0.63	3.31	4.13	Long. in [110]	6.32	$\parallel$	44.6
GaP	0.63			Trans. in [100]	4.13	$\parallel$ or $\perp$ in [010]	24.1
GaAs	1.15	3.37	5.34	Long. in [110]	5.15	$\parallel$	104
GaAs	1.15			Trans. in [100]	3.32	$\parallel$ or $\perp$ in [010]	46.3
TiO_2	0.63	2.58	4.6	Long. in [11–20]	7.86	$\perp$ in [001]	3.93
$LiNbO_3$	0.63	2.20	4.7	Long. in [11–20]	6.57		6.99
YAG	0.63	1.83	4.2	Long. in [100]	8.53	$\parallel$	0.012
YAG	0.63			Long. in [110]	8.60	$\perp$	0.073
YIG	1.15	2.22	5.17	Long. in [100]	7.21	$\perp$	0.33
$LiTaO_3$	0.63	2.18	7.45	Long. in [001]	6.19	$\parallel$	1.37
As_2S_3	0.63	2.61	3.20	Long.	2.6	$\perp$	433
As_2S_3	1.15	2.46		Long.		$\parallel$	347
SF-4	0.63	1.616	3.59	Long.	3.63	$\perp$	4.51
β-ZnS	0.63	2.35	4.10	Long. in [110]	5.51	$\parallel$ in [001]	3.41
β-ZnS	0.63			Trans. in [110]	2.165	$\parallel$ or $\perp$ in [001]	0.57
α-Al_2O_3	0.63	1.76	4.0	Long. in [001]	11.15	$\parallel$ in [11–20]	0.34
CdS	0.63	2.44	4.82	Long. in [11–20]	4.17	$\parallel$	12.1
ADP	0.63	1.58	1.803	Long. in [100]	6.15	$\parallel$ in [010]	2.78
ADP	0.63			Trans. in [100]	1.83	$\parallel$ or in [001]	6.43
KDP	0.63	1.51	2.34	Long. in [100]	5.50	$\parallel$ in [010]	1.91
KDP	0.63			Trans. in [100]		$\parallel$ or $\perp$ in [001]	3.83
H_2O	0.63	1.33	1.0	Long.	1.5		160
Te	10.6	4.8	6.24	Long. in [11–20]	2.2	$\parallel$ in [0001]	4400
$PbMO_4$[14]	0.63	2.4		Long. $\parallel$ c axis	3.75	$\parallel$ or $\perp$	73

[a] The optical-beam direction actually differs from that indicated by the magnitude of the Bragg angle. The polarization is defined as parallel or perpendicular to the scattering plane formed by the acoustic and optical **k** vectors.

Source: Reference 19.

For other materials and at other wavelengths, we can combine the last two equations to obtain a convenient working formula.

$$\frac{I_{\text{diffracted}}}{I_{\text{incident}}} = \sin^2\left(1.4 \frac{0.6328}{\lambda\ \mu m} l\sqrt{M_\omega I_{\text{acoustic}}}\right) \qquad (14.9\text{-}26)$$

where $M_\omega \equiv M_{\text{material}}/M_{H_2O}$ is the diffraction figure of merit of the material relative to water. Values of M and M_ω for some common materials are listed in Tables 14.4 and 14.5.

According to (14.9-22), at small diffraction efficiencies the diffracted light intensity is proportional to the acoustic intensity. This fact is used in acoustic modulation of optical radiation. The information signal is used to modulate the intensity of the acoustic beam. This modulation is then transferred, according to (14.9-22) as intensity modulation onto the diffracted optical beam (Reference 20).

Numerical Example: Scattering $PbMO_4$. Calculate the fraction of 0.6328 μm light that is diffracted under Bragg conditions from a sound wave in $PbMO_4$ with the following characteristics:

Acoustic power = 1 W

Acoustic beam cross section = 1 mm × 1 mm

l = optical path in acoustic beam = 1 mm

M_ω (from Table 14.4) = 0.22

Substituting these data into (14.9-26) yields

$$\frac{I_{\text{diffracted}}}{I_{\text{incident}}} \simeq 37\%$$

14.10 DEFLECTION OF LIGHT BY SOUND

One of the most important applications of acoustooptic interactions is in the deflection of optical beams. This can be achieved by changing the sound frequency while operating near the Bragg-diffraction condition. The situation is depicted in Figure 14.17 and can be understood using Figure 14.18. Let us assume first that the Bragg condition (14.9-10) is satisfied. The momentum vector diagram originally introduced in Figure 14.15a is thus closed and the beam is diffracted along the direction θ as given by (14.9-12). Now let the sound frequency change from ν_s to $\nu_s + \Delta\nu_s$. Since $k_s = 2\pi\nu_s/v_s$, this causes a change of $\Delta k_s = 2\pi(\Delta\nu_s)/v_s$ in the magnitude of the sound wave vector as shown. Since the angle of incidence remains θ and the magnitude of the diffracted $\mathbf{k}_d$ vector is unchanged,[9] its tip is constrained to the circle shown in Figure 14.18. We can no longer close the momentum diagram and thus

[9] The small change in the diffracted wave vector that is attributable to the frequency change is typically about $\Delta k/k \simeq 10^{-7}$ and is neglected.

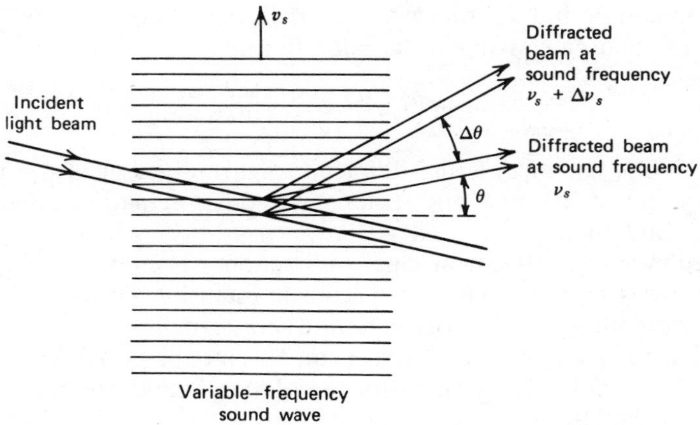

FIGURE 14.17 A change of frequency of the sound wave from ν, to $\nu_s + \Delta\nu_s$ causes a change $\Delta\theta$ in the direction of the diffracted beam according to (14.10-1).

momentum is no longer strictly conserved. The beam will be diffracted along the direction that least violates the momentum conservation.[10] This takes place along the direction OB, causing deflection of the beam by $\Delta\theta$. Recalling that the angles θ and $\Delta\theta$ are all small and that $k_s = 2\pi\nu_s/v_s$ we obtain

$$\Delta\theta = \frac{\Delta k_s}{k} = \frac{\lambda}{n v_s}\Delta\nu_s \tag{14.10-1}$$

so that the deflection angle is proportional to the change of the sound frequency.

As in the case of electrooptic deflection, we are not interested so much in the absolute deflection $\Delta\theta$ as we are in the number of resolvable spots—that is, the factor by which $\Delta\theta$ exceeds the beam divergence angle. If we take the diffraction angle as $\sim\lambda/Dn$, where D is the beam diameter, the number of resolvable spots is thus

$$\begin{aligned} N &= \frac{\Delta\theta}{\theta_{\text{diffraction}}} = \left(\frac{\lambda}{n v_s}\right)\left(\frac{\Delta\nu_s}{\lambda/nD}\right) \\ &= \Delta\nu_s\left(\frac{D}{v_s}\right) = \Delta\nu_s\tau \end{aligned} \tag{14.10-2}$$

where $\tau = D/v_s$ is the transit time for the sound across the optical-beam diameter.

[10] The violation of momentum conservation is equivalent to destructive interference in the diffracted beam, so that the beam intensity will be smaller than under Bragg conditions, where momentum is conserved. The diffracted beam will thus have its maximum value along the direction in which the destructive interference is smallest. This corresponds to the direction that minimizes the momentum mismatch, as shown in Figure 14.18.

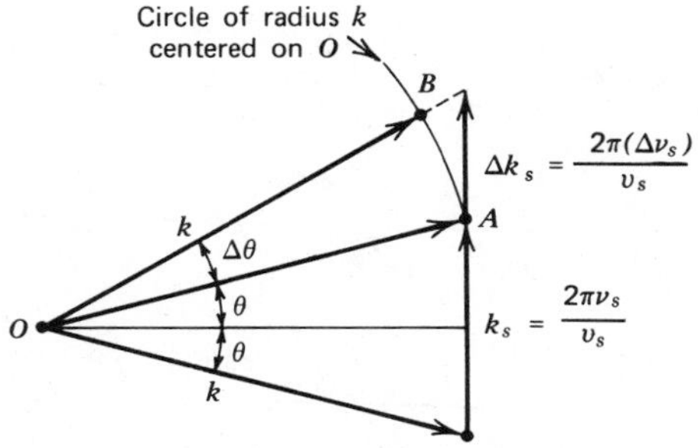

FIGURE 14.18 Momentum diagram, illustrating how the change in sound frequency from ν_s to $\nu_s + \Delta\nu_s$ deflects the diffracted light beam from θ to $\theta + \Delta\theta$.

Numerical Example: Beam Deflection. Consider a deflection system using flint glass and a sound beam that can be varied in frequency from 80 to 120 MHz; thus, $\Delta\nu_s = 40$ MHz. Let the optical beam diameter be $D = 1$ cm. From Table 14.4, we obtain $v_s = 3.1 \times 10^5$ cm/s; therefore, $\tau = D/v_s = 3.23 \times 10^{-6}$ sec and the number of resolvable spots in $N = \Delta\nu_s\tau \simeq 130$.

It is appropriate to recall now that all the results up to this point were derived using the plus sign of (14.9-6). All of the results would still be valid were we to choose, instead, the minus sign so that

$$\omega_i = \omega_d - \omega_s$$
$$\mathbf{k}_i = \mathbf{k}_d - \mathbf{k}_s \tag{14.10-3}$$

A simple consideration of (14.9-5), or basic Doppler reasoning, will show that the scattering described by (14.10-3) will result if the direction of propagation of the sound assumed in Section 14.9 were *reversed*. This situation is depicted in Figure 14.15*c*, *d*.

14.11 BRAGG SCATTERING IN NATURALLY BIREFRINGENT CRYSTALS

The Bragg scattering treated above is described by a vector diagram 14.15*a* involving an isosceles triangle. This is a consequence of the basic "momentum" conservation condition

$$\mathbf{k}_i = \mathbf{k}_d \pm \mathbf{k}_s \tag{14.11-1}$$

when $k_i \simeq k_d = (\omega_d/c)n$. More generally, $n_i \neq n_d$ and, consequently, $k_i \neq k_d$. The relative directions of the incident and diffracted beams are found from the vector triangle $\mathbf{k}_i = \mathbf{k}_d \pm \mathbf{k}_s$. For a given optical frequency and specified directions $\mathbf{k}_i$, $\mathbf{k}_d$, the acoustic wave frequency and direction are thus determined. A simple example, which follows, may best illustrate this point.

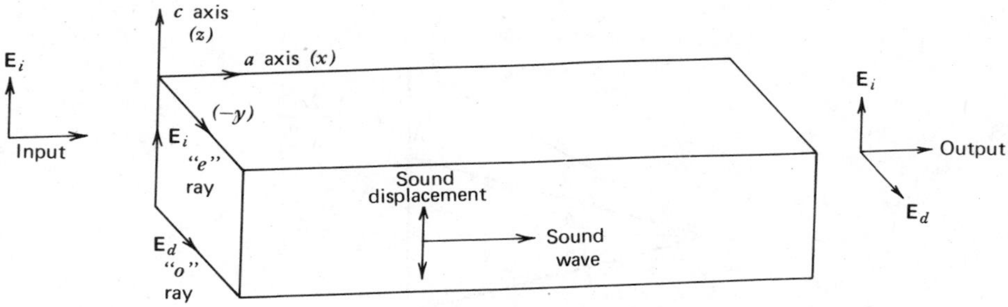

FIGURE 14.19 Collinear coupling between a z-polarized (extraordinary) and a y polarized (ordinary) optical beam by a shear S_{zx} acoustic wave.

Example: Collinear Scattering in $CaMoO_4$ (References 22, 23, and 24). Consider the configuration sketched in Figure 14.19 for coupling by an acoustic wave, two orthogonally polarized optical beams propagating along the x axis of $CaMoO_4$. The crystal possesses a nonvanishing p_{45} photoelastic element that makes it possible, according to (14.9-5), to couple collinearly an ordinary ray and an extraordinary ray as illustrated. The collinear phase matching condition for this case becomes, according to (14.11-1),

$$\frac{\omega}{c}(n_o - n_e) = \frac{\omega_s}{v_s} \tag{14.11-2}$$

and is demonstrated in Figure 14.20. Here, we associate i with the z polarized extraordinary ray, whereas d is the y-polarized ordinary ray. For the two waves to be coupled by the acoustic wave, it is thus necessary, according to (14.9-7), that $p_{zykl} \neq 0$. Since in $CaMoO_4$, $p_{zyzx} = p_{45}$ is finite, the coupling is accomplished by a shear S_{zx} wave propagating along the x direction.

The mathematical description of this configuration is given by (14.9-16) with $r_i = r_d = x$. If we take the input field E_0 as polarized along the z axis and for an acoustic frequency satisfying (14.11-2), the interaction is given by

$$\begin{aligned} E_z(x) &= E_0 \cos \eta x \\ E_y(x) &= iE_0 \sin \eta x \\ \eta &= \frac{\pi l}{\sqrt{2}\,\lambda} \sqrt{\frac{n^6 p_{45}^2}{\rho v_{\text{shear}}^3} I_{\text{acoustic}}} \end{aligned} \tag{14.11-3}$$

The coupling described above can be "turned off" by changing ω_s, since this causes a violation of the momentum conservation condition (14.11-2) (see Problem 14.10). An optical fiber based on this principle is described in References 22 and 23.

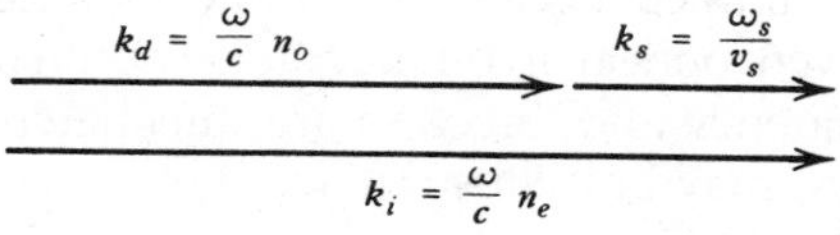

FIGURE 14.20 Collinear phase matching in a birefringent crystal.

References

1. See, for example, J. F. Nye, *Physical Properties of Crystals* (England: Oxford University Press, 1957).
2. Yariv, A., *Introduction to Optical Electronics* (New York: Holt, Rinehart and Winston, 1971).
3. Kaminow, I. P. and E. Turner, "Linear Electrooptic Materials," in *Handbook of Lasers,* Chemical Rubber Co., Cleveland, 1971, Chapter 15. Also, Landolt-Borenstein Numerical Data, "Functional Relationships in Science and Technology," in *Group III: Crystal and Solid State Physics, Vol. 2, Electrooptic Constants* K. H. Hellwege, ed. (Berlin: Springer-Verlag, 1969).
4. Yariv, A., C. A. Mead, and J. V. Parker, "GaAs as an Electrooptic Modulator at 10.6 μ," *IEEE J. Quant. Elect.* **QE-2,** 243 (1966).
5. Kiefer, J. and A. Yariv, "Electrooptic Characteristics of CdTe at 3.39 and 10.6 μ," *Appl. Phys. Letters* **15,** 26 (1969). Also, K. Tada, and M. Aoki, "Linear Electrooptic Effect of ZnTe at 10.6 μ," *Jap. J. Appl. Phys.* **10,** 998 (1971).
6. Namba, S., "Electrooptical Effect of Zincblende," *J. Opt. Soc. Am.* **51,** 76 (1961).
7. Peters, L. C., "Gigacycle Bandwidth Coherent Light Traveling-Wave Phase Modulators," *Proc. IEEE* **51,** 147 (1963).
8. Rigrod, W. W. and I. P. Kaminow, "Wide-Band Microwave Light Modulation," *Proc. IEEE* **51,** 137 (1963).
9. Kaminow, I. P. and J. Lin, "Propagation Characteristics of partially Loaded Two-Conductor Transmission Lines for Broadband Light Modulators," *Proc. IEEE* **51,** 132 (1963).
10. White, R. M. and C. E. Enderby, "Electro-Optical Modulators Employing Intermittent Interaction," *Proc. IEEE* **51,** 214 (1963).
11. Fowler, V. J. and J. Schlafer, "A Survey of Laser Beam Deflection Techniques," *Proc. IEEE* **54,** 1437 (1966).
12. Ninomiya, Y., "Ultrahigh Resolving Electrooptic Prism Array Light Deflectors," *IEEE J. Quant. Elect.* **QE-9,** 791 (1973).
13. See, for example, C. Kittel, *Introduction to Solid State Physics,* 4th ed. (New York: Wiley, 1971).
14. Brillouin, L., "Diffusion de la lumière et des rayons X par un corps transparent homgène," *Ann. Physique* **17,** 88 (1922).
15. Debye, P. and F. W. Sears, "On the Scattering of Light by Supersonic Waves," *Proc. Nat. Acad. Sci. U.S.* **18,** 409 (1932).
16. Dransfeld, K., "Kilomegacycle Ultrasonics," *Sci. Amer.* **208,** 60 (1963).
17. See, for example, Robert Adler, "Interaction Between Light and Sound," *IEEE Spectrum* **4,** 42 (May 1967).
18. Born, M. and E. Wolf, *Principles of Optics* (New York: Pergamon Press, 1965).
19. Dixon, R. W. "Photoelastic Properties of Selected Materials and Their Relevance for Applications to Acoustic Light Modulators and Scanners," *J. Appl. Phys.* **38,** 5149 (1967).
20. Quate, C. F., C. D. W. Wilkinson, and D. K. Winslow, "Interactions of Light and Microwave Sound," *Proc. IEEE* **53,** 1604 (1965). Also, M. G. Cohen, and E. I. Gordon, "Acoustic Beam Probing Using Frequencies," *Bell Syst. Tech. J.* **44,** 693 (1965).
21. Cummings, H. Z and N. Knable, "Single Sideband Modulation of Coherent Light by Bragg Reflection from Acoustical Waves," *Proc. IEEE* **51,** 1246 (1963).
22. Harris, S. E., S. T. K. Nieh, and D. K. Winslow, "Electronically Tunable Acoustooptic Filter," *Appl. Phys. Letters* **15,** 325 (1969).

23. Harris, S. E., S. T. K. Nieh, and R. Feigelson, "$CaMoO_4$ Electronically Tunable Optical Filter," *Appl. Phys. Letters* **17,** 223 (1970).
24. Pinnow, D. A., L. G. Van Uitert, A. W. Warner, and W. A. Bonner, "$PbMO_4$: A Melt Grown Crystal with a High Figure of Merit for Acoustooptic Device Applications," *Appl. Phys. Letters* **15,** 83 (1969).
25. Yariv, A. and P. Yeh, "Optical Waves in Crystals" (New York: Wiley, 1984).

Problems

14.1 Derive the equations of the nine ellipses traced by the optical field vector as shown in Figure 14.3*c* as a function of the retardation Γ.

14.2 Discuss the consequence of the field-independent retardation $(\omega l/c) \times (n_0 - n_e)$ in (14.5-1) on an amplitude modulator such as that shown in Figure 14.4.

14.3 Use the Bessel-function expansion of $\sin(a \sin x)$ to express (14.3-7) in terms of the harmonics of the modulation frequency ω_m. Plot the ratio of the third harmonic $(3\omega_m)$ of the output intensity to the fundamental as a function of Γ_m. What is the maximum allowed Γ_m if this ratio is not to exceed 10^{-2}?

Answer:
$\Gamma m < 0.5$.

14.4 Show that, if a phase-modulated optical wave is incident on a square-law detector, the output consists only of the dc current component.

14.5 Using References 8 and 9, design a partially loaded KDP phase modulator that operates at $\nu_m = 10^9$ Hz and yields a peak phase excursion of $\delta = \pi/3$. What is the modulation power?

14.6 Derive the expression [similar to (14.6-2)] for the modulation power of a transverse $\bar{4}3m$ crystal electrooptic modulator of the type described in the example of Section 14.5.

14.7 (a) Show that if a ray propagates at an angle $\theta (\ll 1)$ to the z axis in the arrangement of Figure 14.4, it exercises a birefringent contribution to the retardation

$$\Delta\Gamma_{\text{birefringent}} = \frac{\omega l}{2c} n_0 \left(\frac{n_0^2}{n_e^2} - 1\right)\theta^2$$

that corresponds to a change in index

$$n_0 - n_e(\theta) = \frac{n_0 \theta^2}{2} \left(\frac{n_0^2}{n_e^2} - 1\right)$$

(b) Derive an approximate expression for the maximum allowable beam spreading angle for which $\Delta\Gamma_{\text{birefringent}}$ does not interfere with the operation of the modulator.

Answer:

$$\theta < \left[\frac{\lambda}{4 n_0 l \left(\frac{n_0^2}{n_e^2} - 1\right)}\right]^{1/2}$$

14.8 Consult the literature (see References 17 and 18, e.g.) and describe the difference between Bragg diffraction and Debye–Sears diffraction. Under what conditions is each observed?

14.9 Bragg's law for diffraction of x rays in crystals is (Reference 13)

$$2d \sin \theta = m \frac{\lambda}{n} \qquad m = 1, 2, 3$$

where n is the index of refraction, d is the distance between equivalent atomic planes, θ is the angle of incidence, and λ is the vacuum wavelength of the diffracted radiation. Bragg diffraction of light from sound (see Figure 14.15) takes place when

$$2\lambda_s \sin \theta = \frac{\lambda}{n}$$

Thus, if we compare it to the X-ray result and take $\lambda_s = d$, only the case of $m = 1$ is allowed. Explain the difference. Why do we not get light diffracted along directions θ corresponding to $m = 2, 3, \ldots$ in the case of scattering from acoustic waves?

Hint:
The diffraction of x rays takes place at discrete atomic planes, which can be idealized as infinitely thin sheets, whereas the sound wave disturbance is sinusoidal.

14.10 Solve the coupled mode equations (14.9-8) for the case $\Delta k = |\mathbf{k}_i - \mathbf{k}_s - \mathbf{k}_d| \neq 0$ and a small input $E_i(0)$. Assume a collinear interaction so that $r_i = r_d = r_s = z$. Show that the maximum fraction of the incident power transferred to the diffracted beam is

$$\frac{\eta^2}{\eta^2 + \left(\frac{\Delta k}{2}\right)^2}$$

It follows that the amount of mismatch Δk that can be tolerated depends on η. Can you give an intuitive explanation for the (implied) violation of momentum conservation?

14.11 Derive (14.9-11) from Doppler arguments and (14.9-12).

CHAPTER 15

Coherent Interactions of a Radiation Field and an Atomic System

15.0 INTRODUCTION

In the treatment, up to this point, of the interaction of radiation and an atomic system, we considered only equilibrium situations. It was assumed that the duration of the interaction was long compared to the inelastic (τ) and elastic (T_2) collision times so that the atomic medium response can be described by means of the susceptibility, χ, as in (8.1-19).

In situations involving intense optical fields and/or long relaxation times, we are often concerned with the atomic response to the field on a time scale *shorter* than the collision times. In such cases, the atomic polarization is *not an explicit function of the instantaneous electric field.*

A number of new phenomena occur in this regime and their study requires some new analytic tools. Some of these phenomena are photon echoes, superradiant states, and self-induced transparency.

Before considering these phenomena, we introduce a formalism, due to Feynman, Vernon, and Hellwarth (Reference 1), which establishes a formal similarity between the response of a two-level atomic system to that of a magnetic spin in a dc magnetic field. This point of view makes it possible to visualize the atomic dipolar behavior in terms of the conceptually simpler spin precession. The "equations of motion" describing the evolution of the atomic system are shown to be identical to those of a gyromagnet in a dc magnetic field.

15.1 VECTOR REPRESENTATION OF THE INTERACTION OF A RADIATION FIELD WITH A TWO-LEVEL ATOMIC SYSTEM (REFERENCE 1)

The Schrödinger equation, when applied to the interaction of a two-level atomic system with an electromagnetic field, can be cast in a simple but rigorous, geometric form. The evolution of the atomic wavefunction can be represented by the motion of a fictitious vector. This formalism is especially useful in physical situations where collision processes can be ignored. This is the case, for example, in the propagation of intense ultrashort pulses. We will use it to treat the problem of optical nutation, spin echoes, and self-induced transparency.

We need to solve the Schrödinger equation

$$\mathcal{H}\psi = i\hbar \frac{\partial \psi}{\partial t} \tag{15.1-1}$$

where

$$\mathcal{H} = \mathcal{H}_0 + V(t) \tag{15.1-2}$$

$V(t)$ is the Hamiltonian representing the interaction of the atomic system with the electromagnetic field. $\mathcal{H}_0$ is the Hamiltonian with zero field.

The wavefunction of an individual system in an ensemble of noninteracting systems can be taken in the form

$$\psi(t) = a(t)u_a + b(t)u_b \tag{15.1-3}$$

where u_a and u_b are the time-independent eigenfunctions of $\mathcal{H}_0$,

$$\mathcal{H}_0 u_a = \frac{\hbar\omega}{2} u_a$$

$$\mathcal{H}_0 u_b = -\frac{\hbar\omega}{2} u_b$$

and the interaction is assumed to involve transitions between states $|a\rangle$ and $|b\rangle$ only, whose energy separation is $\hbar\omega$ (see Figure 15.1).

In general, we need four constants to completely specify $\psi(t)$. These are the real and imaginary parts of $a(t)$ and $b(t)$. Since the absolute phase of $\psi(t)$ has no physical significance, we need only three. These can be taken as the magnitudes of $a(t)$ and $b(t)$ and their relative phase. Alternatively, we can construct three real functions (r_1, r_2, r_3) of a and b that can be viewed as the components of a vector $\mathbf{r}$ in some mathematical space with coordinate systems labeled (1, 2, 3).

$$\begin{aligned} r_1 &= ab^* + ba^* \\ r_2 &= i(ab^* - ba^*) \\ r_3 &= aa^* - bb^* \end{aligned} \tag{15.1-4}$$

so that $|\mathbf{r}|^2 = (|a|^2 + |b|^2)^2 = (\int \psi^*\psi\, dv)^2 = 1$.

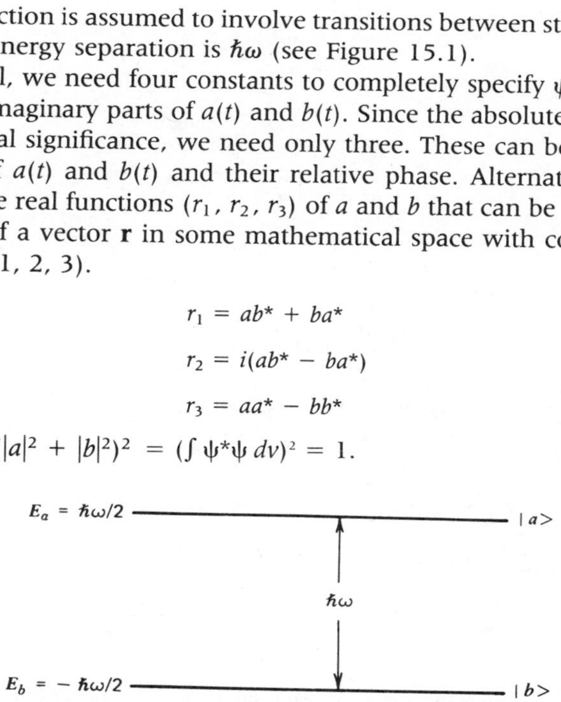

FIGURE 15.1 The two level atomic system used in the vector model (Section 15.1).

In terms of the density matrix defined in Section 8.1, we have

$$\begin{aligned} r_1 &= 2\,\mathrm{Re}(\rho_{21}) \\ r_2 &= -2\,\mathrm{Im}(\rho_{21}) \\ r_3 &= \rho_{22} - \rho_{11} \end{aligned} \tag{15.1-5}$$

The time dependence of **r** can be obtained from the Schrödinger equation (15.1-1) written as

$$(\mathcal{H}_0 + V)(au_a + bu_b) = i\hbar(\dot{a}u_a + \dot{b}u_b) \tag{15.1-6}$$

Multiplying (15.1-6) by u_a^* and integrating over all space lead to

$$\frac{da}{dt} = -\frac{i}{\hbar}\left[a\left(\frac{\hbar\omega}{2} + V_{aa}\right) + bV_{ab}\right] \tag{15.1-7a}$$

and repeating the same procedure with u_b^* gives

$$\frac{db}{dt} = -\frac{i}{\hbar}\left[b\left(-\frac{\hbar\omega}{2} + V_{bb}\right) + aV_{ba}\right] \tag{15.1-7b}$$

In the following, we will limit ourselves to situations where $V_{aa}, V_{bb} \ll \hbar\omega$[1]. If we neglect V_{aa} and V_{bb} in (15.1-7), we can show that the state vector **r** defined by (15.1-4) obeys the simple equation

$$\frac{d\mathbf{r}}{dt} = \boldsymbol{\omega}(t) \times \mathbf{r} \tag{15.1-8}$$

where

$$\begin{aligned} \omega_1 &\equiv (V_{ab} + V_{ba})/\hbar \\ \omega_2 &\equiv i(V_{ab} - V_{ba})/\hbar \\ \omega_3 &\equiv \omega \end{aligned} \tag{15.1-9}$$

As a proof, consider the 1 component of (15.1-8)

$$\frac{dr_1}{dt} = \omega_2 r_3 - \omega_3 r_2 \tag{15.1-10}$$

The left side of (15.1-10) is equal to $\dot{a}b^* + a(\dot{b})^* + \text{c.c.}$ If we use (15.1-7) with $V_{aa} = V_{bb} = 0$, it becomes

$$\frac{dr_1}{dt} = \frac{i}{\hbar}(-\hbar\omega ab^* + V_{ba}^* aa^* - V_{ab} bb^*) + \text{c.c.}$$

Using (15.1-4) and (15.1-9), we obtain

$$\omega_2 r_3 - \omega_3 r_2 = \frac{i}{\hbar}(V_{ab} - V_{ba})(aa^* - bb^*) - i\omega(ab^* - ba^*)$$

[1] Both V_{aa} and V_{bb} are zero in magnetic dipole transitions between the states $m_s = \pm\frac{1}{2}$ of a "spin $\frac{1}{2}$" system, or in electric dipole $\Delta m = \pm 1$ transitions, in a situation where $\mathcal{H}_0$ possesses inversion symmetry so that u_a and u_b have definite parity.

which is the same as the expression for dr_1/dt. The proof for the 2 and 3 components of (15.1-8) is similar.

To proceed further, we need to be more specific about the transition $a \to b$ and the electromagnetic field. Let us consider the important class of dipole transitions involving the selection rule $\Delta m = \pm 1$. Our notation will correspond to electric dipole transitions. The interaction Hamiltonian in this case becomes

$$V = -\mu_x E_x - \mu_y E_y \tag{15.1-11}$$

Defining

$$\begin{aligned} \mu^+ &\equiv \mu_x + i\mu_y \qquad & E^+ &\equiv E_x + iE_y \\ \mu^- &\equiv \mu_x - i\mu_y \qquad & E^- &\equiv E_x - iE_y \end{aligned} \tag{15.1-12}$$

gives

$$V = -\tfrac{1}{2}(\mu^+ E^- + \mu^- E^+) \tag{15.1-13}$$

For $\Delta m = \pm 1$ transitions, we have

$$\begin{aligned} \langle m + 1 | \, \mu^- \, | m \rangle &= 0 \\ \langle m | \, \mu^+ \, | m + 1 \rangle &= 0 \end{aligned} \tag{15.1-14}$$

so that from (15.1-12) and (15.1-13) and letting $u_a \to |m + 1\rangle$, $u_b \to |m\rangle$

$$\begin{aligned} V_{ab} &= -\tfrac{1}{2}\overset{+}{\mu}_{ab}(E_x - iE_y) \\ V_{ba} &= -\tfrac{1}{2}\overset{-}{\mu}_{ba}(E_x + iE_y) \end{aligned} \tag{15.1-15}$$

we are free to choose the phases of u_a and u_b so that $\overset{+}{\mu}_{ab}$ is a real positive number that we designate as 2μ.

$$\overset{+}{\mu}_{ab} = \overset{-}{\mu}_{ba} \equiv 2\mu \qquad \mu = \langle b | \, \mu_x \, | a \rangle \tag{15.1-16}$$

with the equality following from the fact that $\overset{+}{\mu}_{ab} = (\overset{-}{\mu}_{ba})^*$. From (15.1-9) and (15.1-15), we get

$$\begin{aligned} \omega_1(t) &= (V_{ab} + V_{ba})/\hbar = -\frac{2\mu E_x(t)}{\hbar} \\ \omega_2(t) &= i(V_{ab} - V_{ba})/\hbar = -\frac{2\mu E_y(t)}{\hbar} \end{aligned} \tag{15.1-17}$$

so that the vector $\boldsymbol{\omega}$ behaves in the mathematical 1–2 plane exactly as the vector $\mathbf{E}$ does in the physical x–y plane. To attach, similarly, a physical significance to $\mathbf{r}$, consider the expectation value of a transverse (e.g., x) component of the dipole operator,

$$\langle \mu_x \rangle = \tfrac{1}{2}\langle \mu^+ + \mu^- \rangle = \tfrac{1}{2}\int (a^* u_a^* + b^* u_b^*)(\mu^+ + \mu^-)(a u_a + b u_b)\, dv$$

that, using (15.1-14), gives

$$\langle \mu_x \rangle = \mu r_1 \qquad \text{and} \qquad \langle \mu_y \rangle = \mu r_2 \tag{15.1-18}$$

so that the expectation value of the dipole moment operator (which corresponds to the radiating dipole of one atomic system) behaves in the physical x–y plane in the same way as the **r** vector in the fictitious 1–2 plane.

The machinery we have just constructed for treating the problem of the dipole interaction of a two-level atomic system with an electromagnetic field is thus apparent. All we need to do is solve the vector equation

$$\frac{d\mathbf{r}}{dt} = \boldsymbol{\omega}(t) \times \mathbf{r} \tag{15.1-8}$$

for $\mathbf{r}(t)$ where $\boldsymbol{\omega}$ is given by (15.1-17). The transverse dipole moments of a single atomic system are given directly by **r** using (15.1-18). Since the wave function $\psi(t)$ is related uniquely to **r** via (15.1-4), a knowledge of $\mathbf{r}(t)$ is formally equivalent to a complete (in the quantum mechanical sense) specification of the system. The procedure outlined above requires a knowledge of the initial value $\mathbf{r}(0)$, which is equivalent to specifying $\psi(0)$ when solving the Schrödinger equation.

In the following sections, we will study the solution of (15.1-8) and its implications in a few simple cases. Before we do that, however, we may find it instructive to consider the significance of (15.1-8) in the simple case of a spin 1/2 magnetic system. Here, we have $u_a = |1/2\rangle$, $u_b = |-1/2\rangle$ and

$$\boldsymbol{\mu} = \frac{2\beta}{\hbar}\mathbf{S}$$

where β is the Bohr magneton, $\boldsymbol{\mu}$ the magnetic dipole moment, and **S** the spin angular momentum operator. Using (15.1-16), we get

$$2\mu \equiv \mu_{ab}^{+} = \frac{2\beta}{\hbar}\langle\tfrac{1}{2}|\, S^{+}\, |-\tfrac{1}{2}\rangle = 2\beta$$

From (15.1-18), it follows that

$$\langle\mu_x\rangle = \beta r_1 \qquad \langle\mu_y\rangle = \beta r_2$$

In this case, we also have

$$\begin{aligned}\langle\mu_z\rangle &= \langle a u_{1/2} + b u_{-1/2}|\, \frac{2\beta}{\hbar} S_z\, |a u_{1/2} + b u_{-1/2}\rangle \\ &= \beta(aa^* - bb^*) = \beta r_3\end{aligned}$$

so that the identification of $\langle\boldsymbol{\mu}\rangle$ with $\beta\mathbf{r}$ is complete.

Using (15.1-17) and replacing $E_{x,y}$ with the magnetic field component $H_{x,y}$ give

$$\omega_1 = -\frac{2\beta H_x}{\hbar} \qquad \omega_2 = -\frac{2\beta H_x}{\hbar}$$

All that remains is to show that $\omega_3 = -2\beta H_z/\hbar$. This is done by recognizing that the levels' energy separation $\hbar\omega$ is given in this case by $-2\beta H_z$ so that

$$\hbar\omega_3 = \hbar\omega = -2\beta H_z$$

There is, thus, also a complete correspondence between $\boldsymbol{\omega}(t)$ and $-2\beta\mathbf{H}(t)/\hbar$. Using this correspondence as well as $\langle\boldsymbol{\mu}\rangle \to \beta\mathbf{r}$, derived above, in (15.1-8) gives

$$\frac{d\langle\boldsymbol{\mu}\rangle}{dt} = \gamma\langle\boldsymbol{\mu}\rangle \times \mathbf{H} \tag{15.1-19}$$

where $\gamma \equiv 2\beta/\hbar$ is the gyromagnetic ratio. This is the well-known equation of motion of a gyromagnet in a magnetic field.

This simple physical correspondence between r_3 and μ_z does not generally exist in the case of electric dipole transitions. As a matter of fact, in the electric $\Delta m = \pm 1$ case considered here, we have

$$\langle\mu_z\rangle \propto \langle au_a + bu_b|\, z\, |au_a + bu_b\rangle = 0$$

since both $\langle u_a|\, z\, |u_a\rangle$ and $\langle u_b|\, z\, |u_a\rangle = 0$, the first because of the odd parity of the integrand $|u_a|^2 z$ and the second because it involves the integral $\int_0^{2\pi} \exp(i\phi)\, d\phi$. The correspondence between the transverse components of $\mathbf{r}$ and $\boldsymbol{\mu}$ given by (15.1-18) is valid, however, and, as shown below, is extremely useful. The component r_3, in this case, is proportional to the expectation value of the unperturbed Hamiltonian since

$$\begin{aligned}\langle H_0\rangle &= \int \psi^* H_0 \psi\, dv = (aa^* - bb^*)\frac{\hbar\omega}{2} \\ &= r_3 \frac{\hbar\omega}{2}\end{aligned} \tag{15.1-20}$$

Transformation to a Rotating Coordinate System

The solution of the dipolar equation of motion (15.1-8)

$$\frac{d\mathbf{r}}{dt} = \boldsymbol{\omega} \times \mathbf{r}$$

is greatly facilitated by transforming from the stationary (1, 2, 3) coordinate system to one that is rotating about it at a (radian) rate $\boldsymbol{\Omega}$. According to a basic theorem in vector calculus, the rate of change $d\mathbf{r}_R/dt$ of any vector $\mathbf{r}$ as *observed in the rotating system* is related to that observed in the stationary one by

$$\frac{d\mathbf{r}_R}{dt} = \left(\frac{d\mathbf{r}}{dt}\right)_R - \boldsymbol{\Omega} \times \mathbf{r}_R \tag{15.1-21}$$

where the R subscript indicates vector transformation to the rotated system.[2] Applying (15.1-21) to (15.1-8) leads to

$$\frac{d\mathbf{r}_R}{dt} = (\boldsymbol{\omega}_R - \boldsymbol{\Omega}) \times \mathbf{r}_R \tag{15.1-22}$$

[2] If a vector $\mathbf{A} = (A_1, A_2, A_3)$ in the (1, 2, 3) system, then in a system (I, II, III) rotated by an angle Ωt about the "3" axis, it becomes $\mathbf{A}_R = (A_{\rm I}, A_{\rm II}, A_{\rm III})$, where $A_{\rm I} = A_1 \cos \Omega t + A_2 \sin \Omega t$, $A_{\rm II} = -A_1 \sin \Omega t + A_2 \cos \Omega t$, $A_{\rm III} = A_3$. See Problem 15.4 for a related discussion.

The Behavior of $\mathbf{r}(t)$ with No Applied Field

Consider the solution of $\mathbf{r}(t)$ with no radiation field present. In this case, we obtain, from (15.1-17),

$$\omega_1 = 0 \qquad \omega_2 = 0 \qquad \omega_3 = \omega \Rightarrow \boldsymbol{\omega} = \mathbf{a}_3\omega$$

where $\mathbf{a}_3$ is a unit vector along the 3 direction. It is convenient to choose the rotation axis to be parallel to $\mathbf{a}_3$, in this case, designating the axes in the rotating system as I, II, III, we have $\boldsymbol{\Omega} = \mathbf{a}_{\mathrm{III}}\Omega$ and (15.1-22) becomes

$$\frac{d\mathbf{r}_R}{dt} = \mathbf{a}_{\mathrm{III}}(\omega - \Omega) \times \mathbf{r}_R \tag{15.1-23}$$

It may be useful to recall at this point that the choice of the direction and magnitude of $\boldsymbol{\Omega}$ is strictly a matter of convenience, so that if we take $\Omega = \omega$, the solution of (15.1-23) is

$$\mathbf{r}_R = \text{const}$$

and transforming back to the (1, 2, 3) system, we have from Figure 15.2 (for $\Omega = \omega$)

$$\begin{aligned} r_1 &= r_{\mathrm{I}} \cos \omega t - r_{\mathrm{II}} \sin \omega t \\ r_2 &= r_{\mathrm{I}} \sin \omega t + r_{\mathrm{II}} \cos \omega t \\ r_3 &= \sqrt{1 - r_{\mathrm{I}}^2 - r_{\mathrm{II}}^2} = \text{const} \end{aligned} \tag{15.1-24}$$

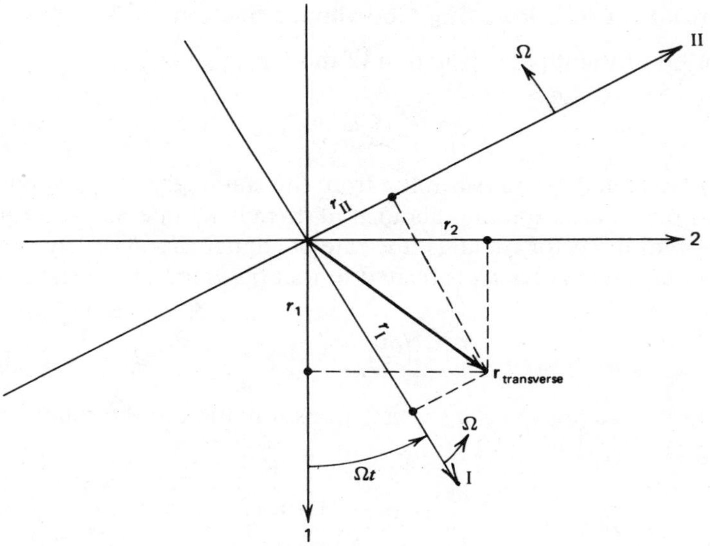

FIGURE 15.2 The relationship between the transverse components of $\mathbf{r}$ in the (1,2) and the rotating (I, II) coordinate system. System I, II, III rotates at a radian rate Ω about the 3 axis (axes 3 and III coincide).

The motion of $\mathbf{r}(t)$ is thus one of precession at a rate ω about the 3 axis at an arbitrary inclination angle so that r_3 and, according to (15.1-20), the energy are constants of the motion. Since $r_3 = (aa^* - bb^*)$, its constancy is just a geometrical representation of the fact that with no applied field, no transitions between $|a\rangle$ and $|b\rangle$ can take place, so that $|a|^2$ and $|b|^2$ are constant.[3]

The components r_{I} and r_{II} can be related simply to the amplitudes U and V of the medium polarization

$$P_x = U \cos \omega t - V \sin \omega t$$

From (15.1-18) and (15.1-24), we obtain

$$P_x = N\langle \mu_x \rangle = N\mu r_{\text{I}} = N\mu(r_{\text{I}} \cos \omega t - r_{\text{II}} \sin \omega t)$$

so that

$$U = N\mu r_{\text{I}} \qquad V = N\mu r_{\text{II}}$$

where N is the density of atoms. If at $t = 0$ there are N_a atoms/m³ in the upper level and N_b in the lower one then

$$U = (N_a - N_b)\mu r_{\text{I}} \qquad V = (N_a - N_b)\mu r_{\text{II}}$$

where $\mathbf{r}_{\text{I,II}}(t)$ is the solution corresponding to an atom that is initially in the upper level $|a\rangle$, that is, $r_{\text{III}}(0) = 1$. The proof of this last statement is left as a problem.

The Behavior of $\mathbf{r}(t)$ in an Applied Field

Consider next the behavior of $\mathbf{r}(t)$ when a circularly polarized electric field

$$E_x = E \cos \omega_0 t$$
$$E_y = E \sin \omega_0 t \tag{15.1-25}$$

is applied in the x, y plane. From (15.1-17), the components of $\boldsymbol{\omega}(t)$ are

$$\omega_1 = -\frac{2\mu}{\hbar} E \cos \omega_0 t$$

$$\omega_2 = -\frac{2\mu}{\hbar} E \sin \omega_0 t \tag{15.1-26}$$

$$\omega_3 = \omega$$

so that, viewed in the 1, 2 plane, $\boldsymbol{\omega}(t)$ is a circularly polarized vector rotating at a rate ω_0 about the 3 axis with a constant magnitude $-2\mu E/\hbar$. In a coordinate system rotating in synchronism with $\boldsymbol{\omega}(t)$ (i.e., $\boldsymbol{\Omega} = \mathbf{a}_{\text{III}}\omega_0$), the vector $\boldsymbol{\omega}(t)$ becomes stationary so that $\boldsymbol{\omega}_R = (-2\mu E/\hbar,\ 0,\ \omega)$ and the equation of

[3] The possibility of spontaneous transition from $|a\rangle$ to $|b\rangle$ is not included in the model due to the classical representation of the electric field in the Hamiltonian (15.1-11).

motion (15.1-22) becomes

$$\frac{d\mathbf{r}_R(\omega)}{dt} = \left[\mathbf{a}_\mathrm{I}\left(-\frac{2\mu E}{\hbar}\right) + \mathbf{a}_\mathrm{III}(\omega - \omega_0)\right] \times \mathbf{r}_R(\omega) \equiv \boldsymbol{\omega}_\mathrm{eff} \times \mathbf{r}_R(\omega) \quad (15.1\text{-}27)$$

The problem has thus been reduced to that of the precession of $\mathbf{r}_R$ about a *stationary* vector $\boldsymbol{\omega}_\mathrm{eff} = \mathbf{a}_\mathrm{I}[-(2\mu E/\hbar)] + \mathbf{a}_\mathrm{III}(\omega - \omega_0)$. If we use the solution (15.1-24) for a stationary $\boldsymbol{\omega}$, the rate of precession is

$$\omega_e = |\boldsymbol{\omega}_\mathrm{eff}| = \sqrt{\left(\frac{2\mu E}{\hbar}\right)^2 + (\omega_0 - \omega)^2} \quad (15.1\text{-}28)$$

This motion is depicted in Figure 15.3, drawn for the initial condition $\mathbf{r}_\mathrm{III}(0) = \mathbf{a}_\mathrm{III}1$, which corresponds, according to (15.1-4), to an atom found initially in the upper state $|a\rangle$. The rotating I direction was chosen to coincide with the projection of ω on the 1–2 plane so that $\omega_\mathrm{II} = 0$.

Using basic trigonometric relations, we obtain from Figure 15.3,

$$r_\mathrm{I} = \frac{\omega_\mathrm{I}(\omega - \omega_0)}{\omega_e^2}(1 - \cos\omega_e t)$$

$$r_\mathrm{II} = -\frac{\omega_\mathrm{I}}{\omega_e}\sin\omega_e t \quad (15.1\text{-}29)$$

$$r_\mathrm{III} = 1 - 2\left(\frac{\omega_\mathrm{I}}{\omega_e}\right)^2 \sin^2\left(\frac{\omega_e t}{2}\right)$$

where $\omega_\mathrm{I} \equiv [-(2\mu E/\hbar)]$ is a negative number.

When we use (15.1-29) as well as the relation $r_\mathrm{III} = |a|^2 - |b|^2$ and the normalization condition, $|a|^2 + |b|^2 = 1$ results in

$$|a|^2 = 1 - \left(\frac{\omega_\mathrm{I}}{\omega_e}\right)^2 \sin^2\left(\frac{\omega_e t}{2}\right)$$

$$|b|^2 = \left(\frac{\omega_\mathrm{I}}{\omega_e}\right)^2 \sin^2\left(\frac{\omega_e t}{2}\right) \quad (15.1\text{-}30)$$

for the probability of finding an atom, initially in the upper state, in states a (upper) and b, respectively.

At resonance $(\omega = \omega_0)$, $\omega_e = \omega_\mathrm{I}$ and

$$|a|^2 = \cos^2\left(\frac{\omega_\mathrm{I} t}{2}\right)$$

$$|b|^2 = \sin^2\left(\frac{\omega_\mathrm{I} t}{2}\right) \quad (15.1\text{-}31)$$

so that a complete population exchange between the upper and lower levels takes place every π/ω_I sec. Occupation probabilities $|a|^2$ and $|b|^2$ are plotted in Figure 15.4 for both the resonance and off-resonance conditions.

In closing, we should notice that in most experimental situations, an

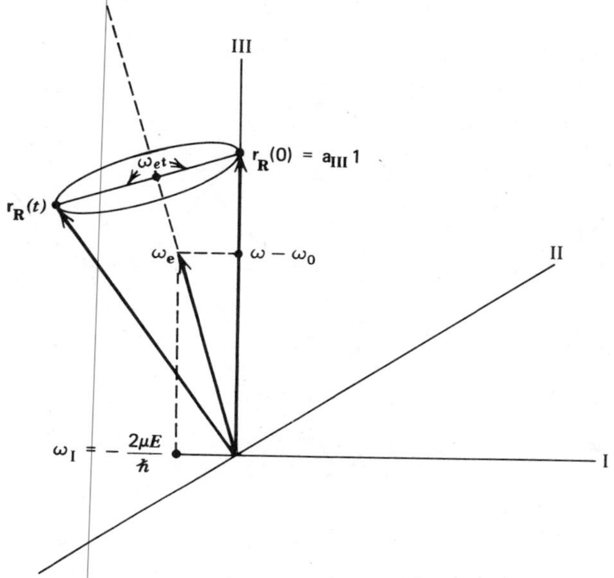

FIGURE 15.3 The motion of the vector $\mathbf{r}_R(t)$ in the rotating coordinate system (I, II, III). The motion consists of a precession at a rate $\omega_e = \sqrt{(\omega - \omega_o)^2 + (2\mu E/\hbar)^2}$ about the vector $\omega_{eff} = \mathbf{a}_I(-2\mu E/\hbar) + \mathbf{a}_{III}(\omega - \omega_0)$. The drawing corresponds to the initial condition $\mathbf{r}_R(0) = \mathbf{a}_{III}1$, that is, the atom is initially in the upper state $|a\rangle$.

atom is subjected to a linearly polarized field

$$E_x = E \cos \omega_0 t \tag{15.1-32}$$

rather than to the circularly polarized field (15.1-25) used in our formalism. The field of (15.1-32) can be resolved into two oppositely (circularly) polarized fields

$$\begin{aligned} E_{x1} &= \frac{E}{2} \cos \omega_0 t \\ E_{y1} &= \frac{E}{2} \sin \omega_0 t \end{aligned} \tag{15.1-33}$$

and

$$\begin{aligned} E_{x2} &= \frac{E}{2} \cos \omega_0 t \\ E_{y2} &= -\frac{E}{2} \sin \omega_0 t \end{aligned} \tag{15.1-34}$$

In a rotating coordinate system that is synchronous with the field (15.1-33), the field of (15.1-34) is seen to rotate at an angular rate of $2\omega_0$ so that to first order, it exerts no average "torque" on $\mathbf{r}$ and can be neglected. The resulting

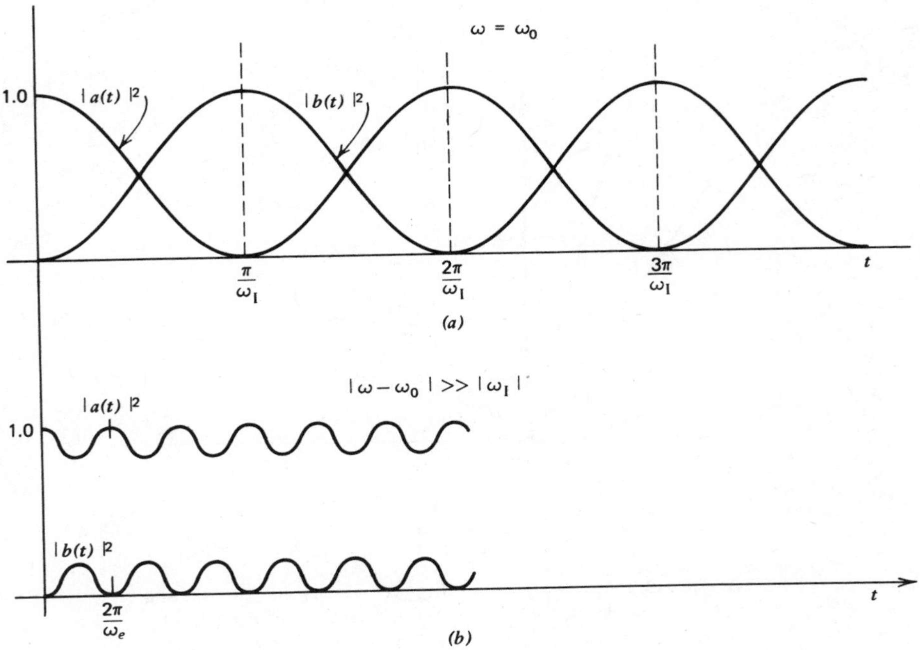

FIGURE 15.4 Oscillation of the occupation probability $|a|$ and $|b|$ when an optical field is applied. (*a*) $\omega = \omega_0$ ("on" resonance). (*b*) $|\omega - \omega_0| \gg |\omega_I|$.

motion of **r** is then the same as that studied above except that we must replace everywhere E by $E/2$.

15.2 SUPERRADIANCE

Consider the precession of the **r** vector in the (I, II, III) space under the influence of an applied circularly polarized field at a frequency $\omega_0 = \omega$ where $\hbar\omega$ is the transition energy. The atom is taken to be initially in the lower state $|b\rangle$, so that $\mathbf{r}_R(0) = -1\mathbf{a}_{III}$. Redrawing Figure 15.3 for this special case, we obtain the situation depicted in Figure 15.5. Let us turn the field off at a time t_0 where $|\omega_I|t_0 = \pi/2$. For $t \geq t_0$,

$$\begin{aligned} r_I &= r_{III} = 0 \\ r_{II} &= -1 \end{aligned} \qquad (15.2\text{-}1)$$

and using (15.1-18) and (15.1-24), we obtain the oscillating atomic dipole moment in real space as

$$\begin{aligned} \mu_x &= \mu r_1 = \mu \sin \omega(t - t_0) \\ \mu_y &= \mu r_2 = -\mu \cos \omega(t - t_0) \end{aligned} \qquad (15.2\text{-}2)$$

This is the largest dipole moment that can be "squeezed" out of one atom. If

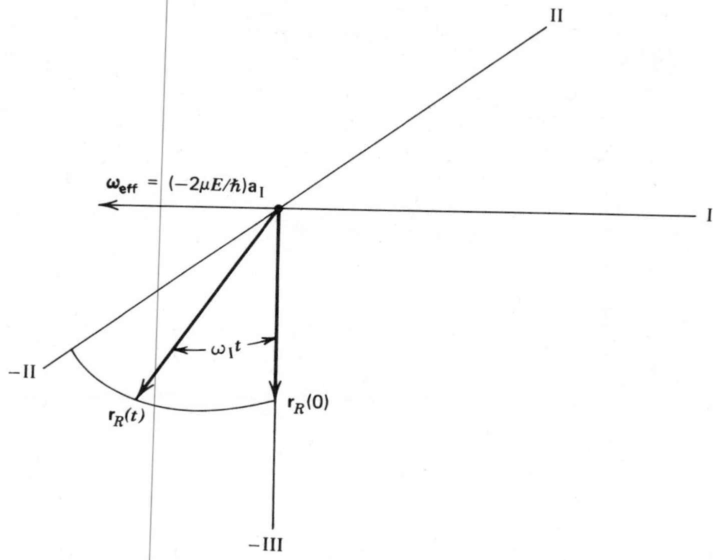

FIGURE 15.5 The motion of $\mathbf{r}_R$ *in the (I, II, III) space in the case of an atom which is in the ground state* $|b\rangle$ at $t = 0$. The tip of the vector $\mathbf{r}_R$ describes a circle in the (II, III) plane. If the field is turned off at $t_0 = \pi/2\omega_\mathrm{I}$ the vector, $\mathbf{r}_R(t{>}t_0)$, is left pointing along the -11 direction.

the initial conditions correspond to N_a atoms per m^3 in state a and N_b atoms per m^3 in state b, the polarization at $t > t_0$ is

$$\begin{aligned} P_x &= (N_b - N_a)\mu \sin \omega(t - t_0) \\ P_y &= -(N_b - N_a)\mu \cos \omega(t - t_0) \end{aligned} \tag{15.2-3}$$

in which *all* the atoms in the ensemble are contributing coherently to a single giant dipole moment. This atomic state was called the superradiant state by Dicke (Reference 2), the term superradiance denoting the increased radiation rate that characterizes the spontaneous decay of this giant dipole.

To get an expression for the spontaneous rate of decay of the dipole described by (15.2-3), consider a sample of volume V_s and an initial density of active atoms N_a and N_b immediately following a "$\pi/2$" pulse. Let the sample be situated within an optical resonator (or some enclosure) with a figure of merit Q and a volume V_c. The power radiated by the dipoles excites the radiation field of the resonator. By equating the power transferred from the polarization to the field [here we use (5.1-13)] to the power dissipated by the enclosure, we obtain

$$\text{Power} = \frac{\omega Q (N_b - N_a)^2 \mu^2 V_s^2}{\varepsilon V_c} \tag{15.2-4}$$

for the initial rate of energy radiation by the atoms. The energy stored initially (at t_0) by the atoms is $\mathscr{E} = (N_b - N_a)\hbar\omega V_s/2$. The resulting initial decay time

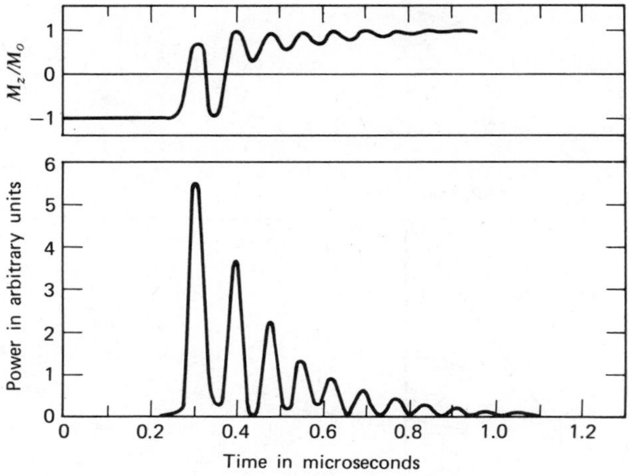

FIGURE 15.6 Superradiant emission of an inverted spin system. Upper trace: magnetization. Lower trace: radiated power. *Source:* Reference 5.

constant is thus

$$\tau = \frac{\mathscr{E}}{\text{Power}} = \frac{\hbar\varepsilon}{2Q(N_b - N_a)\mu^2}\left(\frac{V_c}{V_s}\right) \tag{15.2-5}$$

that, for $N_b \gg N_a$ (i.e., atoms initially in the ground state), gives $\tau \propto (1/N_b)$ indicative of the cooperative nature of the decay.

Superradiant effects were studied extensively in the microwave regions

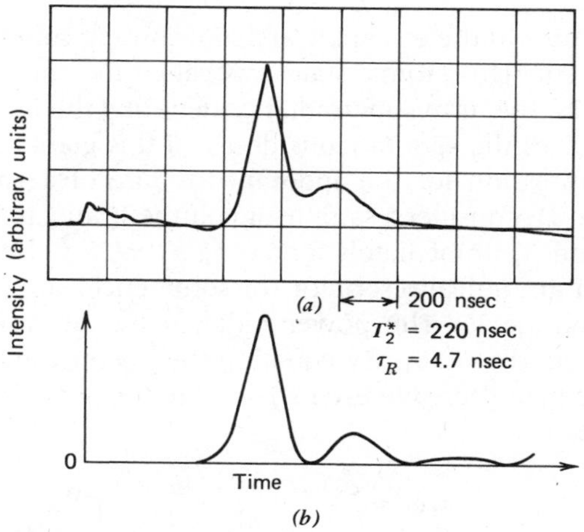

FIGURE 15.7 (*a*) Oscilloscope trace of superradiant pulses from optically pumped HF and (*b*) computer fits. *Source:* Reference 6.

in connection with nonequilibrium spin populations (References 3–5). A discussion of superradiance effects at long infrared wavelength is given in Reference 6. A theoretical superradiant emission pulse and experimental infrared data are shown in Figures 15.6 and 15.7.

15.3 PHOTON ECHOES

A direct manifestation of optical coherent effects is the phenomenon of photon echoes. This effect was first observed and investigated in nuclear magnetic resonance (Reference 7) and, more recently, in experiments involving optical transitions (Reference 8). The basic experiment involves the application of two intense and short optical pulses to an ensemble of resonantly absorbing atoms (or molecules) and observing a radiated pulse (echo) delayed relative to the second pulse by a time equal to the separation of the two exciting pulses.

Consider the sequence of two linearly polarized optical pulses, shown in Figure 15.8, applied to an atomic sample. Let the center frequency of the optical pulses be ω_0. Let us follow the evolution in the rotating frame of the $\mathbf{r}_R$ vectors of three atoms with resonant frequencies $\omega_1 = \omega_0 - \Delta\omega$, $\omega_2 = \omega_0$, and $\omega_3 = \omega_0 + \Delta\omega$, which are initially in the ground state. If the first pulse causes a nutation of $\pi/2(\mu E t_1/\hbar = \pi/2)$, then at t_1 the three $\mathbf{r}$ vectors will be along the $-\text{II}$ direction as shown in Figure 15.9*a*. The behavior between t_1 and t_2 is given by (15.1-27) with $E = 0$

$$\frac{d\mathbf{r}(\omega)}{\partial t} = \mathbf{a}_{\text{III}}(\omega - \omega_0) \times \mathbf{r}(\omega)$$

$\mathbf{r}_1$ will thus precess about the z axis at rate $-\Delta\omega$, $\mathbf{r}_2$ is stationary whereas $\mathbf{r}_3$ rotates at a rate $\Delta\omega$. If we extend this argument to a large ensemble of atoms, we conclude that the fanning in the I, II plane of the individual $\mathbf{r}$ vectors will cause the resultant (superradiant) dipole to disappear within a time $\sim 2\pi/\Delta$ where Δ is the characteristic width of the resonant frequency distribution [i.e., $\Delta \approx (\Delta\omega)_{\text{inhomog}}$]. This disappearance, however, in the case of collisionless inhomogeneous broadening is reversible. To show this, consider what happens when the second pulse is applied. The duration of the pulse is chosen to cause a rotation (about the I axis) of π radians so that $(\mu E/\hbar)(t_3 - t_2) = \pi$. The result at t_3 is shown in Figure 15.9*c*. The relative order of the $\mathbf{r}$ vectors is

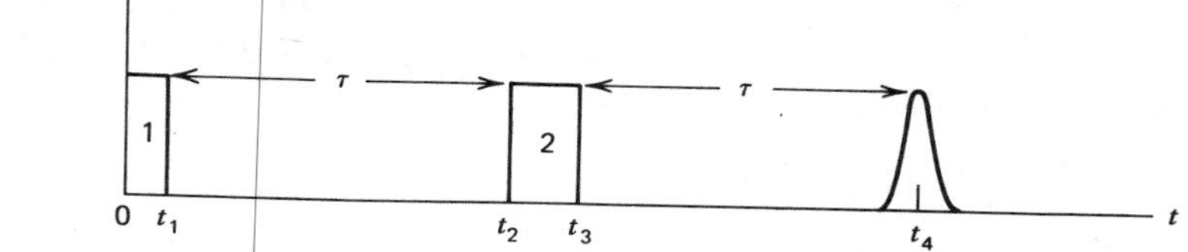

FIGURE 15.8 A $\pi/2$ pulse ① and a π pulse ② applied to an ensemble of absorbing atoms exciting a radiated echo pulse at t_4.

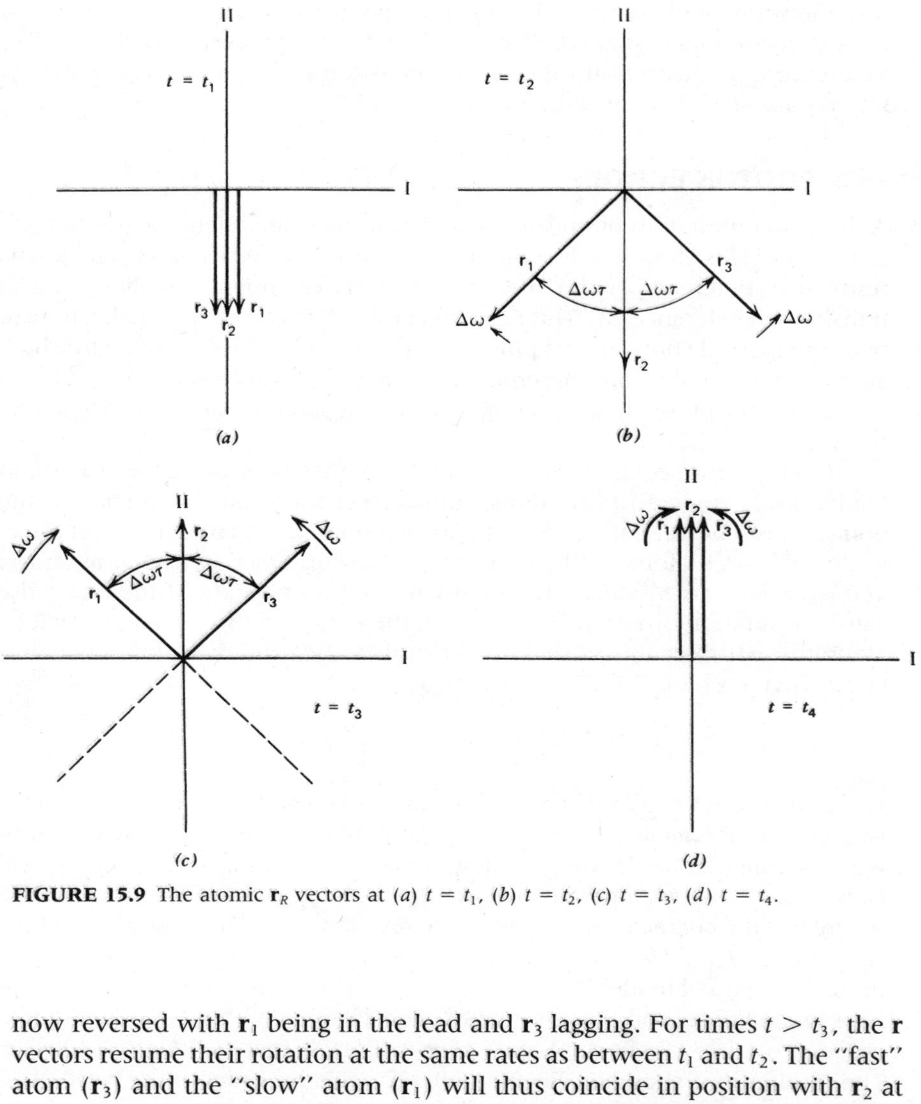

FIGURE 15.9 The atomic $\mathbf{r}_R$ vectors at (a) $t = t_1$, (b) $t = t_2$, (c) $t = t_3$, (d) $t = t_4$.

now reversed with $\mathbf{r}_1$ being in the lead and $\mathbf{r}_3$ lagging. For times $t > t_3$, the $\mathbf{r}$ vectors resume their rotation at the same rates as between t_1 and t_2. The "fast" atom ($\mathbf{r}_3$) and the "slow" atom ($\mathbf{r}_1$) will thus coincide in position with $\mathbf{r}_2$ at $t_4 = t_3 + \tau$. At this point, the resultant transverse dipole moment is a maximum, giving rise to a superradiant echo pulse.

The generalization of the above picture to a distribution of atoms is easily accomplished. Assume N atoms with an inhomogeneous broadening described by a lineshape function $g(\Delta\omega)$ where

$$\int_{-\infty}^{\infty} g(\Delta\omega)\, d(\Delta\omega) = 1 \tag{15.3-1}$$

and where the characteristic width of $g(\Delta\omega)$ is taken as Δ. If we define a *complex* vector $\mathbf{R}$ as the (vectorial) sum of all the individual $\mathbf{r}$ vectors in the I, II plane, then at $t_1 < t < t_2$, we have

$$\mathbf{R}(t) = -iN \int_{-\infty}^{\infty} g(\Delta\omega) e^{i(\Delta\omega)(t-t_1)} \, d(\Delta\omega) \tag{15.3-2}$$

The factor $-i$ in front of (15.3-2) is chosen so that at $t = t_1$, $\mathbf{R}(t_1) = -iN$, that is, it is parallel to the $-$II axis consistent with Figure 15.9*a*. We note that for times t such that $\Delta\omega(t - t_1) > \pi$, $\mathbf{R}(t) \to 0$. At $t = t_2$,

$$\mathbf{R}(t_2) = -iN \int_{-\infty}^{\infty} g(\Delta\omega) e^{i\,\Delta\omega\tau} \, d(\Delta\omega) \tag{15.3-3}$$

The effect of the second "π" pulse is to replace the azimuthal angle $\Delta\omega\tau$ by $\pi - \Delta\omega\tau$ as shown in Figure 15.9*c* so that at t_3,

$$\mathbf{R}(t_3) = -iN \int_{-\infty}^{\infty} g(\Delta\omega) e^{i(\pi-\Delta\omega\tau)} \, d(\Delta\omega) \tag{15.3-4}$$

The atoms will now continue to evolve according to $\exp[i\,\Delta\omega(t - t_3)]$ so that

$$\mathbf{R}(t > t_3) = -iN \int_{-\infty}^{\infty} g(\Delta\omega) e^{i(\pi-\Delta\omega\tau)} e^{i\,\Delta\omega(t-t_3)} \, d(\Delta\omega) \tag{15.3-5}$$

At $t_4 = t_3 + \tau$, we have

$$\mathbf{R}(t_4) = -iN \int_{-\infty}^{\infty} g(\Delta\omega) e^{i\pi} \, d(\Delta\omega) = iN$$

This corresponds to a maximum superradiant superposition of all the individual $\mathbf{r}$ vectors along the II axis as in Figure 15.9*d*. The resulting radiation pulse constitutes the echo at t_4.

Photon echoes were observed in ruby crystals (References 9–10) as well as in gas molecules (Reference 8). Experimental data of photon echoes in $C^{13}H_3F$ are shown in Figure 15.10.

15.4 SELF-INDUCED TRANSPARENCY

Another important manifestation of the coherent interactions of radiation and atomic systems is the phenomenon of self-induced transparency (References 11, 12, and 13). Above a well-defined threshold intensity, short resonant pulses of a given duration will propagate through a *normally absorbing* medium with anomalously low attenuation. This happens when the pulse width is short compared to the relaxation times in the medium and the pulse center frequency is resonant with a two-level normally absorbing transition. After a few classical absorption lengths, the pulse achieves a steady state in which its *width, energy,* and *shape* remain constant. The pulse velocity is greatly reduced from the "normal" velocity of light in the medium. The reduction factor may be as high as a few orders of magnitude.

This is probably the most striking manifestation of the failure of the conventional treatment of propagation phenomena in terms of constitutive medium parameters such as dielectric or magnetic susceptibilities.

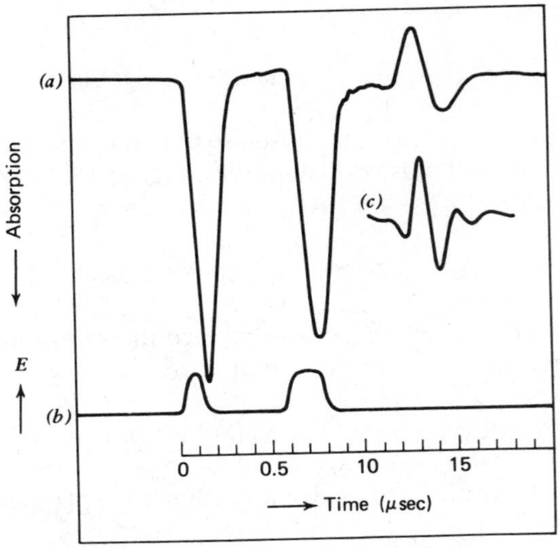

FIGURE 15.10 Photon echo in $C^{13}H_3F$ at 5.2 m torr pressure. (*a*) Optical response to $\pi/2$ and π pulses followed by the photon echo. (*b*) The $\pi/2$ and π Stark pulses. In this experiment, the optical field was left on continuously whereas the electric field dependence of the energy levels (Stark effect) was used to switch the molecules in and out of resonance by means of the electric field pulses (*b*). *Source:* Reference 8.

Our treatment follows the spirit of the original paper by McCall and Hahn (Reference 12) who first predicted and observed this phenomenon. We assume an optical pulse $E(z, t)$ and solve, via the Schrödinger equation, in the density matrix formulation, for the atomic polarization of a two-level system that is induced by this pulse. Next, we reverse the order and solve, through Maxwell's wave equation, for the field that is induced by the polarization. The resulting two sets of relations are made self-consistent. This yields the sought solution for the pulse shape, energy, and velocity.

A flowchart demonstrating the self-consistent approach is shown in Figure 15.11.

In the analysis of this phenomenon that follows, we will rely heavily on the density matrix formalism of Section 8.1.

The scalar field of the propagating pulse and the medium polarization are taken in the form

$$E_x(z, t) = \tfrac{1}{2}\{\varepsilon(z, t)e^{i[k_0 z-\omega_0 t+\phi(z,t)]} + \text{c.c.}\} \tag{15.4-1}$$

$$P_x(z, t) = \tfrac{1}{2}\{[U(z, t) + iV(z, t)]e^{i[(k_0 z-\omega_0 t+\phi(z,t)]} + \text{c.c.}\} \tag{15.4-2}$$

where the inclusion of $\phi(z, t)$ is necessary to accommodate any complex phase evolution. ε, U, and V are real. U and V are the dispersion (in phase) and absorption (quadrature) components of the polarization.

Equations (15.4-1) and (15.4-2) are substituted into the wave equation

$$\frac{\partial^2 E}{\partial z^2} - \frac{n^2}{c^2}\frac{\partial^2 E}{\partial t^2} = \mu_0 \frac{\partial^2 P}{\partial t^2} \tag{15.4-3}$$

where the x subscripts are dropped. By equating the real and imaginary parts of (15.4-3), we obtain two complicated equations. These equations assume simple forms under slowly varying envelope conditions, that is, when $\partial F/\partial t \ll \omega_0 F$, $\partial F/\partial z \ll k_0 F$, $\partial^2 F/\partial t^2 \ll \omega_0 \partial F/\partial t$, $\partial^2 F/\partial z^2 \ll k_0 \partial F/\partial z$ where $F = \varepsilon, U, V, \phi$. The result is

$$\frac{\partial \varepsilon}{\partial z} + \frac{n}{c}\frac{\partial \varepsilon}{\partial t} = -\frac{\omega_0 c \mu_0}{2n} V \tag{15.4-4}$$

$$\varepsilon\left(\frac{\partial \phi}{\partial z} + \frac{n}{c}\frac{\partial \phi}{\partial t}\right) = \frac{\omega_0 c \mu_0}{2n} U \tag{15.4-5}$$

Our next task is to relate the medium polarization U, V to the electric field. We start with the equations of motion for the density matrix (8.1-6) and (8.1-7)

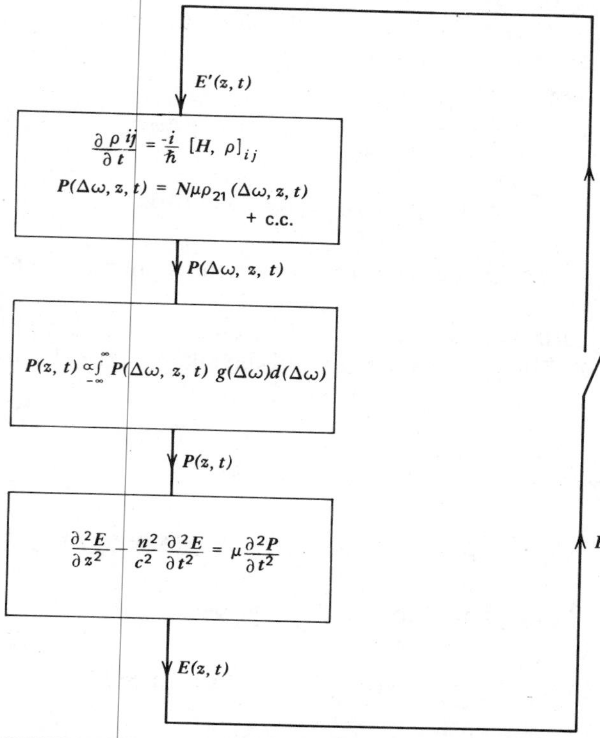

FIGURE 15.11 A flowchart demonstrating the self-consistent analysis for the pulse $E(z, t)$ propagating in a resonant medium.

$$\frac{\partial}{\partial t}(\rho_{11} - \rho_{22}) = \frac{2i\mu}{\hbar} E(\rho_{21} - \rho_{12}) \tag{15.4-6}$$

$$\frac{\partial}{\partial t}\rho_{21} = -i\omega\rho_{21} + \frac{i\mu}{\hbar} E(\rho_{11} - \rho_{22}) \tag{15.4-7}$$

where all the variables are functions of z and t, and μ is the dipole matrix element for the resonant transition

$$\mu \equiv \langle 1|\, \mu_x \,|2\rangle = \langle 2|\, \mu_x \,|1\rangle \tag{15.4-8}$$

and $\hbar\omega = E_2 - E_1$ is the resonant transition energy of the ensemble of atoms represented by the density matrix ρ_{ij}. Equations (15.4-6) and (15.4-7) apply to those atoms in the ensemble whose resonant frequency is ω. We will denote this fact by writing the density matrix as $\rho_{ij}(z, t, \Delta\omega)$ where $\Delta\omega \equiv \omega - \omega_0$, ω_0 being the center frequency of the propagating pulse.

Consider the contribution to the medium polarization from atoms with resonant frequencies $\omega = \omega_0 + \Delta\omega$. From (8.1-5),

$$P(\Delta\omega, z, t) = N\mu[\rho_{21}(\Delta\omega, z, t) + \rho_{12}(\Delta\omega, z, t)] \tag{15.4-9}$$

Now let us return, for a moment, to (15.4-2) for the polarization. Recognizing that $U(z, t)$ and $V(z, t)$ are due to contributions from atoms spanning the whole range of $\Delta\omega$, we can write

$$\begin{aligned} U(z, t) &= \int_{-\infty}^{\infty} u(\Delta\omega, z, t) g(\Delta\omega)\, d(\Delta\omega) \\ V(z, t) &= \int_{-\infty}^{\infty} v(\Delta\omega, z, t) g(\Delta\omega)\, d(\Delta\omega) \end{aligned} \tag{15.4-10}$$

where $g(\Delta\omega)$ is the inhomogeneous broadening normalized lineshape function.

The total polarization is then given by (15.4-2), where U and V have the form of (15.4-10). The polarization can also be taken, according to (15.4-9), as

$$\begin{aligned} P(z, t) &= \int_{-\infty}^{\infty} P(\Delta\omega, z, t) g(\Delta\omega)\, d(\Delta\omega) \\ &= N\mu \int_{-\infty}^{\infty} [\rho_{21}(\Delta\omega, z, t) + \text{c.c.}] g(\Delta\omega)\, d(\Delta\omega) \end{aligned}$$

By equating the last equation to (15.4-2), we get

$$\rho_{21}(\Delta\omega, z, t) = \frac{1}{2N\mu} [u(\Delta\omega, z, t) + iv(\Delta\omega, z, t)] e^{i(k_0 z - \omega_0 t + \phi)} \tag{15.4-11}$$

Substituting (15.4-11) in (15.4-7) and equating real and imaginary parts lead to

$$\frac{\partial u}{\partial t} = v\left(\Delta\omega + \frac{\partial \phi}{\partial t}\right)$$
$$\frac{\partial v}{\partial t} = -u\left(\Delta\omega + \frac{\partial \phi}{\partial t}\right) + \frac{\mu\varepsilon(z, t)}{\hbar} w \tag{15.4-12}$$

where

$$w(\Delta\omega, z, t) \equiv N\mu[\rho_{11}(\Delta\omega, z, t) - \rho_{22}(\Delta\omega, z, t)] \tag{15.4-13}$$

is the population difference (multiplied by μ) per unit volume per unit $\Delta\omega$. Using (15.4-1) and (15.4-11) in (15.4-6) and neglecting nonsynchronous terms oscillating at $2\omega_0$ give

$$\frac{\partial w}{\partial t} = -\frac{\mu}{\hbar}\varepsilon(z, t)v \tag{15.4-14}$$

Equations (15.4-12) and (15.4-14) are the collisionless Bloch equations (Reference 1, Chapter 8) for the transverse polarization amplitudes u and v and the (normalized) population difference w. The addition of phenomenological relaxation terms τ and T_2, as in Section 8.1, leads to

$$\frac{\partial u}{\partial t} = v\left(\Delta\omega + \frac{\partial \phi}{\partial t}\right) - \frac{u}{T_2}$$
$$\frac{\partial v}{\partial t} = -u\left(\Delta\omega + \frac{\partial \phi}{\partial t}\right) + \frac{\mu\varepsilon}{\hbar} w - \frac{v}{T_2} \tag{15.4-15}$$
$$\frac{\partial w}{\partial t} = -\frac{\mu\varepsilon}{\hbar} v - \frac{w - w_0}{\tau}$$

If we use (15.4-10), the field equations (15.4-4, 5) become

$$\frac{\partial \varepsilon}{\partial z} + \frac{n}{c}\frac{\partial \varepsilon}{\partial t} = -\frac{\omega_0 c\mu_0}{2n}\int_{-\infty}^{\infty} v(\Delta\omega, z, t)g(\Delta\omega)\, d(\Delta\omega)$$
$$\varepsilon\left(\frac{\partial \phi}{\partial z} + \frac{n}{c}\frac{\partial \phi}{\partial t}\right) = \frac{\omega_0 c\mu_0}{2n}\int_{-\infty}^{\infty} u(\Delta\omega, z, t)g(\Delta\omega)\, d(\Delta\omega) \tag{15.4-16}$$

The coupled set of Eqs. (15.4-15), together with Eqs. (15.4-16), with the appropriate boundary and temporal conditions describe the behavior of the atomic-radiation system.

Next, let us assume that $g(\Delta\omega)$ is an even function. The following set of statements are, according to (15.4-15) and (15.4-16), self-consistent.

$u(\Delta\omega, z, t)$ is an odd function of $\Delta\omega$

$v(\Delta\omega, z, t)$ is an even function of $\Delta\omega$

$w(\Delta\omega, z, t)$ is an even function of $\Delta\omega$

$\phi(z, t) = 0$

We will also limit our discussion to pulses whose duration is short compared to the relaxation times τ and T_2 so that in (15.4-15), we put $\tau = \infty$, $T_2 = \infty$. The set of equations (15.4-15) and (15.4-16) becomes

$$\begin{aligned}
\frac{\partial u}{\partial t} &= \Delta\omega v \\
\frac{\partial v}{\partial t} &= -\Delta\omega u + \frac{\mu\varepsilon}{\hbar} w \\
\frac{\partial w}{\partial t} &= -\frac{\mu\varepsilon}{\hbar} v \\
\frac{\partial\varepsilon}{\partial z} + \frac{n}{c}\frac{\partial\varepsilon}{\partial t} &= -\frac{\omega_0 c\mu_0}{2n}\int_{-\infty}^{\infty} v(\Delta\omega, z, t)g(\Delta\omega)\, d(\Delta\omega)
\end{aligned} \tag{15.4-17}$$

which are our main working equations.

The first three equations of (15.4-17) can be recognized as the components of the vector relation

$$\frac{\partial \mathbf{r}}{\partial t} = \mathbf{T} \times \mathbf{r} \tag{15.4-18}$$

with

$$\mathbf{r} = \frac{1}{N\mu}(\mathbf{e}_u u + \mathbf{e}_v v + \mathbf{e}_w w) \tag{15.4-19}$$

$$\mathbf{T} = -\mathbf{e}_u\left(\frac{\mu\varepsilon}{\hbar}\right) - \mathbf{e}_w\, \Delta\omega \tag{15.4-20}$$

thus corresponding to a precession of a pseudo vector $\mathbf{r}$ about $\mathbf{T}$ in the fictitious $(\mathbf{e}_u, \mathbf{e}_v, \mathbf{e}_w)$ space. Equation (15.4-18) can be identified with (15.1-8) by making the correspondence $r_u \to r_1$, $r_v \to -r_{11}$, $r_w = -r_3$, which follows directly from the original definitions. Also note that instead of $2E_{x,y}$ in (15.1-17), we have ε in (15.4-20). This difference is due, as pointed out in (15.1-32, 33), to our use here of a linearly polarized electric field instead of a circularly polarized field.

The Area Theorem

Using (15.4-17), we will derive an important result known as the "area" theorem. Before proceeding, we will need to examine the following two results.

I. For those atoms with $\Delta\omega = 0$, the following applies:

$$\begin{aligned}
u(0, z, t) &= 0 \\
v(0, z, t) &= w_0 \sin\theta(z, t) \\
w(0, z, t) &= w_0 \cos\theta(z, t)
\end{aligned} \tag{15.4-21}$$

where

$$\theta(z, t) \equiv \frac{\mu}{\hbar} \int_{-\infty}^{t} \varepsilon(z, t')\, dt' \tag{15.4-22}$$

is, according to (15.4-18, 19, 20), the nutation angle of $\mathbf{r}(\Delta\omega = 0, z, t)$ about the u axis.

II. If $\varepsilon(z, t) = 0$ for $t \geq t_0$, then for $t > t_0$

$$\begin{aligned} u(\Delta\omega, z, t) &= u_0 \cos[\Delta\omega(t - t_0)] + v_0 \sin[\Delta\omega(t - t_0)] \\ v(\Delta\omega, z, t) &= -u_0 \sin[\Delta\omega(t - t_0)] + v_0 \cos[\Delta\omega(t - t_0)] \\ w(\Delta\omega, z, t) &= w(\Delta\omega, z, t_0) \end{aligned} \tag{15.4-23}$$

where

$$u_0 \equiv u(\Delta\omega, z, t_0)$$

$$v_0 \equiv v(\Delta\omega, z, t_0)$$

Proof of I:

For $\Delta\omega = 0$ and assuming $u(0, z, -\infty) = v(0, z, -\infty) = 0$, we have from (15.4-17)

$$\begin{aligned} \frac{\partial u}{\partial t} &= 0 \qquad \text{so that} \qquad u(0, z, t) = 0 \\ \frac{\partial v}{\partial t} &= \frac{\mu}{\hbar} \varepsilon w \\ \frac{\partial w}{\partial t} &= -\frac{\mu}{\hbar} \varepsilon v \end{aligned} \tag{15.4-24}$$

and, consequently,

$$\frac{\partial}{\partial t}(v^2 + w^2) = 2v \frac{\partial v}{\partial t} + 2w \frac{\partial w}{\partial t} = \frac{2\mu}{\hbar} \varepsilon(vw - vw) = 0$$

and

$$v_2 + w^2 = w_0^2 \tag{15.4-25}$$

from the last of (15.4-24) and (15.4-25),

$$\frac{\partial w}{\partial t} = -\frac{\mu}{\hbar} \varepsilon \sqrt{w_0^2 - w^2}$$

$$\frac{dw}{\sqrt{w_0^2 - w^2}} = -\frac{\mu}{\hbar} \varepsilon\, dt$$

Integrating from $-\infty$ to t yields

$$-\cos^{-1} \frac{w}{w_0}\bigg]_{w_0}^{w} = -\cos^{-1} \frac{w}{w_0} = -\frac{\mu}{\hbar} \int_{-\infty}^{t} \varepsilon(z, t')\, dt'$$

so that

$$w(0, z, t) = w_0 \cos\left[\frac{\mu}{\hbar} \int_{-\infty}^{t} \varepsilon(z, t')\, dt'\right] \equiv w_0 \cos \theta(z, t)$$

and

$$v = \sqrt{w_0^2 - w_0^2 \cos^2 \theta} = w_0 \sin \theta(z, t)$$

This completes the proof of I.

Proof of II:
If $\varepsilon(z, t) = 0$ for $t > t_0$, (15.4-17) reduce to

$$\frac{\partial u}{\partial t} = (\Delta\omega)v$$

$$\frac{\partial v}{\partial t} = -(\Delta\omega)u \qquad (15.4\text{-}26)$$

$$\frac{\partial w}{\partial t} = 0$$

In a manner similar to the proof of I, we show that

$$\frac{\partial}{\partial t}(u^2 + v^2) = 0 \Rightarrow u^2 + v^2 = u_0^2 + v_0^2 \qquad v = \sqrt{u_0^2 + v_0^2 - u^2}$$

and from the first of (15.4-26),

$$\frac{du}{\sqrt{u_0^2 + v_0^2 - u^2}} = (\Delta\omega)\, dt$$

that, after integration from t_0 to t, gives

$$-\cos^{-1}\left[\left(\frac{u}{\sqrt{u_0^2 + v_0^2}}\right)\right]_{u_0}^{u} = \Delta\omega(t - t_0)$$

and

$$\frac{u}{\sqrt{u_0^2 + v_0^2}} = \cos\left[-(\Delta\omega)(t - t_0) + \cos^{-1}\frac{u_0}{\sqrt{u_0^2 + v_0^2}}\right]$$

Since

$$\sin\left(\cos^{-1}\frac{u_0}{\sqrt{u_0^2 + v_0^2}}\right) = \frac{v_0}{\sqrt{u_0^2 + v_0^2}}$$

$$u = u_0 \cos[\Delta\omega(t - t_0) + v_0 \sin[\Delta\omega(t - t_0)]$$

and

$$v = \sqrt{u_0^2 + \omega_0^2 - u^2} = v_0 \cos[\Delta\omega(t - t_0)] - u_0 \sin[\Delta\omega(t - t_0)]$$

as stated in (15.4-23).

We can now proceed with the proof of the area theorem. The theorem can be stated as

$$\frac{dA}{dz} = -\frac{\alpha}{2}\sin A \tag{15.4-27}$$

where the pulse area A is defined by

$$A(z) \equiv \lim_{t\to\infty} \theta(z, t) = \frac{\mu}{\hbar}\int_{-\infty}^{\infty} \varepsilon(z, t')\, dt' \tag{15.4-28}$$

$$\alpha = \frac{\omega_0 \pi \mu_0 N \mu^2 c g(0)}{n\hbar} \tag{15.4-29}$$

where it is assumed that when $t \to \infty$, $\varepsilon(z, t) \to 0$ and N is the absorbing atom density.

To prove (15.4-27), we start by taking the derivative of (15.4-28)

$$\frac{dA}{dz} = \lim_{t\to\infty} \frac{\mu}{\hbar}\int_{-\infty}^{t} \frac{\partial}{\partial z}\varepsilon(z, t')\, \mathrm{d}t'$$

Substituting for $\partial\varepsilon/\partial z$ from the last of (15.4-17), we obtain

$$\begin{aligned}\frac{dA}{dz} &= \lim_{t\to\infty} \frac{\mu}{\hbar}\int_{-\infty}^{t} dt' \left[\frac{-\omega_0 c \mu_0}{2n}\int_{-\infty}^{\infty} v(\Delta\omega, z, t') g(\Delta\omega)\, d(\Delta\omega) - \frac{n}{c}\frac{\partial \varepsilon}{\partial t'}\right] \\ &= \lim_{t\to\infty}\left\{\frac{-n\mu}{c\hbar}[\varepsilon(z, t) - \varepsilon(z, -\infty)] - \frac{\omega_0 c \mu_0 \mu}{2n\hbar}\int_{-\infty}^{\infty} d(\Delta\omega) g(\Delta\omega)\int_{-\infty}^{t} dt' v(\Delta\omega, z, t')\right\}\end{aligned} \tag{15.4-30}$$

The term in square brackets is zero since $\varepsilon(z, +\infty) = \varepsilon(z, -\infty) = 0$. We use the first of (15.4-17) to replace $v(\Delta\omega, z, t')$ in the integral. The result is

$$\begin{aligned}\frac{dA}{dz} &= -\frac{\omega_0 c \mu_0 \mu}{2n\hbar}\lim_{t\to\infty}\int_{-\infty}^{\infty} d(\Delta\omega)\frac{g(\Delta\omega)}{\Delta\omega}\int_{-\infty}^{t} dt' \frac{\partial u(\Delta\omega, z, t')}{\partial t'} \\ &= -\frac{\omega_0 c \mu_0 \mu}{2n\hbar}\lim_{t\to\infty}\int_{-\infty}^{\infty} d(\Delta\omega)\frac{g(\Delta\omega)}{\Delta\omega}[u(\Delta\omega, z, t) - u(\Delta\omega, z, -\infty)] \\ &= -\frac{\omega_0 c \mu_0 \mu}{2n\hbar}\lim_{t\to\infty}\int_{-\infty}^{\infty} d(\Delta\omega)\frac{g(\Delta\omega)}{\Delta\omega} u(\Delta\omega, z, t)\end{aligned}$$

since $u(\Delta\omega, z, -\infty) = 0$.

Next, we choose some time t_0 such that for $t \geqslant t_0$, $\varepsilon(z, t) \approx 0$, then use the first of (15.4-23) in the last equation. The result is

$$\frac{dA}{dz} = -\frac{\omega_0 c \mu_0 \mu}{2n\hbar}\lim_{t\to\infty}\int_{-\infty}^{\infty} d(\Delta\omega)\frac{g(\Delta\omega)}{\Delta\omega}\left\{\begin{array}{l} u(\Delta\omega, z, t_0)\cos[\Delta\omega(t - t_0)] \\ + v(\Delta\omega, z, t_0)\sin[\Delta\omega(t - t_0)]\end{array}\right\}$$

Because of the oscillatory nature of $\cos[\Delta\omega(t - t_0)]$ and $\sin[\Delta\omega(t - t_0)]$, in the limit of $t \to \infty$ the contribution to the integral is from a small region near $\Delta\omega = 0$. Since $u(\Delta\omega, z, t_0)$ is an odd function of $\Delta\omega$, we can expand it near

$\Delta\omega = 0$ as $u(\Delta\omega, z, t_0) \approx a_1 \Delta\omega + a_2(\Delta\omega)^3$. The integral containing u is then

$$\lim_{t\to\infty} \int_{-\infty}^{\infty} d(\Delta\omega) \frac{g(\Delta\omega)}{\Delta\omega} u(\Delta\omega, z, t_0) \cos[\Delta\omega(t - t_0)]$$

$$= \lim_{t\to\infty} \frac{g(0)a_1 \sin[\Delta\omega(t - t_0)]}{t - t_0}\Bigg]_{-\infty}^{\infty} = 0$$

Since $v(\Delta\omega, z, t_0)$ is an even function of $\Delta\omega$, we have

$$\frac{dA}{dz} = -\frac{\omega_0 c \mu_0 \mu}{2n\hbar} v(0, z, t_0) g(0) \lim_{t\to\infty} \int_{-\infty}^{\infty} d(\Delta\omega) \frac{\sin[\Delta\omega(t - t_0)]}{\Delta\omega}$$

The last integral is equal to π. Also, from (15.4-21) and (15.4-28),

$$v(0, z, t_0) = w_0 \sin\theta(z, t_0) = w_0 \sin A$$

with these the last expression dA/dz becomes

$$\frac{dA}{dz} = -\frac{\alpha}{2} \sin A$$

$$\alpha = \frac{\omega_0 \pi \mu_0 N \mu^2 c g(0)}{n\hbar}$$

where we used the relation $w_0 = N\mu$ that follows from (15.4-13) for atoms initially in the ground state so that $\rho_{11}(\Delta\omega, z, -\infty) = 1$. The proof of (15.4-27) is thus complete.

Let us now contemplate some of the implications of the area theorem (15.4-27). For weak (small area) pulses, $\sin A \sim A$ so that the solution of (15.4-27) is

$$A(z) = A(0)e^{-(\alpha/2)z}$$

and the pulse energy decays according to

$$\varepsilon^2(z) = \varepsilon^2(0)e^{-\alpha z} \tag{15.4-31}$$

The small-pulse extinction coefficient α, as given by (15.4-29), is the same as that exercised by a weak CW monochromatic signal of frequency ω_0. This last statement can be verified by using (8.3-7) to establish the identity of α of this section and the absorption coefficient given by the negative of (8.4-4).

In the general case of pulses of arbitrary area, it follows, from (15.4-27), that equilibrium solutions $[(dA/dz) = 0]$ are possible for $A = m\pi$, $m = 1, 2, 3 \cdots$. The solutions involving odd m are unstable to small deviations of A, whereas those with $m = 0, 2, 4$ are stable. A pulse starting with a given area will thus approach a value corresponding to the nearest even integer of π. This situation is depicted by Figure 15.12, which also illustrates the evolution of a pulse with $A(0) = 0.9\pi$ and one with $A(0) = 1.1\pi$. Also shown in part (*b*) are computer plots of the pulse shape and area at various z. The qualitative explanation of the observed pulse broadening with distance follows from simple energy considerations and is left to the student.

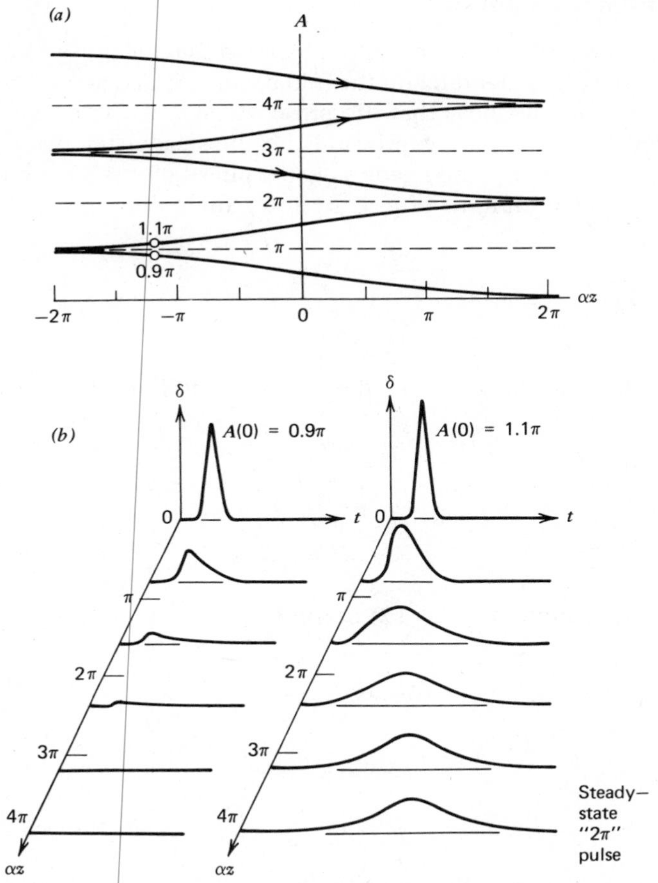

FIGURE 15.12 Pulse area plots of self-induced transparency-area theorem. (*a*) Branch solutions to (15.4-27) are plotted. The entry face of the medium may be at any value of z. For an absorbing medium with $\alpha < 0$, the pulse area evolves in the direction of increasing distance z toward the nearest even multiple of π. Even and odd multiples of π area solutions are, respectively, stable and unstable. (*b*) Computer plots of evolution of input $A(0) = 0.9\pi$ and $A(0) = 1.1\pi$ pulses with distance. The same diagram can be used for the case of an amplifying medium ($\alpha > 0$). The pulse area, in this case, evolves in the direction of decreasing z toward the nearest odd multiple of π. *Source:* Reference 12.

As noted above, $A(z)$ corresponds to the angle of precession of the pseudo vector $\mathbf{r}(\Delta\omega = 0, z)$ so that a steady-state pulse with $A = 2\pi$ causes atoms with $\Delta\omega = 0$ to undergo a complete transition, halfway through the pulse, to the upper state and then back exactly to the ground state. This point will be considered in detail below.

The Steady-State Solutions

In the discussion just concluded, we have shown that the pulse area tends to a constant value that, depending on the initial value, is an even integer of π. In what follows, we will show that the pulse shape and width reach a steady state as well and will obtain expressions for the steady-state values.

Let us assume that in the steady state the pulses propagate with a velocity V. The steady-state solutions of u, v, w, and ε must thus be functions of the variable

$$\gamma = t - \frac{z}{V} \tag{15.4-32}$$

only. The pulse velocity V is to be determined. It follows that for any $f(\gamma)$

$$\frac{\partial f}{\partial t} = \frac{df}{d\gamma}$$

$$\frac{\partial f}{\partial z} = \frac{df}{d\gamma}\left(-\frac{1}{V}\right)$$

The equations of motion (15.4-17) become

$$\begin{aligned}
\frac{du}{d\gamma} &= (\Delta\omega)v \\
\frac{dv}{d\gamma} &= -(\Delta\omega)u + \frac{\mu}{\hbar}\varepsilon w \\
\frac{dw}{d\gamma} &= -\frac{\mu}{\hbar}\varepsilon v \\
\frac{d\varepsilon}{d\gamma}\left(\frac{n}{c} - \frac{1}{V}\right) &= \frac{-\omega_0 c\mu_0}{2n}\int_{-\infty}^{\infty} v(\Delta\omega, \gamma)g(\Delta\omega)\, d(\Delta\omega)
\end{aligned} \tag{15.4-33}$$

The solutions of (15.4-33) are[4]

$$\begin{aligned}
u(\Delta\omega, z, t) &= 2N\mu\frac{(\Delta\omega)\tau}{1 + (\Delta\omega)^2\tau^2}\operatorname{sech}\left(\frac{t - \frac{z}{V}}{\tau}\right) \\
w(\Delta\omega, z, t) &= N\mu - 2N\mu\frac{1}{1 + (\Delta\omega)^2\tau^2}\operatorname{sech}^2\left(\frac{t - \frac{z}{V}}{\tau}\right)
\end{aligned} \tag{15.4-34}$$

[4] The pulse width τ here and in the remainder of the chapter should not be confused with the inversion relaxation time τ used in Chapter 8.

$$v(\Delta\omega, z, t) = -2N\mu \frac{1}{1 + (\Delta\omega)^2\tau^2} \tanh\left(\frac{t - \dfrac{z}{V}}{\tau}\right) \operatorname{sech}\left(\frac{t - \dfrac{z}{V}}{\tau}\right)$$

$$\varepsilon(z, t) = \frac{2\hbar}{\mu\tau} \operatorname{sech}\left(\frac{t - \dfrac{z}{V}}{\tau}\right)$$

where the pulse width τ is *arbitrary*. The pulse velocity V is given by

$$\begin{aligned}\frac{1}{V} &= \frac{n}{c} + \frac{\omega_0 c\mu_0 N\mu^2\tau^2}{2n\hbar} \int_{-\infty}^{\infty} \frac{g(\Delta\omega)}{1 + (\Delta\omega)^2\tau^2} d(\Delta\omega) \\ &= \frac{n}{c} + \frac{\alpha\tau^2}{2\pi g(0)} \int_{-\infty}^{\infty} \frac{g(\Delta\omega)}{1 + (\Delta\omega)^2\tau^2} d(\Delta\omega) \end{aligned} \qquad (15.4\text{-}34a)$$

where α was defined in (15.4-29). The proof of (15.4-34) follows.

Assume a solution for $v(\Delta\omega, \gamma)$ in the form

$$v(\Delta\omega, \gamma) = v(\gamma) f(\Delta\omega) \qquad (15.4\text{-}35)$$

where $v(\gamma)$ and $f(\Delta\omega)$ are to be found and $f(0)$ is taken, arbitrarily, as 1. This determines the form of the other variables. From the first of (15.4-33),

$$\frac{du}{d\gamma} = v(\gamma)[\Delta\omega f(\Delta\omega)]$$

and

$$u(\Delta\omega, \gamma) = \left[\int_{-\infty}^{\gamma} v(\gamma')\, d\gamma'\right] [\Delta\omega f(\Delta\omega)] \equiv u(\gamma) f(\Delta\omega)\, \Delta\omega$$

for $u(\Delta\omega, -\infty) = 0$. Next, from the third of (15.4-33),

$$\frac{dw(\Delta\omega, \gamma)}{d\gamma} = -\frac{\mu}{\hbar} \varepsilon v(\Delta\omega, \gamma) = -\frac{\mu}{\hbar} \varepsilon(\gamma) v(\gamma) f(\Delta\omega)$$

and after integrating from $-\infty$ to γ and using $w(\Delta\omega, -\infty) = N\mu$, we obtain

$$w(\Delta\omega, \gamma) = N\mu - \frac{\mu}{\hbar} f(\Delta\omega) \int_{-\infty}^{\gamma} \varepsilon(\gamma')\, v(\gamma')\, d\gamma \equiv N\mu - w(\gamma) f(\Delta\omega)$$

Finally, from the last of (15.4-33),

$$\begin{aligned}\frac{d\varepsilon(\gamma)}{d\gamma} &= \frac{\omega_0 c\mu_0}{2n\left(\dfrac{1}{V} - \dfrac{n}{c}\right)} v(\gamma) \int_{-\infty}^{\infty} f(\Delta\omega) g(\Delta\omega)\, d(\Delta\omega) \\ &\equiv \frac{\hbar}{N\mu^2\tau^2} v(\gamma)\end{aligned}$$

where

$$\frac{1}{\tau^2} \equiv \frac{\omega_0 c\mu_0 N\mu^2}{2n\hbar\left(\dfrac{1}{V} - \dfrac{n}{c}\right)} \int_{-\infty}^{\infty} f(\Delta\omega) g(\Delta\omega)\, d(\Delta\omega)$$

The equations of motion have thus been reduced to

$$\frac{du(\gamma)}{d\gamma} = v(\gamma)$$

$$\frac{dv(\gamma)}{d\gamma} = \frac{1}{f(\Delta\omega)}\left[-u(\gamma)(\Delta\omega)^2 f(\Delta\omega) + \varepsilon(\gamma)\frac{N\mu^2}{\hbar} - \frac{\mu}{\hbar}\varepsilon(\gamma)w(\gamma)f(\Delta\omega)\right]$$

$$\frac{dw(\gamma)}{d\gamma} = \frac{\mu}{\hbar}\varepsilon(\gamma)v(\gamma) \qquad (15.4\text{-}36)$$

$$\frac{d\varepsilon(\gamma)}{d\gamma} = \frac{\hbar}{N\mu^2\tau^2}v(\gamma)$$

These equations follow from (15.4-33) and the relations

$$u = u(\gamma)\,\Delta\omega f(\Delta\omega)$$

$$v = v(\gamma)f(\Delta\omega) \qquad (15.4\text{-}37)$$

$$w = N\mu - w(\gamma)f(\Delta\omega)$$

derived above.

From the first and last of (15.4-36),

$$\frac{du}{d\gamma} = v(\gamma) = \frac{d}{d\gamma}\left[\frac{N\mu^2\tau^2}{\hbar}\varepsilon(\gamma)\right] \qquad (15.4\text{-}38)$$

$$u(\gamma) = \frac{N\mu^2\tau^2}{\hbar}\varepsilon(\gamma) \qquad (15.4\text{-}39)$$

since $u(-\infty) = \varepsilon(-\infty) = 0$.

Similarly,

$$\frac{dw}{d\gamma} = \frac{\mu}{\hbar}\varepsilon(\gamma)\left(\frac{N\mu^2\tau^2}{\hbar}\right)\frac{d\varepsilon(\gamma)}{d\gamma} = \frac{N\mu^3\tau^2}{2\hbar^2}\frac{d}{d\gamma}[\varepsilon^2(\gamma)] \qquad (15.4\text{-}40)$$

so that

$$w(\gamma) = \frac{N\mu^3\tau^2}{2\hbar^2}\varepsilon^2(\gamma) \qquad \text{since} \qquad w(-\infty) = \varepsilon(-\infty) = 0 \qquad (15.4\text{-}41)$$

and finally, using the expressions just obtained for $u(\gamma)$ and $w(\gamma)$ in the second of (15.4-36) yields

$$\begin{aligned}\frac{dv(\gamma)}{d\gamma} &= -\frac{N\mu^2\tau^2}{\hbar}\varepsilon(\gamma)(\Delta\omega)^2 + \frac{N\mu^2}{\hbar f(\Delta\omega)}\varepsilon(\gamma) - \frac{N\mu^4\tau^2}{2\hbar^3}\varepsilon^3(\gamma)\\ &= -\frac{N\mu^4\tau^2}{2\hbar^3}\varepsilon^3(\gamma) + \frac{N\mu^2}{\hbar}\varepsilon(\gamma)\left[\frac{1}{f(\Delta\omega)} - (\Delta\omega)^2\tau^2\right]\end{aligned} \qquad (15.4\text{-}42)$$

The last expression can hold only when the term within the square brackets is independent of $\Delta\omega$

$$\frac{1}{f(\Delta\omega)} - (\Delta\omega)^2\tau^2 = A = \text{constant}$$

from which

$$f(\Delta\omega) = \frac{1}{1 + (\Delta\omega)^2\tau^2} \tag{15.4-43}$$

in keeping with the arbitrary normalization condition $f(0) = 1$. We can now proceed to solve for $\varepsilon(\gamma)$. From (15.4-13) and (15.4-19) and the initial conditions $u(-\infty) = v(-\infty) = 0$, $\rho_{11}(-\infty) = 1$.

$$u^2 + v^2 + w^2 = w_0^2 = N^2\mu^2$$

from (15.4-37), we have

$$u^2 + v^2 + w^2 = u^2(\Delta\omega)^2 f^2(\Delta\omega) + v^2 f^2(\Delta\omega) + N^2\mu^2 - 2wf(\Delta\omega)N\mu + w^2 f^2(\Delta\omega)$$

equating the last two expressions, then solving for v

$$v = \frac{N\mu^2\tau}{\hbar}\,\varepsilon\,\sqrt{1 - \left(\frac{\mu\tau}{2\hbar}\,\varepsilon\right)^2} \tag{15.4-44}$$

where we used (15.4-39) and (15.4-41) to eliminate u and w. According to the last of (15.4-36),

$$v = \frac{N\mu^2\tau^2}{\hbar}\frac{d\varepsilon}{d\gamma}$$

so that

$$\frac{d\varepsilon}{d\gamma} = \frac{1}{\tau}\,\varepsilon\,\sqrt{1 - \left(\frac{\mu\tau}{2\hbar}\,\varepsilon\right)^2}$$

The last expression can be integrated from some arbitrary γ_0 to γ using

$$\int \frac{dx}{x\sqrt{1 - a^2x^2}} = -\ln\left(\frac{1 + \sqrt{1 - a^2x^2}}{ax}\right)$$

to give

$$\frac{\gamma - \gamma_0}{\tau} = -\ln\left[\frac{1 + \sqrt{1 - \left(\frac{\mu\tau}{2\hbar}\,\varepsilon\right)^2}}{\frac{\mu\tau}{2\hbar}\,\varepsilon}\right]_{\varepsilon(\gamma_0)}^{\varepsilon(\gamma)}$$

or

$$e^{-(\gamma-\gamma_0)/\tau} = \frac{1 + \sqrt{1 - \left(\frac{\mu\tau}{2\hbar}\,\varepsilon\right)^2}}{1 + \sqrt{1 - \left[\frac{\mu\tau}{2\hbar}\,\varepsilon(\gamma_0)\right]^2}}\,\frac{\varepsilon(\gamma_0)}{\varepsilon}$$

so that

$$\frac{e^{-(\gamma/\tau)}\varepsilon}{1 + \sqrt{1 - \left(\frac{\mu\tau}{2\hbar}\,\varepsilon\right)^2}} = \frac{e^{-\gamma_0/\tau}\varepsilon(\gamma_0)}{1 + \sqrt{1 - \left[\frac{\mu\tau}{2\hbar}\,\varepsilon(\gamma_0)\right]^2}}$$

since γ_0 is arbitrary, the right side of the last equality must be independent of γ_0. We designate it as B and rewrite the last relation as

$$Be^{\gamma/\tau}\left[1 + \sqrt{1 - \left(\frac{\mu\tau}{2\hbar}\varepsilon\right)^2}\right] = \varepsilon$$

thus

$$\varepsilon - Be^{\gamma/\tau} = Be^{\gamma/\tau}\sqrt{1 - \left(\frac{\mu\tau}{2\hbar}\varepsilon\right)^2}$$

By squaring the last equation and solving for ε, we get

$$\varepsilon(\gamma) = \frac{2}{\dfrac{1}{B}e^{-\gamma/\tau} + \dfrac{\mu^2\tau^2}{4\hbar^2}Be^{\gamma/\tau}}$$

We define a new constant γ_p by means of

$$B = \frac{2\hbar}{\mu\tau}e^{-\gamma_p/\tau}$$

so that the last expression for $\varepsilon(\gamma)$ becomes

$$\varepsilon(\gamma) = \frac{2\hbar}{\mu\tau}\operatorname{sech}\left(\frac{\gamma - \gamma_p}{\tau}\right)$$

Since γ_p only affects the zero reference of the time scale, which is of no interest in this problem, we can take it as zero. Also using $\gamma \equiv t - z/V$, we rewrite that last expression as

$$\varepsilon(z, t) = \frac{2\hbar}{\mu\tau}\operatorname{sech}\left(\frac{t - z/V}{\tau}\right) \tag{15.4-45}$$

which is the solution stated (15.4-34). The remaining relations of (15.4-34) are derived using (15.4-45) in (15.4-39) to obtain $u(\Delta\omega, z, t)$, in (15.4-41) to get $w(\Delta\omega, z, t)$ and in (15.4-44) for $v(\Delta\omega, z, t)$. This completes the proof.

A plot of the steady-state field envelope $\varepsilon(z, t)$ (15.4-45) is shown in Figure 15.13. The abscissa can be taken as either time at a fixed point along the path, or distance at some fixed time. To be specific, let us take the abscissa coordinate as time. The upper curve then shows the "tipping" history of the pseudovector $\mathbf{r}$ of an atom with $\Delta\omega = 0$ at some point z. The ordinate θ is given by (15.4-22). An atom starting in the ground state (1) is tipped gradually till at (5) its $\mathbf{r}$ vector is in the u–v plane. At point (6), halfway through the pulse, the atom is in the upper state, whereas at (11) it is back to ground state. The net energy exchange with the field is thus zero. An atom with $\Delta\omega \neq 0$ never makes a complete transition to the upper level. This can be seen from the second of (15.4-34) by noting that

$$w\left(\Delta\omega \neq 0, t = \frac{z}{V}\right) > -N\mu$$

The same atom will, however, return exactly to the ground state following the pulse since $w(\Delta\omega, t = \infty, z) = N\mu$.

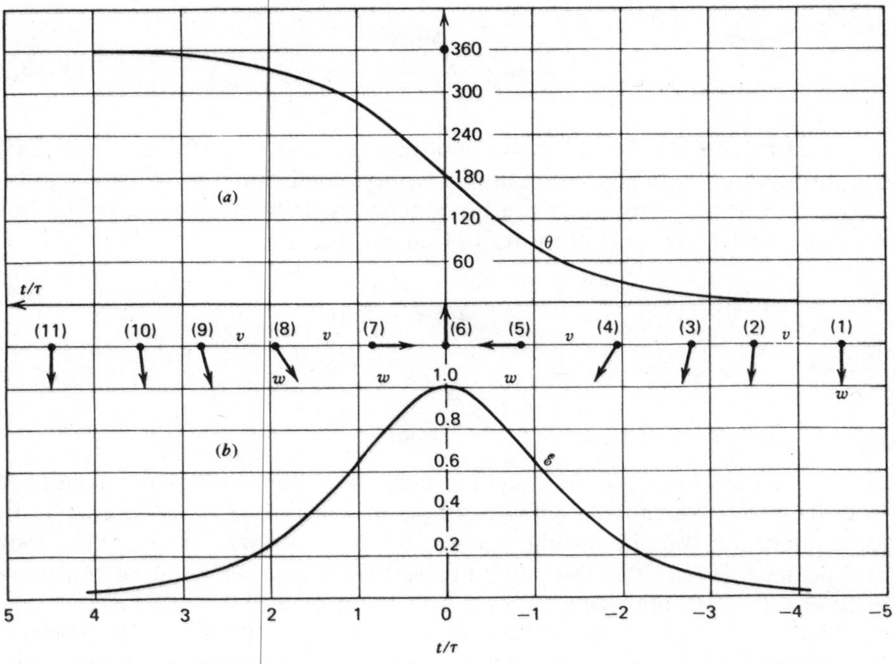

FIGURE 15.13 (*a*) The steady-state pulse shape of a wave propagating in an absorbing medium. (*b*) The tipping angle of an atom (with $\Delta\omega = 0$) initially in the ground state. $\theta = \theta°$ corresponds to the ground state, whereas $\theta = 180°$ corresponds to an atom in the excited state.

It follows that the net energy exchange is zero for all the atoms regardless of $\Delta\omega$, which is consistent with the constant energy content of the steady-state pulse.

The pulse envelope velocity V was derived in (15.4-35) as

$$\frac{1}{V} = \frac{n}{c} + \frac{\alpha\tau^2}{2\pi g(0)} \int_{-\infty}^{\infty} \frac{g(\Delta\omega)}{1 + (\Delta\omega)^2\tau^2}\, d(\Delta\omega) \tag{15.4-46}$$

Consider, for example, a Lorentzian lineshape

$$g(\Delta\omega) = \frac{\Delta\omega_{\text{atomic}}}{2\pi\left[(\Delta\omega)^2 + \left(\dfrac{\Delta\omega_{\text{atomic}}}{2}\right)^2\right]}$$

where $\Delta\omega_{\text{atomic}}$ is the full width at half maximum of $g(\Delta\omega)$.

For the case $\Delta\omega_{\text{atomic}}\tau \gg 1$ (i.e., a broad transition), (15.4-46) becomes

$$\frac{1}{V} = \frac{n}{c} + \frac{\alpha}{2}\tau \tag{15.4-47}$$

whereas for narrow transitions such that $\Delta\omega_{\text{atomic}}\tau \ll 1$, the integration of

(15.4-46) gives

$$\frac{1}{V} = \frac{n}{c} + \frac{\alpha\, \Delta\omega_{\text{atomic}}}{4}\tau^2 \tag{15.4-48}$$

Example. Consider a pulse of a duration $\tau = 5 \times 10^{-9}$ sec propagating through an atomic gas with an absorption coefficient (at the center pulse frequency) $\alpha = 10^4$ m^{-1}. Let the homogeneous linewidth $\Delta\nu_{\text{atomic}} (= (1/2\pi)\, \Delta\omega_{\text{atomic}})$ be 1 MHz. Substitution in (15.4-48) gives

$$\frac{1}{V} = \frac{n}{3 \times 10^8} + \frac{1180}{3 \times 10^8}$$

so that

$$V \approx \frac{c}{1180}$$

and the pulse velocity is reduced by more than three orders of magnitude from its free space value. A good example of self-induced transparency is an experiment utilizing the resonant absorption at $\lambda = 7947.7$ Å in a ^{87}Rb vapor and pulses from a ^{202}Hg laser (Reference 14). Strong self-induced transparency effects were observed with a pulse duration $\tau \sim 7 \times 10^{-9}$ sec that is shorter than the dephasing collision time ($T_2 \sim 55 \times 10^{-9}$ sec) or the spontaneous relaxation times ($\sim 40 \times 10^{-9}$ sec) so that the conclusions of our collisionless theory can be compared with the experiment.

A comparison of the measured energy transmission with the predicted values is shown in Figure 15.14. Note that fractional transmission near 90% is reached in a cell whose transmission for low intensity pulses is exp(−5) ~ 0.7%.

Coherent Pulse Propagation in an Amplifying Medium

The coherent propagation of pulses in an amplifying ($\alpha < 0$) medium are also described by (15.4-27), except that here the inversion $N < 0$ and according to (15.4-29), the sign of α is negative. It follows immediately that stable area solutions of (15.4-27) are now those with $A = \pi, 3\pi, 5\pi$. Figure 15.12*a* and 15.12*b* can still be used except that the pulse evolution along the direction of propagation is now described by moving along the $-z$ direction in the figures. A π pulse (or any odd multiple of π) propagating in an amplifying medium leaves, according to (15.4-21, 22), the initially excited atoms in the lower energy level. The pulse energy thus keeps increasing with distance. Since the pulse area is a constant, the pulse must get progressively narrower. This behavior is shown in Figure 15.12*b* by moving in the $-z$ direction. There exists, thus, a basic difference between the coherent pulse propagation in an absorbing medium that admits a steady-state area, as well as a steady-state shape and that in an amplifying medium where the pulse width keeps decreasing. In the latter case, the pulse energy will eventually increase to the

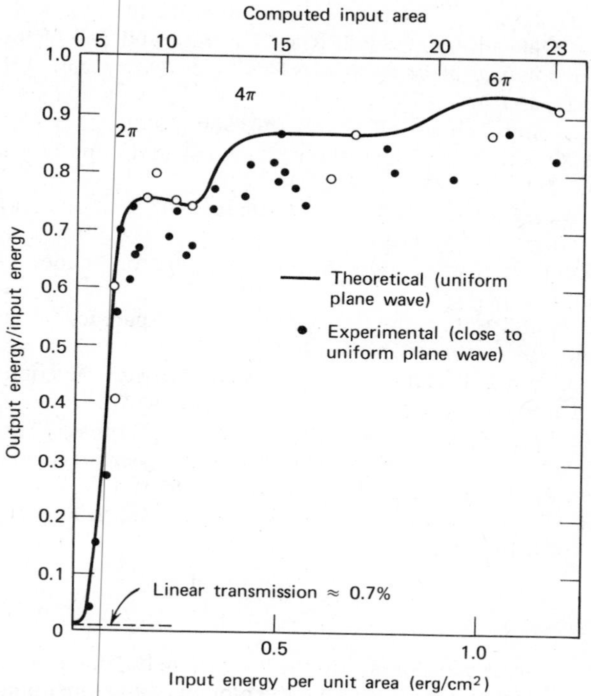

FIGURE 15.14 Self-induced transparency and nonlinear transmission in Rb vapor. Solid curve is a uniform plane-wave computer solution. Solid dots are data taken with 200-μm output aperture to approximate uniform plane wave. Input pulse width $\sim 7 \times 10^{-9}$ sec. The linear low-level transmission through the cell is 0.7%. *Source:* Reference 14.

point where losses can no longer be neglected, which modifies the conclusion reached here (Reference 13).

References

1. Feynman, R. P., F. L. Vernon, Jr., and R. W. Hellwarth, "Geometrical Representation of the Schrödinger Equation for Solving Maser Problems," *J. Appl. Phys.* **28,** 49 (1957).
2. Dicke, R. H., "Coherence in Spontaneous Radiation Processes," *Phys. Rev.* **93,** 99 (1954).
3. Bloembergen, N. and R. V. Pound, "Radiation Damping in Magnetic Resonance Experiments," *Phys. Rev.* **95,** 8 (1954).
4. Feher, G., J. Gordon, E. Buehler, E. Gere, and C. Thurmond, "Spontaneous Emission of Radiation from an Electron Spin System," *Phys. Rev.* **109,** 221 (1958).
5. Yariv, A., "Spontaneous Emission from an Inverted Spin System," *J. Appl. Phys.* **31,** 740 (1960).

6. Herman, I. P., J. C. MacGillivray, N. Skribanowitz, and M. S. Feld, "Self-Induced Emission in Optically Pumped HF Gas: The Rise and Fall of the Superradiant State," in *Proceedings of Vail Conference on Laser Spectroscopy, 1973* (New York: Plenum Press, 1974).
7. Hahn, E. L., "Spin Echoes," *Phys. Rev.* **80,** 580 (1950).
8. Brewer, R. G. and R. L. Shoemaker, "Photon Echoes and Optical Nutation in Molecules," *Phys. Rev.* **27,** 632 (1971).
9. Kurnit, N. A., I. D. Abella, and S. R. Hartman, "Observation of a Photon Echo," *Phys. Rev. Letters* **13,** 567 (1964).
10. Abella, I. D., N. A. Kurnit, and S. R. Hartman, "Photon Echoes," *Phys. Rev.* **141,** 391 (1966).
11. McCall, S. L. and E. L. Hahn, "Self-Induced Transparency by Pulsed Coherent Light," *Phys. Rev. Letters* **18,** 908 (1967).
12. McCall, S. L. and E. L. Hahn, "Pulse-Area Pulse-Energy Description of a Traveling-Wave Laser Amplifier," *Phys. Rev.* **A2,** 861 (1970).
13. Arecchi, F. T., E. Courtens, R. Gilmore, and H. Thomas, "Fundamental and Applied Laser Physics," in *Proceedings of Esfahan Symposium,* M. S. Feld, N. A. Kurnit, and A. Javan, eds. (New York: Wiley, 1973).
14. Slusher, R. E., "Self-Induced Transparency," Vol. XII, in *Progress in Optics* (Amsterdam: North-Holland, 1973).

Problems

15.1 Show that if an atom is initially in the lower state $|b\rangle$, that is, $r_3(0) = -1$, $\mathbf{r}(t)$ is the negative of that which describes the motion of an atom initially in the upper state, $|a\rangle$.

15.2 Derive (15.1-29) and (15.1-30).

15.3 Solve for the induced dipole moment of an ensemble of atoms with a resonant transition $E_a - E_b = \hbar\omega_0$ due to a field $E_x = E_0 \cos(\omega_0 t - \mathbf{k}_0 \cdot \mathbf{r}) + E_1 \cos[(\omega_0 + \Delta)t - \mathbf{k}_1 \cdot \mathbf{r}]$ where $E_1 \ll E_0$ and the atoms are initially in the ground state, $|b\rangle$. Assuming sample dimensions large compared to λ_0, show that the atoms will radiate a wave at $\omega_0 - \Delta$ along $2\mathbf{k}_0 - \mathbf{k}_1$.

15.4 Prove (15.1-21), using instead of $\mathbf{r}(t)$ an arbitrary vector $\mathbf{A}(A_1, A_2, A_3)$.

Hint:
Take

$$\mathbf{A}_R(t) = \begin{vmatrix} \cos \Omega t & \sin \Omega t & 0 \\ -\sin \Omega t & \cos \Omega t & 0 \\ 0 & 0 & 1 \end{vmatrix} \begin{vmatrix} A_1 \\ A_2 \\ A_3 \end{vmatrix}$$

so that

$$\frac{d\mathbf{A}_R(t)}{dt} = \bar{\bar{T}} \frac{d\mathbf{A}}{dt} + \frac{d\bar{\bar{T}}}{dt} \mathbf{A}$$

where $\bar{\bar{T}}$ is the above transformation matrix.

15.5 Show that if

$$\frac{d\mathbf{r}}{dt} = \mathbf{w} \times \mathbf{r}$$

then,

$$\left(\frac{d\mathbf{r}}{dt}\right)_R = \mathbf{w}_R \times \mathbf{r}_R$$

CHAPTER 16

Introduction to Nonlinear Optics—Second-Harmonic Generation

16.0 INTRODUCTION

In Chapter 5 we considered the propagation of electromagnetic radiation in linear media in which the polarization is proportional to the electric field that induces it. In this chapter we consider some of the consequences of the nonlinear dielectric properties of certain classes of crystals in which, in addition to the linear response, polarization is produced that is proportional to the square of the field.

The nonlinear response can give rise to exchange of energy between a number of electromagnetic fields of different frequencies. Three of the most important applications of this phenomenon are (1) second-harmonic generation in which part of the energy of an optical wave of frequency ω propagating through a crystal is converted to that of a wave at 2ω; (2) parametric oscillation in which a strong pump wave at ω_3 causes the simultaneous generation in a nonlinear crystal of radiation at ω_1 and ω_2, where $\omega_3 = \omega_1 + \omega_2$; (3) frequency up-conversion in which a weak signal of a low frequency ω_1 is converted coherently to a signal of a higher frequency ω_3 by mixing with a strong laser field at $\omega_2 = \omega_3 - \omega_1$.

We have already encountered the concept of a nonlinear susceptibility in our study of the electrooptic effect in Section 14.1. This effect was described by means of the tensor r_{jlk} that relates the changes in the indices of refraction (more precisely—the constants of the index ellipsoid) to the applied field according to

$$\Delta\left(\frac{1}{n^2}\right)_{jl} = r_{jlk}E_k$$

An alternative way of describing the linear electrooptic effect would be to relate the complex amplitude of the polarization at the sum frequency $\omega + \Omega$ to the product of the amplitudes of the optical electric field E^{ω} and the low-frequency electric field E^{Ω}

$$P_j^{\omega'=\Omega+\omega} = 2d_{jkl}^{\omega'=\Omega+\omega}E_k^{\Omega}E_l^{\omega} \tag{16.0-1}$$

The tensor $\bar{\bar{d}}$ defined by the last equation is related to the electrooptic tensor $\bar{\bar{r}}$

by the relation

$$d_{jkl}^{\omega'=\Omega+\omega} = -\frac{\varepsilon_j \varepsilon_l}{4\varepsilon_0} r_{jlk}$$

where ε_j and ε_l are the principal dielectric constants along j and l at ω. The derivation of the last relation is identical to that leading to (14.8-8) except for a factor of 1/2 due to the fact that (16.0-1) applies to complex amplitudes.

The optical nonlinearities considered in what follows result when the low frequency at Ω is replaced by a second "optical" frequency, that is, one above the lattice Restrahlen band. At these frequencies, the interatomic lattice motion can no longer be excited and the induced polarization at the sum frequency is due only to nonlinearities of the electronic motion. These nonlinearities are small and their study and utilization had to wait for the advent of the laser with its intense and coherent output.

The first experiment in this field took place in 1961 and involved doubling of the frequency of a ruby laser in a quartz crystal (Reference 1).

16.1 THE NONLINEAR OPTICAL SUSCEPTIBILITY TENSOR

Consider the nonlinear coupling of two optical fields. The first, having its electric field along the j direction, is given by

$$E_j^{\omega_1}(t) = \mathrm{Re}(E_j^{\omega_1} e^{i\omega_1 t}) = \tfrac{1}{2}(E_j^{\omega_1} e^{i\omega_1 t} + \text{c.c.}) \tag{16.1-1}$$

whereas the second field at ω_2 is

$$E_k^{\omega_2}(t) = \mathrm{Re}(E_k^{\omega_2} e^{i\omega_2 t}) \tag{16.1-2}$$

If the medium is nonlinear, then the presence of these field components can give rise to polarizations at frequencies $n\omega_1 + m\omega_2$ where n and m are any integers. If we take the polarization component at $\omega_3 = \omega_1 + \omega_2$ along the i direction as

$$P_i^{\omega_3=\omega_1+\omega_2}(t) = \mathrm{Re}(P_i^{\omega_3} e^{i\omega_3 t}) \tag{16.1-3}$$

the nonlinear susceptibility tensor $d_{ijk}^{\omega_3=\omega_1+\omega_2}$ is defined by the following relations between the complex field amplitudes:

$$P_i^{\omega_3} = 2 d_{ijk}^{\omega_3=\omega_1+\omega_2} E_j^{\omega_1} E_k^{\omega_2} \tag{16.1-4}$$

where we sum over repeated indices.

In a similar manner, we can define the difference frequency susceptibility tensor $d_{ijk}^{\omega_3=\omega_1-\omega_2}$ by

$$P_i^{\omega_3} = 2 d_{ijk}^{\omega_3=\omega_1-\omega_2} E_j^{\omega_1} E_k^{-\omega_2} \tag{16.1-5}$$

where, according to (16.1-1), $E_k^{-\omega_2} = (E_k^{\omega_2})^*$.

Only noncentrosymmetric crystals can possess a nonvanishing d_{ijk} tensor. This follows from the requirement that in a centrosymmetric crystal a reversal of the signs of $E_j^{\omega_1}$ and $E_k^{\omega_2}$ must cause a reversal in the sign of $P_i^{\omega_3=\omega_1+\omega_2}$ and not affect the amplitude. Using (16.1-4), we get

$$d_{ijk}^{\omega_3=\omega_1+\omega_2} E_j^{\omega_1} E_k^{\omega_2} = -d_{ijk}^{\omega_3=\omega_1+\omega_2}(-E_j^{\omega_1})(-E_k^{\omega_2})$$

so that $d_{ijk} \equiv 0$. Lack of an inversion symmetry is also the prerequisite for piezoelectricity so that all piezoelectric crystals can be expected to display second-order ($P \propto E^2$) nonlinear optical properties. The same argument can be used to show that all crystals as well as liquids and gases can display third-order optical nonlinearities.

In most of the nonlinear experiments, the crystal is transparent over a region that includes ω_1, ω_2, and ω_3. This implies a lack of hysteresis in the dependence of **P** on **E**, that is, that **P** is a single-valued function of **E**. We can consequently express the nonlinear polarization by

$$P_i(t) = 2d_{ijk}E_j(t)E_k(t) \tag{16.1-6}$$

where d_{ijk} is *independent of frequency*.

Since no physical significance can be attached to an exchange of E_j and E_k in (16.1-6), it follows that $d_{ijk} = d_{ikj}$. We therefore can replace the subscripts kj by a single symbol according to the piezoelectric contraction

$$xx = 1 \qquad yy = 2 \qquad zz = 3$$

$$yz = zy = 4 \qquad xz = zx = 5 \qquad xy = yx = 6$$

The resulting d_{ij} tensor forms a 3×6 matrix that operates on the $\mathbf{E}^2$ column tensor to yield **P** according to

$$\begin{vmatrix} P_x \\ P_y \\ P_z \end{vmatrix} = \begin{vmatrix} d_{11} & d_{12} & d_{13} & d_{14} & d_{15} & d_{16} \\ d_{21} & d_{22} & d_{23} & d_{24} & d_{25} & d_{26} \\ d_{31} & d_{32} & d_{33} & d_{34} & d_{35} & d_{36} \end{vmatrix} \begin{vmatrix} E_x^2 \\ E_y^2 \\ E_z^2 \\ 2E_zE_y \\ 2E_zE_x \\ 2E_xE_y \end{vmatrix} \tag{16.1-7}$$

where the superfluous (for the lossless case) frequency superscripts in d_{ijk} have been deleted.

The contracted d_{ij} tensor obeys the same symmetry restrictions as the piezoelectric tensor, and in crystals of a given point-group symmetry it has the same form. The tensor forms are given in Table 16.1. In KH_2PO_4 (KDP), for example, which has a $\bar{4}2m$ point-group symmetry, the d_{ij} tensor is given by

$$d_{ij} = \begin{vmatrix} 0 & 0 & 0 & d_{14} & 0 & 0 \\ 0 & 0 & 0 & 0 & d_{14} & 0 \\ 0 & 0 & 0 & 0 & 0 & d_{36} \end{vmatrix} \tag{16.1-8}$$

TABLE 16.1. **The Form of the Nonlinear Optical Tensor d_{ijk} as Defined by (16.1-4)**

Key to Notation

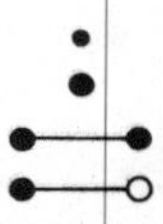

Zero modulus
Nonzero modulus
Equal moduli
Moduli numerically equal, but opposite in sign

Centrosymmetrical Classes
(all moduli vanish)

Noncentrosymmetrical Classes

Triclinic

Class 1

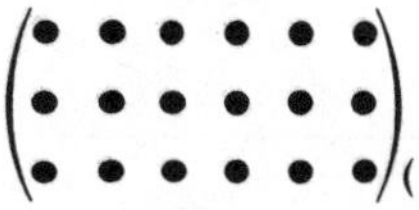
(18)

Monoclinic

Class 2

$2 \| x_2$ (standard orientation)

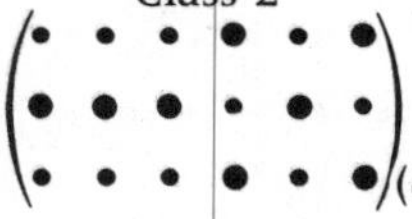
(8)

Class 2

$2 \| x_3$

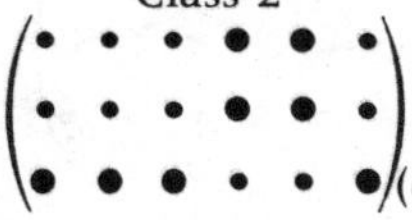
(8)

Class *m*

$m \perp x_2$ (standard orientation)

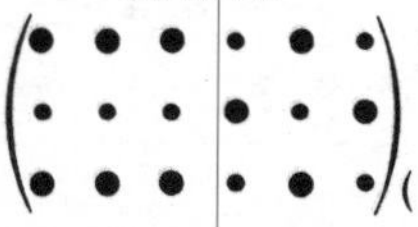
(10)

Class *m*

$m \perp x_3$

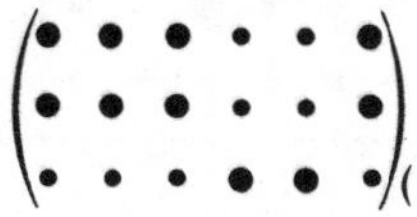
(10)

Orthorhombic

Class 222

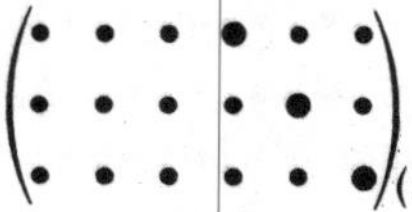
(3)

Class *mm*2

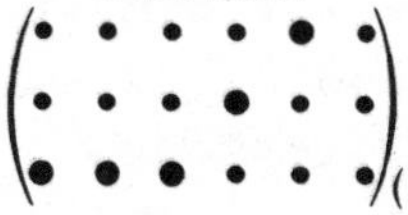
(5)

Tetragonal

Class 4

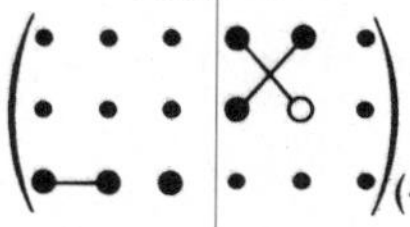
(4)

Class $\bar{4}$

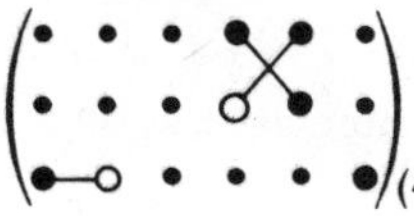
(4)

Class 422

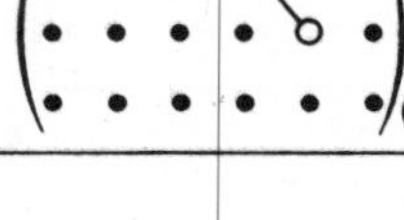
(1)

Class 4*mm*

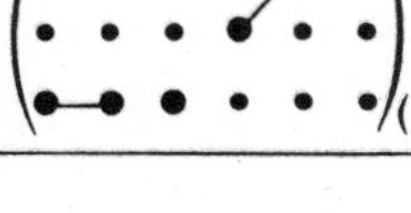
(3)

TABLE 16.1. ***(Continued)***

Tetragonal (continued)

Class $\bar{4}2m$

$2 \| x_1$

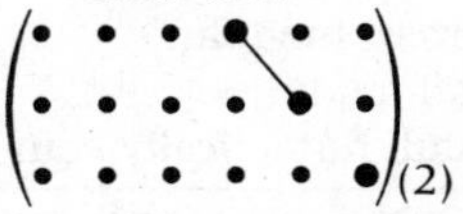

Cubic

Class 432

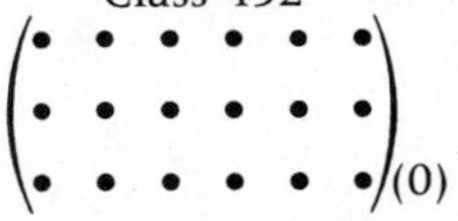

All moduli vanish

Classes $\bar{4}3m$ and 23

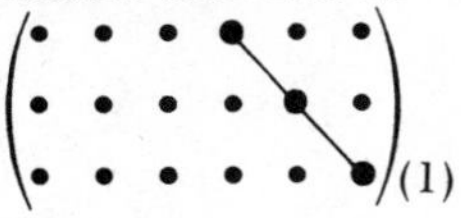

Trigonal

Class 3

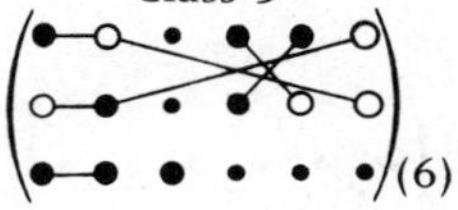

Class 32

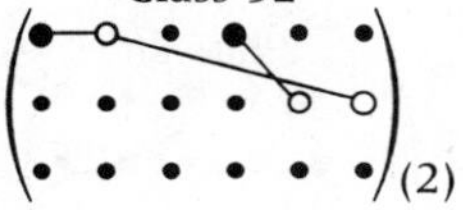

Class $3m$

$m \perp x_1$ (standard orientation)

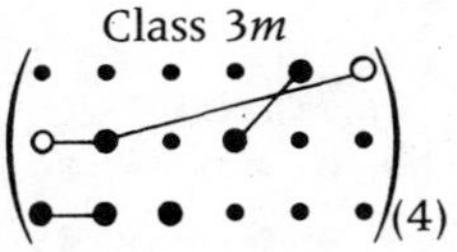

Class $3m$

$m \perp x_2$

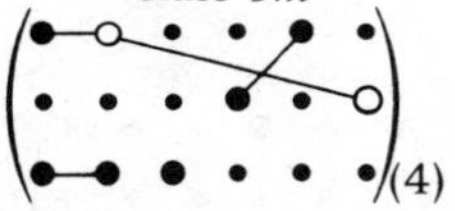

Hexagonal

Class 6

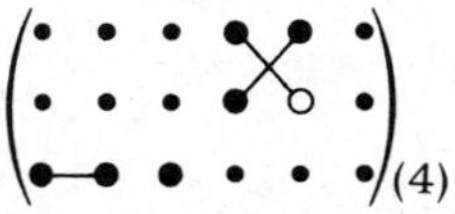

same as class 4

Class $6mm$

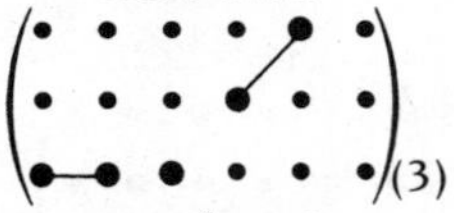

same as class $4mm$

Class 622

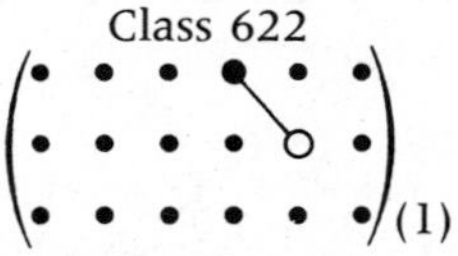

same as class 422

Class $\bar{6}$

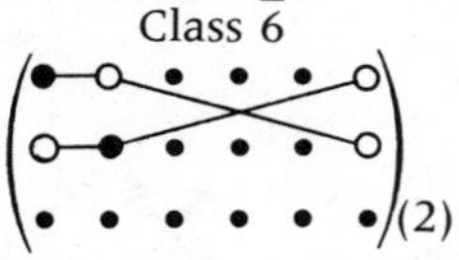

Class $\bar{6}m2$

$m \perp x_1$ (standard orientation)

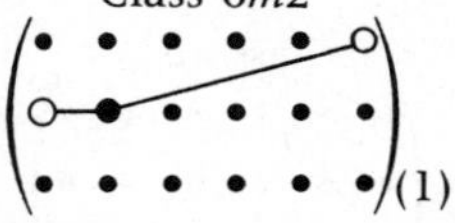

Class $\bar{6}m2$

$m \perp x_2$

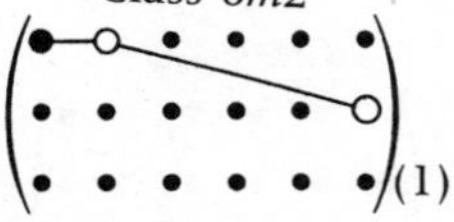

Source: Reference 2.

and the components of the nonlinear polarization are

$$\begin{aligned} P_x &= 2d_{14}E_zE_y \\ P_y &= 2d_{14}E_zE_x \\ P_z &= 2d_{36}E_xE_y \end{aligned} \tag{16.1-9}$$

16.2 THE NONLINEAR FIELD HAMILTONIAN

In addition to the symmetry-imposed restrictions on the number of independent d_{ij} elements, other restrictions apply when the nonlinear polarization is of electronic, rather than ionic, origin and when the crystal is transparent throughout a region that includes all the frequencies involved in the nonlinear process. These conditions were first formulated by Kleinman (Reference 3).

At frequencies sufficiently above ionic resonances (the Restrahlen region), the polarization is due to electronic displacement only, and the ionic contributions are negligible. If, in addition, the frequencies of the mixing fields are well below the electronic absorption region (lossless case), we can take the polarization as a single-valued function of the electric field. This assumption is justified by the treatment of the next section [see (16.3-4) and (16.3-7)] in the limit $\omega \ll \omega_0$, $\gamma = 0$.

When the above conditions are satisfied,

$$\oint_c d(\mathbf{P} \cdot \mathbf{E}) = 0$$

where c is any arbitrary closed path in E_x, E_y, E_z space. Since $d(\mathbf{P} \cdot \mathbf{E}) = \mathbf{P} \cdot d\mathbf{E} + \mathbf{E} \cdot d\mathbf{P}$, it follows that

$$-\oint_c \mathbf{P} \cdot d\mathbf{E} = \oint_c \mathbf{E} \cdot d\mathbf{P} \tag{16.2-1}$$

The right side of (16.2-1) is, according to Section 5.1, the work done by the field on the polarization. Since the medium is lossless, this work over a closed contour is zero. We thus have

$$\oint_c \mathbf{E} \cdot d\mathbf{P} = \text{change in energy} = 0$$

so that from (16.2-1),

$$\oint_c \mathbf{P} \cdot d\mathbf{E} = 0 \tag{16.2-2}$$

If we apply Stokes' theorem,

$$\oint_c \mathbf{P} \cdot d\mathbf{E} = \int_s (\boldsymbol{\nabla}_\mathbf{E} \times \mathbf{P}) \cdot \mathbf{n}\, ds_E$$

leads to

$$\boldsymbol{\nabla}_E \times \mathbf{P} = 0$$

so that there exists an "energy" function $U(\mathbf{E})$ such that

$$\mathbf{P} = -\nabla_E U(\mathbf{E}) \tag{16.2-3}$$

In a nonpolar medium, the lowest power in a Taylor series expansion of $U(\mathbf{E})$ is the second. We can thus expand $U(\mathbf{E})$ as

$$U(\mathbf{E}) = -\frac{\varepsilon_0\chi_{ij}}{2}E_iE_j - \frac{2d_{ijk}}{3}E_iE_jE_k\chi_{ijkl}E_iE_jE_kE_l + \cdots \tag{16.2-4}$$

so that

$$P_i = -\frac{\partial U(\mathbf{E})}{\partial E_i} = \varepsilon_0\chi_{ij}E_j + 2d_{ijk}E_jE_k + 4\chi_{ijkl}E_jE_kE_l + \cdots \tag{16.2-5}$$

It follows that since no physical significance is attached to the order of the electric field components, *all the d_{ijk} coefficients that are related by a rearrangement of the order of the subscripts are equal.* This statement is known as Kleinman's conjecture (Reference 3).

Note that unlike the symmetry condition $d_{ijk} = d_{ikj}$, the Kleinman conjecture applies only to lossless media. Since most nonlinear experiments are carried out in the lossless regime, it is a powerful practical relationship. As an example, consider the coefficient of $E_xE_y^2$ in (16.2-4). It follows that $d_{xyy} = d_{yyx} = d_{yxy}$ or, using the subscript contraction introduced in Section 16.1, we obtain

$$d_{12} = d_{26}$$

Similarly,

$$d_{36} = d_{14}$$

The maximum number of independent d_{ijk} coefficients is thus reduced to 10

$$\begin{matrix} d_{11} & d_{12} & d_{13} & d_{14} & d_{15} & d_{16} \\ d_{16} & d_{22} & d_{23} & d_{24} & d_{14} & d_{12} \\ d_{15} & d_{24} & d_{33} & d_{23} & d_{13} & d_{14} \end{matrix} \tag{16.2-6}$$

In the case of KDP, as an example, whose point group symmetry is $\bar{4}2m$, the number of two independent coefficients (d_{14} and d_{36} in Table 16.1) reduces to one, since $d_{14} = d_{36}$.

16.3 ON THE PHYSICAL ORIGINS OF THE NONLINEAR OPTICAL COEFFICIENTS

Bloembergen (S-1) has used the model of an anharmonic oscillator to discuss the nonlinear optical susceptibility. The same model was used by Garrett and Robinson (Reference 4) to derive an expression for the one-dimensional nonlinear coefficient. We repeat below the essential results of Reference 4 to obtain a numerical estimate for d_{ij}.

The model assumes that the electronic response to a driving electric field can be simulated by that of an electron in an anharmonic potential well. The equation of motion for the electron is then

$$\ddot{X} + \gamma\dot{X} + \omega_0^2 X + DX^2 = \frac{eE_0}{2m}(e^{i\omega t} + e^{-i\omega t}) \tag{16.3-1}$$

where X is the deviation from the potential minimum, mDX^2 is the anharmonic restoring force [corresponding to the term $(m/3)DX^3$ in the potential], the driving electric field is $E_0 \cos \omega t$, and γ is the damping term. Since we are looking for the polarization at 2ω, we assume a solution in the form

$$X = \tfrac{1}{2}(q_1 e^{i\omega t} + q_2 e^{2i\omega t} + \text{c.c.}) \tag{16.3-2}$$

which, when substituted into (16.3-1), yields

$$\begin{aligned} &-\frac{\omega^2}{2}(q_1 e^{i\omega t} + \text{c.c.}) - 2\omega^2(q_2 e^{2i\omega t} + \text{c.c.}) + \frac{i\omega\gamma}{2}(q_1 e^{i\omega t} - \text{c.c.}) \\ &+ i\omega\gamma(q_2 e^{2i\omega t} - \text{c.c.}) + \frac{\omega_0^2}{2}(q_1 e^{i\omega t} + q_2 e^{2i\omega t} + \text{c.c.}) \\ &+ \frac{D}{4}(q_1^2 e^{2i\omega t} + q_2^2 e^{4i\omega t} + q_1 q_1^* + 2q_1 q_2 e^{3i\omega t} \\ &+ 2q_1 q_2^* e^{-i\omega t} + q_2 q_2^* + \text{c.c.}) = \frac{eE_0}{2m}(e^{i\omega t} + \text{c.c.}) \end{aligned} \tag{16.3-3}$$

By equating the coefficients of $e^{i\omega t}$ on both sides of (16.3-3) and assuming that $|Dq_2| \ll [(\omega_0^2 - \omega^2)^2 + \omega^2\gamma^2]^{1/2}$, we obtain

$$q_1 = \left(\frac{eE_0}{m}\right)\frac{1}{(\omega_0^2 - \omega^2) + i\omega\gamma} \tag{16.3-4}$$

The linear susceptibility $\chi_L^{(\omega)}$ is defined by

$$P^{(\omega)}(t) = \frac{\varepsilon_0}{2}(\chi_L^{(\omega)} E_0 e^{i\omega t} + \text{c.c.}) = \frac{Ne}{2}(q_1 e^{i\omega t} + \text{c.c.}) \tag{16.3-5}$$

where N is the number of electrons per unit volume that contribute to P. It follows that

$$\chi_L^{\omega} = \frac{Ne^2}{m\varepsilon_0[(\omega_0^2 - \omega^2) + i\omega\gamma]} \tag{16.3-6}$$

In a similar manner, we proceed to solve for q_2. Equating the multipliers of $e^{2i\omega t}$ on both sides of (16.3-3) leads to

$$q_2(-4\omega^2 + 2i\omega\gamma + \omega_0^2) = -\tfrac{1}{2}Dq_1^2$$

that after we use (16.3-4), becomes

$$q_2 = \frac{-De^2E_0^2}{2m^2[(\omega_0^2 - \omega^2) + i\omega\gamma]^2(\omega_0^2 - 4\omega^2 + 2i\omega\gamma)} \tag{16.3-7}$$

Defining the nonlinear susceptibility $d_{NL}^{(2\omega)}$ by

$$P^{(2\omega)}(t) = \tfrac{1}{2}Ne(q_2 e^{2i\omega t} + \text{c.c.}) = \tfrac{1}{2}[d_{NL}^{(2\omega)} E_0^2 e^{2i\omega t} + \text{c.c.}] \tag{16.3-8}$$

we obtain

$$d_{NL}^{(2\omega)} = \frac{-DNe^3}{2m^2[(\omega_0^2 - \omega^2) + i\omega\gamma]^2[(\omega_0^2 - 4\omega^2) + 2i\omega\gamma]}$$

or, using (16.3-6),

$$d_{NL}^{(2\omega)} = \frac{mD(\chi_L^{(\omega)})^2\chi_L^{2\omega}\varepsilon_0^3}{2N^2\,|e^3|} \tag{16.3-9}$$

which is the desired result. If we define a parameter $\delta^{(2\omega)}$ by

$$\delta^{(2\omega)} = \frac{d_{NL}^{(2\omega)}}{(\chi_L^{(\omega)})^2\chi_L^{2\omega}\varepsilon_0^3} \tag{16.3-10}$$

this parameter has a value

$$\delta^{(2\omega)} = \frac{mD}{2\,|e^3|\,N^2} \tag{16.3-11}$$

This result was first given by Garrett and Robinson (Reference 4). Miller (Reference 6) has observed that the three-dimensional analog of δ, which is defined by

$$\delta_{ijk} = \frac{d_{ijk}^{(2\omega)}}{\chi_{ii}^{2\omega}\chi_{jj}^{\omega}\chi_{kk}^{\omega}\varepsilon_0^3} \tag{16.3-12}$$

is remarkably constant over a large variety of crystals. This fact is evident from Table 16.2, which lists the d_{ijk} and corresponding δ_{ijk} coefficients for several crystals. Although the values of d_{ijk} range over four orders of magnitude (as we go from ADP to Te), the values of δ cluster within a factor of 2 of a mean value $\delta_{\text{mean}} \sim 2 \times 10^9$.

The observed constancy of δ, when examined in light of (16.3-11), suggests that the nonlinear term D is nearly a constant in various materials and that the large variation in the observed values of d_{ijk} reflect the dependence of the latter on the linear susceptibilities.

To obtain an estimate for D, we note that, according to (16.3-1), the anharmonic potential term is

$$V^{(3)} = \frac{m}{3} DX^3 \tag{16.3-13}$$

so that D is zero in centrosymmetric crystals. By assuming a simple model for the potential field of a noncentrosymmetric crystal, we may obtain a numerical estimate for D (and δ). A way to do this is to remove (or add) an ionic charge from one lattice point of a unit cell of a centrosymmetric crystal. The simplest anharmonic potential well in one dimension is, thus, that formed between a charge $2e$ and a charge e a distance r_0 away. Expanding the potential function for this configuration about the potential minimum, we obtain

$$V(X) = \frac{e^2}{4\pi\varepsilon_0 r_0}\left(5.83 + 24.1\frac{X^2}{r_0^2} - 13.3\frac{X^3}{r_0^3}\cdots\right) \tag{16.3-14}$$

where X is the distance from the potential minimum. The first anharmonic

TABLE 16.2. The Nonlinear Optical Coefficients of a Number of Crystals[a]

Crystal	$d_{ijk}^{(2\omega)}$ *in units of* $\frac{1}{9}\times 10^{-22}$ *MKS*	δ_{ijk} *in units of* 10^9
$LiIO_3$	$d_{15} = 4.4$	
$NH_4H_2PO_4$	$d_{36} = 0.45$	3.2
(ADP)	$d_{14} = 0.50 \pm 0.02$	3.2
KH_2PO_4	$d_{36} = 0.45 \pm 0.03$	3.4
(KDP)	$d_{14} = 0.35$	3.4
KD_2PO_4	$d_{36} = 0.42 \pm 0.02$	3.1
	$d_{14} = 0.42 \pm 0.02$	3.1
KH_2ASO_4	$d_{36} = 0.48 \pm 0.03$	2.9
	$d_{14} = 0.51 \pm 0.03$	3.1
Quartz	$d_{11} = 0.37 \pm 0.02$	2.3
$AlPO_4$	$d_{11} = 0.38 \pm 0.03$	2.5
ZnO	$d_{33} = 6.5 \pm 0.2$	4.0
	$d_{31} = 1.95 \pm 0.2$	1.3
	$d_{15} = 2.1 \pm 0.2$	1.5
CdS	$d_{33} = 28.6 \pm 2$	3.8
	$d_{31} = 30 \pm 10$	1.9
	$d_{36} = 33$	2.3
GaP	$d_{14} = 80 \pm 14$	1.5
GaAs	$d_{14} = 72$	2.0
$BaTiO_3$	$d_{33} = 6.4 \pm 0.5$	1.3
	$d_{31} = 18 \pm 2$	3.1
	$d_{15} = 17 \pm 2$	2.9
$LiNbO_3$	$d_{15} = 4.4$	1.4
	$d_{22} = 2.3 \pm 1.0$	0.66
Te	$d_{11} = 517$	0.8
Se	$d_{11} = 130 \pm 30$	5.0
$Ba_2NaNb_5O_{15}$	$d_{33} = 10.4 \pm 0.7$	
	$d_{32} = 7.4 \pm 0.7$	
Ag_3AsS_3	$d_{22} = 22.5$	
(proustite)	$d_{36} = 13.5$	
CdSe	$d_{31} = 22.5 \pm 3$	
$CdGeAs_2$	$d_{36} = 363 \pm 70$	
$AgGaSe_2$	$d_{36} = 27 \pm 3$	
$AgSbS_3$	$d_{36} = 9.5$	
ZnS	$d_{36} = 13$	

[a] Some authors define the nonlinear coefficient d by $P = \varepsilon_0\, dE^2$ rather than by the relation $P = dE^2$ used here. The data of the table were updated using the results of Supplementary Reference 3.

term is $V^{(3)}(X) = (-13.3e^2/4\pi\varepsilon_0 r_0^4)X^3$ that, according to (16.3-13), gives

$$D = \frac{-39.9e^2}{4\pi\varepsilon_0 m r_0^4} \tag{16.3-15}$$

If we use a value of $r_0 \simeq 5$ Å, the corresponding value of δ is derived from

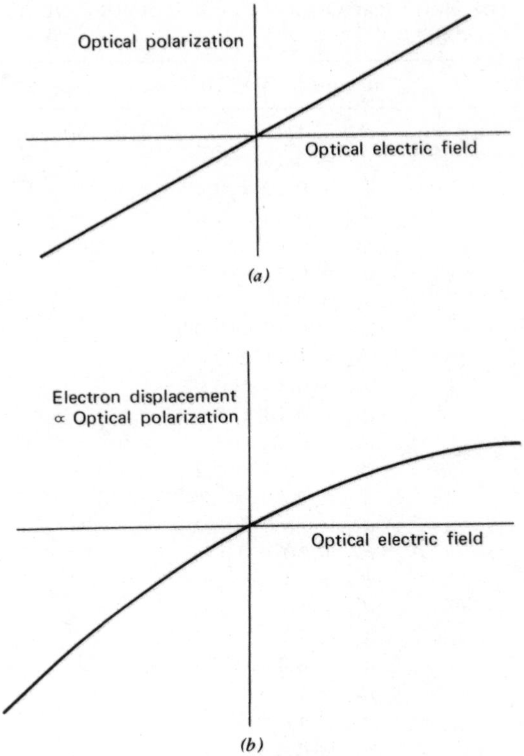

FIGURE 16.1 Relation between induced polarization and the electric field causing it; (*a*) in a linear dielectric and (*b*) in a crystal lacking inversion symmetry in which the electron moves in a potential well of the form (16.3-1).

(16.3-11) using a value of $N \sim 6 \times 10^{28}$ m^{-3}.[1] The result is

$$\delta \sim 3.5 \times 10^{9} \text{ MKS}$$

that is within the range of values spanned in the second column of Table 16.2.

The process of generating a second harmonic polarization by an electron moving in an anharmonic potential well is illustrated in Figures 16.1, 16.2, and 16.3.

A more sophisticated and meaningful approach to the derivation of the nonlinear optical constants is via a quantum mechanical formalism. Such a treatment, at this point, will constitute a considerable diversion and is, consequently, discussed in Appendix 4.

[1] This is a typical value for crystals with two valence electrons per atom.

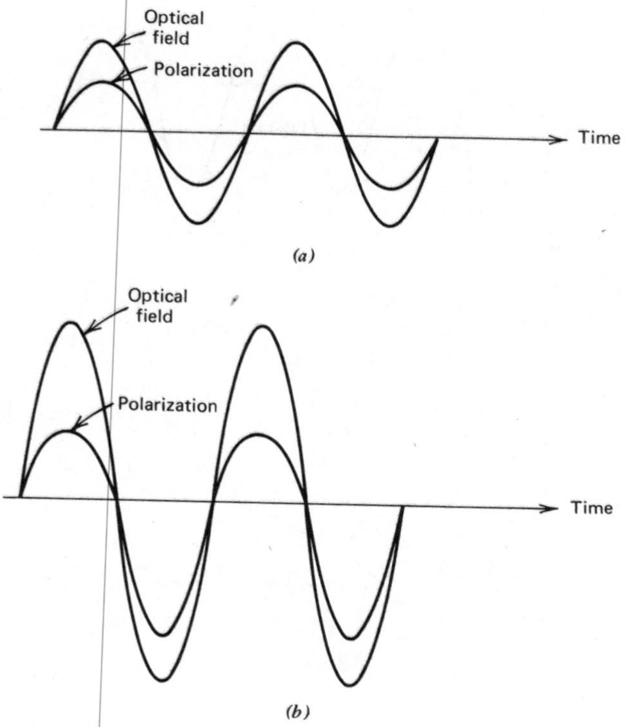

FIGURE 16.2 An applied sinusoidal electric field and the resulting polarization; (*a*) in a linear crystal and (*b*) in a crystal lacking inversion symmetry.

16.4 THE ELECTROMAGNETIC FORMULATION OF THE NONLINEAR INTERACTION

We start with Maxwell equations in a form, which includes the polarization **P**, explicitly

$$\nabla \times \mathbf{H} = \mathbf{i} + \frac{\partial \mathbf{D}}{\partial t} = \mathbf{i} + \frac{\partial}{\partial t}(\varepsilon_0 \mathbf{E} + \mathbf{P})$$
$$\nabla \times \mathbf{E} = -\frac{\partial}{\partial t}(\mu_0 \mathbf{H}) \tag{16.4-1}$$

The polarization **P** is made up of a linear and a nonlinear term

$$\mathbf{P} = \varepsilon_0 \chi_L \mathbf{E} + \mathbf{P}_{NL} \tag{16.4-2}$$

where

$$(P_{NL})_i = 2d_{ijk}E_jE_k \tag{16.4-3}$$

and where the tensor aspect of χ_L is ignored.

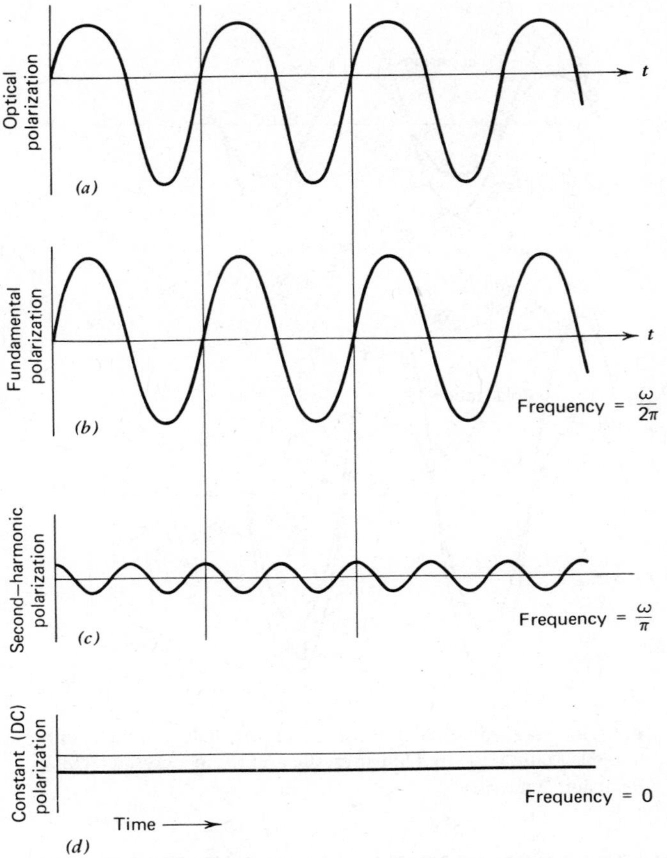

FIGURE 16.3 Analysis of the nonlinear polarization wave (*a*) of Figure 16.2*b* shows that it contains components oscillating at (*b*) the same frequency (ω) as the wave inducing it, (*c*) twice that frequency (2ω), and (*d*) an average (dc) negative component.

The first of (16.4-1) can be written as

$$\nabla \times \mathbf{H} = \sigma \mathbf{E} + \frac{\partial}{\partial t} \varepsilon \mathbf{E} + \frac{\partial \mathbf{P}_{NL}}{\partial t} \tag{16.4-4}$$

where σ is the conductivity and $\varepsilon = \varepsilon_0(1 + \chi_L)$. After taking the curl of both sides of the second of (16.4-1), replacing $\nabla \times \mathbf{H}$ by (16.4-4), and using $\nabla \times \nabla \times \mathbf{E} = \nabla\nabla \cdot \mathbf{E} - \nabla^2 \mathbf{E}$, we get

$$\nabla^2 \mathbf{E} = \mu_0 \sigma \frac{\partial \mathbf{E}}{\partial t} + \mu_0 \varepsilon \frac{\partial^2 \mathbf{E}}{\partial t^2} + \mu_0 \frac{\partial^2}{\partial t^2} \mathbf{P}_{NL} \tag{16.4-5}$$

where $\nabla \cdot \mathbf{E} = 0$.

At this point, we specialize the problem to one dimension by taking $\partial/\partial y = \partial/\partial x = 0$ and denoting the arbitrary direction of propagation as z. We also limit the consideration to three frequencies ω_1, ω_2, and ω_3 and take the corresponding fields to be in the form of traveling plane waves

$$
\begin{aligned}
E_i^{(\omega_1)}(z, t) &= \tfrac{1}{2}[E_{1i}(z)e^{i(\omega_1 t - k_1 z)} + \text{c.c.}] \\
E_k^{(\omega_2)}(z, t) &= \tfrac{1}{2}[E_{2k}(z)e^{i(\omega_2 t - k_2 z)} + \text{c.c.}] \\
E_j^{(\omega_3)}(z, t) &= \tfrac{1}{2}[E_{3j}(z)e^{i(\omega_3 t - k_3 z)} + \text{c.c.}]
\end{aligned}
\tag{16.4-6}
$$

where i, j, k refer to the Cartesian coordinates and can each take on values x and y. Note that for $\mathbf{P}_{NL} = 0$, the solution of (16.4-5) is given by (16.4-6) with $E_{1i}(z)$, $E_{2k}(z)$, and $E_{3j}(z)$ independent of z.

The i component of the nonlinear polarization at $\omega_1 = \omega_3 - \omega_2$, as an example, is given according to (16.4-3) and (16.4-6) as[2]

$$[P_{NL}^{(\omega_1)}(z, t)]_i = d'_{ijk}E_{3j}(z)E_{2k}^*(z)e^{i[(\omega_3-\omega_2)t-(k_3-k_2)z]} + \text{c.c.} \tag{16.4-7}$$

Returning to the wave equation (16.4-5) and taking the ith component (for $\partial/\partial x = \partial/\partial y = 0$) yield

$$\nabla^2 E_i^{(\omega_1)}(z, t) = \frac{\partial^2}{\partial z^2} E_i^{(\omega_1)}(z, t) = \frac{1}{2}\frac{\partial^2}{\partial z^2}[E_{1i}(z)e^{i(\omega_1 t - k_1 z)} + \text{c.c.}] \tag{16.4-8}$$

By carrying out the indicated differentiation and assuming that the variation of the complex field amplitudes with z is small enough so that

$$\frac{dE_{1i}}{dz}k_1 \gg \frac{d^2E_{1i}}{dz^2}$$

we get from (16.4-8),

$$\nabla^2 E_i^{(\omega_1)}(z, t) = -\frac{1}{2}\left[k_1^2E_{1i}(z) + 2ik_1\frac{dE_{1i}(z)}{dz}\right]e^{i(\omega_1 t - k_1 z)} + \text{c.c.} \tag{16.4-9}$$

with similar expressions obtained for $\nabla^2 E_j^{(\omega_3)}(z, t)$ and $\nabla^2 E_k^{(\omega_2)}(z, t)$. Using the last equation in (16.4-5), we may write the wave equation for $E_i^{(\omega_1)}(z, t)$ as

$$
\begin{aligned}
&\left(\frac{k_1^2}{2}E_{1i} + ik_1\frac{dE_{1i}}{dz}\right)e^{i(\omega_1 t - k_1 z)} + \text{c.c.} \\
&= (-i\omega_1\mu_0\sigma + \omega_1^2\mu_0\varepsilon)\left[\frac{E_{1i}}{2}e^{i(\omega_1 t - k_1 z)} + \text{c.c.}\right] - \mu_0\frac{\partial^2}{\partial t^2}[P_{NL}^{\omega_1}(z, t)]_i
\end{aligned}
\tag{16.4-10}
$$

where we used $\partial/\partial t = i\omega_1$. We have also assumed that when the number of interacting frequencies is finite, (16.4-5) must be satisfied separately by each frequency component.

Replacing $[P_{NL}^{\omega_1}(z, t)]_i$ in the last equation by (16.4-7) and recognizing that $\omega_1^2\mu_0\varepsilon = k_1^2$, we obtain

$$ik_1\frac{dE_{1i}}{dz}e^{-ik_1 z} = -\frac{i\omega_1\sigma\mu_0}{2}E_{1i}e^{-ik_1 z} + \mu_0\omega_1^2 d'_{ijk}E_{3j}E_{2k}^*e^{-i(k_3-k_2)z}$$

[2] d'_{ijk} is the $\bar{\bar{d}}$ tensor of (16.1-6) transformed from the crystal coordinate system to that used here to describe the field propagation [see Problem 16.9 and Eq. (16.5-16)].

or after dividing by $ik_1e^{-ik_1z}$ (and allowing σ to be a function of frequency),

$$\frac{dE_{1i}}{dz} = -\frac{\sigma_1}{2}\sqrt{\frac{\mu_0}{\varepsilon_1}}\,E_{1i} - i\omega_1\sqrt{\frac{\mu_0}{\varepsilon_1}}\,d'_{ijk}E_{3j}E^*_{2k}e^{-i(k_3-k_2-k_1)z}$$

and similarly

$$\frac{dE^*_{2k}}{dz} = -\frac{\sigma_2}{2}\sqrt{\frac{\mu_0}{\varepsilon_2}}\,E^*_{2k} + i\omega_2\sqrt{\frac{\mu_0}{\varepsilon_2}}\,d'_{kij}E_{1i}E^*_{3j}e^{-i(k_1-k_3+k_2)z}$$

$$\frac{dE_{3j}}{dz} = -\frac{\sigma_3}{2}\sqrt{\frac{\mu_0}{\varepsilon_3}}\,E_{3j} - i\omega_3\sqrt{\frac{\mu_0}{\varepsilon_3}}\,d'_{jik}E_{1i}E_{2k}e^{-i(k_1+k_2-k_3)z} \qquad (16.4\text{-}11)$$

These equations constitute the main result of this section. We will apply them in the following section and in the next chapter to some specific cases.

16.5 OPTICAL SECOND-HARMONIC GENERATION

The second-harmonic generation experiment that ushered in the field of nonlinear optics was performed by Franken, Hill, Peters, and Weinreich (Reference 1) in 1961. A sketch of the original experiment is shown in Figure 16.4. A ruby laser beam at 6943 Å was focused on the front surface of a crystalline quartz plate. The emergent radiation was examined with a spectrometer and was round to contain radiation at twice the input frequency (i.e., at λ = 3471.5 Å). The conversion efficiency in this first experiment was $\sim 10^{-8}$. The utilization of more efficient materials, higher intensity lasers, and index-matching techniques have resulted, in the last few years, in conversion efficiencies approaching unity. These factors will be discussed later in the section.

Consider next (16.4-11) for a second-harmonic generation (SHG). This is the limiting case of the three-frequency interaction where two of the frequencies, ω_1 and ω_2, are equal, and $\omega_3 = 2\omega_1$. We consequently need to consider only two of (16.4-11), the first (or second) and the last. To further simplify the analysis, and yet retain its validity for the majority of the experimental

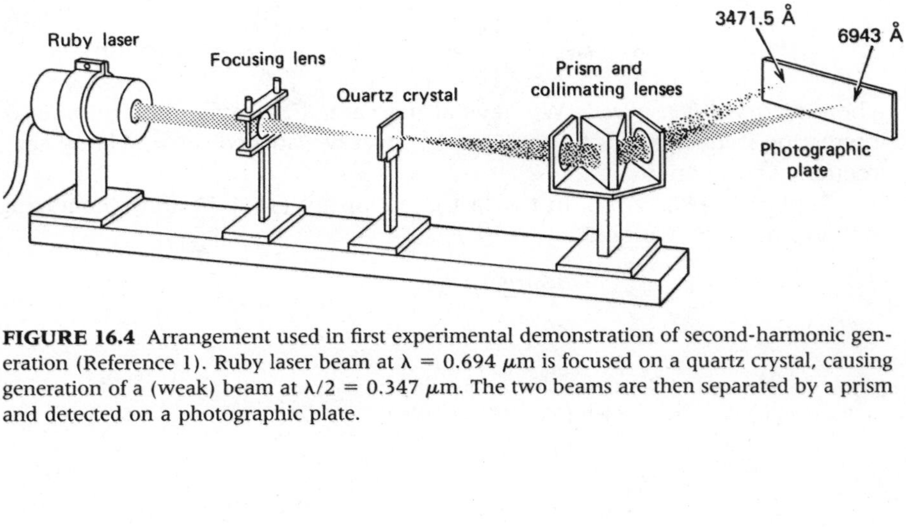

FIGURE 16.4 Arrangement used in first experimental demonstration of second-harmonic generation (Reference 1). Ruby laser beam at λ = 0.694 μm is focused on a quartz crystal, causing generation of a (weak) beam at $\lambda/2$ = 0.347 μm. The two beams are then separated by a prism and detected on a photographic plate.

situations, we assume that the amount of power lost by the input (ω_1) beam (by conversion to $2\omega_1$) is negligible so that $dE_{1i}/dz \simeq 0$ and we need to consider only the last of Eqs. (16.4-11). If the medium is transparent at ω_3, then $\sigma_3 = 0$, and we have

$$\frac{dE_{3j}}{dz} = -i\omega \sqrt{\frac{\mu_0}{\varepsilon}} d'_{jik} E_{1i} E_{1k} e^{i\,\Delta kz} \tag{16.5-1a}$$

where

$$\omega = \omega_1 = \frac{\omega_3}{2}$$

and

$$\Delta k = k_3^{(j)} - k_1^{(i)} - k_1^{(k)} \tag{16.5-1b}$$

and where $k_1^{(i)}$ is the propagation constant for the beam at ω_1, which is polarized along the i direction. The solution of (16.5-1) for $E_{3j}(0) = 0$ (i.e., no second-harmonic input) and for a crystal of length L is

$$E_{3j}(L) = -i\omega\, 2 \sqrt{\frac{\mu_0}{\varepsilon}}\, d'_{jik} E_{1i} E_{1k} \frac{e^{i\,\Delta kL} - 1}{i\,\Delta k} \tag{16.5-2}$$

where factor of 2 accounts for summation over i, k

$$E_{3j}(L) E^*_{3j}(L) = 4\,\frac{\mu_0}{\varepsilon}\, \omega^2 (d'_{jik})^2 E^2_{1i} E^2_{1k} L^2 \frac{\sin^2(\Delta kL/2)}{(\Delta kL/2)^2} \tag{16.5-3}$$

where $\varepsilon \equiv \varepsilon_3$. To obtain an expression for the second-harmonic power output $P^{(2\omega)}$, we use the relation

$$\frac{P^{(2\omega)}}{\text{Area}} = \frac{1}{2} \sqrt{\frac{\varepsilon}{\mu_0}}\, E_{3j} E^*_{3j}$$

Using this last expression in (16.5-3) results in

$$\frac{P^{(2\omega)}}{\text{Area}} = 2 \sqrt{\frac{\mu_0}{\varepsilon}}\, \omega^2 (d'_{ijk})^2 E^2_{1i} E^2_{1k} L^2 \frac{\sin^2(\Delta kL/2)}{(\Delta kL/2)^2} \tag{16.5-4}$$

The conversion efficiency is thus

$$\frac{P^{(2\omega)}}{P^{(\omega)}} = 8 \left(\frac{\mu_0}{\varepsilon_0}\right)^{3/2} \frac{\omega^2 (d'_{ijk})^2 L^2}{n^3} \left(\frac{P^{(\omega)}}{\text{Area}}\right) \frac{\sin^2\left(\frac{\Delta kL}{2}\right)}{(\Delta kL/2)^2} \tag{16.5-5}$$

where we took $\varepsilon_1 \simeq \varepsilon_3 = \varepsilon_0 n^2$.

Phase-Matching in Second-Harmonic Generation

According to (16.5-5), a prerequisite for efficient second-harmonic generation is that $\Delta k = 0$—or, using $\omega_3 = 2\omega$, $\omega_1 = \omega_2 = \omega$, we obtain

$$k^{(2\omega)} = 2k^{(\omega)} \tag{16.5-6}$$

If $\Delta k \neq 0$, the second-harmonic wave generated at some plane (e.g., z_1), having propagated to some other plane (z_2), is not in phase with the second-harmonic wave generated at z_2. This results in the interference described by the factor

$$\frac{\sin^2(\Delta kL/2)}{(\Delta kL/2)^2}$$

in (16.5-5). Two adjacent peaks of this spatial interference pattern are separated by the so-called "coherence length,"

$$l_c = \frac{2\pi}{\Delta k} = \frac{2\pi}{k^{(2\omega)} - 2k^{(\omega)}} \tag{16.5-7}$$

The coherence length l_c is thus a measure of the *maximum crystal length that is useful in producing the second-harmonic power.* Under ordinary circumstances, it may be no larger than 10^{-2} cm. This is because the index of refraction n^ω normally increases with ω so Δk is given by

$$\Delta k = k^{(2\omega)} - 2k^{(\omega)} = \frac{2\omega}{c}(n^{2\omega} - n^\omega) \tag{16.5-8}$$

where we used the relation $k^{(\omega)} = \omega n^\omega/c$. The coherence length is thus

$$l_c = \frac{\pi c}{\omega(n^{2\omega} - n^\omega)} = \frac{\lambda}{2(n^{2\omega} - n^\omega)} \tag{16.5-9}$$

where λ is the free-space wavelength of the fundamental beam. If we take a typical value of $\lambda = 1$ μm and $n(2\omega) - n(\omega) \simeq 10^{-2}$, we get $l_c \simeq 100$ μm. If l_c were to increase from 100 μm to 2 cm, as an example, according to (16.5-4), the second-harmonic power would go up by a factor of 4×10^4.

The technique that is used widely (see References 7 and 8) to satisfy the *phase-matching* requirement, $\Delta k = 0$, takes advantage of the natural birefringence of anisotropic crystals, which was discussed in Chapter 5. If we use the relation $k^{(\omega)} = \omega\sqrt{\mu\varepsilon_0}\, n^\omega$, (16.5-6) becomes

$$n^{2\omega} = n^\omega \tag{16.5-10}$$

so the indices of refraction at the fundamental and second-harmonic frequencies must be equal. In normally dispersive materials, the index of the ordinary wave or the extraordinary wave along a given direction increases with ω, as can be seen from Table 16.3. This makes it impossible to satisfy (16.5-10) when both the ω and 2ω beams are of the same type—that is, when both are extraordinary or ordinary. We can, however, under certain circumstances, satisfy (16.5-10) by using two waves of different type: one extraordinary and one ordinary. To illustrate the point, consider the dependence of the index of refraction of the extraordinary wave in a uniaxial crystal on the angle θ between the propagation direction and the crystal optic (z) axis. It is given by (5.4-2) as

$$\frac{1}{n_e^2(\theta)} = \frac{\cos^2\theta}{n_0^2} + \frac{\sin^2\theta}{n_e^2} \tag{16.5-11}$$

TABLE 16.3. Index of Refraction Dispersion Data of KH_2PO_4

Wavelength, μm	*Index*	
	n_o *(ordinary ray)*	n_e *(extraordinary ray)*
0.2000	1.622630	1.563913
0.3000	1.545570	1.498153
0.4000	1.524481	1.480244
0.5000	1.514928	1.472486
0.6000	1.509274	1.468267
0.7000	1.505235	1.465601
0.8000	1.501924	1.463708
0.9000	1.498930	1.462234
1.0000	1.496044	1.460993
1.1000	1.493147	1.459884
1.2000	1.490169	1.458845
1.3000	1.487064	1.457838
1.4000	1.483803	1.456838
1.5000	1.480363	1.455829
1.6000	1.476729	1.454797
1.7000	1.472890	1.453735
1.8000	1.468834	1.452636
1.9000	1.464555	1.451495
2.0000	1.460044	1.450308

Source: Reference 9.

If $n_e^{2\omega} < n_0^{\omega}$, there exists an angle θ_m at which $n_e^{2\omega}(\theta_m) = n_0^{\omega}$; so if the fundamental beam (at ω) is launched along θ_m as an ordinary ray, the second-harmonic beam will be generated along the *same direction* as an extraordinary ray. The situation is illustrated by Figure 16.5. The angle θ_m is determined by the intersection between the sphere (shown as a circle in the figure) corresponding to the index surface of the ordinary beam at ω with the index ellipsoid (16.5-11) of the extraordinary ray that gives $n_e^{2\omega}(\theta)$. The angle θ_m for negative uniaxial crystals—that is, crystals in which $n_e^{\omega} < n_0^{\omega}$—is that satisfying $n_e^{2\omega}(\theta_m) = n_0^{\omega}$ or, using (16.5-11),

$$\frac{\cos^2\theta_m}{(n_0^{2\omega})^2} + \frac{\sin^2\theta_m}{(n_e^{2\omega})^2} = \frac{1}{(n_0^{\omega})^2} \qquad (16.5\text{-}12)$$

and, solving for θ_m,

$$\sin^2\theta_m = \frac{(n_0^{\omega})^{-2} - (n_0^{2\omega})^{-2}}{(n_e^{2\omega})^{-2} - (n_0^{2\omega})^{-2}} \qquad (16.5\text{-}13)$$

Example: Second-Harmonic Generation in KH_2PO_4(KDP). If we use a fundamental beam derived from a ruby laser ($\lambda = 6943$ Å) and a KDP crystal, the indices of refraction are obtained by extrapolating the data of Table 16.3

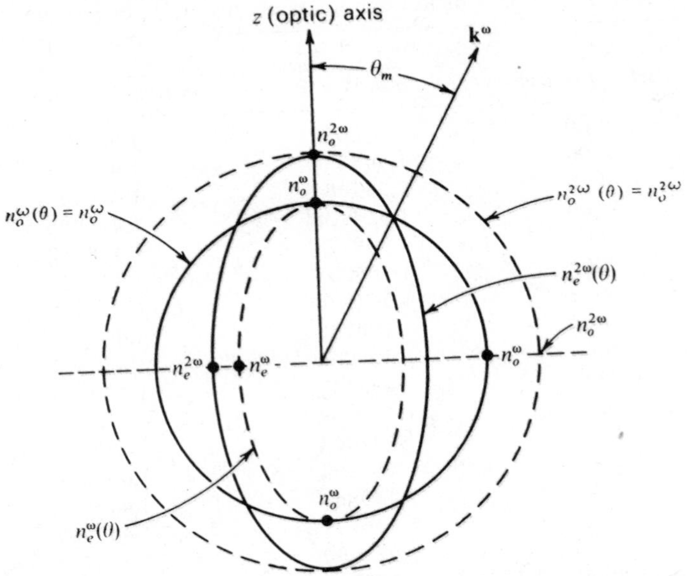

FIGURE 16.5 Normal (index) surfaces for the ordinary and extraordinary rays in a negative ($n_e < n_0$) uniaxial crystal. If $n_e^{2\omega} < n_0^\omega$, the condition $n_e^{2\omega}(\theta) = n_0^\omega$ is satisfied at $\theta = \theta_m$. The eccentricities shown are exaggerated.

$$n_e^\omega = 1.466 \qquad n_e^{2\omega} = 1.487 \qquad n_0^\omega = 1.506 \qquad n_0^{2\omega} = 1.534$$

The matching angle θ_m is then, according to (16.5-2), $\theta_m = 50.4°$.

Another mode of index matching in KDP is one in which the components, i and k of the input beam at ω, do not both belong to ordinary rays, but where one (e.g., k) is an extraordinary ray. The 2ω beam remains an extraordinary ray. Here, the index-matching condition, $\Delta k = 0$, can be written as

$$n_{je}^{2\omega}(\theta) = \tfrac{1}{2}[n_{i0}^{(\omega)} + n_{ke}^{(\omega)}(\theta)] \tag{16.5-14}$$

where j, i, k refer to the axes of the index ellipsoid that are chosen as the reference frame. The index-matching angle θ_m for this case is given by

$$\left[\frac{\cos^2\theta_m}{(n_0^{2\omega})^2} + \frac{\sin^2\theta_m}{(n_e^{2\omega})^2}\right]^{-1/2} = \frac{1}{2}\left\{n_0^\omega + \left[\frac{\cos^2\theta_m}{(n_0^\omega)^2} + \frac{\sin^2\theta_m}{(n_e^\omega)^2}\right]^{-1/2}\right\} \tag{16.5-15}$$

which is a reexpression of (16.5-14) with $n_{je}^{2\omega}(\theta)$ and $n_{ke}^\omega(\theta)$ replaced by their explicit expressions using (16.5-11).

In a SHG experiment, the need to satisfy the index-matching condition (16.5-15) plus the restrictions imposed by the form of the nonlinear optical tensor limit the degrees of freedom that are available in choosing the directions of polarization. In KDP, as an example, the nonlinear polarization is, according to (16.1-9),

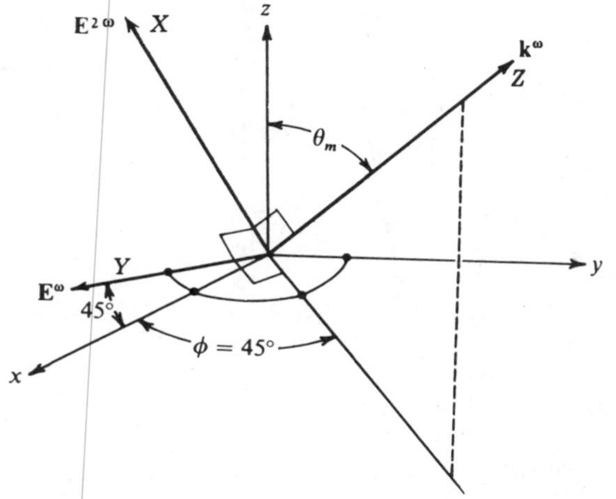

FIGURE 16.6 Second-harmonic generation in KDP. $\mathbf{E}^{\omega}$ is at 45° to the *x-y* axes. The direction of propagation k^{ω} is at angle θ_m to the optic (z) axis. (All vectors passing through the dotted arc are in the *x-y* plane.) The effective d_{ijk} is thus $d_{xyy} = d_{36} \sin \theta_m$ [sec (16.5-16)].

$$
\begin{aligned}
P_x^{2\omega} &= 2d_{14}E_z^{\omega}E_y^{\omega} \\
P_y^{2\omega} &= 2d_{14}E_z^{\omega}E_x^{\omega} \\
P_z^{2\omega} &= 2d_{36}E_x^{\omega}E_y^{\omega},\ P_X^{2\omega} = d_{36}(E^{\omega})^2 \sin \theta_m (\phi = \pi/4)
\end{aligned}
\tag{16.5-16}
$$

In the index-matching scheme that leads to (16.5-13), the fundamental beam is an ordinary ray and, consequently, $E_z^{\omega} = 0$.[3] This dictates that the second-harmonic polarization is given by the last of (16.5-16) and has a z component only. The component of P_z normal to the direction of propagation is equal to $2d_{36}E_x^{\omega}E_y^{\omega} \sin \theta_m$ and is thus maximum for $E_x^{\omega} = E_y^{\omega} = E^{\omega}/\sqrt{2}$, which occurs when the azimuthal angle for $\mathbf{E}^{\omega}$ is $\pi/4$ as shown in Figure 16.6.

According to (16.5-4), the penalty for deviating from the index-matching condition at a fixed L is a reduction of the second-harmonic power output by the factor

$$
\frac{P^{(2\omega)}}{P_{\max}^{(2\omega)}} = \frac{\sin^2(\Delta kL/2)}{(kL/2)^2}
\tag{16.5-17}
$$

Equation (16.5-17) can be checked most easily by varying the angle $\sigma = \theta - \theta_m$ between the index-matching direction and the direction of propagation. For small values of σ, we can approximate $\Delta k(\theta) = k_e^{2\omega}(\theta) - 2k_0^{\omega}$ by

$$
\Delta k(\theta) = 2\beta\sigma
$$

[3] Here, we ignore the fact that, strictly speaking, D_z and not E_z is zero. For $n_e \simeq n_0$, D_z, and E_z are nearly parallel.

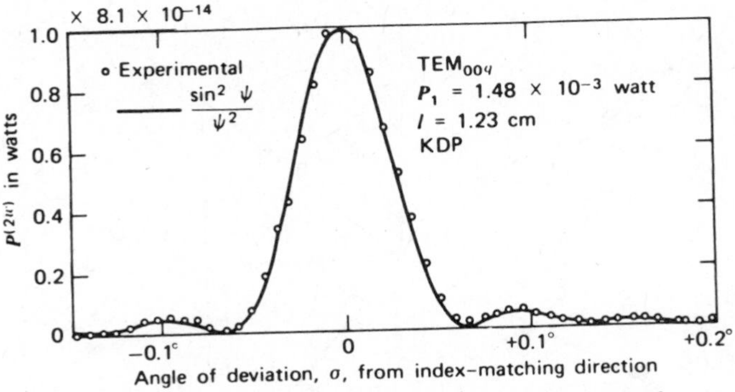

FIGURE 16.7 Variation of second-harmonic power with σ, the angle between the fundamental beam, and the phase-matching direction. *Source:* Reference 10.

where β is a constant that depends on n_0^ω, $n_0^{2\omega}$, and $n_e^{2\omega}$. The factor $(\Delta kL/2)$ in (16.5-17) is thus equal to $\beta\sigma L \equiv \psi$. A plot of the second-harmonic power output as a function of σ is shown in Figure 16.7. The figure also contains the theoretical $\sin^2 \psi/\psi^2$ curve.

An impressive color photograph of second-harmonic generation from ruby light (red) to near ultraviolet (blue in photograph) that uses an ADP (ammonium dihydrogen phosphate) crystal is reproduced next to the title page of this book.

16.6 SECOND-HARMONIC GENERATION WITH A DEPLETED INPUT

In the treatment of second-harmonic generation leading to (16.5-4), it was assumed that the input intensity at ω was not affected by the interaction. This limits the validity of the result to situations where the fraction of the power converted from ω to 2ω is small. In this section we will lift this restriction.

We return to (16.4-11) and define the new field variables A_l by

$$A_l = \sqrt{\frac{n_l}{\omega_l}}\, E_l \qquad l = 1, 2, 3 \tag{16.6-1}$$

so that the intensity at ω_l is

$$I_l = \frac{P_l}{A} = \frac{1}{2}\sqrt{\frac{\varepsilon_0}{\mu_0}}\, n_l\, |E_l|^2 = \frac{1}{2}\sqrt{\frac{\varepsilon_0}{\mu_0}}\, \omega_l\, |A_l|^2 \tag{16.6-2}$$

Since a photon's energy is $\hbar\omega_l$, it follows, from (16.6-2), that $|A_l|^2$ is proportional to the *photon flux* at ω_l, the proportionality constant being independent of frequency.

Equation (16.4-11) can now be written as

$$\frac{dA_1}{dz} = -\tfrac{1}{2}\alpha_1 A_1 - i\kappa A_2^* A_3 e^{-i(\Delta k)z}$$

$$\frac{dA_2^*}{dz} = -\tfrac{1}{2}\alpha_2 A_2^* + i\kappa A_1 A_3^* e^{i(\Delta k)z} \tag{16.6-3}$$

$$\frac{dA_3}{dz} = -\tfrac{1}{2}\alpha_3 A_3 - i\kappa A_1 A_2 e^{i(\Delta k)z}$$

where 1, 2, 3 are the polarization directions of $\mathbf{E}_1$, $\mathbf{E}_2$, $\mathbf{E}_3$.

$$\Delta k \equiv k_3 - (\mathrm{k}_1 + k_2)$$

$$\kappa \equiv d'_{123}\sqrt{\left(\frac{\mu_0}{\varepsilon_0}\right)\frac{\omega_1\omega_2\omega_3}{n_1 n_2 n_3}} \tag{16.6-4}$$

$$\alpha_l \equiv \sigma_l\sqrt{\frac{\mu_0}{\varepsilon_l}} \qquad l = 1, 2, 3$$

The advantage of using the A_l instead of E_l is now apparent since, unlike (16.4-11), relations (16.6-3) involve a single coupling parameter κ.

In the case of second-harmonic generation, $\mathrm{A}_1 = A_2$ and (16.6-3) becomes

$$\frac{dA_1}{dz} = -i\kappa A_3 A_1^* e^{-i\Delta kz}$$

$$\frac{dA_3}{dz} = -i\kappa A_1^2 e^{i\Delta kz} \tag{16.6-5}$$

In the phase-matched case ($\Delta k = 0$), it follows from (16.6-5) that if we choose $A_1(0)$ as a real number, then $\mathrm{A}_1(z)$ is real and (16.6-5) become

$$\frac{dA_1}{dz} = -\kappa A_3' A_1$$

$$\frac{dA_3'}{dz} = \kappa A_1^2 \tag{16.6-6}$$

where $A_3 \equiv -iA_3'$. It follows from (16.6-6) that

$$\frac{d}{dz}(A_1^2 + A_3'^2) = 0$$

so that if we assume no input at ω_3

$$A_1^2 + A_3'^2 = A_1^2(0)$$

or, from (16.6-6),

$$\frac{dA_3'}{dz} = \tfrac{1}{2}\kappa[A_1^2(0) - A_3'^2]$$

and (Reference 12)

$$A_3'(z) = A_1(0)\tan h[\kappa A_1(0)z] \tag{16.6-7}$$

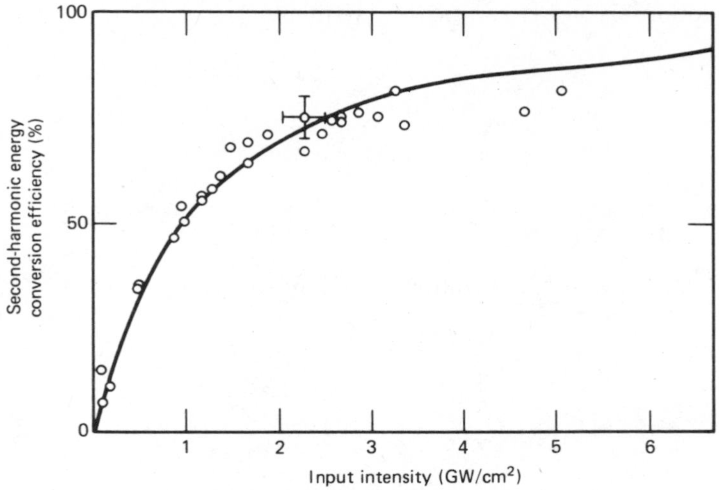

FIGURE 16.8 Frequency doubling energy conversion efficiency. Solid curve: theoretical prediction of Eq. (16.6-8) [recall that $A_1(0) \propto \sqrt{I_{in}}$). The circles correspond to experimental points. *Source:* Supplementary Reference 7.

The conversion efficiency is

$$\frac{P^{(2\omega)}}{P^{(\omega)}} = \frac{|A_3(z)|^2}{|A_1(0)|^2} = \tan h^2[\kappa A_1(0)z] \qquad (16.6\text{-}8)$$

Note that as $\kappa A_1(0)z \rightarrow \infty$, $A_3'(z) \rightarrow A_1(0)$ so that all the input photons can be converted into half (since $A_1 = A_2$) as many output photons (at twice the frequency) and *no more.*

A plot of the conversion efficiency in the no-depletion approximation (16.5-5) and that given by (16.6-8) is shown in Figure 16.8.

16.7 SECOND-HARMONIC GENERATION WITH GAUSSIAN BEAMS

The analysis of second-harmonic generation in Sections 16.5 and 16.6 is based on a plane-wave model. In practice, one uses Gaussian beams and the question arises as to the effect of the diffraction of such beams on nonlinear processes such as second-harmonic generation.

The inverse dependence of the conversion efficiency as given by (16.5-5) on the beam cross-sectional area decrees that the beam be focused onto the nonlinear crystal. A typical situation is sketched in Figure 16.9. Here, $z_0 \equiv (\pi n \omega_0^2/\lambda)$ is, according to (6.6-11), the distance in which the beam cross-sectional area doubles relative to its value at the waist. If the crystal length L is much shorter than z_0, the beam cross section remains essentially a constant within the crystal, and we can use the plane wave result (16.5-3)

$$|E^{(2\omega)}|^2 = 4\,\frac{\mu_0}{\varepsilon}\,\omega^2\, d'_{ijk}\,|E^{(\omega)}|^4\, L^2\, \frac{\sin^2(\Delta kL/2)}{(\Delta kL/2)^2} \qquad (16.7\text{-}1)$$

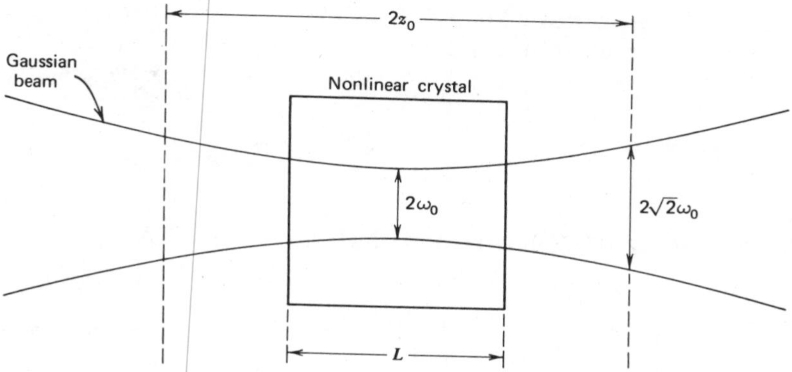

FIGURE 16.9 Gaussian beam focused inside a nonlinear optical crystal.

where

$$E^{(\omega)}(r) \cong E_0 e^{-r^2/\omega_0^2} \tag{16.7-2}$$

Using

$$P^{(\omega)} = \frac{1}{2}\sqrt{\frac{\varepsilon}{\mu_0}} \int_{\text{cross section}} |E^{(\omega)}|^2 \, dx \, dy \cong \sqrt{\frac{\varepsilon}{\mu_0}} E_0^2 \left(\frac{\pi\omega_0^2}{4}\right)$$

as well as (16.7-2), we obtain, by integrating (16.7-1),

$$\frac{P^{(2\omega)}}{P^{(\omega)}} = 8\left(\frac{\mu_0}{\varepsilon_0}\right)^{3/2} \frac{\omega^2 d^2 L^2}{n^3} \left[\frac{P^{(\omega)}}{\pi\omega_0^2}\right] \frac{\sin^2(\Delta kL/2)}{(\Delta kL/2)^2} \tag{16.7-3}$$

where we took $n^\omega \simeq n^{2\omega}$ and $d'_{ijk} \equiv d$. Equation (16.7-3) is identical to (16.5-5) except that it now applies to a Gaussian beam. The L^2 dependence of the conversion efficiency will tempt us to use long crystals. But, according to Figure 16.9, once $L > 2z_0$, the increase of the beam cross section will reduce the conversion efficiency. We thus expect the maximum conversion efficiency to result when $L \simeq 2z_0$ and to be given by (16.7-3) by putting $L = 2z_0 = 2(\pi\omega_0^2 n/\lambda)$. We will refer to this condition as confocal focusing. The result is

$$\left.\frac{P^{(2\omega)}}{P^{(\omega)}}\right|_{\substack{\text{confocal}\\ \text{focusing}}} = \frac{8}{\pi c}\left(\frac{\mu_0}{\varepsilon_0}\right)^{3/2} \frac{\omega^3 \delta^2 L}{n^2} P^{(\omega)} \frac{\sin^2(\Delta kL/2)}{(\Delta kL/2)^2} \tag{16.7-4}$$

An exact analysis [11] shows that optimum conversion results when $L = 5.68z_0$ and the resulting power is approximately 1.2 times that given by (16.7-4).[4]

By comparing (16.7-4) to the plane wave solution (16.5-5), we find that under optimum focusing conditions, the conversion efficiency increases as the crystal length L rather than as L^2.

[4] This is strictly true when the group velocities of the fundamental and second-harmonic beams are parallel, which happens at $\theta_m = 90°$.

Example: Optimum Focusing. Consider second-harmonic conversion under confocal focusing conditions, in KH_2PO_4 from $\lambda = 1\ \mu m$ to $\lambda = 0.5\ \mu m$. Using $L = 1$ cm, $d = d_{36} = 3.6 \times 10^{-24}$ MKS, $n = 1.5$, we obtain, from (16.7-4),

$$\frac{P^{(2\omega)}}{P^{(\omega)}} = 1.72 \times 10^{-4} P^{(\omega)}$$

16.8 INTERNAL SECOND-HARMONIC GENERATION

The development leading to (16.7-3) shows that if an optical beam passes through a nonlinear crystal, an amount of power

$$P^{(2\omega)} = q[P^{(\omega)}]^2 \tag{16.8-1}$$

is converted from ω to 2ω. For the case of a short ($L \ll z_0$) crystal placed at the waist of Gaussian beam, the constant q is

$$q = 8\left(\frac{\mu_0}{\varepsilon_0}\right)^{3/2} \frac{\omega^2\, d^2 L^2}{n^3(\pi\omega_0^2)} \frac{\sin^2(\Delta kL/2)}{(\Delta kL/2)^2} \tag{16.8-2}$$

If a nonlinear crystal is placed within the optical resonator of a laser oscillating at ω, it constitutes, according to (16.8-1), a loss mechanism with a fractional loss per pass of $T_{\text{eff}} = qP^{(\omega)}$.[5]

Consider now a laser with nearly 100% reflectivity at ω pumped so that the unsaturated gain per pass is g_0. Next, we place a nonlinear crystal inside the resonator and gradually increase the coupling constant q. As q is increased gradually, $P(\omega)$—the one-way fundamental power flow inside the laser—decreases. At a value of q such that

$$[qP^{(\omega)}]_{\text{opt}} = \sqrt{g_0 L_i} - L_i \tag{16.8-3}$$

the laser is optimally coupled in the sense discussed in Section 9.3. The total available power of the laser is now converted to 2ω and is given by (9.3-18) as

$$[P^{(2\omega)}]_{\text{opt}} = 2I_s A(\sqrt{g_0} - \sqrt{L_i})^2 \tag{16.8-4}$$

We can generalize the above result by stating that any coupling mechanism characterized by a fractional loss per pass of $T_{\text{eff}} = q[P^{(\omega)}]^n$, $n = 0, 1, 2, \ldots$ can be used to extract the optimum (available) laser power when $q[P^{(\omega)}]^n = \sqrt{g_0 L_i} - L_i$.

To derive an expression for the value of q, which yields optimum coupling, we use (16.8-3), (16.8-4), and (16.8-1) to obtain

$$[P^{(\omega)}]_{\text{opt}} = \frac{[P^{(2\omega)}]_{\text{opt}}}{[qP^{(\omega)}]_{\text{opt}}} = 2I_s A\left[\sqrt{\frac{g_0}{L_i}} - 1\right]$$

so that

$$q_{\text{opt}} = \frac{[qP^{(\omega)}]_{\text{opt}}}{[P^{(\omega)}]_{\text{opt}}} = \frac{\sqrt{g_0 L_i} - L_i}{2I_s A\left[\sqrt{\dfrac{g_0}{L_i}} - 1\right]} = \frac{L_i}{2I_s A} \tag{16.8-5}$$

[5] We use the symbol T to emphasize the fact that as far as the laser is concerned, the crystal is equivalent to a mirror with a transmission T_{eff}.

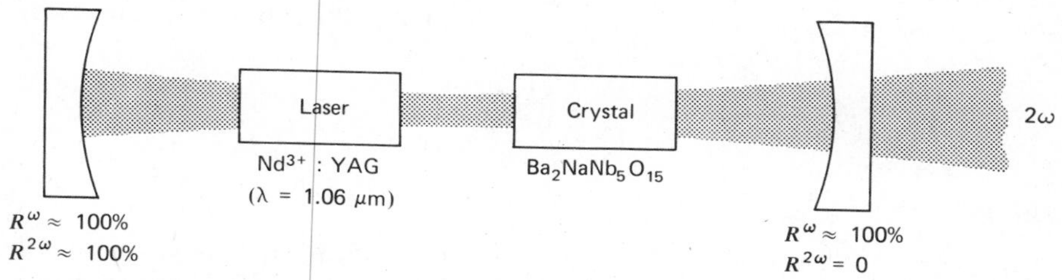

FIGURE 16.10 Typical setup for second-harmonic conversion inside a laser resonator. *Source:* Reference 12.

An experimental arrangement used in internal second-harmonic generation from $\lambda = 1.06\ \mu m$ to $\lambda = 0.53\ \mu m$ is shown in Figure 16.10.

Numerical Example—Internal Second-Harmonic Generation. Consider the problem of designing an internal second-harmonic generator of the type illustrated in Figure 16.10. The Nd^{3+} : YAG laser is assumed to have the following characteristics:

$$\lambda = 1.06\ \mu m = 1.06 \times 10^{-6}\ m$$

$$\Delta\nu = 1.35 \times 10^{11}\ \text{Hz (this is the width of the gain profile)}$$

$$\text{Beam diameter (averaged over entire resonator length)} = 2\ \text{mm}$$

$$L_i = \text{internal loss per pass} = 2 \times 10^{-2}$$

$$n = 1.5$$

The crystal used for second-harmonic generation is $BaNaNb_5O_{15}$, whose second-harmonic coefficient (see Table 16.2) is $d \simeq 1.1 \times 10^{-22}$ MKS units.

Our problem is to calculate the length L of the nonlinear crystal that results in a full conversion of the optimally available fundamental power into the second harmonic at $\lambda = 0.53\ \mu m$. The crystal is assumed to be oriented at the phase-matching condition, so $\Delta k = k^{2\omega} - 2k^{\omega} = 0$.

The optimum coupling parameter is given by (16.8-5) as $q_{opt} = L_i/2I_sA$ where I_s is the saturation intensity defined by (8.7-5). Using the foregoing data in (8.7-5) gives

$$I_sA = 2\ \text{W}$$

that, taking $L_i = 2 \times 10^{-2}$ in (16.8-5) yields

$$q_{opt} = 5 \times 10^{-3}\ (\text{W})^{-1}$$

Next, we use the definition (16.8-2)

$$q = 8\left(\frac{\mu_0}{\varepsilon_0}\right)^{3/2} \frac{\omega^2\, d^2 L^2}{n^3 \pi \omega_0^2}$$

where we put $\Delta k = 0$. We take ω_0, the beam diameter at the crystal, as 70 μm. (The crystal can be placed near a beam waist so the diameter is a minimum.)

Equating the last expression to $q_{opt} = 5 \times 10^{-3}$ using the numerical data given above and solving for the crystal length result in

$$L_{opt} = 0.285 \text{ cm}$$

References

1. Franken, P. A., A. E. Hill, C. W. Peters, and G. Weinreich, "Generation of Optical Harmonics," *Phys. Rev. Letters* **7,** 118 (1961).
2. See, for example, J. F. Nye, *Physical Properties of Crystals* (New York: Oxford, 1957). A difference of a factor X2 in some coefficients is due to a different definition in our case.
3. Kleinman, D. A., "Nonlinear Dielectric Polarization in Optical Media," *Phys. Rev.* **126,** 1977 (1962).
4. Garret, C. G. B. and F. N. H. Robinson, "Miller's Phenomenological Rule for Computing Nonlinear Susceptibilities," *IEEE J. Quant. Elect.* **QE-2,** 328 (1966).
5. Garrett, C. G. B., "Nonlinear Optics, Anharmonic Oscillators and Pyroelectricity," *J. Quant. Elect.* **QE-4,** 70 (1968).
6. Miller, R. C., "Optical Second Harmonic Generation in Piezoelectric Crystals," *Appl. Phys. Letters* **5,** 17 (1964).
7. Maker, P. D., R. W. Terhune, M. Nisenhoff, and C. M. Savage, "Effects of Dispersion and Focusing on the Production of Optical Harmonics," *Phys. Rev. Letters* **8,** 21 (1962).
8. Giordmaine, J. A., "Mixing of Light Beams in Crystals," *Phys. Rev. Letters* **8,** 19 (1962).
9. Zernike, F., Jr., "Refractive Indices of Ammonium Dihydrogen Phosphate and Potassium Dihydrogen Phosphate Between 2000 Å and 1.5 μm," *J. Opt. Soc. Amer.* **54,** 1215 (1964).
10. Ashkin, A., G. D. Boyd, and J. M. Dziedzic, "Observation of Continuous Second Harmonic Generation with Gas Lasers," *Phys. Rev. Letters* **11,** 14 (1963).
11. G. D. Boyd and D. A. Kleinman, "Parametric Interaction of Focused Gaussian Light Beams," *J. Appl. Phys.* **39,** (1968).
12. Armstrong, J. A., N. Bloembergen, J. Ducuing, and P. S. Pershan, "Interactions Between Light Waves in a Nonlinear Dielectric," *Phys. Rev.* **127,** 1918 (1962).
13. Geusic, J. E., H. J. Levinstein, S. Singh, R. G. Smith, and L. G. Van Uitert, "Continuous 0.53 μm Solid-State Source Using $Ba_2NaNb_5O_{15}$," *IEEE J. Quant. Elect.* **QE-4,** 352 (1968).
14. Patel, C. K. N., "Efficient Phase-Matched Harmonic Generation in Tellurium with a CO_2 Laser at 10.6 μ," *Phys. Rev. Letters* **15,** 1027 (1965).
15. Caldwell, R. W., H. Y. Fan, "Optical Properties of Tellurium and Selenium," *Phys. Rev.* **114,** 664 (1959).

Supplementary References

1. Bloembergen, N., *Nonlinear Optics* (New York: W. A. Benjamin, 1965). (Especially for quantum mechanical derivation of nonlinear coefficients.)
2. Levine, B. F., "Magnitude and Dispersion of Kleinman's Forbidden Nonlinear Coefficients," *IEEE J. Quant. Elect.* **QE-9,** 1946 (1973).
3. Levine, B. F. and C. G. Bethea, "Nonlinear Susceptibility of GaP," *Appl. Phys. Letters* **20,** 272 (1972).

4. Kildal, H. and J. C. Mikkelsen, "The Nonlinear Optical Coefficient, Phase Matching and Optical Damage in $AgGaSe_2$," *Optics Comm.* **9,** 315 (1973).
5. Miles, R. B. and S. E. Harris, "Optical Third Harmonic Generation in Alkali Metal Vapors," *IEEE J. Quant. Elect.* **QE-9,** 470 (1973).
6. Hodgson, R. T., P. P. Sorokin, and J. J. Wynne, "Tunable Coherent Vacuum Ultraviolet Generation in Atomic Vapors," *Phys. Rev. Letters* **32,** 343 (1974).
7. Seka, W., S. D. Jacobs, J. E. Rizzo, R. Boni, and R. S. Craxton, "Demonstration of High Efficiency Third Harmonic Conversion of High Power Nd-Glass Laser Radiation," *Optics Comm.* **34,** 469 (1980). Also, Craxton, R. S., "High Efficiency Frequency Tripling Schemes for High Power Nd: Glass Lasers," *IEEE J. Quant. Elect.* **QE-17,** 177 (1981). (Additional articles on doubling and frequency conversion are to be found in the same issue.)
8. Holzrichter, J. F., D. Eimerl, E. V. George, J. B. Trenholme, W. W. Simmons, and J. T. Hunt, *Physics of Laser Fusion,* Vol. III. (Lawrence Livermore National Laboratory, 1982). For availability, see Supplementary Reference 3.
9. George, E. V., ed., *1981 Laser Program Annual Report—Lawrence Livermore Laboratory,* Chapter 7. (Available from the National Technical Information Service, U.S. Department of Commerce, Springfield, Va.)
10. Yariv, A. and P. Yeh, *Optical Waves in Crystals* (New York: Wiley, 1984).

Problems

16.1 Derive the second-harmonic generation index-matching angle for a positive uniaxial crystal (i.e., a crystal with $n_e > n_0$).

Answer:

$$\sin^2 \theta_m = \frac{(n_0^{\omega})^{-2} - (n_0^{2\omega})^{-2}}{(n_0^{\omega})^{-2} - (n_e^{\omega})^{-2}}$$

16.2 Design a second-harmonic generation experiment in Te using an input at $\lambda = 10.6\ \mu m$ (Te belongs to point-group symmetry 32). Find the index-matching angle θ_m and decide on the proper beam polarization and crystal orientation for maximum power output at 5.3 μ. Compare your results with that of Reference 14, footnote 10.

Dispersion of Te

$\lambda(\mu)$	n_0	n_e
4	4.929	6.372
5	4.864	6.316
6	4.838	6.286
7	4.821	6.257
8	4.809	6.253
10	4.796	6.246
12	4.789	6.237
14	4.785	6.230

Source: Data taken from Reference 15.

16.3 Derive (16.3-14).

$$V(X) = \frac{e^2}{4\pi\varepsilon_0 r_0}\left(5.83 + 24.1\frac{X^2}{r_0^2} - 13.3\frac{X^3}{r_0^3} + \cdots\right)$$

Hint:
Recall that X is the deviation from the potential minimum.

16.4 Show that, if applicable, the Kleinman conditions preclude high-frequency (i.e., optical) nonlinear polarization in crystals with point symmetry D_4 and D_6. Would you expect these crystals to possess a linear electrooptic effect? Show that the Kleinman's condition requires that in quartz $d_{123} = 0$.

16.5 Show that Kleinman's conditions can be derived directly from the symmetry condition $d_{ijk} = d_{ikj}$ together with the relation

$$\mathbf{P} = -\nabla_E U$$

16.6 Show that the scalar nonlinear susceptibility $d_{NL}^{(\omega_1+\omega_2)}$ defined by

$$P^{(\omega_1+\omega_2)} = \tfrac{1}{2}[d_{NL}^{(\omega_1+\omega_2)} E_1^{(\omega_1)} E_2^{(\omega_2)} e^{i(\omega_1+\omega_2)t} + \text{c.c.}]$$

is given by

$$d_{NL}^{(\omega_1+\omega_2)} = \frac{-DNe^3}{2m^2[(\omega_0^2-\omega_1^2)+i\omega_1\gamma][(\omega_0^2-\omega_2^2)+i\omega_2\gamma][\omega_0^2-(\omega_1+\omega_2)^2+i(\omega_1+\omega_2)\gamma]}$$

16.7 Show that if θ_m is the phase-matching angle for an ordinary wave at ω and an extraordinary wave at 2ω, then

$$\Delta k(\theta)L\big|_{\theta=\theta_m} = -\frac{2\omega L}{c}\sin(2\theta_m)\frac{(n_e^{2\omega})^{-2}-(n_0^{2\omega})^{-2}}{2(n_0^{\omega})^{-3}}(\theta-\theta_m)$$

16.8 Derive the expression for the phase-matching angle of a parametric amplifier using KDP in which two of the waves are extraordinary, whereas the third is ordinary. Which of the three waves (i.e., signal, idler, or pump) would you choose as ordinary? Can this type of phase matching be accomplished with $\omega_3 =$ 10,000 cm^{-1}, $\omega_1 = \omega_2 = 5000$ cm^{-1}? If so, what is θ_m?

16.9 Show that the coefficients d'_{ijk} in (16.4-11) are related to $d_{\alpha\beta\gamma}$ as defined by (16.1-5) by

$$d'_{ijk} = l_{i\alpha} l_{j\beta} l_{k\gamma} d_{\alpha\beta\gamma} \qquad (1)$$

where i, j, k is the coordinate system used in describing the propagation [see (16.4-6)] and α, β, γ refer to the crystal axes. $l_{i\alpha}$ are the direction cosines. Show that in the case of KDP, (see example on p. 397) the application of (1) results in $d'_{xyy} = d_{36}\sin\theta_m$.

16.10 (a) Show that Eq. (16.1-4) is a consequence of relation (16.1-6).
(b) Show, starting from (16.1-6) that in the case of second harmonic generation ($\omega_1 = \omega_2 \equiv \omega$) Eq. (16.1-4) becomes

$$P_i^{2\omega} = d_{ijk}^{2\omega=\omega+\omega} E_j^{\omega} E_k^{\omega}$$

CHAPTER 17

Parametric Amplification, Oscillation, and Fluorescence

17.0 INTRODUCTION AND LUMPED CIRCUIT ANALOG

The optical nonlinearity responsible for second harmonic generation can also be used to amplify weak optical signals. The basic configuration involves an input "signal" at ω_1 that is incident on a nonlinear optical crystal together with an intense pump wave at ω_3 where $\omega_3 > \omega_1$ (References 1, 2, 3). The amplification of the ω_1 wave is accompanied by a generation of an "idler" wave at $\omega_2 = \omega_3 - \omega_1$.

Microwave devices based on this principle are playing an important role as low-noise amplifiers (References 4, 5).

Optical parametric amplification in its simplest form involves the transfer of power from a "pump" wave at ω_3 to waves at frequencies ω_1 and ω_2, where $\omega_3 = \omega_1 + \omega_2$. It is fundamentally similar to the case of second-harmonic generation treated in Section 16.5. The only difference is in the direction of power flow. In second-harmonic generation, power is fed from the low-frequency optical field at ω to the field at 2ω. In parametric amplification, power flow is from the high-frequency field (ω_3) to the low-frequency fields at ω_1 and ω_2. In the special case where $\omega_1 = \omega_2$ we have the exact reverse of second-harmonic generation. This is the case of the so-called degenerate parametric amplification.

Before we embark on a detailed analysis of the optical case it may be worthwhile to review some of the low-frequency beginnings of parametric oscillation.

Consider a classical nondriven oscillator whose equation of motion is given by

$$\frac{d^2v}{dt^2} + \kappa \frac{dv}{dt} + \omega_0^2 v = 0 \qquad (17.0\text{-}1)$$

The variable v corresond to the excursion of a mass M, which is connected to a spring with a constant $\omega_0^2 M$, or to the voltage across a parallel RLC circuit, in which case $\omega_0^2 = (LC)^{-1}$ and $\kappa = (RC)^{-1}$. The solution of (17.0-1) is

$$v(t) = v(0) \exp\left(-\frac{\kappa t}{2}\right) \exp\left(\pm i \sqrt{\omega_0^2 - \frac{\kappa^2}{4}}\, t\right) \qquad (17.0\text{-}2)$$

that is, a damped sinusoid.

In 1883 Lord Rayleigh [3], investigating parasitic resonances in pipe organs, considered the consequences of the following equation

$$\frac{d^2v}{dt^2} + \kappa\frac{dv}{dt} + (\omega_0^2 + 2\alpha \sin \omega_p t)v = 0 \tag{17.0-3}$$

This equation may describe an oscillator in which an energy storage parameter (mass or spring constant in the mechanical oscillator, L or C in the RLC oscillator) is modulated at a frequency ω_p.

We will consider in what follows the case of an oscillator whose capacitance is modulated according to

$$C = C_0\left(1 - \frac{\Delta C}{C_0}\sin \omega_p t\right) \tag{17.0-4}$$

The equation of the voltage across the RLC circuit is given by (17.0-1) with $\omega_0^2 = (LC)^{-1}$.

Assuming $\Delta C \ll C_0$ (17.0-1) becomes

$$\frac{d^2v}{dt^2} + \kappa\frac{dv}{dt} + \frac{1}{LC_0}\left(1 + \frac{\Delta C}{C_0}\sin \omega_p t\right)v = 0 \tag{17.0-5}$$

which, if we make the identification

$$\omega_0^2 = \frac{1}{LC_0}, \qquad \alpha = \frac{\omega_0^2 \Delta C}{2C_0} \tag{17.0-6}$$

is identical to (8.6-3).

To bring out the essential features of the parametrically driven oscillator described by (17.0-5) assume that

$$v(t) = Re[2ae^{i(\omega t + \phi)}e^{st}] \tag{17.0-7}$$

where s and a are real. We shall inquire under what conditions can the signal $v(t)$ be amplified ($s > 0$) by the parametric pumping. Substituting (17.0-7) in 17.0-5) leads to

$$\begin{aligned} &-(\omega - is)^2 e^{i[(\omega - is)t + \phi]} + i\kappa(\omega - is)e^{i[(\omega - is)t+\phi]} \\ &+ \omega_0^2 e^{i[(\omega - is)t+\phi]} + \text{c.c.} \\ &= -\frac{\omega_0^2 \Delta C}{C_0}\frac{(e^{i\omega_p t} - e^{-i\omega_p t})}{2i}\{e^{i[(\omega - is)t+\phi]} + e^{-i[(\omega + is)t+\phi]}\} \end{aligned} \tag{17.0-8}$$

A necessary condition for the satisfaction of (17.0-8) is that for each exponent on the left side, there be a corresponding equal exponent on the right side. This can only happen if

$$\begin{aligned} \omega_p - \omega &= \omega \qquad (\omega_p = 2\omega) \\ e^{i\phi} &= e^{-i\phi} \qquad (\phi = 0 \text{ or } \pi) \end{aligned} \tag{17.0-9}$$

In words: the pump frequency needs to be twice that of the signal while the pump phase (ϕ) relative to that of the signal $v(t)$ must be 0 or π. This is the so-called degenerate parametric amplifier (or oscillator). Nondegenerate operation, which involves amplification by a pump field at ω_p, simultaneously, of two different frequencies ω_1 and ω_2 where $\omega_p = \omega_1 + \omega_2$, will be discussed in Section 17.2. We can equate the sum of imaginary terms with the same exponential factors on both sides of 17.0-8 obtaining (for $\omega \simeq \omega_0$)

$$2s = \frac{\omega_0 \Delta C}{2C_0} - \kappa \tag{17.0-10}$$

showing that sustained growth ($s > 0$) can occur provided

$$\frac{\omega_0 \Delta C}{2C_0} > \kappa \tag{17.0-11}$$

If we replace the > sign by an equal sign we have the threshold condition for sustained oscillation at $\omega = \omega_p/2$. It is clear that in this case the oscillation signal will possess a phase of 0 or π, since this is the phase corresponding to maximum gain. An electromechanical explanation of the significance of the phase condition in (17.0-9) is given in Ref. 41.

If we equate the sum of the real coefficients in (17.0-8) to zero we obtain

$$\omega^2 = \omega_0^2 + s^2$$

so that at steady state, $s = 0$, $\omega = \omega_0$.

It follows straightforwardly from the above that an input signal, which is in phase quadrature (i.e., shifted by $\pi/2$) with the amplified signal, *will* be attenuated with a temporal rate $2s = -\kappa - \Delta C/2C_0$. This follows analytically from the fact that a phase shift of $\Delta\phi = \pi/2$ in the signal is equivalent to a change of π in the (double frequency) pump phase, which inverts the sign of the right side (17.0-8) and thus of ΔC in (17.0-10).

The degenerate parametric amplifier thus acts as a *phase sensitive amplifier* providing gain to sympathetic signals ($\phi = 0, \pi$) and deamplifying signals with $\phi = \pi/2, 3\pi/2$. This phase sensitivity is at the heart of the recent noise "squeezing" experiments that are discussed in Section 17.9.

17.1 THE BASIC EQUATIONS OF PARAMETRIC AMPLIFICATION

In the nondepleted pump approximation, we take $A_3(z) = A_3(0)$ and write the first two equations of (16.6-3) as

$$\begin{aligned}\frac{dA_1}{dz} &= -\tfrac{1}{2}\alpha_1 A_1 - igA_2^* e^{-i\,1/2\Delta kz}\\ \frac{dA_2^*}{dz} &= -\tfrac{1}{2}\alpha_2 A_2^* + igA_1 e^{i\,1/2\Delta kz}\end{aligned} \tag{17.1-1}$$

$$\begin{aligned}2\Delta k &= k_3 - k_1 - k_2\\ g = \kappa A_3(0) &= 2\sqrt{\left(\frac{\mu_0}{\varepsilon_0}\right)\frac{\omega_1\omega_2}{n_1 n_2}}\, d' E_3(0)\end{aligned} \tag{17.1-2}$$

Let us consider the general case in which both "signal" and "idler" waves with (complex) amplitudes $A_1(0)$ and $A_2(0)$, respectively, are present at the input. The solution of (17.1-1) with $\alpha_i = 0$ (no losses) is

$$A_1(z)e^{i(\Delta kz)} = A_1(0)\left[\cosh(bz) + \frac{i\,\Delta k}{b}\sinh(bz)\right] - i\frac{g}{b}A_2^*(0)\sinh(bz) \qquad (17.1\text{-}3)$$

$$A_2^*(z)e^{-i(\Delta kz)} = A_2^*(0)\left[\cosh(bz) - \frac{i\,\Delta k}{b}\sinh(bz)\right] + i\frac{g}{b}A_1(0)\sinh(bz) \qquad (17.1\text{-}4)$$

where

$$b = \sqrt{g^2 - (\Delta k)^2} \qquad (17.1\text{-}5)$$

$$2\Delta k = k_3 - k_1 - k_2 \qquad (17.1\text{-}6)$$

In the simple case of a parametric amplifier, we have a single input, for example, $A_1(0)$. Putting $A_2(0) = 0$ and considering for simplicity the phase matched case $\Delta k = 0$, we obtain, from (17.1-3),

$$A_1(z) = A_1(0)\cosh\frac{gz}{2}$$
$$A_2^*(z) = iA_1(0)\sinh\frac{gz}{2} \qquad (17.1\text{-}7)$$

The increase in power of the signal (ω_1) and idler (ω_2) waves is at the expense of the pump (ω_3) wave. As a matter of fact, we can show by using (16.6-3) that when $\alpha_i = 0$,

$$-\frac{d}{dz}(A_3A_3^*) = \frac{d}{dz}(A_1A_1^*) = \frac{d}{dz}(A_2A_2^*) \qquad (17.1\text{-}8)$$

Since $A_iA_i^*$ is proportional to the photon flux at ω_i, Eq. (17.1-8) is a statement of the fact that for each photon added to the signal wave (ω_1), one photon is added to the idler wave (ω_2) and one photon is removed from the pump wave (ω_3). Since $\omega_3 = \omega_1 + \omega_2$, energy is conserved. Relation (17.1-8) can be extended by integration to the whole interaction volume, in which case the changes in total power between the input and output planes are related by

$$-\Delta\left(\frac{P_3}{\omega_3}\right) = \Delta\left(\frac{P_1}{\omega_1}\right) = \Delta\left(\frac{P_2}{\omega_2}\right) \qquad (17.1\text{-}9)$$

where P represents beam power. Equation (17.1-9) is known as the Manley–Rowe relation (Reference 6).

Numerical Example: Parametric Amplification. To obtain an appreciation for the magnitude of the gain available in parametric amplification, consider the case of a L_iNbO_3 crystal pumped by a traveling pump. Using the following data:

$$\nu_1 = \nu_2 = 3 \times 10^{14} \text{ Hz} \qquad (\lambda_1 = \lambda_2 = 1 \ \mu\text{m})$$

$$\nu_3 = 6 \times 10^{14} \text{ Hz}$$

$$d_{15} \cong 0.5 \times 10^{-22} \qquad \text{(see Table 16.2)}$$

$$n_1 = n_2 \approx 2.2, \qquad d' = 0.5\, d_{15}$$

$$P_{3/\text{Area}} = 5 \times 10^6 \text{ W/cm}^2, \qquad E_3 = 4.13 \times 10^6 \text{ v/m}$$

in (17.1-2) gives

$$g = 0.2 \text{ cm}^{-1}$$

This shows that even at high densities of pump power, the amount of parametric gain is modest. It is for this reason that the effect is used primarily to obtain oscillation rather than as a means of amplification.

17.2 PARAMETRIC OSCILLATION

In the last section it was shown that a pump wave at ω_3 can provide, via interaction in a nonlinear crystal, simultaneous amplification for optical waves at ω_1 and ω_2 such that $\omega_3 = \omega_1 + \omega_2$. If the nonlinear crystal is placed within an optical resonator that provides resonances for the signal or idler waves (or both), the parametric gain will, at some threshold pumping intensity, cause simultaneous oscillation at both the signal and idler frequencies. The threshold for this oscillation corresponds to the point at which the parametric gain just balances the losses of the signal and idler waves. This is the physical basis of the optical parametric oscillator (References 1, 2, and 3). Its practical importance derives from its ability to convert the power output of a pump laser to a coherent output at the signal and idler frequencies that, as will be shown below, can be tuned continuously over large ranges.

A doubly resonant parametric oscillator, that is, one where both the signal and idler modes resonate (possess high Q), is shown in Figure 17.1.

Before embarking on a rigorous analysis of parametric oscillation, we will consider an extremely simple point of view that is helpful in illustrating the basic nature of the interaction. We start with (17.1-1) and take the fields $A_1(z)$ and $A_2(z)$ as being those of the parametric oscillator sketched in Figure 17.1. The distributed loss constants α_1 and α_2 now account for the mirror losses. This is possible when the loss per pass is small. We then have

$$\alpha_i l = 1 - R_i \qquad (i = 1, 2) \tag{17.2-1}$$

where l is the length of the resonator. In a steady-state oscillation, $dA_1/dz =$

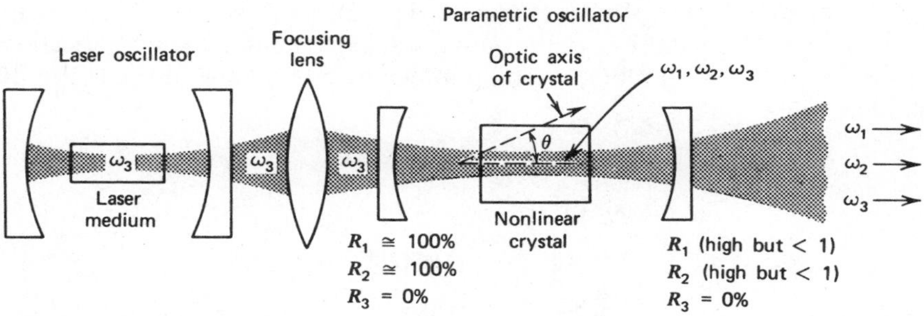

FIGURE 17.1 Schematic diagram of an optical parametric oscillator in which the laser output at ω_3 is used as the pump. The resulting gain gives rise to oscillations at ω_1 and ω_2 (where $\omega_3 = \omega_1 + \omega_2$) in an optical cavity that contains the nonlinear crystal and resonates at ω_1 and ω_2.

$dA_2/dz = 0$ and (17.1-1) becomes

$$-\frac{\alpha_1}{2} A_1 - igA_2^* = 0$$
$$igA_1 - \frac{\alpha_2}{2} A_2^* = 0 \tag{17.2-2}$$

Nontrivial solutions for A_1 and A_2 thus exist if the determinant of the coefficients in (17.2-2) vanishes, that is, when

$$g_t^2 = \alpha_1\alpha_2/4 \tag{17.2-3}$$

We will now consider the same problem on a more rigorous fashion.

Self-Consistent Analysis of Parametric Oscillation

In this section we derive the equations governing steady-state oscillation in parametric oscillators, employing the self-consistent approach previously used in Section 9.1 to describe laser oscillation and in Section 11.3 in the treatment of mode locking.

The basic model is shown in Figure 17.2. We assume, for simplicity, a nonlinear crystal shaped as an optical resonator whose curved ends present reflectances[1] r_1 and r_2, respectively, to the signal and idler fields while being transparent to the pump field. The combined idler-signal field at an arbitrary plane z will be described by the column "vector"

$$\tilde{\mathbf{A}}(z) = \begin{vmatrix} A_1(z)e^{-ik_1z} \\ A_2^*(z)e^{ik_2z} \end{vmatrix} \tag{17.2-4}$$

where $k_i = (\omega_i/c)n_i$. The propagation of $\tilde{A}(z)$ through a nonlinear crystal of

[1] The complex field reflection coefficients are r_1 and r_2. They are related to the mirror reflectivities by $|r_i|^2 = R_i$.

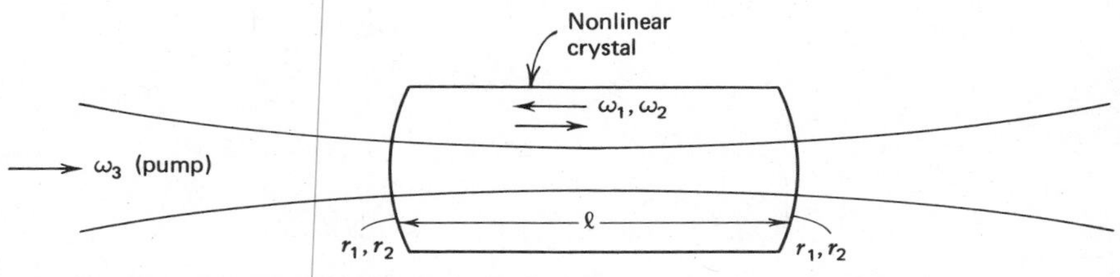

FIGURE 17.2 A crystal parametric oscillator.

length l is described, according to (17.1-34), by a matrix product

$$\tilde{\mathbf{A}}(l) = \begin{vmatrix} e^{-i[k_1+(\Delta k)]l}\left[\cosh(bl) - \dfrac{i\,\Delta k}{b}\sinh(bl)\right] & -ie^{-i[k_1+(\Delta k)]l}\left(\dfrac{g}{b}\right)\sinh(bl) \\ ie^{i[k_2+(\Delta k)]l}\left(\dfrac{g}{b}\right)\sinh(bl) & e^{i[k_2+(\Delta k)]l}\left[\cosh(bl) - \dfrac{i\,\Delta k}{b}\sinh(bl)\right] \end{vmatrix} \tilde{\mathbf{A}}(0) \tag{17.2-5}$$

We now require that the vector $\tilde{\mathbf{A}}(z)$ reproduce itself after one complete round trip inside the resonator. If we use Figure 17.3, this condition is

$$\tilde{\mathbf{A}}_e = \tilde{\mathbf{A}}_a \tag{17.2-6}$$

$\tilde{\mathbf{A}}_e$ is obtained from $\tilde{\mathbf{A}}_a$ by multiplying the latter by four matrices: one accounting for reflection at the left mirror, one for propagation from right to left for which there is no parametric gain, one for reflection at the right mirror, and one matrix, given by (17.2-5), for the pass from left to right.

If we assume a phase-matched $\Delta k = 0$ operation, a complete round trip is thus described by

$$\tilde{\mathbf{A}}_e = \begin{vmatrix} r_1 & 0 \\ 0 & r_2^* \end{vmatrix} \begin{vmatrix} e^{-ik_1l} & 0 \\ 0 & e^{ik_2l} \end{vmatrix} \begin{vmatrix} r_1 & 0 \\ 0 & r_2^* \end{vmatrix} \times \begin{vmatrix} \cosh(gl)e^{-ik_1l} & -i\sinh(gl)e^{-ik_1l} \\ i\sinh(gl)e^{ik_2l} & \cosh(gl)e^{ik_2l} \end{vmatrix} \tilde{\mathbf{A}}_a \tag{17.2-7}$$

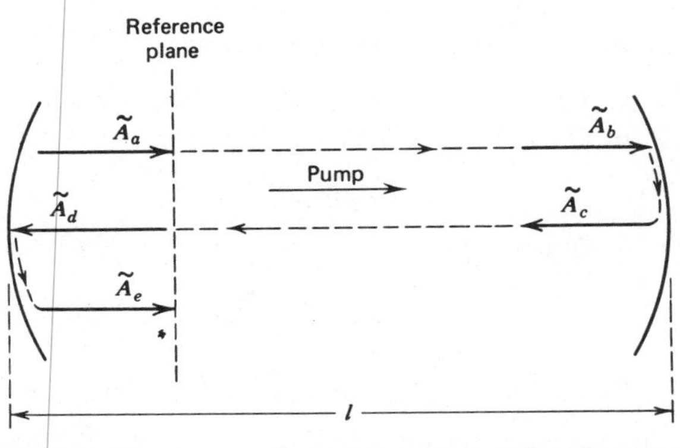

FIGURE 17.3 The round-trip propagation of the idler-signal vector $\tilde{A}$ is used in deriving the oscillation condition (17.2-12).

or

$$\tilde{\mathbf{A}}_e = \tilde{\tilde{\mathbf{M}}}\tilde{\mathbf{A}}_a \tag{17.2-8}$$

where

$$\tilde{\tilde{\mathbf{M}}} = \begin{vmatrix} r_1^2 \cosh(gl)e^{-i2k_1l} & -ir_1^2 \sinh(gl)e^{-i2k_1l} \\ i(r_2^*)^2 \sinh(gl)e^{i2k_2l} & (r_2^*)^2 \cosh(gl)e^{i2k_2l} \end{vmatrix} \tag{17.2-9}$$

The self-consistency condition (17.2-6) can now be written as

$$\tilde{\mathbf{A}}_a = \tilde{\tilde{\mathbf{M}}}\tilde{\mathbf{A}}_a \tag{17.2-10}$$

so that for a nontrivial $\tilde{\mathbf{A}}_a$

$$\det|\tilde{\tilde{\mathbf{M}}} - \tilde{\tilde{\mathbf{I}}}| = 0 \tag{17.2-11}$$

or using (17.2-9) yields

$$[r_1^2 \cosh(gl)e^{-i2k_1l} - 1][(r_2^*)^2 \cosh(gl)e^{i2k_2l} - 1] = r_1^2(r_2^*)^2 \sinh^2(gl)e^{i2(k_2-k_1)l} \tag{17.2-12}$$

This is the *oscillation condition* of the parametric oscillator.

An inspection of (17.2-12) reveals that the minimum threshold gain g_t will result when each of the two factors on the left side of (17.2-12) is a *real positive* number. For this to happen, the following relations must hold:

$$\begin{aligned} -\phi_1 + 2k_1 l &= 2m\pi \\ -\phi_2 + 2k_2 l &= 2s\pi \end{aligned} \tag{17.2-13}$$

where m and s are two integers and

$$(r_{1,2})^2 = R_{1,2} \exp(i2\phi_{1,2}) \tag{17.2-14}$$

Conditions (17.2-13) are equivalent to stating that the signal (ω_1) and idler (ω_2) oscillation frequencies must correspond to two longitudinal modes of the optical resonator.

We will next use (17.2-12) to derive the threshold conditions for two important classes of parametric oscillators.

The Doubly Resonant Parametric Oscillator

In this case, the configuration provides high Q resonances to both ω_1 and ω_2. Using (17.2-13) in (17.2-12) leads to

$$(R_1 + R_2) \cosh(g_t) - R_1 R_2 = 1 \tag{17.2-15}$$

For high-reflectivity mirrors, $R_1, R_2 \approx 1$, $\cosh(g_t) \approx 1 + g_t^2 l^2/2$ and (17.2-15) becomes

$$(g_t l) = \sqrt{(1 - R_1)(1 - R_2)} \tag{17.2-16}$$

If we use (17.1-2) and express the pump field E_3 in terms of the intensity

$$I_3 = \frac{1}{2}\sqrt{\frac{\varepsilon_0 n_3^2}{\mu_0}}\, E_3^2$$

Equation (17.2-16) becomes

$$I_{3t} = 8\left(\frac{\varepsilon_0}{\mu_0}\right)^{3/2} \frac{n_1 n_2 n_3 (1 - R_1)(1 - R_2)}{\omega_1 \omega_2 l^2 d^2} \tag{17.2-17}$$

Example: Parametric Oscillation Threshold. Let us estimate the threshold pump requirement of a parametric oscillator of the kind shown in Figure 17.3 that utilizes a $LiNbO_3$ crystal. We use the following set of parameters:

$$(1 - R_1) = (1 - R_2) = 2 \times 10^{-2} \qquad \text{(i.e., total loss per pass at } \omega_1 \text{ and } \omega_2 = 2\%)$$

$$\lambda_1 = \lambda_2 = 1\ \mu\text{m}$$

$$l(\text{crystal length}) = 1\ \text{cm}$$

$$n_1 = n_2 = n_3 = 1.5$$

$$d_{311}(\text{LiNbO}_3) = 5 \times 10^{-23} \qquad \text{(MKS)}$$

Substitution in (17.2-17) yields

$$I_{3t} = 3.54 \times 10^2\ \text{W/cm}^2$$

This is an easily achievable intensity even on a continuous basis so that the example helps us appreciate the attractiveness of optical parametric oscillation as a means for generating coherent optical radiation at new optical frequencies.

The need to satisfy simultaneously the conditions

$$\omega_3 = \omega_1 + \omega_2$$

and (17.2-13)

$$\frac{\omega_1 n_1 l}{c} = m\pi + \phi_1$$
$$\frac{\omega_2 n_2 l}{c} = s\pi + \phi_2 \tag{17.2-18}$$

places a severe requirement on the stability of the doubly resonant parametric oscillator. To illustrate the point, let us assume that the above conditions are satisfied for some pair ω_1 and ω_2 (ω_3 is fixed). Now if l changes slightly by dl due to vibration or a temperature drift, then the change in ω_1 and ω_2 needed to keep (17.2-18) satisfied is

$$\frac{d\omega_1}{\omega_1} = \frac{d\omega_2}{\omega_2} = -\frac{dl}{l}$$

But if this change takes place, the condition $\omega_3 = \omega_1 + \omega_2$ is no longer satisfied. This situation is depicted in Figure 17.4. From this figure, we find

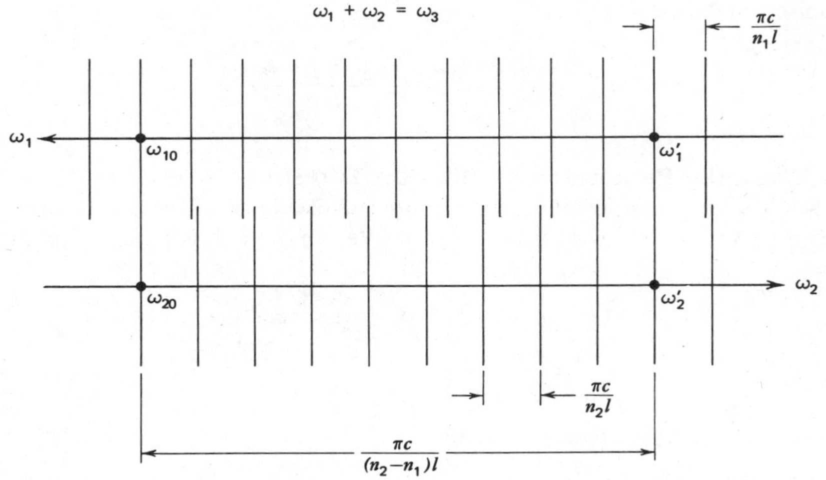

FIGURE 17.4 A construction illustrating how the oscillation phase conditions (17.2-18) can be satisfied simultaneously at both (ω_{10}, ω_{20}) and at (ω_1', ω_2'). Note that the ω_1 scale and ω_2 scale increase in opposite directions so that at any position on this diagram $\omega_1 + \omega_2 = \omega_3$. The vertical lines correspond to the individual longitudinal resonances of the idler and signal frequencies.

that approximately a distance

$$\Delta\omega = \frac{\pi c/l}{n_2 - n_1} \tag{17.2-19}$$

from the first set of frequencies there exists another set (ω_1', ω_2') where the above condition is satisfied. The oscillator will thus react to small changes of length by large frequency variations of the order of $\Delta\omega$. Since $n_2 - n_1$ is typically $\sim 10^{-2} - 10^{-1}$, the frequency fluctuations may correspond to 10 to 100 times the longitudinal frequency mode spacing $\pi c/nl$ (Reference 1).

The Singly Resonant Parametric Oscillator

In the singly resonant oscillator (References 7 and 8), only one of the two frequencies (e.g., ω_1) is reflected back, whereas the idler wave (ω_2) propagates in one direction only. A typical setup is shown in Figure 17.5. In this experiment, one resorts to noncollinear phase matching to separate the signal and idler beams. In this kind of phase matching, each of the three beams propagates in a different direction and the phase matching condition is vectorial

$$\mathbf{k}_3 = \mathbf{k}_1 + \mathbf{k}_2 \tag{17.2-20}$$

in a manner reminescent of the phase matching condition of Bragg scattering of light from an acoustic wave (Section 14.10). The direction of $\mathbf{k}_3$ is that of the input pump beam. The direction of $\mathbf{k}_1$ is fixed by the axis of the resonator. The magnitude of $\mathbf{k}_1$ and the magnitude and direction of $\mathbf{k}_2$ adjust themselves in such a way that both (17.2-20) and the energy conservation condition

$$\omega_3 = \omega_1 + \omega_2$$

are satisfied.

The threshold condition of the singly resonant oscillation is derived from (17.2-12) by putting $r_2 = 0$. The result is

$$r_1^2 \cosh(gl) e^{-i2k_1 l} = 1 \tag{17.2-21}$$

Taking

$$r_1^2 = R_1 e^{i2\phi_1} \tag{17.2-22}$$

we can separate (17.2-21) into a phase condition

$$\phi_1 - k_1 l = m\pi \tag{17.2-23}$$

and an amplitude condition

$$R_1 \cosh(g_t l) = 1 \tag{17.2-24}$$

The phase condition (17.2-23) is the same as (17.2-13) except that no restriction is placed on the phase ϕ_2 of the idler wave so that the frequency-hopping problem of the doubly resonant oscillator discussed above is alleviated.

For $R_1 \approx 1$, the threshold condition (17.2-24) can be rewritten as

$$(g_t l) = \sqrt{2}\sqrt{1 - R_1} \tag{17.2-25}$$

The increase in threshold pumping intensity of the singly resonant case relative to the doubly resonant oscillator with the same value of R_1 is thus

$$\left[\frac{(g_t l)_{\text{singly resonant}}}{(g_t l)_{\text{doubly resonant}}}\right]^2 = \frac{2}{(1 - R_2)} \tag{17.2-26}$$

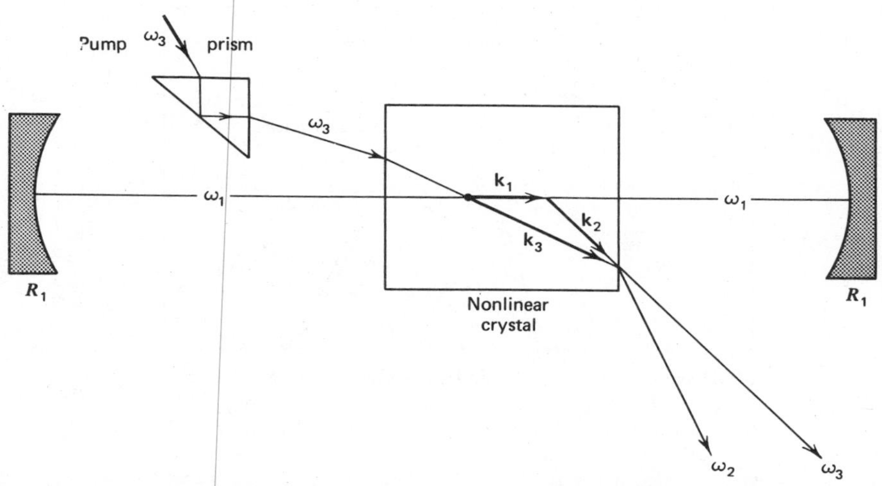

FIGURE 17.5 Singly resonant optical parametric oscillator with noncollinear phase matching. *Source:* Reference 7.

This increase, which for $R_2 \approx 1$ is very large, is not objectionable if sufficient pump power is available to exceed threshold by a considerable factor. Since, like a laser, the excess pumping power goes into the coherent output at ω_1 and ω_2, the high threshold under those conditions does not lead to a sacrifice in efficiency. This point will be made clearer in Section 17.3.

17.3 POWER OUTPUT AND PUMP SATURATION IN PARAMETRIC OSCILLATORS

In the treatment of the laser oscillator in Chapter 9, we found that in the steady state the gain is clamped at its threshold value regardless of the pumping intensity. Increased pumping, which would have otherwise led to a larger inversion (and gain), gives rise to a larger intensity of the laser field that saturates the inversion at its threshold value. A similar phenomenon occurs in parametric oscillation. The pump field E_3 gives rise to amplification with an

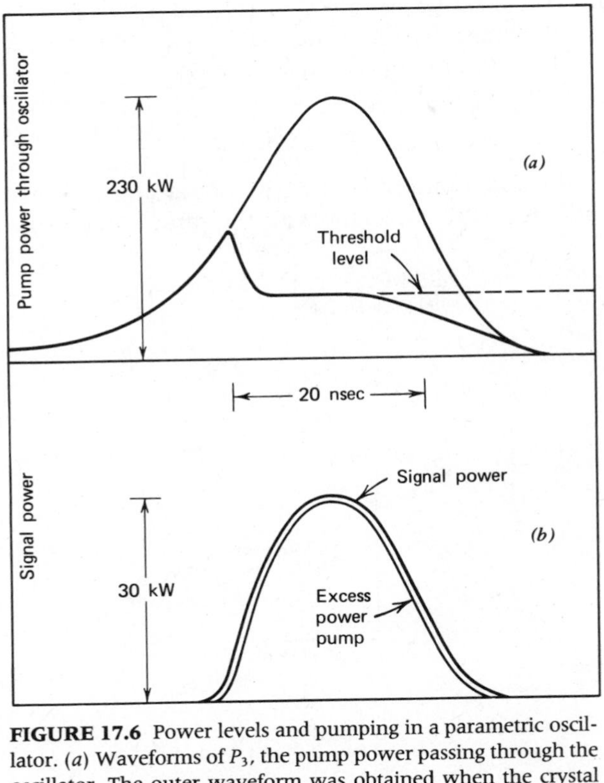

FIGURE 17.6 Power levels and pumping in a parametric oscillator. (*a*) Waveforms of P_3, the pump power passing through the oscillator. The outer waveform was obtained when the crystal was rotated so that oscillation did not occur; the solid waveform was obtained when oscillation took place. (*b*) Signal power and the normalized difference between the waveforms in (*a*). *Source:* Reference 9.

amplification constant given by (17.1-2). At threshold, $g_t^2 l^2 = 4(1 - R_1)(1 - R_2)$. Above threshold, E_3 at steady state must be clamped at its threshold value so that g will not exceed g_t. (If $g > g_t$, the power density at ω_1 and ω_2 must increase with time.) As power is conserved, it follows that any additional pump power input must be diverted into power at the signal and idler fields. Since $\omega_3 = \omega_1 + \omega_2$, it follows that for each input pump photon above threshold, we generate one photon at the signal (ω_1) and one at the idler (ω_2) frequencies, so that

$$\frac{P_1}{\omega_1} = \frac{P_2}{\omega_2} = \frac{(P_3)_t}{\omega_3}\left[\frac{P_3}{(P_3)_t} - 1\right] \tag{17.3-1}$$

The last argument shows that in principle, the parametric oscillator can attain high efficiencies. This requires operation well above threshold, and thus $P_3/(P_3)_t \gg 1$. These considerations are borne out by actual experiments (Reference 9).

Figure 17.6 shows experimental confirmation of the phenomenon of pump saturation.

17.4 FREQUENCY TUNING IN PARAMETRIC OSCILLATION

Unlike the laser, the parametric oscillator does not depend on resonant transitions and can, consequently, be tuned over wide-frequency regions. The pair of idler (ω_1) and signal (ω_2) frequencies that oscillates is that for which the phase-matching condition

$$\omega_3 n_3 = \omega_1 n_1 + \omega_2 n_2 \tag{17.4-1}$$

and the energy conservation condition

$$\omega_3 = \omega_1 + \omega_2 \tag{17.4-2}$$

are satisfied simultaneously. Relation (17.4-1) is, of course, just another way of writing $k_3 = k_1 + k_2$.

In a crystal, the indices n_1, n_2, and n_3 depend, in general, on the crystal orientation (for extraordinary rays), temperature, electric field, and pressure. The control of any of these variables can be used, in view of (17.4-1), for turning the output frequencies ω_1 and ω_2 of the parametric oscillator.

To be specific, let us take up the problem of angle tuning. Consider a parametric oscillator pumped by an extraordinary pump wave at ω_3. ω_1 and ω_2 are ordinary waves. At some crystal orientation θ_0 (the angle between the crystal c axis and the axis of the resonator), oscillation takes place at ω_{10} and ω_{20} where the indices of refraction are n_{10} and n_{20}, respectively. At $\theta = \theta_0$, we have, according to (17.4-1),

$$\omega_3 n_{30}(\theta_0) = \omega_{10} n_{10} + \omega_{20} n_{20}$$

We now rotate the crystal by $\Delta\theta$. This causes the index n_3 to change so that the need to satisfy the phase matching condition (17.4-1) causes ω_1 and

ω_2 to change slightly. The new oscillation takes place with the following changes relative to oscillation at θ_0:

$$\omega_3 \to \omega_3 \qquad \text{(pump frequency is unchanged)}$$

$$n_{30} \to n_{30} + \Delta n_3$$

$$n_{10} \to n_{10} + \Delta n_1$$

$$n_{20} \to n_{20} + \Delta n_2$$

$$\omega_{10} \to \omega_{10} + \Delta\omega_1$$

$$\omega_{20} \to \omega_{20} + \Delta\omega_2$$

$$\Delta\omega_2 = -\Delta\omega_1$$

Since (17.4-1) needs to be satisfied at the new set of frequencies, we have

$$\omega_3(n_{30} + \Delta n_3) = (\omega_{10} + \Delta\omega_1)(n_{10} + \Delta n_1) + (\omega_{20} - \Delta\omega_1)(n_{20} + \Delta n_2)$$

Neglecting second-order terms $\Delta n\, \Delta\omega$ and using (17.4-2), we obtain

$$\Delta\omega_1\big|_{\theta\simeq\theta_0} = \frac{\omega_3\, \Delta n_3 - \omega_{10}\, \Delta n_1 - \omega_{20}\, \Delta n_2}{n_{10} - n_{20}} \tag{17.4-3}$$

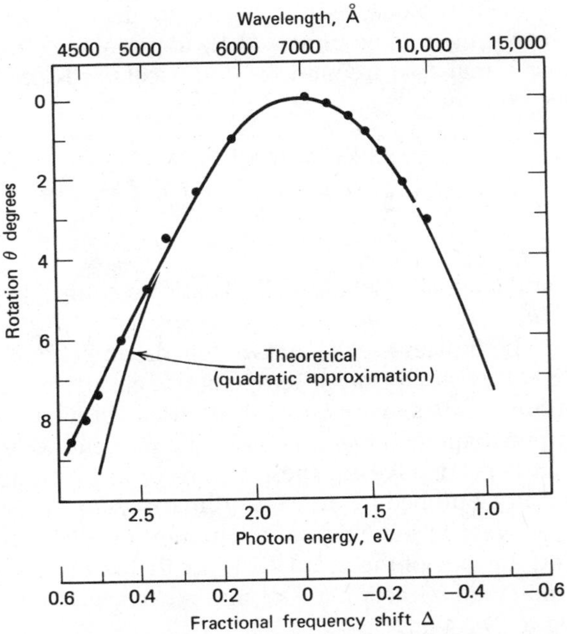

FIGURE 17.7 Dependence of the signal (ω_1) frequency on the angle between the pump propagation direction and the optic axis of the ADP crystal. The angle θ is measured with respect to the angle for which $\omega_1 = \omega_3/2$. $\Delta \equiv (\omega_1 - \omega_3/2)/(\omega_3/2)$. *Source:* Reference 11.

Since the pump is extraordinary, n_3 is a function of θ. The indices n_1 and n_2 of the ordinary rays depend on frequency but not θ. We can thus write

$$\Delta n_1 = \left.\frac{\partial n_1}{\partial \omega}\right|_{\omega_{10}} \Delta\omega_1$$
$$\Delta n_2 = \left.\frac{\partial n_2}{\partial \omega}\right|_{\omega_{20}} \Delta\omega_2 \tag{17.4-4}$$

and

$$\Delta n_3 = \left.\frac{\partial n_3}{\partial \theta}\right|_{\theta_{20}} \Delta\theta \tag{17.4-5}$$

These relations as well as $\Delta\omega_2 = -\Delta\omega_1$ can be used in (17.4-3) to yield

$$\frac{\partial \omega_1}{\partial \theta} = \frac{\omega_3 \dfrac{\partial n_3}{\partial \theta}}{(n_{10} - n_{20}) + \left[\omega_{10}\left(\dfrac{\partial n_1}{\partial \omega}\right) - \omega_{20}\left(\dfrac{\partial n_2}{\partial \omega}\right)\right]} \tag{17.4-6}$$

for the rate of change of the oscillation frequency with respect to the crystal orientation. Using (5.4-2) and the relation $d(1/x^2) = -(2/x^3)\,dx$, we obtain

$$\frac{\partial n_3}{\partial \theta} = -\frac{n_3^3}{2}\sin(2\theta)\left[\left(\frac{1}{n_e^{\omega_3}}\right)^2 - \left(\frac{1}{n_0^{\omega_3}}\right)^2\right]$$

that, when substituted in (17.4-6), gives (Reference 10)

$$\frac{\partial \omega_1}{\partial \theta} = \frac{-\frac{1}{2}\omega_3 n_{30}^3\left[\left(\dfrac{1}{n_e^{\omega_3}}\right)^2 - \left(\dfrac{1}{n_0^{\omega_3}}\right)^2\right]\sin(2\theta)}{(n_{10} - n_{20}) + \left(\omega_{10}\dfrac{\partial n_1}{\partial \omega} - \omega_{20}\dfrac{\partial n_2}{\partial \omega}\right)} \tag{17.4-7}$$

An experimental curve showing the dependence of the signal and idler frequencies on θ in $NH_4H_2PO_4$(ADP) is shown in Figure 17.7. Also shown is a theoretical curve based on a quadratic approximation of (17.4-7), which was plotted using the dispersion (i.e., n vs. ω) data of ADP. An angular tuning curve of a CdSe oscillator (Reference 12) is shown in Figure 17.8.

Reasoning similar to that used to derive the angle-tuning expression (17.4-7) can be applied to determine the dependence of the oscillation frequency on any other physical variable.

17.5 QUANTUM MECHANICAL TREATMENT OF PARAMETRIC INTERACTIONS

For most of the usual applications, the classical treatment of parametric interactions presented above is adequate. It is preferable in many respects to the quantum mechanical treatment in the sense that it handles more simply the phase coherent nature of the interacting fields. Some aspects of the interac-

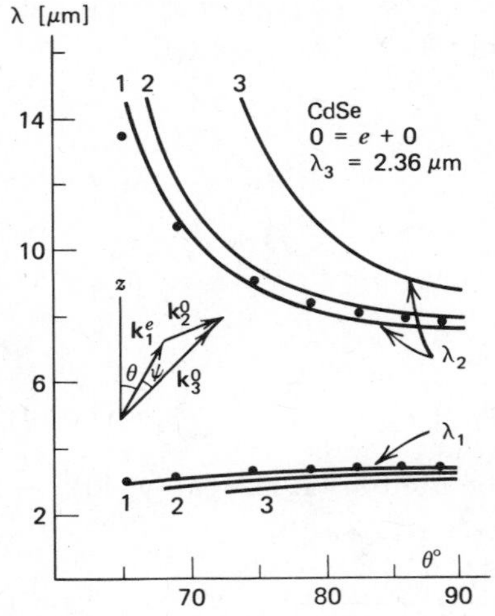

FIGURE 17.8 Angle tuning curves for CdSe with $\lambda_3 = 2.36\ \mu$. 1 = collinear interaction $\psi = 0$; 2, 3, = noncollinear interaction $\psi = 0.5°, 1°$; $\mathbf{k}_3, \mathbf{k}_1, \mathbf{k}_2$ = wave vectors of pump, signal, idler frequencies; θ = angle between signal wave vector and CdSe optic axis; ψ = angle between pumping wave vector and signal 1. *Source:* Reference 12.

tion, however, require the quantum mechanical approach. As an example of these, we may single out the problem of noise fluctuations in parametric processes (Reference 13) and the closely related problem of spontaneous parametric fluorescence.

We consider the case of three interacting fields $\mathbf{E}_1(\mathbf{r}, t)$, $\mathbf{E}_2(\mathbf{r}, t)$, and $\mathbf{E}_3(\mathbf{r}, t)$ at ω_1, ω_2, and ω_3, respectively, where

$$\omega_3 = \omega_1 + \omega_2$$

The field interaction Hamiltonian density is taken as in (16.2-4) in the form of

$$U = -\tfrac{2}{3}\, dE_1E_2E_3 \qquad (17.5\text{-}1)$$

where d is an effective nonlinear coefficient made up of a linear combination of d_{ijk} coefficients depending on the specific choice of crystal symmetry class, orientation, and the polarization of $\mathbf{E}_1$, $\mathbf{E}_2$, and $\mathbf{E}_3$.

Since all of the parametric amplification and oscillation experiments utilize an intense pump, we treat classically E_3 but quantize the signal (ω_1) and idler (ω_2) fields by replacing E_1 and E_2 by their corresponding quantum mechanical operators as given by (5.5-11) and (5.6-14, 15). The result is

$$\begin{aligned}\mathcal{H}' &= \int_V U\, dv = \frac{dB_3\hbar\sqrt{\omega_1\omega_2}}{3\sqrt{\varepsilon_1\varepsilon_2}}\int_V E_1(\mathbf{r})E_2(\mathbf{r})E_3(\mathbf{r})\, dv \cos\omega_3 t(a_1^\dagger - a_1)(a_2^\dagger - a_2) \\ &\equiv s\hbar\cos\omega_3 t(a_1^\dagger - a_1)(a_2^\dagger - a_2)\end{aligned} \qquad (17.5\text{-}2)$$

where we took

$$E_3(\mathbf{r}, t) = B_3E_3(\mathbf{r})\cos\omega_3 t \qquad (17.5\text{-}3)$$

and defined s by

$$s = \frac{dB_3\sqrt{\omega_1\omega_2}}{6\sqrt{\varepsilon_1\varepsilon_2}}\int_V E_1(\mathbf{r})E_2(\mathbf{r})E_3(\mathbf{r})\, dv \qquad (17.5\text{-}4)$$

The parameter s is thus proportional to the pump "amplitude" B_3. The functions $E_1(\mathbf{r})$, $E_2(\mathbf{r})$, and $E_3(\mathbf{r})$ are normalized as in (5.5-10) so that

$$\int_V E_i^2\, dv = 1 \tag{17.5-5}$$

The total Hamiltonian is taken as the sum of the unperturbed ($s = 0$) Hamiltonian (5.6-10) and the interaction Hamiltonian (17.5-2)

$$\mathcal{H} = \sum_{l=1,2} \hbar\omega_l(a_l^\dagger a_l + \tfrac{1}{2}) + 2s\hbar \cos \omega_3 t(a_1^\dagger - a_1)(a_2^\dagger - a_2) \tag{17.5-6}$$

Using the equation of motion (3.7-6) in the Heisenberg representation

$$\frac{dA}{dt} = -\frac{i}{\hbar}[A, \mathcal{H}] \tag{17.5-7}$$

where A is any operator, on $a_1^\dagger$ leads to

$$\begin{aligned} \frac{da_1^\dagger}{dt} &= -\frac{i}{\hbar}\left[a_1^\dagger, \sum_l \hbar\omega_l(a_l^\dagger a_l + \tfrac{1}{2})\right] - \frac{i}{\hbar}[a_1^\dagger, \mathcal{H}'] \\ &= -\frac{i}{\hbar}[a_1^\dagger, \hbar\omega_1 a_1^\dagger a_1] - \frac{i}{\hbar}[a_1^\dagger, 2s\hbar \cos \omega_3 t(a_1^\dagger - a_1)(a_2^\dagger - a_2)] \end{aligned} \tag{17.5-8}$$

From the commutator relations (5.6-8),

$$\begin{aligned} [a_l, a_m] &= [a_l^\dagger, a_m^\dagger] = 0 \\ [a_l, a_m^\dagger] &= \delta_{l,m} \end{aligned} \tag{17.5-9}$$

we obtain from (17.5-8)

$$\frac{da_1^\dagger}{dt} = i\omega_1 a_1^\dagger - \frac{is}{2}(e^{i\omega_3 t} + e^{-i\omega_3 t})(a_2^\dagger - a_2) \tag{17.5-10}$$

In the limit $s = 0$,

$$a_1^\dagger = a_1^\dagger(0)e^{i\omega_1 t}$$

$$a_2 = a_2(0)e^{-i\omega_2 t}$$

so that a synchronous driving term (i.e., a term oscillating at ω_1) can result from the product $e^{i\omega_3 t}a_2(t)$ on the right side of (17.5-10), provided that $\omega_3 = \omega_1 + \omega_2$. Neglecting the nonsynchronous terms results in

$$\frac{da_1^\dagger}{dt} = i\omega_1 a_1^\dagger + isa_2 e^{i\omega_3 t} \tag{17.5-11}$$

and similarly,

$$\frac{da_2}{dt} = -i\omega_2 a_2 - isa_1^\dagger e^{-i\omega_3 t} \tag{17.5-12}$$

The nonsynchronous terms oscillate at $(\omega_3 + \omega_2)$ and thus average out to zero in time intervals exceeding a few optical periods. The solution of (17.5-11) and (17.5-12) is

$$\begin{aligned} a_1^\dagger(t) &= [a_1^\dagger(0) \cosh st + ia_2(0) \sinh st]e^{i\omega_1 t} \\ a_2(t) &= [a_2(0) \cosh st - ia_1^\dagger(0) \sinh st]e^{-i\omega_1 t} \end{aligned} \tag{17.5-13}$$

Since we are interested in the number of quanta at ω_1 and ω_2, we need the operators $a_1^\dagger(t)a_1(t)$ and $a_2^\dagger(t)a_2(t)$. These are obtained directly by multiplying each of (17.5-13) by its Hermitian adjoint. If we use the commutation relation $[a_l(t), a_m^\dagger(t)] = \delta_{lm}$, we obtain

$$\begin{aligned} a_1^\dagger(t)a_1(t) &= a_1^\dagger(0)a_1(0)\cosh^2 st + [1 + a_2^\dagger(0)a_2(0)]\sinh^2 st \\ &\quad - \frac{i}{2}\sinh(2st)[a_1^\dagger(0)a_2^\dagger(0) - a_1(0)a_2(0)] \\ a_2^\dagger(t)a_2(t) &= a_2^\dagger(0)a_2(0)\cosh^2 st + [1 + a_1^\dagger(0)a_1(0)]\sinh^2 st \\ &\quad + \frac{i}{2}\sinh(2st)[a_1^\dagger(0)a_2^\dagger(0) - a_1(0)a_2(0)] \end{aligned} \tag{17.5-14}$$

According to (3.7-3), the expectation value of a quantum mechanical operator is given by

$$\langle A\rangle = \langle\psi(0)|\, A_H(t)\, |\psi(0)\rangle \tag{17.5-15}$$

where $A_H(t)$ is the operator in the Heisenberg representation and $\psi(0)$ is the solution of (the time-dependent) Schrödinger wave equation taken at $t = 0$. It follows that the number of quanta (photons) at ω_1 and ω_2 is

$$\begin{aligned} \langle n_1(t)\rangle &= \langle\psi(0)|\, a_1^\dagger(t)a_1(t)\, |\psi(0)\rangle \\ \langle n_2(t)\rangle &= \langle\psi(0)|\, a_2^\dagger(t)a_2(t)\, |\psi(0)\rangle \end{aligned} \tag{17.5-16}$$

The radiation field at $t = 0$ is assumed to have n_{10} quanta at ω_1 and n_{20} quanta at ω_2. The corresponding wavefunction $\psi(0)$ is thus

$$\psi(0) = |n_{10}, n_{20}\rangle = |n_{10}\rangle|n_{20}\rangle \tag{17.5-17}$$

where $|n_{10}\rangle$, for example, is the harmonic oscillator wavefunction corresponding to an energy $\hbar\omega_1(n_{10} + \frac{1}{2})$ that obeys the relations

$$\begin{aligned} a_1^\dagger|n_{10}\rangle &= (n_{10} + 1)^{1/2}|n_{10} + 1\rangle \\ a_1|n_{10}\rangle &= (n_{10})^{1/2}|n_{10} + 1\rangle \\ a_1^\dagger a_1|n_{10}\rangle &= n_{10}|n_{10}\rangle \end{aligned} \tag{17.5-18}$$

Similar relations apply when the subscript 1 is replaced by 2.

If we use (17.5-17) and (17.5-18) plus the orthonormality condition $\langle n_1 + l|n_1 + k\rangle = \delta_{ik}$ in (17.5-16), we obtain the result

$$\begin{aligned} \langle n_1(t)\rangle &= n_{10}\cosh^2 st + (1 + n_{20})\sinh^2 st \\ \langle n_2(t)\rangle &= n_{20}\cosh^2 st + (1 + n_{10})\sinh^2 st \end{aligned} \tag{17.5-19}$$

Tracing back the derivation of (17.5-19), we find that the term unity in the factors $(1 + n_{20})$ and $(1 + n_{10})$ on the right side is due to the noncommutativity of a^+ and a and is thus of a quantum mechanical origin. These terms correspond to a finite output power for the case of no input, that is, when $n_{10} = n_{20} = 0$, and are thus noise terms.

To obtain an expression for the noise output power of a parametric amplifier, consider the case of a single frequency input at ω_1 so that $n_{20} = 0$. For $st \gg 1$, we replace $\sinh^2 st$ by $K = (1/4)e^{st}$ where K is the gain[2] so that the first of (17.5-19) becomes

$$\langle n_1(t) \rangle = K(n_{10} + 1) \tag{17.5-20}$$

For a more rigorous derivation of (17.5-20), the reader is referred to Problem 17.1 at the end of this chapter and to Reference 13. According to (17.5-20), the number of noise photons at the output is K that, if we assume an integration time $T \sim 1/d\nu$, corresponds to a noise power output of $Kh\nu_1\, d\nu$. The effective input noise is obtained by dividing by the gain K and is

$$N_{\text{input}} = h\nu_1\, d\nu \tag{17.5-21}$$

and is thus the same as that of an ideal laser amplifier (Section 13.1).

17.6 FREQUENCY UP-CONVERSION

Parametric interactions in a crystal can be used to convert a signal from a "low" frequency ω_1 to a "high" frequency ω_3 by mixing it with a strong laser beam at ω_2, where

$$\omega_1 + \omega_2 = \omega_3 \tag{17.6-1}$$

Using the quantum mechanical photon picture described in Section 17.5, we can consider the basic process taking place in frequency up-conversion as one in which a signal (ω_1) photon and a pump (ω_2) photon are annihilated while, simultaneously, one photon at ω_3 is generated (References 13, 15, 16, and 17). Since a photon energy is $\hbar\omega$, conservation of energy dictates that $\omega_3 = \omega_1 + \omega_2$, and the conservation of momentum leads to the relationship

$$\mathbf{k}_3 = \mathbf{k}_1 + \mathbf{k}_2 \tag{17.6-2}$$

between the wave vectors at the three frequencies. This point of view also suggests that the number of output photons at ω_3 cannot exceed the input numbers of photons at ω_1.

The experimental situation is demonstrated by Figure 17.9. The ω_1 and ω_2 beams are combined in a partially transmissive mirror (or prism), so that they traverse together (in near parallelism) the length l of a crystal possessing nonlinear optical characteristics.

The analysis of frequency up-conversion starts with (16.6-3). Assuming negligible depletion of the pump wave A_2 and no losses ($\alpha = 0$) at ω_1 and ω_3, we can write the first and third of these equations as

$$\begin{aligned} \frac{dA_1}{dz} &= -igA_3 \\ \frac{dA_3}{dz} &= -igA_1 \end{aligned} \tag{17.6-3}$$

[2] The fact that our gain is time-dependent should not disturb us. Since there is a one-to-one correspondence between z and t, our results apply to the traveling wave with $t \rightarrow z/c$. The quantum mechanics is handled more easily, however, in the time domain.

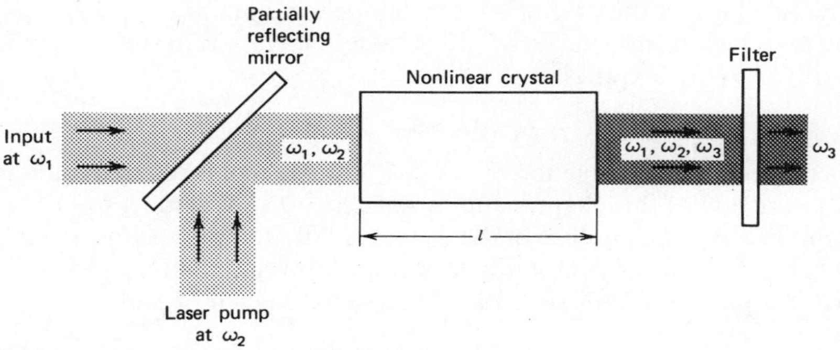

FIGURE 17.9 Parametric up-conversion in which a signal at ω_1 and a strong laser beam at ω_2 combine in a nonlinear crystal to generate a beam at the sum frequency $\omega_3 = \omega_1 + \omega_2$.

where if we use (16.6-4) and (16.6-1) and choose without loss of generality the pump phase as zero so that $A_2(0) = A_2^*(0)$,

$$g \equiv 2\sqrt{\frac{\omega_1\omega_3}{n_1n_3}\left(\frac{\mu_0}{\varepsilon_0}\right)}\, dE_2 \tag{17.6-4}$$

where E_2 is the amplitude of the electric field of the pump laser. If we take the input waves with (complex) amplitudes $A_1(0)$ and $A_3(0)$, the general solution of (17.6-3) is

$$\begin{aligned} A_1(z) &= A_1(0)\cos(gz) - iA_3(0)\sin(gz) \\ A_3(z) &= A_3(0)\cos(gz) - iA_1(0)\sin(gz) \end{aligned} \tag{17.6-5}$$

In the case of a single-(low)-frequency input at ω_1, we have $A_3(0) = 0$. In this case,

$$\begin{aligned} |A_1(z)|^2 &= |A_1(0)|^2\cos^2(gz) \\ |A_3(z)|^2 &= |A_1(0)|^2\sin^2(gz) \end{aligned} \tag{17.6-6}$$

Therefore,

$$|A_1(z)|^2 + |A_3(z)|^2 = |A_1(0)|^2$$

In the discussion following (16.6-2), we pointed out that $|A_l(z)|^2$ is proportional to the photon flux (photons per square meter per second) at ω_l. Using this fact, we may interpret (17.6-6) as stating that the photon flux at ω_1 plus that at ω_3 at any plane z is a constant equal to the input ($z = 0$) flux at ω_1. If we rewrite (17.6-6) in terms of powers, we obtain

$$\begin{aligned} P_1(z) &= P_1(0)\cos^2(gz) \\ P_3(z) &= \frac{\omega_3}{\omega_1} P_1(0)\sin^2(gz) \end{aligned} \tag{17.6-7}$$

In a crystal of length l, the conversion efficiency is thus

$$\frac{P_3(l)}{P_1(0)} = \frac{\omega_3}{\omega_1} \sin^2(gl) \tag{17.6-8}$$

and can have a maximum value of ω_3/ω_1, corresponding to the case in which all the input (ω_1) photons are converted to ω_3 photons.

In most practical situations, the conversion efficiency is small (see the following numerical example) so using $\sin x \simeq x$ for $x \ll 1$, we get

$$\frac{P_3(l)}{P_1(0)} \simeq \left(\frac{\omega_3}{\omega_1} g^2 l^2\right)$$

that, by the use of (17.6-4) and (16.6-2), can be written as

$$\frac{P_3(l)}{P_1(0)} \simeq \frac{2\omega_3^2 l^2 d^2}{n_1 n_2 n_3} \left(\frac{\mu_0}{\varepsilon_0}\right)^{3/2} \left(\frac{P_2}{A}\right) \tag{17.6-9}$$

where A is the cross-sectional area of the interaction region.

Numerical Example: Frequency Up-Conversion. The main practical interest in parametric frequency up-conversion stems from the fact that it offers a means of detecting infrared radiation (a region where detectors are either inefficient, very slow, or require cooling to cryogenic temperatures) by converting the frequency into the visible or near-visible part of the spectrum. The radiation can then be detected by means of efficient and fast detectors such as photomultipliers or photodiodes (References 15 and 16).

As an example of this application, consider the problem of up-converting a 10.6-μm signal, originating in a CO_2 laser, to 0.96 μm by mixing it with the 1.06-μm output of an Nd^{3+} : YAG laser. The nonlinear crystal chosen for this application has to have low losses at 1.06 μm and 10.6 μm, as well as at 0.96 μm. In addition, its birefringence has to be such as to make phase matching possible. The crystal proustite, Ag_3AsS_3 (Reference 18), listed in Table 16.2 meets these requirements.

Using the data

$$\frac{P_{1.06\,\mu\text{m}}}{A} = 10^4\ \text{W/cm}^2 = 10^8\ \text{W/m}^2$$

$$l = 1\ \text{cm}$$

$$n_1 \simeq n_2 \simeq n_3 = 2.6$$

$$d_{\text{eff}} = 1.1 \times 10^{-22}$$ (taken conservatively as a little less than half the value given in Table 16.2 for d_{22})

we obtain, from (17.6-9),

$$\frac{P_{\lambda=0.96\,\mu\text{m}}(l = 1\ \text{cm})}{P_{\lambda=10.6\,\mu\text{m}}} = 2.4 \times 10^{-4}$$

indicating a useful amount of conversion efficiency. Actual experiments in

$LiIO_3$ (Reference 19) using a ruby 0.6943-μm pump resulted in photon conversion efficiencies near 100% at 3.39 μm.

Some General Remarks Concerning Parametric Interactions

We have considered in this chapter two types of interactions: (1) parametric amplification and (2) frequency conversion. In this section we will show how with the aid of a simple but rigorous point of view we may deduce which form of parametric interaction takes place in any given situation.

From the treatment of the last chapter and this one, we know that in a nonlinear dielectric three optical fields at frequencies ω_1, ω_2, and ω_3 can exchange energy. We know, in addition, that this exchange involves, in its most basic form, *one photon* of each of the radiation fields.

These facts enable us to determine the type of interactions that are consistent with the conservation of energy but are not sufficient, in some cases, to determine the direction of power flow. This issue is then settled with the added condition that total photon momentum be conserved. This is just another way of stating the index-matching condition.

To clarify these ideas, let us consider the (one-dimensional) case of an *intense* laser beam at ω_2 plus a weak signal beam at ω_1, which are incident at $z = 0$ on a nonlinear medium. We want to determine the types of possible interactions and the conditions for their occurrence.

If $\omega_1 > \omega_2$, then the restriction of one-photon interaction allows two types of processes to occur initially. One is a process in which a photon at ω_1 is annihilated, while simultaneously photons at ω_2 and ω_3 are created. Quantum mechanically, this process can be described as a transition $|n_1, n_2, n_3\rangle \rightarrow |n_1 - 1, n_2 + 1, n_3 + 1\rangle$. Since the total photon energy is conserved, $\omega_3 = \omega_1 - \omega_2$. The direction of power flow between the ω_1 and the ω_2 and ω_3 fields is illustrated schematically by the arrow diagram of Figure 17.10*a*. Conservation of photon momentum leads to the condition $k_1 = k_2 + k_3$. This process obviously causes a depletion of energy at ω_1 and a corresponding increase at ω_2 and ω_3, as shown in the initial portion, $0 \leqslant gz \leqslant \pi$, of Figure 17.10*c*.

When the radiation field at ω_1 is completely depleted, the process described cannot continue, depending as it does on the existence of ω_1 photons. The reverse process, depicted by Figure 17.10*b*, in which quanta at ω_2 and ω_3 combine to generate a quantum of ω_1, can take place, however. This corresponds to the region $\pi \leqslant gz \leqslant 2\pi$ in Figure 17.10*c*. These processes were described by (17.6-6).

The second possibility is that the initial interaction does not lead to a generation of the difference frequency $\omega_3 = \omega_1 - \omega_2$ but rather to the sum frequency $\omega_3' = \omega_1 + \omega_2$. This is described by Figure 17.11. Since the initial interaction (near $z = 0$) uses up pump photons (at ω_2) as well as ω_1 photons, the pump field (at ω_2) is diminished initially as shown in Figure 17.11*c*. This is the main difference between this case and the difference frequency generation discussed above in which the pump energy increased initially.

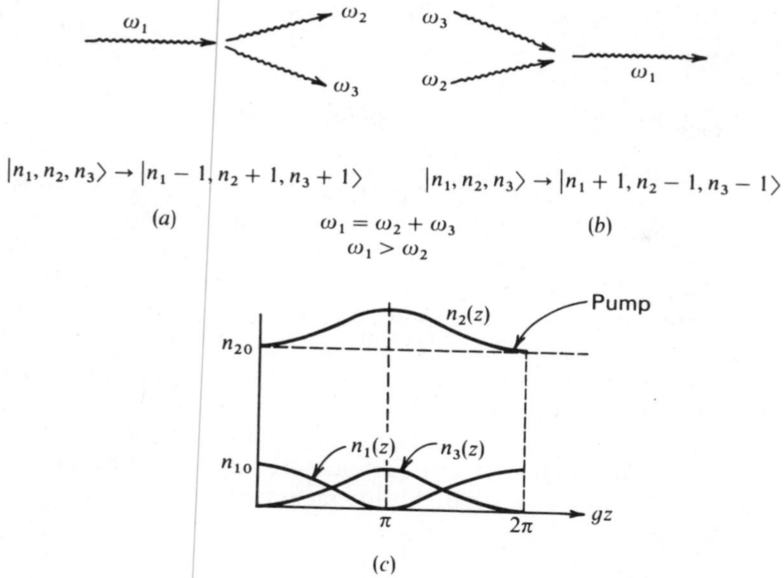

FIGURE 17.10 The process of difference frequency generation where a pump input at ω_2 and a signal at $\omega_1 > \omega_2$ initially produce photons at $\omega_3 = \omega_1 - \omega_2$. The process is reversed further along in the interaction path. The direction of power flow is given in (*a*) and (*b*). The photon densities are shown in (*c*).

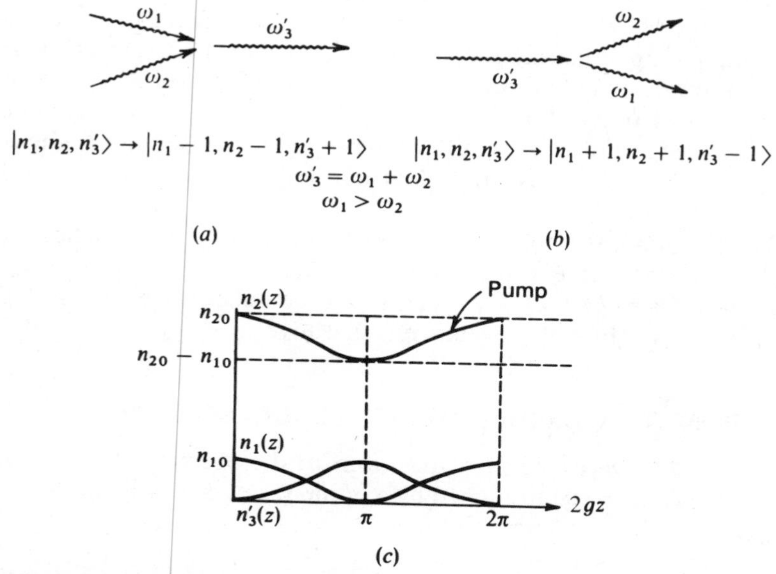

FIGURE 17.11 The process of sum frequency generation where a pump photon at ω_2 and a signal photon at ω_1 combine initially to generate a photon at $\omega_3' = \omega_1 + \omega_2$. The process is reversed further along the interaction path. The direction of power flow is given in (*a*) and (*b*). The photon densities are drawn in (*c*).

Since the input conditions for both of these processes are identical (i.e., an input at ω_2 and at $\omega_1 > \omega_2$), we are justified in inquiring as to the factors that determine which of the two possible interactions takes place. The choice is determined by the index-matching condition. If the condition

$$k_1 = k_2 + k_3$$

is satisfied where $\omega_3 = \omega_1 - \omega_2$, then the difference frequency generation depicted by Figure 17.10 takes place. If, on the other hand, $k_1 + k_2 = k_3'$ where $\omega_3' = \omega_1 + \omega_2$, the sum-frequency generation (of ω_3') as described by Figure 17.11 takes place. The satisfaction of either one of these conditions is achieved by means of the proper crystal orientation as discussed in Section 17.4.

To appreciate why index matching determines the type of interaction, we recall that, locally, the input fields at ω_1 and ω_2 produce (radiating) polarizations both at $\omega_3 = \omega_1 - \omega_2$ and at $\omega_3' = \omega_1 + \omega_2$. The index matching is necessary to ensure cumulative in-phase addition of the radiated power over substantial path lengths, so that the interaction which does not conserve "photon momentum" does not build up.

The Case of $\omega_1 < \omega_2$

Here the frequency of the signal input at ω_1 is smaller than that of the intense pump at ω_2. One possible process is the "splitting" of ω_2 photons into ω_1 and ω_3 photons where $\omega_2 = \omega_1 + \omega_3$. The quantum mechanical transition is $|n_1, n_2, n_3\rangle \rightarrow |n_1 + 1, n_2 - 1, n_3 + 1\rangle$ and is depicted by Figure 17.12. This is the case of the parametric amplifier described in Section 17.1. The growth of n_1 and n_3 continues as long as $n_2 > 0$. Once the pump power at ω_2 is depleted, the direction of power flow is reversed. This reversal shown at point Q is rarely observed in practice. In most of the experimental situations, the condition $n_2(z) \approx$ const, that is, the initial portion of the plot in Figure 17.12 applies.

In concluding, we note that instead of the parametric amplification just described, we may have a sum generation of photons at $\omega_3 = \omega_1 + \omega_2$. This will happen if $k_3 = k_1 + k_2$. This process was described above and is depicted by Figure 17.11. The only difference is that here $\omega_1 < \omega_2$.

17.7 SPONTANEOUS PARAMETRIC FLUORESCENCE

In Section 17.5 we established that even in the absence of any inputs, a nonlinear optical crystal irradiated by a pump (ω_3) wave will emit spontaneously radiation at frequencies ω_1 and ω_2 where

$$\omega_1 + \omega_2 = \omega_3 \tag{17.7-1}$$

and

$$\mathbf{k}_1 + \mathbf{k}_2 \approx \mathbf{k}_3 \tag{17.7-2}$$

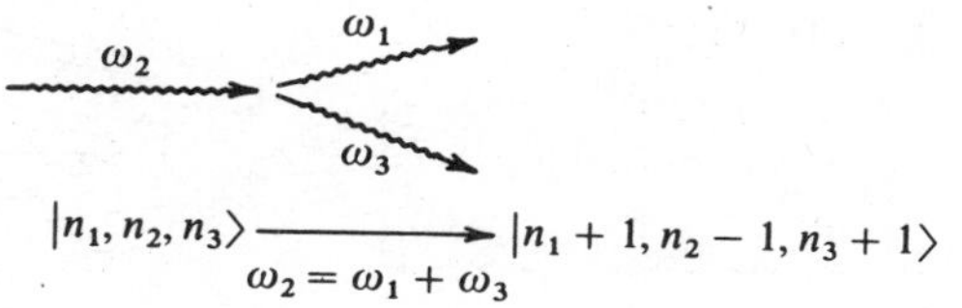

(a)

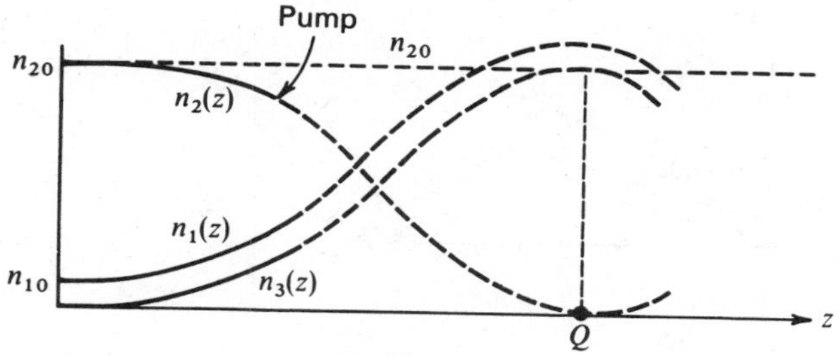

(b)

FIGURE 17.12 Parametric amplification. Here, we have energy conversion from ω_2 to ω_1 and ω_3 where $\omega_2 = \omega_1 + \omega_3$. The photons' densities are shown in (*b*). The dark portions of the curves correspond to the "nondepleted" pump approximation as described by the parametric amplification analysis of Section 17.2 (with ω_2 taken as pump instead of ω_3). The dashed curves are in the nonlinear region.

This phenomenon, which is predicted quantum mechanically but not classically, can still be treated classically by thinking of the zero-point energy of the electromagnetic modes as providing an effective input field with an intensity equivalent to one *photon per mode*. The last statement is based on (17.5-19) reproduced here

$$\langle n_2(t) \rangle = n_{20} \cosh^2(st) + (1 + n_{10}) \sinh^2(st) \tag{17.7-3}$$

The parametric fluorescence at ω_2 is the power emitted with no inputs (i.e., $n_{10} = n_{20} = 0$) and is thus due to the unity in the parenthesis on the right side of (17.7-3). The idler $\langle n_2(t) \rangle$ grows as if driven by one input photon at the signal (ω_1) frequency.

In an actual experiment, there exists a large number of different ω_1, ω_2 pairs that satisfy (17.7-1) and (17.7-2). A measurement of the emitted power near ω_2 (or ω_1) will thus involve a sum of such pairs.

We will, in what follows, derive an expression for the total power near ω_2 measured by a detector in an experiment such as that sketched in Figure 17.13. The procedure we will use is the following. We will first find the total number of modes with frequencies near ω_1 that act as effective inputs for ω_2 modes. We will next put one photon into each ω_1 mode and, using the classical results of Section 17.1, find the corresponding output power in each driven ω_2 mode. We will then sum over all such pairs to get the total output at ω_2.

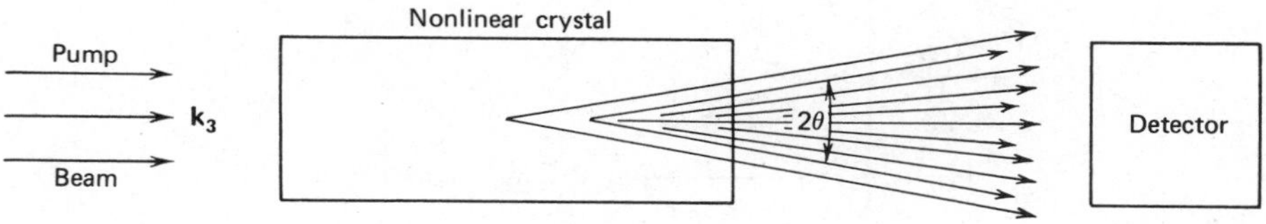

(a)

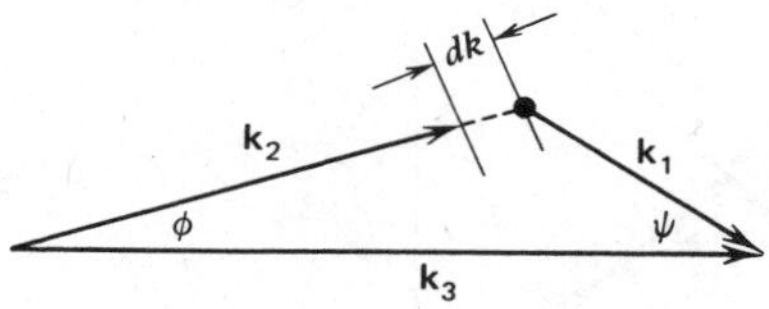

(b)

FIGURE 17.13 (*a*) The geometry used in deriving the parametric fluorescence power. The solid angle subtended by the detector at the crystal is $\Delta\Omega = \pi\theta^2$. ($\theta \ll 1$.) The **k** vectors for signal mode ($\mathbf{k}_1$), an idler mode ($\mathbf{k}_2$), and the pump wave $\mathbf{k}_3$.

Consider the number of signal modes with $\mathbf{k}_1$ vectors contained between ψ and $\psi + d\psi$ with magnitudes between k_1 and $k_1 + dk_1$ as shown in Figure 17.13*b*. The number of such modes is

$$dN_1 = \frac{2\pi k_1^2\, dk_1 \sin\psi\, d\psi}{(2\pi)^3/V} \tag{17.7-4}$$

The numerator of (17.7-4) is equal to the volume in **k** space occupied by the modes. $(2\pi)^3/V$ is the **k** space volume per mode. V is the interaction volume.

The effective input intensity, as far as output near ω_2 is concerned, is obtained by putting one quantum of energy $\hbar\omega_1$ in each of the dN_1 modes and then multiplying by c/Vn_1 to get the corresponding intensity

$$\begin{aligned} dI_1 &= dN_1 \hbar\omega_1 \frac{c}{Vn_1} \\ &= \frac{k_1^2\, dk_1 \psi\, d\psi \hbar\omega_1 c}{(2\pi^2)n_1} \end{aligned} \tag{17.7-5}$$

From (17.1-3), we find that when $A_2(0) = 0$, an idler wave grows according to

$$A_2(l) = -igA_1^*(0)e^{-i(\Delta k)l}\, \frac{\sin\sqrt{(\Delta k)^2 - g^2}\; l}{\sqrt{(\Delta k)^2 - g^2}\; l} \tag{17.7-6}$$

so that the output power near ω_2 due to an input dI_1A (A = area) is

$$dP_2 = (dI_2)A = dI_1\left(\frac{gl}{2}\right)^2\left(\frac{\omega_2}{\omega_1}\right)A\, \frac{\sin^2(\sqrt{(\Delta k)^2 - g^2}\; l)}{[(\Delta k)^2 - g^2](l^2)} \tag{17.7-7}$$

The factor (ω_2/ω_1) is due to the fact that the intensity I_i is related to the photon flux $|A_i|^2$ by $I_i \propto \omega_i\,|A_i|^2$.

Using the following relations:

$$k_1 = \frac{\omega_i}{c}\,n_1$$

$$\psi\,d\psi = \frac{k_2^2}{k_1^2}\,\phi\,d\phi \qquad \text{(follows from Figure 17.13}b\text{)}$$

$$d\omega_2 = -d\omega_1 \qquad (\omega_1 + \omega_2 = \omega_3 = \text{constant})$$

$$g^2 = 4\left(\frac{\mu_0}{\varepsilon_0}\right)\frac{\omega_1\omega_2}{n_1 n_2}\,d^2 E_3^2 \quad \text{[Eq. (17.1-2)]}$$

$$E_3^2 = 2\left(\frac{P_3}{A}\right)\sqrt{\frac{\mu_0}{\varepsilon_0 n_3^2}}$$

in Eqs. (17.7-7) results in

$$dP_2 = \frac{-\hbar\omega_1\omega_2^4 n_2\,d^2}{\pi^2 c^5 n_1 n_3 \varepsilon_0^3}\,l^2 P_3\,d\omega_2\phi\,d\phi\,\operatorname{sinc}^2\!\left(\sqrt{(\Delta k)^2 - g^2}\,l\right) \tag{17.7-8}$$

where sinc $x \equiv \sin x/x$. The total power at P_2 is obtained by integrating (17.7-8) over all frequencies and over the range of angles $0 < \phi < \theta$ accepted by the detector

$$P_2 = -\beta P_3 l^2 \int_{-\infty}^{\infty}\int_0^{\theta} d\omega_2\,d\phi\phi\,\operatorname{sinc}^2\!\left(\sqrt{(\Delta k)^2 - g^2}\,l\right) \tag{17.7-9}$$

where

$$\beta = \frac{\hbar\omega_1\omega_2^4 n_2\,d^2}{\pi^2 c^5 n_1 n_3 \varepsilon_0^3} \tag{17.7-10}$$

In (17.7-9), we considered the factor $\omega_1\omega_2^4$ as a constant since the sinc function acts as a δ function, limiting the effective range of frequencies near ω_2 that contribute to the integral.

To perform the indicated integration, we need to express Δk in (17.7-9) as a function of ω_2 and ϕ. Referring to Figure 17.13b, applying the law of sines, and assuming $\phi \ll 1$, $\psi \ll 1$, $k_3 \approx k_1 + k_2$, give

$$\Delta k \approx \frac{k_2 k_3}{k_1}\frac{\phi^2}{2} + (k_3 - k_2 - k_1) \tag{17.7-11}$$

Next, we designate the pair of frequencies ω_1, ω_2, which is phase-matched in the forward direction ($\phi = 0$) as ω_{10}, ω_{20} and expand k_2 and k_1 as

$$k_2 = k_{20} + \left.\frac{\partial k_2}{\partial\omega_2}\right|_{\omega_{20}}(\omega_2 - \omega_{20})$$

$$k_1 = k_{10} - \left.\frac{\partial k_1}{\partial\omega_1}\right|_{\omega_{10}}(\omega_2 - \omega_{20})$$

where we used $\omega_2 - \omega_{20} = -(\omega_1 - \omega_{10})$. We can now rewrite (17.7-11) as

$$\Delta k = -b(\omega_2 - \omega_{20}) + a\phi^2 \tag{17.7-12}$$

$$a = k_2 k_3 / 2k_1 \qquad b = \left.\frac{\partial k_2}{\partial \omega_2}\right|_{\omega_{20}} - \left.\frac{\partial k_1}{\partial \omega_1}\right|_{\omega_{10}} \tag{17.7-13}$$

with (17.7-12), the total idler power (17.7-9) becomes

$$P_2 = -\beta l^2 P_3 \int_{-\infty}^{\infty} \int_0^{\theta} \phi \, d\phi \, d\omega_2 \operatorname{sinc}^2 \{\sqrt{[a\phi^2 - b(\omega_2 - \omega_{20})]^2 - g^2}\, l\} \tag{17.7-14}$$

Now except for a very narrow range where $a\phi^2 \approx b(\omega_2 - \omega_{20})$, $[a\phi^2 - b(\omega_2 - \omega_{20})]^2 \gg g^2$. Neglecting the term g^2, we can perform the integration, obtaining

$$P_2 = \pi \frac{\beta l P_3}{|b|} \theta^2 \tag{17.7-15}$$

This is the desired result (References 20, 21, and 22).

Experimental Results

It is clear from (17.7-14) that the total measured power near ω_2 contains a continuum of frequencies. To estimate the spectral width of the radiation, we return to (17.7-12)

$$\Delta k = a\phi^2 - b\Delta\omega_2$$

The contribution to P_2 in (17.7-9) due to a given Δk is zero when $\Delta k l \approx 2\pi$ [the first zero of the sinc function in (17.7-9) in the limit $\Delta k \gg g$]. It follows that in a given direction ϕ

$$(\Delta\omega_2)_{\max} \approx \frac{a\phi^2 l - 2\pi}{bl} \tag{17.7-16}$$

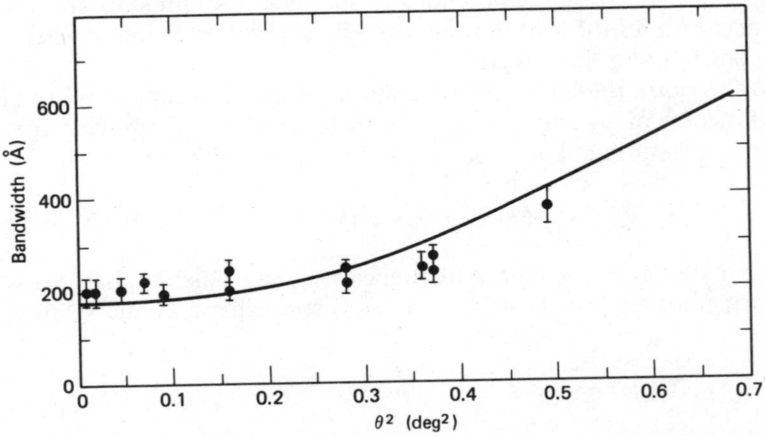

FIGURE 17.14 Signal fluorescence bandwidth as a function of detector acceptance angle. The solid line is the theoretical bandwidth, (17.7-16). *Source:* Reference 23.

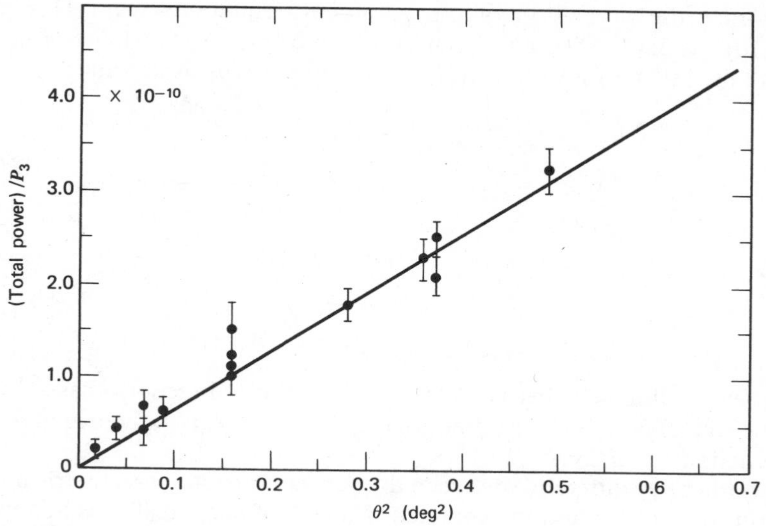

FIGURE 17.15 Total signal power versus detector acceptance angle. The solid line is the theoretical in-crystal fluorescence power per unit pump power calculated using (17.7-15). The experimental data are normalized to the solid theoretical curve at $\theta^2 = 0.16$ deg². *Source:* Reference 23.

so that for small $\theta(\equiv \phi_{\max})$, $\Delta\omega_2 = \pi/|b|$, whereas for $\theta > (2\pi/al)^{1/2}$

$$(\Delta\omega_2) \approx \frac{a\theta^2}{|b|} \tag{17.7-16a}$$

The experimentally observed dependence of $(\Delta\omega_2)$ on θ^2 is demonstrated in Figure 17.14.

The dependence of the total power, P_2, on P_3 and θ^2 is shown in Figure 17.15.

The tuning curve for spontaneous parametric fluorescence, that is, the dependence of λ_2 on crystal orientation, temperature, and other possible factors, is the same as that of the parametric oscillator since in both we need to *satisfy* the condition $\mathbf{k}_3 \approx \mathbf{k}_1 + \mathbf{k}_2$.

Since the parametric fluorescence is not a threshold-dependent phenomenon, it can be used to obtain the oscillator tuning curve when the index dispersion data is not available.

17.8 BACKWARD PARAMETRIC AMPLIFICATION AND OSCILLATION (REFERENCE 27)

Here, we consider the possibility of parametric interactions when the signal and idler travel in opposite directions. To be specific, we choose the signal wave to travel in the $-z$ direction so that

$$\begin{aligned} A_1(z, t) &= A_1(z)e^{i(\omega_1 t + \kappa_1 z)} \\ A_2(z, t) &= A_2(z)e^{i(\omega_2 t - k_2 z)} \end{aligned} \tag{17.8-1}$$

If we trace the derivation of the parametric equations (17.1-1) back to (16.4-10), we find that the only effect here is a change in the sign of k_1. This causes Eqs. (17.1-1) for the normal mode amplitudes to become

$$\frac{dA_1}{dz} = \frac{\alpha_1}{2} A_1 + igA_2^* e^{-i\,\Delta kz}$$
$$\frac{dA_2^*}{dz} = -\frac{\alpha_2}{2} A_2 + igA_1 e^{i\,\Delta kz} \tag{17.8-2}$$

where

$$\omega_3 = \omega_1 + \omega_2$$
$$2\Delta k = k_3 - k_2 + k_1 \tag{17.8-3}$$

Note that the difference between (17.8-2) and the codirectional case (17.1-1) is in the sign of g in the first equation, the (obvious) sign difference in α_1, and in the definition of Δk.

Another key difference between the contradirectional parametric interaction considered here and the codirectional case of Section 17.1 is in applying the boundary conditions. In keeping with the directions of propagation, we need to specify $A_2(0)$ and $A_1(L)$ as shown in Figure 17.16.

We leave it to the student to show that the solution of (17.8-2) in the lossless ($\alpha = 0$) and phase-matched ($\Delta k = 0$) case is

$$A_1(z) = \frac{A_1(L)}{\cos(gL)} \cos(gz) + i\frac{A_2^*(0)}{\cos(gL)} \sin g(z - L) \tag{17.8-4}$$

$$A_2^*(z) = i\frac{A_1(L)}{\cos(gL)} \sin(gz) + \frac{A_2^*(0)}{\cos(gL)} \cos g(z - L) \tag{17.8-5}$$

The output fields are

$$A_1(0) = \frac{A_1(L)}{\cos(gL)} - iA_2^*(0) \tan(gL)$$
$$A_2^*(L) = iA_1(L) \tan(gL) + \frac{A_2^*(0)}{\cos(gL)} \tag{17.8-6}$$

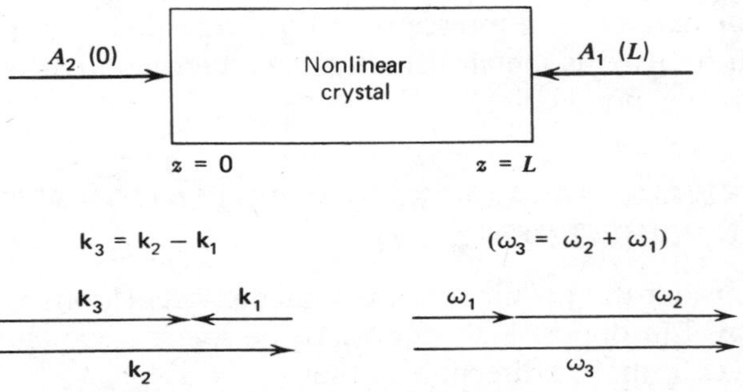

FIGURE 17.16 The boundary conditions and phase matching in contradirectional parametric interaction.

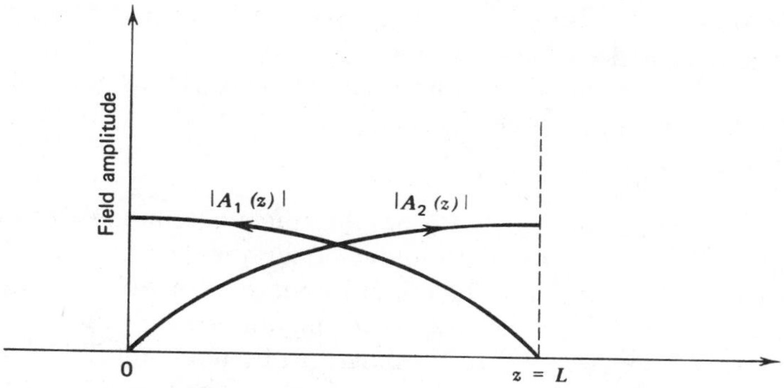

FIGURE 17.17 The signal A_1 and idler A_2 fields in a contradirectional parametric oscillator.

Of special interest is the case

$$gL = \pi/2 \tag{17.8-7}$$

$A_1(0)$ and $A_2^*(L)$ become infinite for a finite input at either end. Stated differently, we obtain finite outputs $A_1(0)$ and $A_2^*(L)$ with no input ($A_1(L) = 0$. $A_2^*(0) = 0$).

In the limit of $gL \to \pi/2$, the field distribution (17.8-4,5) becomes

$$\begin{aligned} A_1(z) &= iA \sin g(z - L) \\ A_2^*(z) &= A \cos g(z - L) \end{aligned} \tag{17.8-8}$$

and are plotted in Figure 17.17.

We note that oscillation can occur here *without mirror feedback*. The feedback is due to the opposite directions of the waves' propagation. This is similar to the principle of operation of backward traveling wave oscillators (References 24 and 25) and distributed feedback lasers (Reference 26), which is discussed in Chapter 19. No parametric oscillators based on this principle have yet been demonstrated (Reference 27). The difficulty is in the lack of a nonlinear optical material with sufficient birefringence to make it possible to satisfy the index matching condition $\Delta k = 0$. This is illustrated in Figure 17.16. To satisfy the vector diagram, it is necessary that $\omega_3 n_3 < \omega_2 n_2$, while $\omega_3 > \omega_2$. It is clear that the eventual operation of such devices will be limited to $\omega_1 \ll \omega_2, \omega_3$.

17.9 SQUEEZED STATES OF THE ELECTROMAGNETIC FIELD

In Section 17.5 we found that the statistical properties of the optical field, for instance, the fluctuations in the number of quanta in a mode, depend on the physical process that was involved in preparing it. This is because the (Heisenberg representation) mode annihilation and creation operators $a(t)$ and $a^\dagger(t)$

which, along with the initial field state, completely determine the statistical properties of the field evolve according to (17.5-7) in a manner that is determined by the field-matter interaction Hamiltonian. It is thus possible to create optical fields whose statistics differ markedly from those of "ordinary" fields such as the coherent states, discussed in Section (5.8), approximated by the output of ordinary lasers.

We find it advantageous to decompose the single-mode electric field—the signal field—of a given mode into two quadrature components with time-dependence $\cos \omega t$ and $\sin \omega t$. The quantum mechanical operators that correspond to the amplitudes of the two quadratures of the field, called *quadrature-phase amplitudes,* will be shown to be noncommuting Hermitian operators; the two quadrature-phase amplitudes thus obey an uncertainty principle. When the field is in a vacuum state or a coherent state, the uncertainty product for the two quadrature amplitudes is a minimum, the uncertainties in the two quadratures being equal. When the field is in a squeezed state, the variance of one quadrature—the *squeezed* quadrature—is less than that of the vacuum, whereas the variance of the other quadrature is increased in accordance with the uncertainty principle.

To make use of the squeezed states in, for example, a communication system, we can modulate the amplitude of the squeezed quadrature component with the contents of some information bearing waveform. On the receiving side of the channel, a phase sensitive detector responding only to the squeezed quadrature component can recover the signal with a signal-to-noise ratio in excess of that corresponding to full shot noise. In what follows we show that the output of a parametric amplifier is a squeezed state.

The electric field for a single-mode electromagnetic field, from (5.5-11) and (5.6-9), is

$$\mathbf{E}_s(\mathbf{r}, t) = -i \left(\frac{\hbar\omega}{2\varepsilon}\right)^{1/2} \mathbf{E}_s(\mathbf{r})[a^\dagger(t) - a(t)] \tag{17.9-1}$$

where the integral over the modal volume

$$\int |E_s(\mathbf{r})|^2 \, dV = 1$$

We consider the degenerate ($\omega_1 = \omega_2$) parametric amplifier. The "pump" field is taken as in (17.5-3)

$$\begin{aligned} \mathbf{E}_p(\mathbf{r}, t) = B_3\mathbf{E}_3(\mathbf{r}) \cos(2\omega t + 2\theta) &= B_3\mathbf{E}_3(\mathbf{r}) \sin(2\omega t + 2\theta + \pi/2) \\ &\equiv B_3\mathbf{E}_3(\mathbf{r}) \sin \phi_p(t) \end{aligned} \tag{17.9-1a}$$

The corresponding field Hamiltonian is obtained from (17.5-6) for the case $a_1 = a_2 \equiv a$

$$\begin{aligned} H &= \hbar\omega(a^\dagger a + \tfrac{1}{2}) + s\hbar \cos(2\omega t + 2\theta)(a^\dagger - a)^2 \\ &= \hbar\omega(a^\dagger a + \tfrac{1}{2}) + s\hbar \sin(2\omega t + \pi/2 + 2\theta)(a^\dagger - a)^2 \end{aligned} \tag{17.9-2}$$

where s is defined by (17.5-4). We can, and we will, without loss of generality choose $\theta = 0$. This fixes the time $t = 0$ as that when the pump field is a maximum. The equations of motion for the operators a and $a^\dagger$ are

$$\frac{da}{dt} = -\frac{i}{\hbar}[a, H] = -i\omega a - 2is\cos 2\omega t(a^\dagger - a) \tag{17.9-3a}$$

$$\frac{da^\dagger}{dt} = -\frac{i}{\hbar}[a^\dagger, H] = i\omega a^\dagger - 2is\cos 2\omega t(a^\dagger - a) \tag{17.9-3b}$$

In the absence of an interaction (i.e., $s = 0$) between the signal and the pump modes (as in free space), the Heisenberg equations (17.9-3a) and (17.9-3b) assume the simple form

$$\frac{da}{dt} = -i\omega a \Rightarrow a(t) = a(0)e^{-i\omega t} \tag{17.9-4a}$$

$$\frac{da^\dagger}{dt} = i\omega a^\dagger \Rightarrow a^\dagger(t) = a^\dagger(0)e^{i\omega t} \tag{17.9-4b}$$

As in Section 17.5, a synchronous driving term results from the product $e^{-2i\omega t}a^\dagger(t)$ in (17.9-4a) and $e^{2i\omega t}a(t)$ in (17.9-4b). By neglecting the nonsynchronous terms in (17.9-3a) and (17.9-3b), we obtain

$$\frac{da}{dt} = -i\omega a - isa^\dagger e^{-2\omega t} \tag{17.9-5a}$$

$$\frac{da^\dagger}{dt} = i\omega a^\dagger + isae^{i2\omega t} \tag{17.9-5b}$$

It is convenient to uncouple (17.9-5a) and (17.9-5b) before attempting a solution. With this in mind, we define the quadrature-phase amplitude operators

$$X_1(t) = \tfrac{1}{2}[a(t)e^{i(\omega t+\pi/4)} + a^\dagger(t)e^{-i(\omega t+\pi/4)}], \tag{17.9-6a}$$

$$X_2(t) = -\frac{i}{2}[a(t)e^{i(\omega t+\pi/4)} - a^\dagger(t)e^{-i(\omega t+\pi/4)}] \tag{17.9-6b}$$

Inverting (17.9-6a) and (17.9-6b), we find that

$$a(t) = [X_1(t) + iX_2(t)]e^{-i(\omega t+\pi/4)} \tag{17.9-7a}$$

$$a^\dagger(t) = [X_1(t) - iX_2(t)]e^{i(\omega t+\pi/4)} \tag{17.9-7b}$$

Substituting (17.9-7a) and (17.9-7b) into (17.9-1), we can write the signal electric field operator as

$$\mathbf{E}_s(\mathbf{r}, t) = \left(\frac{2\hbar\omega}{\varepsilon}\right)^{1/2} \mathbf{E}_s(\mathbf{r})[X_1(t)\sin(\omega t + \pi/4) - X_2(t)\cos(\omega t + \pi/4)] \tag{17.9-8}$$

$X_1(t)$ is the operator corresponding to the amplitude of the "in-phase" component, that is, the one in phase with a field $E_0 \sin[\phi_p(t)/2]$, where $\phi_p(t)$ is defined by (17.9-1a). $X_2(t)$ is thus the amplitude of the "out of phase" amplitude.

By combining (17.9-5a) and (17.9-5b) with (17.9-6a) and (17.9-6b), we obtain the Heisenberg equations of motion:

$$\frac{dX_1}{dt} = sX_1 \tag{17.9-9a}$$

$$\frac{dX_2}{dt} = -sX_2 \tag{17.9-9b}$$

Equations (17.9-9a) and (17.9-9b) have the simple solutions

$$X_1(t) = X_1(0)e^{st} \tag{17.9-10a}$$

$$X_2(t) = X_2(0)e^{-st} \tag{17.9-10b}$$

Classically, we would interpret equations (17.9-10) as stating that a degenerate parametric amplifier will amplify (for $s > 0$) the "in-phase" input field at ω and attenuate the "out-of-phase" component. We recall that here X_1 and X_2 are operators—but the classical conclusion will carry through also in the quantum analysis.

From their definition (17.9-6) it follows that the quadrature-phase amplitude operators X_1 and X_2, unlike the creation and annihilation operators, are Hermitian; they correspond to physically observable quantities and, hence, can be measured. They cannot, however, be measured simultaneously with arbitrary accuracy, since X_1 and X_2 do not commute. From (17.9-6a), (17.9-6b), and the commutation relations (5.6-8), we find that

$$[X_1(t), X_2(t)] = [X_1(0), X_2(0)] = \frac{i}{2} \tag{17.9-11a}$$

Just as the noncommutivity of position and momentum imposes the familiar uncertainty principle [see Eq. (1.2-19)] $\langle \Delta x^2 \rangle^{1/2} \langle \Delta p^2 \rangle^{1/2} \geq \hbar/2$, the commutation relation (17.9-11a) imposes the uncertainty principle

$$\langle \Delta X_1^2(t) \rangle^{1/2} \langle \Delta X_2^2(t) \rangle^{1/2} \geq \tfrac{1}{4} \tag{17.9-11b}$$

Matrix elements involving the operators $X_1(t)$ and $X_2(t)$ can be obtained straightforwardly by using their definition, equations (17.9-6). As an example, taking the input initial state as a coherent state [see (5.8-12)]

$$|\psi_c(0)\rangle = \sum_{n=0}^{\infty} \left(\frac{e^{-\lambda} \lambda^n}{n!} \right)^{1/2} e^{in\phi} |n\rangle$$

we can obtain by using (17.9-6)

$$\begin{aligned}
\langle X_1(t) \rangle &= \langle \psi_c(0) | X_1(t) | \psi_c(0) \rangle \\
&= \tfrac{1}{2} \langle \psi_c(0) | a e^{i(\omega t + \pi/4)} + a^\dagger e^{-i(\omega t + \pi/4)} | \psi_c(0) \rangle \\
\langle X_1^2(t) \rangle &= \tfrac{1}{4} \langle \psi_c(0) | a^\dagger a + a a^\dagger + i a^2 e^{2i\omega t} - i (a^\dagger)^2 e^{-2i\omega t} | \psi_c(0) \rangle
\end{aligned}$$

Following a method similar to that used to arrive at (5.8-20) we can show straightforwardly that for an initial coherent state $|\lambda\rangle$ in free space ($s = 0$)

$$\langle \Delta X_1^2(t) \rangle^{1/2} \equiv \{\langle \lambda | [X_1(t) - \langle X_1(t) \rangle]^2 | \lambda \rangle\}^{1/2}$$

$$= [\langle X_1^2(t) \rangle - \langle X_1(t) \rangle^2]^{1/2} \qquad (17.9\text{-}12a)$$

$$= \tfrac{1}{2}$$

$$\langle \Delta X_2^2(t) \rangle^{1/2} = \tfrac{1}{2} \qquad (17.9\text{-}12b)$$

independent of the initial average number of quanta λ, so that in free space

$$\langle \Delta X_1^2(t) \rangle^{1/2} \langle \Delta X_2^2(t) \rangle^{1/2} = \tfrac{1}{4} \qquad (17.9\text{-}13)$$

which is another manifestation of the minimum uncertainty of the coherent states discussed in Section 5.8 and specifically in (5.8-24). The same result holds true for the vacuum state, since the vacuum state is a coherent state $|\lambda\rangle$ with $\lambda = 0$. The vacuum fluctuations are thus equally divided between the two quadratures. A squeezed state is defined as one that possesses a quadrature component whose fluctuation (noise) $\langle \Delta X_1^2 \rangle$ or $\langle \Delta X_2^2 \rangle$ is reduced below the value of a coherent state, the corresponding orthogonal quadrature having excess noise as a consequence. Mathematically, we say that

$$\langle \Delta X_1^2(t) \rangle^{1/2} = \frac{g}{2} \qquad (17.9\text{-}14a)$$

$$\langle \Delta X_2^2(t) \rangle^{1/2} = \frac{1}{2g} \qquad (17.9\text{-}14b)$$

If $g < 1$, then the quadrature X_1 is squeezed; if $g > 1$, the quadrature X_2 is squeezed.

The output of a degenerate parametric amplifier with a coherent state input is squeezed, as is seen by examining (17.9-10). For such an input, the quadrature-phase uncertainties at the output are

$$\langle \Delta X_1^2(t) \rangle^{1/2} = \langle \Delta X_1^2(0) \rangle^{1/2} e^{st} = \tfrac{1}{2} e^{st} \qquad (17.9\text{-}15a)$$

$$\langle \Delta X_2^2(t) \rangle^{1/2} = \langle \Delta X_2^2(0) \rangle^{1/2} e^{-st} = \tfrac{1}{2} e^{-st} \qquad (17.9\text{-}15b)$$

where we have used (17.9-12a) and (17.9-12b). Here e^{st} plays the role of g in (17.9-14a) and (17.9-14b). If $s < 0$, the quadrature X_1 is squeezed; if $s > 0$, the quadrature X_2 is squeezed. Since the noise is no longer equally distributed between the two quadratures, the noise is *phase-sensitive*. We will show explicitly the phase-sensitivity of the noise when we describe the process used to detect squeezed states.

One might ask what happens to the squeezing as the field propagates from the squeezer to the detector, since there is usually some distance between the two. As we mentioned previously, the quadrature-phase amplitudes are constant operators in free space. Their value beyond the squeezer is the same as that at the output end of the squeezer. Using (17.9-8), the electric field operator at the detector input can be written as

$$\mathbf{E}_S(\mathbf{r}, t) = \left(\frac{2\hbar\omega}{\varepsilon}\right)^{1/2} \mathbf{E}(\mathbf{r})[X_1(t_l)\sin(\omega t + \pi/4) - X_2(t_l)\cos(\omega t + \pi/4)] \qquad (17.9\text{-}16)$$

where $t > t_I$. Here the subscript S denotes that the field to be detected is the parametrically amplified signal at ω. The interaction time t_I is the time required for the field to propagate through the squeezer (in our case a degenerate parametric amplifier).

A balanced homodyne detector,[28–30] shown in Figure 17.18, is preferred for detecting squeezed states. The signal beam $\mathbf{E}_S(\mathbf{r}, t)$ is mixed with a powerful local oscillator (LO) beam $\mathscr{E}_{LO}(\mathbf{r}, t)$ of the same frequency by a 50-50 beamsplitter. The LO can be expressed as the sum of a monochromatic field at frequency ω plus a fluctuation term:

$$\mathscr{E}_{LO}(\mathbf{r}, t) = \left(\frac{2\hbar\omega}{\varepsilon}\right)^{1/2} [\mathbf{E}_{LO}(\mathbf{r}) \cos(\omega t + \phi_{LO}) + \tfrac{1}{2}\, \Delta\mathbf{E}_{LO}(\mathbf{r}, t)] \tag{17.9-17}$$

where, treating henceforth the powerful LO field classically,

$$\langle \Delta\mathbf{E}_{LO}(\mathbf{r}, t)\rangle_{\substack{\text{time}\\\text{average}}} = 0 \tag{17.9-17a}$$

Define

$$\mathbf{E}_{LO}(\mathbf{r}, t) = \left(\frac{2\hbar\omega}{\varepsilon}\right)^{1/2} \mathbf{E}_{LO}(\mathbf{r}) \cos(\omega t + \phi_{LO}) \tag{17.9-18}$$

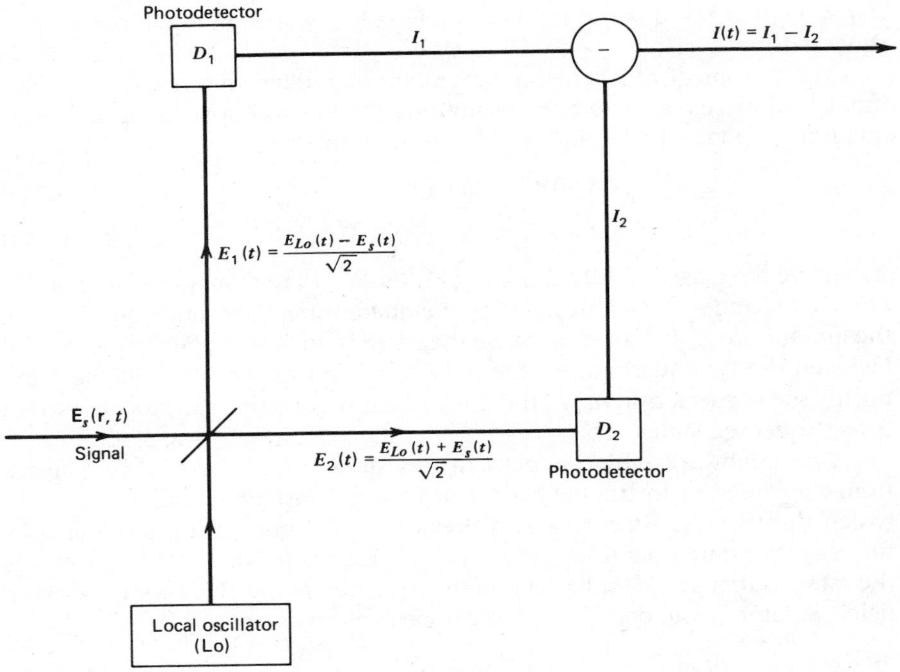

FIGURE 17.18 A balanced homodyne detector.

We will find it useful to decompose the monochromatic and the fluctuation term into their positive ($\propto e^{-i\omega t}$) and negative ($\propto e^{i\omega t}$) frequency components:

$$\mathbf{E}_{\text{LO}}(\mathbf{r}, t) = \mathbf{E}_{\text{LO}}^{(+)}(\mathbf{r}, t) + \mathbf{E}_{\text{LO}}^{(-)}(\mathbf{r}, t) \tag{17.9-19a}$$

where

$$\mathbf{E}_{\text{LO}}^{(+)}(\mathbf{r}, t) = \mathbf{E}_{\text{LO}}^{(-)}(\mathbf{r}, t)^* = \left(\frac{\hbar\omega}{2\varepsilon}\right)^{1/2} \mathbf{E}_{\text{LO}}(\mathbf{r})e^{-i\phi_{\text{LO}}}e^{-i\omega t} \tag{17.9-19b}$$

and

$$\Delta\mathbf{E}_{\text{LO}}(\mathbf{r}, t) = \Delta\mathbf{E}_{\text{LO}}^{(+)}(\mathbf{r}, t) + \Delta\mathbf{E}_{\text{LO}}^{(-)}(\mathbf{r}, t) \tag{17.9-20}$$

The signal field operator can be decomposed in a similar manner. We find

$$\mathbf{E}_{\text{S}}(\mathbf{r}, t) = \mathbf{E}_{\text{S}}^{(+)}(\mathbf{r}, t) + \mathbf{E}_{\text{S}}^{(-)}(\mathbf{r}, t) \tag{17.9-21a}$$

where

$$\begin{aligned}\mathbf{E}_{\text{S}}^{(+)}(\mathbf{r}, t) = [\mathbf{E}_{\text{S}}^{(-)}(\mathbf{r}, t)]^\dagger &= -\left(\frac{\hbar\omega}{2\varepsilon}\right)^{1/2} \mathbf{E}_{\text{S}}(\mathbf{r})[X_2(t_l) - iX_1(t_l)]e^{-i\pi/4}e^{-i\omega t} \\ &\equiv \left(\frac{\hbar\omega}{2\varepsilon}\right)^{1/2} \Delta\mathbf{E}_{\text{S}}^{(+)}(\mathbf{r}, t)\end{aligned} \tag{17.9-21b}$$

The 50-50 beamsplitter mixes the LO and the signal beams, producing the following linear superpositions of the two at its outputs:

$$E_1(t) = \frac{\mathscr{E}_{\text{LO}}(t) - \mathscr{E}_{\text{S}}(t)}{\sqrt{2}} = E_1^{(+)}(t) + E_1^{(-)}(t) \tag{17.9-22a}$$

$$E_2(t) = \frac{\mathscr{E}_{\text{LO}}(t) + \mathscr{E}_{\text{S}}(t)}{\sqrt{2}} = E_2^{(+)}(t) + E_2^{(-)}(t) \tag{17.9-22b}$$

where

$$E_1^{(+)}(t) = [E_1^{(-)}(t)]^\dagger = \frac{1}{\sqrt{2}}\left\{E_{\text{LO}}^{(+)}(t) + \left(\frac{\hbar\omega}{2\varepsilon}\right)^{1/2} [\Delta E_{\text{LO}}^{(+)}(t) - \Delta E_{\text{S}}^{(+)}(t)]\right\} \tag{17.9-23a}$$

$$E_2^{(+)}(t) = [E_2^{(-)}(t)]^\dagger = \frac{1}{\sqrt{2}}\left\{E_{\text{LO}}^{(+)}(t) + \left(\frac{\hbar\omega}{2\varepsilon}\right)^{1/2} [\Delta E_{\text{LO}}^{(+)}(t) + \Delta E_{\text{S}}^{(+)}(t)]\right\} \tag{17.9-23b}$$

The sign difference between eqs. (17.9-22a) and (17.9-22b) is due to energy conservation; the field energy incident at the input ports of the beamsplitter must be equal to the field energy leaving the output ports.[28] We have dropped the vector notation, since we are assuming that the signal and the LO are polarized along the same direction; we retain the field amplitudes only. The electric fields E_1 and E_2 are incident on the photodetectors D_1 and D_2, respectively. The photodetectors respond to the normally ordered part of the incident power flux, $E^{(-)}(t)E^{(+)}(t)$, since the remaining terms, of the type like $E^{(+)2}(t)$ and $E^{(-)2}(t)$, oscillate at optical frequencies, far beyond the bandwidth of any photodetector. An ideal photodetector emits one photoelectron for each incident photon, so the photocurrent operator is obtained by using the classical expression for the current but treating the fields as operators. Taking the electron charge as e and the area of the detector as σ_{det}, we find that the current operators are

$$\begin{aligned} I_1(t) &= \frac{2e\sigma_{\text{det}}}{\hbar\omega}\sqrt{\frac{\varepsilon}{\mu}}\,E_1^{(-)}(t)E_1^{(+)}(t) \\ &= \frac{ec\sigma_{\text{det}}}{2}\left\{\frac{A^2}{V} + \frac{A}{\sqrt{V}}\,[[\Delta E_{\text{LO}}^{(-)}(t) - \Delta E_{\text{S}}^{(-)}(t)]e^{-i(\omega t+\phi_{\text{LO}})} \right. \\ &\quad + [\Delta E_{\text{LO}}^{(+)}(t) - \Delta E_{\text{S}}^{(+)}(t)]e^{i(\omega t+\phi_{\text{LO}})}] \\ &\quad \left. + [\Delta E_{\text{LO}}^{(-)}(t) - \Delta E_{\text{S}}^{(-)}(t)][\Delta E_{\text{LO}}^{(+)}(t) - \Delta E_{\text{S}}^{(+)}(t)]\right\} \end{aligned} \tag{17.9-24a}$$

$$\begin{aligned} I_2(t) &= \frac{2e\sigma_{\text{det}}}{\hbar\omega}\sqrt{\frac{\varepsilon}{\mu}}\,E_2^{(-)}(t)E_2^{(+)}(t) \\ &= \frac{e\sigma_{\text{det}}}{2}\left\{\frac{A^2}{V} + \frac{A}{\sqrt{V}}\,[[\Delta E_{\text{LO}}^{(-)}(t) + \Delta E_{\text{S}}^{(-)}(t)]e^{-i(\omega t+\phi_{\text{LO}})} \right. \\ &\quad + [\Delta E_{\text{LO}}^{(+)}(t) + \Delta E_{\text{S}}^{(+)}(t)]e^{i(\omega t+\phi_{\text{LO}})}] \\ &\quad \left. + [\Delta E_{\text{LO}}^{(-)}(t) + \Delta E_{\text{S}}^{(-)}(t)][\Delta E_{\text{LO}}^{(+)}(t) + \Delta E_{\text{S}}^{(+)}(t)]\right\} \end{aligned} \tag{17.9-24b}$$

where we have replaced $\mathbf{E}_{\text{LO}}(\mathbf{r})$ of (17.9-19b) by $A/\sqrt{V}$, A being a dimensionless amplitude. We will assume that A is large enough that the last term of (17.9-24a) and of (17.9-24b) can be neglected.

A balanced homodyne detector takes the photocurrents from the two detectors and subtracts them coherently:

$$\begin{aligned} I(\phi_{\text{LO}}) &= I_1(t) - I_2(t) \\ &= -\frac{ecA\sigma_{\text{det}}}{\sqrt{V}}\,[\Delta E_{\text{S}}^{(-)}(t)e^{-i(\omega t+\phi_{\text{LO}})} + \Delta E_{\text{S}}^{(+)}(t)e^{i(\omega t+\phi_{\text{LO}})}] \end{aligned} \tag{17.9-25}$$

Let $\phi_{\text{LO}} = \pi/4 + \phi$. Using (17.9-21), we can write (17.9-25) as

$$I(\phi) = \frac{2ecA\sigma_{\text{det}}}{V}\,[X_1(t_I)\sin\phi + X_2(t_I)\cos\phi] \tag{17.9-26}$$

where we have replaced $\mathbf{E}_{\text{S}}(\mathbf{r})$ of (17.9-21) by $1/\sqrt{V}$. Notice that the fluctuations from the local oscillator cancel and make no contribution to the photocurrent.

By changing the LO phase ϕ_{LO}, and thus ϕ, we can, according to (17.9-26), make the current operator $I(\phi)$ proportional to any relative combination of X_1 and X_2. In particular, for $\phi = \pi/2$ (that is, $\phi_{\text{LO}} = 3\pi/4$), the current is proportional to $X_1(t_I)$; for $\phi = 0$ (that is, $\phi_{\text{LO}} = \pi/4$), the current is proportional to $X_2(t_I)$. A more general argument makes this even clearer. We can re-express the signal field (17.9-8) by using trigonometric identities so that it possesses the same argument as the local oscillator field (17.9-17):

$$\mathbf{E}_S(\mathbf{r}, t) = -\left(\frac{2\hbar\omega}{\varepsilon}\right)^{1/2} \mathbf{E}_S(\mathbf{r})\{[X_1(t_I) \sin\phi + X_2(t_I)\cos\phi]\cos(\omega t + \phi_{LO}) - [X_1(t_I)\cos\phi + X_2(t_I)\sin\phi]\sin(\omega t + \phi_{LO})\} \quad (17.9\text{-}27)$$

where $\phi \equiv \phi_{LO} - \pi/4$. The coefficient of the $\cos(\omega t + \phi_{LO})$ term on the right side of (17.9-27) is proportional to the detector current operator $I(\phi)$ in (17.9-26), so that, at any phase ϕ_{LO}, the output of a balanced homodyne detector is proportional to the quadrature amplitude *which is in phase* with the local oscillator.[31]

To illustrate the phase-dependence of $I(\phi)$, we will consider a particular example in greater detail. Suppose that we let our squeezer—a degenerate parametric amplifier—amplify the zero-point vibrations of the electromagnetic field at ω. This corresponds to having no input at all. In this case, $\langle X_1(t_I)\rangle = \langle X_2(t_I)\rangle = 0$, so that

$$\langle \Delta X_1^2(t_I)\rangle = \langle [X_1(t_I) - \langle X_1(t_I)\rangle]^2\rangle = \langle X_1^2(t_I)\rangle \quad (17.9\text{-}28a)$$

$$\langle \Delta X_2^2(t_I)\rangle = \langle [X_2(t_I) - \langle X_2(t_I)\rangle]^2\rangle = \langle X_2^2(t_I)\rangle \quad (17.9\text{-}28b)$$

where

$$\langle X_1^2(0)\rangle = \langle X_2^2(0)\rangle = \tfrac{1}{2} \quad (17.9\text{-}29)$$

Since the expectation values of the quadrature-phase amplitude operators $X_1(t)$ and $X_2(t)$ are zero for the case we are studying it follows from (17.9-26) that, $\langle I(\phi)\rangle = 0$; we will thus not get very far by making direct measurements of the current (operator) $I(\phi)$. To make accessible the second order moments like $\langle X_1^2(t_I)\rangle$ and $\langle X_2^2(t_I)\rangle$, we must square the output current $I(\phi)$ and perform our measurements on the *squared* output current operator $I^2(\phi)$. Clearly we need a square-law detector for the current. A spectrum analyzer measures the power of an input signal $f(t)$ due to frequencies between $\omega - \Delta\omega/2$ and $\omega + \Delta\omega/2$; the power in this interval is given by

$$P_\omega = \int_{\omega-\Delta\omega/2}^{\omega+\Delta\omega/2} W_f(\omega)\, d\omega \quad (17.9\text{-}30)$$

where $W_f(\omega)$ is the *spectral-density function*. From (13.3-4), the spectral density function is given as the Fourier transform of the autocorrelation function $\phi_f(\tau)$. For the present case, the autocorrelation function is independent of τ, since the current $I(\phi)$ is independent of time. From (13.3-3),

$$\phi_I(\tau) = \lim_{T\to\infty} \frac{1}{2T}\int_{-T}^{T} I^2(\phi)\, dt = I^2(\phi) \quad (17.9\text{-}31)$$

The spectral density function is thus

$$W_I(\omega) = \frac{1}{\pi}\int_{-\infty}^{\infty} \phi_I(\tau)e^{-i\omega\tau}\, d\tau = \frac{I^2(\phi)}{\pi}\int_{-\infty}^{\infty} e^{-i\omega\tau}\, d\tau = 2I^2(\phi)\delta(\omega) \quad (17.9\text{-}32)$$

We do not measure $W_I(\omega)$ directly, but rather integrals of $W_I(\omega)$ over small frequency intervals. The interval we are interested in is one about $\omega = 0$; the power in this interval is

$$P_0 = \int_{-\Delta\omega/2}^{\Delta\omega/2} W_I(\omega)\, d\omega = 2I^2(\phi)R \int_{-\Delta\omega/2}^{\Delta\omega/2} \delta(\omega)\, d\omega = 2I^2(\phi)R \qquad (17.9\text{-}33)$$

where R is an effective resistance.

By squaring the output current $I(\phi)$ and choosing the appropriate local oscillator phase, we can measure directly the uncertainties in the quadrature-phase amplitudes or combinations thereof.

$$\left\langle \left(\frac{I(\phi)}{I_0}\right)^2 \right\rangle = \langle X_1^2(t_I)\rangle \sin^2\phi + \tfrac{1}{2}\langle X_1(t_I)X_2(t_I) + X_2(t_I)X_1(t_I)\rangle \sin 2\phi + \langle X_2^2(t_I)\rangle \cos^2\phi \qquad (17.9\text{-}34)$$

$$= \tfrac{1}{4}(e^{2st_I}\sin^2\phi + e^{-2st_I}\cos^2\phi)$$

since $\langle X_1(t_I)X_2(t_I) + X_2(t_I)X_1(t_I)\rangle = 0$ for a vacuum state, and, from (17.9-29), $\langle X_{1,2}^2(0)\rangle = \frac{1}{2}$. Here $I_0 = 2ecA\sigma_{\text{det}}/V$.

If we let $r \equiv \langle (I(\phi)/I_0)^2\rangle^{1/2}$, then we can plot (17.9-34) by using polar coordinates, as in Figure 17.19. The dotted circle in Figure 17.19 shows the rms current $r(\phi)$ for $st_I = 0$, corresponding to the noise in an incident vacuum field. The vacuum noise, often referred to as shot noise, is clearly insensitive to the phase of the local oscillator. Figure 17.19 also shows the corresponding rms current for a *squeezed* vacuum state with $st_I = \ln 4$. We see clearly the reduced fluctuations for $\phi = 0$, where the rms current is equal to the uncer-

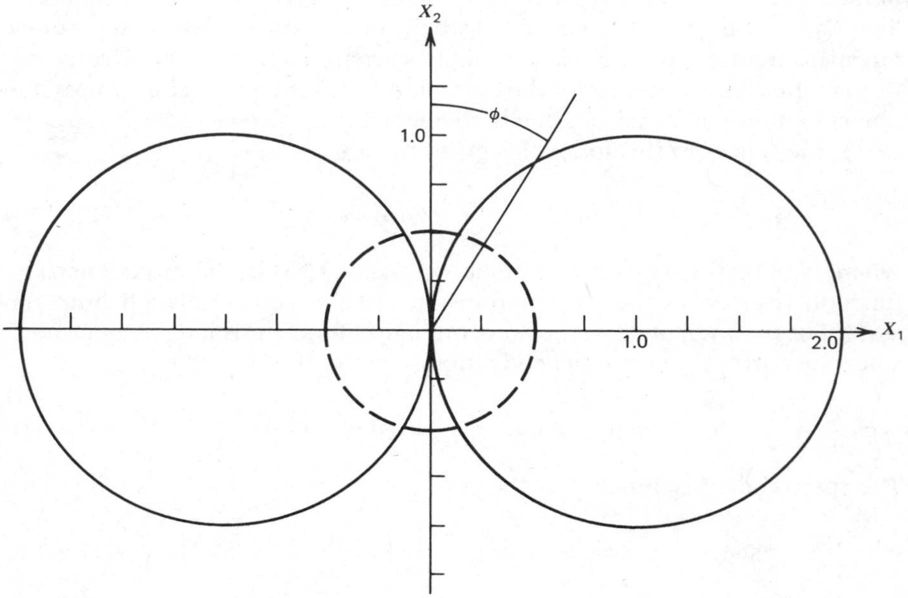

FIGURE 17.19 The rms current as a function of LO phase ϕ. The dashed line is for a vacuum state ($st_I = 0$), and the solid line is for a squeezed vacuum state ($st_I = \ln 4$).

tainty in the squeezed quadrature X_2, and the increased fluctuations for $\phi = \pi/2$, where the rms current is equal to the uncertainty in the amplified quadrature X_1. The phase-sensitivity of the noise in a squeezed vacuum state is obvious; any fluctuation in the phase of the local oscillator near $\phi = 0$ will degrade the observed squeezing by mixing part of the amplified quadrature X_1 with the squeezed quadrature X_2.

A typical experimental setup, which was used to observe squeezing, is shown in Figure 17.20.[32] Here a degenerate parametric amplifier (in an oscillator configuration) acts as the squeezer. The effective length of the nonlinear medium is increased by placing the crystal in an optical cavity, since it allows the light to make multiple passes through the crystal, increasing the interaction time. Notice that the pump also acts as the source for the local oscillator. Figure 17.21 shows some of the results of this experiment. The rms noise voltage $V(\phi)$ is plotted as a function of the local oscillator phase ϕ. Here the noise voltage corresponds to our rms noise current $r(\phi)$ plotted in Figure 17.19. For certain values of ϕ, the noise voltage dips below the vacuum level, given by the dotted line in Figure 17.21.

Until now we have discussed only the parametric amplifier as a generator of squeezed light. There are others, however. Squeezed light has also been generated by forward and backward four-wave mixing in a vapor of sodium atoms,[33,34] and by forward four-wave mixing in an optical fiber.[35] It is hoped that squeezed light will find applications in communications,[36–39] spectroscopy,[40] and high-precision measurements.[41]

To illustrate some of the concepts discussed in this section we might resort to the drawing of Figure 17.20. In (*a*) we represent the expectation value of the amplitude $\langle x_1 \rangle$ of the coherent state electric field x_1 by the horizontal arrow so that $\langle x_2 \rangle = 0$. An actual measurement will discover the tip of the field vector $x_1 + ix_2$ to fall somewhere within the uncertainty circle whose radius according to (17.9-13) is $\langle (\Delta x_1)^2 \rangle^{1/2} = \frac{1}{2}$. The circle thus represents the uncertainty, that is, noise in the measurement of the total field amplitude

$$\sqrt{x_1^2 + x_2^2}$$

If, however, the coherent state is subjected to squeezing, the uncertainty area is distorted as shown. The uncertainty in x_1 decreases while that of x_2 increases, or vice versa (the area remains constant). The vacuum state and the squeezed vacuum state are represented in (*c*) and (*d*), respectively. The situation is similar to that of (*a*) and (*b*) except that here $\langle x_1 \rangle = \langle x_2 \rangle = 0$. Consider, for example, an optical communication system that is based on amplitude modulation. At the receiver the task is to measure the amplitude, say, $x_1(t)$ of the wave. Such a measurement is subject to an uncertainty (noise) of Δx_1. This uncertainty is ~ 1 in the case of the unsqueezed coherent state Figure 17.20*a*, but is reduced to $\Delta x_1 \sim e^{-st}$ if we use as the carrier the x_1 quadrature component of the squeezed state that is represented in 17.20*b* and use a phase coherent detection scheme. The increased resolution, that is, less uncertainty in determining x_1 signifies a higher information data rate, that is, less noise in the channel.

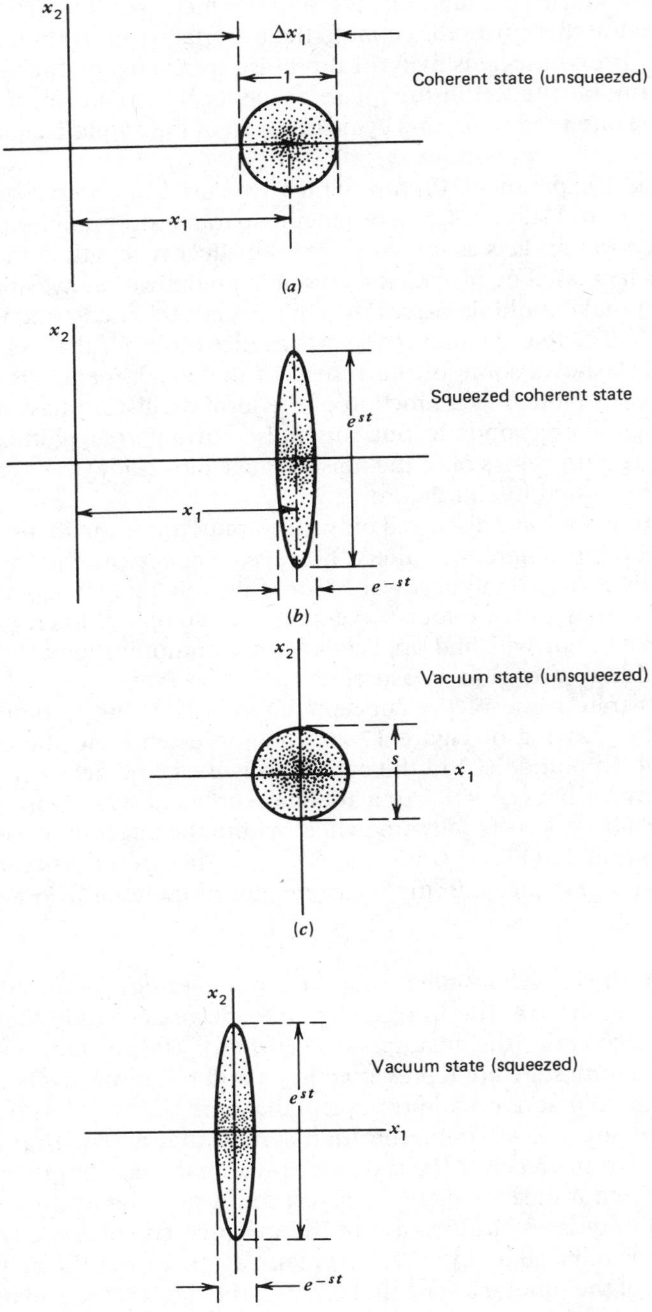

FIGURE 17.20 (*a*) Coherent state (unsqueezed). (*b*) Squeezed coherent state. (*c*) Vacuum state (unsqueezed). (*d*) Squeezed vacuum state.

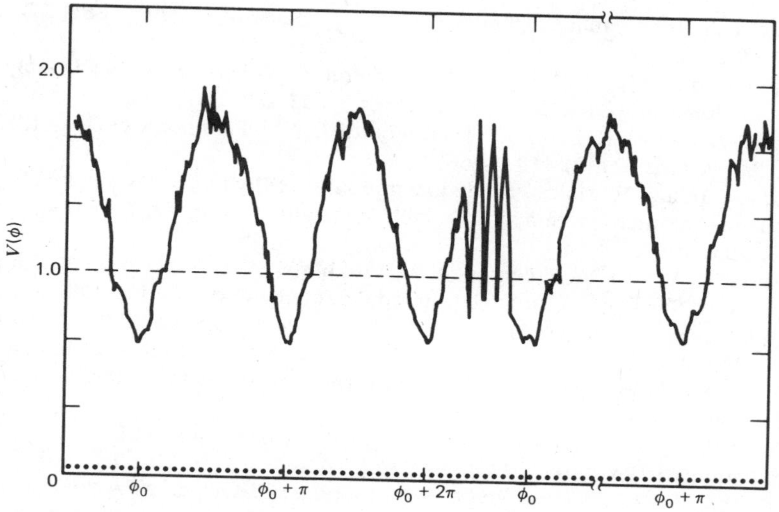

FIGURE 17.21 Measurement of the phase dependence of the quantum fluctuations in a squeezed state produced by parametric down conversion. The phase dependence of the rms noise voltage from a balanced homodyne detector is displayed as a function of local oscillator phase ϕ at fixed analysis frequency (1.8 MHz) and bandwidth (100 KHz) in the spectral distribution of photocurrent fluctuations. With the optical parametric oscillator (0P0) blocked, the vacuum field entering the signal port of the detector produces the noise voltage given by the dashed line with no sensitivity on ϕ. With the 0P0 input present, the dips below the vacuum level represent a 50% reduction in noise power relative to the vacuum noise level. Note that the ordinate is a linear scale in noise voltage. The dotted line is the amplifier noise level. (After Reference 32)

References

1. Giordmaine, J. A. and R. C. Miller, "Tunable Coherent Parametric Oscillation in $LiNbO_3$ at Optical Frequencies," *Phys. Rev. Letters* **14,** 973 (1965).
2. Akhmanov, S. A., A. I. Kovrigin, A. S. Piskarskas, V. V. Fadeev, and R. V. Khokhlov, "Observation of Parametric Amplification in the Optical Range," *Zh. Eksper. Teor. Fiz. Pis'ma* (USSR) **2,** 300 (1965). See, also, S. A. Akhmanov, A. I. Kourigin, V. A. Kolosov, A. S. Piskarskas, V. V. Fadeev, and R. V. Khokhlov, "Tunable Parametric Light Generator with KDP Crystal," *Zh. Eksper. Teor. Fiz. Pis'ma* (USSR) **3,** 372 (1966).
3. Lord Rayleigh, "On maintained vibrations," *Phil. Mag.*, vol. 15, ser. 5, pt. I, p. 229 (1983).
4. Uenohara, M., "Low Noise Amplification," Vol. 23, *Handbuch der Physik* (Berlin: Springer-Verlag, 1962).
5. Yariv, A., *Quantum Electronics,* 1st ed. (New York: Wiley, 1967).
6. Manley, J. M. and H. E. Rowe, "General Energy Relations in Nonlinear Reactances," *Proc. IRE* **47,** 2115 (1959).
7. Falk, J. and J. E. Murray, "Single Cavity Nonlinear Parametric Oscillator," *Appl. Phys. Letters* **14,** 245 (1969).

8. Harris, S. E., "Tunable Optical Parametric Oscillators," *Proc. IEEE* **57,** 2096 (1969).
9. Bjorkholm, J. E., "Efficient Optical Parametric Oscillation Using Doubly and Singly Resonant Cavities," *Appl. Phys. Letters* **13,** 53 (1968).
10. Yariv, A. and W. H. Louisell, "Theory of the Optical Parametric Oscillator," *IEEE J. Quant. Elect.* **QE-2,** 418 (1966).
11. Magde, D. and H. Mahr, "Study in Ammonium Dihydrogen Phosphate of Spontaneous Parametric Interaction Tunable from 4400 to 16000 Å," *Phys. Rev. Letters* **18,** 905 (1967).
12. Davydov, A. A., L. A. Kulevskii, A. M. Prokhorov, A. D. Savel'ev, V. V. Smirnov, and A. V. Shirkov, "A Tunable Infrared Parametric Oscillator in a CdSe Crystal," *Optics Comm.* **9,** 234 (1973). Also, R. L. Herbst and R. L. Byer, "Efficient Parametric Mixing in CdSe," *Appl. Phys. Letters* **19,** 527 (1971); Herbst, R. L. and R. L. Byer, "Singly Resonant CdSe Infrared Parametric Oscillator," *Appl. Phys. Letters* **21,** 189 (1972).
13. Louisell, W. H., A. Yariv, and A. E. Siegman, "Quantum Fluctuations and Noise in Parametric Processes," *Phys. Rev.* **124,** 1646 (1961).
14. Louisell, W. H., *Radiation and Noise in Quantum Electronics* (New York: McGraw Hill, 1964).
15. Johnson, F. M. and J. A. Durado, "Frequency Up-Conversion," *Laser Focus* **3,** 31 (1967).
16. Midwinter, J. E. and J. Warner, "Up-Conversion of Near Infrared to Visible Radiation in Lithium-Meta-Niobate," *J. Appl. Phys.* **38,** 519 (1967).
17. Warner, J., "Photomultiplier Detection of 10.6 μ Radiation Using Optical Up-Conversion in Proustite," *Appl. Phys. Letters* **12,** 222 (1968).
18. Hulme, K. F., O. Jones, P. H. Davies, and M. V. Hobden, "Synthetic Proustite (Ag_3AsS_3): A New Material for Optical Mixing," *Appl. Phys. Letters* **10,** 133 (1967).
19. T. R. Gurski, "High Quantum Efficiency Infrared Up-Conversion," *Appl. Phys. Letters* **23,** 273 (1973).
20. Harris, S. E., M. K. Oshman, and R. L. Byer, "Observation of Tunable Optical Parametric Fluorescence," *Phys. Rev. Letters* **18,** 732 (1967).
21. Byer, R. L. and S. E. Harris, "Power and Bandwidth Spontaneous Parametric Emission," *Phys. Rev.* **168,** 1064 (1968).
22. Giallorenzi, T. G. and C. L. Tang, "Quantum Theory of Spontaneous Parametric Scattering of Intense Light," *Phys. Rev.* **166,** 225 (1968).
23. Pearson, J. E., U. Ganiel, and A. Yariv, "Observations of Parametric Fluorescence and Oscillation in the Infrared," *Applied Optics* **12,** 1165 (1973).
24. Pierce, J. R., *Traveling Wave Tubes* (Princeton, N.J.: Van Nostrand, 1950).
25. Louisell, W. H., *Coupled Modes and Parametric Electronics* (New York: Wiley, 1962).
26. Kogelnik, H. and C. V. Shank, "Coupled Wave Theory of Distributed Feedback Lasers," *J. Appl. Phys.* **43,** 2328 (1972).
27. Harris, S. E., "Proposed Backward Wave Oscillation in the Infrared," *Appl. Phys. Letters* **9,** 114 (1966).
28. Schumaker, B. L., "Noise in Homodyne Detection," *Optics Letters* **9,** 189 (1984).
29. Yuen, H. P. and Shapiro, J. H., "Generation and Detection of Two-Photon Coherent States in Degenerate Four-Wave Mixing," *Optics Letters* **4,** 334 (1979).
30. Yuen, H. P. and Chan, V. W. S., "Noise in Homodyne and Heterodyne Detection," *Optics Letters* **8,** 177 (1983).

31. Caves, C. M. and Schumaker, B. L., "Broadband Squeezing," in *Quantum Optics IV*, edited by J. D. Harvey and D. F. Walls (Springer, Berlin, 1986), pp. 20–30.
32. Wu, L., Kimble, H. J., Hall, J. L., and Wu, H., "Generation of Squeezed States by Parametric Down Conversion," *Phys. Rev. Lett.* **57,** 2520 (1986).
33. Slusher, R. E., Hollberg, L. W., Yurke, B., Mertz, J. C., and Valley, J. F., "Observation of Squeezed States Generated by Four-Wave Mixing in an Optical Cavity," *Phys. Rev. Lett.* **55,** 2409 (1985).
34. Maeda, M. W., Kumar, P., and Shapiro, J. H., "Observation of Squeezed Noise Produced by Forward Four-Wave Mixing in Sodium Vapor," *Optics Letters* **12,** 161 (1987).
35. Shelby, R. M., Levenson, M. D., Perlmutter, S. H., DeVoe, R. G., and Walls, D. F., "Broad-Band Parametric Deamplification of Quantum Noise in an Optical Fiber," *Phys. Rev. Lett.* **57,** 691 (1986).
36. Yuen, H. P. and Shapiro, J. H., "Optical Communication with Two-Photon Coherent States—Part I: Quantum-State Propagation and Quantum-Noise Reduction," *IEEE Trans. on Inf. Theory* IT–**24,** 657 (1978).
37. Shapiro, J. H., Yuen, H. P., and Machado Mata, J. A., "Optical Communication with Two-Photon Coherent States—Part II: Photoemissive Detection and Structured Receiver Performance," *IEEE Trans. on Inf. Theory* IT–**25,** 179 (1979).
38. Yuen, H. P. and Shapiro, J. H., "Optical Communication with Two-Photon Coherent States—Part III: Quantum Measurements Realizable with Photoemissive Detectors," *IEEE Trans. on Inf. Theory* IT–**26,** 78 (1980).
39. Yamamoto, Y. and Haus, H. A., "Preparation, Measurement and Information Capacity of Optical Quantum States," *Rev. Mod. Phys.* **58,** 1001 (1986).
40. Yurke, B. and Whittaker, E. A., "Squeezed-State-Enhanced Frequency-Modulation Spectroscopy," *Optic Letters* **12,** 236 (1987).
41. Caves, C. M., "Quantum-Mechanical Noise in an Interferometer," *Phys. Rev. D* **23,** 1693 (1981).
42. Special Issue on Squeezed States, *J. Optical Society of Am. B.*, Vol. 4, no. 10 (October 1987). Contains up-to-date review articles on squeezing.
43. Yariv, A., *Optical Electronics*, Holt, Rinehart and Winston, 3rd Ed., New York, 1985, p. 255.

Problems

17.1 In Section 17.5 we calculated $\langle n_1(t)\rangle$ using $\psi(0) = |n_{10}, n_{20}\rangle$. By specifying the number of quanta exactly we have destroyed, according to the third of (1.2-22), all the phase-information since $\Delta\phi = \infty$. This can be remedied by using for $\psi(0)$ a minimum-uncertainty Poisson packet (Reference 13)

$$\psi(0) = \sum_{\Sigma_{10}, n_{20}} [p(n_{10})p(n_{20})]^{1/2} e^{-i(n_{10}\phi_1 + n_{20}\phi_2)} |n_{10}, n_{20}\rangle$$

where the p's are the Poisson distribution functions so that

$$p(n_{i0}) = \frac{\exp(-\overline{n}_{i0})\overline{n}_{i0}^{\,n_{i0}}}{(n_{i0})!}$$

and where $\bar{n}_{i0}$ is the average number of quanta at ω_i at $t = 0$. Show that with this choice of $\psi(0)$ and in the limit $st \gg 1$

$$\langle \bar{n}_1(t) \rangle = \langle \psi(0) | a_1^+(t) a_1(t) | \psi(0) \rangle$$
$$= [1 + \bar{n}_{10} + \bar{n}_{20} + 2(\overline{n_{10} n_{20}})^{1/2} \sin(\phi_1 + \phi_2)] K$$

where K is the gain $\sim(\frac{1}{4})e^{st}$. Note that for $\bar{n}_{20} = 0$, the result is identical with (17.5-20).

17.2 (a) Describe quantum mechanically, in a manner based on Section 17.5, the operation of a frequency up-converter.
(b) Describe its noise behavior, that is, what is the output when $n_{10} = 0$.

17.3 Calculate the quantum mechanical variance $\langle (\Delta n_1^2) \rangle \equiv \langle (n_1 - \bar{n}_1)^2 \rangle$ for the case of the parametric amplifier that is initially in the state $\psi(0) = |n_{10}, n_{20}\rangle$.

17.4 Calculate the variance $\langle (\Delta n_3)^2 \rangle \equiv \langle (n_3 - \bar{n}_3)^2 \rangle$ of a frequency up-converter initially in the state $|n_1 = n_0, n_3 = 0\rangle$. Here, $\omega_3 = \omega_1 + \omega_2$ where 2 indicates the pump field and 3 the up-converted signal.

17.5 Derive the equations describing a backward-wave frequency up-converter assuming a pump wave at ω_2 and a single input $A_1(0)$ at ω_1. Compare qualitatively to the backward wave amplifier.

17.6 Derive (17.7-15).

CHAPTER 18

Third-Order Optical Nonlinearities—Stimulated Raman and Brillouin Scattering

18.0 INTRODUCTION

In this chapter we will consider third-order nonlinear optical processes—that is, phenomena due to an induced material polarization that is proportional to the third power of the electric field. The various phenomena that fit under this umbrella include the optical and dc Kerr effect, self-focusing, third-harmonic generation, stimulated Brillouin scattering, stimulated Raman scattering and optical phase conjugation.

18.1 THE NONLINEAR CONSTANTS

We distinguish between two classes of nonlinear third-order processes. The first group is due to local nonlinearities reflecting the nonlinear response of atoms or molecules to electric fields. In these cases, we may legitimately use the third-order nonlinear susceptibility $\chi^{(3)}$ of the medium. A typical member of this group is the phenomenon of third-harmonic generation. In the second category we have phenomena, which, although still involving the triple product of the electric field, are no longer describable in terms of local susceptibilities. This is due to the fact that, as in the case of Brillouin scattering, they involve extended interactions. We will consider first the phenomenology of the first group.

If we start with the energy function (16.2-4)

$$U(\mathbf{E}) = -\frac{\varepsilon_0}{2}\chi_{ij}E_iE_j - \frac{2}{3}d_{ijk}E_iE_jE_k - \chi_{ijkl}E_iE_jE_kE_l \qquad (18.1\text{-}1)$$

we obtain

$$P_i = -\frac{\partial U(\mathbf{E})}{\partial E_i} = \varepsilon_0\chi_{ij}E_j + 2d_{ijk}E_jE_k + 4\chi_{ijkl}E_jE_kE_l \qquad (18.1\text{-}2)$$

where, we recall, the expression (18.1-2) is valid only when the response is instantaneous [see the discussion leading to (16.2-4)] so that

$$P_i(t) = P_i[E(t)]$$

When this is the case, we obtain

$$\chi_{ijkl} = \chi_{ikjl} = \chi_{ilkj} \quad \text{etc.}$$

since the permutation of $ijkl$ involves merely a rearrangement of the corresponding fields in (18.1-1). By using the reasoning employed in connection with d_{ijk} in Chapter 16, we find that $\chi_{ijkl} \neq 0$ in any medium including centrosymmetric crystals and isotropic homogeneous media.

Next, we consider the case when the third-order nonlinear response is due to the presence of a number (from one to three) of optical waves. First, we will take up the case of frequency addition $\omega_4 = \omega_1 + \omega_2 + \omega_3$ of the fields

$$\mathbf{E}_1(\mathbf{r}, t) = \tfrac{1}{2}\mathbf{E}_1^{(\omega_1)}(\mathbf{r})e^{i\omega_1 t} + \text{c.c.}$$

$$\mathbf{E}_2(\mathbf{r}, t) = \tfrac{1}{2}\mathbf{E}_2^{(\omega_2)}(\mathbf{r})e^{i\omega_2 t} + \text{c.c.}$$

$$\mathbf{E}_3(\mathbf{r}, t) = \tfrac{1}{2}\mathbf{E}_3^{(\omega_3)}(\mathbf{r})e^{i\omega_3 t} + \text{c.c.}$$

If we take the total field $\mathbf{E}$ in (18.1-2) as the sum of $\mathbf{E}_1$, $\mathbf{E}_2$, and $\mathbf{E}_3$ then, limiting ourselves to the third-order polarization at $\omega_4 = \omega_1 + \omega_2 + \omega_3$

$$\begin{aligned} P_1(t) = 4\chi_{ijkl} &\times \tfrac{1}{2}\,[E_j^{(\omega_2)}e^{i\omega_1 t} + E_j^{(\omega_2)}e^{i\omega_2 t} + E_j^{(\omega_3)}e^{i\omega_3 t} + \text{c.c.}] \\ &\times \tfrac{1}{2}\,[j \text{ replaced by } k] \\ &\times \tfrac{1}{2}\,[j \text{ replaced by } l] \end{aligned} \tag{18.1-3}$$

Consider, for example, the contribution to P_i due to the specific triplet $E_s^{(\omega 1)}E_t^{(\omega 1)}E_u^{(\omega 3)}$. Recalling that in (18.1-3) we sum over all repeated indices,

$$\begin{aligned} P_i^{(\omega_4=\omega_1+\omega_2+\omega_3)} &= \frac{4}{8}\,(\chi_{istu} + \chi_{isut} + \chi_{itsu} + \chi_{iust} + \chi_{iuts} \\ &\quad + \chi_{itus})\, E_s^{(\omega_1)}E_t^{(\omega_2)}E_u^{(\omega_3)}e^{i(\omega_1+\omega_2+\omega_3)t} + \text{c.c.} \end{aligned}$$

If we operate in the instantaneous response regime, all the χ's are equal and

$$\begin{aligned} P_i^{(\omega_4=\omega_1+\omega_2+\omega_3)} &= \frac{4 \times 6}{8}\,\chi_{istu}E_s^{(\omega_1)}E_t^{(\omega_2)}E_u^{(\omega_3)}e^{i(\omega_1+\omega_2+\omega_3)t} + \text{c.c.} \\ &= R_e[P_i^{(\omega_4=\omega_1+\omega_2+\omega_3)}e^{i(\omega_1+\omega_2+\omega_3)t}] \end{aligned} \tag{18.1-4}$$

where the complex amplitude of the polarization is given by

$$P_i^{(\omega_4=\omega_1+\omega_2+\omega_3)} = 6\chi_{istu}E_s^{(\omega_1)}E_t^{(\omega_2)}E_u^{(\omega_3)} \tag{18.1-5}$$

It is important to realize here that the χ of (18.1-5) is the *same* as that of (18.1-2). The different numerical factor (six instead of four) merely reflects the fact that the experimental situation considered here is that of frequency addition. We will generalize (18.1-5) to the general non-instantaneous case, that is, the frequencies involved may exceed or approach those of resonant transitions in the medium, and write

$$P_i^{(\omega_4=\omega_1+\omega_2\pm\omega_3)} = 6\chi_{istu}^{(\omega_4=\omega_1+\omega_2\pm\omega_3)}E_s^{(\omega_1)}E_t^{(\omega_2)}E_u^{(\pm\omega_3)} \tag{18.1-6}$$

and the χ tensor element is now a function of the particular frequencies involved, their ordering, and *signs*. Also

$$E_u^{(-\omega_3)} \equiv [E_u^{(\omega_3)}]^*$$

$\chi_{istu}^{(\omega_4=\omega_1\pm\omega_2\pm\omega_3)}$ is invariant to a permutation of the subscripts provided the frequencies are permuted similarly. For example,

$$\chi_{istu}^{(\omega_4=\omega_1-\omega_2+\omega_3)} = \chi_{isut}^{(\omega_4=\omega_1+\omega_3-\omega_2)} \qquad (18.1\text{-}7)$$

since in both cases the electric field component at $(+\omega_3)$ is applied along the u axis and that at $(-\omega_2)$ along t. Note that the sign preceding the frequency tells us whether that frequency is "added" or "subtracted" in the physical phenomenon considered.

In an analogous fashion we find that, for example, in the case of third-harmonic generation the appropriate relation is

$$P_i^{(3\omega=\omega+\omega+\omega)} = \chi_{ijkl}^{(3\omega=\omega+\omega+\omega)} E_j^{(\omega)} E_k^{(\omega)} E_l^{(\omega)} \qquad (18.1\text{-}7)$$

or, in general,

$$P_i^{(\omega_4=\omega_1+\omega_2+\omega_3)} = D\chi_{ijkl}^{(\omega_4=\omega_1+\omega_2+\omega_3)} E_j^{\omega_1} E_k^{\omega_2} E_l^{\omega_3} \qquad (18.1\text{-}8)$$

$$D = \begin{cases} 1 & \omega_1,\ \omega_2, \text{ and } \omega_3 \text{ are all equal} \\ 3 & \text{two of } \omega_1,\ \omega_2, \text{ and } \omega_3 \text{ are equal} \\ 6 & \text{none of } \omega_1,\ \omega_2, \text{ and } \omega_3 \text{ are equal} \end{cases}$$

When we apply this rule, ω and $(-\omega)$ are to be considered as two different frequencies so that, as an example,

$$P_i^{(\omega=\omega+\omega-\omega)} = 3\chi_{ijkl}^{(\omega=\omega+\omega-\omega)} E_j^{\omega} E_k^{\omega} E_l^{-\omega} \qquad (18.1\text{-}9)$$

The nonlinear constant χ_{ijkl} of atoms and molecules can be evaluated using a variety of techniques. We will derive in what follows the expressions for the nonlinear susceptibilities for third-harmonic generation (THG) and atomic Raman scattering. In the first of these cases, we apply the Feynman diagram technique of Chapter 3. We assume a collection of atoms each with an energy level diagram as shown in Figure 18.1*a*. We will assume that $\omega_{da} \approx$

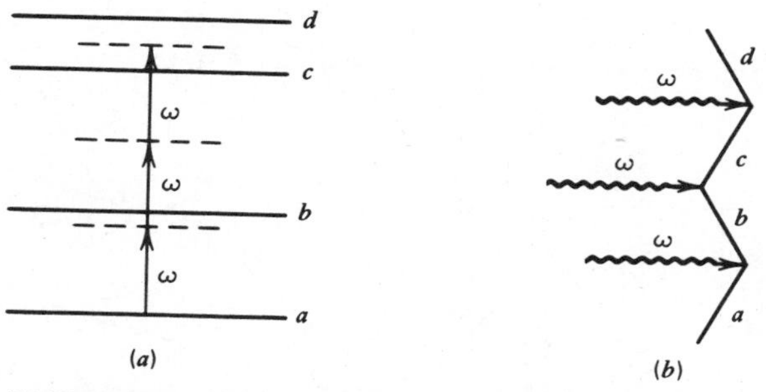

FIGURE 18.1 (*a*) The energy level diagram of an atom (molecule) responsible for third-harmonic generation. (*b*) The dominant Feynman diagram used in the calculation of $\chi^{(3\omega=\omega+\omega+\omega)}$.

3ω so that the main contribution comes from $\psi^{(3)}_{-\omega,-\omega,-\omega}$, as illustrated in Figure 18.1*b*.

$$P_i = N\langle\mu_i\rangle = N\langle\psi(t)|\mu_i|\psi(t)\rangle \approx N\langle\psi^{(3)}_{-\omega,-\omega,-\omega}|\mu_i|\psi^{(0)}\rangle \quad (18.1\text{-}10)$$

where N is the atom density and

$$\psi^{(3)}_{-\omega,-\omega,-\omega} = \frac{1}{8\hbar^3}\frac{(\mu_j)_{ab}(\mu_k)_{bc}(\mu_l)_{cd}}{(\omega_{ba}-\omega)(\omega_{ca}-2\omega-i\gamma_c)(\omega_d-3\omega-i\gamma_d)}|d\rangle$$

$$\cdot\, E_j^{*(\omega)}E_k^{*(\omega)}E_l^{*(\omega)}e^{-i(\omega+\omega+\omega)t}$$

where γ_c and γ_d are the natural linewidths of the respective transitions. We did not include γ_a since we assumed that $\omega_{ba} - \omega \gg \gamma_a$. Using $\psi^{(3)}$ in the expression for P_i leads to

$$\chi_{ijkl}^{(3\omega=\omega+\omega+\omega)} = \frac{N}{4\hbar^3}\frac{(\mu_i)_{da}(\mu_j)_{ba}(\mu_k)_{cb}(\mu_l)_{dc}}{(\omega_{ba}-\omega)(\omega_{ca}-2\omega+i\gamma_c)(\omega_d-3\omega+i\gamma_d)} \quad (18.1\text{-}11)$$

A case of special interest is that of two-photon resonance $\omega_{ca} = 2\omega$. In this case, (18.1-11) reduces to

$$\chi_{ijkl}^{(3\omega=\omega+\omega+\omega)} = -\frac{iN}{4\hbar^3}\frac{(\mu_i)_{da}(\mu_j)_{ba}(\mu_k)_{cb}(\mu_l)_{dc}}{(\omega_{ba}-\omega)\gamma_c(\omega_d-3\omega+i\gamma_d)} \quad (18.1\text{-}12)$$

As a second example, we will derive the nonlinear susceptibility characterizing the process, illustrated in Figure 18.2. Here, an atom absorbs a photon ω_1 and emits one at ω_2, winding up in a state $|s\rangle$ that is above the starting state $|n\rangle$. The induced electric dipole moment at ω_2 due to an atom initially in level n is

$$\begin{aligned}\langle\mu_i\rangle &= \langle\psi^{(1)}_{-\omega_1}(t)|\mu_i|\psi^{(2)}_{-\omega_1,\omega_2}\rangle \\ &= \frac{1}{8\hbar^3}\sum_{m,g,s}\frac{(\mu_1)_{mn}(\mu_i)_{ms}(\mu_1)_{gn}(\mu_2)_{sg}E_1^*E_1E_2e^{i(\omega_1-\omega_1+\omega_2)t}}{(\omega_{mn}-\omega_1)(\omega_{gn}-\omega_1)(\omega_{sn}-\omega_1+\omega_2-i\gamma)}\end{aligned} \quad (18.1\text{-}13)$$

We will assume that some level m satisfies the condition $\omega_{mn} \approx \omega_1$ and that, in

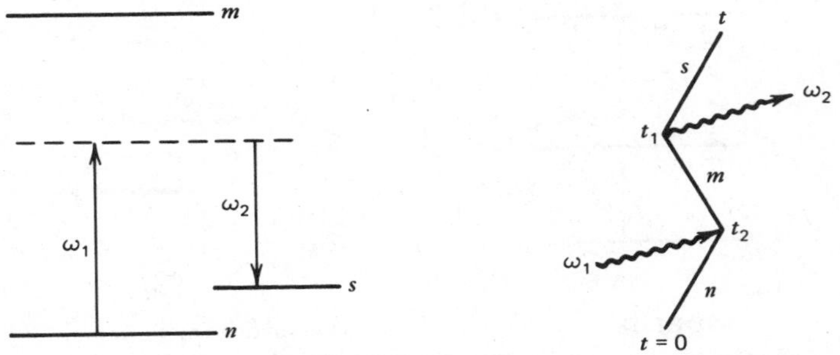

FIGURE 18.2 Energy levels and Feynman diagrams involved in Raman scattering.

addition, for some level s,

$$\omega_{sn} \approx \omega_1 - \omega_2 \tag{18.1-14}$$

(This approximate condition is exactly true in stimulated Raman scattering, in which case the field at ω_2 is not applied externally, but is generated by the interaction. The condition $\omega_{sn} = \omega_1 - \omega_2$ in this case results in maximum gain at ω_2.) Under these conditions,

$$\chi_{i,1,1,2}^{(\omega_2=\omega_1-\omega_1+\omega_2)} = \frac{(N_n - N_s)(\mu_1)_{mn}^2(\mu_i)_{ms}(\mu_2)sm}{4\hbar^3(\omega_{mn} - \omega_1)^2(\omega_{sn} - \omega_1 + \omega_2 - i\gamma)} \tag{18.1-15}$$

It will be left as an exercise to show [also see the development leading to (18.4-22)] that a weak field at ω_2 will exercise an exponential gain constant due to the interaction considered here.

$$g_{\text{Raman}} = \frac{\pi\omega_2(\mu_1)_{mn}^2(\mu_2)_{ms}^2(N_n - N_s)I_1}{2\hbar^3c^2\varepsilon_0^2 n_1 n_2(\omega_{mn} - \omega_1)^2} \cdot \frac{(\gamma/\pi)}{[\omega_{sn} - (\omega_1 - \omega_2)]^2 + \gamma^2} \tag{18.1-16}$$

where n_1 and n_2 are the indices of refraction at ω_1 and ω_2, respectively. I_1 is the intensity of the field at ω_1. This gain is referred to as that of "stimulated electronic raman scattering," since the levels involved are all electronic. This gain can thus be observed most readily in atomic vapors. Since the amplified radiation is at $\omega_2 = \omega_1 - \omega_{sn}$, a change of pumping frequency ω_1 can lead to the generation of tunable wavelength radiation. (See Supplementary Reference 14.)

The reader who wishes to pursue further the subject of the diagrammatic approach in the calculation of nonlinear susceptibilities should consult Supplemental References 4 and 5.

Using the definition of $\chi^{(3)}$ used in this chapter, we can convert between the MKS and CGS systems of units

$$\chi_{MKS}^{(3)} = \frac{\chi_{CGS}^{(3)}}{3^4 \times 10^{17}} = 1.23456 \times 10^{19}\chi_{CGS}^{(3)}$$

18.2 MOLECULAR RAMAN SCATTERING

Ordinary Raman spectroscopy is used mostly as a tool for studying the vibrational energy levels of molecules and of lattice optical branch vibrations in crystals (Reference 1). A cell containing the sample liquid (or gas) or the crystal to be studied is irradiated with a narrow band optical wave. A spectral analysis of the scattered radiation reveals the existence of frequencies that are shifted down by increments equal to vibrational frequencies of the material irradiated. This type of scattering is referred to as Stokes scattering.

Frequencies equal to the sum of the incident wave frequency and the vibrational frequencies are also present in the scattered radiation. This is the so-called anti-Stokes scattering, and its intensity is usually orders of magnitude below that of the Stokes radiation.

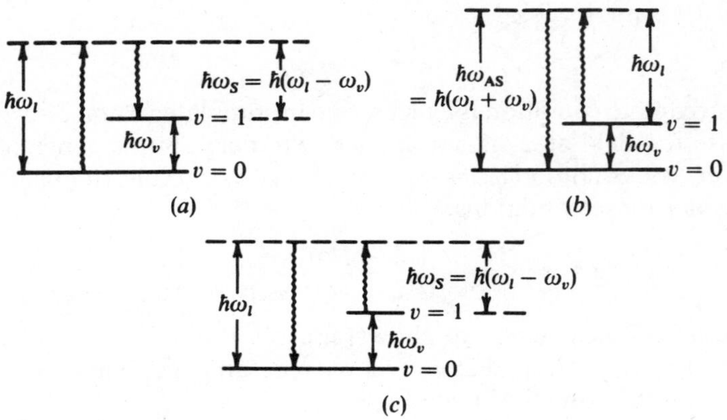

FIGURE 18.3 (*a*) A Stokes scattering in which a laser photon at ω_l is absorbed, while a Stokes $(\omega_l - \omega_v)$ photon is created along with a vibrational $(v = 1)$ quantum. (*b*) An anti-Stokes scattering in which a laser photon at ω_l and a vibrational (ω_v) quantum are absorbed, while a photon at $\omega_l + \omega_v$ is created. (*c*) A process in which the presence of laser radiation at ω_l stimulates the absorption of Stokes photons at $\omega_l - \omega_v$, that is, the reverse of (*a*).

The two types of scattering events are illustrated by Figure 18.3. In Figure 18.3*a* the molecule is initially in the ground state, $v = 0$. An incident photon at ω_l is absorbed, whereas simultaneously a Stokes photon at $\omega_s = \omega_l - \omega_v$ is emitted. To conserve energy, the molecule is excited to the vibrational level $v = 1$ of energy, $\hbar\omega_v$. If, on the other hand, the molecule is initially in the

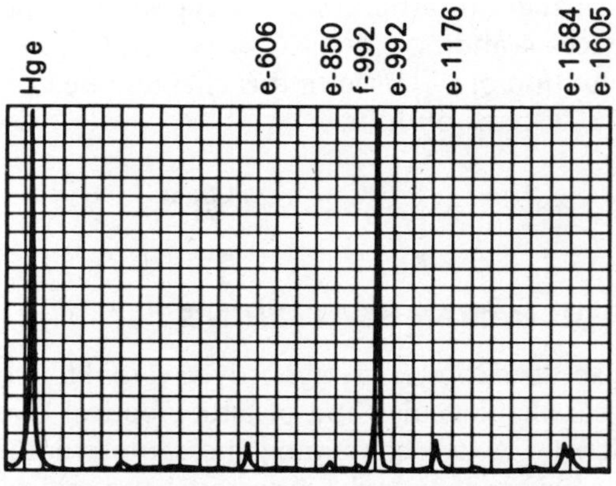

FIGURE 18.4 The spectrum of the Stokes radiation scattered by benzene. The line on the extreme left is the (greatly attenuated) exciting Hg *e* line at 22,938 cm^{-1} (4358 Å) that is the strongest line in the visible emission spectrum of Hg. The numbers on top give the downward shift in frequency (in cm^{-1}), thus corresponding to the vibrational frequencies. They are preceded by letters indicating the parent Hg line. The *f*-992 line is due to the *f* line of Hg at 22,995 cm^{-1}.

excited ($v = 1$) state as shown in Figure 18.3*b*, the scattered anti-Stokes photon is of frequency $\omega_{AS} = \omega_l + \omega_v$. Since anti-Stokes emission depends on the molecule being excited initially, its intensity compared to the Stokes emission is down by a factor of $e^{-h\nu_v/kT}$. The reverse process in which a Stokes photon is absorbed is shown in Figure 18.3*c*.

Raman spectroscopy has been performed until recently with intense incoherent radiation sources, the sources used most often being some of the intense mercury lines. A typical Raman (Stokes) spectrum revealing the vibrational frequencies of benzene is shown in Figure 18.4.

Recently, coherent laser sources have replaced the Hg lamp in Raman spectroscopy (Reference 5).

The vibrational frequencies of a given atomic group vary little from compound to compound, as can be seen from Tables 18.1 and 18.2.

Quantum Mechanical Description of Raman Scattering

A complete quantum mechanical treatment of Raman scattering is of limited usefulness because the matrix elements involved are not known and in most practical cases are too hard to calculate. It can be used, however, as a rough estimate of the scattering cross sections and to point out the dependence of the latter on near coincidence between the incident photon energy and the energies of electronic levels of the scattering system (Reference 3).

Our main concern in this chapter is in phase-coherent stimulated Raman scattering. For this purpose, we find it advantageous to use the semiphenomenological model of Placzek (Reference 4).

For a given electronic configuration, the potential energy of a molecule as a function of a normal vibrational coordinate X (in a simple case, such as a H_2 molecule, X is simply the interatomic separation) can be expressed as

$$V(X) = aX^2 + bX^3 + \cdots \tag{18.2-1}$$

TABLE 18.1. Characteristic Stretching Vibrations of Atomic Groups in Molecules

Frequency (cm^{-1})	*Vibrating Group*	*Type of Compound*
445–550	S—S	aliphatic disulfides
490–522	C—I	aliphatic compound
510–594	C—Br	aliphatic compound
570–650	C—Cl	aliphatic compound
600–700	C—SH	mercaptans
630–705	C—S	aliphatic compound
700–1100	C—C	aliphatic compound
750–850	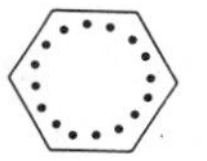	para derivatives of benzene

TABLE 18.1. *(Continued)*

Frequency (cm^{-1})	*Vibrating Group*	*Type of Compound*
884–899	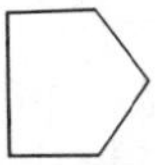	cyclopentane and mono derivatives
939–1005		cyclobutane and derivatives
990–1050	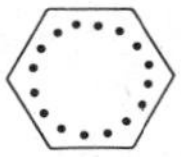	benzene and mono- to tri-substrate benzenes
1020–1075	C—O—C	aliphatic compound
1085–1125	C—OH	aliphatic compound
1120–1130	C=C=O	aliphatic compound
1188–1207		cyclopropane and derivatives
≈1190	SO_2	aliphatic compound
1216–1230	—S=O	aliphatic compound
≈1340	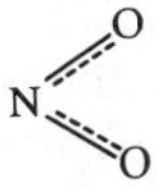	aromatic compound
≈1380	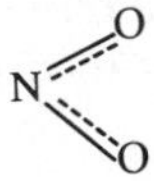	aliphatic compound
≈1380	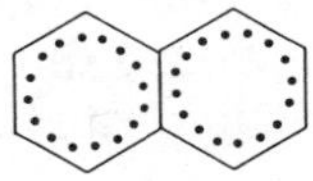	naphthalene and derivatives
1590–1610	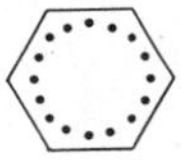	benzene derivatives
1610–1640	N=O	aliphatic compound
1620–1680	C=C	aliphatic compound
≈1630	C=N	aromatic compound
1654–1670	C=N	aliphatic compound
1650–1820	C=O	aliphatic compound
1695–1715	C=O	aromatic compound
1974–2260	C≡C	aliphatic compound
2150–2245	C≡N	nitriles
≈2570	S—H	aliphatic compound
2800–3000	C—H	aliphatic compound
3000–3200	C—H	aromatic compound
3150–3650	O—H	aliphatic compound
3300–3400	N—H	aliphatic compound
4160	H—H	H_2

Source: Reference 6.

TABLE 18.2 Influence of Various Radicals on the Frequency of the Symmetric NO_2 Vibration

Substance	*Vibrational Frequency* (cm^{-1})
1-chloro, 3-nitrobenzene 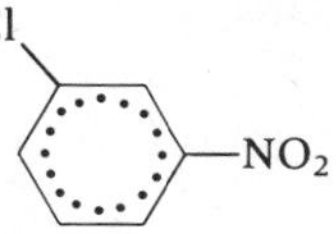	1353
nitrobenzene 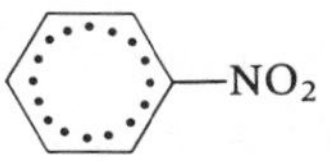	1345
p-nitrotoluene 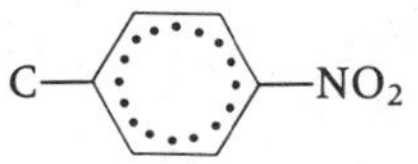	1340
o-nitrophenol 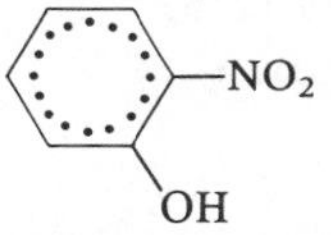	1322*

* This large shift is due to intramolecular hydrogen bonding.
Source: Reference 6.

where $X = 0$ is the equilibrium position. We have already considered, in Chapter 2, the solution of Schrödinger equation for the harmonic oscillator where $V(X) = \frac{1}{2}kX^2$. This is the simplest form that $V(X)$ can have and the energy solutions correspond to a spectrum of equispaced vibrational levels with energies $E_v = \hbar\omega(v + \frac{1}{2})$. In general, the higher-order terms of (18.2-1) cannot be neglected, and the situation is considerably more complicated.

The induced electronic dipole moment of a molecule is taken as $\mu_i = \varepsilon_0 \alpha E$ where α is the molecular polarizability and E the electric field. If the molecule were rigid, we could consider α to be a constant. In a vibrating molecule α is clearly a function of the normal coordinate of vibration X. The first two terms in the series expansion of $\alpha(X)$ are taken as[1] $\alpha(X) = \alpha_0 + (\partial\alpha/\partial X)_0 X$ where $(\partial\alpha/\partial X)_0$ is referred to as the differential polarizability. In addition, an asymmetric molecule may possess a permanent dipole moment μ_p, which is a function of X. We take the first two terms as $\mu_p(X) = \mu_p^0 +$

[1] In a real molecule we must deal with the tensor quantity $\partial\alpha_{ij}/\partial X_k$. In our treatment, we ignore the tensor aspect.

$(\partial\mu_p/\partial X)_0 X$. When studying transitions between vibrational levels that are induced by a radiation field, we consider the perturbation Hamiltonian

$$\mathcal{H}' = -\boldsymbol{\mu} \cdot \mathbf{E} = -\left[\mu_p^0 + \left(\frac{\partial\mu_p}{\partial X}\right)_0 X + \varepsilon_0\alpha_0 E + \varepsilon_0\left(\frac{\partial\alpha}{\partial X}\right)_0 XE\right] E \quad (18.2\text{-}2)$$

where E is the electric field component parallel to $\boldsymbol{\mu}$. The first and third terms in the square brackets are independent of X and, consequently, cannot cause transitions between adjacent vibrational levels since the eigenfunctions $\psi_v(X)$ are orthogonal to each other. The second term gives rise to direct infrared absorption at ω_v. The term that gives rise to Raman scattering is the last one

$$\mathcal{H}'_{\text{Raman}} = -\left(\frac{\partial\alpha}{\partial X}\right)_0 \varepsilon_0 X E^2 \quad (18.2\text{-}3)$$

To demonstrate this point, we consider an electric field composed of two frequencies

$$E = E_l \cos \omega_l t + E_s \cos \omega_s t \quad (18.2\text{-}4)$$

Using (5.6-15) and omitting the multiplying constants, we find that the electric field can be written as

$$E \propto (\omega_l)^{1/2}(a_l^\dagger - a_l) + (\omega_s)^{1/2}(a_s^\dagger - a_s) \quad (18.2\text{-}5)$$

where the $a_v^\dagger$'s and a's are the photon creation and annihilation operators, respectively. In a similar manner we can, using (2.2-25), expand X as

$$X = \propto (a_v^\dagger + a_v) \quad (18.2\text{-}6)$$

where $a_v^\dagger$ and a_v are the harmonic oscillator creation and annihilation operators. The process of emission of a Stokes photon illustrated by Figure 18.3*a* is clearly due to the term $a_l a_s^\dagger a_v^\dagger$, and its rate is therefore proportional to

$$W_{\text{emiss.}} \propto |\langle \underbrace{n_l - 1, n_s + 1, 1}_{\text{Final}} | a_l a_s^\dagger a_v^\dagger | \underbrace{n_l, n_s, 0}_{\text{Initial}} \rangle|^2 = n_l(n_s + 1) \quad (18.2\text{-}7)$$

where n_s and n_l are the number of quanta (photons) in the Stokes and laser radiation modes, respectively. The inverse process illustrated by Figure 18.3*c*, in which a laser (ω_l) photon is emitted while a photon at ω_s is absorbed, has a rate proportional to

$$W_{\text{abs}} \propto |\langle \underbrace{n_l + 1, n_s - 1, 0}_{\text{Final}} | a_l^\dagger a_s a_v | \underbrace{n_l, n_s, 1}_{\text{Initial}} \rangle|^2 = (n_l + 1) n_s \quad (18.2\text{-}8)$$

thus resulting in a deexcitation of the molecule from state $v = 1$ to $v = 0$.

Let us consider next the process of Raman scattering into a *single* Stokes mode ω_s from a laser mode ω_l. We have, from (18.2-7) and (18.2-8),

$$\frac{dn_s}{dt} = DP_a n_l(n_s + 1) - DP_b n_s(n_l + 1) \quad (18.2\text{-}9)$$

where D is a constant to be determined, whereas P_a and P_b are the respective probabilities of finding the molecule in the (ground) state $v = 0$ and in the state $v = 1$. Since the number of photons is conserved.

$$\frac{dn_l}{dt} = -\frac{dn_s}{dt}$$

In spontaneous Raman scattering $\langle n_s \rangle \ll 1$, so that the growth of the Stokes beam is given by

$$\frac{dn_s}{dt} = -\frac{dn_l}{dt} = DP_a n_l \qquad (18.2\text{-}10)$$

An "observer" traveling with the velocity of light $c/n(\nu_l)$ sees the exciting photon density decay as

$$\frac{dn_l}{dz} = \frac{dn_l}{dt}\frac{dt}{dz} = -\frac{Dn(\nu_l)}{c} P_a n_l \qquad (18.2\text{-}11)$$

so that

$$n_l(z) = n_l(0)\, e^{-[Dn(\nu_l)P_a/c]z} = n_l(0)\, e^{-\beta z} \qquad (18.2\text{-}12)$$

where $\beta = [DP_a n(\nu_l)]/c$ and $n(\nu_l)$ is the index of refraction at ν_l.

In actual scattering experiments, the interaction of the incident beam is not with a single Stokes mode but rather with the totality of continuum modes centered at the Stokes frequency that lie within the natural linewidth $\Delta\nu$ of the transition. The cross-section σ per unit volume for Stokes scattering (which is the same as the exponential absorption coefficient) is the product of β and the number of such modes

$$\sigma = \beta\left[\frac{8\pi\nu_s^2 n^3(\nu_s)}{c^3}\Delta\nu V\right] = \frac{8\pi\nu_s^2 Dn^3(\nu_s)\, n(\nu_l)\, P_a V\, \Delta\nu}{c^4} \qquad (18.2\text{-}13)$$

where $V =$ mode volume.

In practice, one does not measure σ but instead the differential cross section per unit volume $[d\sigma/d\Omega(\theta, \phi)]_V$. It is defined by means of Figure 18.5.

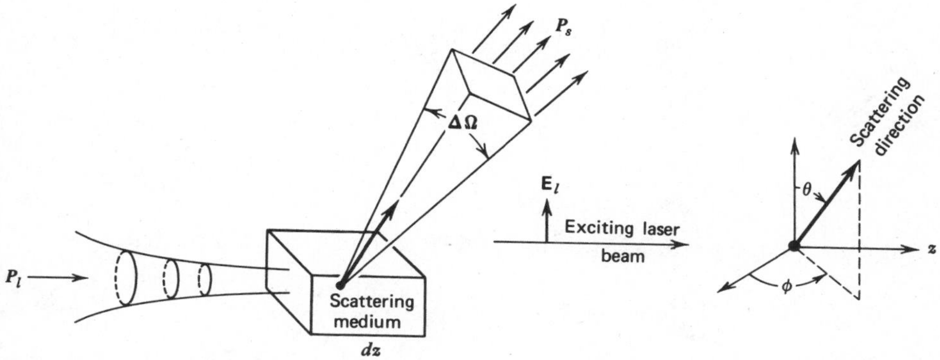

FIGURE 18.5 The scattering geometry used to measure the scattering cross section $[d\sigma/d\Omega(\theta, \phi)]_v$. The *total* scattered power P_s into a solid angle $\Delta\Omega$ from an element of length dz is related to the *total* incident power P_l by $P_s = P_l[d\sigma/d\Omega(\theta, \phi)]_v\, \Delta\Omega\, dz$.

The *total* power P_s scattered in a distance dz into a solid angle $\Delta\Omega$ in a direction θ, ϕ is

$$P_s = P_l \left[\frac{d\sigma}{d\Omega}(\theta, \phi)\right]_V \Delta\Omega\, dz \tag{18.2-14}$$

where P_l is the *total* incident power. Since for each scattered photon one photon is subtracted from the incident beam, it follows that

$$\frac{dP_l}{dz} = -\frac{\omega_l}{\omega_s} P_l \int_0^{4\pi} \left[\frac{d\sigma}{d\Omega}(\theta, \phi)\right]_V d\Omega = -\sigma P_l$$

so that the attenuation coefficient of the incident beam is

$$\sigma = \frac{\omega_l}{\omega_s} \int_0^{4\pi} \left[\frac{d\sigma}{d\Omega}(\theta, \phi)\right]_V d\Omega \tag{18.2-15}$$

For the case of dipolar scattering, we have

$$\frac{d\sigma}{d\Omega}(\theta, \phi) = \frac{d\sigma}{d\Omega}(\theta = 90°) \sin^2\theta$$

where θ and ϕ are defined in Figure 18.5.

$$\int_0^{4\pi} \frac{d\sigma}{d\Omega}(\theta, \phi)\, d\Omega = \int_0^{2\pi} d\phi \int_0^{\pi} d\theta \frac{d\sigma}{d\Omega}(\theta = 90°) \sin^3\theta$$

$$= 4\pi \left(\frac{2}{3}\right) \frac{d\sigma}{d\Omega}(\theta = 90°) \tag{18.2-16}$$

so that

$$\sigma = 4\pi \left(\frac{2}{3}\right) \frac{\omega_l}{\omega_s} \left[\frac{d\sigma}{d\Omega}(\theta = 90°)\right]_V \tag{18.2-17}$$

In practice, it is customary to use the differential scattering cross section per molecule. If the density of molecules is N,

$$\left[\frac{d\sigma}{d\Omega}(\theta = 90°)\right]_V = N \left[\frac{d\sigma}{d\Omega}(\theta = 90°)\right]_{\text{molec}}$$

and (18.2-17) becomes

$$\sigma = 4\pi \left(\frac{2}{3}\right) \frac{\omega_l}{\omega_s} N \left[\frac{d\sigma}{d\Omega}(90°, \phi)\right]_{\text{molec}} \tag{18.2-18}$$

But σ as given by (18.2-18) is the same as that of (18.1-13) so that

$$\left[\frac{d\sigma}{d\Omega}(90°, \phi)\right]_{\text{molec}} = \frac{3\nu_s^3 n^3(\nu_s)\, n(\nu_l)\, DV\, \Delta\nu P_a}{\nu_l N c^4} \tag{18.2-19}$$

We have thus related the quantum mechanical rate constant D to the experimentally measurable scattering cross section.

18.3 STIMULATED MOLECULAR RAMAN SCATTERING (REFERENCE 7)

Here, we consider the stimulated terms in (18.2-9), that is, the terms proportional to n_s. These, taken alone, give

$$\frac{dn_s}{dt} = D(P_a - P_b)\, n_l n_s \tag{18.3-1}$$

or

$$\frac{dn_s}{dz} = \frac{dn_s}{dt} \times \frac{dt}{dz} = \frac{Dn(\nu_s)}{c} (P_a - P_b)\, n_l n_s \tag{18.3-2}$$

so that the photon density (or intensity) at ν_s grows exponentially with distance according to

$$I_s(z) = I_s(0)\, e^{g_s z}$$

where

$$g_s = \frac{Dn(\nu_s)}{c} (P_a - P_b)\, n_l \tag{18.3-3}$$

Using (18.2-19) for D leads to

$$g_s = \frac{d\sigma}{d\Omega}(\theta = 90°)_{\text{molec}} \frac{Nc^2[1 - e^{-h(\nu_l - \nu_s)/kT}]}{3h\nu_s^3 n^2(\nu_s)\, \Delta\nu} I_l \tag{18.3-4}$$

where we used

$$\frac{n_l}{V} = \frac{n(\nu_l)}{h\nu_l c} I_l$$

and assumed an equilibrium temperature T so that

$$\frac{P_a - P_b}{P_a} = 1 - e^{-h(\nu_l - \nu_s)/kT}$$

Equation (18.3-4) states that if a medium possesses a nonvanishing Raman scattering cross section, then, in the presence of a laser beam with intensity I_l, it will amplify radiation at a frequency $\nu_s = \nu_l - \nu_v$. The exponential amplifi-

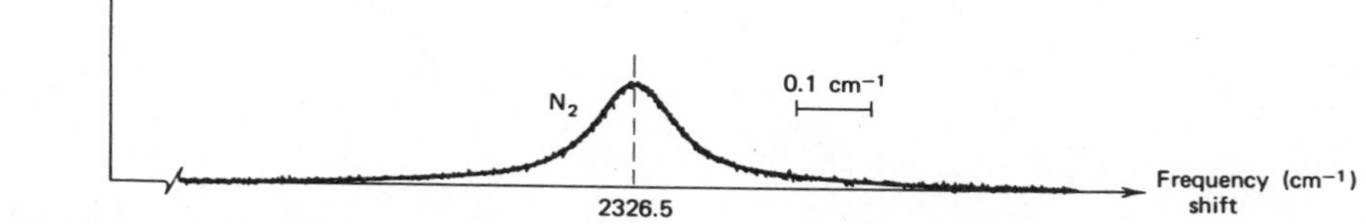

FIGURE 18.6 A spectral plot of the spontaneous Raman emission from liquid N_2. *Source:* Reference 9.

cation constant g, is proportional to the product of the cross section and the laser intensity.

If one performs a spectral analysis on the radiation scattered into a direction (θ, ϕ) one observes a narrow band with a width $\Delta\nu$ centered on ν_s. The normalized lineshape function of the scattered radiation $S(\nu)$ describes, as it does in ordinary laser amplification, the dependence of the Raman (Stokes) gain g_s on ν_s. We can thus replace $(\Delta\nu)^{-1}$ in (18.3-4) by $S(\nu)$, obtaining

$$g_s(\nu) = \left[\frac{d\sigma}{d\Omega}(\theta = 90°)\right]_{\text{molec}} \frac{Nc^2[1 - e^{-h(\nu_l-\nu_s)/kT}]}{3h\nu_s^3 n^2(\nu_s)} I_l S(\nu) \qquad (18.3\text{-}5)$$

where $\int_{-\infty}^{\infty} S(\nu)\, d\nu = 1$.

A typical spontaneous Raman scattering lineshape function $S(\nu)$ is shown in Figure 18.6.

Some cross-section data on Raman scattering for various molecular liquids and gases are given in Tables 18.3, 18.4, and 18.5.

Numerical Example: Raman Gain in CS_2. Here, we calculate the gain in CS_2, which is experienced by Stokes radiation in the presence of a

TABLE 18.3 Raman Scattering Cross Sections per Molecule of Some Liquids

Raman Lines	*Wavelength of the Exciting Light* (Å)	*Raman Scattering Cross Section:* $(d\sigma/d\Omega)\|$ (10^{-29} cm^2 molecule^{-1} · sr^{-1})
C_6H_6	6328	0.800 ± 0.029
992 cm^{-1}	5145	2.57 ± 0.08
Benzene	4880	3.25 ± 0.10
$C_6H_5CH_3$	6328	0.353 ± 0.013
1002 cm^{-1}	5145	1.39 ± 0.05
Chlorobenzene	4880	1.83 ± 0.06
$C_6H_5NO_2$	6328	1.57 ± 0.06
1345 cm^{-1}	5145	9.00 ± 0.29
Nitrobenzene	4880	10.3 ± 0.4
	6943	0.755
CS_2	6328	0.950 ± 0.034
656 cm^{-1}	5145	3.27 ± 0.10
	4880	4.35 ± 0.13
CCl_4	6328	0.628 ± 0.023
459 cm^{-1}	5145	1.78 ± 0.06
	4880	2.25 ± 0.07

Source: Reference 10.

TABLE 18.4 The Ratio of the Raman Cross Section per Molecule to That of N_2. The Value of N_2 is $(d\sigma/d\Omega)_{N_2} = (4.3 \pm 0.3) \times 10^{-31}$ cm²/sr-molec. at $\lambda_l = 4880$ Å.

Gas	*Vibrational Frequency* (cm^{-1})	$\frac{d\sigma}{d\Omega}$
N_2	2331	1.0
O_2	1556	1.3
H_2 (sum)	4161	2.4
$H_2(Q(1))$	4161	1.6
CO	2145	1.0
NO	1877	0.27
$CO_2(\nu_1)$	1388	1.4
$CO_2(2\nu_2)$	1286	0.89
$N_2O(\nu_1)$	1285	2.2
$N_2O(\nu_3)$	2224	0.51
$SO_2(\nu_1)$	1151	5.2
$SO_2(\nu_2)$	519	0.12
$H_2S(\nu_1)$	2611	6.4
$NH_3(\nu_1)$	3334	5.0
$ND_3(\nu_1)$	2420	3.0
$CH_4(\nu_1)$	2914	6.0
$C_2H_6(\nu_3)$	993	1.6
$C_6H_6(\nu_2)$	992	7.0

Source: Reference 11.

ruby laser beam with an intensity I_l. We have

$$T = 300°\text{K}$$

$$\nu_l = 4.32 \times 10^{14} \text{ Hz } (14{,}400 \text{ cm}^{-1})$$

$$\nu_v = 656 \ (\text{cm}^{-1})$$

$$\nu_s = 4.123 \times 10^{14} \text{ Hz } (13{,}744 \text{ cm}^{-1})$$

$$\Delta\nu = 1.5 \times 10^{10} \text{ Hz } (0.5 \text{ cm}^{-1})$$

$$n \simeq 1.63$$

$$N = 1.64 \times 10^{22} \text{ cm}^{-3}, \quad \left(\frac{d\sigma}{d\Omega}\right)_{\text{molec}} = 6.8 \times 10^{-30} \text{ cm}^2/\text{sr-molec.}$$ (from Table 18.3 corrected for later wavelength.)

Using these data in (18.3-4) gives

$$g_{\max}(\text{cm}^{-1}) \approx 0.02 I_l \left(\frac{\text{MW}}{\text{cm}^2}\right)$$

TABLE 18.5 Frequency Shift ν_v, and Scattering Cross Section $N(d\sigma/d\Omega)$ of Spontaneous Raman Scattering; N is the Number of Molecules per cm^3; Steady-State Gain Factor g_s/I_l of Stimulated Raman Scattering in Different Substances (λ_l = 6943 Å)

Substance	*Frequency Shift* ν_v (cm^{-1})	*Linewidth* $\Delta\nu$ (cm^{-1})	*Cross Section* $N\ d\sigma/d\Omega_\parallel \times 10^8$ (cm^{-1} ster^{-1})	*Gain Factor*[a] g_s/I_l *in Units of* 10^{-3} (cm/MW)	*Temp.* T[°K]
Liquid O_2	1552	0.117	0.48 ± 0.14	14.5 ± 4	
				16 ± 5	
Liquid N_2	2326.5	0.067	0.29 ± 0.09	17 ± 5	
				16 ± 5	
Benzene	992	2.15	3.06	2.8	300
		2.3	3.3	3	
			4.1	3.8	
CS_2	655.6	0.50	7.55	24	300
Nitrobenzene	1345	6.6	6.4	2.1	300
			7.9	2.6	
Bromobenzene	1000	1.9	1.5	1.5	300
Chlorobenzene	1002	1.6	1.5	1.9	300
Toluene	1003	1.94	1.1	1.2	300
$LiNbO_3$	256	23	381	E.9	300
	258	7	262	28.7	80
	637	20	231	9.4	300
	643	16	231	12.6	80
Li^6NbO_3	256			17.8	300
	266			35.6	80
	637			9.4	300
	643			12.6	80
$Ba_2NaNb_5O_{15}$	650			6.7	300
	655			18.9	80
$LiTaO_3$	201	22	238	4.4	300
	215	12	167	10	80
Li^6TaO_3	600			4.3	300
	608			7.9	80
SiO_2	467			0.8	300
				0.8	300
				0.6	300
H_2-gas	4155			1.5 (P > 10 atm)	300

[a] To obtain the gain constant g_s(cm^{-1}) at ν_s, multiply by ν_s/ν_l and by the intensity in MW/cm^2.
Source: Reference 12.

Raman Oscillation

Consider a Raman-active medium contained within an optical enclosure of length L that possesses an effective reflectivity R to the Stokes radiation. In the presence of a laser field, the Stokes field will exercise a line center gain as given by (18.3-4). If the gain is sufficient to compensate for the round-trip losses, oscillation at ν_s can result. The necessary condition is thus

$$Re^{g_s L} = 1$$

that, if we use (18.3-5) and take the spontaneous Raman linewidth as $\Delta\nu = S(\nu_0)^{-1}$, gives

$$(I_l)_t = \frac{3h\nu_s^3\,\Delta\nu n^2(\nu_s)\left(-\frac{1}{L}\ln R\right)}{\left(\frac{d\sigma}{d\Omega}\right)_{\text{molec}} Nc^2[1 - e^{-h(\nu_l-\nu_s)/kT}]} \tag{18.3-6}$$

for the threshold intensity.

Using the CS_2 data of the last section, a value of $R \sim 10^{-3}$ and $L = 5$ cm, yields

$$(I_l)_t = 67 \text{ MW/cm}^2$$

Raman oscillation is used in practice to convert the output of some of the common pulsed lasers (e.g., Nd^{3+}: glass at 1.06 μm) to coherent outputs at frequencies shifted by the values listed in the second column of Table 18.5.

18.4 ELECTROMAGNETIC TREATMENT OF STIMULATED RAMAN SCATTERING

The treatment of the preceding section is based on the concept of transition rates. The electromagnetic field quantities appear only as photon densities n_l and n_s so that all phase information is absent. It is still possible to predict the onset of stimulated Raman scattering, as in (18.3-6), but many important features are lost. Since in the region of interest the occupation numbers satisfy $n_l, n_s \gg 1$, we can employ a classical analysis instead of the quantum treatment of the preceding section.

In stimulated Raman scattering experiments it is found that the output consists, simultaneously, of an appreciable number of Stokes frequencies at $(\omega_l - \omega_v)$, $(\omega_l - 2\omega_v)$, . . . , and of anti-Stokes frequencies at $(\omega_l + \omega_v)$, $(\omega_l + 2\omega_v)$, ... To understand the origin of these frequencies we refer to Figure 18.3. The process of Stokes emission shown in Figure 18.3*a* causes the population of the vibrational $v = 1$ level to build up. Once this happens, radiation at $\omega_{AS} = \omega_l + \omega_v$ can be emitted as shown in Figure 18.3*b*. The Stokes (ω_S) and anti-Stokes (ω_{AS}) fields can, next, act as input radiation, thus generating $\omega_S - \omega_v = \omega_l - 2\omega_v$ and $\omega_{AS} + \omega_v = \omega_l + 2\omega_v$, respectively, and so on.

A complete solution of this problem is not feasible since it requires a simultaneous solution of Maxwell equations and the vibration equation that involve all the frequency components of interest (Reference 3). In order to illustrate the basic principles involved in Raman scattering, we take advantage of the fact that only the Stokes radiation $(\omega_l - \omega_v)$ can be amplified initially. The growth of all the other frequency components depends either on the presence of molecules in the state $v = 1$, as in the generation of anti-Stokes components, or on the presence of first-order Stokes radiation, which is necessary to generate the second-order Stokes radiation. We can consequently derive the condition for net gain, or oscillation, at the first Stokes frequency, $\omega_s = \omega_l - \omega_v$, while neglecting all the frequency components except, of course, the laser field at ω_l.

The model used in the analysis is as follows: the Raman medium is taken as consisting of N harmonic oscillators per unit volume, each oscillator representing one molecule. The oscillators are *independent* of each other so that the ensemble of oscillators cannot support a wavemotion with a nonvanishing group velocity. Each oscillator is characterized by its position z (the analysis is one-dimensional so that $\partial/\partial x = \partial/\partial y = 0$) and normal vibrational coordinate $X(z, t)$. The equation of motion for a single oscillator is then

$$\frac{d^2X(z, t)}{dt^2} + \gamma\frac{dX}{dt} + \omega_v^2 X = \frac{F(z, t)}{m} \tag{18.4-1}$$

where γ is the damping constant chosen so that the observed spontaneous Raman scattering linewidth is $\Delta\nu = \gamma/2\pi$, ω_v is the (undamped) resonance frequency, m is the mass, and $F(z, t)$ is the driving force.

The driving term can be derived by considering the electromagnetic energy in the presence of the molecules. The electrostatic stored energy density is

$$\mathscr{E} = \tfrac{1}{2}\varepsilon\mathbf{E}\cdot\mathbf{E}$$

that, if we use

$$\varepsilon = \varepsilon_0(1 + N\alpha) = \varepsilon_0\left\{1 + N\left[\alpha_0 + \left(\frac{\partial\alpha}{\partial X}\right)_0 X\right]\right\} \tag{18.4-2}$$

can be written as

$$\mathscr{E} = \tfrac{1}{2}\varepsilon_0\left\{1 + N\left[\alpha_0 + \left(\frac{\partial\alpha}{\partial X}\right)_0 X\right]\right\}\overline{\mathbf{E}\cdot\mathbf{E}} \tag{18.4-3}$$

The force per unit volume of polarizable material is $\partial\mathscr{E}/\partial X$ that, after dividing by N, gives the force per oscillator as

$$F(z, t) = \tfrac{1}{2}\varepsilon_0\left(\frac{\partial\alpha}{\partial X}\right)_0\overline{\mathbf{E}\cdot\mathbf{E}} \tag{18.4-4}$$

where the overbar indicates averaging over a few optical periods since the molecules cannot respond to optical frequencies. This shows that because of the nonvanishing differential polarizability, $(\partial\alpha/\partial X)_0$, the molecular vibration can be driven by the optical electric field.

Our next problem is to show how the field-induced excitation of molecular vibration $X(z, t)$ reacts back on the electromagnetic fields. The molecular vibration at ω_v causes, according to (18.4-2), a modulation of the dielectric constant ε at ω_v. This leads to phase modulation of any radiation field present, thus creating sidebands separated by ω_v. Stated differently, a modulation of ε at ω_v, caused by molecular vibrations, can lead to energy exchange between electromagnetic fields separated in frequency by multiples of ω_v, such as, for example, the laser (ω_l) and the Stokes ($\omega_s = \omega_l - \omega_v$) fields.

The total field is taken as the sum of the Stokes (ω_1) and laser field (ω_2)

$$\mathbf{E}(z, t) = \tfrac{1}{2}\hat{\mathbf{e}}_1 E_1(z) e^{i\omega_1 t} + \tfrac{1}{2}\hat{\mathbf{e}}_2 E_2(z) e^{i\omega_2 t} + \text{c.c} \tag{18.4-5}$$

so that

$$\overline{\mathbf{E} \cdot \mathbf{E}} \equiv \hat{\mathbf{e}}_1 \cdot \hat{\mathbf{e}}_2 \tfrac{1}{4} E_2(z) E_1^*(z) e^{i(\omega_2-\omega_1)t} + \text{c.c.} \tag{18.4-6}$$

Substituting (18.4-6) in (18.4-4) and then in the molecular equation of motion (18.4-1) gives

$$\tfrac{1}{2}(\omega_v^2 - \omega^2 + i\omega\gamma) X(z) e^{i\omega t} = \frac{\hat{\mathbf{e}}_1 \cdot \hat{\mathbf{e}}_2 \varepsilon_0}{8m} \left(\frac{\partial \alpha}{\partial X}\right)_0 E_2 E_1^* e^{i(\omega_2-\omega_1)t} \tag{18.4-7}$$

where

$$X(z, t) = \tfrac{1}{2} X(z) e^{i\omega t} + \text{c.c.} \tag{18.4-8}$$

It follows from (18.4-7) that the molecular vibration is driven at a frequency $\omega = \omega_2 - \omega_1$ with a complex amplitude

$$X(z) = \frac{\varepsilon_0 \left(\dfrac{\partial \alpha}{\partial X}\right)_0 E_2(z) E_1^*(z) \hat{\mathbf{e}}_1 \cdot \hat{\mathbf{e}}_2}{4m[\omega_v^2 - (\omega_2 - \omega_1)^2 + i(\omega_2 - \omega_1)\gamma]} \tag{18.4-9}$$

The polarization induced in the molecules by the field at ω_1 is

$$P = \varepsilon_0 N \alpha(z, t) E(z, t) = \varepsilon_0 N \left[\alpha_0 + \left(\frac{\partial \alpha}{\partial X}\right)_0 X(z, t)\right] E(z, t) \tag{18.4-10}$$

Our concern here is with the nonlinear polarization term that is proportional to the product XE. If we use (18.4-5) and (18.4-9) in (18.4-10), it becomes

$$P_{\text{NL}}(z, t) = \tfrac{1}{4}\varepsilon_0 N \left(\frac{\partial \alpha}{\partial X}\right)_0 \left\{ \frac{\varepsilon_0 \left(\dfrac{\partial \alpha}{\partial X}\right)_0 \hat{\mathbf{e}}_1 \cdot \hat{\mathbf{e}}_2 E_2 E_1^* e^{i(\omega_2-\omega_1)t}}{4m[\omega_v^2 - (\omega_2 - \omega_1)^2 + i(\omega_2 - \omega_1)\gamma]} + \text{c.c.} \right\} \times [E_1(z) e^{i\omega_1 t} + E_2(z) e^{i\omega_2 t} + \text{c.c.}] \tag{18.4-11}$$

If we multiply the two terms in (18.4-11), we get polarizations oscillating at ω_1, ω_2, $2\omega_1 - \omega_2$, and $2\omega_2 - \omega_1$. Let us concentrate first on the ω_1 term.

$$P_{\text{NL}}^{(\omega_1)}(z, t) = \tfrac{1}{2} P_{\text{NL}}^{(\omega_1)}(z) e^{i\omega_1 t} + \text{c.c.} \tag{18.4-12}$$

where

$$P_{\text{NL}}^{(\omega_1)}(z) = \frac{\varepsilon_0^2 N \left(\dfrac{\partial \alpha}{\partial X}\right)_0^2 |E_2|^2 \hat{\mathbf{e}}_1 \cdot \hat{\mathbf{e}}_2}{8m[\omega_v^2 - (\omega_2 - \omega_1)^2 - i(\omega_2 - \omega_1)\gamma]} E_1(z) \tag{18.4-13}$$

The coefficient relating an induced polarization to the inducing field is the susceptibility. From (18.4-13), we can define a complex Raman nonlinear susceptibility through the relation

$$P_{\text{NL}}^{(\omega_1)}(z) = \varepsilon_0 \chi_{\text{Raman}}(\omega_1) \, |E_2(z)|^2 \, E_1(z) \tag{18.4-14}$$

so that for $\hat{\mathbf{e}}_1 \parallel \hat{\mathbf{e}}_2$

$$\chi_{\text{Raman}}(\omega_1) = \frac{\varepsilon_0 N \left(\frac{\partial \alpha}{\partial X}\right)_0^2}{8m[\omega_v^2 - (\omega_2 - \omega_1)^2 - i(\omega_2 - \omega_1)\gamma]} \tag{18.4-15}$$

More generally, we can characterize the effect of induced molecular vibration by means of a fourth-rank tensor

$$P_i^{(\omega_i = \omega_j - \omega_k + \omega_l)} = \chi_{ijkl}^{(\omega_i = \omega_j - \omega_k + \omega_l)} E_j^{(\omega_j)} E_k^{(\omega_k)*} E_l^{(\omega_l)} \tag{18.4-16}$$

so that (18.4-14) is but a special case where $\omega_j = \omega_k = \omega_2$, $\omega_i = \omega_l = \omega_1$.

Returning to (18.4-15), we define

$$\chi_{\text{Raman}}(\omega_1) = \chi'_{\text{Raman}}(\omega_1) - i\chi''_{\text{Raman}}(\omega_1) \tag{18.4-17}$$

where

$$\chi'_{\text{Raman}}(\omega_1) \simeq \frac{\varepsilon_0 N \left(\frac{\partial \alpha}{\partial X}\right)_0^2 [\omega_v - (\omega_2 - \omega_1)]}{16m\omega_v\{[\omega_v - (\omega_2 - \omega_1)]^2 + \gamma^2/4\}} \tag{18.4-18}$$

and

$$\chi''_{\text{Raman}}(\omega_1) \simeq \frac{-\varepsilon_0 N \left(\frac{\partial \alpha}{\partial X}\right)_0^2 (\gamma/2)}{16m\omega_v\{[\omega_v - (\omega_2 - \omega_1)]^2 + \gamma^2/4\}} \tag{18.4-19}$$

where the approximation applies to the high Q case $\gamma \ll \omega_v$ that is usually the case (typically $\gamma \leqslant 10^{-2}\,\omega_v$).

The nonlinear Raman susceptibility is thus Lorentzian as is its linear counterpart discussed in Section 8.1. It is plotted in Figure 18.7.

The presence of a Raman polarization (18.4-14) at ω_1 can be accounted for by modifying the propagation constant as in (8.3-4) from k_1 to

$$\begin{aligned} k_1' &= k_1 \left[1 + \frac{\chi_{\text{Raman}}(\omega_1)}{2n_1^2} |E_2|^2\right] \\ &= k_1 \left\{1 + \frac{|E_2|^2}{2n_1^2} [\chi'_{\text{Raman}}(\omega_1) - i\chi''_{\text{Raman}}(\omega_1)]\right\} \end{aligned} \tag{18.4-20}$$

so that

$$\begin{aligned} E^{(\omega_1)}(z) = E^{(\omega_1)}(0) \exp\Bigg\{ &-ik_1 z\left[1 + \frac{|E_2|^2 \chi'_{\text{Raman}}(\omega_1)}{2n_1^2}\right] \\ &- k_1 z \frac{|E_2|^2 \chi''_{\text{Raman}}(\omega_1)}{2n_1^2}\Bigg\} \end{aligned} \tag{18.4-21}$$

The exponential gain coefficient is thus

$$g(\omega_1) = -\frac{k_1}{2n_1^2} |E_2|^2 \chi''_{\text{Raman}}(\omega_1) \tag{18.4-22}$$

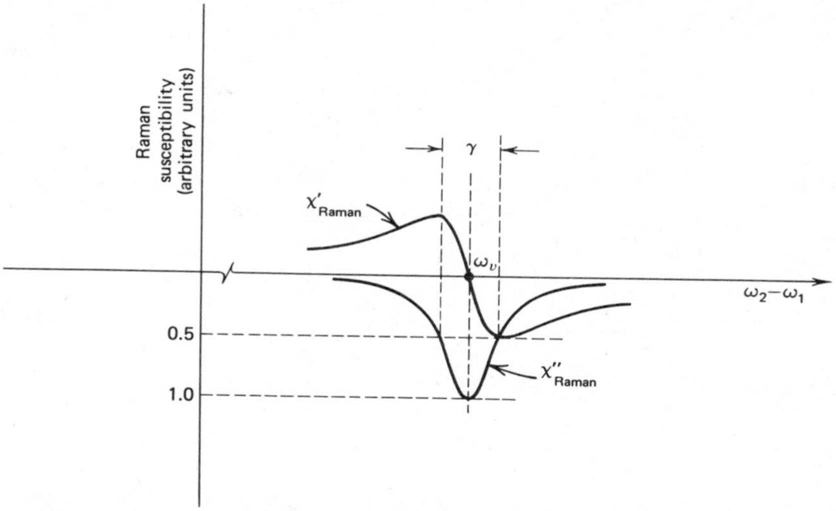

FIGURE 18.7 The in-phase (χ'_{Raman}) and quadrature (χ''_{Raman}) components of the Raman nonlinear susceptibility as a function of the Stokes frequency ω_1. (ω_1 increases from right to left.)

and is positive since $\chi''_{\text{Raman}}(\omega_1) < 0$. If we use (18.4-19), $g(\omega_1)$ is given as

$$g(\omega_1) = \frac{k_1 \varepsilon_0 \left(\dfrac{\partial \alpha}{\partial X}\right)_0^2 N\gamma |E_2|^2}{32 n_1^2 m \omega_v \{[\omega_v - (\omega_2 - \omega_1)]^2 + \gamma^2/4\}} \tag{18.4-23}$$

By comparing (18.4-23) to (18.3-5), we can identify the normalized Raman lineshape as

$$S(\nu_1) = \frac{\gamma/2\pi}{[\nu_v - (\nu_2 - \nu_1)]^2 + \left(\dfrac{\gamma}{4\pi}\right)^2} \tag{18.4-24}$$

18.5 ANTI-STOKES SCATTERING

The considerations of Section 18.2 show that anti-Stokes radiation at $\omega_3 \simeq \omega_2 + \omega_v$ can be generated by Raman transitions originating in the excited ($v = 1$) vibrational state, as in Figure 18.3*b*. To treat the problem electromagnetically, let us consider the ω_3 polarization induced in the Raman medium due to an electric field

$$E(z, t) = \tfrac{1}{2}[E_1(z)\, e^{i\omega_1 t} + E_2(z)\, e^{i\omega_2 t} + E_3(z)\, e^{i\omega_3 t} + \text{c.c.}] \tag{18.5-1}$$

where $\omega_3 - \omega_2 = \omega_2 - \omega_1$.

First, we obtain a term due to the driving of the molecular vibration by the product $E_3 E_2^*$. This term is analogous to (18.4-11) and is derived in an identical manner. We can thus obtain the polarization by modifying (18.4-13), recalling that now E_3 is the high-frequency field and E_2 the low one. We thus replace E_2 by E_3 and E_1 by E_2, $\omega_2 - \omega_1$ by $\omega_3 - \omega_2$. The result for $\hat{\mathbf{e}}_1 = \hat{\mathbf{e}}_2$ is

$$P_{\mathrm{NL}}^{(\omega_3)}(z) = \frac{\varepsilon_0^2 N \left(\dfrac{\partial \alpha}{\partial X}\right)_0^2 |E_2|^2}{8m[\omega_v^2 - (\omega_3 - \omega_2)^2 + i(\omega_3 - \omega_2)\gamma]} E_3(z) \qquad (18.5\text{-}2)$$

The important difference between (18.5-2) and (18.4-13) is the *opposite sign* of the imaginary term. This difference translates into an opposite sign of $\chi''(\omega_3)$ relative to $\chi''(\omega_1)$

$$\chi''_{\mathrm{Raman}}(\omega_3) = \frac{\varepsilon_0 N \left(\dfrac{\partial \alpha}{\partial X}\right)_0^2 (\gamma/2)}{16m\omega_v\{[\omega_v - (\omega_3 - \omega_2)]^2 + \gamma^2/4\}} \qquad (18.5\text{-}3)$$

so that the gain constant exercised by the anti-Stokes wave (ω_3) is

$$g(\omega_3) = -\frac{k_1}{2n_3^2}|E_2|^2 \chi''_{\mathrm{Raman}}(\omega_3) < 0 \qquad (18.5\text{-}4)$$

and the wave is *attenuated.* We thus reach the conclusion that if one were to introduce anti-Stokes ($\omega_3 = \omega_2 + \omega_v$) radiation into a Raman active medium in the presence of an ω_2 wave and in the absence of Stokes ($\omega_1 = \omega_2 - \omega_v$) radiation, it would *attenuate*.

There exists, however, another source of polarization at ω_3. It is obtained by taking the term

$$P_{\mathrm{NL}}^{(\omega_3)} \propto E_2 E_2 E_1^* \exp[i(2\omega_2 - \omega_1)t] \qquad (18.5\text{-}5)$$

in (18.4-11). This term does not involve E_3 and can be viewed as the upper sideband $[(\omega_2 + (\omega_2 - \omega_1)]$ due to a modulation of the dielectric constant "seen" by ω_2 at the driven molecular frequency $(\omega_2 - \omega_1)$. This term acts as a source of radiation at ω_3.

If we insert the spatial dependence into the polarization of (18.5-5), we find that

$$P_{\mathrm{NL}}^{(\omega_3)}(z) \propto E_2 E_2 E_1^* e^{-i(2\mathbf{k}_2 - \mathbf{k}_1)\cdot \mathbf{r}} \qquad (18.5\text{-}6)$$

This term will generate a field at ω_3 with a spatial dependence $E_2 e^{-\mathbf{k}_3 \cdot \mathbf{r}}$ such that[2]

$$\mathbf{k}_3 = 2\mathbf{k}_2 - \mathbf{k}_1 \qquad (18.5\text{-}7)$$

Anti-Stokes radiation will thus be emitted in any direction $\mathbf{k}_3$ that satisfies (18.5-7). The resulting direction $\mathbf{k}_3$ of the emitted anti-Stokes beam is shown in Figure 18.8. (It should be recalled that in an isotropic medium, the magnitudes of $\mathbf{k}_1$, $\mathbf{k}_2$, and $\mathbf{k}_3$ are determined by their respective frequencies and are $|\mathbf{k}_i| = \omega_i n_i/c$, where n_i is the index of refraction at ω_i and is determined by the intersection of the $\mathbf{k}_1$ locus and that of $\mathbf{k}_3$.) This is the reason why anti-Stokes radiation is emitted in the form of a conical shell with a half-apex

[2] This can be ascertained by using $P_{\mathrm{NL}}^{(\omega_3)}$ as the source term in the wave equation (14.9-1).

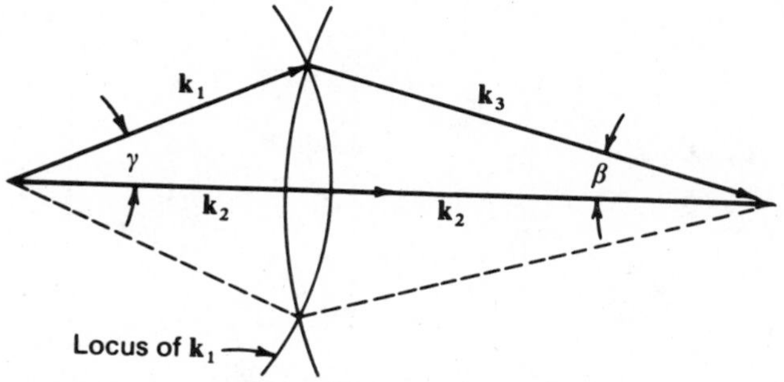

FIGURE 18.8 A construction for finding the direction of propagation $\mathbf{k}_3$ of the anti-Stokes radiation.

angle β about the laser propagation direction (Reference 13). The calculation of the cone angle is discussed in a problem at the end of the chapter.

The "real-life" situation in stimulated Raman emission is considerably more complicated than that portrayed above. In addition to the existence of higher-order Stokes and anti-Stokes radiation that was mentioned earlier, it is found that, as an example, the direction of the emitted anti-Stokes radiation deviates because of "trapping" from that predicted by (18.5-7). For a consideration of some of these effects, the reader is referred to References 14–18.

A color photograph, courtesy of R. W. Terhune (Reference 19), which shows a multiplicity of Stokes and anti-Stokes rings generated by a ruby laser beam in benzene, is reproduced next to the title page of this book.

18.6 STIMULATED BRILLOUIN SCATTERING

The scattering of light from thermally excited acoustic waves was considered as early as 1922 by Brillouin (Reference 20). The phenomenon of stimulated Brillouin scattering in which the acoustic wave that scatters the optical beam is produced by the optical beam itself was discovered in 1964 (Reference 21). It was found that when an intense laser beam of frequency ω_2 passed through a crystal (sapphire or quartz in the original experiment), a coherent acoustic wave at a frequency ω_s was produced within the crystal while, simultaneously, an optical beam at a frequency $\omega_2 - \omega_s$ was generated. Both the acoustic and scattered optical beams were emitted along specific directions, and their generation occurred only above a well-defined input threshold value.

A schematic diagram of the experimental arrangement used in the first experiment is shown in Figure 18.9.

18.7 A CLASSICAL TREATMENT OF BRILLOUIN SCATTERING

The presence of a time-varying electric field in a liquid (or crystal) gives rise to a time-varying electrostrictive strain and is thus capable of driving acoustic

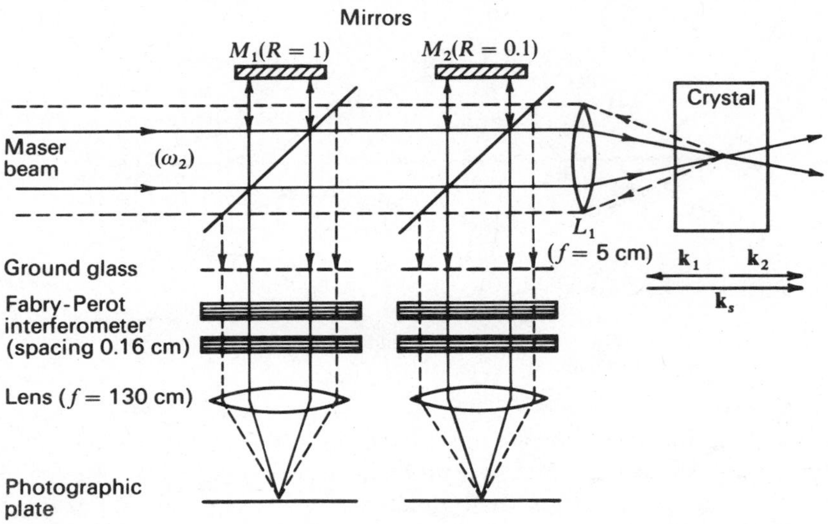

FIGURE 18.9 A schematic diagram of the experimental arrangement used to detect stimulated Brillouin scattering. The presence of a scattered optical beam at $\omega_2 - \omega_s$ is detected by the Fabry–Perot interferometer and causes additional "rings" to appear in the focal plane where they are photographed. *Source:* Reference 21.

waves in the medium. The presence of an acoustic wave, on the other hand, modulates the optical dielectric constant and thus can cause an exchange of energy between electromagnetic waves whose frequencies differ by an amount equal to the acoustical frequency. The effect is thus analogous to stimulated Raman scattering with acoustic waves playing the role of the molecular vibrations.

In order to derive the equation of motion for the sound wave (see Figure 18.10), consider a differential volume $dx\, dy\, dz$ inside a fluid subjected to an electric field E. Let the deviation of a point x from its equilibrium position be $u(x, t)$ so that the one-dimensional strain is $\partial u/\partial x$. We introduce, phenomenologically, a constant γ that describes the change in the optical dielectric constant induced by the strain through the relation

$$\delta\varepsilon = -\gamma \frac{\partial u}{\partial x} \tag{18.7-1}$$

so that the presence of strain changes are stored electrostatic energy density by $-\frac{1}{2}\gamma(\partial u/\partial x)E^2$.

A change in stored energy that is accompanied by strain implies the

FIGURE 18.10 A differential volume of unit cross section ($\Delta z\, \Delta y = 1$) and length Δx used to derive the equation of motion of an electrostrictively driven sound wave.

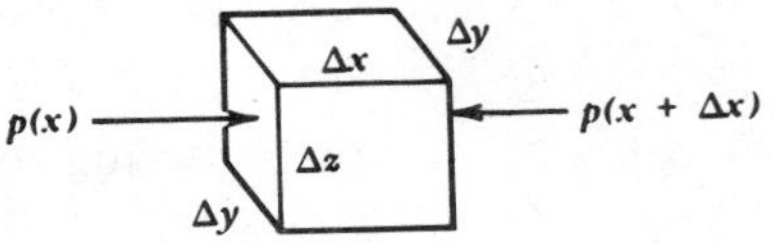

existence of a pressure. This pressure p is found by equating the work $p(\partial u/\partial x)$ done while straining a unit volume (here we take $\Delta x\, \Delta y\, \Delta z = 1$) to the change $-\frac{1}{2}\gamma(\partial u/\partial x)E^2$ of the energy density. This results in

$$p = -\tfrac{1}{2}\gamma E^2 \tag{18.7-2}$$

The net electrostrictive force in the positive x direction acting on a unit volume is thus

$$F_{\text{per unit volume}} = -\frac{\partial p}{\partial t} = \frac{\gamma}{2}\frac{\partial}{\partial x}E^2$$

The equation of motion for $u(x, t)$ is thus

$$-\eta\frac{\partial u}{\partial t} + T\frac{\partial^2 u}{\partial x^2} + \frac{\gamma}{2}\frac{\partial}{\partial x}E^2 = \rho\frac{\partial^2 u}{\partial t^2} \tag{18.7-3}$$

where η is a dissipation constant accounting phenomenologically for acoustic losses, whereas T and ρ are the elastic constant (bulk modulus) and the mass density, respectively.[3]

Next, we assume that the acoustic field and the two electric fields are in the form of plane waves traveling in arbitrary directions and take them in the form

$$\begin{aligned} E_1(\mathbf{r}, t) &= \tfrac{1}{2}E_1(r_1)e^{i(\omega_1 t - \mathbf{k}_1\cdot\mathbf{r})} + \text{c.c.} \\ E_2(\mathbf{r}, t) &= \tfrac{1}{2}E_2(r_2)e^{i(\omega_2 t - \mathbf{k}_2\cdot\mathbf{r})} + \text{c.c.} \\ u(\mathbf{r}, t) &= \tfrac{1}{2}u_s(r_s)e^{i(\omega_s t - \mathbf{k}_s\cdot\mathbf{r})} + \text{c.c.} \end{aligned} \tag{18.7-4}$$

where r_1, r_2, and r_s are the algebraic distances measured along the respective directions of propagation $\mathbf{k}_1$, $\mathbf{k}_2$, and $\mathbf{k}_s$, so that $\mathbf{r}_i = (\mathbf{k}_i \cdot \mathbf{r}_i/k_i)$.

Returning to the wave equation (18.7-3), we use the last of equations (18.7-4), and replacing x by r_s, we obtain

$$\frac{\partial^2 u}{\partial r_s^2} = -\frac{1}{2}\left(k_s^2 u_s + 2ik_s\frac{du_s}{dr_s} - \frac{d^2u_s}{dr_s^2}\right)e^{i(\omega_s t - \mathbf{k}_s\cdot\mathbf{r})} + \text{c.c.}$$

so that (18.7-3) can be written as

$$\begin{aligned} &\left[(-i\eta\omega_s + \rho\omega_s^2)u_s - T\left(k_s^2 u_s + 2ik_s\frac{du_s}{dr_s}\right)\right]e^{i(\omega_s t - \mathbf{k}_s\cdot\mathbf{r})} + \text{c.c.} \\ &\quad = -\frac{\gamma}{8}\frac{\partial}{\partial r_s}\{E_2(r_2)E_1^*(r_1)e^{i[(\omega_2-\omega_1)t - (\mathbf{k}_2-\mathbf{k}_1)\cdot\mathbf{r}]} + \text{c.c.}\} \end{aligned} \tag{18.7-5}$$

where we assumed

$$k_s^2 u_s,\ k_s\frac{du_s}{dr_s} \gg \frac{d^2u_s}{dr_s^2}$$

It follows from (18.7-5) that

$$\omega_s = \omega_2 - \omega_1 \tag{18.7-6}$$

[3] $T = (1/\rho)(d\rho/dp)$ where p is the pressure.

and

$$\mathbf{k}_s = \mathbf{k}_2 - \mathbf{k}_1$$

With these substitutions, the right side of (18.7-5) can be written as

$$-\frac{\gamma}{8}\left[\frac{d}{dr_s}(E_2E_1^*) - ik_sE_2E_1^*\right]e^{i(\omega_s t - \mathbf{k}_s\cdot\mathbf{r})} + \text{c.c.}$$

and the wave equation (18.7-5) as

$$2ik_sv_s^2\frac{du_s(r_s)}{dr_s} + \left(k_s^2v_s^2 - \omega_s^2 + \frac{i\eta\omega_s}{\rho}\right)u_s(r_s) = -\frac{i\gamma k_s}{8\rho}E_2(r_2)E_1^*(r_1) \qquad (18.7\text{-}7)$$

where we assumed $|\partial/\partial r_s(E_2E_1^*)| \ll |k_sE_2E_1^*|$ and used the relation $T/\rho = v_s^2$, v_s being the free propagation velocity of acoustic waves in the medium.

The Electromagnetic Wave Equation

We start with the wave equation

$$\nabla^2E_i(\mathbf{r}, t) = \mu\varepsilon\frac{\partial^2}{\partial t^2}E_i(\mathbf{r}, t) + \mu\frac{\partial^2}{\partial t^2}(P_{\text{NL}})_i \qquad (18.7\text{-}8)$$

where $(P_{NL})_i$ is the ith component of the nonlinear polarization that acts as a source term for $E_i(\mathbf{r}, t)$. Using the first of (18.7-4), we obtain

$$\nabla^2E_1(\mathbf{r}, t) = -\tfrac{1}{2}[k_1^2E_1(r_1) + 2i\mathbf{k}_1\cdot\nabla E_1(r_1) - \nabla^2E_1(r_1)]e^{i(\omega_1 t - \mathbf{k}_1\cdot\mathbf{r})} + \text{c.c.} \qquad (18.7\text{-}9)$$

that, after substituting into (18.7-8) with $i = 1$, neglecting the $\nabla^2E_1(r_1)$ term, and recalling that $\mathbf{k}_1\cdot\nabla E_1(r_1) = k_1(dE_1/dr_1)$, yields

$$\left[k_1\frac{dE_1(r_1)}{dr_1}\right]e^{i(\omega_1 t - \mathbf{k}_1\cdot\mathbf{r})} + \text{c.c.} = i\mu\frac{\partial^2}{\partial t^2}(P_{\text{NL}})_i \qquad (18.7\text{-}10)$$

The nonlinear polarization term that appears in (18.7-10) is the additional polarization caused by the acoustic wave and is, consequently, given by $(P_{NL})_i = (\delta\varepsilon)E$ that, if we use (18.7-1), is equal to

$$(P_{\text{NL}})_i = -\gamma E(\mathbf{r}, t)\frac{\partial u(\mathbf{r}, t)}{\partial r_s} \qquad (18.7\text{-}11)$$

According to (18.7-4), the product $E(\partial u/\partial r_s)$ contains terms with exponential time factors $i(\pm\omega_s \pm \omega_1)t$ and $i(\pm\omega_s \pm \omega_2)t$. Only the terms involving $\pm i(\omega_2 - \omega_s) = \pm i\omega_1$, however, can act as synchronous driving terms on the right side of (18.7-10) so that it can be written as

$$k_1\frac{dE_1}{dr_1}e^{i(\omega_1 t - \mathbf{k}_1\cdot\mathbf{r})} = \frac{i\mu}{4}\frac{\partial^2}{\partial t^2}\left\{-\gamma E_2e^{i(\omega_2 t - \mathbf{k}_2\cdot\mathbf{r})}\frac{\partial}{\partial r_s}[u_s^*e^{-i(\omega_s t - \mathbf{k}_s\cdot\mathbf{r})}]\right\}$$

or

$$k_1\frac{dE_1}{dr_1} = \frac{i\omega_1^2\gamma\mu}{4}E_2\left(ik_su_s^* + \frac{du_s^*}{dr_s}\right)$$

where we used (18.7-6). For the case when $|du_s/dr_s| \ll |k_s u_s|$, the wave equation becomes

$$\frac{dE_1}{dr_1} = -\frac{\omega_1^2 \gamma \mu k_s}{4k_1} E_2 u_s^* - \frac{\alpha E_1}{2} \tag{18.7-12}$$

where the dissipation term $-(\alpha E_1/2)$ was added to account for the losses of the medium at ω_1 that, up to this point, have been neglected.[4]

A completely analogous treatment leads to the relation

$$\frac{dE_2}{dr_2} = -\frac{\omega_2^2 \gamma \mu k_s}{4k_s} E_1 u_s - \frac{\alpha E_2}{2} \tag{18.7-13}$$

for the wave at $\omega_2 = \omega_1 + \omega_s$.

Relations (18.7-7), (18.7-12), and (18.7-13) form a set of coupled equations involving the acoustic variable $u_s(r_s)$ and the field amplitudes $E_1(r_1)$ and $E_2(r_2)$. The solution of these equations under certain conditions is discussed in the next section.

Stimulated Brillouin Scattering

Here, we consider the stimulated Brillouin scattering described in the introductory paragraph. In this case, the application of a sufficiently intense optical field at ω_2 causes a simultaneous generation of an optical beam at ω_1 and of an acoustic wave at $\omega_s = \omega_2 - \omega_1$. The analysis of simplified by limiting it to the case in which the amount of power drained off from the pump field at ω_2 due to the oscillation at ω_1 and ω_s is small compared to the input power. Under these conditions, we may take $E_2(r_2) = \text{const}$ and confine ourselves to the solution of (18.7-7) and (18.7-12). In the first of these equations, we put $\omega_s = k_s v_s$, that is, assume that the acoustic dispersion is the same as in free and lossless propagation. The result is

$$\frac{du_s}{dr_s} = -\frac{\eta}{2\rho v_s} u_s - \frac{\gamma}{16\rho v_s^2} E_2 E_1^* \tag{18.7-14}$$

The equation for the ω_1 beam (18.7-12) is rewritten as

$$\frac{dE_1^*}{dr_1} = -\frac{\alpha E_1^*}{2} - \frac{\gamma k_1 k_s}{4\varepsilon_1} E_2^* u_s \tag{18.7-15}$$

The variables r_1 and r_s, it should be recalled, are the distances as measured along the arbitrary directions of propagations, $\mathbf{k}_1$ and $\mathbf{k}_s$, of the optical and acoustic waves, respectively. The difficulty of having two variables r_1 and r_s in the coupled equations (18.7-14) and (18.7-15) can be removed by transforming to the coordinate ξ measured along the bisectrix, as shown in Figure 18.11. Using the relation $r_s = r_1 = \xi \cos\theta = q$, we can rewrite (18.7-14) and

[4] Had we carried along the finite conductivity σ of the medium in the derivation, the dissipation term would be given by $\alpha = \sigma\sqrt{\mu_0/\varepsilon}$.

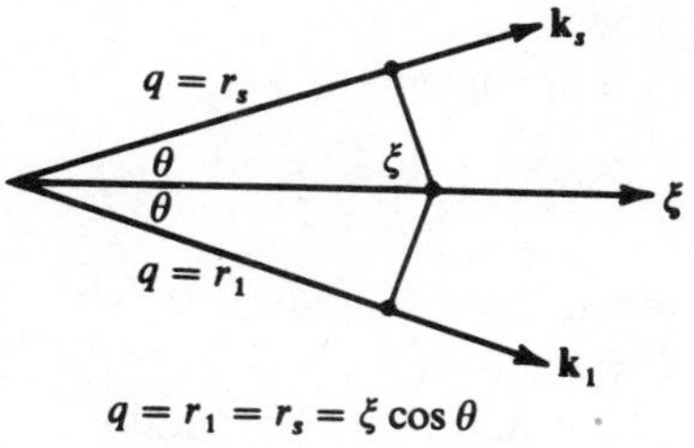

FIGURE 18.11 A one-to-one correspondence between the distance r_s measured normal to the acoustic wavefront and r_1 measured normal to the optical (ω_1) wavefront.

(18.7-15) as

$$\begin{aligned} \frac{du_s}{dq} &= -\frac{\eta}{2\rho v_x} u_s - \frac{\gamma}{16\rho v_s^2} E_2 E_1^* \\ \frac{dE_1^*}{dq} &= -\frac{\alpha E_1^*}{2} - \frac{\gamma k_1 k_s}{4\varepsilon_1} E_2^* u_s \end{aligned} \tag{18.7-16}$$

These equations describe the growth, or decay, of the acoustic displacement u_s and the electric field E_1 as a function of the distance q, as measured along either one of the two directions of propagation.

Assuming an exponential growth rate, we take

$$\begin{aligned} u_s(q) &= u_s^0 e^{gq} \\ E_1^*(q) &= (E_1^0)^* e^{gq} \end{aligned} \tag{18.7-17}$$

and solve the determinantal equation resulting upon substitution in (18.7-16) for the exponential growth factor g. The result is

$$g = -\tfrac{1}{4}(\alpha_s + \alpha) + \frac{1}{4}\sqrt{(\alpha_s + \alpha)^2 - 4\left(\alpha_s \alpha - \frac{k_1 k_s \gamma^2 |E_2|^2}{16\rho\varepsilon_1 v_s^2}\right)} \tag{18.7-18}$$

where the acoustic attenuation constant is $\alpha_s = \eta/\rho v_s$. The exponential gain constant g thus increases with the acoustic frequency, $\omega_s = k_s v_s$. The propagation vector $\mathbf{k}_s$ is determined by (18.7-6). Since $\omega_s \ll \omega_2$, we have $\omega_2 \approx \omega_1$ and in isotropic media $k_2 \approx k_1$. The vector relationship (18.7-6), $\mathbf{k}_2 - \mathbf{k}_1 = \mathbf{k}_s$, thus becomes identical to that for Bragg scattering as given in Figure 14.15 and shown again in Figure 18.12. It follows that

$$k_s = 2k_2 \sin\theta \tag{18.7-19}$$

so that maximum gain obtains for the case of backward scattering $\theta = \pi/2$ where $k_s = 2k_2$ and the resulting forward acoustic wave has a frequency

$$(\omega_s)_{\max} = 2\omega_2 \frac{v_s n_2}{c} \tag{18.7-20}$$

When the exponential growth constant g is positive, thermally excited acoustic waves propagating along $\mathbf{k}_s$ and zero-field optical waves at ω_1 traveling along $\mathbf{k}_1$ will be amplified simultaneously, according to (18.7-17). This will cause large enhancement of the respective powers along these two directions.

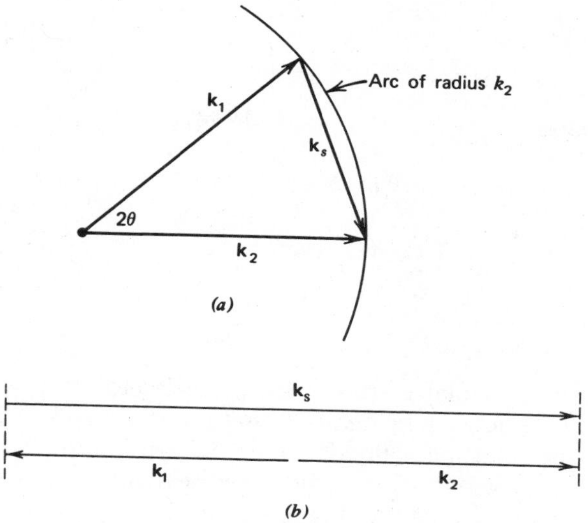

FIGURE 18.12 The vector relationship $\mathbf{k}_2 - \mathbf{k}_1 = \mathbf{k}_s$ for stimulated Brillouin scattering in an isotropic medium ($k_2 \cong k_1$). (*a*) For an arbitrary angle θ. (*b*) For backward scattering ($\theta = \pi/2$).

The condition $g \geq 0$ for stimulated Brillouin scattering occurs, according to (18.7-18), when

$$|E_2|^2 \geq \frac{16T\varepsilon_1\alpha_s\alpha}{\gamma^2 k_1 k_s} \tag{18.7-21}$$

where we used $v_s^2 = T/\rho$.

If we choose to express the acoustic and optical attenuation by the decay distances (i.e., the distances in which the amplitudes would normally decrease by a factor e^{-1}) $L_s = 2/\alpha_s$ and $L_1 = 2/\alpha$, and express $|E_2|^2$ in terms of intensity, we obtain

$$I_2 > \frac{32cT\varepsilon_0^2 n_1^2 n_2}{\gamma^2 k_1 k_s L_1 L_s} \tag{18.7-22}$$

for the threshold condition for the onset of stimulated Brillouin scattering.

Numerical Example: Stimulated Brillouin Scattering. As an estimate of the threshold power for stimulated Brillouin scattering, consider the following order of magnitude calculation based on quartz. Let

$T = 5 \times 10^{10}$ newtons/m^2	This is a typical value for the bulk modulus of solids.
$\gamma \sim \varepsilon_0 \sim 10^{-11}$ (MKS)	This is a typical value for the electrostrictive coefficient. See, for example, J. Stratton, *Electromagnetic Theory* (New York: McGraw-Hill, 1941), p. 151, and recall that $\gamma = \rho(d\varepsilon/d\rho)$.

$\lambda_2 \sim \lambda_1 = 1\ \mu\text{m}$

$L_1 = 1\ \text{m}$

$k_1 L_1 = \dfrac{2\pi L_1}{\lambda_1} = 2\pi \times 10^6$

$\omega_s \sim 2\omega_2 \dfrac{v_s n}{c} \sim 2\pi(6 \times 10^9)$ — Estimate based on $v_s = 3 \times 10^3$ m/sec and $\lambda_2 \sim 1\ \mu$m.

$L_s = 10^{-1}$ cm — Estimate based on typical data for quartz and sapphire (References 22, 23, and 24).

Using the above data in (18.7-22), we get

$$\left(\frac{\text{Power}}{\text{Area}}\right)_{\text{threshold}} \sim 2 \times 10^7\ \text{W/cm}^2$$

This power level is available from giant pulse lasers. In most liquids, the phenomenon of beam trapping discussed in Section 18.8 gives rise to intensities exceeding the threshold value for stimulated Brillouin scattering even at moderate input powers. As a result, the experimentally observed threshold in most materials is that of beam trapping (Reference 14).

A case of special interest is that where $\mathbf{k}_1 \cdot \mathbf{k}_2 < 0$. This happens when the angle 2θ between the incident (ω_2) and the scattered optical beam at ω_1 exceeds $\pi/2$. What makes this case physically different is that at any point along the propagation direction of, for example, the scattered (ω_1) optical beam, the rate of growth of the field E_1 is influenced by the values of E_1 at points lying ahead along the propagation direction $\mathbf{k}_1$. This *feedback* is provided by the sound beam by virtue of its opposite direction of propagation. In a similar manner, the scattered optical beam at ω_1 provides feedback to the sound beam. An accurate treatment of this situation requires a consideration of the boundary conditions as in Section 17.8 (Reference 25). An alternative approach is to treat the scattering of light and sound, not by the traveling wave approach as done above, but by assuming all the interacting modes to be resonant, that is, to exist inside suitable optical and acoustical resonators (Reference 26).

18.8 SELF-FOCUSING OF OPTICAL BEAMS

According to the previous discussions, stimulated Raman scattering can occur in a material only when the optical field intensity exceeds a certain "threshold" value, given by (18.3-6); experimentally, however, the measured threshold intensities required in laser beams are often significantly lower. This discrepancy, which can be as large as ~100 in some liquids, is caused by the phenomenon of beam self-focusing (References 27, 28, 29, 30). It is found that, when the beam power exceeds a critical value, the beam diameter continually contracts as it propagates through the material, ultimately forming an intense "focus" after a certain distance. Near the focal point, the power density usually exceeds that required for stimulated Raman scattering, so that the measured threshold is that of self-focusing and not of Raman scattering.

The self-focusing effect is due to an optical dielectric constant that changes when an electric field is applied. This can be described by the relative dielectric constant

$$\varepsilon_{\text{total}} = \varepsilon + \varepsilon_2\langle \mathbf{E} \cdot \mathbf{E} \rangle \qquad (18.8\text{-}1)$$

where $\langle \mathbf{E} \cdot \mathbf{E} \rangle$ is taken as the time average of the square of the optical field (so that it is equal to one-half of the amplitude squared). It can be seen that this type of dielectric constant results from a polarization of the form

$$\mathbf{P} = \varepsilon_0 \chi \mathbf{E} + \varepsilon_0 \chi_2 \langle \mathbf{E} \cdot \mathbf{E} \rangle \mathbf{E}$$

where

$$\varepsilon = 1 + \chi \qquad \text{and} \qquad \varepsilon_2 = \chi_2$$

The nonlinear polarization, $\varepsilon_0 \chi_2 \langle \mathbf{E} \cdot \mathbf{E} \rangle \mathbf{E}$, can be produced by a number of effects. For example, the (linear) electronic polarization of an atom or molecule can be derived quantum mechani.ally, using first-order perturbation theory (Reference 31); by extending the analysis to third-order perturbations, a quantum mechanical formula for χ_2 can be obtained (Reference 32). We can roughly estimate ε_2 without such an analysis by using the following arguments: the perturbation of the atom's electronic "cloud" will become "strong" when the (time-averaged) electrostatic energy of the applied field $\frac{1}{2}\varepsilon_0 \varepsilon \langle \mathbf{E} \cdot \mathbf{E} \rangle V$ (V is the volume of the atom) is comparable to the energy of the electronic state itself, $\hbar\omega_0$ (here $\hbar\omega_0$ can be taken as the energy of the first excited state; ω_0 typically lies in the ultraviolet for transparent materials). When this "strong" electronic distortion holds, the "nonlinear" polarization will be comparable to the "linear" so that we can estimate

$$\frac{\chi_2 \langle \mathbf{E} \cdot \mathbf{E} \rangle \mathbf{E}}{\chi E} \sim \frac{\frac{1}{2}\varepsilon_0 \varepsilon \langle \mathbf{E} \cdot \mathbf{E} \rangle V}{\hbar\omega_0}$$

This gives

$$\varepsilon_2 = \chi_2 \sim \frac{\varepsilon_0 n^2 (n^2 - 1)}{2\hbar\omega_0 N} \qquad (18.8\text{-}2)$$

where $n^2 = \varepsilon$ and $N \simeq 1/V$ is the number density of the atoms. For typical values (glass) $n \sim 1.52$, $\lambda \sim 2800$ Å, $N \sim 2.7 \times 10^{28}$ m^{-3}, we obtain the order of magnitude estimate $\varepsilon_2 \sim 7 \times 10^{-22}$.

In liquids containing anisotropic molecules, the "orientational Kerr effect" often produces the dominant contribution to ε_2. Consider, for simplicity, a molecule that has rotational symmetry about one axis; we assume that $\alpha_\parallel$, the molecular polarizability along that axis, is different from $\alpha_\perp$, the value at right angles to it. If this axis lies at an angle θ with respect to an applied electric field **E**, then the induced dipole moment **p** changes the molecule's potential energy by

$$\begin{aligned} U(\theta) &= -\tfrac{1}{2}\mathbf{p} \cdot \mathbf{E} \\ &= -\tfrac{1}{2}(\alpha_\parallel |\mathbf{E}| \cos\theta \mathbf{e}_\parallel + \alpha_\perp |\mathbf{E}| \sin\theta \mathbf{e}_\perp) \cdot (|\mathbf{E}| \cos\theta \mathbf{e}_\parallel + |\mathbf{E}| \sin\theta \mathbf{e}_\perp) \\ &= -\tfrac{1}{2}\mathbf{E} \cdot \mathbf{E}(\alpha_\parallel \cos^2\theta + \alpha_\perp \sin^2\theta) \end{aligned}$$

where $\mathbf{e}_\parallel$ and $\mathbf{e}_\perp$ are unit vectors parallel to the molecule's axis of cylindrical symmetry and at right angles to it, respectively. For $\alpha_\parallel > \alpha_\perp$, this interaction tends to align the molecular axis along the field ($\theta \to 0$) where $U(\theta)$ has a minimum. In thermal equilibrium, the molecules will have an angular distribution given by

$$f(\theta) = \frac{e^{-U(\theta)/kT}}{\int_0^\pi e^{-U(\theta)/kT} 2\pi \sin\theta \, d\theta} \simeq \frac{1}{4\pi}\left[1 + \frac{\langle \mathbf{E}\cdot\mathbf{E}\rangle}{2kT}\left(\frac{\alpha_\parallel - \alpha_\perp}{3}\right)(3\cos^2\theta - 1)\right]$$

We have taken $\alpha\langle \mathbf{E}\cdot\mathbf{E}\rangle/kT \ll 1$; also, $\langle \mathbf{E}\cdot\mathbf{E}\rangle$ is used since the molecules cannot reorient themselves at optical frequencies. Finally, given this distribution, the net polarization will be along the original field direction, with a magnitude of

$$|P| = N\int_0^\pi (|\mathbf{E}|\,\alpha_\parallel \cos^2\theta + |\mathbf{E}|\,\alpha_\perp \sin^2\theta) f(\theta) 2\pi \sin\theta \, d\theta$$

$$= |\mathbf{E}|\left(\frac{2\alpha_\perp + \alpha_\parallel}{3}\right) N + |\mathbf{E}|\langle \mathbf{E}\cdot\mathbf{E}\rangle \frac{2N(\alpha_\parallel - \alpha_\perp)^2}{45kT}$$

Thus,

$$\varepsilon = 1 + \frac{(2\alpha_\perp + \alpha_\parallel)}{3\varepsilon_0} N \qquad \text{and} \qquad \varepsilon_2 = \frac{2N(\alpha_\parallel - \alpha_\perp)^2}{45\varepsilon_0 kT} = \frac{2}{5}\frac{(n^2-1)^2}{kTN}\left(\frac{\alpha_\parallel - \alpha_\perp}{\alpha_\parallel + 2\alpha_\perp}\right)^2$$

A comparison with equation (18.8-2) shows that this effect can be much larger than the electronic nonlinearity (at room temperature $\hbar\omega_0/kT \sim 170$ for $\lambda \sim 2800$ Å).

Physically, a nonlinearity of the form (18.8-1) manifests itself as an "intensity dependent refractive index,"

$$n_{\text{total}} = \sqrt{\varepsilon_{\text{total}}} \simeq n + n_2\langle \mathbf{E}\cdot\mathbf{E}\rangle$$

with $n_2 = \frac{1}{2}(\varepsilon_2/n)$. This causes beam self-focusing because the central (intense) portion of a beam "sees" a higher index of refraction than the outer edges; thus, since the "optical length" (i.e., phase delay) is larger near the beam axis, a positive "lens" is formed in the nonlinear material by the beam itself.

To analyze the phenomenon of beam self-focusing, we assume that ε ($= n^2$) and ε_2 are known and write Maxwell's equations as

$$\nabla \times \mathbf{H} = \varepsilon_0 \frac{\partial}{\partial t}[(\varepsilon + \varepsilon_2\langle \mathbf{E}\cdot\mathbf{E}\rangle)\mathbf{E}]$$

$$\nabla \times \mathbf{E} = -\mu \frac{\partial \mathbf{H}}{\partial t}$$

Taking the curl of $\nabla \times \mathbf{E}$ and substituting for $\nabla \times \mathbf{H}$, we obtain the wave equation[5]

$$\nabla^2\mathbf{E} - \frac{n^2}{c^2}\frac{\partial^2\mathbf{E}}{\partial t^2} - \frac{2nn_2}{c^2}\frac{\partial^2}{\partial t^2}(\langle \mathbf{E}\cdot\mathbf{E}\rangle\mathbf{E}) = 0 \qquad (18.8\text{-}3)$$

[5] Here, as in (6.5-3), we neglect the term involving $\nabla \cdot \mathbf{E}$.

We now consider a beam propagating in the z direction and polarized in the x direction, which we write in the form

$$\mathbf{E} = \tfrac{1}{2}[E(\mathbf{r})e^{i(\omega t - kz)} + \text{c.c.}]\hat{\mathbf{e}}_x$$

where $k = n\omega/c$. We assume that $E(\mathbf{r})$ varies slowly in z compared with e^{-ikz}, so that

$$\frac{\partial^2}{\partial z^2}(Ee^{-ikz}) \simeq e^{-ikz}\left(-k^2E - 2ik\frac{\partial E}{\partial z}\right)$$

With this substitution and the relation $\langle \mathbf{E}\cdot\mathbf{E}\rangle = |E|^2/2$, (18.8-3) becomes

$$\nabla_T^2 E - 2ik\frac{\partial E}{\partial z} + \frac{n_2 k^2}{n}|E|^2 E = 0 \qquad (18.8\text{-}4)$$

where we have defined

$$\nabla_T^2 = \frac{\partial^2}{\partial x^2} + \frac{\partial^2}{\partial y^2}$$

When no nonlinearity is present ($n_2 = 0$), (18.8-4) correctly describes linear beam propagation (in a transparent medium). In particular, the beam solutions given in Section 6.9 form a complete set of Gaussian modes, with the "lowest order" beam shape having the form (6.6-14). Ignoring overall phase factors, we will consider a fundamental Gaussian beam entering the material at $z = 0$ as shown in Figure 18.13.

Consider the Gaussian beam solution of (6.6-7). Let us rewrite it for the beam shown in Figure 18.13 by replacing z by $z - z_{\text{min}}$

$$\psi(x, y, z) = \exp\left\{-i\left[-i\ln\left(1 + \frac{z - z_{\text{min}}}{q_0}\right) + \frac{kr^2}{2(q_0 + z - z_{\text{min}})}\right]\right\}$$

$$= \frac{1}{\sqrt{1 + \dfrac{4(z - z_{\text{min}})^2}{k^2\omega_0^4}}}\exp\left\{i\tan^{-1}\frac{2(z - z_{\text{min}})}{k\omega_0^2} - \frac{r^2}{\omega_0^2}\left[1 - i\frac{2(z - z_{\text{min}})}{k\omega_0^2}\right]^{-1}\right\} \qquad (18.8\text{-}5)$$

where we used $q_0 = ik\omega_0^2/2$.

The input beam at $z = 0$ is thus of the form

$$E(x, y, 0) = E_0\exp\left[-\frac{r^2}{\omega_0^2}\left(1 + i\frac{2z_{\text{min}}}{k\omega_0^2}\right)^{-1}\right]$$

$$= E_0\exp\left[-\frac{r^2}{\omega_0^2}\frac{1 - i\dfrac{2z_{\text{min}}}{k\omega_0^2}}{1 + \left(\dfrac{2z_{\text{min}}}{k\omega_0^2}\right)^2}\right] \qquad (18.8\text{-}6)$$

but, from (6.6-11),

$$\omega^2(z) = \omega_0^2\left\{1 + \left[\frac{2(z - z_{\text{min}})}{k\omega_0^2}\right]^2\right\}$$

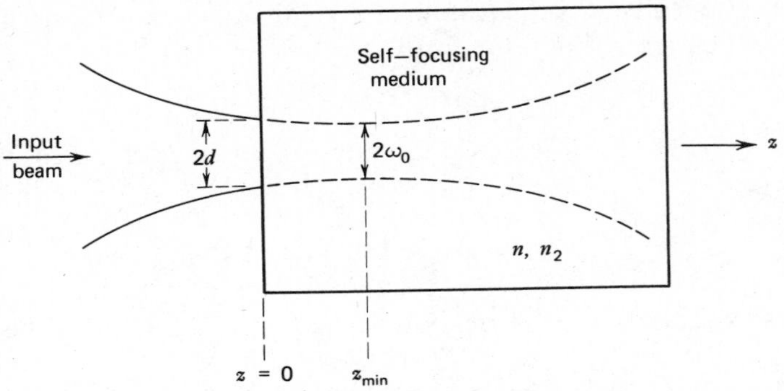

FIGURE 18.13 A Gaussian beam entering a slab of material that has a dielectric constant $\varepsilon = n^2 + nn_2 |E|^2$. The dashed curve gives the beam radius without self-focusing ($n_2 = 0$).

so that (18.8-6) can be written as

$$E(x, y, 0) = E_0 \exp\left[-\frac{r^2}{d^2}\left(1 - i\frac{2z_{\min}}{k\omega_0^2}\right)\right] \tag{18.8-7}$$

$$d^2 \equiv \omega^2(0) = \omega_0^2\left[1 + \left(\frac{2z_{\min}}{k\omega_0^2}\right)^2\right] \tag{18.8-8}$$

The input beam at $z = 0$ can thus be defined by its radius d and the distance to the waist, $z_{\min}$. We will greatly simplify the following analysis by introducing a focusing parameter

$$\theta \equiv \frac{2z_{\min}}{k\omega_0^2} \tag{18.8-9}$$

so that, from (18.8-7) and (18.8-8),

$$E(x, y, 0) = E_0 e^{-r^2(1-i\theta)/d^2} \tag{18.8-10}$$

$$z_{\min} = \frac{kd^2}{2}\frac{\theta}{1 + \theta^2} \tag{18.8-11}$$

$$\omega_0 = \frac{d}{(1 + \theta^2)^{1/2}} \tag{18.8-12}$$

Note that for $\theta = 0$ the beam waist is at $z = 0$. If $\theta > 0$, the input beam at $z = 0$ is converging, while for $\theta < 0$ the beam is diverging.

When the nonlinearity is present ($n_2 \neq 0$), general solutions of (18.8-4) must be found numerically. Our analysis of the problem will consider only the *initial* focusing behavior of a circularly symmetrical beam, with the "input" beam shape given in (18.8-10). By repeated use of (18.8-4), the "intensity" $|E|^2$ of the beam is found to obey the following:

$$\frac{\partial |E|^2}{\partial z} = E\frac{\partial E^*}{\partial z} + E^*\frac{\partial E}{\partial z} = \frac{i}{2k}(E\nabla_T^2 E^* - E^*\nabla_T^2 E)$$

$$\frac{\partial^2 |E|^2}{\partial z^2} = \frac{1}{4k^2}\Big\{(\nabla_T^2 E)(\nabla_T^2 E^*) - E\nabla_T^2(\nabla_T^2 E^*) + \frac{n_2 k^2}{n} E[|E|^2\nabla_T^2 E^* - \nabla_T^2(|E|^2 E^*)] + \text{c.c.}\Big\} \qquad (18.8\text{-}13)$$

To obtain a feeling for the overall beam behavior, we can now find the expansion of the beam's intensity on the axis near $z = 0$. We express $|E(0, 0, z)|^2$ in a Taylor expansion, keeping only the first three terms and using (18.8-10) to obtain the transverse derivatives.

$$|E(x = y = 0)|^2 \simeq E_0^2\left[1 + (4\theta)\frac{z}{kd^2} + \frac{z^2}{k^2 d^4}\left(-4 + 12\theta^2 + \frac{2n_2 k^2 d^2}{n}E_0^2\right) + \cdots\right]$$

The inverse of this function forms a rough approximation to the area of the beam. Keeping the first three terms in the expansion $(1 + x)^{-1} \simeq 1 - x + x^2 + \cdots$ gives

$$a(z) \sim \frac{1}{|E(x = y = 0)|^2} \simeq a(0)\left[1 - 4\theta\frac{z}{kd^2} + \frac{z^2}{k^2 d^4}\left(4 + 4\theta^2 - \frac{2n_2 k^2 d^2}{n}E_0^2\right) + \cdots\right] \qquad (18.8\text{-}14)$$

By assuming that beam focusing occurs where this area vanishes, we can find the strength of the nonlinearity $n_2(E_0^2)_c$ required for this effect. Setting the quadratic inside the brackets of (18.8-14) equal to zero, we obtain the distance from the input plane to the self-focusing point

$$z_f = \frac{kd^2}{2}\frac{1}{\left(\sqrt{\frac{P}{P_c} - 1} + \theta\right)} \qquad (18.8\text{-}15)$$

where the total input beam power P is given by

$$P = \frac{\pi\varepsilon_0 c n d^2}{4}E_0^2$$

and

$$P_c = \frac{\pi\varepsilon_0 c^3}{2n_2\omega^2} \qquad (18.8\text{-}16)$$

According to (18.8-15), if the input beam is initially converging ($\theta > 0$), it will focus catastrophically at z_f [$a(z_f) \to 0$] provided its total power exceeds P_c. The critical power P_c is independent of the initial degree of convergence (i.e., of θ) and of the initial beam diameter d.

If the beam is initially divergent ($\theta < 0$), the critical power for self-focusing is

$$P_{\text{critical}}(\theta < 0) = P_c(1 + \theta^2) \qquad (18.8\text{-}17)$$

These approximate results agree (within a factor of 4) with numerical results based directly on (18.8-4) and (18.8-10).

Numerical Example: Self-Focusing in Carbon Disulfide (CS_2). Using

$$(n_2)_{\text{MKS}} = \tfrac{1}{9} \times 10^{-8}(n_2)_{\text{esu}}$$

$$(n_2)_{\text{esu}} \simeq 10^{-11} \text{ in } CS_2 \quad \text{(see References 27, 28)}$$

$$\lambda \sim 10^{-6}\text{ m} \qquad (\omega \sim 1.9 \times 10^{15}\text{ sec}^{-1})$$

In (18.8-16), we get

$$P_c \sim \times 10^4 \text{ W}$$

so that, as observed, self-focusing occurs even at moderately high power levels. In addition to the anamalous stimulated Raman scattering threshold, there are various ways in which beam focusing is observed experimentally. In

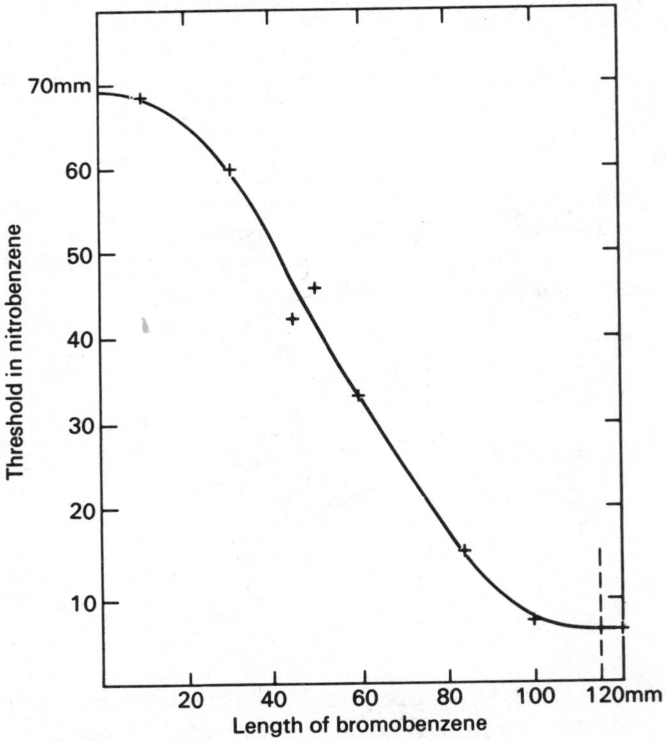

FIGURE 18.14 The threshold length for stimulated Stokes production in a nitrobenzene cell as a function of the length of a cell filled with bromobenzene placed immediately in front. The vertical dashed line indicates the threshold for Stokes production in bromobenzene. *Source:* Reference 33.

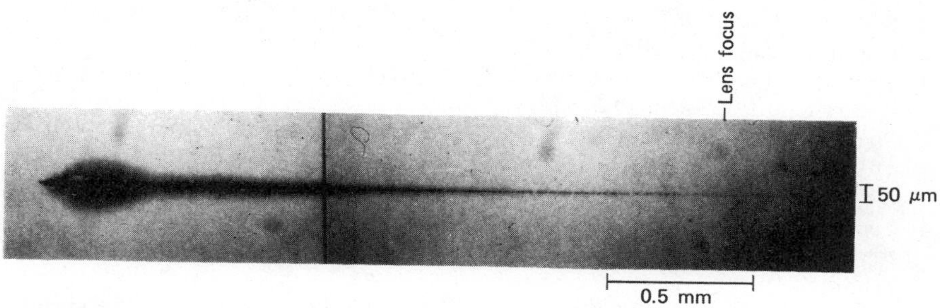

FIGURE 18.15 Damage filament due to self focusing of a Ruby laser beam in sapphire. *Source:* Reference 34.

one experiment, that of Lallemand and Bloembergen (Reference 33), the laser beam first passed through a cell containing bromobenzene and then entered a cell containing nitrobenzene. The threshold length (for stimulated Raman scattering) in the nitrobenzene cell was plotted as a function of the length of the bromobenzene cell, with the laser power held constant. The data, reproduced in Figure 18.14, show that due to partial focusing in the first cell, which acts as a converging lens, the collapse of the beam—hence the observed threshold—in the nitrobenzene cell, which is placed behind it, is accomplished in a shorter distance.

When self-focusing occurs in a solid, the extreme power densities present at the focal point can cause physical damage (Reference 34) to the material, as illustrated in Figure 18.15. This phenomenon is of great concern to experimentalists working with very high-power laser pulses since such damage can occur within the laser source itself.

Self-focusing and self-trapping of low-power (~20 mW) CW laser beams have recently been observed (Reference 35). The trapped beam is that of a dye laser tuned to the immediate vicinity of the sodium D_2 line ($^2S_{1/2} - {}^2P_{3/2}$) at 5890 Å. The focusing medium is sodium vapor.

The intensity-dependent index change is due to the partial saturation of the atomic resonant susceptibility by the laser beam. For a fundamental Gaussian beam and at $\nu > \nu_0$ (ν_0 is the center frequency), the index is larger at the beam center than at the wings and self-focusing results. The observed low critical power is due to the fact that, in this case, n_2 is inversely proportional to the product of the saturation intensity and the Doppler width, both of which are extremely small. At $\nu < \nu_0$ the sign of n_2 is reversed, and the beam is defocused in passing through the cell.

For an extensive bibliography on this subject, the student is referred to Supplementary Reference 3.

References

1. See, for example, G. Herzberg, *Molecular Spectra and Molecular Structure,* (Princeton, N.J.: Van Nostrand, 1961).

2. Brandmüller, J. and H. Moser, *Einfuhrung in die Raman Spectroscopie* (Darmstadt: Dietrich Steinkopff Verlag, 1962).
3. Shen, Y. R. and N. Bloembergen, "Theory of Stimulated Brillouin and Raman Scattering," *Phys. Rev.* **137,** A1787 (1965).
4. Placzek, G., *Handbuch der Radiologie VI* (Leipzig: Akademische Verlagsgesellschaft, 1934) Teil II, p. 205 (English translation, Lawrence Radiation Laboratory, Berkeley, Calif.).
5. Leite, R. C. C. and S. P. S. Porto, "Continuous Photoelectric Recording of the Raman Effect in Liquids Excited by He-Ne Red Laser," *J. Opt. Soc. Amer.* **54,** 981 (1964).
6. Eckhardt, G., "Selection of Raman Laser Materials," *J. Quant. Elect.* **2,** 1 (1966).
7. Eckhardt, G., R. W. Hellwarth, F. J. McClung, S. E. Schwarz, D. Weiner, and E. J. Woodbury, "Stimulated Raman Scattering from Organic Liquids," *Phys. Rev. Letters* **9,** 455 (1962).
8. Hellwarth, R. W., "Theory of Stimulated Raman Scattering," *Phys. Rev.* **130,** 1850 (1963).
9. Clements, W. R. L. and B. P. Stoicheff, "Raman Linewidths for Stimulated Threshold and Gain Calculations," *Appl. Phys. Letters* **12,** 246 (1968).
10. Kato, Y. and H. Takuma, "Experimental Study on the Wavelength Dependence of the Raman Scattering Cross Section," *J. Chem. Phys.* **54,** 5398 (1971).
11. Fenner, W. R., H. A. Hyatt, J. M. Kellman, and S. P. S. Porto, "Raman Cross Sections of Some Simple Gases," *J. Opt. Soc.* **63,** 73 (1973).
12. Kaiser, W. and M. Maier, *Stimulated Rayleigh, Brillouin and Raman Spectroscopy Laser Handbook,* F. T. Arecchi and E. O. Schulz-Dubois, eds. (Amsterdam: North Holland, 1972), p. 1077.
13. Terhune, R. W., "Nonlinear Optics," *Bull. Am. Phys. Soc.* **8,** 359 (1969).
14. Chiao, R. Y., E. Garmire, and C. H. Townes, "Self-Trapping of Optical Beams," *Phys. Rev. Letters* **13,** 479 (1964).
15. Kelley, P. L., "Self-Focusing of Optical Beams," *Phys. Rev. Letters* **15,** 1005 (1965).
16. Askaryan, G. A., "Effects of the Gradient of a Strong Electromagnetic Beam on Electrons and Atoms," *Soviet Phys. JETP* (Trans.) **15,** 1088 (1962).
17. Talanov, V. I., "Self-Focusing of Electromagnetic Waves in a Nonlinear Medium," *Radio Physics* (Trans.) **7,** 254 (1964).
18. Akhmanov, S. A., R. V. Kokhlov, and A. P. Sukhorukov, "Self-Focusing, Self-Defocusing and Self-Modulation of Laser Beams," in *Laser Handbook,* F. T. Arecchi and E. O. Schulz-Dubois, eds. (Amsterdam: North Holland, 1972), p. 1151.
19. Courtesy of R. W. Terhune.
20. Brillouin, L., "Diffusion de la lumière et des rayons par un corps transparent homogène," *Ann. Phys.* **17,** 88 (1922).
21. Chiao, R. Y., C. H. Townes, and B. P. Stoicheff, "Stimulated Brillouin Scattering and Coherent Generation of Intense Hypersonic Waves," *Phys. Rev. Letters* **12,** 592 (1964).
22. Pomerantz M., "Temperature Dependence of Microwave Phonon Attenuation," *Phys. Rev.* **139,** 501 (1965).
23. Bommel, H. E. and K. Dransfeld, "Attenuation of Hypersonic Waves in Quartz," *Phys. Rev. Letters* **2,** 298 (1959).
24. Wilson, R. A., H. J. Shaw, and D. K. Winston, "Measurement of Microwave Acoustic Attenuation in Sapphire and Rutile Using Nickel-Film Transducers, *J. Appl. Phys.* **36,** 3269 (1965).

25. Chiao, R. Y., "Brillouin Scattering and Coherent Phonon Generation," Ph.D. Thesis, M.I.T. Cambridge, Mass. (1965).
26. Yariv, A., "Quantum Theory for the Interaction of Light and Hypersound," *J. Quant. Elect.* **QE-1,** 28 (1965).
27. Chiao, R. Y., E. Garmire, and C. H. Townes, "Self-Trapping of Optical Beams," *Phys. Rev. Letters* **13,** 479 (1964).
28. Kelley, P. L., "Self-Focusing of Optical Beams," *Phys. Rev. Letters* **15,** 1005 (1965).
29. Askaryan, G., "Effects of the Gradient of a Strong Electromagnetic Beam on Electrons and Atoms," *Sovient Phys. LETP* (Trans.) **15,** 1088 (1962).
30. Talanov, I., "Propagation of a Short Electromagnetic Pulse in an Active Medium," *Radio Phys.* (Trans.) **7,** 254 (1964).
31. Yariv, A., *Quantum Electronics,* 1st ed., (New York: Wiley, 1967), p. 90.
32. Armstrong, J. A., N. Bloembergen, J. Ducuing, and P. S. Pershan, "Interactions Between Light Waves in Nonlinear Media," *Phys. Rev.* **127,** 1918 (1962).
33. Lallemand, P. and N. Bloembergen, "Self-Focusing of Laser Beams and Stimulated Raman Gains in Liquids," *Phys. Rev. Letters* **15,** 1010 (1965).
34. Guiliano, C. R. and J. H. Marburger, "Observations of Moving Self-Foci in Sapphire," *Phys. Rev. Letters* **27,** 905 (1971).
35. Bjorkholm, J. E. and A. Ashkin, "CW Self-Focusing and Self-Trapping of Light in Sodium Vapor," *Phys. Rev.* **32,** 129 (1974).

Supplemental References

1. Von der Linde, D., L. Laubereau, and W. Kaiser, "Molecular Vibrations in Liquids; Direct Measurement of the Molecular Dephasing Time; Determination of the Shape of Picosecond Light Pulses," *Phys. Rev. Letters* **26,** 954 (1971).
2. Mooradian, A., *Raman Spectroscopy of Solids, Laser Handbook,* Vol. 2 (Amsterdam: North Holland, 1972), p. 1409.
3. Akhmanov, S. A., R. V. Kokhlov, and A. P. Sukhorukov, "Self-Focusing, Self-Defocusing and Self-Modulation of Laser Beams," in *Laser Handbook,* F. T. Arecchi and E. O. Schultz-Dubois, eds. (Amsterdam: North-Holland, 1972), p. 1151.
4. Yariv, A., "The Application of Feynman Diagrams and the Time Evolution Operator to Nonlinear Operator to Nonlinear Optics," *IEEE J. Quant. Elect.* **QE-13,** 943 (1977).
5. Prior, Y, "Perturbative and Diagrammatic Technique," *IEEE J. Quant. Elect.* **QE-20,** 37 (1984).
6. OwYoung, A., "Absolute Determination of $\chi^{(3)}$ via Two-Beam Nonlinear Interferometry," *Opt. Comm.* **16,** 266 (1976).
7. Rabinowitz, P., A. Stein, R. Brickman, and A. Kaldor, "Efficient Tunable H_2 Raman Laser," *Appl. Phys. Letters* **35,** 739 (1979). See also "Stimulated Rotational Raman Scattering from Para-H_2 Pumped by a CO_2 TEA Laser," *Opt. Letters* **3,** 147 (1978).
8. Harris, S. E., "Proposal for a 207-Å Laser in Lithium," *Opt. Letters* **5,** 1 (1980).
9. Harris, S. E., "Spontaneous Anti-Stokes Scattering as a High-Resolution and Picosecond-Time-Scale VUV Light Source," *Appl. Phys. Letters* **31,** 498 (1977).
10. Zych, L. J., J. Lukasic, J. F. Young and S. E. Harris, "Laser Induced Two-Photon Black Body Radiation in the VUV," *Phys. Rev. Letters* **40,** 1493 (1978).
11. Lin, C. and M. A. Bosch, "Large-Stokes-Shift Stimulated Four-Photon Mixing in Optical Fibers," *Appl. Phys. Letters* **38,** 479 (1981).

12. Shen, Y. R., *The Principles of Nonlinear Optics* (New York: Wiley-Interscience, 1984).
13. Shubert, M. and B. Wilhelmi, *Nonlinear Optics and Quantum Electronics*, (New York: Wiley-Interscience, 1986).
14. Harris, L. A. and J. N. Levinos, "Generation of Nanosecond Infrared Pulses Tunable from 2.8μm to 16μm by Efficient Stimulated Electronic Raman Scattering," *Appl. Opt.* **26,** 396 (1987).

Problems

18.1 Justify the relation of $\gamma = 2\pi\,\Delta\nu$ where γ is that appearing in (18.4-1) and $\Delta\nu$ is the linewidth for spontaneous Raman scattering.

18.2 (a) Assume that a molecular liquid is completely transparent in a certain frequency region. Show that while the liquid is irradiated with an intense laser beam, its transparency is no longer uniform across the frequency spectrum. Specifically, show that it can now absorb radiation at $\omega_1 = \omega_l + \omega_v$ and has negative absorption at $\omega_2 = \omega_l - \omega_v$.

Hint:
Use the reasoning of Section 18.2 with the aid of diagrams such as those of Figure 18.3.

(b) Discuss how you would use the combination of a white light source and a laser beam for determining vibrational levels. [See W. J. Jones and B. P. Stoicheff, *Phys. Rev. Letters* **13,** 657 (1964).]

(c) Derive the expression for the absorption coefficient at $\omega = \omega_l + \omega_v$ in the presence of the laser radiation. Use the known laser intensity and the differential scattering cross section $d\sigma/d\Omega$.

18.3 Show why, in a normally dispersive and isotropic medium, the condition $2\mathbf{k}_2 = \mathbf{k}_1 + \mathbf{k}_3$ cannot be satisfied for propagation along a single direction.

18.4 (a) Show that the peak frequency of the stimulated Raman emission is at $\omega_1 = \omega_2 - \omega_v + (\gamma^2/4\omega_v)$.

(b) What is the shift of the excited vibrational frequency relative to ω_v?

18.5 Show that the anti-Stokes cone angle β in Figure 18.8 is given by

$$\beta = \left\{\frac{1}{n}\frac{\omega_1}{\omega_3}\left[n_3 - n_1 + \frac{\omega_2 - \omega_1}{\omega_2}(n_3 + n_1 - 2n_0)\right]\right\}^{1/2}$$

where the subscripts 1, 2, 3 refer, respectively, to the Stokes, laser, and anti-Stokes frequencies, and the n's are the indices of refraction.

18.6 Derive equations (18.8-13), (18.8-14), and (18.8-15).

18.7 Use the formalism of Appendix 4 to obtain a quantum mechanical expression for x_2 as defined by Eq (18.8-1). Compare the result to (18.8-2).

18.8 Obtain, starting with (18.1-15), the expression for the gain constant g_{Raman} exercised by the signal at ω_2 due to the presence of a field at ω_1 with an intensity I_1 (W/m^2).

18.9 Calculate in a manner similar to that leading to (18.1-15) the nonlinear susceptibility in the expression

$$P^{(\omega_3=\omega_1-\omega_1+\omega_3)} = \chi_{\text{anti-Stokes}}^{(\omega_3=\omega_1-\omega_1+\omega_3)} E_1^{(\omega_1)}[E_1^{(\omega_1)}]^* E^{(\omega_3)}$$

due to the process in which an atom initially in an excited state $|s\rangle$ (see the figure below) absorbs a photon at ω_1 and emits one at $\omega_3 = \omega_1 + \omega_{sn}$. Assume that one Feynman diagram dominates.

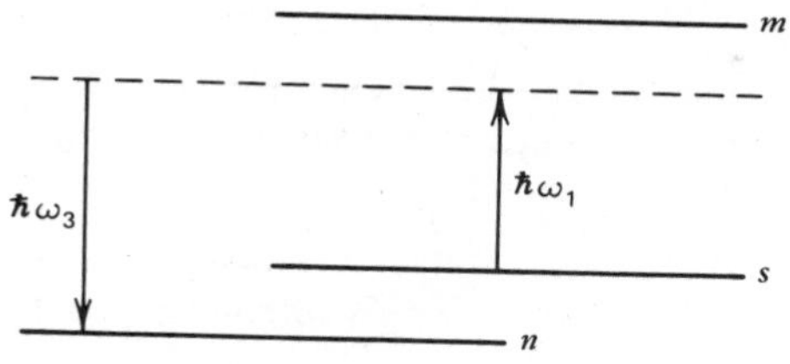

18.10 (a) Obtain an expression for the gain exercised by a "weak" beam at ν' due to a "strong" pump beam at ν in an inverted two-level ($|2\rangle$ and $|1\rangle$) atomic system.

(b) Taking $\hbar\nu = E_2 - E_1$ as being resonant with the transition, plot $\gamma(\nu)$ for $I_2 = 0.25 I_s$, where I_s is the saturation intensity at ν.

Show that the constant ε_2 of (18.8-1) is proportional to the susceptibility $\chi^{(\omega=\omega-\omega+\omega)}$. What is the proportionality constant?

18.11 (a) Calculate the power transfer per unit volume from the field at ω_3 to that at ω_2 in the presence of a field at ω_1 due to the process illustrated in the figure below.

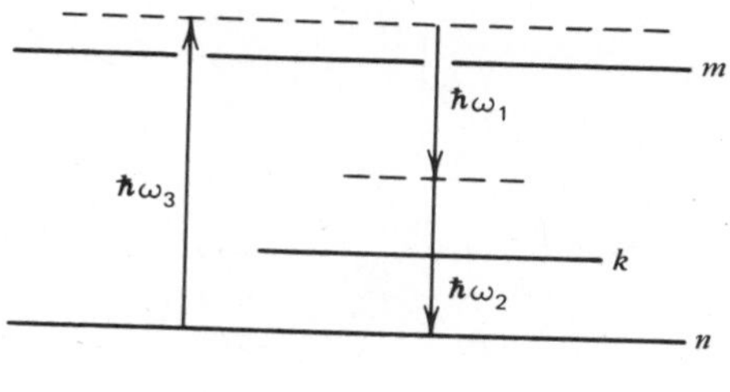

Assume

$$\omega_3 \approx \omega_{mn}$$

$$\omega_2 \approx \omega_{kn}$$

The population of level n is N(atoms/m^3). The other levels are not populated.

Answer:

$$\frac{P_{3\to 2}}{\text{Vol.}} = \frac{N\omega_2 I_1 I_2 I_3}{4\hbar^5 c^3 \varepsilon_0^3 n_1 n_2 n_3} \frac{(\mu_3)_{mn}^2(\mu_1)_{km}^2(\mu_2)_{kn}^2}{\left[(\omega_{mn}-\omega_3)^2+\gamma_{mn}^2\right]\left[(\omega_{kn}-\omega_2)^2+\gamma_{kn}^2\right]\gamma_{nn}}$$

where I's are intensities and n's indices of refraction.

Clue:
Show that the largest contribution to the induced atomic dipole at ω_2 is from

$$\langle \mu(\omega_2) \rangle = \langle \psi^{(2)}_{-\omega_3,\omega_1} | \mu_2 | \psi^{(3)}_{-\omega_3,\omega_1,\omega_2} \rangle$$

(b) Obtain an expression for the nonlinear susceptibility

$$\chi^{(\omega_2=-\omega_3+\omega_3-\omega_1+\omega_1+\omega_2)}_{ijklmn}$$

of an atomic system with an energy level scheme and frequencies as shown in the figure. Consider only resonant terms.

(c) Using the expression (5.1-13)

$$\text{Average power per unit volume} = \overline{\mathbf{E}(\mathbf{r}, t) \cdot \frac{\partial \mathbf{P}(\mathbf{r}, t)}{\partial t}}$$

obtain the power flow $P_{3\to 2}/\text{Vol.}$ from ω_3 to ω_2 using the result of (b). Compare the final result to that stated in (a).

CHAPTER 19

Phase-Conjugate Optics and Photorefractive Beam Coupling

19.0 INTRODUCTION

Phase-conjugate optics is the name given to a field in optics that involves the use of nonlinear optical techniques for an ever-increasing list of technological and scientific applications. The most important of these involve the correction of propagation distortion and the processing of pictorial information (References 1 and 2). We will start by defining optical-phase conjugation and demonstrating some of its basic properties.

19.1 PROPAGATION THROUGH A DISTORTING MEDIUM

Consider as an example the problem of an optical beam given in scalar form by

$$\begin{aligned} E_1(\mathbf{r}, t) &= \mathrm{Re}[\psi(\mathbf{r})e^{i(\omega t - kz)}] \\ &= \mathrm{Re}[A_1(\mathbf{r})e^{i\omega t}] \end{aligned} \tag{19.1-1}$$

propagating through a linear lossless distorting medium essentially in the z direction (from left to right). The dependence of ψ on $\mathbf{r}$ reflects spatial modulation by information, the effects of distortion, and diffraction. If in some region of space near z_0, we somehow generate a field $E_2(\mathbf{r}, t)$ that is described locally by

$$\begin{aligned} E_2(\mathbf{r}, t) &= \mathrm{Re}[\psi^*(\mathbf{r})e^{i(\omega t + kz)}] \\ &= \mathrm{Re}[A_2(\mathbf{r})e^{i\omega t}] \end{aligned} \tag{19.1-2}$$

then

$$A_2(\mathbf{r}) = A_1^*(\mathbf{r}) \qquad \text{for all} \qquad z < z_0$$

The field $E_2(\mathbf{r}, t)$ will be called in this book the *phase conjugate* of $E_1(\mathbf{r}, t)$. We note that to obtain E_2 from E_1, we take the complex conjugate of the *spatial part only,* leaving the factor $e^{i\omega t}$ intact. (This is equivalent to leaving the spatial part alone, but reversing the sign of t. That is, the field E_2 is related to E_1 by "time reversal.")

To appreciate the practical consequences of phase conjugation, consider a field E_1 propagating from left to right through a distorting medium as shown

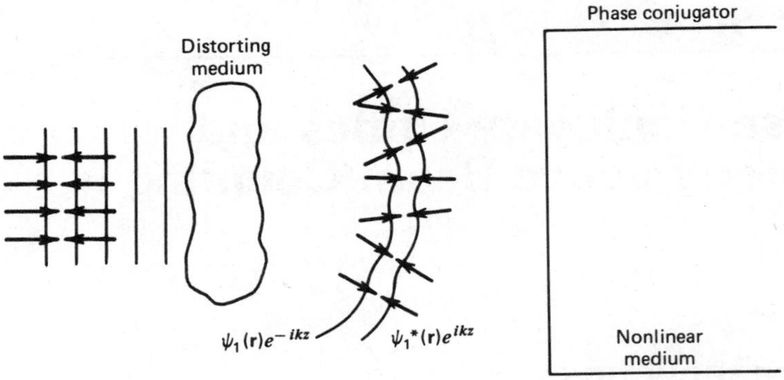

FIGURE 19.1 Complex conjugation of an input field $\psi_1(\mathbf{r})^{-ikZ}$.

in Figure 19.1. The electric permittivity (dielectric constant) of the composite medium is $\varepsilon(\mathbf{r})$, a real function accounting for the presence of passive linear components such as lenses, wedges, and/or distorting media. The scalar forward-propagating $(+z)$ beam is taken as

$$E_1(\mathbf{r}, t) = \text{Re}[\psi_1(\mathbf{r})e^{+i(\omega t - kz)}] \tag{19.1-3}$$

The scalar wave equation obeyed by this wave is taken as

$$\nabla^2\psi_1 + [\omega^2\mu\varepsilon(\mathbf{r}) - k^2]\psi_1 - 2ik\frac{\partial\psi_1}{\partial z} = 0 \tag{19.1-4}$$

(This form is valid as long as the fractional variation of $\varepsilon(\mathbf{r})$ in one optical wavelength is small compared to unity.)

Let us next, as a purely mathematical operation, consider the complex conjugate of Eq. (19.1-4) which, assuming a *real* $\varepsilon(\mathbf{r})$, becomes

$$\nabla^2\psi_1^* + [\omega^2\mu\varepsilon(\mathbf{r}) - k^2]\psi_1^* + 2ik\frac{\partial\psi_1^*}{\partial z} = 0 \tag{19.1-5}$$

This is also the wave equation obeyed by a wave form

$$E_2(\mathbf{r}, t) = \text{Re}[\psi_2(\mathbf{r})e^{+i(\omega t + kz)}] \tag{19.1-6}$$

propagating in the $-z$ direction with a complex amplitude

$$\psi_2(\mathbf{r})e^{ikz} \propto [\psi_1(\mathbf{r})e^{-ikz}]^* \tag{19.1-7}$$

If the dielectric constant $\varepsilon(\mathbf{r})$ in (19.1-4,5) represents some optical distortion in the path, then the above proof shows that the reflected beam is "healed" of the distortion and regains its original state. What we have shown is that a wave E_2, whose complex amplitude is everywhere the complex conjugate of E_1, satisfies the same wave equation as that obeyed by E_1. (This is true for an arbitrary wave in any lossless medium with a real ε; see Problem 19.1).

The practical corollary of the formal proof given above is the following: consider a complex monochromatic beam E_1 that propagates from some plane A to another plane B through a medium with a real $\varepsilon(\mathbf{r})$. If in the vicinity of B, we generate, by some means, a wave E_2 whose complex amplitude over an area that contains the incident beam is the complex conjugate of E_1 (within a multiplicative constant), then E_2 will propagate backward and *remain everywhere the complex conjugate of E_1. Its wavefronts everywhere thus coincide with those of E_1* and is therefore, *everywhere,* the phase conjugate of the incident wave $E_1(\mathbf{r}, t)$. A vector extension of the compensation theorem is taken up in Problem 19.5.

19.2 IMAGE TRANSMISSION IN FIBERS

The second example we choose to illustrate the power of phase conjugation is long-distance image transmission in optical waveguides (References 3 and 4). The problem of phase conjugation in an optical waveguide, say a fiber, is illustrated in Figure 19.2. Let the incident field at the input plane of the guide be

$$f_1(x, y, z = 0, t) = \sum_{m=0}^{N} \sum_{n=0}^{N} A_{mn}E_{mn}(x, y)e^{i\omega t} \tag{19.2-1}$$

where E_{mn} is the spatial mode function m, n of the particular waveguide used. The summation occurs over the discrete spectrum of guided modes, whose number is N^2. The output field at the fiber's end is obtained by propagating each mode a distance L and then summing

$$f_2(x, y, L, t) = \sum_{m} \sum_{n} A_{mn}E_{mn}(x, y) \exp[i(\omega t - \beta_{mn}L)] \tag{19.2-2}$$

where β_{mn} is the propagation constant of mode m, n and where we have neglected the possibility of mode-dependent losses. In general, the picture field (19.2-2) is different from the input field (19.2-1), due to the phase factors $\beta_{mn}L$. In practice, this corresponds to a scrambling of the initially launched spatial information. If, however, we can generate a complex conjugate of the output field (19.2-2),

$$f_3(x, y, L, t) = \sum_{m} \sum_{n} A^*_{mn}E^*_{mn}(x, y) \exp[i(\omega t + \beta_{mn}L)] \tag{19.2-3}$$

and launch it into a section of length L of a fiber identical in length and all

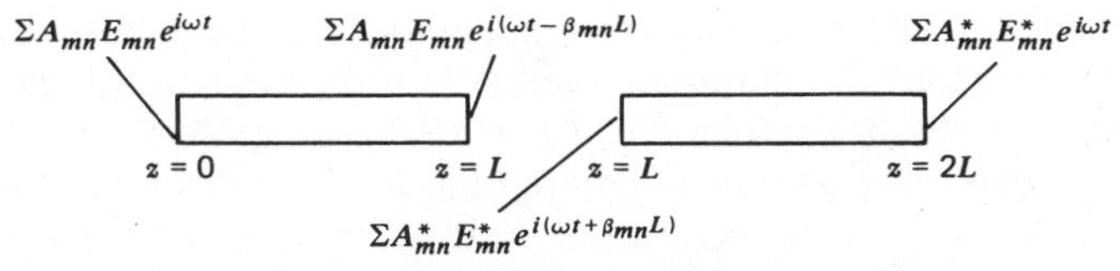

FIGURE 19.2 Compensation (and restoration for image loss) by modal dispersion in a dielectric waveguide using phase conjugation.

other characteristics to the first one, the result will be an output field

$$\begin{aligned} f_4(x, y, 2L, t) &= \sum_m \sum_n A^*_{mn} E^*_{mn}(x, y) \exp[i(\omega t + \beta_{mn} L - \beta_{mn} L)] \\ &= \sum_m \sum_n A^*_{mn} E^*_{mn}(x, y) e^{i\omega t} \end{aligned} \tag{19.2-4}$$

Except for the complex conjugation, the field f_4 at $z = 2L$ is the same as the input field f_1, and the spatial information carried by the wave has thus been recovered.

The practical realization of this scheme requires two sections of identical near-perfect fibers that have negligible intermodal scattering and negligible loss for all the modes. A similar scheme using a single section of fiber was demonstrated experimentally (see Reference 5). In this scheme, the phase-conjugate signal was generated via the degenerate four-wave mixing (see the next section) in a barium titanate ($BaTiO_3$) crystal and then retraversed the same fiber. This results in a reconstruction of the input image.

19.3 THEORY OF PHASE CONJUGATION BY FOUR-WAVE MIXING

In this section we present the basic theory of phase conjugation by four-wave mixing. Here, we depend on the third-order nonlinearity that, according to Chapter 18, can be represented using

$$P_i = 4\chi_{ijkl} E_j E_k E_l \tag{19.3-1}$$

The complex amplitude of the induced polarization at the frequency $\omega_1 = \omega_2 + \omega_3 - \omega_4$ is related to the electric field amplitudes according to (18.1-6) by

$$\begin{aligned} P_i(\omega_1 &= \omega_2 + \omega_3 - \omega_4) \\ &= 6\chi_{ijkl}(-\omega_1, \omega_2, \omega_3, -\omega_4) E_j(\omega_2) E_k(\omega_3) E_l^*(\omega_4) \\ \chi_{ijkl}&(-\omega_1, \omega_2, \omega_3, -\omega_4) \equiv \chi_{ijkl}^{(\omega_1=\omega_2+\omega_3-\omega_4)} \end{aligned} \tag{19.3-2}$$

In the case of a material medium with an instantaneous polarization response, the element χ_{ijkl} in Eq. (19.3-2) is numerically equal to that in Eq. (19.3-1). The factor 6 results from the number of different ways in which the combination $E_j(\omega_2) E_k(\omega_3) E_l^*(\omega_4)$ arises in Eq. (19.3-1). Unlike the second-order coefficient d_{ijk}, which is nonvanishing only in noncentrosymmetric crystals, symmetry allows the coefficient χ_{ijkl} to be nonvanishing in any medium, including isotropic materials (gases, liquids, glasses) as well as cubic crystals, as discussed in Chapter 18. Table 19.1 lists some χ_{ijkl}.

The basic experimental arrangement of four-wave mixing is illustrated in Figure 19.3. The nonlinear medium is pumped by two counter propagating beams E_1 and E_2 of frequency ω. In what follows, we use the subscripts to label the wave and not to indicate its polarization. The polarizations are

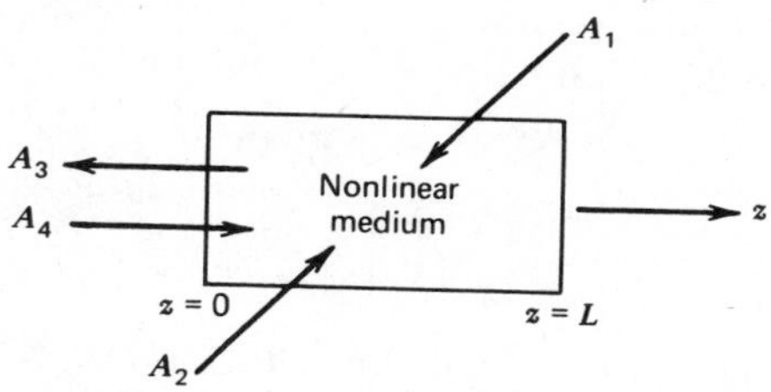

FIGURE 19.3 The basic experimental arrangement for phase conjugation by four-wave mixing.

assumed fixed. The waves are taken as

$$E_{1,2} = A_{1,2}(\mathbf{r}) e^{i(\omega t - \mathbf{k}_{1,2} \cdot \mathbf{r})} \tag{19.3-3}$$

Present in the medium are also two other waves E_3 and E_4 that are propagating essentially in opposition to each other along some arbitrary direction z. Their complex amplitudes are $A_3(\mathbf{r})$ and $A_4(\mathbf{r})$. The waves E_3 and E_4 are not, in general, plane waves, and we consider E_4 as some kind of input wave and E_3 as the output. If the frequency ω_4 of wave 4 is ω (i.e., the same as that of the pump waves 1 and 2), then wave 3 will be generated with $\omega_3 = \omega$ and its complex amplitude everywhere will be the complex conjugate of that of E_4. We take the input and output waves as

$$E_4 = A_4(z) e^{i(\omega t - kz)}$$

$$E_3 = A_3(z) e^{i(\omega t + kz)}$$

The four waves are assumed to be polarized along a given set of directions, and we will treat the interaction using scalar symbols.

TABLE 19.1 $\chi_{xyyx}(-\omega, \omega, \omega, -\omega)$ **of Some Materials at** $\lambda = 694$ **nm**

Material	n	$\chi_{xyyx}(-\omega, \omega, \omega, -\omega)$ (10^{-34} *MKS*)
CS_2	1.612	441
CCl_4	1.454	6.2
Fused quartz	1.455	1.5
YAG	1.829	7.41
Benzene	1.493	68.9
LSO glass	1.505	2.26
ED-4 glass	1.557	2.8
SF-7 glass	1.631	9.8
BS-7 glass	1.513	2.26
LaSF-7 glass	1.91	12.4

Source: Reference 6.

The four waves 1, 2, 3, 4 are coupled by means of the nonlinear polarization (19.3-2). Consider, as an example, the nonlinear polarization term due to the mixing of waves 1, 2, and 4

$$\begin{aligned} P_{\text{NL}}^{(\omega=\omega+\omega-\omega)}(\mathbf{r},t) &= \tfrac{1}{2}\chi^{(3)}A_1A_2A_4^* e^{i[(\omega+\omega-\omega)t-(\mathbf{k}_1+\mathbf{k}_2)\cdot\mathbf{r}+kz]} \\ &= \tfrac{1}{2}\chi^{(3)}A_1A_2A_4^* e^{ir(\omega t+kz)} \end{aligned} \tag{19.3-4}$$

since $\mathbf{k}_1 + \mathbf{k}_2 = 0$. Note by comparison with Eq. (19.3-2) that $\chi^{(3)} \equiv 6\chi_{ijkl}$. Equation (19.3-4) corresponds to a polarization wave with a frequency ω and a wave vector $\mathbf{k} = -\hat{\mathbf{e}}_z k$ and will thus excite a wave E_3 of the form

$$E_3 = A_3(z)e^{i(\omega t+kz)}$$

Once generated, this new wave E_3 will mix with waves E_1, E_2 to generate a polarization

$$P_{\text{NL}}^{(\omega=\omega+\omega-\omega)}(z,t) = \tfrac{1}{2}\chi^{(3)}A_1A_2A_3^* e^{i(\omega t-kz)} \tag{19.3-5}$$

which has the same frequency and propagation vector as E_4, so that it will interact strongly with this wave. This is how the interaction and power exchange of waves 3 and 4 are mediated via the pump waves 1 and 2. In treating this interaction mathematically, we will assume that the intensities of the pump waves E_1 and E_2 are so strong as to be little affected by the power exchange, so that A_1 and A_2 will be taken as constants. We will also assume that $|A_1| = |A_2|$, so that the slowing down (optical Kerr effect) of each pump wave by the other is the same and thus can be neglected (since the propagation constants are subtracted). We therefore need to consider only the propagation of waves E_3 and E_4.

We start with the basic coupling equation of a wave to a nonlinear polarization. In the case of the backward wave E_3, we replace k_1 in (16.4-10) by $-k$, assume no losses ($\sigma = 0$), and by recalling that $k_1^2 = \omega_1^2\mu_0\varepsilon$, obtain

$$-ik\frac{dA_3}{dz}e^{i(\omega t+kz)} = -\mu_0\frac{\partial^2}{\partial t^2}[P_{\text{NL}}^{(\omega)}(z,t)] \tag{19.3-6}$$

Using Eq. (19.3-4) for P_{NL} leads to

$$\frac{dA_3}{dz} = i\frac{\omega}{2}\sqrt{\frac{\mu_0}{\varepsilon}}\chi^{(3)}A_1A_2A_4^* \tag{19.3-7}$$

and similarly,

$$\frac{dA_4^*}{dz} = i\frac{\omega}{2}\sqrt{\frac{\mu_0}{\varepsilon}}\chi^{(3)}A_1^*A_2^*A_3$$

where we have used $k = \omega\sqrt{\mu_0\varepsilon}$. The last two equations can be rewritten as

$$\begin{aligned} \frac{dA_3}{dz} &= i\kappa^*A_4^* \\ \frac{dA_4^*}{dz} &= i\kappa A_3 \end{aligned} \tag{19.3-8}$$

where

$$\kappa^* = \frac{\omega}{2}\sqrt{\frac{\mu_0}{\varepsilon}}\chi^{(3)}A_1A_2 \tag{19.3-8a}$$

Most of the research literature in phase-conjugate optics uses the CGS system of units. Our derivation is based on MKS units. The Eqs. (19.3-8), which are the basic starting point for describing many applications, remain unchanged in CGS units (esu) except that the definition of the coupling constant is

$$(\kappa^*)_{\text{esu}} = \frac{2\pi\omega}{cn}\chi^{(3)}A_1A_2 \tag{19.3-9}$$

where, of course, all the physical quantities $\chi^{(3)}$, $A_{1,2}$ must be expressed in esu. To be able to move back and forth between the two systems of units, it is useful to remind ourselves that

$$1 \text{ statvolt/cm} = 2.99793 \times 10^4 \text{ V/m}$$

and

$$\chi^{(3)}_{\text{MKS}} = \frac{4\pi\varepsilon_0}{9 \times 10^8}\chi^{(3)}_{\text{esu}} = \frac{\chi^{(3)}_{\text{esu}}}{8.1 \times 10^{18}} \tag{19.3-10}$$

If we specify the complex amplitudes $A_3(L)$ and $A_4(0)$ of the two weak waves at their respective input planes ($z = L$, $z = 0$), the solution of Eqs. (19.3-8) is

$$\begin{aligned} A_3(z) &= \frac{\cos|\kappa|z}{\cos|\kappa|L}A_3(L) + i\frac{\kappa^* \sin|\kappa|(z-L)}{|\kappa|\cos|\kappa|L}A_4^*(0) \\ A_4^*(z) &= i\frac{|\kappa|\sin|\kappa|z}{\kappa^*\cos|\kappa|L}A_3(L) + \frac{\cos|\kappa|(z-L)}{\cos|\kappa|L}A_4^*(0) \end{aligned} \tag{19.3-11}$$

Most of the experimental situations involve a single input $A_4(0)$ at $z = 0$. In this case, the reflected wave at the input ($z = 0$) is

$$A_3(0) = -i\left(\frac{\kappa^*}{|\kappa|}\tan|\kappa|L\right)A_4^*(0) \tag{19.3-12}$$

while at the output ($z = L$)

$$A_4^*(L) = \frac{A_4^*(0)}{\cos|\kappa|L} \tag{19.3-13}$$

First, we note that the reflected wave $A_3(0)$ is proportional to $A_4^*(0)$, that is, to the complex conjugate of the input wave $A_4(0)$. It follows directly from the linearity of the Eqs. (19.3-8) that if the monochromatic input field E_4 is not a plane wave but possesses complex wavefronts so that

$$E_4 = \text{Re}[\psi(\mathbf{r})e^{i(\omega t - kz)}]$$

[where $\psi(\mathbf{r})$ is some complex function], then the reflected field E_3 is of the form

$$E_3(\mathbf{r})_{z<0} = \mathrm{Re}\left[-i\left(\frac{\kappa^*}{|\kappa|}\tan|\kappa|L\right)\psi^*(\mathbf{r})e^{i(\omega t+kz)}\right] \qquad (19.3\text{-}14)$$

so that the analysis of the Eqs. (19.3-8) applies not only to plane-wave inputs but to waves with arbitrary wavefronts.

We have thus established that the four-wave mixing geometry of Figure 19.3 performs the function of phase conjugation demanded of the magical mirror of Figure 19.1 and that the reflected wave E_3 will unravel in its backward propagation any distortion suffered by the incident wave E_4.

Although our main purpose was to demonstrate phase conjugation, the formal solution leading to Eq. (19.3-12) reveals some new features. We note that, for example,

$$\tfrac{1}{4}\pi < |\kappa|L < \tfrac{3}{4}\pi \qquad (19.3\text{-}15)$$

we have

$$|A_3(0)| > |A_4(0)|$$

and the phase-conjugate mirror acts as a (reflection) amplifier. The power gain of the amplifier is

$$\left|\frac{A_3(0)}{A_4(0)}\right|^2 = \tan^2|\kappa|L \qquad (19.3\text{-}16)$$

It also follows from Eq. (19.3-13) that for a single input $A_4(0)$ the transmitted wave $A_4(L)$ is given by

$$A_4(L) = \frac{1}{\cos|\kappa|L}A_4(0) \qquad (19.3\text{-}17)$$

and is thus always bigger than $A_4(0)$, that is, the device functions as a transmission amplifier for all values of κ at any pumping level.

A quantum-mechanical analysis of the four-wave interaction reveals that the amplification of the forward wave A_4 and the generation of the backward wave A_3 come at the expense of the pump waves A_1 and A_2. For each photon added to A_3, one photon must be added to A_4 while, simultaneously, a photon is subtracted from each of A_1 and A_2.

The field distribution inside the medium under the high-amplification condition $|\kappa|L \lesssim \frac{1}{2}\pi$ is shown in Figure 19.4.

If the pump intensity (i.e., the product $|A_1A_2|$) is increased until $|\kappa|L = \frac{1}{2}\pi$, then according to Eqs. (19.3-16) and (19.3-17),

$$\left|\frac{A_3(0)}{A_4(0)}\right| = \infty, \qquad \frac{A_4(L)}{A_4(0)} = \infty \qquad (19.3\text{-}17a)$$

The corresponding intensity distribution is shown in Figure 19.5. Equation (19.3-17a) can be interpreted as a situation in which a vanishingly small input $A_4(0)$ gives rise to finite outputs $A_4(L)$ and $A_3(0)$ from both sides of the interaction volume. A device that emits radiation without an input is, by

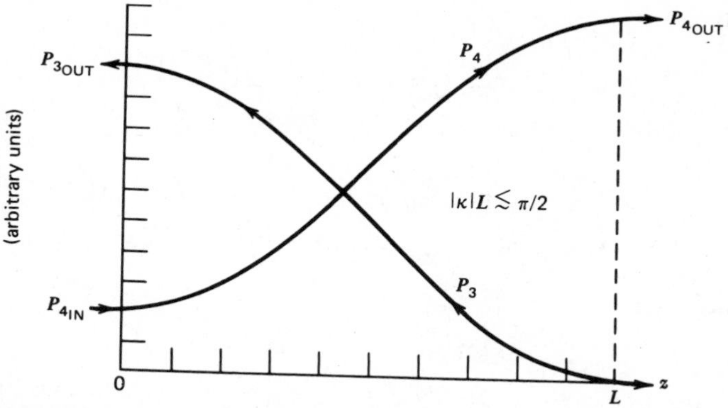

FIGURE 19.4 The intensity distribution inside the interaction region corresponding to the amplifier case $\pi/4 < |\kappa|L < \pi/2$.

definition, an oscillator. We thus established theoretically the possibility of mirrorless oscillation.

Numerical Estimate of the Coupling Constant

Most practical situations require that we achieve [see (19.3-11)] the condition $\kappa L \sim 1$. Three classes of material have been used to date: (1) polar liquids with highly anisotropic molecules such as CS_2; (2) atomic vapors such as mercury operated on, or very near to, an optical resonance; (3) photorefractive crystals, which will be discussed in some detail in Section 19.8. Here, we will evaluate the coupling constant and the requisite power levels in the case of CS_2.

Using Eq. (19.3-8a) and

$$A_1 = A_2 = \sqrt{\frac{2I}{\varepsilon_0 nc}}$$

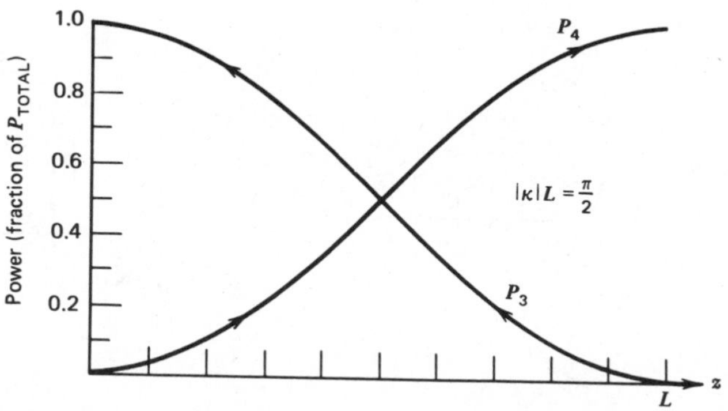

FIGURE 19.5 The intensity distribution inside the interaction region when the oscillation condition $|\kappa L| = \pi/2$ is satisfied.

we obtain

$$\kappa_{\text{MKS}} = \frac{2\pi}{n\varepsilon_0\lambda}\sqrt{\frac{\mu}{\varepsilon}}\,\chi^{(3)}I_1 = 2.68 \times 10^{14}\frac{\chi^{(3)}}{n^2\lambda}I_1$$

Using

$$\chi^{(3)}_{CS_2} = 6\chi_{yxxy}(-\omega, \omega, \omega, -\omega) = 2.03 \times 10^{-31}\ \text{MKS}$$

$$\lambda = 10^{-6}\ \text{m}, \qquad n = 1.62$$

we obtain

$$\kappa(\text{m}^{-1}) = 0.21 I_1\ (\text{MW/cm}^2)$$

To get values of $\kappa L \sim 1$ with, say, $L = 10$ cm requires intensities in the range of tens of megawatts per cm^2, which are available only from pulsed lasers. The requisite intensities can be reduced by orders of magnitude if one takes advantage of the large values of $\chi^{(3)}$ that result when the applied frequency is near that of some transition in a gas or atomic vapor (Reference 7) or by using photorefractive crystals that will be discussed in Section 19.8.

Some Experiments Demonstrating the Basic Features of Phase Conjugation

A good introduction to the practice of phase-conjugate optics would be (1) the verification of the basic phase-conjugation equations (19.3-14) and (2) the demonstration that the backward traveling wave A_3 is indeed the phase-conjugate replica of the incident A_4. The latter can be achieved by introducing some distortion or passive optical element, say a lens, and seeing if A_3 in its reverse propagation assumes everywhere the same wavefronts as A_4, that is, is "healed" of the distortion.

An experiment verifying both of the above points is illustrated in Figure 19.6. The collimated pump beam A_1 ($\lambda = 0.694\ \mu$m) is polarized in the plane of the figure (↑) and originates in a Q-switched ruby laser. The counter-

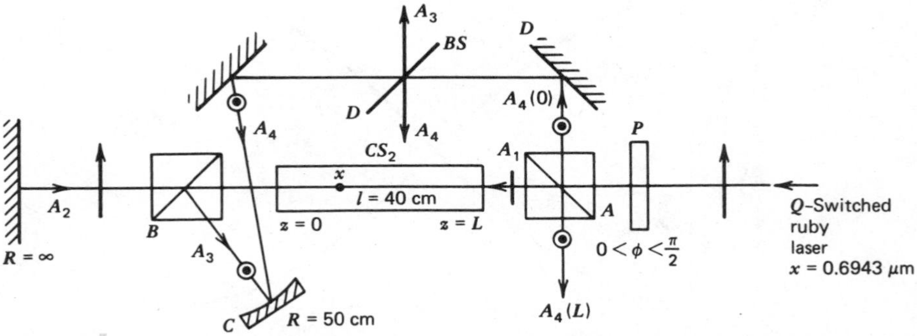

FIGURE 19.6 The experimental arrangement of a basic phase conjugation experiment. *Source:* Reference 8.

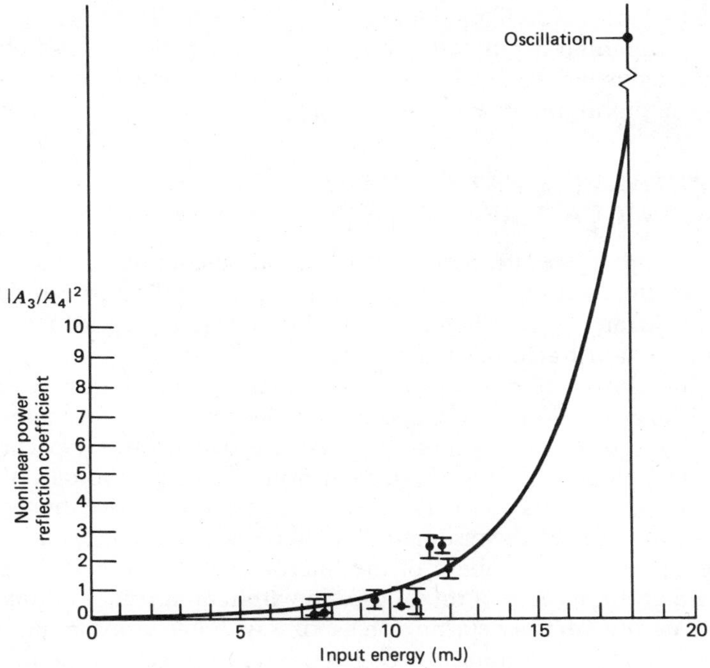

FIGURE 19.7 A plot of the reflection power coefficient versus pump pulse energy in (mJ) for the experimental setup of Figure 19.4. Data points [.] include oscillation ($|A_3/A_4|^2 = \infty$). The solid curve is the least-square fit to $R = \tan^2 (CI)$. *Source:* Reference 8.

propagating pump beam A_2 is generated by reflection from a mirror on the extreme left. A small part of the incident ($\uparrow$) energy is converted to the orthogonal O polarization by the birefringent plate P. This radiation is separated spatially from the main beam by polarizing prism A to yield the input probing beam A_4. A spherical mirror C focuses the collimated beam A_4 to point x in the CS_2 cell. This focusing constitutes the "distortion." (One person's focusing is another person's distortion.)

The reflected beam A_3 generated by the mixing of A_1, A_2, and A_4 is O-polarized and can be photographed and/or read off beam splitter D.

The measurement of the reflection coefficient $|A_3/A_4|^2$ for various pump intensities is shown in Figure 19.7. The solid curve is a plot of the theoretical expression (19.3-16)

$$\left|\frac{A_3(0)}{A_4(0)}\right|^2 = \tan^2 |\kappa| L$$

The verification of the phase-conjugate relationship between A_3 and A_4 is provided by the fact that after reflection from C, beam 3 is collimated. Beam 3 thus emanates from point source x, which is the very point on which beam 4 is converging. It follows that the wavefronts of these two beams are identical.

The prediction of oscillation for $|\kappa|L = \pi/2$ is verified (see Figure 19.5) by eliminating the input beam $A_4(0)$, say by removing mirror D, and observing the simultaneous emergence of O-polarized beams 3 and 4 with further increase of the pump intensity.

19.4 OPTICAL RESONATORS WITH PHASE-CONJUGATE REFLECTORS

One of the more interesting consequences of phase-conjugate optics involves optical resonators where one of the two conventional reflectors is replaced by a phase-conjugate mirror (Reference 9), hence referred to as a PCM. The situation is demonstrated in Figure 19.8.

Consider some arbitrary transverse Gaussian beam with transverse quantum numbers m, n (of the type considered in Section 6.9). Let the phase shift of this beam due to propagation between the two mirrors (spacing l) be $\phi_l(m, n)$. The phase shift upon reflection from the conventional reflector is taken as ϕ_R, whereas that of the PCM is α. We shall now derive the resonance frequency condition for the resonator in light of the reasoning used in Section 7.2 by requiring that the phase of the internally shuttling beam reproduce itself after a given number of round trips to within an integer multiple of 2π. We designate the arbitrary starting phase of, say, a left-propagating beam at some plane A as ϕ_1. Without loss of generality, we take A to lie just to the right of the reflector.

The phases of the beam at the various stages are

$$\phi_2 = \phi_1 + \phi_R$$

$$\phi_3 = \phi_2 + \phi_l(m, n) = \phi_1 + \phi_R + \phi_l(m, n)$$

$$\phi_4 = -\phi_3 + \alpha = -(\phi_1 + \phi_R + \phi_l(m, n)) + \alpha$$

Notice the sign inversion in going from ϕ_3 to ϕ_4 due to phase conjugation

$$\phi_5 = \phi_4 + \phi_l(m, n) = -\phi_1 - \phi_R + \alpha$$

$$\phi_6 = \phi_5 + \phi_R = -\phi_1 + \alpha$$

$$\phi_7 = \phi_6 + \phi_l(m, n) = -\phi_1 + \alpha + \phi_l(m, n) \tag{19.4-1}$$

$$\phi_8 = -\phi_7 + \alpha = \phi_1 - \phi_l(m, n)$$

$$\phi_9 = \phi_8 + \phi_l(m, n) = \phi_1$$

The self-consistent condition $\phi_9 = \phi_1$ is thus satisfied automatically. The phase-conjugate resonator has a resonance at the frequency of the pump beams. (This follows since no allowance was made for a frequency shift upon reflection from the PCM, which would be the case if the resonant mode frequency did not equal that of the pump.) *The resonance condition is satisfied independently of the length l of the resonator or the transverse order (m, n) of the Gaussian beam.* This requires *two* complete round trips. By tracing the arrows

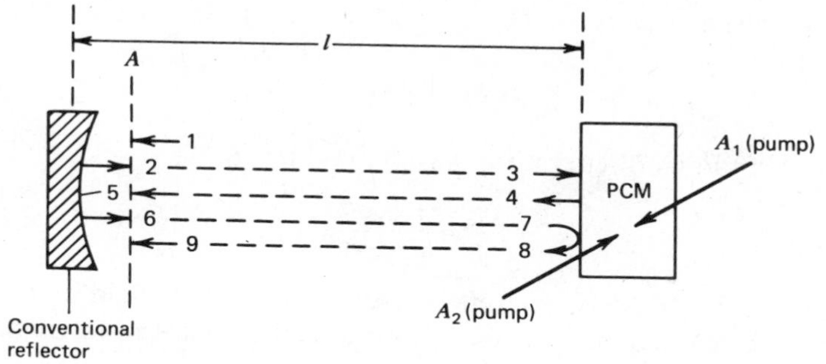

FIGURE 19.8 An optical resonator formed between a conventional reflector and a phase-conjugate mirror (PCM).

of Figure 19.9, we can verify that the radius of curvature of the Gaussian beam will also reproduce itself after two round trips. It follows that the *phase-conjugate resonator is stable* (in the sense defined in Section 7.2) *regardless of the radius of curvature R of the mirror and the mirror spacing l.*

19.5 THE *ABCD* FORMALISM OF PHASE-CONJUGATE OPTICAL RESONATORS

In this section we extend the *ABCD* Gaussian mode formalism of Section 5.7 to the case of a phase-conjugate resonator. The analysis follows closely that of Reference 9.

The ABCD Matrix of a Phase-Conjugate Mirror

Consider a Gaussian field E_i propagating along the z axis, to be incident upon the PCM. We thus have from (6.6-14)

$$E_i = \mathscr{E}_i(\mathbf{r}) \exp\left[i\left(\omega t - kz - \frac{kr^2}{2\rho}\right) - \frac{r^2}{w^2}\right] \tag{19.5-1}$$

where $\mathscr{E}(\mathbf{r})$ is the complex amplitude of E_i, and ρ and w are the radius of curvature and the spot size of the incident field, respectively. This field can

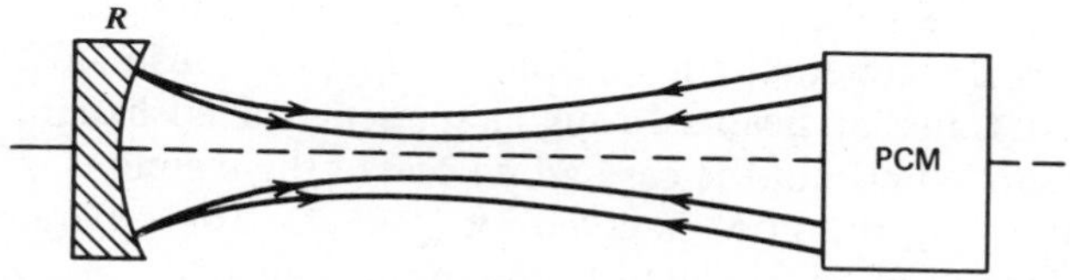

FIGURE 19.9 A self-consistent beam solution inside a phase-conjugate resonator reproduces its waveform curvature after two round trips.

also be written as

$$E_i = \mathscr{E}_i(\mathbf{r}) \exp\left[i\left(\omega t - kz - \frac{kr^2}{2q_i}\right)\right] \tag{19.5-2}$$

The complex radius of curvature q_i is defined by (6.7-15)

$$\frac{1}{q_i} = \frac{1}{\rho} - \frac{i\lambda}{\pi w^2} \tag{19.5-3}$$

The effect of the PCM is to "reflect" such an incident field as to yield its conjugate replica, leaving the wavefront and the spot size unchanged. The reflected field is thus

$$E_r \propto \mathscr{E}_i^*(\mathbf{r}) \exp\left[i\left(\omega t + kz + \frac{kr^2}{2\rho}\right) - \frac{r^2}{w^2}\right] \tag{19.5-4}$$

which can also be expressed as

$$E_r \propto \mathscr{E}_i^*(\mathbf{r}) \exp\left[i\left(\omega t + kz - \frac{kr^2}{2q_r}\right)\right] \tag{19.5-5}$$

The reflected field complex radius of curvature is, according to (19.5-3) and (19.5-4),

$$\frac{1}{q_r} = -\frac{1}{\rho} - \frac{i\lambda}{\pi w^2} = -\frac{1}{q_i^*} \tag{19.5-6}$$

An observer traveling with the beam will thus find the spot size unchanged after phase-conjugate reflection, but will see an opposite sign for the curvature of the wavefront.

If we introduce the ray matrix formalism of Section 6.4, the effect of the PCM can be represented by the matrix

$$\mathbf{M} = \begin{pmatrix} A & B \\ C & D \end{pmatrix} = \begin{pmatrix} 1 & 0 \\ 0 & -1 \end{pmatrix} \tag{19.5-7}$$

with the output and input q parameters related by

$$q_r = \frac{Aq_i^* + B}{Cq_i^* + D} \tag{19.5-8}$$

Note the *complex conjugation of* q_i, as opposed to the conventional formalism [Equation (6.7-6)], where the input field is not conjugated. We note that this matrix also describes the reflection of rays from the conjugate mirror.

It follows directly that the ordinary *ABCD* formalism for treating the propagation of Gaussian beams through a sequence of lenslike media (Section 6.7) can be applied also in the case when one of the elements is a PCM. The matrix representing the PCM is given by (19.5-7). The q parameter at any plane *following* the PCM is related to the input q by

$$q_{\text{out}} = \frac{A_T q_i^* + B_T}{C_T q_i^* + D_T} \tag{19.5-9}$$

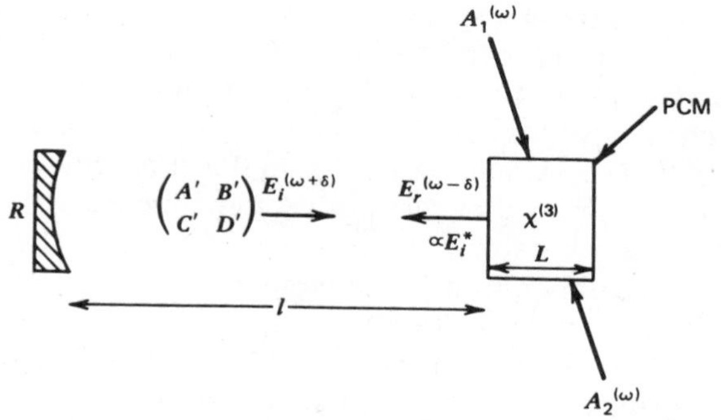

FIGURE 19.10 The phase-conjugate resonator (PCR). This general resonator is formed by placing some arbitrary optical components, represented collectively by an equivalent $A'B'C'D'$ ray matrix, between a "real" mirror (of radius R) at one end and a phase-conjugate mirror (PCM) at the other end. In the case of degenerate modes, which is considered here, $\delta = 0$.

where the subscript T implies that the matrix elements correspond to that of the resultant matrix for the given sequence of optical elements, including that of the PCM. Since all the matrices are assumed to be real, the complex conjugation of q_i imposed by (19.5-9) can be performed at any plane.

Consider next the situation sketched in Figure 19.10. The resonator is bounded on one end by a mirror having a radius of curvature R, containing arbitrary intracavity optical components described collectively by an $A'B'C'D'$ matrix $\mathbf{M}'$ for optical propagation from left to right and again by an $A''B''C''D''$ matrix $\mathbf{M}''$ for propagation from right to left. The resonator is bounded on the other end by a PCM. In order to investigate the stability criterion for such a resonator, we apply the standard self-consistent condition, whereby we require that the complex radius of curvature of the beam reproduce itself after two round trips. Choosing a reference plane arbitrarily to the immediate right of the real mirror, we trace a beam that propagates to the right and obtain, after one round trip, the following matrix product:

$$\begin{aligned} \mathbf{M}_1 = \begin{pmatrix} A_1 & B_1 \\ C_1 & D_1 \end{pmatrix} &= \begin{pmatrix} 1 & 0 \\ \dfrac{-2}{R} & 1 \end{pmatrix} \begin{pmatrix} A'' & B'' \\ C'' & D'' \end{pmatrix} \begin{pmatrix} 1 & 0 \\ 0 & -1 \end{pmatrix} \begin{pmatrix} A' & B' \\ C' & D' \end{pmatrix} \\ &= \begin{pmatrix} 1 & 0 \\ \dfrac{-2}{R} & 1 \end{pmatrix} \begin{pmatrix} 1 & 0 \\ 0 & -1 \end{pmatrix} \end{aligned} \qquad (19.5\text{-}10)$$

where we have used the relation

$$\mathbf{M}'\mathbf{M}\mathbf{M}' = \mathbf{M} \qquad (19.5\text{-}11)$$

which can be shown straightforwardly, using the reciprocity property of the group of optical elements represented by $\mathbf{M}'$ (or $\mathbf{M}''$), where $\mathbf{M}$ is given by (19.5-7). Equation (19.5-10) is merely a reaffirmation of the fact that an arbitrary sequence of passive and lossless optical elements followed by a PCM is equivalent to the PCM alone. This is due to the "time reversal" occurring at the PCM and the reciprocity of the passive components.

We have already established in Section 19.4 that the self-consistent phase condition of a mode in a phase-conjugate resonator is satisfied automatically after two round trips. The *ABCD* matrix describing the effect of two round trips on the complex beam radius is $\mathbf{M}_2 = (\mathbf{M}_1)^2$, where $\mathbf{M}_1$, given by (19.5-10), is the single round-trip Gaussian beam evolution matrix

$$\mathbf{M}_2 = (\mathbf{M}_1)^2 = \mathbf{I} \tag{19.5-12}$$

where $\mathbf{I}$ is the identity matrix. It follows that *any* Gaussian beam (i.e., one with an arbitrary waist location, waist size, and transverse mode order m, n) is a proper mode solution as far as shape reproducibility (after two round trips) is concerned. This, coupled with the above demonstration (19.4-1) concerning the phase condition, completes the proof that any arbitrary Gaussian beam with a frequency equal to that of the pump beams is a proper mode solution of a phase-conjugate resonator independent of the resonator length and the radius of curvature of its one spherical mirror.

For an additional discussion of this topic including the problem of modes whose resonant frequencies differ from that of the pump waves, one should consult References 9–11.

19.6 SOME PRACTICAL APPLICATIONS OF PHASE CONJUGATION

Dynamic Distortion Correction within a Laser Resonator

One of the more interesting practical applications of phase-conjugate optics is dynamic real-time correction of distortion in optical resonators. The situation is depicted in Figure 19.11. A laser oscillator comprises a gain medium, a

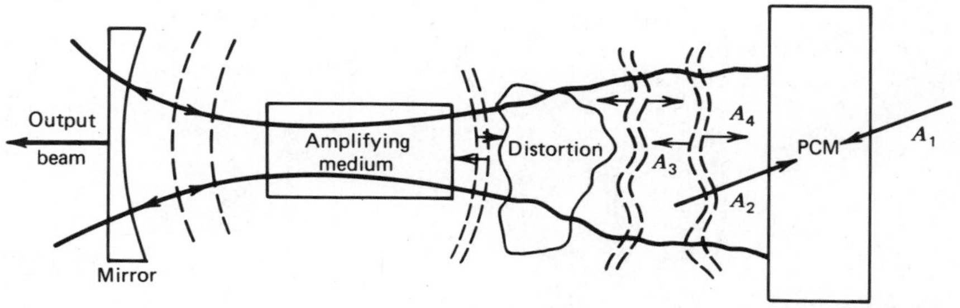

FIGURE 19.11 A phase-conjugate reflector compensates in real time for a time-varying distortion inside an optical resonator.

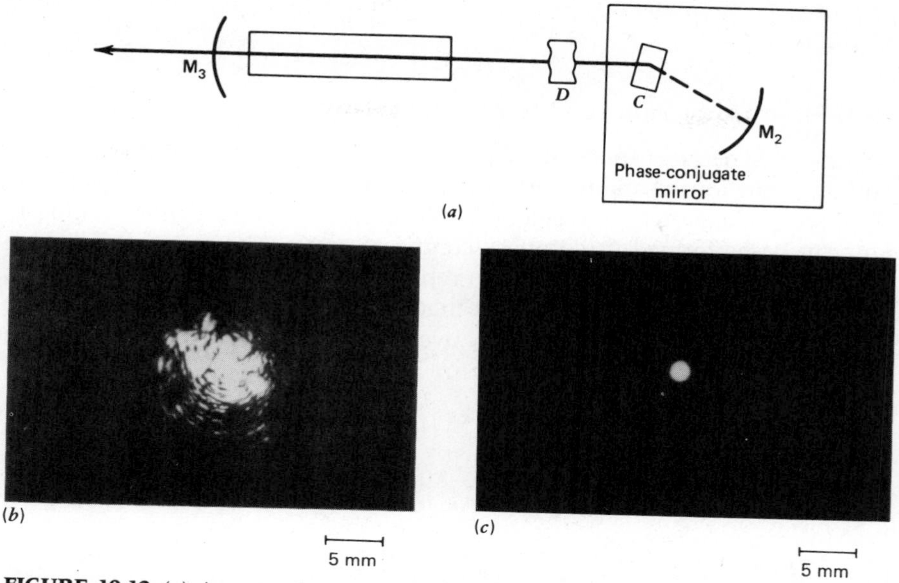

FIGURE 19.12 (*a*) An argon laser gain tube with a distortion *D*, a (photorefractive) phase-conjugating crystal *C*, and a feedback mirror $\mathbf{M}_2$. (*b*) The highly degraded output beam from $\mathbf{M}_3$ when *C* is replaced by a conventional mirror. (*c*) The beam regains its diffraction-limited shape in the presence of the distortion when the phase-conjugate reflector of configuration (*a*) is used. *Source:* Reference 12.

mirror, a phase conjugating mirror, and a distortion. The distortion may be due to the gain medium itself or to "bad" optics. Let us assume, for a moment, that the wave incident on the distortion from the left corresponds to a perfect Gaussian beam whose radius of curvature at the left mirror matches that of the mirror. The beam is distorted in passage through the distortion, but after reflection from the PCM and the reverse propagation through the distortion it regains, according to the distortion correction theorem of Section 19.1, its original undistorted form with the reflected (left-going) wavefronts coinciding in space with those of the right-going beam. It follows immediately by repeating the above scenario that the situation depicted in Figure 19.11 is self-reproducing and self-consistent (if not necessarily unique). It should thus be possible to extract the full available power of a laser oscillator in the form of an ideal Gaussian beam, that is, the output on the left side of Figure 19.11 in the presence of considerable and even time-varying distortion inside the resonator corresponds to that of a Gaussian beam.

An experimental demonstration (Reference 12) of a laser oscillator with dynamic phase-conjugate distortion correction is illustrated in Figure 19.12. The phase-conjugate mirror utilizes a crystal of barium titanate.[1] The gain

[1] The use of photorefractive crystals for phase-conjugate reflectors without pump beams will be discussed in Section 19.8.

medium is a commercial argon laser tube and the distortion is an acid-etched glass flat.

Real-Time Holography and Image Processing

The analogy between phase-conjugate optics and holography is interesting both from the formal and the practical points of view and suggests that nearly all of the applications envisaged or demonstrated with conventional holography can be performed using phase-conjugate optics. The main attraction in the use of phase-conjugate optics to replace the conventional holography is the real-time aspect of the former that obviates the need to develop the hologram. To appreciate this analogy, we use the expression (19.3-14) for the reflectivity of a phase-conjugate mirror

$$A_3(0) = -i\left(\frac{\kappa^*}{|\kappa|}\tan|\kappa|L\right)A_4^*(0)$$

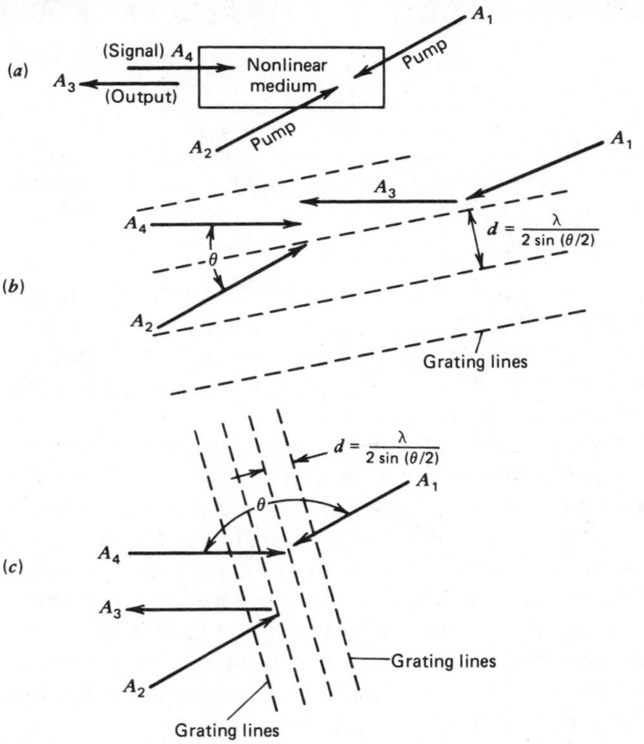

FIGURE 19.13 (*a*) The conventional geometry of phase-conjugate optics. (*b*) Beams 2 and 4 interfere to form a grating $A_2A_4^*$. Beam 1 is Bragg-diffracted from the grating to yield the output phase-conjugate (to A_4) beam $A_3 \;\alpha\; (A_2A_4^*)A_1$. (*c*) Beams 1 and 4 interfere to form a grating $A_1A_4^*$. Beam 2 is Bragg-diffracted from the grating to yield the phase-conjugate beam $A_3 \;\alpha\; (A_1A_4^*)A_2$.

To simplify the discussion, consider the case of small reflectivity $|A_3(0)/A_4(0)|^2 \ll 1$. The last relation simplifies to

$$A_3(0) = -i\kappa^* A_4^*(0) L$$

where L is the thickness of the phase-conjugating medium. Using (19.3-8*a*), we obtain

$$A_3(0) = -i\frac{\omega}{2}\sqrt{\frac{\mu}{\varepsilon}}\,\chi^{(3)} L A_1 (A_2 A_4^*) \tag{19.6-1}$$

$$= -i\frac{\omega}{2}\sqrt{\frac{\mu}{\varepsilon}}\,\chi^{(3)} L (A_1 A_4^*) A_2 \tag{19.6-2}$$

The placement of the brackets in (19.6-1) is to suggest that we may view the process of phage conjugation as the reflection of beam A_1 from the *stationary* holographic grating formed by the interference of A_2 and A_4. This situation is depicted in Figure 19.13*b*. We may, likewise, using the group of (19.6-2), view the process as the reflection of A_2 from the grating formed by beams A_1 and A_4. This situation is depicted in Figure 19.13*c*.

It should be emphasized here that the grating point of view used above is employed mostly for pedagogic reasons and contains no new physics. Both sets of "gratings" are represented automatically in the electromagnetic formulation of phase conjunction in Section 19.3.

The multiplication property [see (19.6-2)]

$$A_3 \propto A_1 A_2 A_4^*$$

of the phase-conjugate optical configuration of Figure 19.13*a* is the basis for numerous "real-time holographic applications" such as imaging through distortion and correlation and convolution in real time of optical images (References 13–15). These and other applications of phase-conjugate optics are described in detail in Reference 15. An account and references to the major contributions to this field by the Russian school are given in Reference 16.

19.7 OPTICAL PHASE CONJUGATION BY STIMULATED NONLINEAR SCATTERING

Historically, the first observation of distortion correction following phase conjugation was performed by Zeldovich et al. (Reference 17) using the experimental configuration depicted in Figure 19.14. A ruby laser beam, which was distorted by passing through a roughened plate of glass, served as a pump for stimulated Brillouin scattering in pressurized methane gas. The backward propagating stimulated beam passed in reverse through the distorting plate and emerged from it with an undistorted wavefront similar to that of the pumping beam before impinging on the distorting plate.

A similar experiment was performed by Wang and Giuliano (Reference 18) who verified the distortion compensation by accurate measurements of the beam divergence before and after the compensation. The results of this experiment are shown in Figure 19.15.

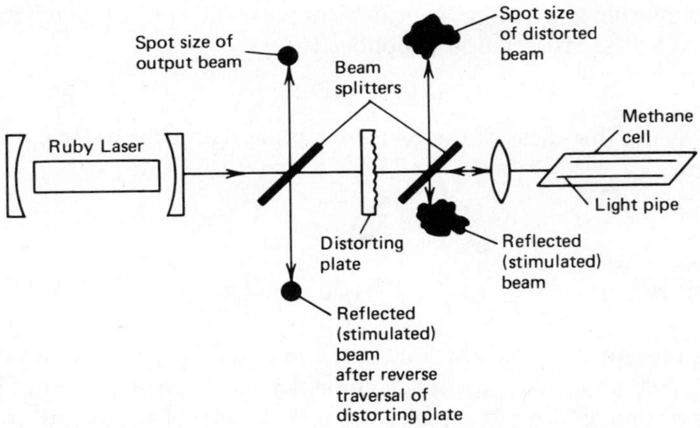

FIGURE 19.14 Experimental setup used by Zeldovich *et al.* for correction of phase distortion by using stimulated Brillouin scattering. *Source:* Reference 17.

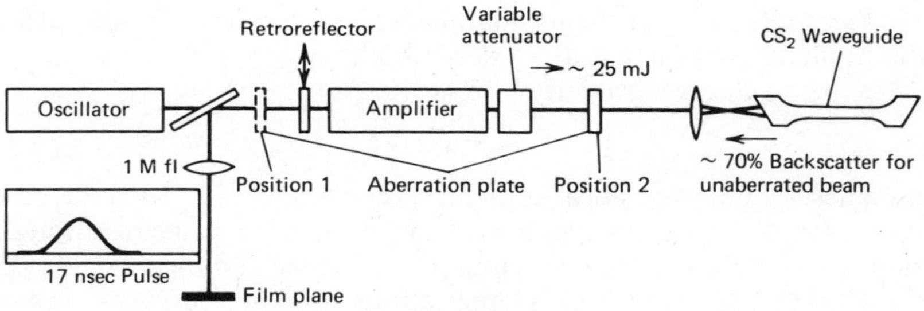

(*a*)

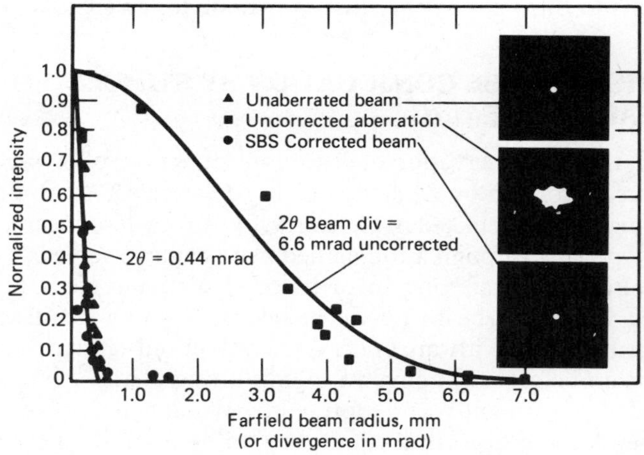

(*b*)

FIGURE 19.15 (*a*) The setup used in the experiment of Wang and Giuliano. (*b*) A radial intensity plot of the uncorrected and corrected beam. *Source:* Reference 18.

In this experiment, the phase conjugation occurs in the process of stimulation with the distorted beam acting as a pump. The explanation of why the stimulated (reflected) beam is the conjugate of the input beam is more subtle than that of the direct four-wave mixing schemes described above.

The basic explanation that was advanced by Zeldovich et al. (Reference 17) is that the phase-conjugate replica of the pump wave is the field configuration that experiences the *largest* spatial gain coefficient and is thus favored in the stimulation process.

The theoretical proof of this last statement proceeds as follows. Consider the wave equations obeyed by the backward propagating "signal" beam E_1 and the forward pump beam E_2 inside the nonlinear medium.

Let

$$\begin{aligned} E_2(\mathbf{r}, t) &= \tfrac{1}{2}\mathcal{E}_2(\mathbf{r}_\perp, z)e^{+i(\omega_2 t - k_2 z)} \\ E_1(\mathbf{r}, t) &= \tfrac{1}{2}\mathcal{E}_1(\mathbf{r}_\perp, z)e^{+i(\omega_1 t + k_1 z)} \end{aligned} \tag{19.7-1}$$

where z is the nominal direction of propagation of the "pump" beam. The deviations of both beams from ideal plane wave behavior are represented by the $\mathbf{r}_\perp$ dependence of complex amplitudes $\mathcal{E}_1$ and $\mathcal{E}_2$. The basic configuration is that used already to analyze stimulated Brillouin scattering in Section 18.7. We can following the derivation of 18.7-15 with $du_s/dr_s << \eta u_s/\rho v_s$, obtain

$$\frac{\partial \mathcal{E}_1}{\partial z} - \frac{i}{2k_1}\nabla_\perp^2 \mathcal{E}_1 + \frac{1}{2}A|\mathcal{E}_2(\mathbf{r}_\perp, z)|^2 \mathcal{E}_1 = 0 \tag{19.7-2a}$$

$$\frac{\partial \mathcal{E}_2}{\partial z} + \frac{i}{2k_2}\nabla_\perp^2 \mathcal{E}_2 + \frac{1}{2}A|\mathcal{E}_1(\mathbf{r}_\perp, z)|^2 \mathcal{E}_2 = 0 \tag{19.7-2b}$$

where (see Ref. 1 for details)

$$A = \frac{\gamma^2 k^2}{4\varepsilon\rho\alpha_S v_S^2} = \frac{\pi^2 \varepsilon_0 n^8 p^2}{\rho \alpha_S v_S^2 \lambda_0^2} \tag{19.7-3}$$

where v_S = velocity of sound, α_S = sound absorption coefficient, ρ = mass density, and p = photoelastic constant. Since $\omega_2 - \omega_1 \ll \omega_2, \omega_1$, we will put $k_2 = k_1 \equiv k$ in (19.7-2). Consider a system of orthonormal function $f_k(\mathbf{r}_\perp, z)$ such that

$$\frac{\partial f_k}{\partial z} - \frac{i}{2k}\nabla_\perp^2 f_k = 0 \tag{19.7-4}$$

and

$$\int f_i^*(\mathbf{r}_\perp, z) f_k(\mathbf{r}_\perp, z)\, d\mathbf{r}_\perp = \delta_{ik} \tag{19.7-5}$$

Choose the function f_0^* such that it coincides within a constant B with the input (pump) field

$$\mathcal{E}_2(\mathbf{r}_\perp, z) = Bf_0^*(\mathbf{r}_\perp, z) \tag{19.7-6}$$

The remaining members of the set f_k are generated starting from the orthogonality relation (19.7-5). The sought field $\mathcal{E}_1(\mathbf{r}_\perp, z)$ is expanded in the form of

$$\mathcal{E}_1(\mathbf{r}_\perp, z) = \sum_{k=0}^{\infty} C_k(z) f_k(\mathbf{r}_\perp, z) \tag{19.7-7}$$

Substituting (19.7-7) in (19.7-2*a*) yields

$$\sum_k \left(\frac{dC_k}{dz} f_k + C_k \frac{\partial f_k}{\partial z} - i \frac{C_k}{2k} \nabla_t^2 f_k + \frac{1}{2} AB^2 |f_0|^2 C_k f_k \right) = 0 \qquad (19.7\text{-}8)$$

The sum of the second and third terms is zero in virtue of (19.7-4). We next multiply (19.7-8) by $f_i^*(\mathbf{r}_\perp, z)$ and integrate over $\mathbf{r}_\perp$. Using (19.7-5) leads to

$$\frac{dC_i}{dz} + \frac{1}{2} \sum_{k=0}^{\infty} g_{ik}(z)\, C_k(z) = 0 \qquad (19.7\text{-}9)$$

with

$$g_{ik}(z) = AB^2 \int d\mathbf{r}_\perp |f_0(\mathbf{r}_\perp, z)|^2 f_i^*(\mathbf{r}_\perp, z) f_k(\mathbf{r}_\perp, z) \qquad (19.7\text{-}10)$$

If the intensity $|f_0(\mathbf{r}_\perp, z)|^2$ of the laser field fluctuates strongly as a function of $\mathbf{r}_\perp$, then it follows from (19.7-10) that, in general, the partial overlap of maxima and minima of $|f_0(\mathbf{r}_\perp, z)|^2$ and $f_i^*(\mathbf{r}_\perp, z) f_k(\mathbf{r}_\perp, z)$ as well as the complex nature of these functions will cause g_{ik} to be a small number except for g_{oo}, since

$$g_{oo} = AB^2 \int d\mathbf{r}_\perp |f_0(\mathbf{r}_\perp, z)|^4 \qquad (19.7\text{-}11)$$

and no cancellation occurs. It follows that under these conditions, the coefficient C_0 grows (with distance) more rapidly than the other C_i so that after a sufficient distance, the field ξ_1 is given, according to (19.7-7),

$$\boldsymbol{\varepsilon}_1(\mathbf{r}_\perp, z) = \sum_{k=0}^{\infty} C_k(z) f_k(\mathbf{r}_\perp, z) \simeq C_0(z) f_0(\mathbf{r}_\perp, z) = \frac{C_0(z)}{B^*} \boldsymbol{\varepsilon}_2^*(z) \qquad (19.7\text{-}12)$$

that is, the *Stokes field generated by stimulated Brillouin* (*back*) *scattering is the complex conjugate of the incident laser field* and is thus in the proper form to correct in its backward travel for the phase distortions undergone by the laser field.

We note that if $f_0(\mathbf{r}_\perp, z)$ is not a strongly fluctuating function of $\mathbf{r}_\perp$, then g_{oo} and g_{ii} will be of the same order of magnitude and the preferential growth of C_0, which leads to phase conjugation, will not take place. The introduction of additional phase aberration in front of the Brillouin cell may thus improve the phase conjugation.

In the above derivation, no initial boundary for $C_i(z)$ was specified, since no input exists. It is assumed that the input fields are due to zero point vibrations of the electromagnetic field at the Stokes' frequency.

The area of phase conjugation by stimulated Brillouin and Raman scattering is playing an increasing role in high-laser-power applications (Reference 2) in which the high intensities that are required are not a problem.

19.8 BEAM COUPLING AND PHASE CONJUGATION BY THE PHOTOREFRACTIVE EFFECT

Many nonlinear operations involving beam coupling, phase conjugation, and a variety of image processing experiments make use of the photorefractive effect. The basic phenomenon involves light-induced changes of the index of refraction of certain crystals (Reference 19). These changes make it possible to record in the volume of a crystal the intensity pattern of two (or more) interferring optical beams as a corresponding index grating (Reference 20). This "hologram" is, in turn, used to affect the propagation of some other beam or of the same beams that wrote it (Reference 21). More specifically, the phenomenon involves the excitation of charge carriers within the crystal volume by the optical interference pattern as shown in Figure 19.16. These charge carriers redistribute themselves through diffusion and drift, thus leading to a spatial charge distribution ρ_{sc}. The resulting electric field E_{sc} operates through the electrooptic effect to produce a refractive index grating Δn. It follows that only noncentrosymmetric crystals can be photorefractive. Two main crystal classes have been used to date. (1) Oxygen octahedra ferroelectrics such as barium titanate ($BaTiO_3$), barium sodium niobate ($Ba_2NaNb_5O_{15}$), strontium barium niobate ($Sr_{1-x}Ba_xNb_2O_6$), and potassium niobate ($KNbO_3$). Many of these crystals have the perovskite structure (References 23 and 24). (2) A second class includes BSO and BGO.

Most theoretical analyses of the photorefractive effect follow closely the original pioneering work of Kukhtarev (Reference 25). The basic features of his analysis were borne out in experiments by Huignard et al. (Reference 27).

A photorefractive crystal is illuminated by an intensity distribution

$$I(x) = I_0 + [I_1 e^{i(\Omega t - Kx)} + \text{c.c.}] \tag{19.8-1}$$

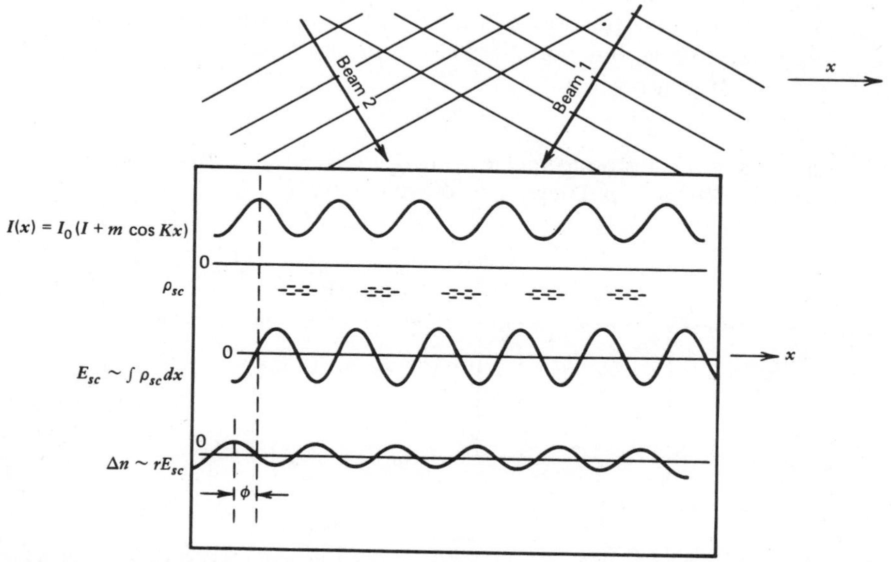

FIGURE 19.16 The photorefractive mechanism. Two laser beams intersect, forming an interference pattern. Charge is excited preferentially in regions of high intensity and migrates to regions of low intensity. The electric field associated with the resultant space charge operates through the electroopic effect to produce a refractive index grating.

due to the interference of two beams

$$\begin{aligned} \mathbf{E}_1(\mathbf{r}, t) &= \hat{\mathbf{e}}_1 A_1(\mathbf{r}) e^{i(\omega_1 t - \mathbf{k}_1 \cdot \mathbf{r})} + \text{c.c.} \\ \mathbf{E}_2(\mathbf{r}, t) &= \hat{\mathbf{e}}_2 A_2(\mathbf{r}) e^{i(\omega_2 t - \mathbf{k}_2 \cdot \mathbf{r})} + \text{c.c.} \end{aligned} \tag{19.8-1a}$$

so that

$$\begin{aligned} K &= |\mathbf{k}_2 - \mathbf{k}_1|, \ \hat{e}_x = \frac{\mathbf{k}_2 - \mathbf{k}_1}{|\mathbf{k}_2 - \mathbf{k}_1|} \\ \Omega &= \omega_2 - \omega_1 \end{aligned} \tag{19.8-2}$$

$$I_0 \equiv |A_1|^2 + |A_2|^2, \qquad I_1 \equiv \hat{\mathbf{e}}_1 \cdot \hat{\mathbf{e}}_2 A_1 A_2^* \tag{19.8-2a}$$

Charge (electrons or holes) is excited by the periodic intensity distribution (19.8-1) into a conduction band, where it migrates by diffusion and drift. The excited charge carriers are trapped preferentially in regions of low optical intensity thus leading to charge separation. The positive and negative charge sheets are separated by half the intensity modulation period, that is, by π/K. A periodic space charge is thus set up that, in turn, modulates the refractive index via the electrooptic effect. Since the band gap of common PR materials is in the ultraviolet and holographic recording takes place in the visible, the charge cannot be excited directly from the valence band to the conduction band, rendering the absorption coefficient of the material to the recording light sufficiently small so that volume holograms can be recorded. An energy level diagram of a typical material is shown in Figure 19.17.

A large concentration $N_D \sim 10^{18} - 10^{19}\ \text{cm}^{-3}$ of impurity atoms (or ions) might be found in one of two valence states. In one valence state, these atoms (or ions) with concentration N_D^0 act as donors with an energy level inside the gap. A much smaller fraction of the donor atoms are to be found in an oxidized state, that is, have one less electron, and their concentration is denoted by N_D^+. It follows that

$$N_D = N_D^0 + N_D^+$$

In addition a second atomic species called an acceptor with a concentration N_A act as an electron trap. These "traps" are always full.

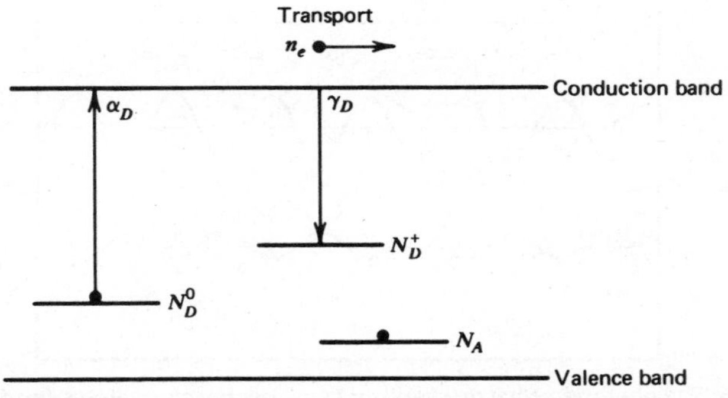

FIGURE 19.17 The donor and acceptor levels of an impurity atom in a photorefractive crystal.

An electron in our model can be either bound to a donor atom where it is immobile or excited to the conduction band where it is "free" to roam with a mobility μ under the influence of any electric field E_x (internal or externally applied). The concentration of electrons in the conduction band is n_e. The three species listed above coexist in the presence of the interaction with the optical field (19.8-1) as well as diffusion, drift, and trapping processes. The process indicated in Figure 19.17 converts the donor (N_D^0) atom into an electron trap (N_D^+) while the trap, having gained an electron, becomes an N_D^0 atom so that the spatial average of each species remains constant while the local concentration may vary. The acceptor atoms N_A are fully occupied by electrons at all times so that in the dark $\langle N_D^+ \rangle = N_A$ where $\langle\rangle$ stands for spatial averaging. The above mechanism can operate with holes as well or with both types of charge carriers simultaneously. To keep our treatment generically simple we will limit our discussions to the case where the mobile carriers are electrons. The relevant rate equation is (see Reference 25)

$$\frac{\partial N_D^+(x, t)}{\partial t} = \left(\frac{\alpha_D}{h\nu}\right) I(x)(N_D - N_D^+) - \gamma_D n_e N_D^+ \tag{19.8-3}$$

where α_D is the absorption cross section of the $N_D^0 = N_D - N_D^+$ donor state atoms. γ_D is the recombination coefficient of a free electron at an N_D^+ site. γ_D is related to the commonly used recombination cross-section σ_D by $\gamma_D = \sigma_D v_{th}$, where v_{th} is the mean thermal velocity of the free electrons. The electrons given off by the donor atoms (in the process of becoming ionized) can be either present as mobile charge carriers in the conduction band (n_e), or be trapped by the N_D^+ traps thereby converting each filled trap to an N_D^0 atom (ion), or be *permanently* trapped by N_A acceptor atoms. In the dark $n_e = 0$ and $\langle N_D^+ \rangle = N_A$.

The current density $J_x(\text{A/m}^2)$ is the sum of a drift and diffusion terms

$$J_x = \mu e n_e E_x + eD\frac{\partial n_e}{\partial x} = \mu e n_e E_x + k_B T \mu \frac{\partial n_e}{\partial x} \tag{19.8-4}$$

where we used the Einstein relation $eD = k_B T \mu$ relating the electron diffusion coefficient D to the Boltzmann constant k_B, the mobility μ, and the temperature T. The current continuity relation $\mathbf{\nabla} \cdot \mathbf{J} = -\partial\rho/\partial t$ becomes

$$\frac{\partial J_x}{\partial x} = -e\frac{\partial}{\partial t}(N_D^+ - n_e) \tag{19.8-5}$$

and the Gauss relation

$$\frac{\partial E_x}{\partial x} = \rho/\varepsilon = \frac{e(N_D^+ - n_e - N_A)}{\varepsilon} \tag{19.8-6}$$

We will first solve for $n(x, t)$, $\rho(x, t)$, and $E_x(x, t)$ by assuming small spatial modulation that can be represented by the first two harmonics ($n = 0, n = 1$) of the spatial Fourier amplitudes

$$N_D^+(x, t) = D_0 + [D_1 e^{-iKx} + \text{c.c.}] \tag{19.8-7}$$

$$n_e(x, t) = n_{e0} + [n_1 e^{-iKx} + \text{c.c.}] \tag{19.8-8}$$

$$E_x(x, t) = E_0 + [E_1^{sc} e^{-iKx} + \text{c.c.}] \tag{19.8-9}$$

Since the crystal is charge-balanced, it follows that

$$\langle \rho \rangle = e\langle N_D^+ - n_e - N_A \rangle \equiv D_0 - n_{e0} - N_A = 0$$

where $\langle\rangle$ denotes averaging over x. In addition, E_0, the average value of the internal field E_x, is equal to the externally applied electric field if one exists.

The following approximations that are justified by the actual numerical values in real crystals (Reference 26) are made

$$\underset{\sim 10^{19}\ \text{cm}^{-3}}{\underset{\downarrow}{N_D}} \gg \underset{\sim 10^{16}\ \text{cm}^{-3}}{\underset{\downarrow}{N_A}} \gg \underset{10^{13}\ \text{cm}^{-3}}{\underset{\downarrow}{n_{e0}}}, \qquad D_1 \gg n_1$$

We substitute (19.8-4) into (19.8-5) and eliminate n and E using (19.8-3) and (19.8-6). We take advantage of the inequalities and neglect the product of second-order terms.

The result is

$$E_1^{sc} = -i\frac{I_1}{I_0}\frac{E_N(E_0 + iE_D)(e^{i\Omega t} - e^{-t/\tau})}{[E_0 - \Omega t_0(E_D + E_\mu)] + i(E_N + E_D + \Omega t_0 E_0)}$$

$$\Omega \equiv \omega_2 - \omega_1, \qquad E_N = \frac{eN_A}{\varepsilon K}, \qquad E_\mu = \frac{\gamma_D N_A}{\mu K} \tag{19.8-10}$$

$$E_0 = \text{externally applied field}, \qquad E_D = \frac{k_B T K}{e}$$

$$t_0 = \frac{N_A h\nu}{\alpha_D N_D I_0}, \qquad \tau = t_0\frac{E_0 + i(E_D + E_\mu)}{E_0 + i(E_N + E_D)}$$

The steady-state response is obtained at $t \gg \tau$, at which time the transient term $\exp(-t/\tau)$ is zero. Under typical conditions using a grating period $2\pi K^{-1} = 2\mu\text{m}$ and the above estimates of N_D, N_D^+, and N_A, we estimate in $B_aT_iO_3$

$$E_D \sim 800\ \text{V/cm}, \qquad E_N \sim 75\ \text{V/cm}$$

If no applied field is present $E_0 = 0$, the steady-state internal field of (19.8-10) is

$$E_1^{sc} \simeq -i\left(\frac{I_1}{I_0}\right)\frac{1}{1 + i\Omega\tau}\frac{E_D}{(1 + E_D/E_N)}$$
$$\tau = t_0\frac{1 + E_\mu/E_D}{1 + E_N/E_D} \tag{19.8-11}$$

From the above numerical estimates, we expect internal fields of the order of magnitude of $\sim 10^2$V/cm and time constants τ that are inversely proportional to the intensity I_0. Experimental evidence (Reference 27) supports the basic qualitative features predicted by (19.8-10) and (19.8-11) such as the dependence of E_1 (or Δn) on the grating vector K and I_1/I_0. The quantitative agreement is satisfactory, depending as it does on detailed atomic data of absorption and recombination cross sections. Calculated values of $E_1 < 10^2$ V/cm are in the region of the experimental observations. The calculated values of τ are shorter by a factor of 10–100 from those measured in most ferroelectric crystals.

Two of the most important features of the analysis are the dependence of the internal field E_1^{sc} on the grating period λ_g and the dependence of τ on intensity. If we use the above definitions of E_D and E_N, we can rewrite E_1 of (19.8-10) as

$$E_1^{sc} = -i\left(\frac{I_1}{I_0}\right)\left(\frac{1}{1+i\Omega\tau}\right)\frac{(k_BTK/e)}{1+\dfrac{\varepsilon k_BTK^2}{e^2N_A}} \qquad (19.18\text{-}11a)$$

so that in the limits of $K \to 0$ ($\lambda_g \to \infty$) and $K \to \infty$ ($\lambda_g \to 0$), $E_1^{sc} \to 0$. Experimental evidence illustrating the dependence of E_1 on the grating period is shown in Figure 19.18, which shows an experimental plot of a quantity proportional to the internal field as a function of the grating period $\lambda_g = 2\pi K^{-1}$. Figure 19.19 shows the measured dependence of τ on the intensity I_0.

A detailed comparison of E_1^{sc} vs. K and τ vs. I_0 was first reported by Huignard et al. (Reference 27).

The factor $(-i)$ in (19.8-11) represents a quarter wavelength ($\lambda_g/4$) shift of the electric field (E_1^{sc}) grating relative to that of the optical intensity $I(x)$. This spatial shift causes power transfer between the two optical beams whose interference gives rise to the intensity distribution $I(x)$. This power transfer,

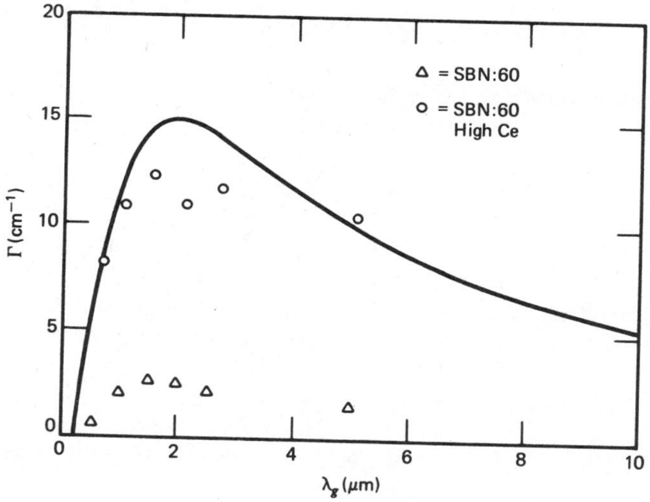

FIGURE 19.18 Two-beam coupling coefficient versus grating wavelength for $E_0 = 0$ V/cm. The coupling coefficient Γ is proportional to the internal field E_1 of Eq. (19.8-11). *Source:* Reference 26.

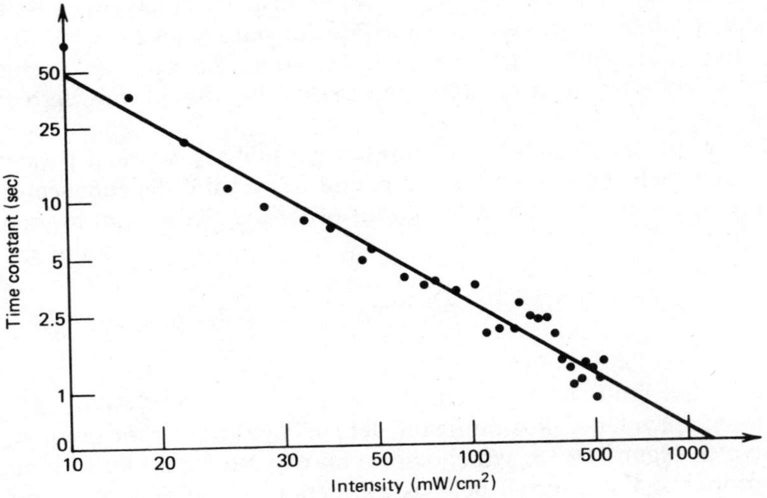

FIGURE 19.19 Photorefractive response time of the $BaTiO_3$ crystal versus intensity for $\lambda = .605\ \mu m$, $\lambda_g = 1.4\ \mu m$. *Source:* Reference 26.

which will be studied next, is at the heart of most of the applications of photorefractive media.

The Grating Formation

The change in the index of refraction due to an electric field E_k is given by

$$\Delta\left(\frac{1}{n^2}\right)_{ij} = r_{ijk}E_k$$

where r_{ijk} is the electrooptic tensor element. In any specific situation, this leads to a change in the index of refraction (14.1-12)

$$\Delta n \equiv -\frac{1}{2}n_0^3 r_{\text{eff}}E$$

where E is the *low-frequency* electric field, and r_{eff} is some linear combination of electrooptic tensor elements that depends on the crystal orientation and the field direction. If we replace in the last equation the electric field by relation (19.8-9), we obtain for the spatial dependence of the index of refraction

$$n(x, t) \equiv n_0 + \frac{1}{2}\left[\frac{n_1 e^{-i\phi}I_1}{I_0}e^{-i(\Omega t - Kx)} + \text{c.c.}\right] \qquad (19.8\text{-}12)$$

where $I_0 = |A_1|^2 + |A_2|^2$, $I_1 = \hat{\mathbf{e}}_1 \cdot \hat{\mathbf{e}}_2 A_1 A_2^*$ so that n_1 is defined as

$$\frac{n_1 e^{-i\phi}I_1}{I_0}\left(= \frac{n_1 e^{-i\phi}\hat{\mathbf{e}}_1 \cdot \hat{\mathbf{e}}_2 A_1 A_2^*}{I_0}\right) = -n_0^3 r_{\text{eff}}E_1^{sc} \qquad (19.8\text{-}13)$$

is the amplitude of the index (sinusoidal) grating induced by the spatial space charge field. We shall take n_1 as real so that ϕ is the phase shift between the intensity pattern and that of the index $n(x)$. The quantity n_1 is thus the index modulation amplitude that results when $I_1 = I_0$. From (19.8-10) and (19.8-11) in the case $\omega_1 = \omega_2$ ($\Omega = 0$)

$$n_1 e^{-i\phi} \simeq r_{\text{eff}} n_0^3 \frac{iE_N(E_0 + iE_D)}{E_0 + i(E_N + E_D)} \qquad (19.8\text{-}14)$$

We note from (19.8-12) and (19.8-14) that the index modulation *does not depend on the absolute intensity,* but only on the spatial modulation index $I_1/I_0 \equiv |\hat{\mathbf{e}}_1 \cdot \hat{\mathbf{e}}_2 A_1 A_2^*|/I_0$. This reflects the fact that the role of the optical field is only to redistribute the electronic charge so that the maximum space charge field, and Δn, are limited by the total available charge. The optical intensity does determine, however, how fast the steady state is achieved. To illustrate this point, consider the maximum electric field that can be obtained by distributing a limitless supply of free electrons with a sinusoidal distribution sin Kx. At equilibrium, the diffusion current and the drift current must balance each other.

$$Ne\mu E = eD \frac{\partial}{\partial x} (N \sin Kx) \qquad (19.8\text{-}15)$$

where N is the free charge density. Solving for E, we obtain

$$E = \frac{DK}{\mu} \cos Kx = \frac{k_B T K}{e} \cos Kx \qquad (19.8\text{-}16)$$

where we used the Einstein relation $eD = k_B T \mu$. The maximum field is thus $E_d \equiv k_B T K/e$ and is *independent of the donor density* N. Note that E_d is the limiting value of E_1 as given by (19.8-10) when $E_0 = 0$ and $E_N \gg E_D$ (i.e., infinite available charge).

Refractive Two-Beam Coupling

The index grating registered by beams E_1 and E_2 in a photorefractive medium causes power transfer by coupling E_1 and E_2 to each other. To understand this coupling, we will solve the wave equation of the combined field $E = E_1 + E_2$ but take the index of refraction $n(r)$ of the medium, self-consistently, as that produced by the interference of E_1 and E_2 as in (19.8-12).

We start with the wave equation (6.5-3)

$$\nabla^2 \mathbf{E} + \omega^2 \mu \varepsilon(\mathbf{r}) \mathbf{E} = 0, \qquad \varepsilon(\mathbf{r}) = \varepsilon_0 n^2(\mathbf{r}) \qquad (19.8\text{-}17)$$

Replacing $n(r)$ in (19.8-17) by (19.8-12), assuming $n_1 \ll n_0$ in (19.8-12), and putting $\omega^2 \mu \varepsilon_0 = \omega^2/c^2$ lead to

$$\nabla^2 \mathbf{E} + \frac{\omega^2}{c^2}\left[n_0^2 + \frac{n_0 n_1 e^{-i\phi} \hat{\mathbf{e}}_1 \cdot \hat{\mathbf{e}}_2 A_1 A_2^*}{I_0} e^{i(\Omega t - Kx)} + \text{c.c.}\right] \mathbf{E} = 0 \qquad (19.8\text{-}18)$$

The field $\mathbf{E}$ is the sum of the two fields $\mathbf{E}_1$ and $\mathbf{E}_2$ that "write" the grating as given by (19.8-1*a*). If we substitute $\mathbf{E}$ into (19.8-18) and replace for simplicity the vector **E** by a scalar E (this requires the use of the proper electrooptic coefficient r in Eq. (19.8-13)], the result is

$$\begin{aligned}&\left[-2ik_1\frac{dA_1}{dr_1}-\frac{d^2}{dr_1^2}A_1(r_1)-k_1^2A_1(r_1)e^{i(\omega_1t-k_1r_1)}+\text{c.c.}\right]\\&+\left[-2ik_2\frac{dA_2}{dr_2}-\frac{d^2}{dr_2^2}A_2(r_2)-k_2^2A_2(r_2)e^{i(\omega_2t-k_2r_2)}+\text{c.c.}\right]\\&+\frac{\omega^2}{c^2}\left[n_0^2+(n_0n_1e^{-i\phi}\frac{\hat{\mathbf{e}}_1\cdot\hat{\mathbf{e}}_2A_1A_2^*}{I_0}e^{i[(\omega_1-\omega_2)t-(k_1r_1-k_2r_2)]}+\text{c.c.}\right]\\&\times\left[A_1(r_1)\varepsilon^{i(\omega_1t-k_1r_1)}+A_2(r_2)e^{i(\omega_2t-k_2r_2)}+\text{c.c.}\right]=0\end{aligned}\tag{19.8-19}$$

Recognizing that $k_{1,2} = \frac{\omega_{1,2}n_0}{c}$, neglecting the second derivative terms compared to those involving the first derivatives (this is the slowly varying amplitude approximation), and equating separately synchronous terms with the same exponential factors lead to the coupled wave equations

$$\begin{aligned}\cos\theta_1\frac{dA_1}{dz}&=-\frac{\alpha}{2}A_1+i\frac{\pi n_1}{\lambda}e^{-i\phi}\frac{|A_2|^2}{I_0}\hat{\mathbf{e}}_1\cdot\hat{\mathbf{e}}_2A_1\\\cos\theta_2\frac{dA_2}{dz}&=-\frac{\alpha}{2}A_2+i\frac{\pi n_1^*}{\lambda}e^{+i\phi}\frac{|A_1|^2}{I_0}\hat{\mathbf{e}}_1\cdot\hat{\mathbf{e}}_2A_2\end{aligned}\tag{19.8-20}$$

where θ_1 and θ_2 are the angles between $\mathbf{k}_1$ and $\mathbf{k}_2$, respectively, and the normal to the crystal input face, a plane defined as $z = 0$. The loss term α was added phenomenologically to account for absorption in the crystal.

Before considering some exact consequences of Eq. 19.8-20, we might contemplate some qualitative features. Using (19-8-14) in the limit $E_n \gg E_D$, $\phi = -\pi/2$ and we can recast (19.8-20) in the form of

$$\begin{aligned}\cos\theta_1\frac{dA_1}{dz}&=-\frac{\alpha}{2}A_1+\left[\frac{\pi n_0^3}{\lambda}r_{\text{eff}}(\hat{\mathbf{e}}_1\cdot\hat{\mathbf{e}}_2)\frac{E_D|A_2|^2}{I_0}\right]A_1\\\cos\theta_2\frac{dA_2}{dz}&=-\frac{\alpha}{2}A_2-\left[\frac{\pi n_0^3}{\lambda}r_{\text{eff}}(\hat{\mathbf{e}}_1\cdot\hat{\mathbf{e}}_2)\frac{E_D|A_1|^2}{I_0}\right]A_2\end{aligned}\tag{19.8-21}$$

which for $r_{\text{eff}} > 0$ indicates the growth of A_1 at the expense of A_2 with an initial exponential growth constant

$$\gamma=\left[\frac{\pi n_0^3}{2\lambda}r_{\text{eff}}(\hat{\mathbf{e}}_1\cdot\hat{\mathbf{e}}_2)E_D\frac{|A_2|^2}{I_0}-\frac{\alpha}{2}\right]\tag{19.8-22}$$

The direction of power flow depends on the sign of $r_{\text{eff}}E_D$ and thus can be reversed by inverting the crystal orientation. Defining normalized intensities as $\mathcal{I}_1 = |A_1|^2$, $\mathcal{I}_2 = |A_2|^2$, we obtain directly from (19.8-20)

$$\begin{aligned}\cos\theta_1\frac{d\mathcal{I}_1}{dz}&=-\alpha\mathcal{I}_1-\frac{2\pi n_1}{\lambda}\hat{\mathbf{e}}_1\cdot\hat{\mathbf{e}}_2\sin\phi\mathcal{I}_1\mathcal{I}_2\\\cos\theta_2\frac{d\mathcal{I}_2}{dz}&=-\alpha\mathcal{I}_2+\frac{2\pi n_1}{\lambda}\hat{\mathbf{e}}_1\cdot\hat{\mathbf{e}}_2\sin\phi\mathcal{I}_1\mathcal{I}_2\end{aligned}\tag{19.8-23}$$

A similar analysis for ordinary nonlinear transparent materials that are characterized by a real $\chi^{(3)}$ will lead to $\phi = 0$ [$I(x)$ and $n(x)$ "in step"] and no power transfer takes place. In a photorefractive material, on the other hand, where $n(x)$ is given by (19.8-12,14), $\phi = -\pi/2$ and the power transfer is maximum. By adding the last two equations, we obtain

$$\frac{d}{dz}(\mathcal{I}_1 \cos\theta_1 + \mathcal{I}_2 \cos\theta_2) = -\alpha(\mathcal{I}_1 + \mathcal{I}_2)$$

which in the case of $\alpha = 0$ amounts to conservation of total power.

Two-Beam Coupling—Symmetric Geometry

In the case of $\theta_1 = -\theta_2 \equiv \theta$ illustrated in Figure 19.20, Eqs. (19.8-23) can be sloved exactly. If we define $J_{1,2} \equiv \mathcal{I}_{1,2}\, e^{\alpha r}$, $r_1 = r_2 \equiv r = z/\cos\theta$ is the distance measured along the beams' propagation directions

$$\begin{aligned} \frac{dJ_1}{dr} &= -2\Gamma\frac{J_1 J_2}{J_t} = -2\Gamma\frac{J_1(J_t - J_1)}{J_t} \\ \frac{dJ_2}{dr} &= 2\Gamma\frac{J_1 J_2}{J_t} = 2\Gamma\frac{J_2(J_t - J_2)}{J_t} \end{aligned} \qquad (19.8\text{-}24)$$

$$2\Gamma = \frac{2\pi n_1}{\lambda}\hat{\mathbf{e}}_1 \cdot \hat{\mathbf{e}}_2 \sin\phi \qquad (19.8\text{-}25)$$

These equations can be integrated directly. The result, when expressed in terms of the original variables, is (see Reference 28)

$$\begin{aligned} \mathcal{I}_1(r) &= \mathcal{I}_1(O)e^{-\alpha r}\frac{\mathcal{I}_1(O) + \mathcal{I}_2(O)}{\mathcal{I}_1(O) + \mathcal{I}_2(O)e^{2\Gamma r}} \\ \mathcal{I}_2(r) &= \mathcal{I}_2(O)e^{-\alpha r}\frac{\mathcal{I}_1(O) + \mathcal{I}_2(O)}{\mathcal{I}_1(O)e^{-2\Gamma r} + \mathcal{I}_2(O)} \end{aligned} \qquad (19.8\text{-}26)$$

In the case of $\mathcal{I}_2(0) \ll \mathcal{I}_1(0)e^{-2\Gamma r}$, the last equation becomes

$$\mathcal{I}_2(r) = \mathcal{I}_2(0)e^{(2\Gamma - \alpha)r} \qquad (19.8\text{-}27)$$

This predicted power exchange has been first observed by Staebler and Amodei in 1972 (Reference 21).

Numerical Example: Two-Beam Coupling in $BaTiO_3$. $BaTiO_3$ is an electrooptic crystal with a perovskite structure that is ferroelectric at room

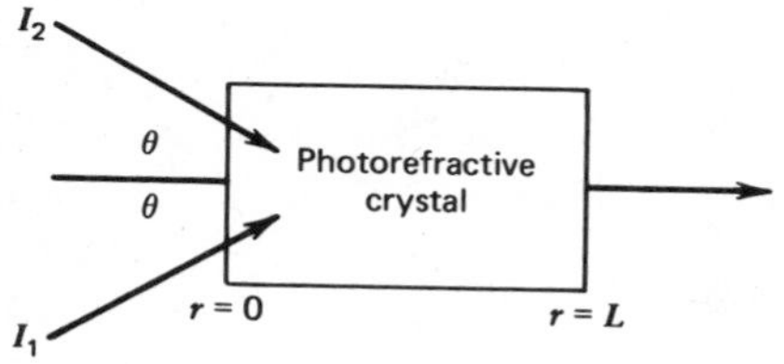

FIGURE 19.20 The symmetric two-beam coupling configuration.

temperature. This causes it to possess an extremely large electrooptic coefficient. The particular coefficient that comes into play in two-beam coupling is $r_{42} = 8.2 \times 10^{-10}$ m/V. Using the following data $n_0 = 2.5$, $r_{\text{eff}} \approx r_{42}/2$, $E_D \approx$ 800 V/cm, $\lambda = 0.5$ μm, $\phi = \pi/2$, $\hat{\mathbf{e}}_1 = \hat{\mathbf{e}}_2$, and using (19.8-14), we obtain

$$2\Gamma = \frac{2\pi n_1}{\lambda} \sin \phi \approx \frac{2\pi}{\lambda} n_0^3 r_{\text{eff}} E_D \approx 128 \text{ cm}^{-1}$$

This is a very large coupling constant and, according to (19.8-27), we should expect to see very considerable power transfer, even in lengths $l \sim 1$ mm of crystals, which is indeed the case. As a matter of fact, the two-beam amplification of radiation scattered from internal imperfections leads to a strong diversion of power from the main beam into a broad fan of light that is characteristic of experiments with photorefractive crystals. This large gain can be used to pump a new class of oscillators, as will be discussed next.

Passive Phase Conjugation

The large amplification obtainable by two-beam coupling photorefractive crystals can be used to "pump" a new class of optical oscillators and these, in turn, can perform a variety of tasks (Reference 29), including passive phase-conjugate reflectors, that is, phase-conjugating mirrors that do not require pumping beams (References 31 and 32). To illustrate this principle, consider the configuration of Figure 19.21.

An input beam 4 provides gain, by two-beam coupling, to beam 1 in a photorefractive crystal placed inside a two-mirror optical resonator. If this gain is sufficient to overcome the crystal and mirror losses, an oscillation optical field builds up inside the resonator. The two resulting beams 1 and 2 then act as the conventional pump beams as in the canonical four-wave phase configuration geometry of Figure 19.3, resulting in a reflected beam 3 that is the phase-conjugate replica of the input beam 4.

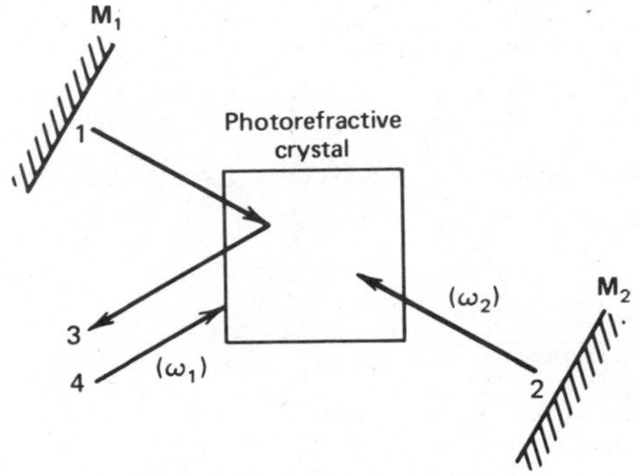

FIGURE 19.21 A passive phase-conjugate mirror geometry.

Other examples of photorefractively pumped optical oscillators and their applications are included in Reference 29.

We will conclude this section by pointing out that the coupled wave equations (19.8-20) were derived from the outset for the general case when the two interacting beams E_1 and E_2 are not necessarily of the same frequency. The frequency difference (Ω) dependence of E_1^{sc} in (19.8-10) translates to a frequency dependence of $n_1 \exp(-i\phi)$, which in the case $E_0 = 0$ can be expressed as

$$n_1 e^{-i\phi}(\omega_1 \neq \omega_2) = \frac{(n_1 e^{-i\phi})_{\omega_2=\omega_1}}{1 + i(\omega_2 - \omega_1)\tau} \qquad (19.8\text{-}28)$$

This implies directly that the oscillation in photorefractively pumped oscillators is not necessarily at the same frequency as that of the pump beam. This follows from the fact that for a given input frequency ω_1, the amplification constant Γ experienced by a beam at ω_2 is, according to (19.8-25), proportional to n_1 so that the gain bandwidth is, according to (19.8-28), $\sim \tau^{-1}$ (~ 1 s in most experimental situations). The oscillation frequency at ω_2 (see Figure 19.21) is thus able to seek whatever value (within the gain bandwidth of τ^{-1} Hz) is required to satisfy the resonator (M_1M_2) phase conditions. The resulting frequency shift ($\omega_2 - \omega_1$) has been observed (References 29–31) in accordance with (19.8-28).

A summary of recent applications of photorefractive beam coupling is given in Reference 29.

An exploration of the basic similarity between photorefractive two-beam coupling and that of stimulated Brillouin scattering (SBS) is assigned as Problem 19.11. In both phenomena, electromagnetic power is transferred coherently between two optical beams of slightly different frequencies and the role in SBS of an acoustic wave mediating the interaction is played in the photorefractive case by a moving $n(x, t)$ grating [Eq. (19.8-12)].

References

1. Yariv, A., "Phase Conjugate Optics and Real-Time Holography," *IEEE J. Quant. Elect.* **QE-14,** 650 (1978).
2. Fisher, R. A., ed., *Optical Phase Conjugation* (New York: Academic Press, 1983).
3. Yariv, A., "Three-Dimensional Pictorial Transmission in Optical Fibers," *Appl. Phys. Letters* **28,** 88 (1976).
4. Yariv, A., "On Transmission and Recovery of Three-Dimensional Image Information in Optical Waveguides," *J. Opt. Soc. Amer.* **66,** 301 (1976).
5. Dunning, G. J. and R. C. Lind, "Demonstration of Pictorial Transmission Through Fibers by Optical Phase Conjugation," *Opt. Letters* **7,** 558 (1982).
6. Hellwarth, R. W., "Third-Order Optical Susceptibilities of Liquids and Solids," *Proc. Quant. Elect.* **5,** 1–68 (1977).
7. Abrams, R. L., J. F. Lam, R. C. Lind, D. G. Steele, and P. F. Liao, "Phase Conjugation by Resonant Degenerate Four-Wave Mixing," in *Optical Phase Conjugation,* R. A. Fisher, ed. (New York: Academic Press, 1983), Chapter 8.
8. Pepper, D. M., D. Fekete, and A. Yariv, "Observation of Amplified Phase Conju-

gate Reflection and Optical Parametric Oscillation by Degenerate Four Wave Mixing in a Transparent Medium," *Appl. Phys. Letters* **33,** 41 (1978).

9. Auyeung, J., D. Fekete, D. M. Pepper, and A. Yariv, "A Theoretical and Experimental Investigation of the Modes of Optical Resonators with Phase-Conjugate Mirrors," *IEEE J. Quant. Elect.* **QE-15,** 1180 (1979).
10. Beldyugin, I. M. and E. M. Zemskov, "Theory of Resonators with Wave-Front Reversing Mirrors," *Sov. J. Quant. Elect.* **9,** 1198 (1979).
11. Siegman, A. E., P. Belanger, and A. Hardy, "Optical Resonators Using Phase-Conjugate Mirrors," in *Optical Phase Conjugation,* R. A. Fischer, ed. (New York: Academic Press, 1983).
12. Cronin-Golomb, M., B. Fischer, J. Nilsen, J. O. White, and A. Yariv, "Laser with Dynamic Holographic Intracavity Distortion Correction Capability," *Appl. Phys. Letters* **41,** 220 (1982).
13. Yariv, A. and T. L. Koch, "One Way Coherent Imaging Through a Distorting Medium Using Four-Wave Mixing," *Opt. Letters* **7,** 113 (1982).
14. White, J. O. and A. Yariv, "Real Time Image Processing via Four-Wave Mixing in a Photorefractive Medium," *Appl. Phys. Letters,* **37,** 5, (1980).
15. O'Meara, T., D. M. Pepper, and J. O. White, "Applications of Nonlinear Optical Phase Conjugation," in *Optical Phase Conjugation,* R. A. Fisher, ed. (New York: Academic Press, 1983), Chapter 14.
16. Zeldovich, B. Y., N. F. Pilipetsky, and V. V. Shkunov, "Principles of Phase Conjugation," (Vol. 42, Optical Science series, (Berlin Springer-Verlag, 1985).
17. Zeldovich, B. Y., V. I. Popovichev, V. V. Ragulskii, and F. S. Faisullov, "Connection Between the Wave Fronts of the Reflected and Exciting Light in Stimulated Mandel'shtam-Brillouin Scattering," *Sov. Phys. JETP* **15,** 109 (1972).
18. Wang, V. and C. R. Giuliano, "Correction of Phase Aberrations via Stimulated Brillouin Scattering," presented at the 1977 IEEE OSA Conference on Laser Engineering and Applications, Washington, D.C., June 1977, Paper 17.6; also, *Opt. Lett.* **2,** 4 (1978).
19. Ashkin, A., G. D. Boyd, J. M. Dziedzic, R. G. Smith, A. A. Ballman, J. J. Levinstein, and K. Nassau, "Optically-Induced Refractive Index Inhomogeneities in $LiNbO_3$ and $LiTaO_3$," *Appl. Phys. Letters* **9,** 72 (1966).
20. Chen, F. S., J. T. Lamacchia, and D. B. Fraser, "Holographic Storage in Lithium Niobate," *Appl. Phys. Letters* **13,** 223 (1968).
21. Staebler, J. J. and D. L. Amodei, "Coupled-Wave Analysis of Holographic Storage in $LiNbO_3$," *J. Appl. Phys.* **43,** 1042 (1972); also, *RCA Rev.* **32,** 185 (1971).
22. Vahey, D. W., "A Nonlinear Coupled-Wave Theory of Holographic Storage in Ferroelectric Materials," *J. Appl. Phys.* **46,** 3510 (1975).
23. Glass, A. M., D. van der Linde, and T. J. Negran, "High-Voltage Bulk Photovoltaic Effect and the Photorefractive Process in $LiNbO_3$," *Appl. Phys. Letters* **25,** 233 (1974).
24. Neurganokar, R. R. and W. K. Cory, "Progress in Photorefractive Tungsten Bronze Crystals," *J. Opt. Soc. Amer.* **B3,** 274 (1986).
25. Kukhtaerev, N. V., V. B. Markov, S. G. Odulov, M. S. Soskin, and V. L. Vinetsky, "Holographic Storage in Electrooptic Crystals," Part I: Steady State, Part II: Beam Coupling Light Amplification," *Ferroelectrics* **22,** 949 (1979).
26. Rakuljic, G., "Photorefractive Properties and Applications of $BaTiO_3$ and Tungsten Bronze Ferroelectrics," Ph.D. Thesis, California Institute of Technology, Pasadena, Calif., 1987.
27. Huignard, J. P., J. P. Herriau, and G. Rivet, "Progress in Photorefractive Tungsten Bronze Crystals," *Opt. Letters* **5,** 102 (1980).

28. Cronin-Golomb, M., "Large Nonlinearities in Four Wave Mixing in Photorefractive Crystals and Applications in Optical Phase Conjugation," Ph.D. Thesis, California Institute of Technology, Pasadena, Calif., 1983.
29. Kwong, S. K., M. Cronin-Golomb, and A. Yariv, "Oscillation with Photorefractive Gain," *IEEE J. Quant. Elect.* **QE-22,** 1508 (1986).
30. Yariv, A. and S. K. Kwong, "Theory of Laser Oscillators with Photorefractive Gain," *Opt. Letters* **10,** 455 (1985).
31. Cronin-Golomb, M., B. Fischer, J. O. White, and A. Yariv, "Passive (Self-Pumped) Phase Conjugate Mirror: Theoretical and Experimental Investigations," *Appl. Phys. Letters* **41,** 689 (1982).
32. Feinberg, J., "Self Pumped Continuous Wave Phase Conjugator Using Internal Reflection," *Opt. Letters,* **7,** 486 (1982).

Problems

19.1 Consider a medium whose permittivity depends on the field intensity according to

$$\varepsilon(\mathbf{r}) = \varepsilon_0(\mathbf{r}) + \varepsilon_2(\mathbf{r})|E|^2$$

where E is the total complex amplitude of the monochromatic field at point $\mathbf{r}$. Show that the wave equation

$$\nabla^2 E - \frac{\mu}{c^2}[\varepsilon_0(\mathbf{r}) + \varepsilon_2(\mathbf{r})|E|^2]\frac{\partial^2 E}{\partial t^2} = 0 \quad (1)$$

is satisfied by

$$E = [\psi(\mathbf{r})e^{-ikz} + af(\mathbf{r})e^{ikz}]e^{i\omega t} + \text{c.c.} \quad (2)$$

provided

$$f(r) = \psi^*(r)$$

$$|a| = 1$$

and that this is true for any lossless medium as long as ε is real. In the proof we discard spatially nonsynchronous terms with $e^{\pm i3kz}$ dependence. If you have completed the proof, you have succeeded in showing that a phase-conjugate wave can compensate by backward propagation not only for static index inhomogeneities $[\varepsilon_0(\mathbf{r})]$, but also for intensity-dependent distorions $[\varepsilon_2|E|^2]$.

19.2 Consider the canonical phase-conjugation experimental arrangement of the figure below. Show that the threshold condition for oscillation in a direction normal to the plane of the mirror is

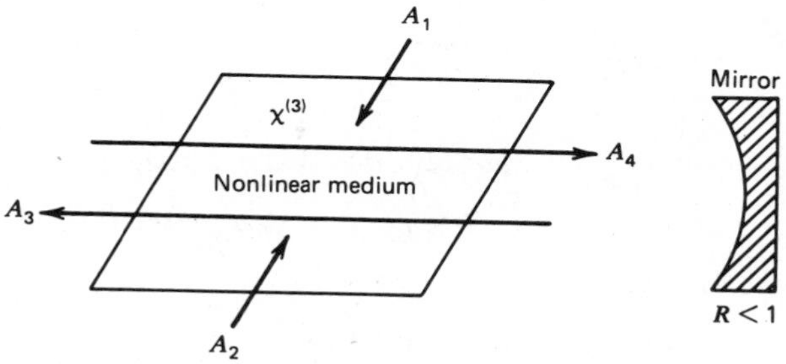

$$|\kappa|L = \tan^{-1}\frac{1}{r}$$

where $R = |\mathbf{r}|^2$ is the reflectivity of the mirror.

19.3 Show that if the nonlinear medium used in the experimental arrangement of Figure 19.3 is lossy, then the coupled mode equations (19.3-8) become

$$\frac{dA_3}{dz} = i\kappa^* e^{-\alpha L/2} A_4^* + \frac{\alpha}{2} A_3$$

$$\frac{dA_4^*}{dz} = i\kappa e^{-\alpha L/2} A_3 - \frac{\alpha}{2} A_4^*$$

where α is the intensity absorption coefficient and

$$\kappa = \frac{\omega}{2}\sqrt{\frac{\mu_0}{\varepsilon}}\,\chi^{(3)} A_1^*(L) A_2^*(0)$$

Solve these equations subject to the boundary conditions $A_3(L) = 0$ and a given $A_4(0)$. Show that the reflection coefficient is

$$r = \frac{A_3(0)}{A_4^*(0)} = \frac{-2i\kappa^* e^{-\alpha L/2}\tan(\kappa_{\text{eff}}L)}{\alpha\tan(\kappa_{\text{eff}}L) + 2\kappa_{\text{eff}}}$$

with

$$\kappa_{\text{eff}} = \sqrt{|\kappa|^2 e^{-\alpha L} - \left(\frac{\alpha}{2}\right)^2}$$

19.4 Solve the coupling equations (19.3-8) for the case when the frequency of the weak input A_4 is not the same as that of the pump wave (i.e., $\omega_1 = \omega_2$, $\omega_4 \neq \omega_1$). Plot the frequency response

$$|r| = \left|\frac{A_3(\omega + \delta)}{A_4^*(\omega - \delta)}\right|$$

as a function of the frequency offset δ. Here, ω is the pump frequency and $\omega - \delta$ is the input frequency.

Caution:
Watch for the phase factor $\mathbf{k}_1 + \mathbf{k}_2 - (\mathbf{k}_3 + \mathbf{k}_4)$ in

$$(\mathbf{k}_1 + \mathbf{k}_2)\cdot\mathbf{r} + (k_{3z} + k_{4z})z$$

which for $\delta \neq 0$ is no longer zero.

19.5 ***Conjugate Waves***
Consider the electric-field vector of a monochromatic radiation

$$\mathbf{E}_1(\mathbf{r}, t) = \mathbf{E}(\mathbf{r})e^{i\omega t}$$

where the amplitude function $\mathbf{E}(\mathbf{r})$ satisfies the equation

$$\nabla \times (\nabla \times \mathbf{E}) - \omega^2\mu\varepsilon\mathbf{E} = 0$$

(a) Show that the conjugate wave

$$\mathbf{E}_2(\mathbf{r}, t) = \mathbf{E}^*(\mathbf{r})e^{i\omega t}$$

also obeys the same equation provided $\mu\varepsilon$ is a real quantity (i.e., in a lossless and non magnetic medium).

(b) Let

$$\mathbf{E}(\mathbf{r}) = \mathbf{A}(\mathbf{r})e^{i\phi(\mathbf{r})}$$

where $\mathbf{A}$ and ϕ are real functions of $\mathbf{r}$. Show that the wavefronts (surfaces of constant phase) of the waves $\mathbf{E}_1$ and $\mathbf{E}_2$ are given, respectively, by

$$\phi(\mathbf{r}) + \omega t = C_1$$

$$-\phi(\mathbf{r}) + \omega t = C_2$$

where C_1 and C_2 are constants. Since C_1 and C_2 are arbitrary, these wavefronts can be simply represented at any instant t by

$$\phi(\mathbf{r}) = \text{constant}$$

In other words, $\mathbf{E}_1(\mathbf{r}, t)$ and $\mathbf{E}_2(\mathbf{r}, t)$ have exactly the same wavefronts in the entire space $\mathbf{r}$.

(c) We now consider the propagation of a wavefront across space during a time interval $(t, t + dt)$. Show that the wavefront displacements for the waves E_1 and E_2 are given, respectively, by

$$d\mathbf{r}_1 \cdot \nabla\phi(\mathbf{r}) + \omega\, dt = 0$$

and

$$-d\mathbf{r}_2 \cdot \nabla\phi(\mathbf{r}) + \omega\, dt = 0$$

This shows that the motion of the wavefront of the conjugate wave is the opposite of that of the original wave.

(d) Let

$$\mathbf{E}(\mathbf{r}) = \int \mathbf{B}(\mathbf{k})e^{i\mathbf{k}\cdot\mathbf{r}}d^3k$$

Show that each Fourier component of the conjugate wave is propagating exactly opposite to the corresponding component of the original wave.

19.6 ***Polarization States of Conjugate Wave***

Let

$$\mathbf{E}_1(z, t) = \mathbf{E}e^{i(\omega t - kz)}$$

$$\mathbf{E}_2(z, t) = \mathbf{E}^*e^{i(\omega t + kz)}$$

$$\mathbf{E}_3(z, t) = \mathbf{E}e^{i(\omega t + kz)}$$

where $\mathbf{E}$ is a complex constant and k is the propagation constant.

(a) Show that if $\mathbf{E}_1$ is linearly polarized, so are $\mathbf{E}_2$ and $\mathbf{E}_3$.

(b) Show that if $\mathbf{E}_1$ is right-handed circularly polarized, so is its conjugate wave $\mathbf{E}_2$. The wave $\mathbf{E}_3$ is now left-handed circularly polarized. The physical implications are the following: The spin of a photon is reversed upon reflection from a conjugator-mirror, whereas the helicity (spin component along the propagation direction) remains unchanged.

19.7 ***Self-Induced Index Modulation***

Consider the propagation of an intense optical beam in a medium possessing a strong Kerr effect.

(a) Show that the complex amplitude of the polarization can be written

$$P_i^\omega = \varepsilon_0 \chi_{ij} E_j^\omega + 3\chi_{ijkl} E_j^\omega E_k^{\omega *} E_l^\omega$$

where E_j^ω is the complex amplitude of the electric field.

(b) Assume that the electric field is polarized along the x direction. Show that the index of refraction associated with the beam can be written

$$n = n_0 + \tfrac{1}{2} n_2 \mathbf{E}^* \cdot \mathbf{E}$$

and show that

$$n_2 = \frac{3\chi_{111}}{\varepsilon_0 n_0}$$

This n_2 is responsible for the self-focusing of laser beams in certain media and is discussed in Section 18.7

19.8 (a) Ignoring the vector nature of the electromagnetic fields, show, using the basic reasoning of Section 16.2, that the interaction Hamiltonian due to $\chi^{(3)}$ in a nonlinear medium can be taken as

$$H' = S \int_{\text{volume of medium}} \prod_{i=1}^{4} \chi^{(3)} (a_i^\dagger - a_i) E_i(\mathbf{r}) e^{-\mathbf{k}_i \cdot \mathbf{r}} \, d^3\mathbf{r}$$

where S is some constant and $i = 1, 2, 3, 4$ refer to the four waves of the canonical phase-conjugation geometry of Figure 19.3. The four waves are of the same frequency.

(b) Show that the basic physical phenomenon involved in phase conjugation involves a simultaneous generation of one photon apiece in waves 3 and 4 and an annihilation of one photon from each of waves 1 and 2.

19.9 Show that a field with a spatial (periodic) amplitude

$$E_N = \frac{e\langle N_D^+ \rangle}{\varepsilon K}$$

would result from a periodic spatial modulation of the crystal charge density distribution with an amplitude $e\langle N_d^+ \rangle$ and a period $2\pi K^{-1}$. Discuss this result in the context of Eq. (19.8-11a).

19.10 (a) Express the photorefractive result of (19.8-12) and (19.8-28)

$$\Delta n(x, t) = \frac{1}{2} \left\{ \frac{n_1 e^{-i\phi} A_1 A_2^* e^{i(\Omega t - Kx)}}{[|A_1|^2 + |A_2|^2][1 + i(\omega_2 - \omega_1)\tau]} \right\}$$

Using the terminology of nonlinear optics of Chapter 18 that relates the induced polarization to the product of optical fields.

Clue:
Use Polarization $\propto \Delta n E$.

(b) Compare the effective third-order photorefractive nonlinear susceptibility of (a) to that of Brillouin scattering discussed in Chapter 18. Can you see any common features between the two phenomena.

19.11 (a) Consider the power transfer $P_{2\to 1}$ via Brillouin-mediated scattering from a wave

$$E_2(\mathbf{r}, t) = E_2^{(\omega_2)}(\mathbf{r}_\perp)e^{i(\omega_2 t - k_2 z)}$$

to a wave at ω_1 ($<\omega_2$)

$$E_1(\mathbf{r}, t) = E_1^{(\omega_1)}(\mathbf{r}_\perp)e^{i(\omega_1 t + k_1 z)}$$

The two waves co-exist inside a volume V of a nonliner material. Show that

$$P_{2\to 1} = \frac{\chi_I}{2}\omega_1 \int |E_1^{(\omega_1)}|^2\, |E_2^{(\omega_2)}|^2\, dV$$

where χ_I is the imaginary part of the Brillouin nonlinear susceptibility.

(b) If $E_1^{(\omega_1)}$ were to be built up from a noise input, reason out what spatial distribution of $E_1^{(\omega_1)}(\mathbf{r})$ will most effectively "milk" (the spatially modulated) $E_2^{(\omega_2)}$ and be amplified.

Answer:

$$E_1^{(\omega_1)}(\mathbf{r}) \propto [E_2^{(\omega_2)}(\mathbf{r})]^*$$

C H A P T E R 20

Q-Switching and Mode Locking of Lasers

20.0 INTRODUCTION

In the analysis of Chapters 9 and 10 we considered lasers operating in a continuous (CW) fashion. Some of the most important applications of lasers, however, involve pulsed operation. In the pulsed mode, the pump energy can be concentrated into extremely short time durations, thereby increasing the peak power. This is of key importance in numerous industrial applications such as machining and welding with lasers, as well as in scientific applications. In the latter category, the use of extremely short pulses makes it possible to probe very short-lived ($\sim 10^{-14}$ sec) transient phenomena.

In the rest of this chapter we will consider in detail the techniques of *Q*-switching and mode locking used to generate short laser pulses.

20.1 Q-SWITCHING

The technique of *Q*-switching (References 1, 2, and 3) is used extensively to obtain intense and short bursts of oscillation from lasers. The principle of the technique is as follows: the quality factor *Q* of the optical resonator of the laser is degraded (i.e., the losses are increased) during the pumping so that the gain can build up to a very high value and yet not exceed the oscillation threshold value. During this stage the atomic system acts as an energy storage mechanism. When the inversion reaches its peak, the *Q* is restored abruptly to its high value. The gain is now well above the (lowered) oscillation threshold. This causes an extremely rapid buildup of the oscillation field and a simultaneous exhaustion of the inversion by stimulated transitions. This process converts most of the pump energy stored by the excited state atoms into photons, which are now inside the optical resonator. These proceed to bounce back and forth between the reflectors with a fraction $(1 - R)$ "escaping" from the resonator each pass. This causes a decay of the pulse with a characteristic time constant (the "photon lifetime") given in (7.4-2) as

$$t_c \simeq \frac{n_0 l}{c[\alpha l - \ln \sqrt{R_1 R_2}]}$$

Both experiment and theory indicate that the total evolution of the giant laser pulse as described above is typically completed in $\sim 2 \times 10^{-8}$ sec. We will, consequently, neglect the effect of population relaxation and pumping that take place during the pulse. We will also assume that the switching of the Q from the low to the high value is accomplished instantaneously.

The laser is characterized by the following variables: the total number of photons in the optical resonator ϕ, the mode volume V, the total inversion $n \equiv [N_2 - (g_2/g_1) N_1] V$, and the decay time constant for photons in the *passive* resonator t_c. The exponential gain constant γ is proportional to n. The radiation intensity I thus grows with distance as $I(z) = I_0 \exp(\gamma z)$ so that $dI/dz = \gamma I$. An observer traveling with the wave velocity will see it grow at a rate

$$\frac{dI}{dt} = \frac{dI}{dz}\frac{dz}{dt} = \gamma \frac{c}{n_0} I$$

and thus, the temporal exponential growth constant is $\gamma c/n_0$. If the laser rod is of length L whereas the resonator length is l, then only a fraction L/l of the photons is undergoing amplification at any one time and the average growth constant is $\gamma c/n_0 (L/l)$. Following Reference 4, we write

$$\frac{d\phi}{dt} = \phi\left(\frac{\gamma c L}{n_0 l} - \frac{1}{t_c}\right) \tag{20.1-1}$$

where $-\phi/t_c$ is the decrease in the number of resonator photons per unit time due to incidental resonator losses and to the output coupling. Defining a dimensionless time $\tau = t/t_c$, we obtain, upon multiplying (20.1-1) by t_c,

$$\frac{d\phi}{d\tau} = \phi\left[\left(\frac{\gamma}{n_0 l/cLt_c}\right) - 1\right] = \phi\left(\frac{\gamma}{\gamma_t} - 1\right)$$

where $\gamma_t = (n_0 l/cLt_c)$ is the minimum value of the gain constant at which oscillation (i.e., $d\phi/d\tau = 0$) can be sustained. Since, according to (8.4-4), γ is proportional to the inversion n, the last equation can also be written as

$$\frac{d\phi}{d\tau} = \phi\left(\frac{n}{n_t} - 1\right) \tag{20.1-2}$$

where $n_t = \Delta N_t V$ is the total inversion at threshold as given by (9.1-16).

The term $\phi(n/n_t)$ in (20.1-2) gives the number of photons generated by induced emission per unit of normalized time. Since each generated photon results from a single transition, it corresponds to a decrease of $\Delta n = -2$ in the total inversion. We can thus write directly

$$\frac{dn}{d\tau} = -2\phi \frac{n}{n_t} \tag{20.1-3}$$

The coupled pair of equations (20.1-2) and (20.1-3) describes the evolution of ϕ and n. It can be solved easily by numerical techniques. Before we proceed to give the results of such a calculation, we will consider some of the consequences that can be deduced analytically.

Dividing (20.1-2) by (20.1-3) results in

$$\frac{d\phi}{dn} = \frac{n_t}{2n} - \frac{1}{2}$$

and, by integration,

$$\phi - \phi_i = \frac{1}{2}\left[n_t \ln \frac{n}{n_i} - (n - n_i)\right]$$

Assuming that ϕ_i, the initial number of photons in the cavity, is negligible, we obtain

$$\phi = \frac{1}{2}\left[n_t \ln \frac{n}{n_i} - (n - n_i)\right] \tag{20.1-4}$$

for the relation between the number of photons ϕ and the inversion n at any moment. At $t \gg t_c$, the photon density ϕ will be zero so that setting $\phi = 0$ in (20.1-4) results in the following expression for the final inversion n_f:

$$\frac{n_f}{n_i} = \exp\left(\frac{n_f - n_i}{n_t}\right) \tag{20.1-5}$$

This equation is of the form $(x/a) = \exp(x - a)$, where $x = n_f/n_t$ and $a = n_i/n_t$, so that it can be solved graphically (or numerically) for n_f/n_i as a function of n_i/n_t. The result is shown in Figure 20.1. We notice that the fraction of the energy originally stored in the inversion that is converted into laser oscillation energy is $(n_i - n_f)/n_i$ and that it tends to unity as n_i/n_t increases.

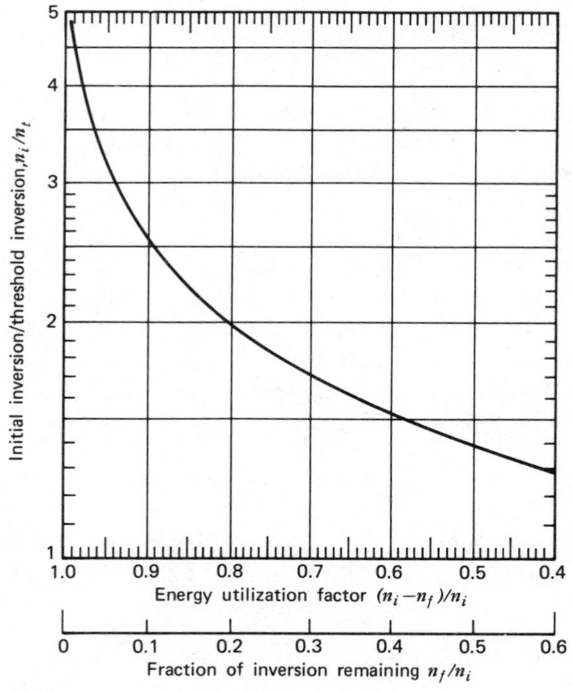

FIGURE 20.1 Energy utilization factor $(n_i - n_f)/n_i$ and inversion remaining after the giant pulse. *Source:* Reference 4.

The instantaneous power output of the laser is given by $P = \phi h\nu/t_c$ or, if we use (20.1-4), by

$$P = \frac{h\nu}{2t_c}\left[n_t \ln \frac{n}{n_i} - (n - n_i)\right] \tag{20.1-6}$$

Of special interest to us is the peak power output. Setting $\partial P/\partial n = 0$, we find that maximum power occurs when $n = n_t$. Putting $n = n_t$ in (20.1-6) gives

$$P_{\max} = \frac{h\nu}{2t_c}\left[n_t \ln \frac{n_t}{n_i} - (n_t - n_i)\right] \tag{20.1-7}$$

for the peak power. If the initial inversion is well in excess of the (high Q) threshold value (i.e., $n_i \gg n_t$), we obtain from (20.1-7)

$$(P_{\max})_{n_i \gg n_t} \approx \frac{n_i h\nu}{2t_c} \tag{20.1-8}$$

Since the power P at any moment is related to the number of photons by $P = \phi h\nu/t_c$, it follows, from (20.1-8), that the maximum number of stored photons inside the resonator is $n_i/2$. This can be explained by the fact that if $n_i \gg n_t$, the buildup of the pulse to its peak value occurs in a time short compared to t_c so that at the peak of the pulse, when $n = n_t$, most of the photons that were generated by stimulated emission are still present in the resonator. Moreover, since $n_i \gg n_t$, the number of these photons $(n_i - n_t)/2$ is very nearly $n_i/2$.

A typical numerical solution of (20.1-2) and (20.1-3) is given in Figure 20.2.

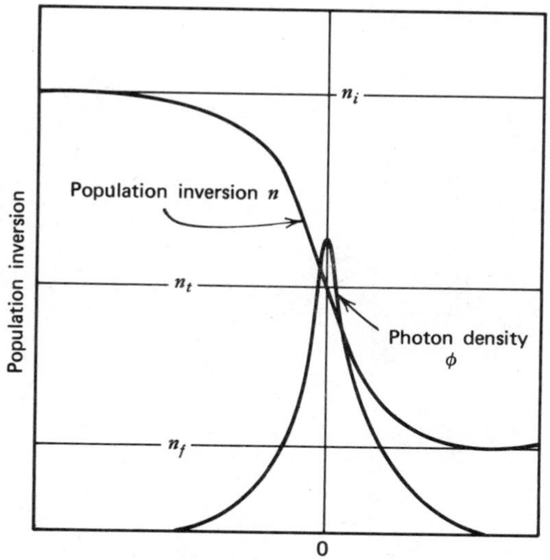

FIGURE 20.2 Inversion and photon density during a giant pulse. *Source:* Reference 4.

To initiate the pulse, we need, according to (20.1-2) and (20.1-3), to have $\phi_i \neq 0$. Otherwise, the solution is trivial ($\phi = 0, n = n_i$). The appropriate value of ϕ_i is usually estimated on the basis of the number of spontaneously emitted photons within the acceptance solid angle of the laser mode at $t = 0$. We also notice, as discussed above, that the photon density, hence the power, reaches a peak when $n = n_t$. The energy stored in the cavity ($\propto \phi$) at this point is maximum, so stimulated transitions from the upper to the lower laser levels continue to reduce the inversion to a final value $n_f < n_t$.

Numerical solutions of (20.1-2) and (20.1-3) corresponding to different initial inversions n_i/n_t are shown in Figure 20.3. We notice that for $n_i \gg n_t$ the

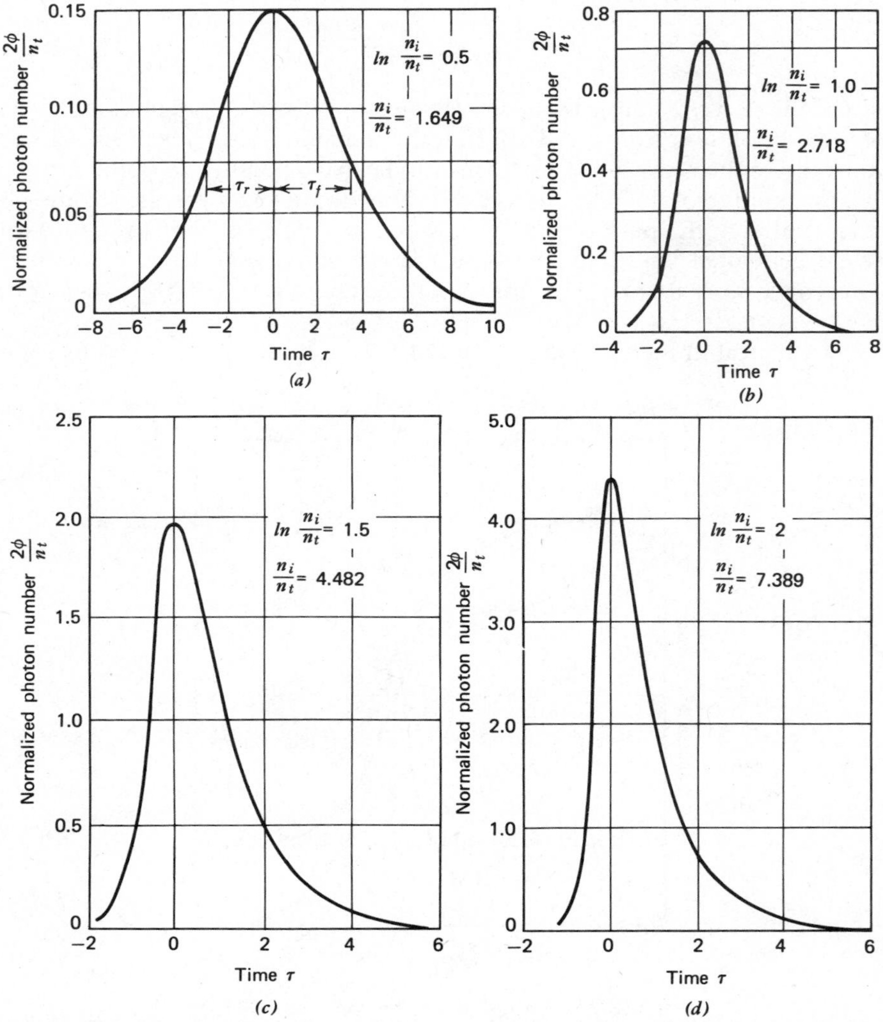

FIGURE 20.3 Photon number versus time in central region of giant pulse. Time is measured in units of photon lifetime. *Source:* Reference 4.

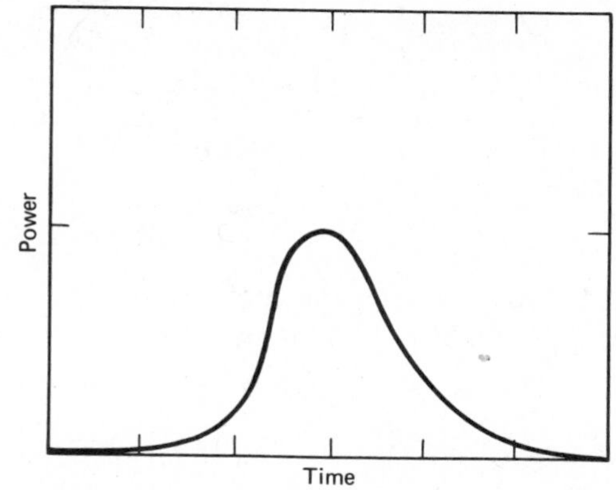

FIGURE 20.4 An oscilloscope trace of a *Q*-switched pulse in a ruby laser. Time scale is 20 nsec per division.

rise time becomes short compared to t_c, but the fall time approaches a value nearly equal to t_c. The reason is that the process of stimulated emission is essentially over at the peak of the pulse ($\tau = 0$) and the observed output is due to the free decay of the photons in the resonator.

In Figure 20.4 we show an actual oscilloscope trace of a giant pulse. Giant laser pulses are used extensively in applications that depend on their extremely high peak powers and short duration. These applications include experiments in nonlinear optics, ranging, material machining and drilling, initiation of chemical reactions, and plasma diagnostics.

Numerical Example: Giant Pulse Ruby Laser. Consider the case of dark ruby with a chromium ion density of $N = 1.58 \times 10^{20}$ cm^{-3}. Its absorption coefficient is $\alpha \simeq 2$ cm^{-1} (at 300°K). Other assumed characteristics are

$$l = \text{length of ruby rod} = 10 \text{ cm}$$

$$A = \text{cross-sectional area of mode} = 1 \text{ cm}^2$$

$$(1 - R) = \text{fractional intensity loss per pass}^1 = 20\%$$

Since, according to (8.4-4), the exponential loss coefficient is proportional to $(g_2/g_1)N_1 - N_2$, we have

$$\alpha(\text{cm}^{-1}) = 2\,\frac{N_1 - N_2(g_1/g_2)}{1.58 \times 10^{20}} \tag{20.1-9}$$

Thus, at room temperature, when $N_2 = 0$, $N_1 = 1.58 \times 10^{20}$ cm^{-3} and $\alpha = 2$ cm^{-1} as observed. The expression for gain coefficient follows directly from (20.1-9):

[1] We express the loss in terms of an effective reflectivity even though it is due to a number of factors.

$$\gamma(\text{cm}^{-1}) = \frac{2\left(\frac{g_1}{g_2} N_2 - N_1\right)}{1.58 \times 10^{20}} = \frac{2\frac{g_1}{g_2} n}{1.58 \times 10^{20} V} \tag{20.1-10}$$

where

$$n \equiv \left(N_2 - N_1 \frac{g_2}{g_1}\right) V$$

is the total inversion and $V = AL$ is the crystal volume in cubic centimeters.

Threshold is achieved when the net gain per pass is unity. This happens when

$$e^{\gamma_t l} R = 1 \qquad \text{or} \qquad \gamma_t = -\frac{1}{l} \ln R \tag{20.1-11}$$

where the subscript t indicates the threshold value.

Using (20.1-10) in the threshold condition (20.1-11) plus the appropriate data from above and $g_1/g_2 = 2$ gives

$$n_t = 8.8 \times 10^{18} \tag{20.1-12}$$

Assuming that the initial inversion is $n_i = 1.64 n_t = 1.44 \times 10^{19}$, we find from (20.1-8) that the peak power is approximately

$$P_{\text{max}} = \frac{n_i h\nu}{2t_c} \tag{20.1-13}$$

where $t_c \cong nl/c(1 - R) \simeq 2.5 \times 10^{-9}$ sec.

Substituting the foregoing data in (20.1-8) gives a peak power

$$P_{\text{max}} = 8 \times 10^8 \text{ W}$$

The total pulse energy is

$$\mathscr{E} \sim \frac{n_i h\nu}{2} \sim 2\text{J}$$

whereas the pulse duration (see Figure 20.3) $\simeq 7t_c \simeq 17.5 \times 10^{-9}$ sec.

Methods of Q-Switching

Some of the schemes used in Q-switching are the following:

1. Mounting one of the two end reflectors on a rotating shaft so that the optical losses are extremely high except for the brief interval in each rotation cycle in which the mirrors are nearly parallel.
2. The inclusion of a saturable absorber (bleachable dye, e.g.) in the optical resonator (see Reference 5). The absorber whose opacity decreases (saturates) with increasing optical intensity prevents rapid inversion depletion due to buildup of oscillation by presenting a high loss to the early stages of oscillation, during which the slowly increasing intensity is not high enough to saturate the absorption. As the intensity increases the loss de-

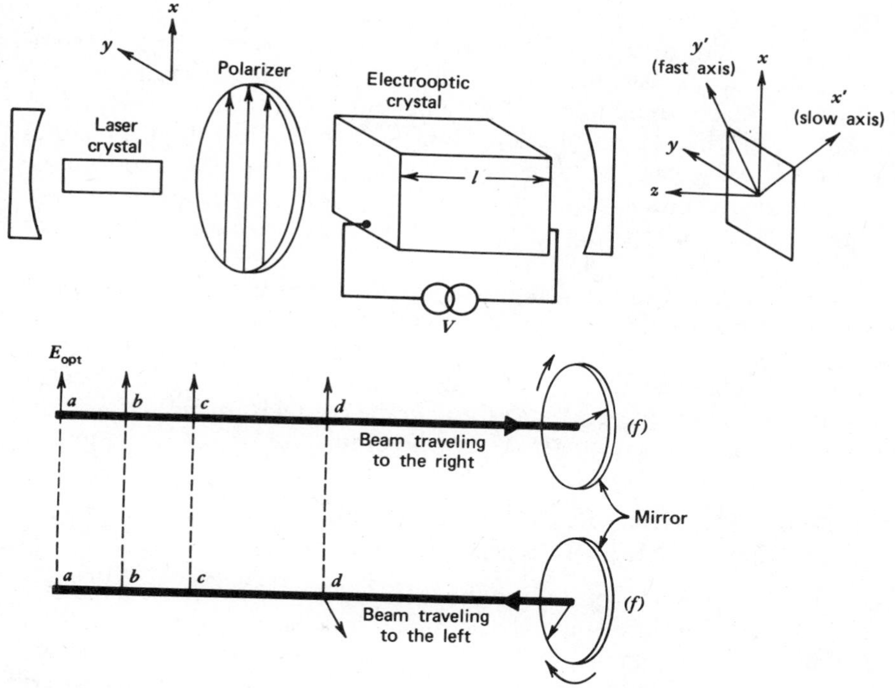

For beam traveling to right:

At point d,

$$\left.\begin{aligned} E_{x'} &= \frac{E}{\sqrt{2}}\cos \omega t \\ E_{y'} &= \frac{E}{\sqrt{2}}\cos \omega t \end{aligned}\right\}$$ The optical field is linearly polarized with its electric field vector parallel to x

At point f,

$$\left.\begin{aligned} E_{x'} &= \frac{E}{\sqrt{2}}\cos\left(\omega t - kl - \frac{\pi}{2}\right) \\ E_{y'} &= \frac{E}{\sqrt{2}}\cos(\omega t - kl) \end{aligned}\right\}$$ Circularly polarized

For beam traveling to left:

At point f,

$$\left.\begin{aligned} E_{x'} &= -\frac{E}{\sqrt{2}}\cos\left(\omega t - kl - \frac{\pi}{2}\right) \\ E_{y'} &= -\frac{E}{\sqrt{2}}\cos(\omega t - kl) \end{aligned}\right\}$$ Circularly polarized

At point d,

$$\left.\begin{aligned} E_{x'} &= -\frac{E}{\sqrt{2}}\cos(\omega t - 2kl - \pi) \\ E_{y'} &= -\frac{E}{\sqrt{2}}\cos(\omega t - 2kl) \end{aligned}\right\}$$ Linearly polarized along y

FIGURE 20.5 Electrooptic crystal used as voltage-controlled gate in Q-switching a laser.

creases, and the effect is similar, but not as abrupt, as that of a sudden increase of Q.

3. The use of an electrooptic crystal (or liquid Kerr cell) as a voltage-controlled gate inside the optical resonator. It provides a precise control over the losses (Q). Its operation is illustrated by Figure 20.5 and is discussed in some detail in the following. The control of the phase delay in the electrooptic crystal by the applied voltage is discussed in detail in Chapter 14.

During the pumping of the laser by the light from a flashlamp, a voltage is applied to the electrooptic crystal of such magnitude as to introduce a $\pi/2$ relative phase shift (retardation) between the two mutually orthogonal components (x' and y') that make up the linearly polarized (x) laser field. On exiting from the electrooptic crystal at point f, the light traveling to the right is circularly polarized. After reflection from the right mirror, the light passes once more through the crystal. The additional retardation of $\pi/2$ adds to the earlier one to give a total retardation of π, thus causing the emerging beam at d to be linearly polarized along y and consequently to be blocked by the polarizer.

It follows that with the voltage on, the losses are high, so oscillation is prevented. The Q-switching is timed to coincide with the point at which the inversion reaches its peak and is achieved by a removal of the voltage applied to the electrooptic crystal. This reduces the retardation to zero so that the state of polarization of the wave passing through the crystal is unaffected, and the Q regains its high value associated with the ordinary losses of the system.

20.2 MODE LOCKING IN INHOMOGENEOUSLY BROADENED LASER SYSTEMS (References 6–8)

The technique of mode locking has resulted in one of the most important ways in which lasers are employed. It makes it possible to generate intense laser pulses with durations as short as 10^{-14} sec. It is, from the pedagogic point of view, an astounding manifestation of the coherence properties of laser radiation.

We will first describe the phenomenon of mode locking in inhomogeneously broadened lasers in somewhat simple terms that emphasize the physical processes. A more rigorous mathematical description is given last. The phenomenon of mode locking in homogeneously broadened lasers is treated in the next section.

In an inhomogeneous laser system, atoms with different transition energies are independent of each other. In the presence of a strong field, the saturation effects are local and manifest themselves as a "hole" in the gain curve as discussed in Section 8.7. In a homogeneously broadened laser, the whole gain profile saturates as in Figure 8.4*a*. One consequence of this difference is that in the ideal homogeneous case only one laser mode, the one with the lowest threshold, can oscillate as demonstrated in Figure 20.6*a*. This is due to the fact that the gain at the oscillating mode frequency (ν_0) is clamped at a value equal to the losses as discussed in Section 9.3. Since the homogeneous line can tolerate no "holes," the gain exercised by the other modes is below the threshold value.

Now consider what happens as we gradually increase the gain in an inhomogeneously broadened laser. The situation is depicted in Figure 20.6*d*. Once threshold is reached as in curve B, the gain at ν_0 remains clamped at the threshold value. There is no reason, however, why the gain at other frequencies should not increase with further pumping. This gain is due to atoms that do not communicate with those contributing to the gain at ν_0. Further pump-

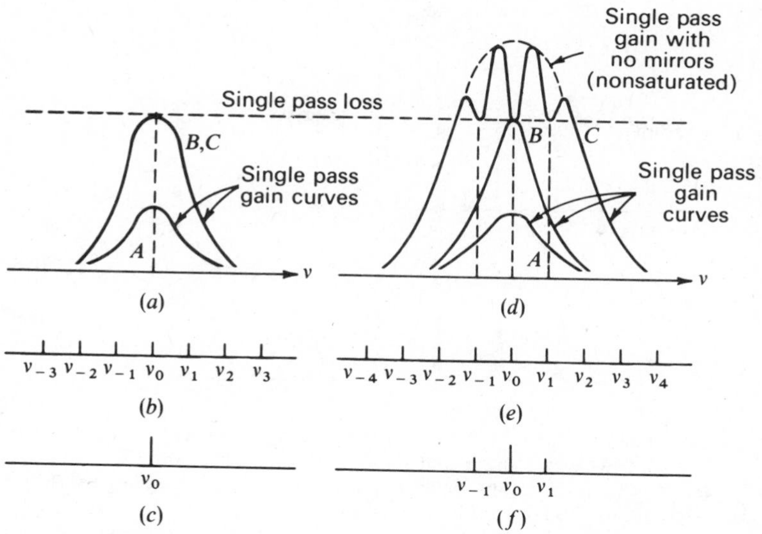

FIGURE 20.6 (*a*) Single-pass gain curves for a homogeneous atomic system (*A*—below threshold; *B*—at threshold; *C*—well above threshold). (*b*) Mode spectrum of optical resonator. (*c*) Oscillation spectrum (only one mode oscillates). (*d*) Single-pass gain curves for an inhomogeneous atomic system (*A*—below threshold; *B*—at threshold; *C*—well above threshold). (*e*) Mode spectrum of optical resonator. (*f*) Oscillation spectrum for pumping level *C*, showing three oscillating modes.

ing will thus lead to oscillation at additional longitudinal-mode frequencies as shown in curve *C*. Since the gain at each oscillating frequency is clamped, the gain profile curve acquires "holes" at the oscillation frequencies.

A plot of the output frequency spectrum showing the multimode oscillation of a He–Ne 0.6328 μm laser is shown in Figure 20.7.

Mode Locking—Qualitative Introduction

We have argued above that in an inhomogeneously broadened laser, oscillation can take place at a number of frequencies, which are separated, according to (7.3-4), by[2]

$$\omega_{q+1} - \omega_q = \frac{\pi c}{l} \equiv \omega$$

Now consider the total optical electric field resulting from such multimode oscillation at some arbitrary point, say one of the mirrors, in the optical

[2] In the following we assume for simplicity that the index of refraction is unity. For cases where $n \neq 1$, c should be replaced everywhere by c/n.

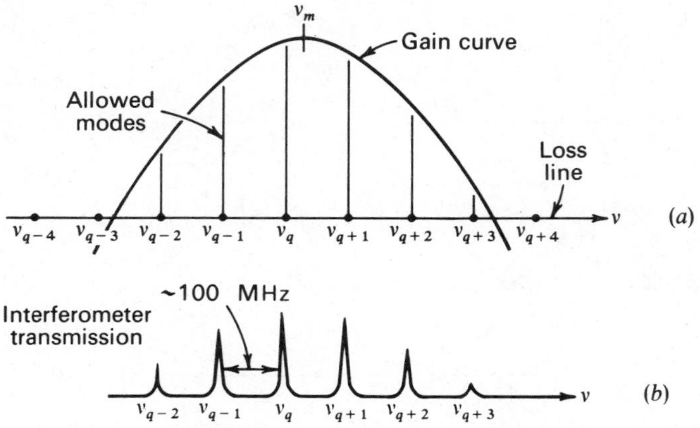

FIGURE 20.7 (*a*) Inhomogeneously broadened Doppler gain curve of the 6328 Å Ne transition and position of allowed longitudinal mode frequencies. (*b*) Intensity versus frequency profile of an oscillating He-Ne laser. Six modes have sufficient gain to oscillate. *Source:* Reference 9.

resonator. It can be taken, if we use complex notation, as

$$E(t) = \sum_n E_n e^{i[(\omega_0 + n\omega)t + \phi_n]} \tag{20.2-1}$$

where the summation is extended over the oscillating modes and ω_0 is chosen arbitrarily as a reference frequency. Symbol ϕ_n is the phase of the nth mode. One property of (20.2-1) is that $E(t)$ is periodic in $T \equiv 2\pi/\omega = 2l/c$, which is the round-trip transit time inside the resonator.

$$\begin{aligned} E(t + T) &= \sum_n E_n \exp\left\{i\left[(\omega_0 + n\omega)\left(t + \frac{2\pi}{\omega}\right) + \phi_n\right]\right\} \\ &= \sum_n E_n \exp\{i[(\omega_0 + n\omega)t + \phi_n]\} \exp\left\{i\left[2\pi\left(\frac{\omega_0}{\omega} + n\right)\right]\right\} \\ &= E(t) \end{aligned} \tag{20.2-2}$$

since ω_0/ω is an integer ($\omega_0 = m\pi c/l$).

Note that the periodic property of $E(t)$ depends on the fact that the phases ϕ_n are fixed. In typical lasers, the phases ϕ_n are likely to vary randomly with time. This causes the intensity of the laser output to fluctuate randomly[3] and greatly reduces its usefulness for many applications where temporal coherence is important.

There are two ways in which this problem can be attacked. The first is to make it possible for the laser to oscillate at a single frequency only, so that mode interference is eliminated. This can be achieved in a variety of ways,

[3] It should be noted that this fluctuation takes place because of random interference between modes and not because of intensity fluctuations of individual modes.

including shortening the resonator length l, thus increasing the mode spacing ($\omega = \pi c/l$) to a point where only one mode falls within the gain linewidth. The second approach is to *force* the phases ϕ_n to *maintain* their relative values. This is the so-called "mode locking" (Reference 6) technique, which (as shown previously) causes the oscillation intensity to consist of a periodic train with a period of $T = 2l/c = 2\pi/\omega$.

One of the most useful forms of mode locking results when the phases ϕ_n are made equal to zero. To simplify the analysis of this case, assume that there are N oscillating modes with equal amplitudes. Taking $E_n = 1$ and $\phi_n = 0$ in (20.2-1) gives

$$E(t) = \sum_{-(N-1)/2}^{(N-1)/2} e^{i(\omega_0 + n\omega)t} \tag{20.2-3}$$

$$= e^{i\omega_0 t} \frac{\sin(N\omega t/2)}{\sin(\omega t/2)} \tag{20.2-4}$$

The average[4] laser power output is proportional to $E(t)E^*(t)$

$$P(t) \propto \frac{\sin^2(N\omega t/2)}{\sin^2(\omega t/2)} \tag{20.2-5}$$

Some of the analytic properties of $P(t)$ are immediately apparent (Reference 6).

1. The power is emitted in a form of a train of pulses with a period $T = 2\pi/\omega = 2l/c$.
2. The peak power $P(sT)$ (for $s = 1, 2, 3, \ldots$) is equal to N times the average power, where N is the number of modes locked together.
3. The peak field amplitude is equal to N times the amplitude of a single mode.
4. The individual pulse width, defined as the time from the peak to the first zero, is $\tau = T/N$. The number of oscillating modes can be estimated by $N \simeq \Delta\omega/\omega$—that is, the ratio of the transition lineshape width $\Delta\omega$ to the frequency spacing ω between modes. Using this relation, as well as $T = 2\pi/\omega$ in $\tau = T/N$, we obtain

$$\tau \sim \frac{2\pi}{\Delta\omega} = \frac{1}{\Delta\nu} \tag{20.2-6}$$

Thus, the length of the mode-locked pulses is approximately the inverse of the gain lineswidth.

A theoretical plot of $\sqrt{P(t)}$ as given by (20.2-5) for the case of five modes ($N = 5$) is shown in Figure 20.8. The ordinate may also be considered as being proportional to the instantaneous field amplitude.

The foregoing discussion was limited to the consideration of mode locking as a function of time. It is clear, however, that since the solution of

[4] The averaging is performed over a time that is long compared with the optical period $2\pi/\omega_0$, but short compared with the modulation period $2\pi/\omega$.

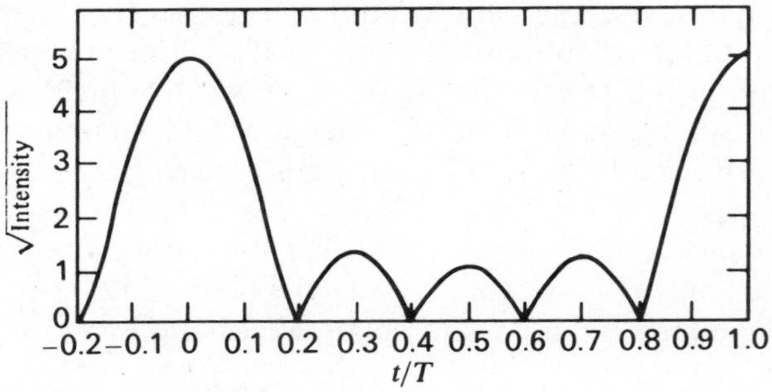

FIGURE 20.8 Theoretical plot of optical field amplitude $[\sqrt{P(t)} \propto \sin(N\omega t/2)\ \sin(\omega t/2)]$ resulting from phase locking of five ($N = 5$) equal-amplitude modes separated from each other by a frequency interval $\omega = 2\pi/T$.

Maxwell's equation in the cavity involves traveling waves (a standing wave can be considered as the sum of two waves traveling in opposite directions), mode locking causes the oscillation energy of the laser to be condensed into a *packet* that travels back and forth between the mirrors with the velocity of light c. The pulsation period $T = 2l/c$ corresponds simply to the time interval between two successive arrivals of the pulse at the mirror. The spatial length of the pulse L_p must correspond to its time duration multiplied by its velocity c. Using $\tau = T/N$, we obtain

$$L_p \sim c\tau = \frac{cT}{N} = \frac{2\pi c}{\omega N} = \frac{2l}{N} \tag{20.2-7}$$

We can verify the last result by taking the basic resonator mode as being proportional to $\sin k_n z \sin \omega_n t$; the total optical field is then

$$E(z, t) = \sum_{n=-(N-1)/2}^{(N-1)/2} \sin\left[(m+n)\frac{\pi z}{l}\right] \sin\left[(m+n)\frac{\pi c}{l}t\right] \tag{20.2-8}$$

where $\omega_n = (m + n)(\pi c/l)$, $k_n = \omega_n/c$, and m is the integer corresponding to the central mode. We can rewrite (20.2-8) as

$$E(z, t) = \frac{1}{2} \sum_{n=-(N-1)/2}^{(N-1)/2} \left\{\cos\left[(m+n)\frac{\pi}{l}(z - ct)\right] - \cos\left[(m+n)\frac{\pi}{l}(z + ct)\right]\right\} \tag{20.2-9}$$

which can be shown to have the spatial and temporal properties described previously. Figure 20.9 shows a spatial plot of (20.2-9) at some time t.

Methods of Mode Locking

In the preceding discussion we considered the consequences of fixing the phases of the longitudinal modes of a laser. Mode locking can be achieved by

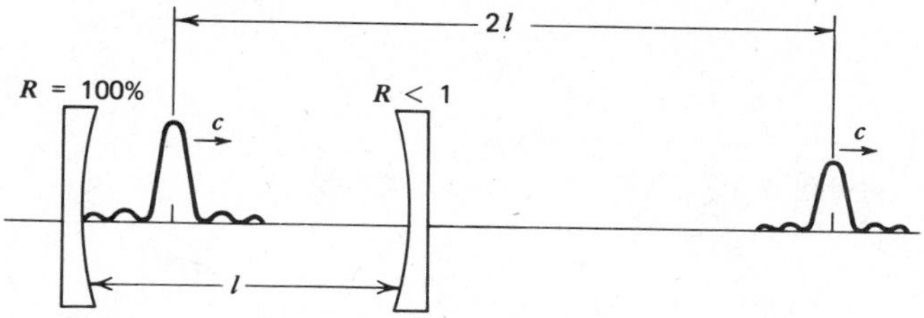

FIGURE 20.9 Traveling pulse of energy resulting from the mode locking of N laser modes; based on (20.2-9).

modulating the losses (or gain) of the laser at a radian frequency $\omega = \pi c/l$, which is equal to the intermode frequency spacing. The theoretical proof is given later; however, a good plausibility argument can be made as follows: as a form of loss modulation, consider a thin shutter inserted inside the laser resonator. Let the shutter be closed (high optical loss) most of the time except for brief periodic openings for a duration of τ_{open} every $T = 2\pi/\omega$ sec. This situation is illustrated by Figure 20.10. A single mode laser wll not oscillate in this case because of the high losses (we assume that τ_{open} is too short to allow the oscillation to build up during each opening). The same applies to multimode oscillation with arbitrary phases. There is one exception, however, if the phases were locked as in (20.2-3); the energy distribution inside the resonator would correspond to that shown in Figure 20.9 and would consist of a narrow ($L_p \simeq 2l/N$) traveling pulse. If this pulse should arrive at the shutter's position when it is open, and if the pulse (temporal) length τ is short compared to the opening time τ_{open}, the mode-locked pulse will be "unaware" of the shutter's existence and, consequently, *will not be attenuated by it.* We may thus reach the conclusion that loss modulation causes mode locking through some kind of "survival of the fittest" mechanisms. In reality, the periodic shutter chops off any intensity tails acquired by the mode-locked pulses due to a "wandering" of the phases from their ideal ($\phi_n = 0$) values. This has the effect of continuously restoring the phases.

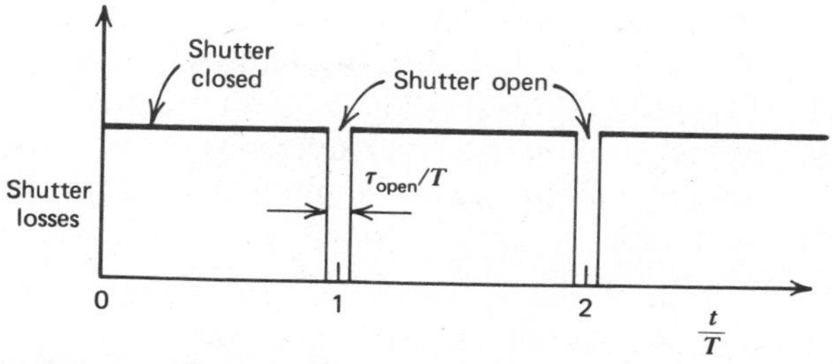

FIGURE 20.10 Periodic losses introduced by a shutter to induce mode locking. The presence of these losses favors the choice of mode phases that results in a pulse passing through the shutter during open intervals—that is, mode locking.

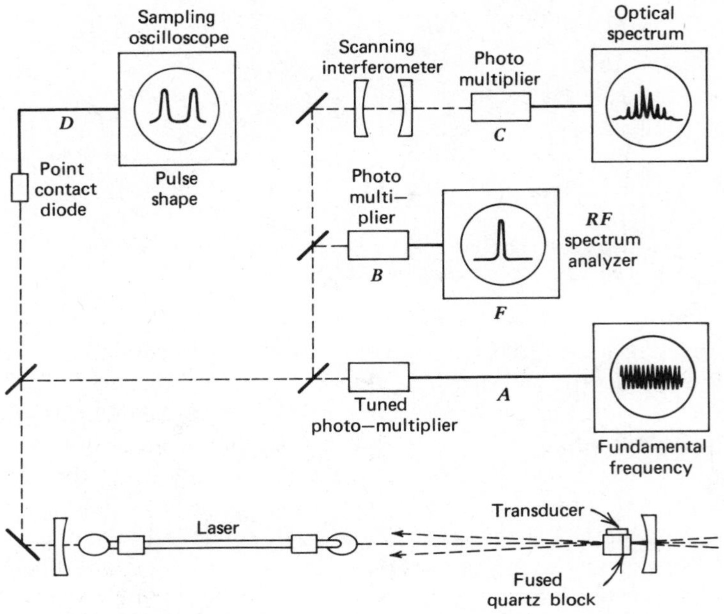

FIGURE 20.11 Experimental setup for laser mode locking by acoustic (Bragg) loss modulation. Parts *A*, *B*, *C*, and *D* of the experimental setup are designed to display the fundamental component of the intensity modulation, the power spectrum of the intensity modulation, the power spectrum of the optical field $E(t)$, and the optical intensity, respectively. *Source:* Reference 8.

An experimental setup used to mode lock a He-Ne laser is shown in Figure 20.11; the periodic loss (Reference 8) is introduced by Bragg diffraction (see Chapter 14) of a portion of the laser intensity by a standing acoustic wave. The loss, which is proportional to the acoustic intensity, is thus modulated at twice the acoustic frequency.

Figure 20.12 shows the pulses resulting from mode locking a He-Ne laser.

Mode locking occurs spontaneously in some lasers if the optical path contains a saturable absorber (an absorber whose opacity decreases with increasing optical intensity). This method is used to induce mode locking in the high-power-pulsed solid-state lasers (References 5, 10, and 11).

Table 20.1 lists some of the lasers commonly used in mode locking and the observed pulse durations.

Mode Locking by Loss Modulation—Theoretical Derivation

The theoretical approach (Reference 6) consists of solving Maxwell's equations for an oscillating resonator for the case where the losses are modulated.

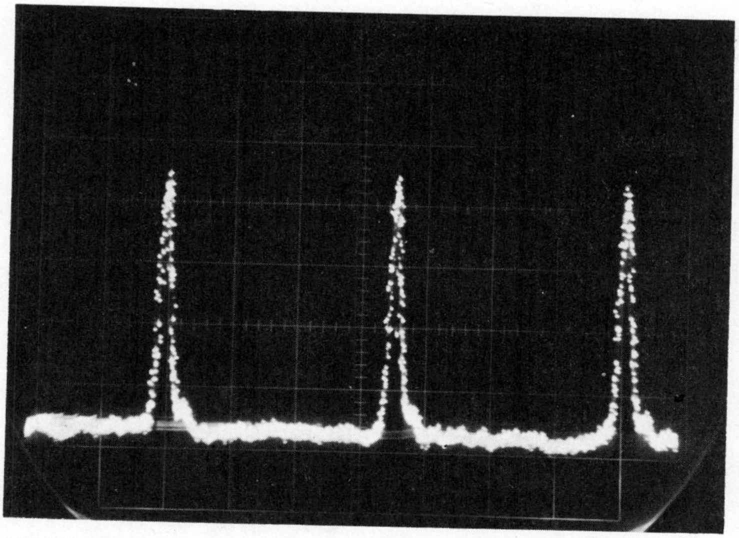

FIGURE 20.12 Pulse train from mode-locked 6118 Å He-Ne laser. Horizontal scale: 5 nsec/div. *Source:* Reference 12.

We find immediately that the proper modes given in Chapter 5 no longer satisfy Maxwell's equations. The new solutions can be expressed, however, as expansions in terms of the old modes. The coefficients of this expansion that are now related to each other in amplitude as well as in phase are the complex amplitudes of the "locked" modes.

We will account for the time varying "gate" of Fig. 20.10 by a generalized time and space dependent loss term, which is represented mathematically by assuming that the medium inside the resonator possesses a conduc-

TABLE 20.1. Some Laser Systems, Their Gain Linewidth $\Delta\nu$, and the Length of Their Pulses in the Mode-Locked Operation

Laser Medium	*$\Delta\nu$ Hz*	*$(\Delta\nu)^{-1}$ Sec*	*Observed Pulse Duration, Sec*
He-Ne (0.6328 μm) CW	1.5×10^9	6.66×10^{-10}	6×10^{-10}
Nd: YAG (1.06 μm) CW	1.2×10^{10}	8.34×10^{-11}	7.6×10^{-11}
Ruby (0.6943 μm) pulsed	6×10^{10}	1.66×10^{-11}	1.2×10^{-11}
Nd^{3+}: glass (1.06 μm) pulsed	3×10^{12}	3.33×10^{-13}	4×10^{-13}
Rhodamine 6G (~0.6 μm)	10^{13}	10^{-13}	10^{-13}

tivity $\sigma(\mathbf{r}, t)$. We write Maxwell's equations as

$$\nabla \times \mathbf{H} = \sigma(\mathbf{r}, t)\mathbf{E} + \varepsilon \frac{\partial \mathbf{E}}{\partial t}$$
$$\nabla \times \mathbf{E} = -\mu \frac{\partial \mathbf{H}}{\partial t} \qquad (20.2\text{-}10)$$

By taking the curl of the second equation, using the vector identity $\nabla \times \nabla \times \mathbf{A} = -\nabla^2 \mathbf{A} + \nabla(\nabla \cdot \mathbf{A})$, and assuming a charge free medium so that $\nabla \cdot \mathbf{E} = 1/\varepsilon\ (\nabla \cdot \mathbf{D}) = 0$, we obtain from (20.2-10)

$$\nabla^2 \mathbf{E} - \mu\sigma \frac{\partial \mathbf{E}}{\partial t} - \mu \mathbf{E} \frac{\partial \sigma}{\partial t} - \mu\varepsilon \frac{\partial^2 \mathbf{E}}{\partial t^2} = 0$$

In the last equation we take advantage of the fact that the variation of σ is "slow" compared to that of $\mathbf{E}$ so that $\mathbf{E}(\partial\sigma/\partial t) \ll \sigma(\partial \mathbf{E}/\partial t)$ and will be neglected. The last equation is then rewritten as

$$\nabla^2 \mathbf{E} - \mu\sigma(\mathbf{r}, t) \frac{\partial \mathbf{E}}{\partial t} - \mu\varepsilon \frac{\partial^2 \mathbf{E}}{\partial t^2} = 0 \qquad (20.2\text{-}11)$$

A convenient set of eigenmodes in which to express the field $\mathbf{E}$ is the mode solutions of the "cold" ($\sigma = 0$) resonator. These obey

$$\nabla^2 \mathbf{E}_a(\mathbf{r}) + \omega_a^2 \mu\varepsilon \mathbf{E}_a(\mathbf{r}) = 0 \qquad (20.2\text{-}11a)$$

where ω_a is the resonant frequency of the ath mode. The modes $\mathbf{E}_a$ satisfy the resonator boundary conditions and form a complete orthogonal set. The arbitrary scale factor of $\mathbf{E}_a$ is chosen, arbitrarily, to satisfy (5.5-10)†

$$\int_{V_{\text{resonator}}} \mathbf{E}_a \cdot \mathbf{E}_b \, dV = \delta_{ab} \qquad (20.2\text{-}12)$$

In the presence of a time-varying conductivity $\sigma(\mathbf{r}, t)$ the modes $\mathbf{E}_a(\mathbf{r})$ $\exp(i\omega_a t)$ do not satisfy the wave equation (20.2-11) individually. We can, however, take advantage of the completeness of the set $\mathbf{E}_a(\mathbf{r})$ and expand the field $\mathbf{E}(\mathbf{r}, t)$

$$\mathbf{E}(\mathbf{r}, t) = \sum_s \mathbf{A}_s(t) \mathbf{E}_s(\mathbf{r}) e^{i\omega_s t} \qquad (20.2\text{-}13)$$

Substituting (20.2-13) in (20.2-11) gives

$$\sum_s \left\{ \mathbf{A}_s(\nabla^2 \mathbf{E}_s + \omega_s^2 \mu\varepsilon \mathbf{E}_s) - (\mu\sigma + 2i\omega_s\mu\varepsilon) \frac{dA_s}{dt} \mathbf{E}_s - i\omega_s\mu\sigma A_s \mathbf{E}_s \right\} e^{i\omega_s t} = 0 \qquad (20.2\text{-}14)$$

The sum of the first two terms is zero according to (20.2-11a). We also take advantage of $\sigma/\varepsilon \lll \omega_s$ (ε/σ is the photon lifetime, that is, the decay time constant of the resonator which in an oscillating laser approaches infinity so

† For a plane wave like mode in a resonator with a length ℓ we have $\mathbf{E}_a = \hat{x}\sqrt{2/V} \sin\left(\frac{a\pi}{\ell} z\right)$, a = any integer, V is the volume of the resonator.

that $\sigma/\varepsilon \to 0$, ω_s is $\sim 2 \times 10^{15}$ at ($\lambda = 1\ \mu$m)). The result is

$$\sum_s \mathbf{E}_s \frac{dA_s}{dt} e^{i\omega_s t} = -\sum_s A_s \frac{\sigma(\mathbf{r}, t)}{2\varepsilon} \mathbf{E}_s e^{i\omega_s t} \tag{20.2-15}$$

Since $\sigma(\mathbf{r}, t)$ represents in our model both the time-independent average loss (or gain) as well as the periodic loss (or gain) modulation that causes mode locking, we will take it as

$$\sigma(\mathbf{r}, t) = \sigma_0 + \sigma_m \cos(\omega_m t) f(\mathbf{r}) \tag{20.2-16}$$

We use this form in (20.2-15), then dot multiply both sides by $\mathbf{E}_a(\mathbf{r})$ and integrate over the resonator volume which leads, taking advantage of (20.2-12), to

$$\begin{aligned} e^{i\omega_a t} \frac{dA_a}{dt} &= -e^{i\omega_a t} \frac{\sigma_0}{2\varepsilon} A_s - \sum_s S_{a,s} \frac{\sigma_m \cos \omega_m t}{2\varepsilon} A_s e^{i\omega_s t} \\ S_{s,a} &= \int_{V_{\text{resonator}}} f(\mathbf{r}) \mathbf{E}_s \cdot \mathbf{E}_a \, dV \end{aligned} \tag{20.2-17}$$

It follows that for $\sigma_m = 0$, that is, no modulation, $A_a = A_a(0) \exp(-\sigma_0 t/2\varepsilon)$ so that ε/σ_0 is the photon decay time. In an oscillating laser the losses are balanced by the gain, and the photon lifetime is nearly infinite (exactly infinite if we neglect spontaneous emission input to the oscillating mode) so that $\sigma_0 = 0$ and (20.2-17) becomes

$$\frac{dA_a}{dt} = -\sum_s \frac{S_{a,s}\sigma_m}{4\varepsilon} A_s (e^{i\omega_m t} + e^{-i\omega_m t}) e^{i(\omega_s - \omega_a)t} \tag{20.2-18}$$

For sustained (synchronous) cumulative interaction the sum of the terms $(\omega_s - \omega_a \pm \omega_m)$ must be near zero, that is, the modulation frequency must be nearly equal to the difference $(\omega_s - \omega_a)$ of two "cold" resonant frequencies. This agrees with the intuitive model of mode locking depicted in Fig. 20.10. In practice one chooses, most often, ω_m to lie close to the difference of two neighboring "cold" resonator frequencies. We thus define a detuning parameter as the difference between ω_m and the intermode frequency difference Ω.

$$\begin{aligned} &\Delta \equiv \omega_{s+1} - \omega_s - \omega_m \equiv \Omega - \omega_m \\ &\Omega \equiv \omega_{s+1} - \omega_s \text{ is the intermode frequency difference} \end{aligned} \tag{20.2-19}$$

The only two synchronous terms on the right side of (20.2-18), that is, the terms that satisfy the condition $\omega_s - \omega_a \approx \pm\omega_m$ are $s = a + 1$ and $s = a - 1$. Limiting the right side of (20.2-18) to these terms gives

$$\frac{dA_a}{dt} = \kappa A_{a+1} e^{i\Delta t} + \kappa A_{a-1} e^{i\Delta t} + \kappa A_{a+1} e^{i\Delta t} \tag{20.2-20}$$

$$\kappa_{a,a+1} = \kappa_{a,a-1} \equiv \kappa = \frac{S_{a,a+1}\sigma_m}{4\varepsilon} \tag{20.2-21}$$

The effect of synchronous loss (or gain) modulation is thus reduced to a very simple coupled mode equation. The coupling between adjacent modes is represented by the parameter κ which, according to (20.2-21) is proportional to the loss modulation σ_m. Another noteworthy feature is that if the loss modulation were uniform in space, that is, $f(\mathbf{r})$ = constant, then according to (20.2-17) and (20.2-12) $\kappa = 0$ and no mode coupling can take place. This agrees with the intuitive explanation of mode locking using the concept of a localized (nonuniform) gate in Fig. 20.10. The solution of (20.2-20) is facilitated by defining

$$\begin{aligned} c_a(t) &= -ie^{ia\,\Delta t}e^{-ia\pi/2}A_a(t) \\ A_a(t) &= ic_a(t)e^{-ia\,\Delta t}e^{+ia\pi/2} \end{aligned} \tag{20.2-22}$$

so that (20.2-20) becomes

$$i\frac{dc_a}{dt} + a\,\Delta c_a = -\kappa c_{a+1} + \kappa c_{a-1} \tag{20.2-23}$$

This recurrence equation has a steady state ($dc_a/dt = 0$) solution

$$c_a = I_a\left(\frac{\kappa}{\Delta}\right) \tag{20.2-24}$$

where I_a is the hyperbolic Bessel function of order a. For $\kappa/\Delta \gg 1$, $I_a(\kappa/\Delta) \to (2\pi\kappa/\Delta)^{-1/2}$ and the original mode amplitudes become

$$A_a(t) = i(2\pi\kappa/\Delta)^{-1/2}e^{-ia(\Delta t-\pi/2)}$$

The total resonator field (20.2-13) is

$$\begin{aligned} \mathbf{E}(\mathbf{r}, t) &= \sum_s A_s(t)\mathbf{E}_s(\mathbf{r})e^{i\omega_s t} \\ &= \sum_s i(2\pi\kappa/\Delta)^{-1/2}e^{i(\omega_0+s\omega_m)t}e^{is\pi/2}\mathbf{E}_s(\mathbf{r}) \end{aligned} \tag{20.2-25}$$

where we put $\omega_s = \omega_0 + s\Omega$ and made use of the definition (20.2-19).

We thus find that the resonator field is no longer composed of individual modes but must be expressed as a superposition of modes with *fixed amplitudes, phases,* and *frequencies,* that is, a mode-locked field.

We also note (in 20.2-25) that the intermode frequency difference of the locked modes does not have the "cold" resonator value Ω but is equal to the modulation frequency ω_m. Effective mode locking, however, requires that $\omega_m \approx \Omega$, since this leads to $\kappa/\Delta \gg 1$, a condition that was shown above to give rise to equal values of the mode amplitudes $I_s(\kappa/\Delta) = (2\pi\kappa/\Delta)^{-1/2}$.

The solution (20.2-25) is identical to the form assumed in (20.2-3) for the ideal mode-locked field. The phase factor $\exp(is\pi/2)$ corresponds to a mere shift in the time origin.

Additional references on mode locking, which include the gain dependence on the frequency, are listed at the end of this chapter (Refs. 13, 14).

20.3 MODE LOCKING IN HOMOGENEOUSLY BROADENED LASER SYSTEMS

The analysis of mode locking in inhomogeneous laser systems assumed that the role of internal modulation was that of locking together the phases of modes that, in the absence of modulation, oscillate with random phases. In the case of homogeneous broadening, only one mode can normally oscillate. Experiments, however, reveal that mode locking leads to short pulses in a manner quite similar to that described in Section 20.2. One way to reconcile the two points of view and the experiments is to realize that *in the presence of internal modulation,* power is transferred continuously from the high gain mode to those of lower gain (i.e., those that would not normally oscillate). This power can be viewed simply as that of the sidebands at $(\omega_0 \pm n\omega)$ of the mode at ω_0 created by a modulation at ω. Armed with this understanding, we see that the physical phenomenon is not one of mode locking but of mode generation. The net result, however, is that of a large number of oscillating modes with equal frequency spacing and fixed phases, as in the inhomogeneous case, leading to ultrashort pulses.

The analytical solution to this case (References 15 and 16) follows an approach used originally to analyze short pulses in traveling wave microwave oscillators (Reference 17).

Referring to Figure 20.13, we consider an optical resonator with mirror reflectivities R_1 and R_2 that contains, in addition to the gain medium, a periodically modulated loss cell. The method of solution is to follow one pulse through a complete round trip through the resonator and to require that the resulting pulse reproduce itself. The temporal pulse shape at each stage is assumed to be Gaussian.

Before proceeding, we need to characterize the effect of the gain medium and the loss cell on a traveling Gaussian pulse.

The Transfer Function of the Gain Medium

Assume that an optical pulse with a field $E_{\text{in}}(t)$ is incident on an amplifying optical medium of length l. If we take the Fourier transform of $E_{\text{in}}(t)$ as $E_{\text{in}}(\omega)$, the amplifier can be characterized by a transfer function $g(\omega)$ where

$$E_{\text{out}}(\omega) = E_{\text{in}}(\omega)g(\omega) \qquad (20.3\text{-}1)$$

is the Fourier transform of the output field. Equation (20.3-1) is a linear relationship and applies only in the limit of negligible saturation $4\Omega^2 T_2 \tau \ll 1$.

Using (8.2-4) and (8.1-19), we have

$$\begin{aligned} g(\omega) &= \exp\left\{-ikl\left[1 + \frac{1}{2n^2}(\chi' - i\chi'')\right]\right\} \\ &= \exp\left\{-ikl - \frac{kl\mu^2 T_2\,\Delta N_0}{2n^2\varepsilon_0\hbar}\left[\frac{1}{1 + i(\omega - \omega_0)T_2}\right]\right\} \\ &\simeq \exp\left\{-ikl + \frac{\gamma_{\max}l}{2}\left[1 - i(\omega - \omega_0)T_2 - (\omega - \omega_0)^2T_2^2\right]\right\} \end{aligned}$$

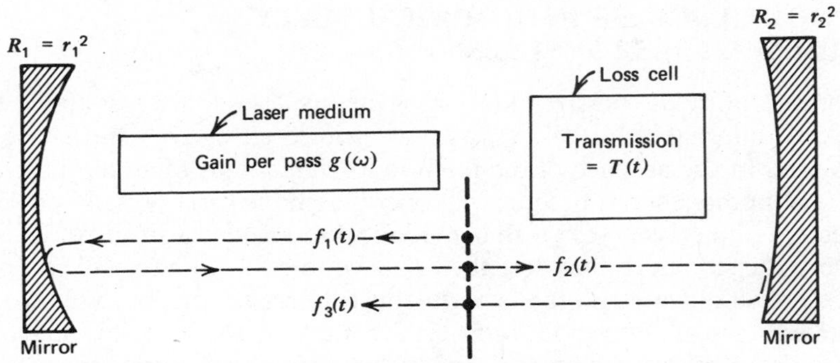

FIGURE 20.13 The experimental arrangement assumed in the theoretical analysis of mode locking in homogeneously broadened lasers.

where the approximation is good for $(\omega - \omega_0)T_2 \ll 1$, and where we recall $\Delta N < 0$ for gain. Since the pulse is making two passes through the cell, we take

$$\frac{E_{\text{out}}(\omega)}{E_{\text{in}}(\omega)} = [(g(\omega)]^2 = \exp\{-i2kl + \gamma_{\max} l[1 - i(\omega - \omega_0)T_2 - (\omega - \omega_0)^2 T_2^2]\}$$

The imaginary terms in the exponent correspond to a time delay (due to the finite group velocity of the pulse) of

$$\tau_d = \frac{2l}{c} + l\gamma_{\max} T_2$$

We are only considering here the effect on the pulse shape so that, ignoring the imaginary term,[6] we obtain

$$[g(\omega)]^2 = e^{\gamma_{\max} l[1-(\omega-\omega_0)^2 T_2^2]} \tag{20.3-1a}$$

The Transfer Function of the Loss Cell

Here, we need to express the effect of the loss cell on the pulse in the time domain.

Assume that the double pass amplitude transmission factor $T(t)$ of the loss cell is given by

$$E_{\text{out}}(t) = E_{\text{in}}(t)T(t) = E_{\text{in}}(t) \exp[-2\delta_l^2 \sin^2(\pi \, \Delta\nu_{\text{axial}} t)] \tag{20.3-2}$$

where $\Delta\nu_{\text{axial}}$, the longitudinal mode spacing, is given by

$$\Delta\nu_{\text{axial}} = \frac{c}{2l_c}$$

where l_c is the effective optical length of the resonator. The transmission peaks are thus separated by $2l_c/c$ sec so that a mode-locked pulse can pass through

[6] The finite propagation delay affects the round-trip pulse propagation time that must be equal to the period of the loss modulation.

the cell on successive trips with minimum loss. Since the pulses pass through the cell centered on the point of maximum transmission, we approximate (20.3-2) by

$$E_{\text{out}}(t) = E_{\text{in}}(t)T(t) = E_{\text{in}}(t)\exp[-2\delta_l^2(\pi\,\Delta\nu_{\text{axial}}t)^2] \tag{20.3-3}$$

We can view the form of (20.3-3) as the prescribed transmission function of the cell. The form, however, is suggested by physical considerations. In the case of an electrooptic shutter with a retardation (see Section 14.3) $\Gamma(t) = \Gamma_m \sin\omega_m t$, the transmission factor is $T(t) = \cos^2[\Gamma(t)/2]$. Near the transmission peaks, $\Gamma(t) \ll 1$ and $T(t)$ is given by

$$T(t) \simeq \exp[-\tfrac{1}{4}(\Gamma_m^2\omega_m^2 t^2)] = \exp[-2\delta_l^2(\pi\,\Delta\nu_{\text{axial}}t)^2]$$

where $\omega_m = \pi\,\Delta\nu_{\text{axial}}$ and $\Gamma_m = 2\sqrt{2}\,\delta_l$.

We now return to the main analysis. The starting pulse $f_1(t)$ in Figure 20.13 is taken as

$$f_1(t) = Ae^{-\alpha_1 t^2}e^{i(\omega_0 t + \beta_1 t^2)} \tag{20.3-4}$$

corresponding to a "chirped" frequency

$$\omega(t) = \omega_0 + 2\beta_1 t \tag{20.3-5}$$

Its Fourier transform is

$$\begin{aligned} F_1(\omega) &= \frac{1}{2\pi}\int_{-\infty}^{\infty} f_1(t)e^{-i\omega t}\,dt \\ &= \frac{A}{2}\sqrt{\frac{1}{\pi(\alpha_1 - i\beta_1)}}\exp[-(\omega-\omega_0)^2/4(\alpha_1 - i\beta_1)] \end{aligned} \tag{20.3-6}$$

A double pass through the amplifier and one mirror reflection (r_1) are accounted for by multiplying $F_1(\omega)$ by the transfer factor $[g(\omega)]^2 r_1$

$$\begin{aligned} F_2(\omega) &= F_1(\omega)[g(\omega)]^2 r_1 \\ &= \frac{r_1 A}{2}e^{g_0}\sqrt{\frac{1}{\pi(\alpha_1 - i\beta_1)}}\exp\left\{[-(\omega-\omega_0)^2]\left[\frac{1}{4(\alpha_1 - i\beta_1)} + g_0 T_2^2\right]\right\} \end{aligned} \tag{20.3-7}$$

where $g_0 \equiv \gamma_{\max} l$ and $[g(\omega)]^2$ is given by (20.3-1a). Transforming back to the time domain, we get

$$\begin{aligned} f_2(t) &= \int_{-\infty}^{\infty} F_2(\omega)e^{i\omega t}\,d\omega \\ &= \frac{r_1 A e^{g_0}}{2\pi}\sqrt{\frac{\pi}{\alpha_1 - i\beta_1}}\,e^{-\omega_0^2 Q}\sqrt{\frac{\pi}{Q}}\exp[-(2i\omega_0 Q - t)^2/4Q] \end{aligned} \tag{20.3-8}$$

where

$$Q \equiv \frac{1}{4(\alpha_1 - i\beta_1)} + g_0 T_2^2 \tag{20.3-9}$$

A reflection from mirror 2 and a passage through the loss cell lead, according to (20.3-3), to

$$
\begin{aligned}
f_3(t) &= r_2 f_2(t) e^{-2\delta_l^2 \pi^2 (\Delta\nu_{\text{axial}})^2 t^2} \\
&= \frac{r_1 r_2 A e^{g_0}}{2} \sqrt{\frac{1}{(\alpha_1 - i\beta_1)Q}}\, e^{i\omega_0 t} e^{-[2\delta_l^2 (\pi\,\Delta\nu_{\text{axial}})^2 + (1/4Q)]t^2}
\end{aligned}
\tag{20.3-10}
$$

For self-consistency, we require that $f_3(t)$ be a replica of $f_1(t)$. We thus equate the exponent of (20.3-10) to that of (20.3-4)

$$
\begin{aligned}
\alpha_1 &= 2\delta_l^2 (\pi\,\Delta\nu_{\text{axial}})^2 + \text{Re}\left(\frac{1}{4Q}\right) \\
\beta_1 &= -\text{Im}\left(\frac{1}{4Q}\right)
\end{aligned}
\tag{20.3-11}
$$

If we use (20.3-9), the second of (20.3-11) gives

$$
\beta_1 = \frac{\beta_1}{(1 + 4g_0 T_2^2 \alpha_1)^2 + (4g_0 T_2^2 \beta_1)^2}
$$

so that a self-consistent solution requires that

$$
\beta_1 = 0
$$

that is, no chirp. With $\beta_1 = 0$, the first of (20.3-11) becomes

$$
2\delta_l^2 (\pi\,\Delta\nu_{\text{axial}})^2 + \frac{\alpha_1}{(1 + 4g_0 T_2^2 \alpha_1)} = \alpha_1 \tag{20.3-12}
$$

that, if we assume

$$
4g_0 T_2^2 \alpha_1 \ll 1 \tag{20.3-13}
$$

results in

$$
\alpha_1 = \left(\frac{\delta_l^2}{2g_0}\right)^{1/2} \frac{\pi\,\Delta\nu_{\text{axial}}}{T_2}
$$

The pulse width at the half-intensity points is from (20.3-4)

$$
\tau_p = (2\ln 2)^{1/2} \alpha_1^{-1/2}
$$

so that the self-consistent pulse has a width

$$
\tau_p = \frac{(2\ln 2)^{1/2}}{\pi} \left(\frac{2g_0}{\delta_l^2}\right)^{1/4} \left(\frac{1}{\Delta\nu_{\text{axial}}\,\Delta\nu}\right)^{1/2} \tag{20.3-14}
$$

where $\Delta\nu \equiv (\pi T_2)^{-1}$. The condition (20.3-13) can now be interpreted as requiring that $\tau_p \gg 2\sqrt{g_0}\,T_2$, which is true in most cases.

An experimental setup demonstrating mode locking in a pressure-broadened CO_2 laser is sketched in Figure 20.14*a*. The inverse square root dependence of τ_p on $\Delta\nu$ is displayed by the data of Figure 20.14*b*.

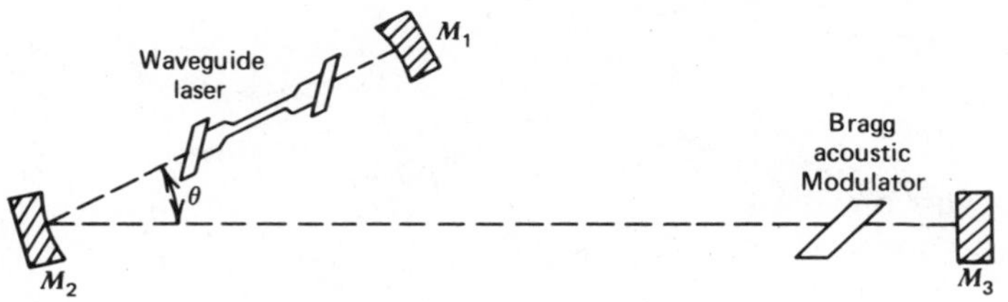

FIGURE 20.14*a* A schematic drawing of the mode-locking experiment in a high pressure CO_2 laser. *Source:* Reference 18.

Mode Locking by Phase Modulation

Mode locking can be induced by internal phase, rather than loss, modulation. This is usually done by using an electrooptic crystal inside the resonator oriented in the basic manner of Figure 14.6, such that the passing wave undergoes a phase delay proportional to the instantaneous electric field across the crystal. The frequency of the modulating signal is equal, as in the loss modulation case, to the inverse of the round-trip delay time, that is, to the longitudinal intermode frequency separation.

The analysis for the case of the inhomogeneous laser is similar to that of Section 20.2 and leads to similar results (Reference 6).

In the homogeneous laser case, one employs an analysis similar to that of Section 20.3, except that the transfer function through the modulation cell is

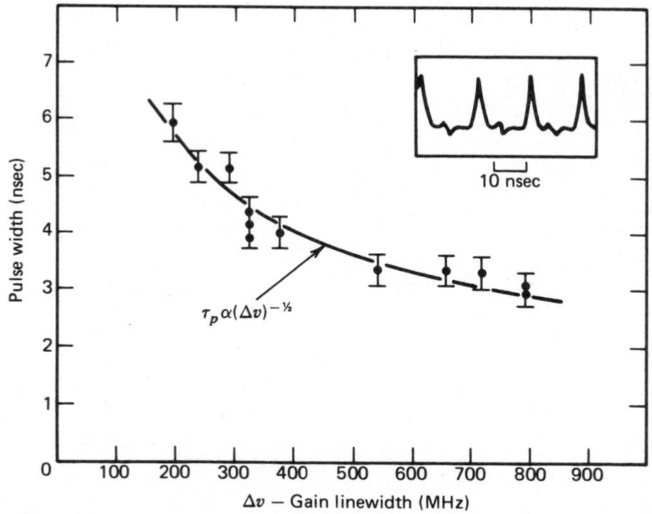

FIGURE 20.14*b* The dependence of the pulse width on the gain linewidth $\Delta\nu$ that is controlled by varying the pressure ($\Delta\nu = 8 \times 10^8$ at 150 torr). *Source:* Reference 18.

taken, instead of (20.3-2), as

$$E_{\text{out}}(t) = E_i(t)\exp(-i2\delta_\phi \cos 2\pi\Delta\nu_{\text{axial}}t) \tag{20.3-15}$$

For pulses passing near the extrema of the phase excursion, we can approximate the last equation as

$$E_{\text{out}}(t) = E_i(t)\exp(\mp i2\delta_\phi \pm i\delta_\phi 4\pi^2\Delta\nu_{\text{axial}}^2 t^2) \tag{20.3-16}$$

An analysis identical to that leading to (20.3-14) yields (Reference 16)

$$\tau_p = \frac{(2\ln 2)^{1/2}}{\pi}\left(\frac{2g_0}{\delta_\phi}\right)^{1/4}\left(\frac{1}{\Delta\nu_{\text{axial}}\,\Delta\nu}\right)^{1/2} \tag{20.3-17}$$

In this case, self-consistency leads to a chirped pulse with

$$\beta = \pm\alpha = \pm\pi^2\,\Delta\nu_{\text{axial}}\,\Delta\nu\,\sqrt{\frac{\delta_\phi}{2g_0}} \tag{20.3-18}$$

The upper and lower signs in (20.3-16) and (20.3-18) correspond to two possible solutions, one passing through the cell near the maximum of the phase excursion and the other near its minimum.

We note that (20.3-17) is similar to the loss modulation result (20.3-14) except for the fact that δ_ϕ appears instead of δ_l^2. This difference can be traced to a difference between (20.3-2) and (20.3-15). The choice in both cases is such that δ *corresponds* to the *retardation* induced by the electrooptic crystal.

The shortest mode-locked pulses to date were obtained in passively locked dye lasers employing Rhodamine 6G as the gain medium. The mode

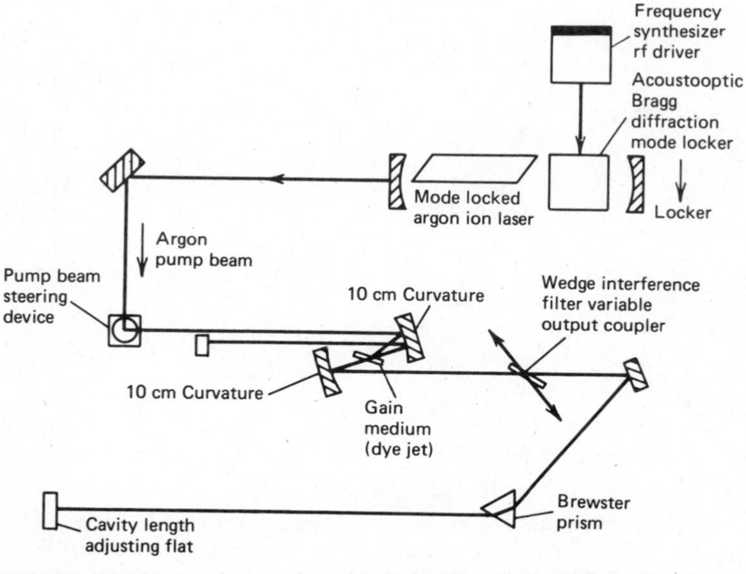

FIGURE 20.15*a* Synchronously mode-locked dye laser configuration. *Source:* T. L. Koch, The California Institute of Technology, Pasadena, CA.

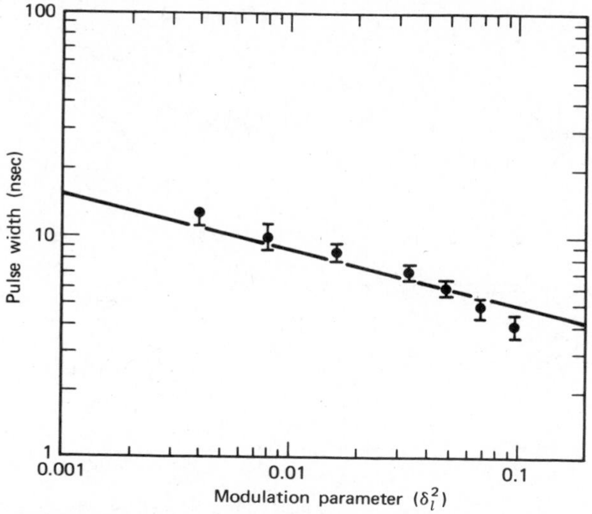

FIGURE 20.15*b* The mode-locked pulse width as a function of the modulation parameter δ_l^2. *Source:* Reference 18.

locking is due to saturable absorption in a second dye cell present in the optical cavity.

A sketch of a synchronously mode-locked dye laser configuration is shown in Figure 20.15.

The shortest pulses obtained to date are 80×10^{-15}s. These pulses have been narrowed down further to 8×10^{-15}s by the use of nonlinear optical techniques [Supplementary Reference 3].

Ultrashort mode-locked pulses are now used in an ever-widening circle of applications involving the measurement and study of short-lived molecular and electronic phenomena. The use of ultrashort optical pulses has led to an improvement of the temporal resolution of such experiments by more than three orders of magnitude. For a description of many of these applications as well as of the many methods used to measure the pulse duration, the reader should consult Supplementary References 4–6.

Another important method of Q-switching applies to semiconductor lasers (Chapter 11). Since the gain constant of the active region is determined by a bias current applied through an electrode, a current signal applied to such an electrode can cause Q-switching in a manner similar to that described above. By separating the current electrode shown in Figure 20.16 into two segments, it is possible to use one segment to provide the necessary dc bias current that supplies the laser gain while using the second electrode independently, to control the loss (gain). The experimental laser demonstrated in Figure 20.16 emitted Q-switched pulses as short as 18 ps at rates of up to 5 GHz.

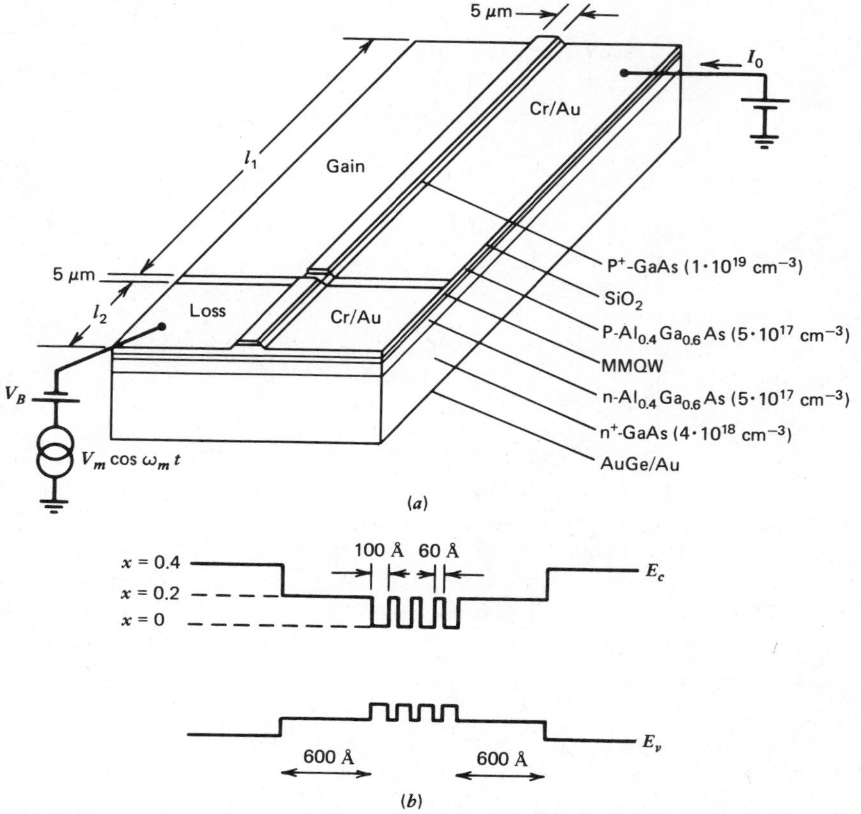

FIGURE 20.16 (*a*) Perspective view of the two-segment quantum well laser. The lengths of the gain section l_1 and the modulator section l_2 are 250 and 50 μm, respectively. (*b*) Expanded band gap diagram of the MQW active layer.

20.4 RELAXATION OSCILLATION IN LASERS

Relaxation oscillation of the intensity has been observed in most types of lasers (References 19 and 20). This oscillation takes place characteristically with a period that is considerably longer than the cavity decay time t_c, or the resonator round-trip time $2nl/c$. Typical values range between 0.1 μs to 10 μs.

The basic physical mechanism is an interplay between the oscillation field in the resonator and the atomic inversion (Reference 19). An increase in the field intensity causes a reduction in the inversion due to the increased rate of stimulated transitions. This causes a reduction in the gain that tends to decrease the field intensity.

In the mathematical modeling of this phenomenon, we assume an ideal homogeneously broadened laser referring to Figure 9.4. We also assume that the lower-level population N_1 is negligible (i.e., $W_i^{-1} \gg t_1 \ll t_2$) and take the inversion density $N \equiv N_2 - N_1(g_2/g_1) \cong N_2$. The pumping rate into level 2 (atoms/m^3 − sec) is R and the lifetime, due to all causes except stimulated emission, of atoms in level 2 is τ. Taking the induced transition rate per atom as W_i, we have

$$\frac{dN}{dt} = R - W_i N - \frac{N}{\tau} \tag{20.4-1}$$

The transition rate W_i is, according to (8.3-10), proportional to field intensity I and hence to the photon density q in the optical resonator. We can, consequently, rewrite (20.4-1) as

$$\frac{dN}{dt} = R - qBN - \frac{N}{\tau} \tag{20.4-2}$$

where B is a proportionality constant defined by $W_i = Bq$. Since qBN is also the rate (m^{-3} − $\sec^{-1}$) at which photons are generated, we have

$$\frac{dq}{dt} = qBN - \frac{q}{t_c} \tag{20.4-3}$$

where t_c is the decay time constant for photons in the optical resonator as discussed in Section 7.4. Equations (20.4-2) and (20.4-3) describe the interplay between the photon density q and the inversion N.

First, we notice that in equilibrium, $dq/dt = dN/dt = 0$, the following relations are satisfied:

$$\begin{aligned} N_0 &= \frac{1}{Bt_c} \\ q_0 &= \frac{RBt_c - \dfrac{1}{\tau}}{B} \end{aligned} \tag{20.4-4}$$

From (20.4-4), it follows that when $R = (Bt_c\tau)^{-1}$, $q_0 = 0$. We denote this threshold pumping rate by R_t and define the pumping factor $r \equiv R/R_t$[7], so that the second of (20.4-4) can also be written as

$$q_0 = \frac{(r-1)}{B\tau} \tag{20.4-5}$$

Next, we consider the behavior of small perturbations from equilibrium. We take

$$N(t) = N_0 + N_1(t) \qquad N_1 \ll N_0$$

and

$$q(t) = q_0 + q_1(t) \qquad q_1 \ll q_0$$

Substituting these relations in (20.4-2) and (20.4-3) and making use of

[7] r is equal to the ratio of the unsaturated ($q = 0$) gain to the saturated gain (the saturated gain is the actual gain "seen" by the laser field and is equal to the loss).

(20.4-4), we obtain

$$\frac{dN_1}{dt} = -RBt_cN_1 - \frac{q_1}{t_c} \tag{20.4-6}$$

$$\frac{dq_1}{dt} = \left(RBt_c - \frac{1}{\tau}\right) N_1 \tag{20.4-7}$$

Taking the derivative of (20.4-7), substituting (20.4-6) for dN_1/dt, and using (20.4-4) lead to

$$\frac{d^2q_1}{dt^2} + RBt_c \frac{dq_1}{dt} + \left(RB - \frac{1}{\tau t_c}\right) q_1 = 0 \tag{20.4-8}$$

or, in terms of the pumping factor $r = RBt_c\tau$ introduced above,

$$\frac{d^2q_1}{dt^2} + \frac{r}{\tau}\frac{dq_1}{dt} + \frac{1}{\tau t_c}(r-1)q_1 = 0 \tag{20.4-9}$$

This is the differential equation describing a damped harmonic oscillator so that assuming a solution $q \propto e^{pt}$, we obtain

$$p^2 + \frac{r}{\tau}p + \frac{1}{\tau t_c}(r-1) = 0$$

with the solutions

$$p(\pm) = -\alpha \pm i\omega_m$$

$$\alpha = \frac{r}{2\tau}, \qquad \omega_m = \sqrt{\frac{1}{t_c\tau}(r-1) - \left(\frac{r}{2\tau}\right)^2} \tag{20.4-10}$$

$$\simeq \sqrt{\frac{1}{t_c\tau}(r-1)} \qquad \frac{1}{t_c\tau}(r-1) \gg \left(\frac{r}{2\tau}\right)^2$$

so that $q_1(t) \propto e^{-\alpha t} \sin \omega_m t$. The predicted perturbation in the power output (which is proportional to the number of photons q) is thus a damped sinusoid, with the damping rate α and the oscillation frequency ω_m increasing with excess pumping.

Although some lasers display the damped sinusoidal perturbation of intensity described above, in many other laser systems the perturbation is undamped. An example of the first is illustrated in Figure 20.17, which shows the "spiking" output of a $CaWO_4:Nd^{3+}$ laser.

Numerical Example: Relaxation Oscillation. Consider the case shown in Figure 20.17 with the following parameters:

$$\tau = 1.6 \times 10^{-4} \text{ sec}$$

$$t_c \simeq 10^{-8} \text{ sec}$$

$$r \simeq 2$$

that if we use (20.4-10) gives $T_m \equiv 2\pi/\omega_m \simeq 8 \times 10^{-6}$ sec.

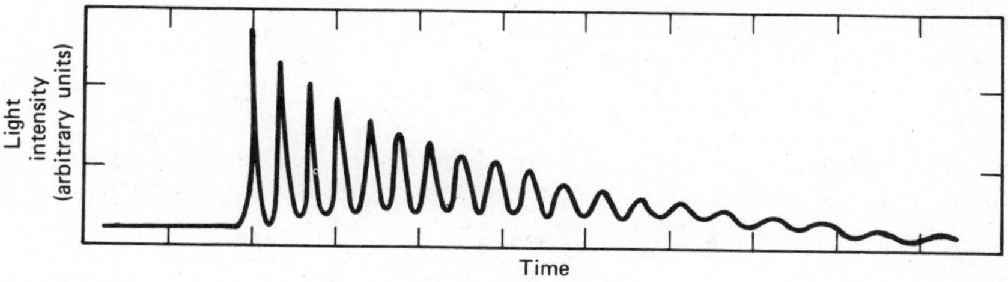

FIGURE 20.17 Intensity relaxation oscillation in a $CaWO_4 : Nd^{3+}$ laser at 1.06 μm. Horizontal scale = 20 μsec/div. *Source:* Reference 20.

The undamped relaxation oscillation observed in many cases can be understood, at least qualitatively, by considering (20.4-9). As it stands, the equation is identical in form to that describing a damped nondriven harmonic oscillator or, equivalently, a resonant *RLC* circuit. Persistent (i.e., nondamped) oscillation is possible when the "oscillator" is driven. In this case, the driving function will replace the zero on the right side of (20.4-9). One such driving mechanism may be due to time variation of the pumping rate *R*. In this case, we may take the pumping in the form

$$R = R_0 + R_1(t) \tag{20.4-11}$$

where R_0 is the average pumping and $R_1(t)$ is the deviation.

Retracing the steps leading to (20.4-6), but using (20.4-11), we find that the inversion equation is now

$$\frac{dN_1}{dt} = R_1 - R_0 B t_c N_1 - \frac{q_1}{t_c}$$

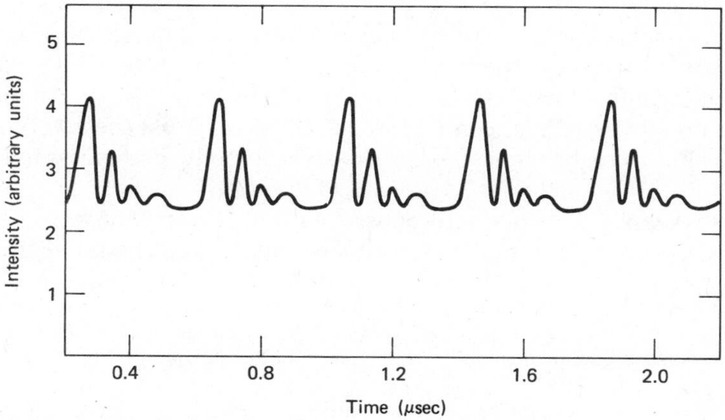

FIGURE 20.18 Intensity relaxation oscillation in a xenon 3.51-μm laser. *Source:* Reference 21.

and that (20.4-9) takes the form

$$\frac{d^2q_1}{dt^2} + \frac{r}{\tau}\frac{dq_1}{dt} + \frac{1}{\tau t_c}(r-1)q_1 = \frac{1}{\tau}(r-1)R_1 \qquad (20.4\text{-}12)$$

Taking the Fourier transform of both sides of (20.4-12), defining $Q(\omega)$ and $R(\omega)$ as the transforms of $q(t)$ and $R_1(t)$, respectively, and then solving for $Q(\omega)$ give

$$Q(\omega) = \frac{-\frac{1}{\tau}(r-1)R(\omega)}{\omega^2 - i\frac{r}{\tau}\omega - \frac{1}{\tau t_c}(r-1)} = \frac{-\frac{1}{\tau}(r-1)R(\omega)}{(\omega - \omega_m - i\alpha)(\omega + \omega_m - i\alpha)} \qquad (20.4\text{-}13)$$

$$\omega_m = \sqrt{\frac{1}{t_c\tau}(r-1) - \left(\frac{r}{2\tau}\right)^2} \cong \sqrt{\frac{1}{t_c\tau}(r-1)} \qquad \frac{1}{t_c\tau}(r-1) \gg \left(\frac{r}{2\tau}\right)^2 \qquad (20.4\text{-}14)$$

$$\alpha = \frac{r}{2\tau} \qquad (20.4\text{-}15)$$

where we notice that ω_m and α correspond to the oscillation frequency and damping rate, respectively, of the transient case as given by (20.4-10). If we assume that the spectrum $R(\omega)$ of the driving function $R(t)$ is uniform (i.e., like "white" noise) near $\omega \simeq \omega_m$, we may expect the intensity spectrum $Q(\omega)$ to have a peak near $\omega = \omega_m$ with a width $\Delta\omega \simeq 2\alpha \equiv r/\tau$. In addition, if $\Delta\omega \ll \omega_m$, we may expect the intensity fluctuation $q(t)$ as observed in the time domain to be modulated at a frequency ω_m[8] since for frequencies $\omega \simeq \omega_m$, $Q(\omega)$ is a maximum.

These conclusions are verified in experiments on different laser systems. In Figure 20.18 we show the intensity fluctuations of a xenon 3.51 μm. The corresponding intensity spectrum $Q(\omega)$ is shown in Figure 20.19. An increase in the pumping strength is seen (Figure 20.20) to lead to a spectral broadening and a shift to higher frequencies consistent with the discussion following (20.4-15).

The resonant nature of the response $Q(\omega)/R(\omega)$ is manifest in any experiment in which some laser parameter is modulated at frequencies near $\omega_m/2\pi$.

[8] To verify this statement, assume that $R(t)$ is approximated by a superposition of uncorrelated sinusoids $R(t) \propto \sum_n a_n e^{i\omega_n t}$ and using $R(\omega) \propto \int_{-\infty}^{\infty} R(t)e^{-i\omega t}\,dt$, we get $R(\omega) \propto \sum_n a_n\,\delta(\omega - \omega_n)$. From the inverse transform relation $q(t) \propto \int_{-\infty}^{\infty} Q(\omega)e^{i\omega t}\,d\omega$ and (20.4-13), we get

$$q(t) \propto \sum_n \frac{a_n e^{i\omega_n t}}{(\omega_n - \omega_m - i\alpha)(\omega_n + \omega_m - i\alpha)}$$

so that in the limit $\omega_m \gg \alpha$, $q(t)$ is a quasi-sinusoidal oscillation with a frequency ω_m.

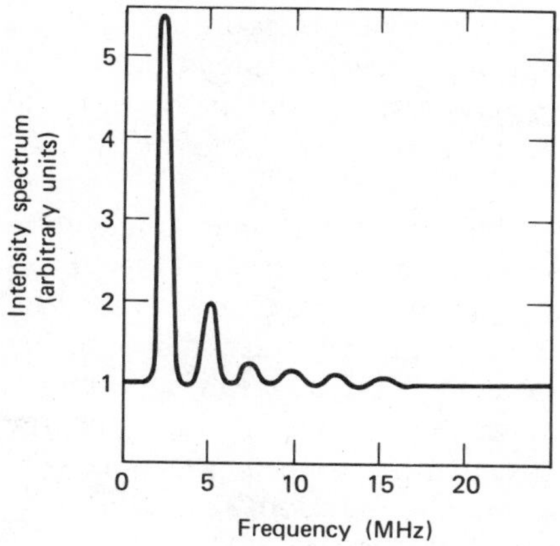

FIGURE 20.19 The intensity fluctuation spectrum of the laser output shown in Figure 20.18. *Source:* Reference 21.

Internal coupling modulation schemes of lasers (References 22 and 23) at these frequencies are usually beset by severe disortion (Reference 24).

20.5 PASSIVE MODE LOCKING

In the discussion following (20.2-9), it was pointed out that mode locking can be caused by incorporating a thin periodic shutter in the optical path. Such "shutters" (loss cells) are shown in Figures 20.11 and 20.13. In these cases, an external signal is required to activate the loss cells, and the term "active mode locking" is often used to describe the resulting laser oscillation.

The effect of a periodic gate can also be provided by the insertion of a saturable absorber in the optical path (References 25 and 26). This is a material, usually a dilute solution of an organic dye, or an appropriate molecular gas, whose absorption at the laser wavelength decreases with increasing intensity. The saturable absorber will clearly "encourage" the laser to oscillate

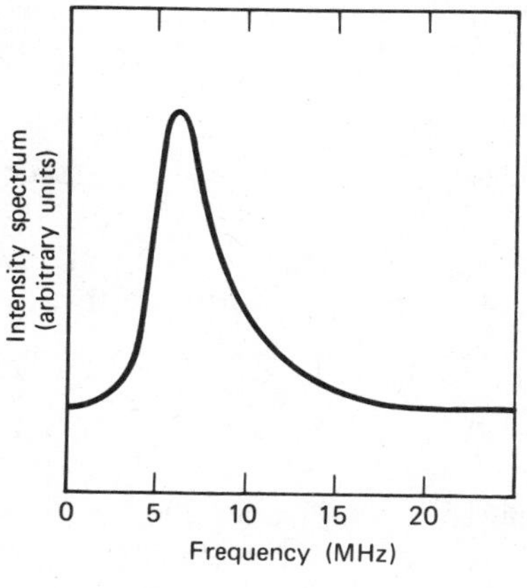

FIGURE 20.20 Same as Figure 20.19 except at increased pumping. *Source:* Reference 21.

in a pulsed fashion since this mode of oscillation will undergo smaller losses (References 27 and 28) than one in which the energy is spread more uniformly. This causes the largest of the blackbody intensity fluctuations in the optical resonator to grow, once the pumping begins, at a far larger (exponential) rate than other intensity peaks. The result is a pulsation with a period equal to the round-trip transit time $2nL/c$. This form of "passive" mode locking is especially useful in pulsed lasers where the short duration of the pumping pulse ($\tau < 10^{-3}$ sec) makes the active form of mode locking impractical.

The above arguments suggest that the presence of a saturable absorber with a recovery time of s^{-1} sec will induce mode locking, resulting in pulses with a duration $\tau \sim s^{-1}$ sec. This is due to the fact that the total energy absorbed from a pulse of a given energy becomes independent of τ once $s\tau \ll 1$, since the dye cannot "recover" during the pulse duration. This conclusion follows from an analysis of a simple model (Reference 28) in which the absorbing transition takes place in a two-level system described by

$$\frac{dN_1}{dt} = R(N_2 - N_1) + sN_2 \qquad N_a = N_1 + N_2 \tag{20.5-1}$$

Here, N_1 and N_2 are the molecular densities of the ground (absorbing) state and the excited state of the absorber. Symbol s is the spontaneous rate per molecule (sec^{-1}) for a $2 \rightarrow 1$ transition, R is the rate for a stimulated $2 \rightleftarrows 1$ transition. N_a is the total molecular density.

We assume a rectangular optical pulse containing E photons per unit area and of duration τ. The decrease in the photon flux E/τ (photons $-$ m^{-2} $-$ sec^{-1}) with distance is equal to the net number of absorptive transitions per unit time per unit volume

$$\frac{d}{dx}\left(\frac{E}{\tau}\right) = -\frac{\sigma_a E}{\tau}(N_1 - N_2) \tag{20.5-2}$$

where we used the relation $R = \sigma_a E/\tau$, σ_a being the absorption cross section (m^2) per ground-state molecule. If at $t = 0$, $N_1 = N_a$ (i.e., the molecules are all in their ground state), (20.5-1) and (20.5-2) can be solved to yield

$$-\frac{d}{dx}\left(\frac{E}{\tau}\right) = \frac{N_a\sigma_a E}{\tau\left(2\dfrac{\sigma_a E}{\tau} + s\right)}\left[s + 2\frac{\sigma_a E}{\tau}e^{-(2\sigma_a E/\tau + s)t}\right] \tag{20.5-3}$$

If we assume that the light is not greatly reduced in intensity in its passage through the cell (this will apply if the total unsaturated absorption per pass is less than, e.g., 35%, we may integrate (20.5-3) over the length L_a of the absorbing cell to obtain the total number E_a of photons per unit area lost by the pulse in passing through the cell.

$$\frac{E_a}{E} = \frac{N_a\sigma_a s\tau L_a}{2\sigma_a E + s\tau} + \frac{2N_a\sigma_a^2 E L_a}{(2\sigma_a E + s\tau)^2}\left[1 - e^{-(2\sigma_a E + s\tau)}\right] \tag{20.5-4}$$

The fractional energy absorption E_a/E given by the last equation is plotted in Figure 20.21 as a function of the pulse energy E with $s\tau$ as a parameter.

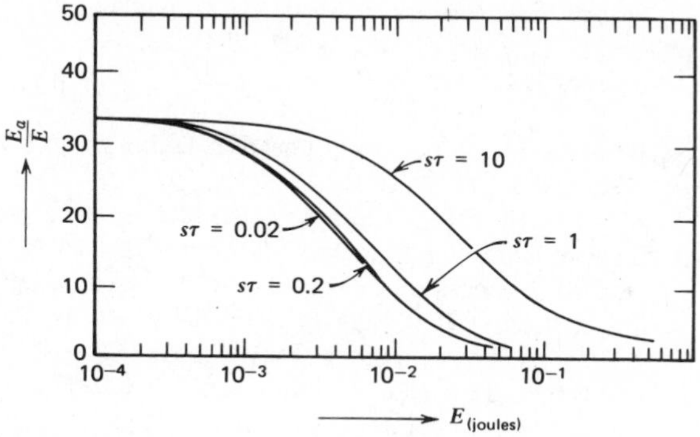

FIGURE 20.21 Fractional absorption of pulse energy in a two-level molecular system $N_a = 6 \times 10^{22}\ m^{-3}$, $L_a = 10^{-3}$ m, $\sigma_a = 0.58 \times 10^{-20}\ m^2$. The parameter $s\tau$ is equal to the ratio of the pulse duration to the molecular relaxation time. *Source:* Reference 29.

We find that pulse durations much shorter than the molecular relaxation rate, that is, $s\tau \ll 1$, do not lead to a significant reduction of losses. This causes the pulse duration to tend toward a value of $\tau \sim s^{-1}$, provided the gain linewidth $\Delta\omega_{homog} \gg s$ so that all the pulse spectral components can be amplified by the laser medium.

References

1. Hellwarth, R. W., "Control of Fluorescent Pulsations," in *Advances in Quantum Electronics,* J. R. Singer, ed. (New York: Columbia University Press, 1961), p. 334.
2. McClung, F. J. and R. W. Hellwarth, *J. Appl. Phys.* **33,** 828 (1962).
3. Hellwarth, R. W., "*Q* Modulation of Lasers," in *Lasers,* Vol. 1, A. K. Levine, ed. (New York: Marcel Dekker, 1966), p. 253.
4. Wagner, W. G. and B. A. Lengyel, "Evolution of the Giant Pulse in a Laser," *J. Appl. Phys.* **34,** 2042 (1963).
5. Mocker, H. and R. J. Collins, "Mode Competition and Self-Locking Effects in a *Q*-Switched Ruby Laser," *Appl. Phys. Letters* **7,** 270 (1965).
6. Yariv, A., "Internal Modulation in Multimode Laser Oscillators," *J. Appl. Phys.* **36,** 388 (1965).
7. DiDomenico, M., Jr., "Small Signal Analysis of Internal Modulation of lasers," *J. Appl. Phys.* **35,** 2870 (1964).
8. Hargrove, L. E., R. L. Fork, and M. A. Pollack, "Locking of He-Ne Laser Modes Induced by Synchronous Intracavity Modulation," *Appl. Phys. Letters* **5,** 4 (1964).
9. Fork, R. L., D. R. Herriott, and H. Kogelnik, "A Scanning Spherical Mirror Interferometer for Spectral Analysis of Laser Radiation," *Appl. Optics* **3,** 1471 (1964).
10. DeMaria, A. J., W. H. Glenn, M. J. Brienza, and M. E. Mack, "Picosecond Laser Pulses," *Proc. IEEE* **57,** 2 (1969).
11. DeMaria, A. J., "Mode Locking," *Electronics,* 12 (1968); see also, P. W. Smith, "Mode Locking of Lasers," *Proc. IEEE* **58,** 1342 (1970).

12. Teng, T. C., R. Gerlach, and Y. H. Pao, "Mode Locking of a 6118 Å Laser by Use of an Ne Discharge Cell," *J. Quant. Elect.* **QE-9,** 784 (1973).
13. H. Haus, "A Theory of Forced Mode Locking," *IEEE J. Quant. Elec.* **QE-11,** 323 (1975).
14. J. AuYeung, "Theory of Mode Locking of a Laser Diode in an External Cavity," *IEEE J. Quant. Elec.* **QE-17,** 398 (1981).
15. Siegman, A. E. and D. J. Kuizenga, "Simple Analytic Expressions for AM and FM Mode Locked Pulses in Homogeneous Lasers," *Appl. Phys. Letters* **14,** 181 (1969).
16. Kuizenga, D. J. and A. E. Siegman, "FM and AM Mode Locking of the Homogeneous Laser: Part I, Theory; Part II, Experiment," *J. Quant. Elect.* **QE-6,** 694 (1970).
17. Cutler, C. C., "The Regenerative Pulse Generator," *Proc. IRE* **43,** 140 (1955).
18. Smith, P. W., T. J. Bridges, and E. G. Burkhardt, "Mode Locked High Pressure CO_2 Laser," *Appl. Phys. Letters* **21,** 470 (1972).
19. For additional references on relaxation oscillation, the reader should consult: Birnbaum, G., *Optical Masers* (New York: Academic Press, 1964), p. 191; Evtuhov, V., "Pulsed Ruby Lasers" in *Lasers,* A. K. Levine, ed. (New York: Marcel Dekker, 1966), p. 76; Dunsmuir, R. J., "Theory of Relaxation Oscillation of Optical Masers," *J. Elec. Control* **10,** 453 (1961); Statz, H., and G. DeMars, *Quantum Electronics* (New York: Columbia University Press, 1960), p. 530.
20. Johnson, L. F. "Solid State Lasers," in *Lasers,* A. K. Levine, ed. (New York: Marcel Dekker, 1966), p. 174.
21. Casperson, L. and A. Yariv, "The Time Behavior and Spectra of Relaxation Oscillation in a High Gain Laser," *J. Quant. Elect.* **QE-8,** 69 (1972).
22. Gürs, K. and R. Muller, "Internal Modulation of Optical Masers," *Phys. Letters* **5,** 179 (1963).
23. Kiefer, J. E., T. A. Nussmeier, and F. E. Goodwyn, "Intracavity CdTe Modulators for CO_2 Lasers," *IEEE J. Quant. Elect.* **QE-8,** 173 (1972).
24. Yariv, A., T. Nussmeier, and J. E. Kiefer, "Frequency Response of Intracavity Laser Coupling Modulation," *IEEE J. Quant. Elect.* **QE-9,** 594 (1973).
25. Mocker, H. and R. J. Collins, "Mode Competition and Self-Locking Effects in a *Q*-Switched Ruby laser," *Appl. Phys. Letters* **7,** 270 (1965).
26. DeMaria, A. J., D. A. Stetser, and H. Heynau, "Self Mode-Locking of Lasers with Saturable Absorbers," *Appl. Phys. Letters* **8,** 174 (1966).
27. Schwarz, S. E. and T. Y. Tan, "Wave Interactions in Saturable Absorbers," *Appl. Phys. Letters* **10,** 4 (1967).
28. Garmire, E. M. and A. Yariv, "Laser Mode Locking with Saturable Absorbers," *IEEE J. Quant. Elect.* **QE-3,** 222 (1967).
29. Cirkel, H. J. and F. P. Schafer, "Passive Non-Reciprocal Element for Traveling Wave Ring Lasers," *Optics Comm.* **5,** 183 (1972).

Supplementary References

1. Von der Linde, D., "Mode Locked Lasers and Ultrashort Light Pulses," *Appl. Phys.* **2,** 281 (1973).
2. Arakawa, Y., A. Larson, J. Paslaski, and A. Yariv, "Active *Q*-Switching in GaAs/AlGaAs Multiquantum Well Laser with an Intracavity Monolithic Loss Modulator," *Appl. Phys. Letters* **48,** 561 (1986).
3. Fork, R. L., B. J. Greene, and C. V. Shank, *Phys. Letters* **38,** 671 (1981).

4. Shapiro, S. L., ed., *Ultrashort Light Pulses,* Topics in Applied Physics, Vol. 18 (Berlin-New York: Springer-Verlag, 1977).
5. Hochstrasser, R. M., W. Kaiser, and C. V. Shank, eds., *Picosecond Phenomena II,* Series in Chemical Physics, Vol. 14 (Berlin-New York: Springer-Verlag, 1980).
6. Eisenthal, K. B., R. M. Hochstrasser, W. Kaiser, and A. Laubereau, eds., *Picosecond Phenomena III,* Series in Chemical Physics, Vol. 23 (Berlin-New York: Springer-Verlag, 1982).

Problems

20.1 (a) Describe qualitatively what one may expect to see (measure) in parts A, B, C, and D of the mode-locking experiment sketched in Figure 20.11. The reader may find it useful to read first the section on photomultipliers in Reference 7, Chapter 8.

(b) What is the effect of mode locking on the intensity of the beat signal (at $\nu = c/2l$) displayed by the RF spectrum analyzer F? Assume N equal amplitude modes spaced by ν whose phases before mode locking are random.

Answer:
Mode locking increases the beat signal power by N.

20.2 Analyze the case of mode locking when the dielectric constant ε, rather than the losses (σ), is modulated at the intermode separation frequency $c/2l$.

(a) In the case of an inhomogeneous gain broadening as in Section 20.2.

(b) In the case of a homogeneous broadening as in Section 20.3.

C H A P T E R 21

Noise and Spectra of Laser Amplifiers and Oscillators

21.0 INTRODUCTION

The topic of noise in lasers is important for both basic theoretical and practical considerations. In the case of laser oscillators, measured properties, such as spectral linewidth and intensity fluctuations, are often, and especially in semiconductor lasers, the reflection of basic quantum mechanical limitations so that their understanding is a prerequisite to designing less noisy lasers. In the case of laser amplifiers, the "noisy" spontaneous emission places a lower bound on the signals that can be amplified without getting lost in the noise. This chapter will start with basic considerations and will derive the essential noise parameters and spectral features of laser amplifiers and oscillators. The topic of this chapter, which used to be the sole purview of the theoretician, has assumed recently large practical significance. The spectral purity and power fluctuations of semiconductor lasers, for example, have become the factors limiting the data transmission capacity of optical fiber links and the rotation sensing resolution of laser gyroscopes.

21.1 NOISE IN LASER AMPLIFIERS

Here, we will consider a traveling-wave laser amplifier that consists of a medium with an inverted population. The radiation to be amplified is made to make a single pass through the material. Noise radiation emitted spontaneously by the upper laser population intermingles with the "signal," and the fraction of the noise that eventually is intercepted by the detector, along with the signal, leads to a degradation of the "signal-to-noise" power ratio (SNR). It is clear from this short description that a treatment of noise in laser amplifiers must include the "spatial filtering" aspect that determines what fraction of the noise (and signal) is collected. This point was recognized in the first laser paper of Schawlow and Townes (Reference 1). The present treatment follows closely that of Reference 2.

To determine the characteristics of the noise radiated by a laser amplifier of length l, we assume the simple model shown in Figure 21.1. The laser medium with an inversion density $N_2 - N_1(g_2/g_1)$ is distributed uniformly between the planes $z = 0$ and $z = l$.

Light of frequency ν passing through this medium at an angle θ with

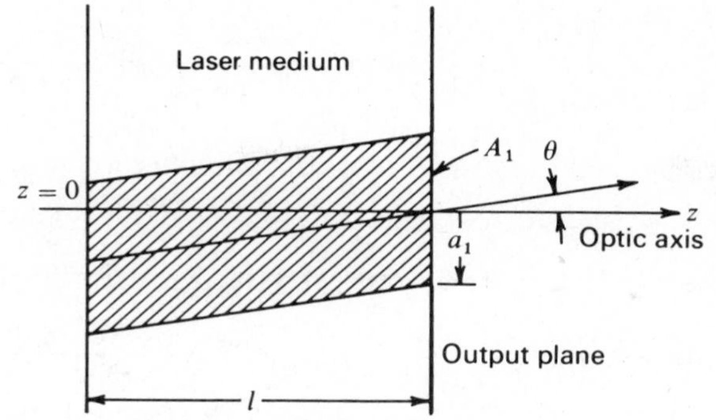

FIGURE 21.1 Amplifier model with aperture stop in output plane.

respect to the optic axis z is amplified as $e^{\gamma z/\cos\theta}$. The expression for the power gain constant γ is, according to (8.4-4),

$$\gamma(\nu) = \frac{[N_2 - N_1(g_2/g_1)]c^2 g(\nu)}{8\pi n^2 \nu^2 t_{\text{spont}}} \tag{21.1-1}$$

As shown in Figure 21.1, our amplifier model includes an absorbing screen in the output plane with a round hole of radius a_1 and area $A_1 = a_1^2\pi$ to provide for passage of the signal. In the following, we propose to determine the amount of noise due to spontaneous emission in the laser medium that is radiated through this hole.

Note that we use neither the model of a rectangular box enclosing the active medium nor the model of an enclosing waveguide. The answer is obtained with the help of fundamental formulas and some bookkeeping.

The amount of noise power emitted spontaneously by an element dV of the amplifying medium at frequencies between ν and $\nu + d\nu$ into a solid angle $d\Omega$ is

$$dN = h\nu \frac{N_2 g(\nu)}{t_{\text{spont}}} \frac{d\Omega}{4\pi} d\nu\, dV \tag{21.1-2}$$

This result can be verified by showing that $\iiint dN = N_2 h\nu V/t_{\text{spont}}$. We assume that the propagation of noise can be described in terms of geometrical optics. Noise emitted at a coordinate z and propagating at an angle θ with respect to the z axis is amplified by a factor $e^{\gamma(l-z)/\cos\theta}$ until it reaches the output hole or is intercepted by the screen. For a given θ and a polar angle ϕ, only noise originating from within the shaded cylinder shown in Figure 21.1 can escape through the output hole (into the differential element $d\Omega$ of solid angle). If we sum the contributions of all volume elements within the cylinder and take account of the proper amplification factor, we obtain the noise power $N(\theta)$ radiated through the hole A_1 at an angle θ into $d\Omega$

$$N(\theta) = A_1 \int_0^l dz \frac{dN}{dV} \cdot e^{\gamma(l-z)/\cos\theta} \tag{21.1-3}$$

The evaluation of this integral yields

$$N(\theta) = A_1 \cos\theta \, \frac{[G(\theta) - 1]h\nu N_2 g(\nu)}{\gamma t_{\text{spont}}} \frac{d\Omega}{4\pi} \, d\nu \tag{21.1-4}$$

where we have defined the gain $G(\theta)$ of the amplifier as

$$G(\theta) = e^{\gamma l/\cos\theta} \tag{21.1-5}$$

Substituting the expression for γ from (21.1-1), we arrive finally at

$$N(\theta) = 2h\nu \, d\nu \, \frac{N_2}{N_2 - N_1(g_2/g_1)} [G(\nu) - 1] \frac{A_1 n^2 \cos\theta \, d\Omega}{\lambda^2} \tag{21.1-6}$$

Equation (21.1-6) can be expected to hold for any atomic medium with populations N_2 and N_1 and not exclusively for the inverted population considered here. Consider, for example, a medium at a (positive) temperature T. To render the medium "black," we take $l = \infty$ so that it becomes perfectly absorbing. Since for an absorber $\gamma < 0$, $G = e^{\gamma l/\cos\theta} = 0$. If we use this last result and the relation $(N_1/N_2)(g_2/g_1) = e^{h\nu/kT}$ (which applies because of the thermal equilibrium condition), Equation (21.1-6) becomes

$$N(\theta) = \frac{2h\nu \, d\nu}{e^{h\nu/kT} - 1} \frac{A_1 n^2 \cos\theta \, d\Omega}{\lambda^2} \tag{21.1-7}$$

Equation (21.1-7) is the classical expression for the power radiated by a "black" surface of area A_1 and temperature T into a solid angle $d\Omega$ at θ. Equation (21.1-6) can thus be considered a generalization of the result of blackbody radiation theory to media of finite extent and nonthermal equilibrium populations.

The quantity

$$N_0 = h\nu \, d\nu \, \frac{N_2}{N_2 - N_1(g_2/g_1)} [G(\nu) - 1] \tag{21.1-8}$$

appearing in (21.1-6) can be recognized as what is usually called the amount of "noise per mode," and it can be shown that it is the amount of noise within a bandwidth $d\nu$ radiated into a solid angle that can be associated with a single mode (see Appendix 2).

The noise power emitted into a cone of very small half-apex angle θ' corresponding to a solid angle $\Omega = \pi\theta'^2$ is proportional to Ω and can be obtained from (21.1-6) as

$$N_\Omega = 2N_0 \frac{A_1 \Omega n^2}{\lambda^2} \tag{21.1-9}$$

for cases in which the angle between the cone axis and z axis is sufficiently small. The above formulas include noise contributions with random polarization. A linear polarizer can be used to reduce the noise by a factor of 2.

The derivation leading to (21.1-9) is also valid if the hole is not round, but of any other shape, as long as its area is A_1. Next, we consider the problem of the minimum value of Ω.

We take a signal beam with a Gaussian transverse field variation and a minimum beam radius (spot size) a_M. A beam of this kind is naturally obtained, as discussed in Chapter 6, from a laser with a spherical mirror resonator oscillating in the fundamental mode. The beam spread due to diffraction is described by a half-apex angle (6.6-18)

$$\theta_{\text{beam}} = \frac{\lambda}{\pi n a_M} \tag{21.1-10}$$

that corresponds to a solid angle

$$\Omega_{\text{beam}} = \pi\theta_{\text{beam}}^2 = \frac{\lambda^2}{\pi n^2 a_M^2} \tag{21.1-11}$$

To achieve the best transmission through the output aperture of the amplifier, the light beam should be so injected into the amplifier that it reaches its minimum radius a_M in the plane of the output screen as shown in Figure 21.2. If the beam radius is equal to the radius of the hole, the signal will pass through essentially undisturbed. In this case we have, from (21.1-11),

$$\Omega_{\text{beam}} = \frac{\lambda^2}{n^2 A_1} \tag{21.1-12}$$

for the solid angle occupied by the output beam.

Having obtained in (21.1-9) an expression for the noise radiated by the laser medium into a solid angle Ω, we can combine this knowledge with that of the propagation characteristics of the optical beam and find the amount of noise added to the signal.

Assume that the signal is intercepted by a detector placed along the z axis far enough from the detector so that "far-field" conditions apply. In order to intercept substantially all of the signal, the detector must subtend, according to (21.1-12), a solid angle $\Omega = \Omega_{\text{beam}} = \lambda^2/n^2A_1$ at the amplifier. The amount of noise detected, along with the signal, is then, from (21.1-9),

$$N = N_0$$

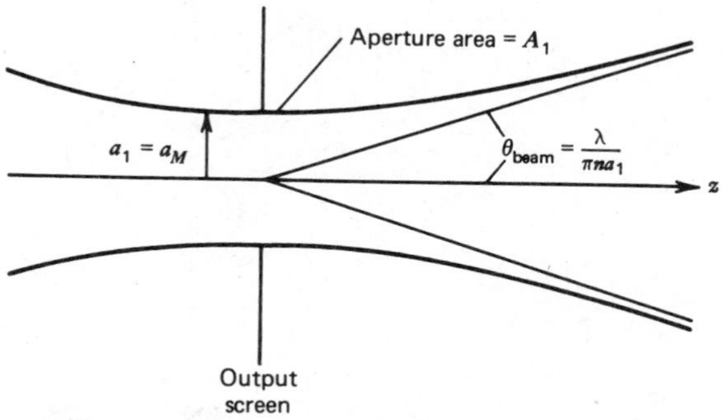

FIGURE 21.2 Gaussian beam with minimum in the output plane of the amplifier.

where it is assumed that half of the noise power is removed by using a linear polarizer in front of the detector.

We have just established the fact that N_0 is the minimum noise power received by a detector subject to the constraint that it be large enough to intercept (essentially) all of the power in a diffraction-limited Gaussian optical beam. This is accomplished by choosing the screen position and the diameter of its aperture as in Figure 21.2. Also, the detector receiving solid angle is chosen equal to λ^2/n^2A_1 where A_1 is the aperture area. If we choose to refer the minimum noise power N_0 to the amplifier input, we get

$$N_{\text{input}} = \frac{N_0}{G} = h\nu \, d\nu \frac{N_2}{N_2 - N_1(g_2/g_1)} \frac{G-1}{G} \tag{21.1-13}$$

One important conclusion resulting from (21.1-13) is that (all other factors being equal) a three-level laser where $N_1 \approx N_2$ is "noisier" than a four-level laser (where $N_1 \ll N_2$) by a factor $N_2/[N_2 - N_1(g_2/g_1)]$. As an example, in the case of ruby with the data of Section 10.2 this factor is ~50. Physically, this factor reflects the increase in the population N_2 of the upper level of a three-level laser relative to a four-level one (having the same gain) and the fact that the noise power is proportional to N_2.

Far-field conditions do not always obtain in the laboratory, but if we insert a lens behind the output hole, the far field is projected into the focal plane of the lens. A screen in the focal plane can be used to select radiation emitted into a given solid angle, as shown in Figure 21.3. A hole of radius a_2 in this screen accepts radiation that, in the far field, would have occupied a half-apex angle θ given by

$$\theta = \tan^{-1}\frac{a_2}{f} \approx \frac{a_2}{f} \tag{21.1-14}$$

where f is the focal length of the lens. To reduce the noise power passing

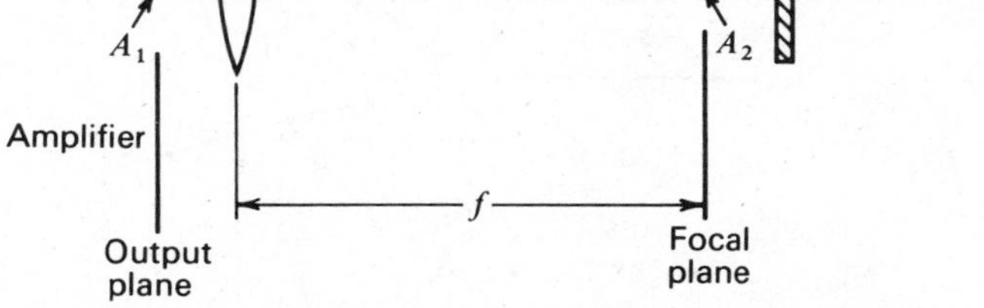

FIGURE 21.3 Noise reduction structure with iris in focal plane.

through the hole A_2 to N_0, we need $\theta = \lambda/\pi n a_1$ so that

$$a_2 = \frac{f\lambda}{\pi n a_1} \tag{21.1-15}$$

This is also the value that the radius of the signal beam assumes in the focal plane. From (21.1-15), we derive the relation

$$\frac{n^2 A_1 A_2}{f^2 \lambda^2} = 1 \tag{21.1-16}$$

between λ, f, A_1, and the area $A_2 = \pi a_2^2$ of the hole in the focal plane, which has to be satisfied for a signal-to-noise ratio of about S_0/N_0 where S_0 is the input signal power.

To arrive at this estimate, we have assumed that a beam with a Gaussian field distribution can pass undisturbed through an aperture stop with a radius that is equal to the radius (spot size) of the beam. Actually, the stop intercepts 13.5% of the incident beam power, as is easily computed. Furthermore, we have neglected all diffraction effects that will result from chopping off the smooth field distribution of the signal beam.

Actual measurements of the noise emitted by laser amplifiers are discussed in References 3 and 4.

Noise due to the Particle Nature of Light

In the discussion just concluded, we found that the amount of noise power that intermingles at the output of a laser amplifier with the signal and cannot be separated from it is

$$N_0 = h\nu\, d\nu \frac{N_2}{N_2 - N_1(g_2/g_1)} [G(\nu) - 1]$$

in a four-level laser ($N_1 \ll N_2$) with high gain $N_0 \approx h\nu\, d\nu G(\nu)$. A signal input power giving rise to the same amount of output is

$$P_{\min} = \frac{N_0}{G(\nu)} = h\nu\, d\nu \tag{21.1-17}$$

and is used as a measure of the minimum power that can be detected.[1] The origin of this power is, as shown above, the spontaneous emission of the laser transition.

Now assume that the task of measuring the signal power will be performed not by first amplifying the signal with a laser amplifier, but instead by counting the number of photoelectrons emitted from a photoemissive surface on which the signal is incident. The power P of the optical wave can be written as

$$P = \bar{N} h\nu \tag{21.1-18}$$

[1] The same power is often referred to as the noise equivalent power (NEP) and is defined as the input signal giving rise to a signal to noise power ratio of unity at the output.

where $\bar{N}$ is the average number of photons arriving at the photocathode per second. Next, assume a hypothetical noiseless photoemissive surface in which exactly one electron is produced for each η^{-1} incident photons. The measurement of P is performed by counting the number of electrons produced during an observation period T and then averaging the result over a large number of similar observations.

The average number of electrons emitted per observation period T is

$$\bar{N}_e = \bar{N}T\eta \tag{21.1-19}$$

that, if we assume perfect randomness in the arrival, is equal to the mean-square fluctuation[2]

$$\overline{(\Delta N_e)} \equiv \overline{(N_e - \bar{N}_e)^2} = \bar{N}T\eta$$

Taking the minimum detectable number of quanta as that for which the root-mean-square (rms) fluctuation equals the average value, we get

$$(\bar{N}_{\min} T\eta)^{1/2} = \bar{N}_{\min} T\eta$$

or

$$\bar{N}_{\min} = \frac{1}{T\eta} \tag{21.1-20}$$

If we convert the last result to power by multiplying it by $h\nu$ and recall that $T^{-1} \simeq d\nu$, where $d\nu$ is the bandwidth of the system, we get

$$P_{\min} = \frac{h\nu\, d\nu}{\eta} \tag{21.1-21}$$

that, for $\eta = 1$, is in agreement with (21.1-17).

A detection limit identical to (21.1-21) is also obtained when we consider the noise origin as being that of the shot noise of the photo emitted electrons (Reference 5).

[2] This follows from the assumption that the photon arrival is perfectly random, so the probability of having N photons arriving in a given time interval is given by the Poisson law

$$p(N) = (\bar{N})^N e^{-N}/N!$$

The mean-square fluctuation is given by

$$\overline{(\Delta N)^2} = \sum_{N=0}^{\infty} p(N)(N - \bar{N})^2 = \bar{N} \tag{21.1-19a}$$

where

$$\bar{N} = \sum_{0}^{\infty} Np(N)$$

is the average N. Relation (21.1-19a) holds, for example, for the coherent field ψ_c (5.8-12) which describes the output of lasers.

21.2 SPONTANEOUS EMISSION NOISE IN LASER OSCILLATORS

Another type of noise that plays an important role in quantum electronics is that of spontaneous emission in laser oscillators. According to Section 9.2, we can represent a laser oscillator by an RLC circuit, as shown in Figure 21.4. The presence of the laser medium with negative loss (i.e., gain) is accounted for by including a negative conductance $-G_m$, whereas the ordinary loss mechanisms described in Section 7.4 are represented by the positive conductance G_0. The noise generator associated with the losses G_0 is

$$\overline{i_N^2} = \frac{4\hbar\omega G_0(\Delta\omega/2\pi)}{e^{\hbar\omega/kT} - 1} \tag{21.2-1}$$

and it accounts for the thermal Johnson noise (References 6 and 7) in a bandwidth $\Delta\omega$. T is the actual temperature of the losses. Spontaneous emission is represented by a similar expression[3]

$$\overline{(i_N^2)}_{\substack{\text{spont.}\\ \text{emission}}} = \frac{4\hbar\omega(-G_m)(\Delta\omega/2\pi)}{e^{\hbar\omega/kT_m} - 1} \tag{21.2-2}$$

where the term $(-G_m)$ represents negative losses and T_m is a temperature determined by the population ratio according to

$$\left(\frac{g_1}{g_2}\right)\frac{N_2}{N_1} = e^{-\hbar\omega/kT_m} \tag{21.2-3}$$

Since in an inverted (amplifying) medium $N_2 > (g_2/g_1)N_1$, $T_m < 0$ and $\overline{(i_N^2)}$ in (21.2-2) is positive definite. The justification of (21.2-2) as a representation of spontaneous emission noise is provided by the discussion preceding (21.1-7). Here, we may note that since $G_m \propto N_2 - (g_2/g_1)N_1$, $\overline{(i_N^2)}$ in (21.2-2) can be written, if we use (21.2-3), as[4]

$$\overline{(i_N^2)}_{\substack{\text{spont.}\\ \text{emission}}} \propto \frac{-4\hbar\omega\,\Delta\omega\left(N_2 - \dfrac{g_2}{g_1}N_1\right)}{\left(\dfrac{g_2}{g_1}\right)(N_1/N_2) - 1} = 4\hbar\omega\Delta\omega N_2 \tag{21.2-4}$$

and is thus proportional to N_2. This makes sense, since spontaneous emission power is due to $2 \rightarrow 1$ transitions and should consequently be proportional to N_2.

Returning to the equivalent circuit, we see that its quality factor Q is

[3] The 2π factor appearing in the denominators of $\overline{i_N^2}$ is due to the fact that here we use $\overline{i_N^2}(\omega)$ instead of $\overline{i_N^2}(\nu)$ with

$$\overline{i_N^2}(\omega)\Delta\omega = \overline{i_N^2}(\nu)\Delta\nu$$

[4] The proportionality of G_m to $N_2 - (g_2/g_1)N_1$ can be justified by noting that in the equivalent circuit (Figure 21.4), the stimulated emission power is given by V^2G_m where V is the voltage. If we use the field approach, this power is proportional to $E^2[N_2 - (g_2/g_1)N_1]$ where E is the field amplitude. Since V is proportional to E, G_m is proportional to $N_2 - (g_2/g_1)N_1$.

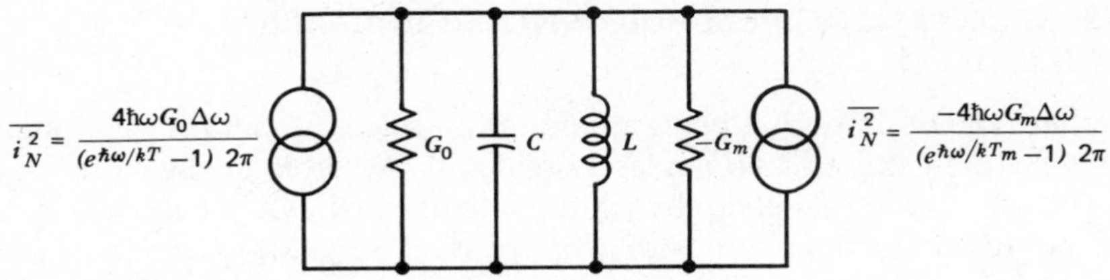

FIGURE 21.4 Equivalent circuit of a laser oscillator.

given by

$$Q^{-1} = \frac{G_0 - G_m}{\omega_0 C} = \frac{1}{Q_0} - \frac{1}{Q_m} \tag{21.2-5}$$

where $\omega_0^2 = (LC)^{-1}$. The circuit impedance is

$$\begin{aligned} Z(\omega) &= \frac{1}{(G_0 - G_m) + (1/i\omega L) + i\omega C} \\ &= \frac{i\omega}{C} \frac{1}{(i\omega\omega_0/Q) + (\omega_0^2 - \omega^2)} \end{aligned} \tag{21.2-6}$$

so the voltage across this impedance due to a current source with a complex amplitude $I(\omega)$ is

$$V(\omega) = \frac{i}{C} \frac{I(\omega)}{[(\omega_0^2 - \omega^2)/\omega] + (i\omega_0/Q)} \tag{21.2-7}$$

that, near $\omega = \omega_0$, becomes

$$\overline{|V(\omega)|^2} = \frac{1}{4C^2} \frac{\overline{|I(\omega)|^2}}{(\omega_0 - \omega)^2 + (\omega_0^2/4Q^2)} \tag{21.2-8}$$

The current sources driving the resonant circuit are those shown in Figure 21.4; since they are not correlated, we may take $|I(\omega)|^2$ as the sum of their mean-square values

$$\overline{|I(\omega)|^2} = 4\hbar\omega \left[\frac{G_m N_2}{N_2 - (g_2/g_1)N_1} + \frac{G_0}{e^{\hbar\omega/kT} - 1} \right] \frac{d\omega}{2\pi} \tag{21.2-9}$$

where in the first term inside the square brackets, we used (21.2-3). In the optical region, $\lambda = 1\ \mu$m, for example, and for $T = 300°$K, we have $\hbar\omega/kT \simeq 50$; thus, since near oscillation $G_m \simeq G_0$, we may neglect the thermal (Johnson) noise term in (21.2-9), thereby obtaining

$$\overline{|V(\omega)|^2}_{\omega \simeq \omega_0} = \frac{\hbar G_m}{2\pi C^2} \left[\frac{N_2}{N_2 - (g_2/g_1)N_1} \right] \left[\frac{\omega\, d\omega}{(\omega_0 - \omega)^2 + (\omega_0^2/4Q^2)} \right] \tag{21.2-10}$$

Equation (21.2-10) represents the spectral distribution of the power output. It should apply to operation below threshold as well as above threshold. In both cases, the full spectral width of the mode field is

$$\Delta\nu_{\text{laser}} = \frac{\nu_0}{Q} \tag{21.2-11}$$

Using (21.2-5), we rewrite (21.2-11) as

$$\Delta\nu_{\text{laser}} = \frac{1}{2\pi}\left(\frac{G_0 - G_m}{C}\right) = \frac{G_0}{2\pi C}\left(1 - \frac{G_m}{G_0}\right) \tag{21.2-12}$$

The negative conductance G_m is the one exercised by the field; that is, it must be taken as the saturated value so that, if we follow (8.7-4), it can be written as

$$G_m = \frac{G_{m0}}{1 + \dfrac{I}{I_S}} \tag{21.2-13}$$

for a homogeneously broadened laser medium. I is the intensity; G_{m0} is the unsaturated value of G_m.

We will now consider separately the two cases: below threshold and above threshold.

Linewidth Below Threshold

Here, $I \ll I_S$ and $G_m \approx G_{m0}$. Using (21.2-12) and the expression (7.4-6) for the passive ($G_m = 0$) resonator linewidth

$$\Delta\nu_{1/2} = \frac{\nu_0}{Q_c} = \frac{1}{2\pi}\frac{G_0}{C} \tag{21.2-14}$$

leads to

$$\Delta\nu_{\substack{\text{mode below threshold} \\ (G_{m0} < G_0)}} = \Delta\nu_{1/2}\left(1 - \frac{G_{m0}}{G_0}\right) \tag{21.2-15}$$

This shows how $\Delta\nu_{\text{mode}}$ narrows from the passive ($G_{m0} = 0$) cavity value of $\Delta\nu_{1/2}$ as threshold ($G_{m0} = G_0$) is approached.

Laser Linewidth Above Threshold

The linewidth above threshold is still given by (21.2-11) or, equivalently, by (21.2-12). In practice, however, this expression is not useful since G_m is nearly equal to G_0 and $\Delta\nu_{\text{laser}}$ is thus proportional [see (21.2-5)] to the difference of two nearly equal quantities, neither of which is known with high enough accuracy. We can avoid this difficulty by showing that Q is related to the laser power output, and thus $\Delta\nu_{\text{laser}}$ may be expressed in terms of the power.

The total emitted power above threshold is, according to (21.2-10),

$$\begin{aligned} P_e &= G_0 \int_0^\infty \frac{\overline{|V(\omega)|^2}}{d\omega}\, d\omega \\ &= \frac{\hbar G_m G_0}{2\pi C^2}\left(\frac{N_{2t}}{N_{2t} - N_{1t}\dfrac{g_2}{g_1}}\right)\int_0^\infty \frac{\omega\, d\omega}{(\omega_0 - \omega)^2 + (\omega_0/2Q)^2} \end{aligned} \tag{21.2-16}$$

Since the integrand peaks sharply near $\omega \simeq \omega_0$, we may replace ω in the numerator of (21.2-16) by ω_0 and, after integration, obtain

$$P_e = \frac{\hbar G_m G_0 Q}{C^2}\left[\frac{N_{2t}}{N_{2t} - (g_2/g_1)N_{1t}}\right] \tag{21.2-17}$$

which is the desired result linking P to Q. In a laser oscillator, the gain very nearly equals the loss or, in our notation, $G_m \simeq G_0$. Using this result in (21.2-17), we obtain

$$Q = \frac{C^2}{\hbar G_0^2}\left[\frac{N_{2t} - (g_2/g_1)N_{1t}}{N_{2t}}\right] P_e$$

that, when substituted in (21.2-11), yields

$$\Delta\nu_{\text{laser}} = \frac{2\pi h\nu_0(\Delta\nu_{1/2})^2}{P_e}\left[\frac{N_{2t}}{N_{2t} - (g_2/g_1)N_{1t}}\right] \tag{21.2-18}$$

where $\Delta\nu_{1/2}$ is the full width of the passive cavity resonance as given by (21.2-14).

Numerical Example. Consider a He–Ne laser oscillator with the following characteristics:

$$\nu = 4.74 \times 10^{14}\ \text{Hz} \qquad (\lambda = 6328\ \text{Å})$$

$$l = 30\ \text{cm}$$

$$\alpha l(\text{loss}) = 1\%\ \text{per pass}$$

$$P_e = 1\ \text{mW}$$

$$N_2 \gg N_1$$

$$n = 1$$

These numbers are typical of low-power laboratory He-Ne lasers. From (7.4-6), we get

$$\Delta\nu_{1/2} = \frac{1}{2\pi t_c} = \frac{c\alpha l}{2\pi l} \simeq 1.5 \times 10^6$$

Using the foregoing data in (21.2-18) gives

$$\Delta\nu_{\text{laser}} \simeq 1.5 \times 10^{-3}\ \text{Hz}$$

for the spectral width of the laser output. We must emphasize, however, that $\Delta\nu$ as given by (21.2-18) represents a theoretical limit and does not necessarily correspond to the value commonly observed in the laboratory. The output of operational lasers is broadened mostly by thermal and acoustic fluctuations in the optical resonator length, which cause the resonance frequencies to shift about rapidly. An experimental determination of the limiting $\Delta\nu_{\text{laser}}$ requires great care in acoustic isolation and thermal stabilization; see References 8 and 9. An observed value of $\Delta\nu \sim 10^3$ Hz reported in Reference 8 in a 1-mW He-Ne laser is still limited by vibrations and thermal fluctuations.

The expression most often given for $\Delta\nu_{\text{laser}}$ is in the form of (21.2-18). It may be interesting, however, to express P_e in (21.2-18) in terms of the pumping parameters. We use (9.3-26), the correspondence,

$$\frac{g_0}{L_i + T} = \frac{G_{m0}}{G_0}$$

as well as footnote 2, Chapter 9, for ΔN_t. The result is

$$\Delta\nu_{\text{laser}} = \frac{(\Delta\nu_{1/2}/\Delta\nu)\phi c\lambda^2 \left[\dfrac{N_{2t}}{N_{2t} - N_{1t}(g_2/g_1)}\right]}{8\pi V_m n^3 \left(\dfrac{G_{m0}}{G_0} - 1\right)} \tag{21.2-19}$$

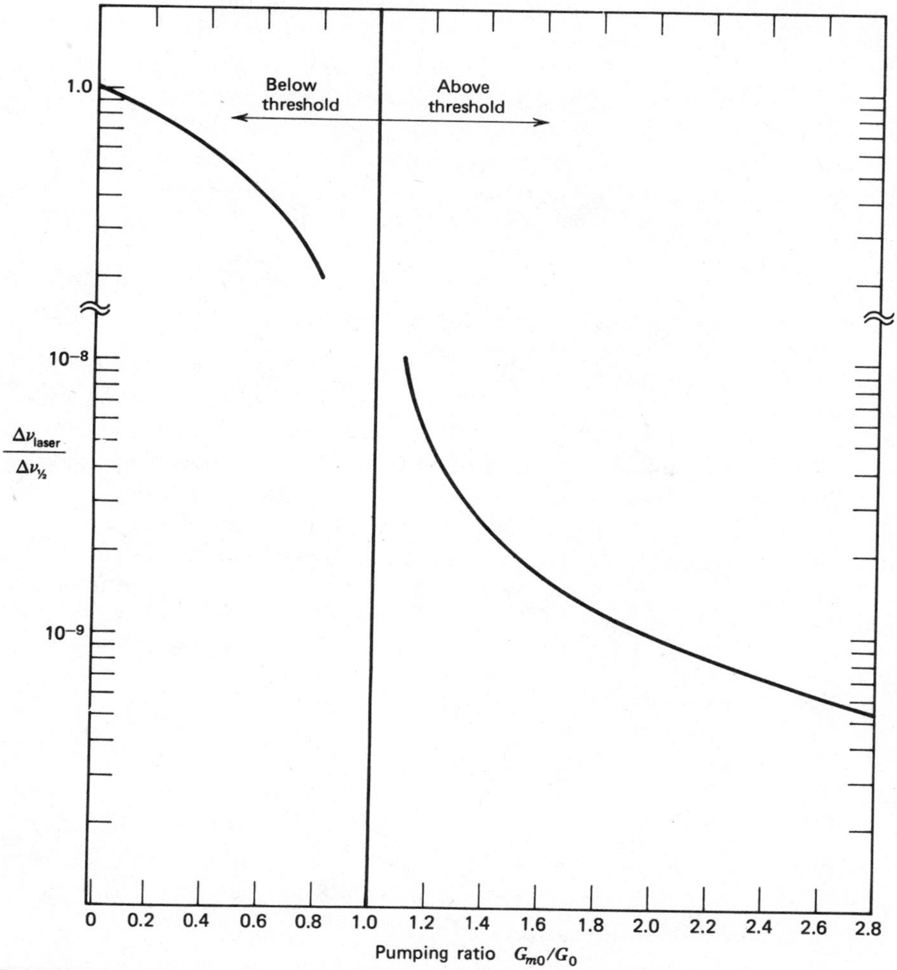

FIGURE 21.5 The laser mode linewidth below threshold [Eq. (21.2-15)] and above threshold [Eq. (21.2-19)]. The data used in the plot correspond to the He-Ne laser example of Section 21.2. Note the break in the ordinate scale.

where we recall $\phi \equiv t_2/t_{\text{spont}}$, $\Delta\nu_{1/2} = (2\pi t_c)^{-1}$, and $\Delta\nu$ is the linewidth of the gain transition.

The functional dependence of the linewidth both above (21.2-19) and below (21.2-15) threshold on pumping is contained in the factor $(G_{m0}/G_0 - 1)$. The big difference between the two regimes, however, is in the numerical factor. If we use the above cited data of the He-Ne laser and for a mode volume of 2 cm^3, $\Delta\nu = 1.5 \times 10^9$ Hz, (21.2-19) becomes

$$\Delta\nu_{\substack{\text{laser}\\(G_{m0}>G_0)}} \approx \Delta\nu_{1/2} \frac{10^{-9}}{\left(\dfrac{G_{m0}}{G_0} - 1\right)} \left(\frac{N_{2t}}{N_{2t} - N_{1t}\dfrac{g_2}{g_1}}\right) \tag{21.2-20}$$

Comparing the last result to (21.2-15), we conclude that typical laser linewidths above threshold are smaller by a factor of $10^{-8} - 10^{-9}$ than below threshold. This difference is intimately related to the big difference in the power output as indicated in Figure 9.7.

The variation of mode linewidth below threshold ($G_{m0} < G_0$) and above it is illustrated in Figure 21.5. The numerical data are based on the above example of a He-Ne 0.6328-μm laser.

Figure 21.6 shows the output spectrum of a $Pb_{0.88}Sn_{0.12}Te$ injection laser at 10.6 μm (Reference 10). The narrowing of the spectrum from $\Delta\nu_{\text{laser}} = 1.74$ MHz to $\Delta\nu_{\text{laser}} = 0.75$ MHz is consistent with the inverse dependence on P_e predicted by Eq. (21.2-18).

21.3 SOME MATHEMATICAL BACKGROUND

The discussion just concluded is too simple to explain some of the more subtle aspects of the laser spectrum especially the phase-amplitude coupling which takes place in semiconductor lasers. Its main virtue, and the reason for its inclusion, are its simplicity and the insight it affords to the concept of a laser as a noise-driven resonant circuit.

In the following, we will expand on this idea in a more realistic model that takes into account the dynamic aspects of the saturation behavior of the laser medium. Before starting, we will list some of the main mathematical tools that will be employed.

The Fourier Integral Relationships

$$F(\omega) = \frac{1}{2\pi}\int_{-\infty}^{\infty} f(t)e^{-i\omega t}\,dt \tag{21.3-1}$$

$$f(t) = \int_{-\infty}^{\infty} F(\omega)e^{i\omega t}\,d\omega \tag{21.3-2}$$

The Autocorrelation Function of $f(t)$

$$\phi_f(\tau) = \langle f(t+\tau)f^*(t)\rangle = \lim_{T\to\infty}\frac{1}{2T}\int_{-T}^{T} f(t+\tau)f^*(t)\,dt \tag{21.3-3}$$

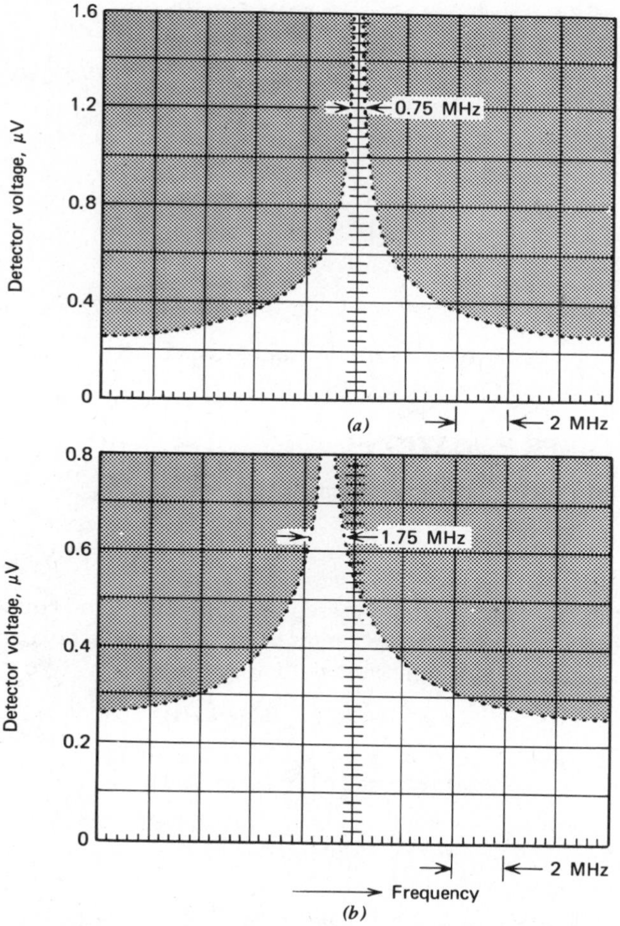

FIGURE 21.6 Spectrum analyzer display of the beat note between a low-power diode laser mode and the $P(16)$ transition of a CO_2 gas laser, corresponding to diode laser current of 865 mA in (a) and 845 mA in (b). Center frequency is 92 MHz. The dotted curves correspond to a Lorentzian lineshape, with the indicated half-power linewidths. *Source:* Reference 10.

where $\langle\ \rangle$ indicates an ensemble average and $f(t)$ is assumed to be a stationary function.

The Spectral Density Function

$$W_f(\omega) = \frac{1}{\pi}\int_{-\infty}^{\infty} \phi_f(\tau)e^{-i\omega\tau}\,d\tau \qquad (21.3\text{-}4)$$

$W_f(\omega)\,d\omega$ is the "power" of $f(t)$ due to frequencies between ω and ω +

$d\omega$. This interpretation is based on the relationship

$$\lim_{T\to\infty} \frac{1}{2T}\int_{-T}^{T} |f(t)|^2\, dt = \frac{1}{2}\int_{-\infty}^{\infty} W_f(\omega)\, d\omega = \frac{\pi}{T}\int_{-\infty}^{\infty} |F(\omega)|^2\, d\omega \tag{21.3-5}$$

It follows from (21.3-4) that

$$\phi_f(\tau) = \frac{1}{2}\int_{-\infty}^{\infty} W_f(\omega) e^{i\omega\tau}\, d\omega \tag{21.3-6}$$

and from (21.3-3),

$$\phi_f(\tau) = \phi_f^*(-\tau) \tag{21.3-7}$$

Equation (21.3-4) is known as the Wiener–Kintchine theorem.

21.4 THE LASER EQUATIONS

The optical electric field inside the laser resonator obeys

$$\nabla^2\mathbf{e}(\mathbf{r}, t) - \mu\sigma\frac{\partial \mathbf{e}}{\partial t} - \mu\varepsilon\frac{\partial^2 \mathbf{e}}{\partial t^2} = \mu\frac{\partial^2}{\partial t^2}[\mathbf{P} + \mathbf{p}] \tag{21.4-1}$$

where σ is the medium conductivity representing losses and coupling through mirrors. $\boldsymbol{P}$ = polarization due to the laser transition; $\boldsymbol{p}$ = polarization source representing spontaneous emission = "Langevin source." We expand

$$\begin{aligned} \mathbf{e}(\mathbf{r}, t) &= \text{Re}\left[\sum_m E_m(t)\mathbf{e}_m(\mathbf{r})\right] \\ \mathbf{P}(\mathbf{r}, t) &= \text{Re}\left[\sum_m P_m(t)\mathbf{e}_m(\mathbf{r})\right] \\ \mathbf{p}(\mathbf{r}, t) &= \text{Re}\left[\sum_m p_m(t)\mathbf{e}_m(\mathbf{r})\right] \end{aligned} \tag{21.4-2}$$

where $\mathbf{e}_m(\mathbf{r})$, the spatial mode function, satisfies

$$\nabla^2\mathbf{e}_m(\mathbf{r}) + \omega_m^2\mu\varepsilon\mathbf{e}_m(\mathbf{r}) = 0 \tag{21.4-3}$$

plus appropriate boundary conditions of the resonator. The quantity $E_n(t)\mathbf{e}_n(\mathbf{r})$ is thus the (complex) analytic signal of the (real) electric field of the nth mode. The spatial mode function $\mathbf{e}_n(\mathbf{r})$ is orthonormalizable

$$\int_{\text{mode volume}} \mathbf{e}_n \cdot \mathbf{e}_m\, dv = \delta_{nm} V \tag{21.4-3a}$$

where V is the volume of the optical resonator. Substituting (21.4-2) into (21.4-1), taking advantage of (21.4-3), then taking the inner (dot) product with $\mathbf{e}_n^*$ and using (21.4-3a) result in

$$\ddot{E}_n + \frac{1}{\tau_p}\dot{E}_n + \omega_n^2 E_n = -\frac{1}{\varepsilon}(\ddot{P}_n + \ddot{p}_n) \tag{21.4-4}$$

where the dots denote differentiation with respect to time. ω_n is thus the resonance frequency of the nth mode in a lossless and non driven resonator.

$\tau_p = \varepsilon/\sigma$ is the photon lifetime in the passive resonator, that is, without the atomic medium. We will relate P_n to E_n by (Also—See prob. 1)

$$P_n(t) = \varepsilon_0[X^{(1)} + X^{(3)}|E_n|^2]\, E_n(t) \tag{21.4-5}$$

The complex coefficients $X^{(1)}$, $X^{(3)}$ are related to the optical susceptibilities $\chi^{(1)}$ and $\chi^{(3)}$ introduced in Chapter 18 by[5]

$$X^{(1)} = \chi^{(1)} \int_{\text{atomic medium}} (\mathbf{e}_n \cdot \mathbf{e}_n)\, dV \approx \chi^{(1)} \frac{l}{L}$$

$$\varepsilon_0 X^{(3)} = \chi^{(3)} \int_{\text{atomic medium}} (\mathbf{e}_n \cdot \mathbf{e}_n)^2\, dV \approx \chi^{(3)} \frac{3l}{2V}$$

where the approximate equalities hold for a plane wave resonator of length L, a crystal length l, and a mode volume V. Next, we express the optical field as

$$E_n = [A_0 + \delta(t)]e^{i[\omega_m t + \phi(t)]} \tag{21.4-6}$$

and the Langevin source as

$$\frac{-\ddot{p}_n}{\varepsilon} = \Delta(t)e^{i[\omega_m t + \phi(t)]} \tag{21.4-7}$$

A_0 is the average field amplitude, $\delta(t)$ the (real) amplitude deviation, $\phi(t)$ the instantaneous phase, and $\Delta(t)$ a random function representing the spontaneous emission. ω_m is the laser (average) frequency. We also assume

$$\delta(t) \ll A_0, \qquad \langle\Delta(t)\rangle = \langle\phi(t)\rangle = \langle\delta(t)\rangle = 0$$

where $\langle\ \rangle$ stands for time or ensemble averaging. We also take the time variation of $\delta(t)$, $\phi(t)$, and $\Delta(t)$ to be slow compared to the optical term $\exp(i\omega_m t)$. Substitution of (21.4-6,7) in (21.4-5) leads to

$$\begin{aligned} &2i\omega_m(\dot{\delta} + iA_0\dot{\phi}) + \frac{3A_0^2X^{(3)}}{n^2}(2i\omega_m\dot{\delta} - \omega_m^2\delta) \\ &+ \left[\omega_n^2 - \omega_m^2 + i\frac{\omega_m}{\tau_p} - \left(X^{(1)} + A_0^2X^{(3)}\right)\frac{\omega_m^2}{n^2}\right] A_0 = \Delta(t) \end{aligned} \tag{21.4-8}$$

where $\varepsilon \equiv \varepsilon_0 n^2$ so that n is the optical index of refraction and products of small terms, that is, $\dot{\delta}\dot{\phi}$ have been neglected. We first equate the sum of all the time-independent terms (21.4-8) to zero, which results in the steady-state solutions

$$A_0^2 = \frac{X_i^{(1)} - \dfrac{n^2}{\omega_m\tau_p}}{-X_i^{(3)}} \tag{21.4-9}$$

Equation (21.4-9) describes the dependence of the average power output ($\propto A_0^2$) on the unsaturated gain term $X_i^{(1)}$ and the saturation parameter $X_i^{(3)}$.

[5] $\chi^{(1)}$ and $\chi^{(3)}$ are the first order and third order complex optical susceptibilities, respectively, of the gain medium. $\chi^{(3)}|E|^2$ represents the effect of saturation.

$$\omega_m^2 = \frac{\omega_n^2}{1 + \left(\frac{X_r^{(1)} + X_r^{(3)} A_0^2}{n^2}\right)} \qquad (21.4\text{-}10)$$

where we took

$$X^{(1,3)} = X_r^{(1,3)} + iX_i^{(1,3)}$$

The remaining terms in (21.4-8) are grouped into real and imaginary terms leading to

$$(1 + 2A_0^2 X_r^{(3)}/n^2)\dot{\delta} - \frac{A_0^2 \omega_m X_i^{(3)}}{n^2} \delta = \frac{\Delta_i(t)}{2\omega_m} \qquad (21.4\text{-}11)$$

$$A_0 \dot{\phi} + \frac{3X_i^{(3)} A_0^2}{n^2} \dot{\delta} + \left(\frac{3}{2}\right) \frac{\omega_m X_r^{(3)} A_0^2}{n^2} \delta = -\frac{\Delta_r(t)}{2\omega_m} \qquad (21.4\text{-}12)$$

where

$$\Delta(t) = \Delta_r(t) + i\Delta_i(t) \qquad (21.4\text{-}13)$$

and use has been made of (21.4-9,10). Equations (21.4-11,12) are our basic results. The term $2X_i^{(3)} A_0^2 \dot{\delta}/n^2$ is very small compared to the other terms and is usually neglected. We note that the term with $X_r^{(3)}$ *couples amplitude fluctuation* (δ) *to phase fluctuation* ($\dot{\phi}$).

21.5 THE LASER SPECTRA

The differential equations (21.4-11,12) will be solved to obtain spectra of some laser field parameters. Of special interest are the spectra (spectral density functions) of the laser field instantaneous $E(t)$, the instantaneous frequency $\omega_m + \dot{\phi}(t)$, and the instantaneous power $[A_0 + \delta(t)]^2$. The various spectra and a method for obtaining them experimentally are illustrated in Figure 21.7.

We start with (21.4-11) and using the inequality $A_0^2 X_r^{(3)}/n^2 \ll 1$, valid for all reasonable fields A_0, obtain

$$\dot{\delta} + \omega_1 \delta = \frac{\Delta_i(t)}{2\omega_m} \qquad (21.5\text{-}1)$$

$$(\omega_1 \equiv -X_i^{(3)} A_0^2 \omega_m / n^2) \qquad (21.5\text{-}2)$$

whose Laplace transform is

$$-\delta(0) + p\tilde{\delta}(p) + \omega_1 \tilde{\delta}(p) = \frac{\tilde{\Delta}_i(p)}{2\omega_m} \qquad (21.5\text{-}3)$$

$$\tilde{\delta}(p) \equiv \int_0^\infty \delta(t) e^{-pt}\, dt \qquad (21.5\text{-}4)$$

From (21.4-5) and taking $\delta(0) = 0$,

$$\tilde{\delta}(p) = \frac{\tilde{\Delta}_i(p)}{2\omega_m(p + \omega_1)} \qquad (21.5\text{-}5)$$

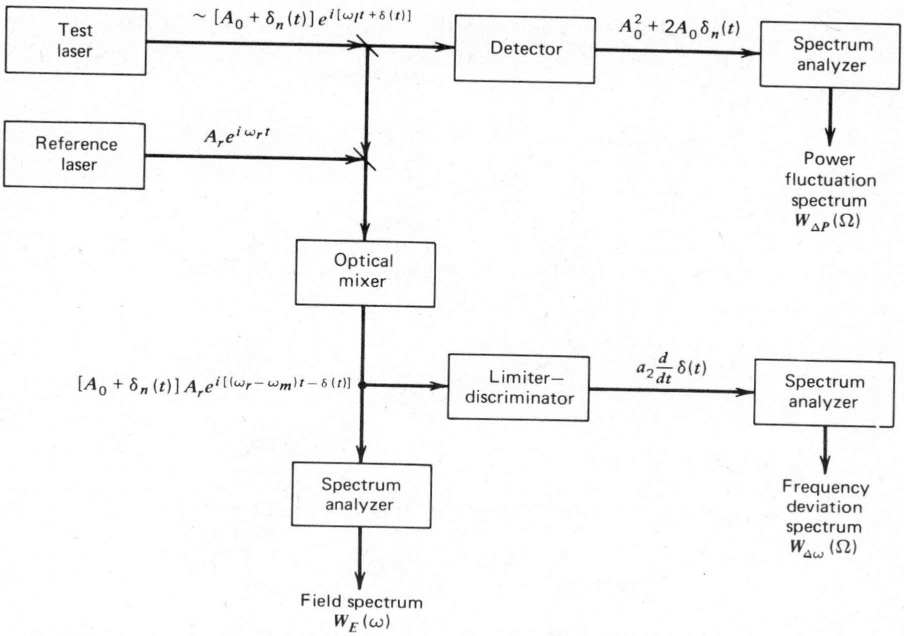

FIGURE 21.7 A block diagram demonstrating a typical arrangement for measuring the various noise spectra of a laser oscillation field.

We take the inverse Laplace transform of (21.5-5) using

$$f(t) = L^{-1}[\tilde{f}(p)] = \frac{1}{2\pi i}\int_{-i\infty+\tau_0}^{i\infty+\tau_0} \tilde{f}(p)\, e^{pt}\, dp; \; Re\; t > 0$$

and the result

$$L^{-1}[\tilde{f}_1(p)\tilde{f}_2(p)] = \int_0^t f_1(\lambda) f_2(t-\lambda)\, d\lambda \equiv f_1(t) * f_2(t)$$

taking

$$\tilde{f}_1(p) = \frac{\tilde{\Delta}_i(p)}{2\omega_m} \qquad \tilde{f}_2(p) = \frac{1}{p + \omega_1}$$

The result is

$$\delta(t) = \frac{1}{2\omega_m}\int_0^t \Delta_i(\lambda) e^{-\omega_1(t-\lambda)}\, d\lambda \tag{21.5-6}$$

Equation (21.5-6) is a formal solution for $\delta(t)$. The driving term $\Delta_i(\lambda)$, however, is random so that the best we can do is solve for statistical aspects of $\delta(t)$ in terms of those of $\Delta(t)$. We start with the autocorrelation function of $\delta(t)$.

$$\langle \delta(t+\tau)\delta(t)\rangle = \frac{1}{4\omega_m^2}\left\langle \int_0^{t+\tau} d\lambda_1 \int_0^t d\lambda_2 \Delta_i(\lambda_1)\Delta_i(\lambda_2) e^{-\omega_1(t+\tau-\lambda_1)} e^{-\omega_1(t-\lambda_2)} \right\rangle \tag{21.5-7}$$

We will take the random Langevin driving term that represents spontaneous emission as one having zero autocorrelation (instant amnesia or "white noise")

$$\langle \Delta_i(\lambda_1)\Delta_i(\lambda_2)\rangle = \langle \Delta_r(\lambda_1)\Delta_r(\lambda_2)\rangle = QD(\lambda_1 - \lambda_2)$$
$$\langle \Delta_i(\lambda_1)\Delta_r(\lambda_2)\rangle = 0 \tag{21.5-8}$$
$$D(x) \equiv \text{``}\delta\text{'' function of } x$$

and Q is a constant related to the "power" of $\Delta(t)$ and will be determined below. Using (21.5-8) in (21.5-7) gives

$$\langle \delta(t+\tau)\delta(t)\rangle = \frac{Qe^{-\omega_1|\tau|}}{8\omega_m^2\omega_1}(1 - e^{-2\omega_1 t}) \xrightarrow[t\to\infty]{} \frac{Qe^{-\omega_1|\tau|}}{8\omega_m^2\omega_1} \tag{21.5-9}$$

The Solution for ϕ

We start with (21.4-12), neglecting the small second term

$$A_0\dot{\phi} + \left(\frac{3}{2}\right)\frac{\omega_m X_r^{(3)} A_0^2}{n^2}\delta = -\frac{\Delta_r(t)}{2\omega_m} \tag{21.5-9a}$$

After Laplace transformation and using (21.5-5), we obtain

$$\tilde{\phi}(p) = -\left(\frac{3}{4}\right)\frac{X_r^{(3)} A_0 \tilde{\Delta}_i(p)}{n^2 p(p+\omega_1)} - \frac{\tilde{\Delta}_r(p)}{2A_0\omega_m p}$$

An inverse transformation results in

$$\Delta\phi(t) \equiv \phi(t) - \phi(0) = -\left(\frac{3}{4}\right)\frac{X_r^{(3)} A_0}{n^2\omega_1}\left[\int_0^t \Delta_i(\lambda)\,d\lambda - \int_0^t \Delta_i(\lambda)e^{-\omega_1(t-\lambda)}\,d\lambda\right]$$
$$- \frac{1}{2A_0\omega_m}\int_0^t \Delta_r(\lambda)\,d\lambda \tag{21.5-10}$$

and with the help of (21.5-8) and (21.5-2)

$$\langle \Delta\phi(t_1)\Delta\phi(t_2)\rangle = \frac{Q}{4\omega_m^2 A_0^2}\left\{1 + \left(\frac{X_r^{(3)}}{X_i^{(3)}}\right)^2\right\}\begin{Bmatrix} t_1 & t_1 < t_2 \\ t_2 & t_2 < t_1 \end{Bmatrix}$$

which we rewrite as

$$\langle \Delta\phi(t_1)\Delta\phi(t_2)\rangle = \frac{Q}{4\omega_m A_0^2}(1+\alpha^2)\begin{Bmatrix} t_1 & t_1 < t_2 \\ t_2 & t_2 < t_1 \end{Bmatrix} \tag{21.5-11}$$

$$\alpha \equiv \frac{X_r^{(3)}}{X_i^{(3)}} = \frac{\chi_r^{(3)}}{\chi_i^{(3)}} \tag{21.5-12}$$

The Frequency Spectrum

If the physical quantity of interest here is the instantaneous laser frequency

$$\omega(t) = \omega_m + \dot{\phi}(t)$$

then defining the frequency deviation yields

$$\Delta\omega(t) \equiv \omega(t) - \omega_m = \dot{\phi}(t)$$

$$\langle \Delta\omega(t_1)\Delta\omega(t_2)\rangle = \langle \dot{\phi}(t_1)\dot{\phi}(t_2)\rangle$$

and from (21.5-9) and (21.5-9a)

$$W_{\Delta\omega}(\Omega) = \frac{1}{\pi}\int_{-\infty}^{\infty} \langle \dot{\phi}(t)\dot{\phi}(t+\tau)\rangle e^{-i\Omega\tau}\, d\tau = \frac{Q}{4\pi\omega_m^2 A_0^2}\left(1 + \frac{\alpha^2}{1 + \left(\frac{\Omega}{\omega_1}\right)^2}\right) \quad (21.5\text{-}13)$$

The parameter Q can be determined using energy balance arguments (see Appendix A.6)

$$Q = \frac{4\hbar\omega_n^3\eta}{\varepsilon V\tau_p} \quad (21.5\text{-}14)$$

where V is the mode volume, and η, the inversion factor,

$$\eta \equiv \left(\frac{N_2}{N_2 - N_1}\right)_{\text{threshold}}$$

accounts for the fact that although gain is proportional to the population inversion $N_2 - N_1$, the spontaneous emission noise power is proportional to the upper laser level population N_2. With the above definition, we obtain

$$W_{\Delta\omega}(\Omega) = \frac{(\hbar\omega_m/\tau_p)\eta}{\pi\varepsilon V A_0^2}\left(\frac{\alpha^2}{1 + \left(\frac{\Omega}{\omega_1}\right)^2}\right) \quad (21.5\text{-}15)$$

realizing that the laser power is $P_0 \equiv \langle P\rangle = \varepsilon A_0^2 V/\tau_p$, we can rewrite

$$W_{\Delta\omega}(\Omega) = \frac{\eta}{\pi\tau_p^2}\left(\frac{\hbar\omega_m}{P_0}\right)\left(\frac{\alpha^2}{1 + \left(\frac{\Omega}{\omega_1}\right)^2}\right) \quad (21.5\text{-}16)$$

We notice the dependence of the noise power on the parameter α. This parameter dominates every aspect of the laser frequency (or phase) noise and it will be discussed separately further on. The basic features of $W_{\Delta\omega}(\Omega)$ were studied experimentally by Mannes and Siegman (Reference 13) and are in agreement with (21.5-6) except for "excess" noise power at $\Omega < 10^3$ Hz whose origin is not understood.

The Power Fluctuations Spectrum

The output of a detector that is receiving the laser output is a current proportional instantaneously to the laser power output (subject to the limited response time of the detector). A spectral analysis of this current will thus be proportional to the spectral density function of the laser power. This situation is illustrated in Figure 21.7. From the relationship

$$P = \frac{\varepsilon E_n{}^2(t)\, V}{\tau_p}, \langle P\rangle = \frac{\varepsilon A_0^2 V/}{\tau_p} \quad (21.5\text{-}17)$$

for the total power emitted by the atomic population and from (21.4-6)

$$\Delta P \equiv P - \langle P \rangle = \frac{2\varepsilon V A_0 \delta(t)}{\tau_p} \tag{21.5-18}$$

we can write

$$W_{\Delta P}(\Omega) = \frac{1}{\pi}\int_{-\infty}^{\infty} \langle \Delta P(t)\Delta P(t+\tau)\rangle e^{-i\Omega\tau}\, d\tau = \frac{4}{\pi}\frac{\varepsilon^2 V^2 A_0^2}{\tau_p^2}\int_{-\infty}^{\infty}\langle \delta(t)\delta(t+\tau)\rangle e^{-i\Omega\tau}\, d\tau \tag{21.5-19}$$

which, using (21.5-9), gives

$$W_{\Delta P}(\Omega) = \left(\frac{2}{\pi}\right)\frac{\varepsilon \Sigma A_0^2 \hbar\omega_m \eta}{\tau_p^3 \omega_1}\int_{-\infty}^{\infty} e^{-\omega_1|\tau| - i\Omega\tau}\, d\tau = \left(\frac{4}{\pi}\right)\frac{\hbar\omega_m \eta\varepsilon V A_0^2}{\tau_p^3}\left(\frac{1}{\Omega^2+\omega_1^2}\right)$$

where $\eta \equiv (N_2/(N_2 - N_1))_t$. From (2.15-17) and $\langle E^2(t)\rangle = A_0^2$, $P_e \equiv \langle P \rangle$,

$$W_{\Delta P}(\Omega) = \left(\frac{4}{\pi}\right)\frac{\hbar\omega_m \eta P_e}{\tau_p^2}\left(\frac{1}{\Omega^2+\omega_1^2}\right) \tag{21.5-20}$$

If we recall the definition (21.5-2) of ω_1

$$\omega_1 \equiv -X_i^{(3)} A_0^2 \omega_m / n^2$$

it follows that $W_{\Delta p}(0) \propto \langle P \rangle^{-1}$.

The Field Spectrum $W_E(\omega)$

This is the most commonly discussed laser spectrum. It is the spectrum displayed by a high-resolution Fabry–Perot etalon that scans through the laser line. In practice, the finesse of such etalons is not sufficiently high to resolve the narrow laser lines. Other methods, to be discussed below, are employed.

We will solve for the spectral density $W_E(\omega)$ of the field using the Wiener–Kintchine theorem so that our first task is to obtain the field autocorrelation function $C_E(\tau)$ of the field amplitude E_n [see (21.4-6)] which, hereafter, will be designated as E.

$$\begin{aligned} C_E(\tau) &= \langle E(t)E(t+\tau)\rangle \\ &= \frac{1}{4}\langle [E(t) + E^*(t)][E(t+\tau) + E^*(t+\tau)]\rangle \\ &= \frac{A_0^2}{4}\langle e^{-i\omega_m\tau + i[\phi(t)-\phi(t+\tau)]} + \text{c.c.}\rangle \\ &= [I(\tau) + I^*(\tau)] \end{aligned} \tag{21.5-21}$$

$$I(\tau) \equiv \frac{A_0^2}{4} e^{-i\omega_m\tau}\langle e^{i\Delta\phi(t,\tau)}\rangle \tag{21.5-22}$$

$$\Delta\phi(t,\tau) \equiv \phi(t) - \phi(t+\tau) \tag{21.5-23}$$

where we neglected the amplitude fluctuation $\delta(t)$ that plays but a minor role in the field spectrum. The expectation value in (21.5-22) can be evaluated statistically. The variable $\Delta\phi$ reflects the phase excursion due to many independent spontaneous emission events and thus, according to the central limit theorem, obeys Gaussian statistics with a probability function

$$g(\Delta\phi) = \frac{1}{\sqrt{2\pi\langle(\Delta\phi)^2\rangle}} e^{-\frac{(\Delta\phi)^2}{2\langle(\Delta\phi)^2\rangle}}$$

which, when used in (21.5-22), leads to

$$\langle e^{i\Delta\phi(t,\tau)}\rangle = \int_{-\infty}^{\infty} g(\Delta\phi)e^{i\Delta\phi}d(\Delta\phi) = e^{-1/2\langle[\Delta\phi(t,\tau)]^2\rangle} \tag{21.5-24}$$

It is clear that for stationary processes $\langle[\Delta\phi(t, \tau)]^2\rangle$ does not depend on the starting time t. Using (21.5-11) and putting $t_1 = t_2 = \tau$, we get

$$\langle[\Delta\phi(t, \tau)]^2\rangle = \frac{Q}{4\omega_m^2 A_0^2}(1 + \alpha^2)|\tau| \tag{21.5-25}$$

so that from (21.5-22) and (21.5-23),

$$I(\tau) = \left(\frac{A_0^2}{4}\right) e^{-i\omega_m\tau} e^{-\frac{Q(1+\alpha^2)}{8\omega_m^2 A_0^2}|\tau|} \tag{21.5-26}$$

$$C_E(\tau) = \frac{A_0^2}{2} e^{-\frac{Q(1+\alpha^2)}{8\omega_m^2 A_0^2}|\tau|} \cos \omega_m\tau \tag{21.5-27}$$

If we apply the Wiener–Kintchine theorem (21.3-4)

$$W_E(\omega) = \frac{1}{\pi}\int_{-\infty}^{\infty} C_E(\tau)e^{-i\omega\tau}\, d\tau$$

which for $\omega \approx \omega_0$ can be approximated as

$$W_E(\omega) \simeq \frac{A_0^2}{4\pi} \frac{Q(1 + \alpha^2)/4\omega_m^2 A_0^2}{\left[\frac{Q(1 + \alpha^2)}{8\omega_m^2 A_0^2}\right]^2 + (\omega - \omega_m)^2} \tag{21.5-28}$$

The laser field spectrum is thus a Lorentzian distribution with a full width at half maximum

$$(\Delta\omega)_{\text{laser}} = \frac{Q(1 + \alpha^2)}{4\omega_m^2 A_0^2}$$

If we use the definition (A.5-5) of Q and the relations $\langle P\rangle = \varepsilon A_0^2 V/\tau_p$, $(\Delta\nu)_{1/2} \equiv (2\pi\tau_p)^{-1}$, Eq. (21.5-27) becomes

$$(\Delta\nu)_{\text{laser}} = \frac{2\pi\hbar\nu_m\eta(\Delta\nu_{1/2})^2}{\langle P\rangle}(1 + \alpha^2) \tag{21.5-29}$$

where $\Delta\nu_{1/2}$ is the passive width of the optical resonator, that is, the width with no contribution to gain or loss from the resonant atomic transition.

21.6 THE LASER SPECTRA EXPERIMENTS

The laser linewidth as given by (21.5-29) exceeds the Schawlow–Townes result (21.2-28) by the factor $(1 + \alpha^2)$ where

$$\alpha = \frac{\chi_r^{(3)}}{\chi_i^{(3)}} \tag{21.6-1}$$

From the definition (21.4-6), we can rewrite α as

$$\alpha = (\partial\chi_r/\partial|E_n|^2)/(\partial\chi_i/\partial|E_n|^2) = \frac{\partial\chi_r/\partial N}{\partial\chi_i/\partial N} \tag{21.6-2}$$

where N is the inversion density and $\chi = \chi_r + i\chi_i = \chi^{(1)} + \chi^{(3)}|E_n|^2$. This factor ($\alpha$) accounts for the effect of the modulation of the index of refraction of the gain medium by the inversion density fluctuations caused by spontaneous transitions. In most gas and solid-state lasers, the lasing gain spectrum $\chi_i(\omega)$ is due to a single atomic transition and is thus symmetric. Since lasing takes place very near the gain maximum, it follows from the Kroning–Kramers relations (A.1-1) that χ_r is zero and that $\partial\chi_r/\partial N$ is zero as shown in Figure 21.8*a*. In a semiconductor laser, the situation is different. Due to the Fermi dirac statistics and the resulting distribution of states, the gain curve depicted is not symmetric. Near the peak of the gain curve, χ_r and $\partial\chi_r/\partial N$ are not zero as shown in Figure 21.8*b* and thus $\alpha \neq 0$.

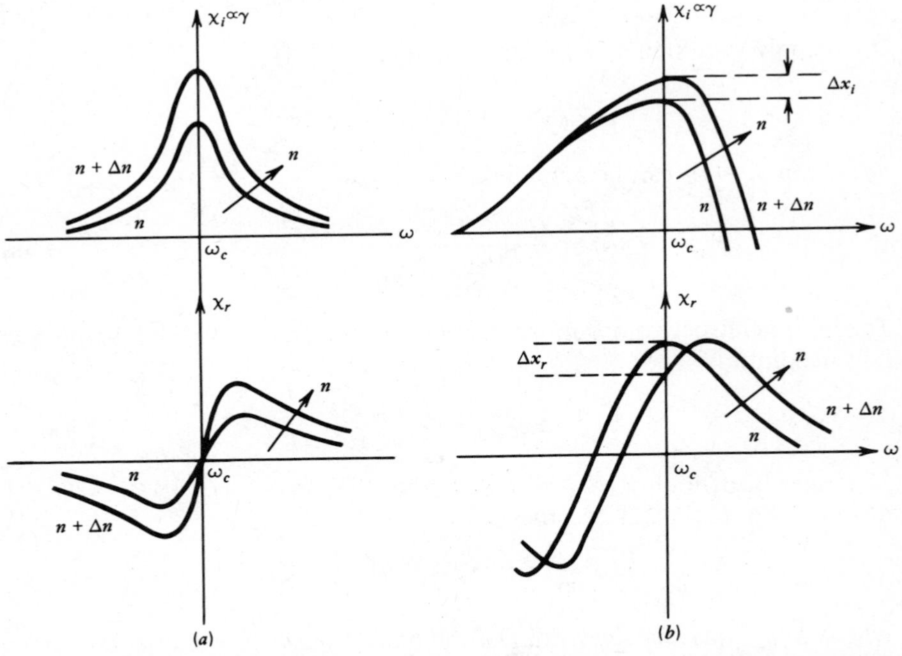

FIGURE 21.8 The gain (χ_i) and dispersion (χ_r) curves in (*a*) typical solid state and gas lasers; (*b*) semiconductor lasers.

Numerical Example

Consider a GaAs/GaAlAs semiconductor laser with the following characteristics:

$$\nu = 3.53 \times 10^{14}\ \text{s}^{-1} (\lambda_0 = 8500\ \text{Å})$$

$$l = 3 \times 10^{-2}\ \text{cm}$$

$$P_0 = 1\ \text{mW}$$

$$R = 31\%\ \text{(Fresnel reflection at GaAs/air interface)}$$

$$n = 3.5$$

$$(\Delta\nu)_{1/2} = \frac{(1 - R)c}{2\pi n l} \sim 3 \times 10^{10} (\tau_p \sim 5 \times 10^{-12} s)$$

If we take $\alpha = 0$, we obtain

$$(\Delta\nu)_{\text{laser}} \sim 4.5 \times 10^6\ \text{Hz}$$

We can trace the origin of the nearly nine orders of magnitude difference between the predicted linewidths of the GaAs/GaAlAs laser and the He-Ne laser (Sec. 21.2) to the much shorter photon lifetime τ_p in the semiconductor laser. This is due to its small length and reduced facet reflectivity. Semiconductor lasers operated in external, long, cavity display linewidths in the KHz regime (Reference 15).

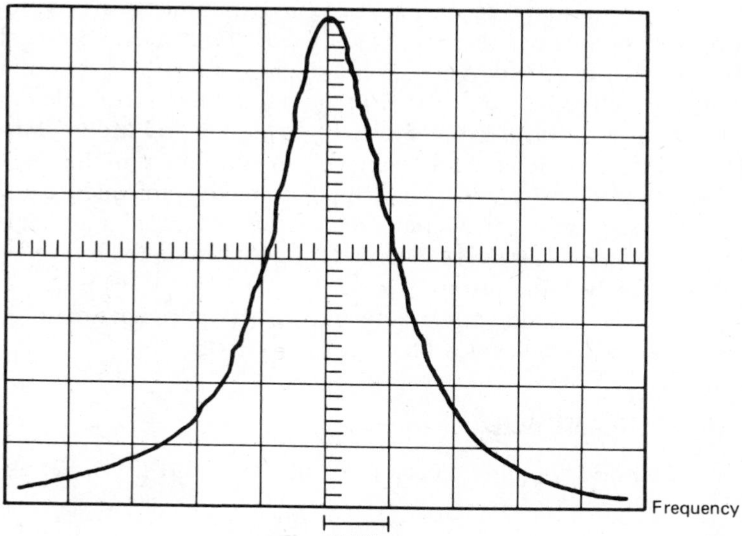

FIGURE 21.9 An experimentally measured Lorentzian spectral profile of the output of a GaAs laser. *Source:* Reference 14.

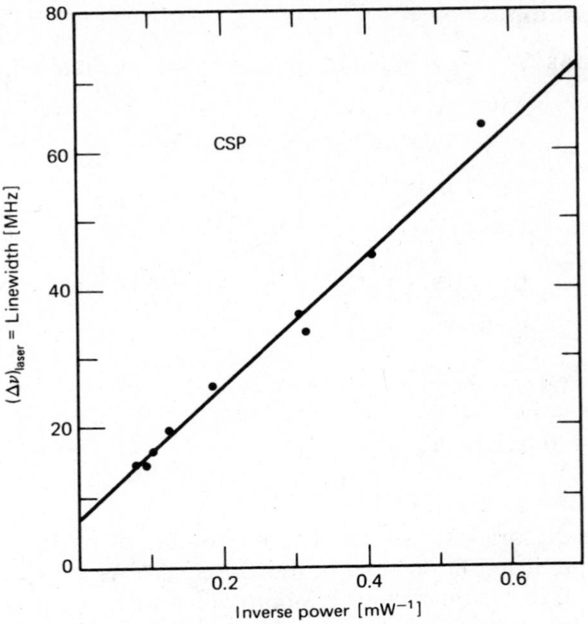

FIGURE 21.10 The measured $(\Delta\nu)_{\text{laser}}$ vs. inverse power dependence in a commercial (Hitachi) Channelled Substrate laser. *Source:* Reference 14.

Measured line shapes of the spectra of semiconductor lasers display the predicted Lorentzian shape (see Figure 21.9). Experiments also verify the inverse power dependence predicted by (21.5-29). Typical experimental data are shown in Figure 21.10. We notice, however, that the experimentally measured values of $(\Delta\nu)_{\text{laser}}$ are ~10–30 times larger than those calculated in the above example. It was this observation by Fleming and Mooradian (Reference 15) that led (see References 16 and 17) to elucidation of the role of the coupling between intensity and phase fluctuations in semiconductor lasers. Such coupling had been considered earlier (Reference 18) but neglected, justifiably, in the case of gas lasers. This coupling is represented by the factor α in (21.5-28). It follows that values of $\alpha \sim 3$–4, that is, $(1 + \alpha^2) \sim 10$–20, are needed to reconcile the difference between theory and experiment. In the next section, we will discuss this point in some detail.

21.7 THE α PARAMETER

The phase-amplitude coupling constant α was defined in (21.5-12) as

$$\alpha = \frac{\chi_r^{(3)}}{\chi_i^{(3)}} = \left(\frac{\partial \chi_r}{\partial N}\right) \Big/ \left(\frac{\partial \chi_i}{\partial N}\right) \qquad (21.7\text{-}1)$$

where N is the carrier density in the active region of the semiconductor laser.

(Other types of lasers are not considered here since in that case, $\alpha \approx 0$.) From the theoretical derivation, Eq. (11.2-13), of the complex susceptibility $\chi(\nu)$ of a semiconductor material, we can write

$$\chi(\nu,N) = \chi_r(\nu,N) + i\chi_i(\nu,N) = \int_0^\infty dEg(N,E)\left[\frac{(E-\hbar\nu)+i\hbar/T_2}{(E-\hbar\nu)^2+(\hbar/T_2)^2}\right] \quad (21.7\text{-}2)$$

Here, ν is the lasing frequency; E, the integration variable, is the transition energy; T_2, the dephasing time (see Section 8.1) due to electron-electron and electron-phonon "collisions" and $g(n, E)$, is the gain envelope function

$$g(N,E) \propto |x_{cv}|^2[f_c(N,E) - f_v(N,E)]\rho_r(E)$$

where x_{cv} is the coordinate matrix element connecting the upper (conduction band) and lower (valence band) states involved in the transition; f_c and f_v are, respectively, the conduction and valence occupation (Fermi) functions evaluated at the transition energy E. They depend on the carrier density n though the quasi-Fermi levels E_{F_c} and E_{f_v} discussed in Section 11.1. From (21.7-2) and (21.7-1), we obtain

$$\alpha(\nu,N) = \frac{\displaystyle\int_0^\infty dE\, g'(N,E)\frac{(E-\hbar\nu)}{(E-\hbar\nu)^2+(\hbar/T_2)^2}}{\displaystyle\int_0^\infty dE\, g'(N,E)\frac{\hbar/T_2}{(E-\hbar\nu)^2+(\hbar/T_2)^2}} \quad (21.7\text{-}3)$$

where

$$g' \equiv \frac{\partial g}{\partial N}$$

A plot of α as a function of the lasing frequency with the carrier density n as a parameter based on (21.7-3) is shown in Figure 21.11. The inset shows the calculated gain profile. The resulting values of α fall within the shaded region in the figure and are in the range of 2–3.

Direct measurements of α were performed (Reference 19) by a deliberate current modulation of a semiconductor laser and then measuring the resulting intensity (AM) spectrum and phase-modulation (FM) spectrum of the laser light. We recall that the increase in laser noise represented by α is due to the fact that the random modulation of the carrier density by spontaneous emission must also modulate the index of refraction and thus the frequency of the semiconductor medium. It is thus not surprising that a deliberate modulation of the carriers by a radio frequency current should have a similar effect. It was shown (Reference 19) that in this case, the α parameter is given by

$$\alpha = -2\,\frac{\text{FM modulation index of laser light}}{\text{AM modulation index of laser light}} \quad (21.7\text{-}4)$$

Values of the α parameter measured by this technique agree with the values obtained from the noise measurements. The description of the α parameter given above suggests that any laser in which $\alpha \neq 0$ should have a wavelength (frequency) modulation attendant on any population modula-

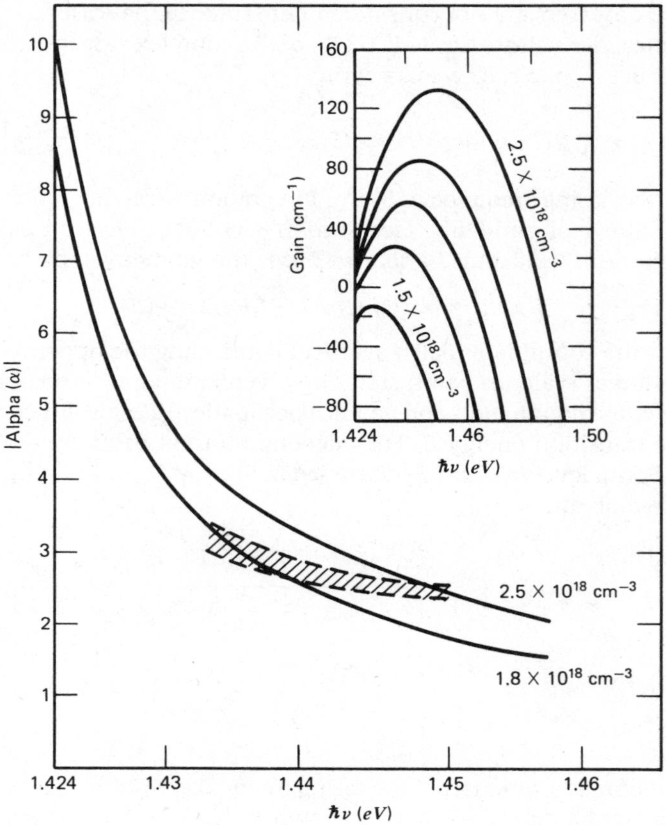

FIGURE 21.11 A plot of the parameter α based on (21.7-3). The inset shows the calculated gain vs. ν plots (*Source:* Reference 18). A plot of the theoretical dependence of α on the frequency of operation at two carrier densities. The shaded region shows the typical zone of operation as determined from the gain vs. plot of the inset.

tion. This causes undesirable wavelength chirping in current pulsed semiconductor lasers (Reference 20).

21.8 THE MEASUREMENT OF $(\Delta\nu)_{\text{laser}}$

The analytic signal of the optical field is given, according to (21.4-6), by

$$E(t) = [A_0 + \delta(t)]e^{i[\omega_m t + \phi(t)]} \tag{21.8-1}$$

In the discussion in Sections 21.4 and 21.5, we argued that the finite width of the laser is due mainly to the factor $exp[i\phi(t)]$, so that the *width of the laser spectrum is equal to* that of $exp[i\phi(t)]$, the two spectra being merely displaced by ω_m.

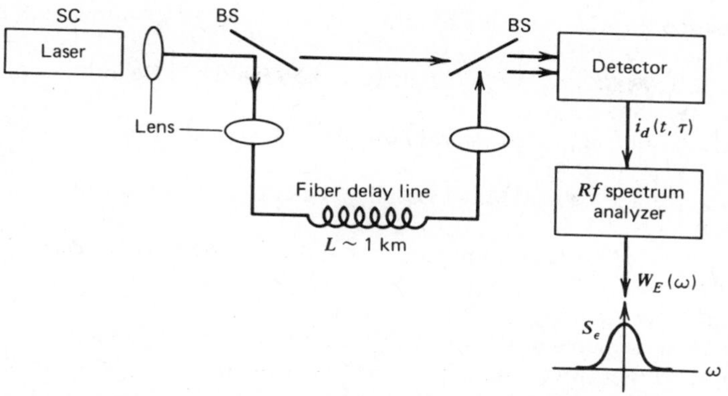

FIGURE 21.12 An interferometric arrangement employing a fiber delay for obtaining the spectrum $W_E(\omega)$ of the laser field. *Source:* Reference 21.

One experimental means for obtaining the spectrum is shown in Figure 21.12. The laser beam is split into two halves that then recombine at a detector with a relative time delay τ. At the detector, the total field is thus

$$E_d(t) = \tfrac{1}{2}A_0\{e^{i[\omega_m t+\phi(t)]} + e^{i[\omega_m(t-\tau)+\phi(t-\tau)]}\} \tag{21.8-2}$$

where $\delta(t)$ was neglected. The output current of the ("square law") detector is proportional to $E_d(t)E_d^*(t)$, which from (21.8-2) gives

$$\begin{aligned} i_d(t, \tau) &\propto e^{-i\omega_m\tau}e^{i\Delta\phi(t,\tau)} + \text{c.c.} \\ \Delta\phi(t, \tau) &\equiv \phi(t) - \phi(t-\tau) \end{aligned} \tag{21.8-3}$$

since τ is a constant in the experiment. The factor $exp(-i\omega_m\tau)$ does not affect the spectrum and will be ignored. Our aim is to obtain an output current from the detector that is proportional to $exp\{i[\phi(t)]\}$ and not $exp\{i[\Delta\phi(t, \tau)]\}$, but if the time delay τ exceeds the coherence time of the laser, that is, if

$$\tau \gg \tau_c \approx \frac{1}{\pi(\Delta\nu)_{\text{laser}}} \tag{21.8-4}$$

then $\phi(t)$ and $\phi(t-\tau)$ are uncorrelated and the statistics of $\phi(t)$ and $\Delta\phi(t, \tau)$ are identical. A spectral analysis of i_d, as shown in Figure 21.12, thus yields directly the laser spectrum $W_E(\omega)$ so that $(\Delta\nu)_{\text{laser}}$ can be read off the oscilloscope. In lasers with $(\Delta\nu)_{\text{laser}} < 10^6$ Hz, condition (21.8-4) translates to a path delay exceeding a few hundred meters.

References

1. Schawlow, L. and C. H. Townes, *Phys. Rev.* **112,** 1940 (1958); also, C. H. Townes, "Some Applications of Optical and Infrared Masers," in *Advances in Quantum Electronics,* J. R. Singer, ed. (New York: Columbia University Press, 1961).
2. Kogelnik, H. and A. Yariv, "Noise and Schemes for Its Reduction in Laser Amplifiers," *Proc. IEEE* **52,** 165 (1964).

3. Paananen, R. A., "Noise Measurement in an He-Ne Laser Amplifier," *Appl. Phys. Letters* **4,** 149 (1964).
4. Klüver, J. W., "Laser Amplifier Noise at 3.5 Microns in Helium-Xenon," *J. Appl. Phys.* **37,** 2987 (1966).
5. Yariv, A., *Introduction to Optical Electronics* (New York: Holt, Rinehart and Winston, 1971), p. 278.
6. Gordon, E. I., "Optical Maser Oscillators and Noise," *Bell System Tech. J.* **43,** 507 (1964).
7. Grivet, P. A. and A. Blaquière, *Optical Masers* (New York: Polytechnic Press, 1963), p. 69.
8. Jaseja, T. J., A. Javan, and C. H. Townes, "Frequency Stability of He-Ne Masers and Measurements of Length," *Phys. Rev. Letters* **10,** 165 (1963).
9. Egorov, Y. P., "Measurements of Natural Line Width of the Emission of a Gas Laser with Coupled Modes," *JETP Letters* **8,** 320 (1968).
10. Hinkley, E. D. and C. Freed, "Direct Observation of the Lorentzian Lineshape as Limited by Quantum Phase Noise in a Laser Above Threshold," *Phys. Rev. Letters* **23,** 277 (1969).
11. Yariv, A. and W. Caton, "Frequency, Intensity, and Field Fluctuations in Laser Oscillators," *IEEE J. Quant. Elect.* **QE-10,** 509 (1974).
12. See, for example, discussion in T. K. Caughey, "Response of Van der Pol's Oscillator to Random Excitation," *ASME J. of Appl. Mech.* **81,** 345 (1959).
13. Mannes, K. R. and A. E. Siegman, "Observations of Quantum Phase Fluctuations in Infrared Gas Lasers," *Phys. Rev.* **4,** 373 (1971).
14. Vahala, K. "Dynamic and Spectral Properties of Semiconductor Lasers." Ph.D. Thesis, California Institute of Technology, Pasadena, Calif., 1985.
15. Fleming, M. and A. Mooradian, "Fundamental Line Broadening of Single Mode (GaAl)As Diode Lasers," *Appl. Phys. Letters* **38,** 511 (1981).
16. Henry, C. H., "Theory of the Linewidth of Semiconductor Lasers," *IEEE J. Quant. Elec.* **QE-18,** 259 (1982).
17. Vahala, K. and A. Yariv, "Semiclassical Theory of Noise in Semiconductor Lasers—Part I," *IEEE. J. Quant. Elect.* **QE-19,** 1096 (1983).
18. Lax, M., "Classical ν Noise in Self-Sustained Oscillators," *Phys. Rev.* **160,** 290 (1967); also, Vahala, K., L. C. Chiu, S. Margalit, and A. Yariv, "On the Linewidth Factor α in Semiconductor Lasers" *Appl. Phys. Letters* **42,** 631 (1983).
19. Harder, C., K. Vahala, and A. Yariv, "Measurement of the Linewidth Enhancement Factor α of Semiconductor Lasers," *Appl. Phys. Letters* **42,** 328 (1983).
20. Koch, T. L. and R. A. Linke, "Effect of Nonlinear Gain Reduction on Semiconductor Laser Wavelength Chirping," *Appl. Phys. Letters* **48,** 613 (1986).
21. Yamamoto, Y., T. Mukai, and S. Saito, "Quantum Phase Noise and Linewidth of a Semiconductor Laser," *Elect. Letters* **17,** 327 (1981); also, Okoshi, T., K. Kikuchi, and A. Nakayma, "Novel Method for High Resolution Measurement of Laser Output Spectrum," *Elec. Letters* **6,** 630 (1980).

Problems

21.1 Show that starting with the macroscopic relations

$$P = \chi^{(1)}E + \chi^{(3)}\langle|E|^2\rangle E$$

relating the induced polarization to the complex field amplitude E, we can use the normal mode expansion of (21.4-2) to obtain (21.4-6).

21.2 Derive (21.4-8).

21.3 Assume a laser amplifier employing a gain medium with a width of $\Delta\nu \sim 10^9$ Hz. The laser is used as a preamplifier in an infrared radiometer "looking" at a background with a mean temperature $T = 20°\text{K}$. What is the smallest increment ΔT that can be measured?

21.4 Using References 19 and 20 at the end of this chapter and the articles mentioned in them, describe the connection between wavelength chirping in modulated semiconductor lasers and the α parameter.

21.5 (A simple-minded question) Explain why a spectral analysis of the output current of a detector illuminated by a laser beam does not yield information about the field spectrum $W_E(\omega)$ of the laser.

21.6 In a rotating ring ("gyro") laser the two counter propagating modes "see" different path lengths thus oscillating at different frequencies.

(a) Relate the frequency offset $W_{CCW} - W_{CW}$ to the angular rotation rate Ω of the laser.

(b) What is the minimum Ω that can be sensed by the laser gyroscope in the presence of quantum frequency fluctuations as given by (21.5-16)?

CHAPTER 22

Guided Wave Optics—Propagation in Optical Fibers

22.0 INTRODUCTION

In this chapter we take up a number of topics that involve propagation of optical modes in dielectric films with thicknesses comparable to the wavelength.

The ability to generate, guide, modulate, and detect light in such thin film configurations (References 1, 2, 3) opens up new possibilities for monolithic "optical circuits" (Reference 4)—an endeavor going under the name of "integrated optics" (Reference 5).

We will first consider the basic problem of TE and TM mode propagation in slab dielectric waveguides. A coupled mode formalism is then developed to describe situations in which the normal power flow of the TE and TM mode is perturbed by some agency.

The coupled mode formalism is applied in analyzing a number of important applications. These include (a) periodic (corrugated) optical waveguides and filters, (b) distributed feedback lasers, (c) electrooptic mode coupling, and (d) laser arrays. The chapter will conclude with a discussion of pulse dispersion and spreading effects in optical fibers.

22.1 THE WAVEGUIDE MODES

A prerequisite to an understanding of guided wave interactions is a knowledge of the properties of the guided modes. A mode of a dielectric waveguide at a (radian) frequency ω is a solution of the wave equation (6.5-3) that, if we put $k^2(\mathbf{r}) \equiv k^2 n^2(\mathbf{r})$, can be written as

$$\nabla^2 \mathbf{E}(\mathbf{r}) + k^2 n^2(\mathbf{r}) \mathbf{E}(\mathbf{r}) = 0 \qquad (22.1\text{-}1)$$

The solutions are subject to the continuity of the tangential components of $\mathbf{E}$ and $\mathbf{H}$ at the dielectric interfaces. In (19.1-1) the form of the field is taken as

$$\mathbf{E}(\mathbf{r}, t) = \mathbf{E}(\mathbf{r})\, e^{i[\omega t - \phi(\mathbf{r})]} \qquad (22.1\text{-}2)$$

$k \equiv \omega/c$, and $n(\mathbf{r})$, the index of refraction, is related to the dielectric constant $\varepsilon(\mathbf{r})$ by $n^2(\mathbf{r}) \equiv \varepsilon(\mathbf{r})/\varepsilon_0$. Limiting ourselves to waves with phase fronts normal to the waveguide axis, z, we have $\phi(\mathbf{r}) = \beta z$ and (22.1-1) becomes

$$\left(\frac{\partial^2}{\partial x^2} + \frac{\partial^2}{\partial y^2}\right) \mathbf{E}(x, y) + [k^2 n^2(\mathbf{r}) - \beta^2]\mathbf{E}(x, y) = 0 \qquad (22.1\text{-}3)$$

The basic features of the behavior of dielectric waveguide can be extracted from a planar model in which no variation exists in one (e.g., y) dimension. Channel waveguides, in which the waveguide dimensions are finite in both the x and y directions, approach the behavior of the planar guide when one dimension is considerably larger than the other (References 6 and 7). Even when this is not the case, most of the phenomena of interest are only modified in a simple quantitative way when going from a planar to a channel waveguide. Because of this and of the immense mathematical simplification that results, we will limit most of the following treatment to planar waveguides such as the one shown in Figure 22.1.

Putting $\partial/\partial y = 0$ in (22.1-3) and writing it separately for regions 1, 2, 3 yield

Region 1

$$\frac{\partial^2}{\partial x^2} E(x, y) + (k^2 n_1^2 - \beta^2) E(x, y) = 0 \qquad (22.1\text{-}4a)$$

Region 2

$$\frac{\partial^2}{\partial x^2} E(x, y) + (k^2 n_2^2 - \beta^2) E(x, y) = 0 \qquad (22.1\text{-}4b)$$

Region 3

$$\frac{\partial^2}{\partial x^2} E(x, y) + (k^2 n_3^2 - \beta^2) E(x, y) = 0 \qquad (22.1\text{-}4c)$$

where $E(x, y)$ is a cartesian component of $\mathbf{E}(x, y)$. Before embarking on a formal solution of (22.1-4), we may learn a great deal about the physical nature of the solutions by simple arguments. Let us consider the nature of the solutions as a function of the propagation constant β at a *fixed* frequency ω. Let us assume that $n_2 > n_3 > n_1$. For $\beta > kn_2$ [i.e., region (a) in Figure 22.2], it

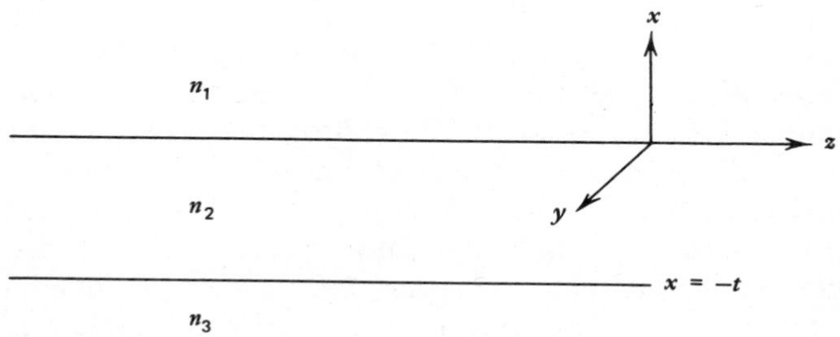

FIGURE 22.1 A slab ($\partial/\partial y = 0$) dielectric waveguide.

follows directly from (22.1-4) that $(1/E)(\partial^2 E/\partial x^2) > 0$ everywhere, and $E(x)$ is exponential in all three layers 1, 2, 3 of the waveguides. Because of the need to match both $E(x)$ and its derivatives (see Section 22.2) at the two interfaces, the resulting field distribution is as shown in Figure 22.2*a*. The field increases without bound away from the waveguide so that the solution is not *physically realizable* and thus does not correspond to a real wave.

For $kn_3 < \beta < kn_2$, as in points (*b*) and (*c*), it follows from (22.1-4) that the solution is sinusoidal in region 2, since $(1/E)(\partial^2 E/\partial x^2) < 0$, but is exponential in regions 1 and 3. This makes it possible to have a solution $E(x)$ that satisfies the boundary conditions while *decaying* exponentially in regions 1 and 3. These solutions are shown in Figures 22.2*b* and 22.2*c*. The energy carried by these modes is confined to the vicinity of the guiding layer 2, and we will, consequently, refer to them as confined, or guided modes. From the above discussion, it follows that a necessary condition for their existence is that $kn_1, kn_3 < \beta < kn_2$ so that confined modes are possible only when $n_2 > n_1, n_3$; that is, the inner layer possesses the highest index of refraction.

Solutions of (22.1-4) for $kn_1 < \beta < kn_3$ (*d*) correspond, according to (22.1-4), to exponential behavior in region 1 and to sinusoidal behavior in regions 2 and 3 as illustrated in Figure 22.2*d*. We will refer to these modes as substrate radiation modes. For $0 < \beta < kn_1$ as in (*e*), the solution for $E(x)$ becomes sinusoidal in all three regions. These are the so-called radiation modes of the waveguides.

A solution of (22.1-4) subject to the boundary conditions at the interfaces given in the next section shows that although in regions (*d*) and (*e*) β is a continuous variable, the values of allowed β in the propagation regime $kn_3 < \beta < kn_2$ are *discrete*. The number of modes depends on the width t, the frequency, and the indices of refraction n_1, n_2, n_3. At a given wavelength the number of guided modes increases from 0 with increasing t. At some t, the mode TE_0 becomes confined. Further increases in t will allow TE_1 to exist as well, and so on.

A useful point of view is one of viewing the wave propagation in the inner layer 2 as that of a plane wave propagating at some angle θ to the horizontal axis and undergoing a series of total internal reflections at the interface 2-1 and 2-3. This is based on (22.1-4*b*). Assuming $E \propto \sin(hx + \alpha) \exp(-i\beta z)$, we obtain

$$\beta^2 + h^2 = k^2 n_2^2 \tag{22.1-5}$$

The resulting right-angle triangles with sides β, h, and kn_2 are shown in Figure 22.2. Note that since the frequency is constant, $kn_2 \equiv (\omega/c)n_2$ is the same for cases (*b*), (*c*), (*d*), and (*e*). The propagation can thus be considered formally as that of a plane wave along the direction of the hypotenuse with a *constant* propagation constant kn_2. As β decreases, θ increases until, at $\beta = kn_3$, the wave ceases to be totally internally reflected at the interface 3-2. The condition for exponential decay in region 3 $\beta = kn_3$ is identified, by writing $\beta = kn_2 \cos\theta$, with the geometrical optics condition for the onset of total internal reflection.

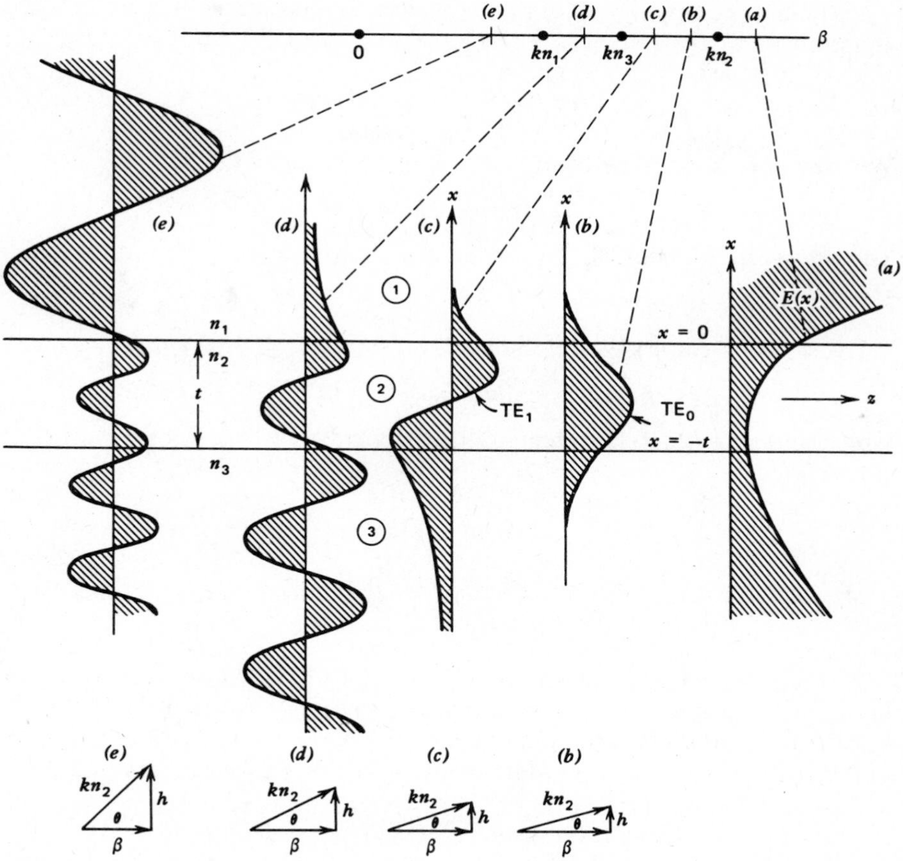

FIGURE 22.2 Top: the different regimes *a, b, c, d, e* of the propagation constant β of the waveguide shown in Figure 22.1. Middle: the field distributions corresponding to the different value of β. Bottom: the propagation triangles corresponding to the different propagation regimes.

22.2 MODE CHARACTERISTICS OF THE PLANAR WAVEGUIDE

TE Modes

Consider the dielectric waveguide sketched in Figure 22.1. It consists of a film of thickness t and index of refraction n_2 sandwiched between media with indices n_1 and n_3. If we take $\partial/\partial y = 0$, this guide can, in the general case, support a finite number of confined TE modes with field components E_y, H_x, and H_z and TM modes with components H_y, E_x, E_z. The "radiation" modes of this structure, which are not confined to the inner layer, are not treated here but are important in considering other problems such as grating couplers and radiation losses (Supplemental Reference 1).

The field component E_y of the TE modes, as an example, obeys the wave equation

$$\nabla^2 E_y = \frac{n_i^2}{c^2}\frac{\partial^2 E_y}{\partial t^2} \qquad i = 1, 2, 3 \tag{22.2-1}$$

We take $E_y(x, z, t)$ in the form

$$E_y(x, z, t) = \mathscr{E}_y(x)\, e^{i(\omega t - \beta z)} \tag{22.2-2}$$

The transverse function $\mathscr{E}_y(x)$ is taken as

$$\mathscr{E}_y = \begin{cases} C \exp(-qx) & 0 \leqslant x < \infty \\ C[\cos(hx) - (q/h)\sin(hx)] & -t \leqslant x \leqslant 0 \\ C[\cos(ht) + (q/h)\sin(ht)]\exp[p(x+t)] & -\infty < x \leqslant -t \end{cases} \tag{22.2-3}$$

which applying (22.2-1) to regions 1, 2, 3 yields

$$\begin{aligned} h &= (n_2^2 k^2 - \beta^2)^{1/2} \\ q &= (\beta^2 - n_1^2 k^2)^{1/2} \\ p &= (\beta^2 - n_3^2 k^2)^{1/2} \\ k &\equiv \omega/c \end{aligned} \tag{22.2-4}$$

The acceptable solutions for $\mathscr{E}_y$ and $\mathscr{H}_z = (i/\omega\mu)(\partial\mathscr{E}_y/\partial x)$ are continuous at both $x = 0$ and $x = -t$. The particular choice of coefficients in (22.2-3) satisfies the continuity condition of $\mathscr{E}_y$ at $x = 0$, $x = -t$, and $\partial\mathscr{E}_y/\partial x$ at $x = 0$. By imposing the continuity condition on $\partial\mathscr{E}_y/\partial x$ at $x = -t$, we get, from (22.2-3),

$$h\sin(ht) - q\cos(ht) = p\left[\cos(ht) + \frac{q}{h}\sin(ht)\right]$$

or

$$\tan(ht) = \frac{q + p}{h\left(1 - \dfrac{pq}{h^2}\right)} \tag{22.2-5}$$

In the symmetric case ($n_1 = n_3$) the field (22.2-3) must be odd or even about $x = -t/2$. This requires, according to (22.2-3), that $(1 + q/h)$ or $(1 - q/h)$ be zero, respectively. Since $p = q$, for this case the denominator of (22.2-5) becomes zero. This special case is treated in References 8 and 9. The last equation in conjunction with (22.2-4) is used to obtain the eigenvalues β for the confined TE modes. An example of such a solution is shown in Figure 22.3.

The constant C appearing in (22.2-3) is arbitrary, yet for many applications, especially those in which propagation and exchange of power involve more than one mode, it is advantageous to define C in such a way that it is related to total power in the mode. This point will become clear in Section 22.3. We choose C so that the field $\mathscr{E}_y(x)$ in (22.2-3) corresponds to a power flow of *one* watt (per unit width in y direction) in the mode. A mode for which

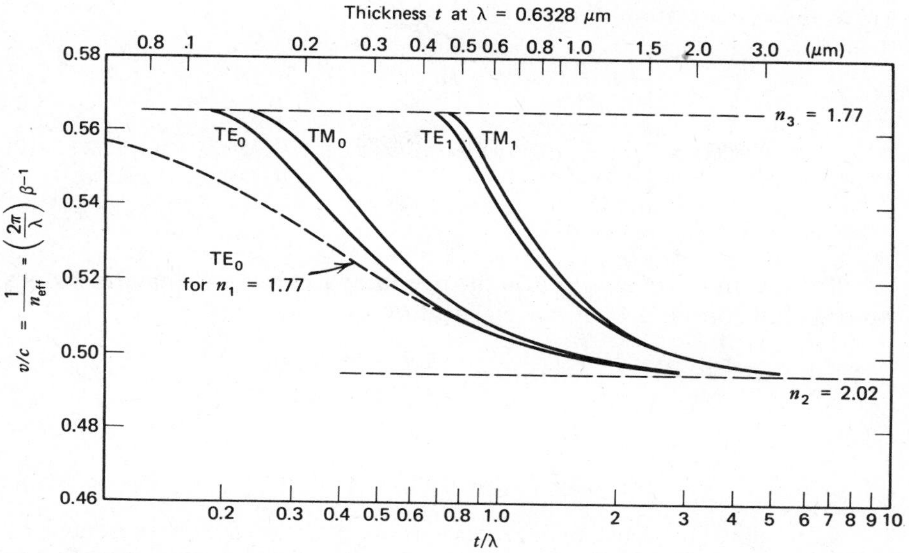

FIGURE 22.3 Dispersion curves for the confined modes of ZnO on sapphire waveguide, $n_1 = 1$. (*Source:* Reference 10.)

$E_y = A\mathscr{E}_y(x)$ will thus correspond to a power flow of $|A|^2$ W/m. The normalization condition becomes

$$-\frac{1}{2}\int_{-\infty}^{\infty} E_y H_x^* \, dx = \frac{\beta_m}{2\omega\mu}\int_{-\infty}^{\infty} [\mathscr{E}_y^{(m)}(x)]^2 \, dx = 1 \tag{22.2-6}$$

where the symbol m denotes the mth confined TE mode corresponding to the mth eigenvalue of (22.2-5) and $H_x = -i(\omega\mu)^{-1}\, \partial E_y/\partial z$.

Using (22.2-3) in (22.2-6) leads to

$$C_m = 2h_m \left[\frac{\omega\mu}{|\beta_m|\left(t + \dfrac{1}{q_m} + \dfrac{1}{p_m}\right)(h_m^2 + q_m^2)}\right]^{1/2} \tag{22.2-7}$$

Since the modes $\mathscr{E}_y^{(m)}$ are orthogonal (see Problem 22.9), we have

$$\int_{-\infty}^{\infty} \mathscr{E}_y^{(l)}\mathscr{E}_y^{(m)} \, dx = \frac{2\omega\mu}{\beta_m}\,\delta_{l,m} \tag{22.2-8}$$

TM Modes

The field components are

$$\begin{aligned} H_y(x, z, t) &= \mathscr{H}_y(x)e^{i(\omega t - i\beta z)} \\ E_x(x, z, t) &= \frac{i}{\omega\varepsilon}\frac{\partial H_y}{\partial z} = \frac{\beta}{\omega\varepsilon}\mathscr{H}_y(x)e^{i(\omega t - \beta z)} \\ E_z(x, z, t) &= -\frac{i}{\omega\varepsilon}\frac{\partial H_y}{\partial x} \end{aligned} \tag{22.2-9}$$

The transverse function $\mathcal{H}_y(x)$ is taken as

$$\mathcal{H}_y(x) = \begin{cases} -C\left[\dfrac{h}{\bar{q}}\cos(ht) + \sin(ht)\right] e^{p(x+t)} & x < -t \\ C\left[-\dfrac{h}{\bar{q}}\cos(hx) + \sin(hx)\right] & -t < x < 0 \\ -\dfrac{h}{\bar{q}}\, Ce^{-qx} & x > 0 \end{cases} \qquad (22.2\text{-}10)$$

The continuity of H_y and E_z at the two interfaces requires that the various propagation constants obey the eigenvalue equation

$$\tan(ht) = \frac{h(\bar{p} + \bar{q})}{h^2 - \bar{p}\bar{q}} \qquad (22.2\text{-}11)$$

where

$$\bar{p} \equiv \frac{n_2^2}{n_3^2}\, p \qquad \bar{q} = \frac{n_2^2}{n_1^2}\, q$$

The normalization constant C is chosen so that the field represented by (22.2-9) and (22.2-10) carries one watt per unit width in the y direction

$$\frac{1}{2}\int_{-\infty}^{\infty} H_y E_x^* \, dx = \frac{\beta}{2\omega}\int_{-\infty}^{\infty} \frac{\mathcal{H}_y^2(x)}{\varepsilon(x)}\, dx = 1$$

or, if we use $n_i^2 \equiv \varepsilon_i/\varepsilon_0$,

$$\int_{-\infty}^{\infty} \frac{[\mathcal{H}_y^{(m)}(x)]^2}{n^2(x)}\, dx = \frac{2\omega\varepsilon_0}{\beta_m} \qquad (22.2\text{-}12)$$

Carrying out the integration using (22.2-10) gives

$$C_m = 2\sqrt{\frac{\omega\varepsilon_0}{\beta_m t_{\text{eff}}}}$$

$$t_{\text{eff}} \equiv \frac{\bar{q}^2 + h^2}{\bar{q}^2}\left(\frac{t}{n_2^2} + \frac{q^2 + h^2}{\bar{q}^2 + h^2}\frac{1}{n_1^2 q} + \frac{p^2 + h^2}{\bar{p}^2 + h^2}\frac{1}{n_3^2 p}\right) \qquad (22.2\text{-}13)$$

The general properties of the TE and TM mode solutions are illustrated in Figure 22.3. In general, a mode becomes confined above a certain (cutoff) value of t/λ. At the cutoff value, $p = 0$ and the mode extends to $x = -\infty$. For increasing values of t/λ, $p > 0$ and the mode becomes increasingly confined to layer 2. This is reflected in the effective mode index $2\pi/(\beta\lambda)$ that at cutoff is equal to n_3, and that for t/λ approaches n_2. In a symmetric waveguide ($n_1 = n_3$), the lowest-order modes TE_0 and TM_0 have no cutoff and are confined for all values of t/λ. The selective excitation of waveguide modes by means of prism couplers and a determination of their propagation constants β_m are described in Reference 11.

22.3 COUPLING BETWEEN GUIDED MODES

In Section 22.2 we obtained solutions for the confined modes supported by a slab dielectric waveguide such as that shown in Figure 22.1. An increasingly large number of experiments and devices involves coupling between such modes (Reference 12). The coupling can, in most cases, be represented as a distributed perturbation polarization source. Two typical examples involve TM-to-TE mode conversion by the electrooptic or acoustooptic effect (References 13 and 14) or coupling of forward to backward modes by means of a corrugation in one of the waveguides interfaces (Reference 15).

We start with the wave equation

$$\nabla^2 \mathbf{E}(\mathbf{r}, t) = \mu\varepsilon_0 \frac{\partial^2 \mathbf{E}}{\partial t^2} + \mu \frac{\partial^2}{\partial t^2} \mathbf{P}(\mathbf{r}, t) \tag{22.3-1}$$

The total medium polarization can be taken as the sum

$$\mathbf{P}(\mathbf{r}, t) = \mathbf{P}_0(\mathbf{r}, t) + \mathbf{P}_{\text{pert}}(\mathbf{r}, t) \tag{22.3-2}$$

where

$$\mathbf{P}_0(\mathbf{r}, t) = [\varepsilon(\mathbf{r}) - \varepsilon_0]\mathbf{E}(\mathbf{r}, t) \tag{22.3-3}$$

is the polarization induced by $\mathbf{E}(\mathbf{r}, t)$ in the *unperturbed* waveguide whose dielectric constant is $\varepsilon(\mathbf{r})$. The perturbation polarization $\mathbf{P}_{\text{pert}}(\mathbf{r}, t)$ is then defined by (22.3-2). Using (22.3-2) and (22.3-3) in (22.3-1) gives

$$\nabla^2 E_y - \mu\varepsilon(\mathbf{r}) \frac{\partial^2}{\partial t^2} E_y = \mu \frac{\partial^2}{\partial t^2} [P_{\text{pert}}(\mathbf{r}, t)]_y \tag{22.3-4}$$

and similar expressions for E_x and E_z.

Ignoring the possibility of coupling to the continuum of radiation modes, we expand the total field in the "perturbed" waveguide as

$$E_y(\mathbf{r}, t) = \frac{1}{2} \sum_m A_m(z)\mathscr{E}_y^{(m)}(x) e^{i(\omega t - \beta_m z)} + \text{c.c.} \tag{22.3-5}$$

where m indicates the mth eigenmode of (22.2-5), which satisfies

$$\left(\frac{\partial^2}{\partial x^2} - \beta_m^2\right) \mathscr{E}_y^{(m)}(\mathbf{r}) + \omega^2\mu\varepsilon(\mathbf{r})\mathscr{E}_y^{(m)}(\mathbf{r}) = 0 \tag{22.3-6}$$

where $\varepsilon(\mathbf{r}) = \varepsilon_0 n^2(\mathbf{r})$.

Substitution of (22.3-5) in (22.3-4) leads to

$$\begin{aligned} e^{i\omega t} \sum_m \Bigg\{ \frac{A_m}{2} \left[-\beta_m^2 \mathscr{E}_y^{(m)} + \frac{\partial^2 \mathscr{E}_y^{(m)}}{\partial x^2} + \omega^2\mu\varepsilon(\mathbf{r})\mathscr{E}_y^{(m)} \right] e^{-i\beta_m z} \\ + \frac{1}{2}\left(-2i\beta_m \frac{dA_m}{dz} + \frac{d^2 A_m}{dz^2}\right) \mathscr{E}_y^{(m)} e^{-i\beta_m z} \Bigg\} + \text{c.c.} \\ = \mu \frac{\partial^2}{\partial t^2} [P_{\text{pert}}(\mathbf{r}, t)]_y \end{aligned} \tag{22.3-7}$$

First, we note that in view of (22.3-6) the sum of the first three terms in (22.3-7) is zero. We assume "slow" variation so that

$$\left|\frac{d^2 A_m}{dz^2}\right| \ll \beta_m \left|\frac{dA_m}{dz}\right|$$

and obtain from (22.3-7)

$$\sum_m - i\beta_m \frac{dA_m}{dz} \mathscr{E}_y^{(m)} e^{i(\omega t - \beta_m z)} + \text{c.c.} = \mu \frac{\partial^2}{\partial t^2} [P_{\text{pert}}(\mathbf{r}, t)]_y \tag{22.3-8}$$

We take the product of (22.3-8) with $\mathscr{E}_y^{(s)}(x)$ and integrate from $-\infty$ to ∞. The result, if we use (22.2-8), is

$$\frac{dA_s^{(-)}}{dz} e^{i(\omega t + \beta_s z)} - \frac{dA_s^{(+)}}{dz} e^{i(\omega t - \beta_s z)} - \text{c.c.} = -\frac{i}{2\omega} \frac{\partial^2}{\partial t^2} \int_{-\infty}^{\infty} [P_{\text{pert}}(\mathbf{r}, t)]_y \mathscr{E}_y^{(s)}(x)\, dx \tag{22.3-9}$$

where we recall that the summation over m in (22.3-8) contains two terms involving $\mathscr{E}_y^{(m)}(x)$ for each value of m, one designated as $(-)$ traveling in the $-z$ direction, and the $(+)$ term traveling in the $+z$ direction.

Equation (22.3-9) can be used to treat a large variety of mode interactions (Reference 12). Some important examples are considered in the following sections.

22.4 THE PERIODIC WAVEGUIDE—DISTRIBUTED FEEDBACK LASERS

Consider a periodic dielectric waveguide in which the periodicity is due to a corrugation of one of the interfaces as shown in Figure 22.4.

The corrugation is described by the dielectric perturbation $\Delta\varepsilon(\mathbf{r}) \equiv \varepsilon_0 \Delta n^2(\mathbf{r})$ such that the total dielectric constant is

$$\varepsilon'(\mathbf{r}) = \varepsilon(\mathbf{r}) + \Delta\varepsilon(\mathbf{r})$$

The perturbation polarization is from (22.3-2) and (22.3-3)

$$\mathbf{P}_{\text{pert}} = \Delta\varepsilon(\mathbf{r})\mathbf{E}(\mathbf{r}, t) = \Delta n^2(\mathbf{r})\varepsilon_0 \mathbf{E}(\mathbf{r}, t) \tag{22.4-1}$$

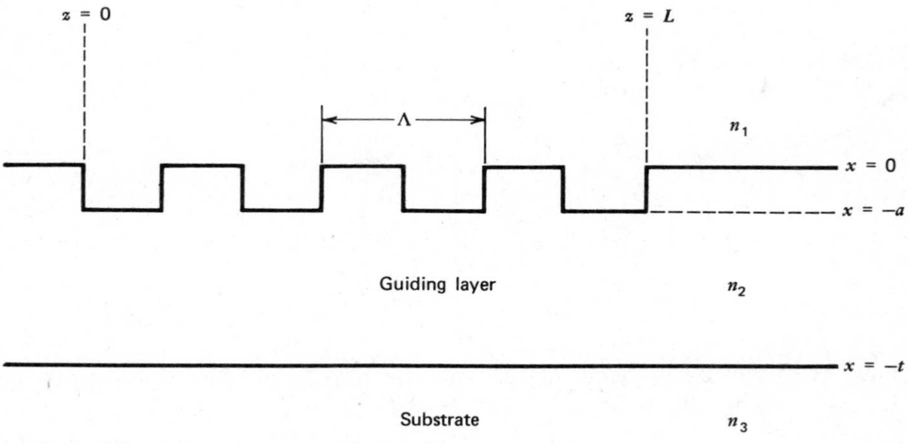

FIGURE 22.4 A corrugated periodic waveguide.

Since $\Delta n^2(\mathbf{r})$ is a scalar, it follows, from (22.3-4), that the corrugation couples only TE to TE modes and TM to TM but not TE to TM.

To be specific, consider TE mode propagation. Using (22.3-5) in (22.4-1) gives

$$[P_{\text{pert}}(\mathbf{r}, t)]_y = \frac{\Delta n^2(\mathbf{r})\varepsilon_0}{2} \sum_m [A_m \mathscr{E}_y^{(m)}(x) e^{i(\omega t - \beta_m z)} + \text{c.c.}] \tag{22.4-2}$$

that, when used in (22.3-9), leads to

$$\begin{aligned} &\frac{dA_s^{(-)}}{dz} e^{i(\omega t + \beta_s z)} - \frac{dA_s^{(+)}}{dz} e^{i(\omega t - \beta_s z)} - \text{c.c.} \\ &= -\frac{i\varepsilon_0}{4\omega}\frac{\partial^2}{\partial t^2} \sum_m \left[A_m \int_{-\infty}^{\infty} \Delta n^2(x, z) \mathscr{E}_y^{(m)}(x) \mathscr{E}_y^{(s)}(x)\, dx e^{i(\omega t - \beta_m z)} + \text{c.c.} \right] \end{aligned} \tag{22.4-3}$$

The two terms on the left side of (22.4-3) can be influenced only by terms on the right side that possess the same z phase dependence, that is, $\exp(i\beta_s z)$ or $\exp(-i\beta_s z)$, that is, are "phase matched." The contribution of nonphase matched terms averages to zero over a few phase beat lengths. We recognize that in order for the mth mode to couple to the sth mode, for example, it is necessary that the product of the perturbation and the mth mode phase factor

$$\Delta n^2(x, z) \exp(\mp i\beta_m z)$$

in (22.4-3) contain a term that is proportional to $\exp(-i\beta_s z)$ or $\exp(+i\beta_s z)$. In the first case, the forward $A_s^{(+)}$ mode is driven synchronously by the perturbation, whereas in the second case, it is the backward $A_s^{(-)}$ mode. The choice of which modes couple is thus determined by the z dependence of $\Delta n^2(x, z)$. To be specific, assume that the period Λ of $\Delta n^2(x, z)$ is chosen so that $l\pi/\Lambda \approx \beta_s$ for an integer l. We can expand $\Delta n^2(x, z)$ as

$$\Delta n^2(x, z) = \Delta n^2(x) \sum_{q=-\infty}^{\infty} a_q e^{i(2q\pi/\Lambda)z}$$

The right side of (22.4-3) now contains a term ($q = l$, $m = s$) proportional to $A_s^{(+)} \exp[i(2l\pi/\Lambda - \beta_s)z]$. But

$$\frac{2l\pi}{\Lambda} - \beta_s \approx \beta_s \tag{22.4-4}$$

so that this term is capable of driving synchronously the amplitude $A_s^{(-)} \exp(i\beta_s z)$ on the left side of (22.4-3) with the result

$$\frac{dA_s^{(-)}}{dz} = \frac{i\omega\varepsilon_0}{4} A_s^{(+)} \int_{-\infty}^{\infty} \Delta n^2(x) [\mathscr{E}_y^{(s)}(x)]^2\, dx a_l e^{i[(2l\pi/\Lambda) - 2\beta_s]z} \tag{22.4-5}$$

The coupling between the backward mode $A_s^{(-)}$ and the forward $A_s^{(+)}$ one by the lth harmonic of $\Delta n^2(x, z)$ can thus be described by

$$\frac{dA_s^{(-)}}{dz} = \kappa A_s^{(+)} e^{-i2(\Delta\beta)z} \tag{22.4-6}$$

and reciprocally

$$\frac{dA_s^{(+)}}{dz} = \kappa^* A_s^{(-)} e^{i2(\Delta\beta)z}$$

where

$$\kappa = \frac{i\omega\varepsilon_0 a_l}{4} \int_{-\infty}^{\infty} \Delta n^2(x)[\mathscr{E}_y^{(s)}(x)]^2\, dx \tag{22.4-7}$$

$$\Delta\beta \equiv \beta_s - \frac{l\pi}{\Lambda} \equiv \beta_s - \beta_0 \tag{22.4-8}$$

We note that the total power carried by both modes is conserved since

$$\frac{d}{dz}[|A_s^{(-)}|^2 - |A_s^{(+)}|^2] = 0 \tag{22.4-9}$$

Let us return to the specific "square wave" corrugation of Figure 22.4. In this case,

$$\begin{aligned}\Delta n^2(x, z) &= \Delta n^2(x)\left[\frac{1}{2} + \frac{2}{\pi}\left(\sin \eta z + \frac{1}{3}\sin 3\eta z + \cdots\right)\right] \\ &= \Delta n^2(x)\sum_l a_l e^{i\eta l z}\end{aligned} \tag{22.4-10}$$

where

$$\Delta n^2(x) = \begin{cases} n_2^2 - n_1^2 & -a \leqslant x \leqslant 0 \\ 0 & \text{elsewhere}\end{cases} \tag{22.4-11}$$

$$\eta \equiv 2\pi/\Lambda$$

so that

$$a_l = \begin{cases} \dfrac{i}{\pi l} & l \text{ odd} \\ 0 & 0 \text{ even}\end{cases} \qquad a_0 = \tfrac{1}{2}$$

and for l odd

$$\kappa = \frac{-\omega\varepsilon_0}{4\pi l}\int_{-\infty}^{\infty} \Delta n^2(x)[\mathscr{E}_y^{(s)}(x)]^2\, dx \tag{22.4-12}$$

In practice, Λ is chosen so that, for some particular l, $\Delta\beta \approx 0$. We note that for $\Delta\beta = 0$

$$\Lambda = l\frac{\lambda_g^{(s)}}{2} \tag{22.4-13}$$

where $\lambda_g^{(s)} = 2\pi/\beta_s$ is the guide wavelength of the sth mode.

We can now use the field expansion (22.2-3) plus (22.4-11) to perform the integration of (22.4-12).

$$\begin{aligned}\int_{-\infty}^{\infty} \Delta n^2(x)[\mathscr{E}_y^{(s)}(x)]^2\, dx &= (n_2^2 - n_1^2)\int_{-a}^{0}[\mathscr{E}_y^{(s)}(x)]^2\, dx \\ &= (n_2^2 - n_1^2)C_s^2 \int_{-a}^{0}\left[\cos(h_s x) - \frac{q_s}{h_s}\sin(h_s x)\right]^2 dx\end{aligned} \tag{22.4-14}$$

Although the integral can be calculated exactly using (22.2-3) and (22.2-5), an especially simple result ensues if we consider that operation is sufficiently above the propagation cutoff, so that

$$\begin{aligned} & h_s \to \pi s/t \qquad s = 1, 2, \ldots = \text{transverse mode number} \\ & \frac{q_s}{h_s} \approx (n_2^2 - n_1^2)^{1/2} \left(\frac{2t}{s\lambda}\right) \\ & \beta_s \approx n_2 k \end{aligned} \tag{22.4-15}$$

These results can be verified using (22.2-4) and (22.2-5). In addition, since $q_s \gg h_s$, we have, from (22.2-7),

$$C_s^2 = \frac{4h_s^2\omega\mu}{\beta_s t q_s^2} \tag{22.4-16}$$

in the well-confined regime, and for $h_s a \ll 1$, the integral of (22.4-14) becomes

$$(n_2^2 - n_1^2) \int_{-a}^{0} [\mathscr{E}_y^{(s)}(x)]^2 \, dx = (n_2^2 - n_1^2) \frac{4\pi^2\omega\mu}{3n_2 k} \left(\frac{a}{t}\right)^3 \left(1 + \frac{3}{q_s a} + \frac{3}{q_s^2 a^2}\right)$$

and, if we use (22.4-15),

$$\kappa_s \approx \frac{2\pi^2 s^2}{3l\lambda} \frac{(n_2^2 - n_1^2)}{n_2} \left(\frac{a}{t}\right)^3 \left[1 + \frac{3}{2\pi} \frac{\lambda/a}{(n_2^2 - n_1^2)^{1/2}} + \frac{3}{4\pi^2} \frac{(\lambda/a)^2}{(n_2^2 - n_1^2)}\right] \tag{22.4-17}$$

The problem has thus been reduced to a pair of coupled differential equations (22.4-6) and an expression (22.4-17) for the coupling constant.

22.5 THE COUPLED-MODE SOLUTIONS

Let us return to the coupled mode equations (22.4-6). For simplicity, let us put $A_s^{(-)} \equiv A$, $A_s^{(+)} \equiv B$ and write them as

$$\begin{aligned} \frac{dA}{dz} &= \kappa_{ab} B e^{-i2(\Delta\beta)z} \qquad && \Delta\beta = \beta - \beta_0 \\ \frac{dB}{dz} &= \kappa_{ab}^* A e^{+i2(\Delta\beta)z} \qquad && \beta_0 = l\pi/\Lambda, \ l = 1, 2, 3, \ldots \end{aligned} \tag{22.5-1}$$

Consider a corrugated section of length L as in Figure 22.5. A wave with an amplitude $B(0)$ is incident from the left on the corrugated section.

The solution of (22.5-1) for this case subject to $A(L) = 0$ is

$$\begin{aligned} A(z)e^{i\beta z} &= B(0) \frac{i\kappa_{ab} e^{i\beta_0 z}}{-\Delta\beta \sinh(SL) + iS\cosh(SL)} \sinh[S(z - L)] \\ B(z)e^{-i\beta z} &= B(0) \frac{e^{-i\beta_0 z}}{-\Delta\beta \sinh(SL) + iS\cosh(SL)} \\ &\quad \cdot \{\Delta\beta \sinh[S(z - L)] + iS \cosh[S(z - L)]\} \end{aligned} \tag{22.5-2}$$

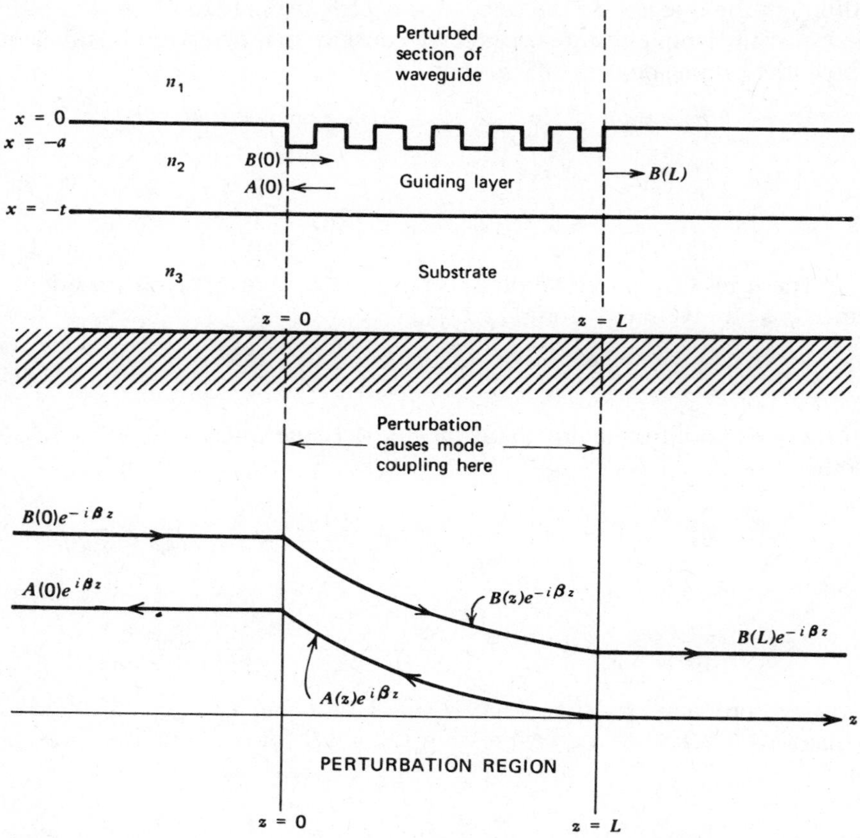

FIGURE 22.5 Upper: A corrugated section of a dielectric waveguide. Lower: The incident and reflected fields.

where

$$S = \sqrt{\kappa^2 - (\Delta\beta)^2}, \qquad \kappa \equiv |\kappa_{ab}| \tag{22.5-3}$$

If the fields at $z > L$ and $z < 0$ are taken as in Figure 22.5, it follows from (22.5-2) that the total field $A(z)\exp(i\beta z) + B(z)\exp(-i\beta z)$ as well as its derivative are continuous at $z = 0$ and $z = L$,

$$A(z) = B(0)\left(\frac{\kappa_{ab}}{\kappa}\right)\frac{\sinh[\kappa(z - L)]}{\cosh(\kappa L)}$$

$$B(z) = B(0)\frac{\cosh[\kappa(z - L)]}{\cosh(\kappa L)} \tag{22.5-4}$$

A plot of the mode powers $|B(z)|^2$ and $|A(z)|^2$ for this case is shown in Figure 22.5. For sufficiently large arguments of the cosh and sinh functions in (22.5-4), the incident mode power drops off exponentially along the perturbation region. This behavior, however, is due not to absorption but to reflection of power into the backward traveling mode A.

From (22.3-5) and (22.5-2), we find that the z-dependent part of the wave solutions in the periodic waveguide are exponentials with propagation constants

$$\beta' = \beta_0 \pm iS = l\frac{\pi}{\Lambda} \pm i\sqrt{\kappa^2 - [\beta(\omega) - \beta_0]^2} \qquad (22.5\text{-}5)$$

where we used $\Delta\beta \equiv \beta - \beta_0$, $\beta_0 \equiv l\pi/\Lambda$.

We note that for a range of frequencies such that $\Delta\beta(\omega) < \kappa$, β' has an imaginary part. This is the so called "forbidden" region in which the evanescence behavior shown in Figure 22.5 occurs and that is formally analogous to the energy gap in semiconductors where the periodic crystal potential causes the electron propagation constants to become complex. Note that for each value of l, $l = 1, 2, 3, \ldots$, there exists a gap whose center frequency ω_{0l} satisfies $\beta(\omega_{0l}) = l\pi/\Lambda$. The exceptions are values of l for which κ is zero. Returning to (22.5-5) and approximating $\beta(\omega)$ near the Bragg value $(\pi l/\Lambda)$ by $\beta(\omega) \approx (\omega/c)n_{\text{eff}}$ where n_{eff} is an effective index of refraction, we have

$$\beta' \cong \frac{l\pi}{\Lambda} \pm i\left[\kappa^2 - \left(\frac{n_{\text{eff}}}{c}\right)^2 (\omega - \omega_0)^2\right]^{1/2} \qquad (22.5\text{-}6)$$

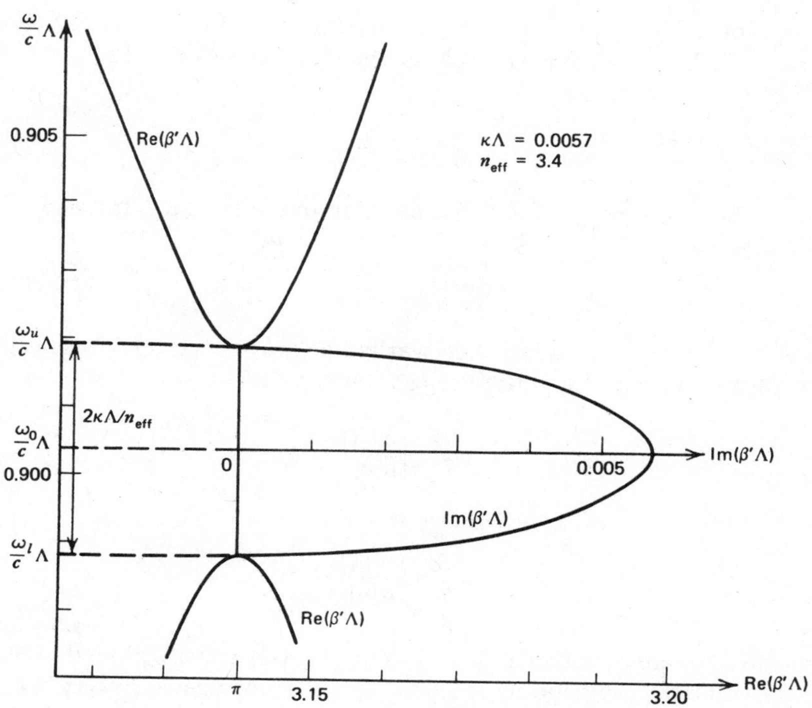

FIGURE 22.6 The dependence of the real and imaginary parts of the mode propagation constant β' of the modes in a periodic waveguide. At frequencies $\omega_l < \omega < \omega_v$, Im$(\beta') \neq 0$ and the modes are evanscent. At these frequencies, Re $\beta' = \pi/\Lambda$.

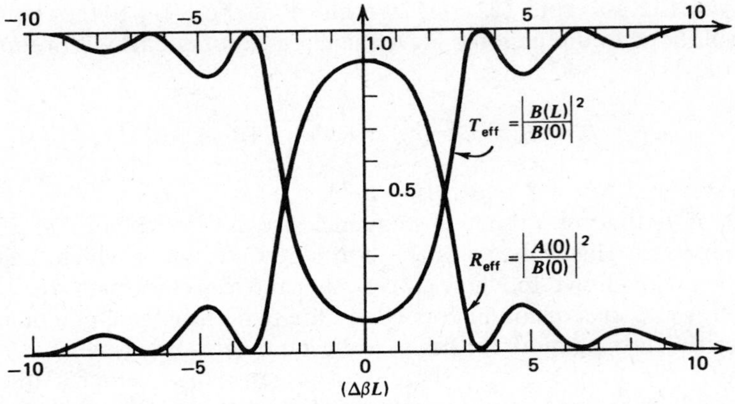

FIGURE 22.7 The transmission and reflection characteristics of a corrugated section of length L as a function of the detuning $\Delta\beta L \approx [(\omega - \omega_0)L/c]n_{eff}$. ($\kappa L =$ 1.84.)

where ω_0, the midgap frequency, is the value of ω for which the unperturbed β is equal to $\beta_0 \equiv l\pi/\Lambda$.

A plot of Re β' and Im β' versus ω, (for $l = 1$) based on (22.5-6), is shown in Figure 22.6. We note that the height of the "forbidden" frequency zone is

$$(\Delta\omega)_{gap} = \frac{2\kappa c}{n_{eff}} \tag{22.5-7}$$

where κ is according to (22.4-17) a function of the integer l. It follows from (22.5-6) that

$$(\mathrm{Im}\,\beta')_{max} = \kappa \tag{22.5-8}$$

A short section of a corrugated waveguide thus acts as a high-reflectivity mirror for frequencies near the Bragg value ω_0. The transmission

$$T_{eff} = \left|\frac{B(L)}{B(0)}\right|^2$$

and reflection

$$R_{eff} = \left|\frac{A(0)}{B(0)}\right|^2$$

of such a filter are obtainable directly from (22.5-2) and are plotted in Figure 22.7. Actual transmission characteristics of a corrugated waveguide are shown in Figure 22.8. The forbidden frequency gap can be observed experimentally in the output spectrum of distributed feedback lasers as discussed at the end of chapter 22.

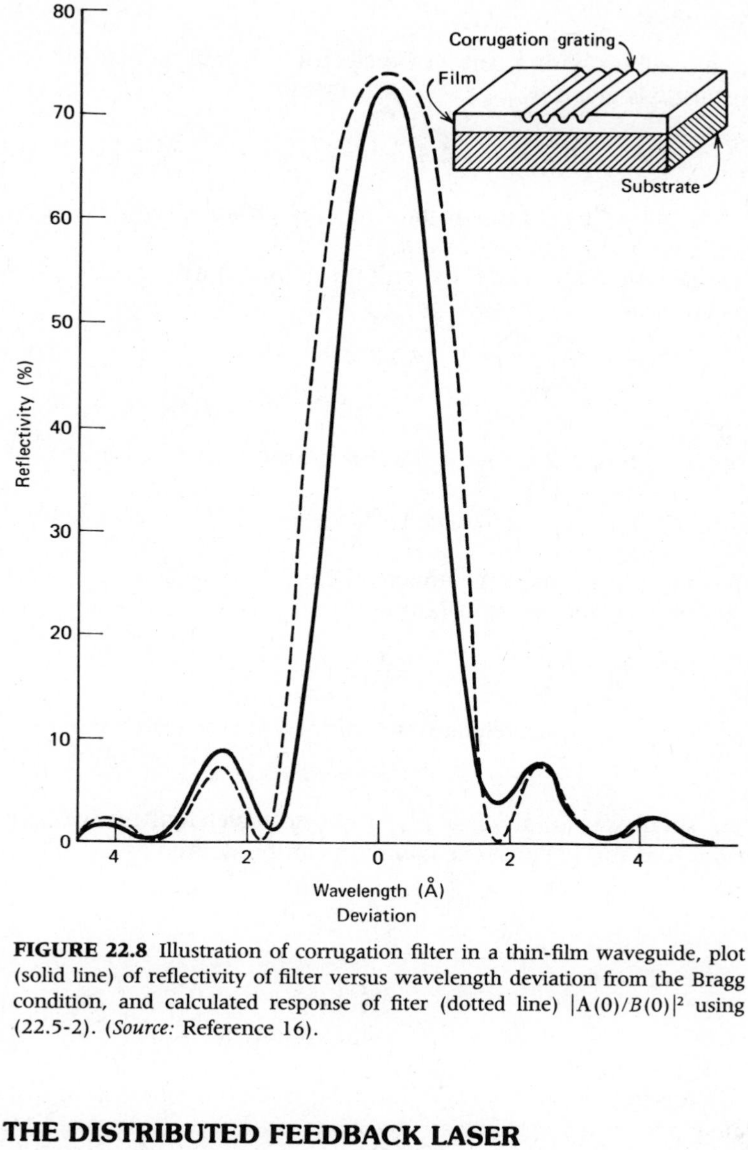

FIGURE 22.8 Illustration of corrugation filter in a thin-film waveguide, plot (solid line) of reflectivity of filter versus wavelength deviation from the Bragg condition, and calculated response of fiter (dotted line) $|A(0)/B(0)|^2$ using (22.5-2). (*Source:* Reference 16).

22.6 THE DISTRIBUTED FEEDBACK LASER

If a periodic medium is provided with sufficient gain at frequencies near the Bragg frequency ω_0 (where $\beta \approx l\pi/\Lambda$), oscillation can result without the benefit of end reflectors. The feedback is now provided by the continuous coherent backscattering from the periodic perturbation. We will consider in what follows two generic cases: (1) the bulk properties of a medium are perturbed periodically (Reference 17). (2) The boundary of a waveguide laser is perturbed periodically (Reference 18). Both cases will be found to lead to the same set of equations.

Bulk Periodicity

Consider a medium with a complex dielectric constant so that the propagation constant k is given by

$$k^2 = \omega^2\mu\varepsilon = \omega^2\mu(\varepsilon_r + i\varepsilon_i) = k_0^2 n^2(z)\left[1 + i\,\frac{2\gamma(z)}{k_0 n}\right] \tag{22.6-1}$$

so that k_0 is the propagation constant in vacuum and (for $\gamma \ll k_0$), γ is the amplitude exponential gain constant.[1]

In a case where the index $n(z)$ and the gain $\gamma(z)$ are harmonic functions of z, we can write

$$\begin{aligned} n(z) &= n + n_1 \cos 2\beta_0 z \\ \gamma(z) &= \gamma + \gamma_1 \cos 2\beta_0 z \end{aligned} \tag{22.6-2}$$

If we use (22.6-2) in (22.6-1) and limiting ourselves to the case $n_1 \ll n$, $\gamma_1 \ll \gamma$

$$k^2(z) = k_0^2 n^2 + i2k_0 n\gamma + 4k_0 n\left(\frac{\pi n_1}{\lambda} + i\,\frac{\gamma_1}{2}\right)\cos 2\beta_0 z$$

the propagation constant in the unperturbed case ($\gamma_1 = 0$, $n_1 = 0$) is $\beta = k_0 n$. If, in addition, we define a constant κ by

$$\kappa = \frac{\pi n_1}{\lambda} + i\,\frac{\gamma_1}{2} \tag{22.6-3}$$

where λ is the vacuum wavelength, we can rewrite the expression for $k^2(z)$ as

$$k^2(z) = \beta^2 + i2\beta\gamma + 4\beta\kappa\cos(2\beta_0 z) \tag{22.6-4}$$

For a small fractional variation of k^2 per wavelength, it was shown in Section 6.5 that the scalar wave equation can be written as

$$\frac{d^2E}{dz^2} + k^2(z)E = 0 \tag{22.6-5}$$

or if we use (22.6-4),

$$\frac{d^2E}{dz^2} + [\beta^2 + i2\beta\gamma + 4\beta\kappa\cos(2\beta_0 z)]E = 0$$

In the discussion following (22.4-3), it was pointed out that a spatially modulated parameter varying as $\cos 2\beta_0 z$ can couple a forward-traveling wave $\exp(-i\beta z)$ and a backward $\exp(i\beta z)$ wave, provided $\beta_0 \cong \beta$. When this (Bragg) condition is nearly satisfied, it is impossible to describe the field $E(z)$ by a single traveling wave, but to a high degree of approximation, we can take it as a linear superposition of both oppositely traveling waves

$$E(z) = A'(z)e^{i\beta' z} + B'(z)e^{-i\beta' z} \tag{22.6-6}$$

[1] γ used here is thus one-half of that appearing in (8.4-3).

where β' is the propagation constant of the uncoupled ($\kappa = 0$) waves

$$\beta'^2 = \beta^2 + i2\beta\gamma$$
$$(\beta' \approx \beta + i\gamma, \gamma \ll \beta) \tag{22.6-7}$$

Using (22.6-6) and

$$\frac{d^2}{dz^2}[A'(z)e^{i\beta' z}] = -\left(\beta'^2 A' - 2i\beta' \frac{dA'}{dz} - \frac{d^2A'}{dz^2}\right) e^{i\beta' z}$$

in (22.6-6), assuming "slow" variation so that $d^2A'/dz^2 \ll \beta' \, dA'/dz$, gives

$$i\beta' \frac{dA'}{dz} e^{i\beta' z} - i\beta' \frac{dB'}{dz} e^{-i\beta' z} = -\beta\kappa e^{i(2\beta_0 - \beta')z}B' - \beta\kappa e^{-i(2\beta_0 - \beta')z}A'$$
$$-\beta\kappa e^{i(2\beta_0 + \beta')z}A' - \beta\kappa e^{-i(2\beta_0 + \beta')z}B' \tag{22.6-8}$$

For operation near the Bragg condition, $\beta_0 \approx \beta$, we can equate terms with nearly equal phase variation (i.e., synchronous), thus ignoring the last two terms in (22.6-8). We obtain

$$\frac{dA'}{dz} = i\kappa B' e^{-i2(\beta' - \beta_0)z} = i\kappa B' e^{-i2(\Delta\beta + i\gamma)z}$$
$$\frac{dB'}{dz} = -i\kappa A' e^{+i2(\beta' - \beta_0)z} = -i\kappa A' e^{i2(\Delta\beta + i\gamma)z} \tag{22.6-9}$$
$$\Delta\beta \equiv \beta - \beta_0$$

We will next derive a similar set of equations to describe a corrugated waveguide laser.

Corrugated Waveguide Distributed Feedback Laser

The case of a passive corrugated waveguide is described by (22.5-1). If the guiding medium possesses gain, we simply need to modify these equations by adding gain terms so that when $\kappa = 0$, the two independent solutions, $A(z)$ and $B(z)$, correspond to exponentially growing waves along the $-z$ and $+z$ directions, respectively. We thus replace (22.5-1) by

$$\frac{dA}{dz} = \kappa_{ab} B e^{-i2(\Delta\beta)z} - \gamma A$$
$$\frac{dB}{dz} = \kappa_{ab}^* A e^{i2(\Delta\beta)z} + \gamma B \tag{22.6-10}$$

where κ is given by (22.4-12) and γ is the exponential gain constant of the medium. If we define $A'(z)$ and $B'(z)$ by

$$A(z) = A'(z)e^{-\gamma z}$$
$$B(z) = B'(z)e^{\gamma z} \tag{22.6-11}$$

Equations (22.6-10) become

$$\frac{dA'}{dz} = \kappa_{ab} B' e^{-i2(\Delta\beta + i\gamma)z}$$
$$\frac{dB'}{dz} = \kappa_{ab}^{*} A' e^{+i2(\Delta\beta + i\gamma)z} \tag{22.6-12}$$

and are thus in a form identical to that of (22.6-9) derived for the case of a bulk periodic medium with index modulation.

Equations (22.6-12) become identical to (22.5-1), provided we replace

$$\Delta\beta \rightarrow \Delta\beta + i\gamma \tag{22.6-13}$$

With this substitution, we can then use (22.5-2) to obtain directly the solution for the total complex field $E(z) = B'(z) \exp[(-i\beta + \gamma)z] + A'(z) \exp[(i\beta - \gamma)z]$ within the periodic section of length L of the waveguide. If we assume an input incident field of $B(0)$ at $z = 0$, the solutions of (22.6-12) for the "forward" wave $B'(z) \exp[(-i\beta + \gamma)z]$ and the "backward" wave $A'(z) \exp[(i\beta - \gamma)z]$ are

$$B'(z)e^{[(-i\beta+\gamma)z]} = B(0) \frac{e^{-i\beta_0 z}\{(\gamma - i\,\Delta\beta)\sinh[S(L-z)] - S\cosh[S(L-z)]\}}{(\gamma - i\,\Delta\beta)\sinh(SL) - S\cosh(SL)} \tag{22.6-14}$$

$$A'(z)e^{[(i\beta-\gamma)z]} = B(0) \frac{\kappa_{ab} e^{i\beta_0 z} \sinh[S(L-z)]}{(\gamma - i\,\Delta\beta)\sinh(SL) - S\cosh(SL)} \tag{22.6-14a}$$

where $S^2 = \kappa^2 + (\gamma - i\,\Delta\beta)^2$, $\kappa^2 = |\kappa_{ab}|^2$.

The fact that S now is complex makes for a qualitative difference between the behavior of the passive periodic guide (22.5-2) and the periodic guide with gain (22.6-14). To demonstrate this difference, consider the case when the condition

$$(\gamma - i\,\Delta\beta)\sinh(SL) = S\cosh(SL) \tag{22.6-15}$$

is satisfied. It follows from (22.6-14) that both the reflectance, $E_r(0)/E_i(0)$, and the transmittance, $E_i(L)/E_i(0)$, become infinite. The device acts as an oscillator since it yields finite output fields $E_r(0)$ and $E_i(L)$ with no input ($E_i(0) = 0$). Condition (22.6-15) is thus the oscillation condition for a distributed feedback laser (Reference 17). For the case of $\gamma = 0$, it follows, from (22.5-2), that $|E_i(L)/E_i(0)| < 1$ and $|E_R(0)/E_i(0)| < 1$ as appropriate to a passive device with no internal gain.

For frequencies very near the Bragg frequency $\omega_0(\Delta\beta = 0)$ and for sufficiently high gain so that (22.6-15) is nearly satisfied, the guide acts as a high gain amplifier. The amplified output is available either in reflection with a ("voltage") gain

$$\frac{E_r(0)}{E_i(0)} = \frac{\kappa_{ab}\sinh(SL)}{(\gamma - i\,\Delta\beta)\sinh(SL) - S\cosh(SL)} \tag{22.6-16}$$

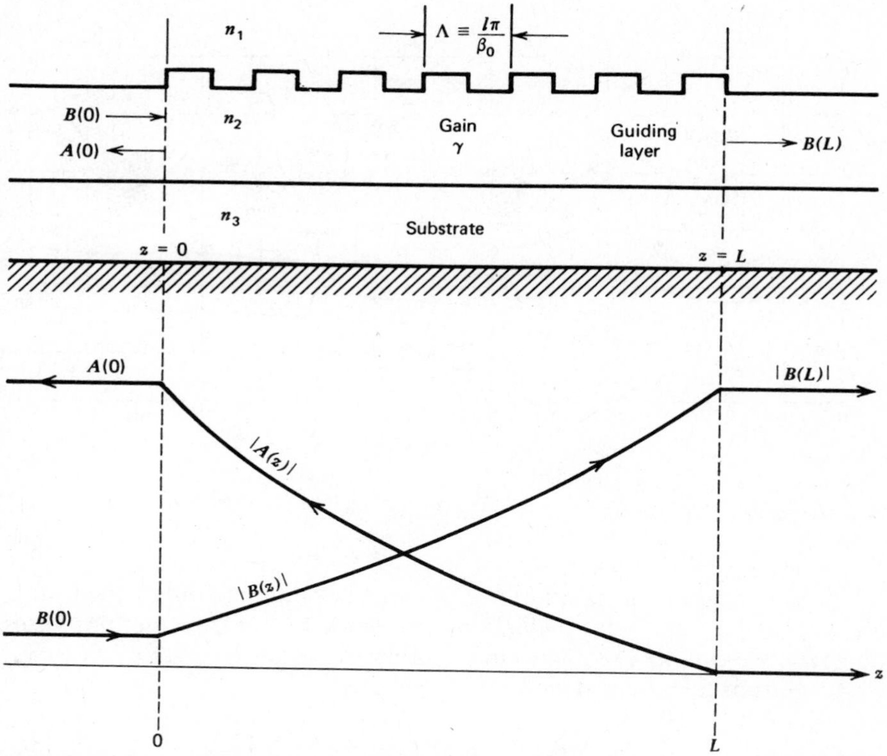

FIGURE 22.9 The incident and reflected fields inside an amplifying periodic waveguide near the Bragg condition $\beta \approx \frac{l\pi}{\Lambda}$.

or in transmission with a gain

$$\frac{E_i(L)}{E_i(0)} = \frac{-Se^{-i\beta_0 L}}{(\gamma - i\,\Delta\beta)\sinh(SL) - S\cosh(SL)} \tag{22.6-17}$$

The behavior of the incident and reflected field for a high gain case is sketched in Figure 22.9. Note the qualitative difference between this case and the passive one depicted in Figure 22.5.

The reflection gain, $|E_r(0)/E_i(0)|^2$, and the transmission gain, $|E_i(L)/E_i(0)|^2$, are plotted in Figures 22.10*a* and 22.10*b*, respectively, as a function of $\Delta\beta$ and γ. Each plot contains four infinite gain singularities at which the oscillation condition (22.6-15) is satisfied. These are four of the longitudinal laser modes.

The Oscillation Condition

The oscillation condition (22.6-15) can be written as

$$\frac{S - (\gamma - i\,\Delta\beta)}{S + (\gamma - i\,\Delta\beta)}\, e^{2SL} = -1 \tag{22.6-18}$$

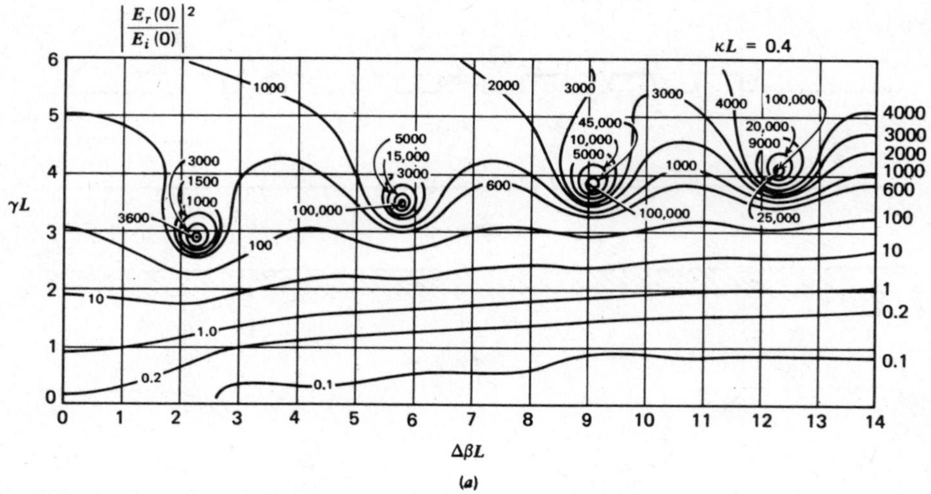

FIGURE 22.10a Reflection Gain Contours in the $\Delta\beta L - \gamma L$ plane.

In general, one has to resort to a numerical solution to obtain the threshold values of $\Delta\beta$ and γ for oscillation (Reference 17). In some limiting cases, however, we can obtain approximate solutions. In the high gain $\gamma \gg \kappa$ case, we have from the above definition of S^2

$$S \approx -(\gamma - i\,\Delta\beta)\left[1 + \frac{\kappa^2}{2(\gamma - i\,\Delta\beta)^2}\right]$$

so that

$$S - (\gamma - i\,\Delta\beta) = -2(\gamma - i\,\Delta\beta)$$

$$S + (\gamma - i\,\Delta\beta) = \frac{-\kappa^2}{2(\gamma - i\,\Delta\beta)}$$

and (22.6-18) becomes

$$\frac{+4(\gamma - i\,\Delta\beta)^2}{\kappa^2}\, e^{2SL} = -1 \qquad (22.6\text{-}19)$$

Equating the phases on both sides of (22.6-19) results in

$$2\tan^{-1}\frac{(\Delta\beta)_m}{\gamma_m} - 2(\Delta\beta)_m L + \frac{(\Delta\beta)_m L\kappa^2}{\gamma_m^2 + (\Delta\beta)_m^2} = (2m+1)\pi \qquad (22.6\text{-}20)$$

$$m = 0, \pm 1, \pm 2, \ldots$$

In the limit $\gamma_m \gg (\Delta\beta)_m, \kappa$, the oscillating mode frequencies are given by

$$(\Delta\beta)_m L \approx -(m + \tfrac{1}{2})\pi$$

The two lowest-order modes correspond to $m = 0$ ($\Delta\beta = -\pi/2L$) and $m = -1(\Delta\beta = +\pi/2L)$. If we use $\Delta\beta \equiv \beta - \beta_0 \approx (\omega - \omega_0)n_{\text{eff}}/c$, the corre-

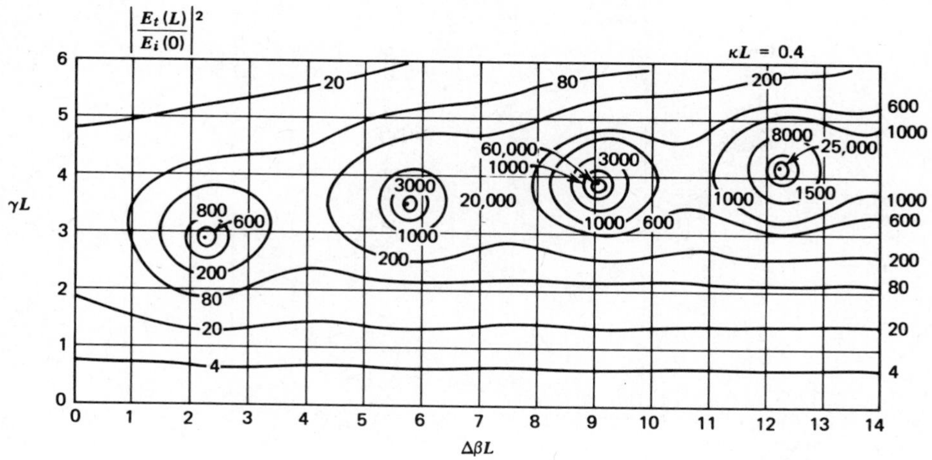

(*b*)

FIGURE 22.10b Transmission Gain Contours in the $\Delta\beta L - \gamma L$ plane.

sponding oscillation frequencies are

$$\omega_{m=0} = \omega_0 - \frac{\pi c}{2n_{\text{eff}}L}$$
$$\omega_{m=-1} = \omega_0 + \frac{\pi c}{2n_{\text{eff}}L} \tag{22.6-21}$$

From the threshold condition (22.6-18), it follows that the threshold gain depends only on the magnitude of $\Delta\beta$ so that the lowest-order modes $m = 0$ and $m = -1$ possess identical thresholds.

In the limit $\gamma_m \gg (\Delta\beta)_m, \kappa$, the oscillating mode frequencies are given by

$$(\Delta\beta_m)L \cong -(m + \tfrac{1}{2})\pi \tag{22.6-22}$$

The mode frequency spacing is

$$\omega_m - \omega_{m-1} \simeq \frac{\pi c}{n_{\text{eff}}L} \tag{22.6-23}$$

and is approximately the same as in a two-reflector resonator of length L.

The threshold gain value γ_m is obtained from the amplitude equality in (22.6-19).

$$\frac{e^{2\gamma_m L}}{\gamma_m^2 + (\Delta\beta)_m^2} = \frac{4}{\kappa^2} \tag{22.6-24}$$

indicating an increase in threshold with increasing mode number m. This is also evident from the numerical gain plots (Figure 22.10)

Figure 22.11 shows a theoretical plot of the threshold gain based on (22.6-18) as well as the experimental data of a GaAlAs corrugated laser.

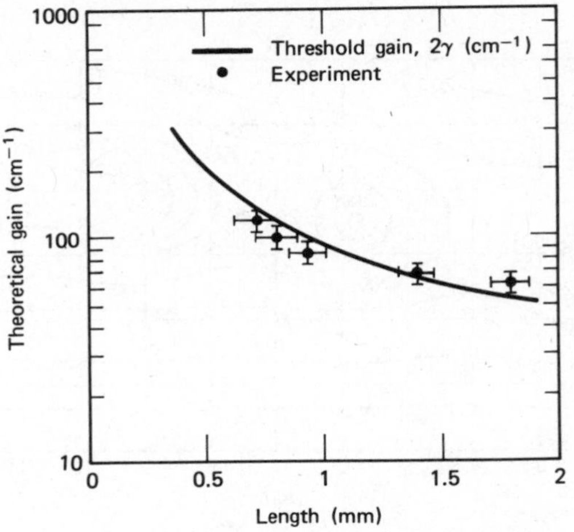

FIGURE 22.11 Threshold gain of a distributed-feedback laser. (*Source:* Reference 19.)

The distributed feedback laser thus has a strong built-in discrimination against all modes except those with $m = \pm 1$ that are equispaced about the Bragg frequency ω_0.

Distributed feedback semiconductor lasers that incorporate a corrugated interface in their epitaxial layered structure have assumed great importance in

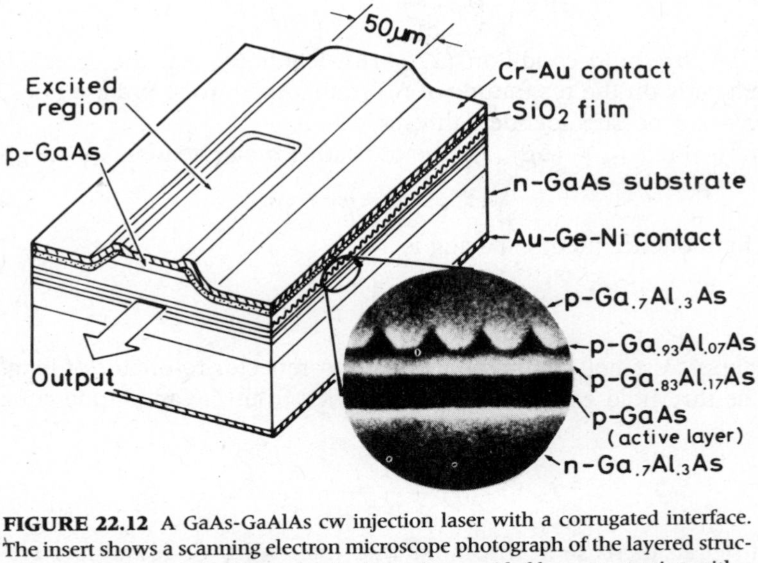

FIGURE 22.12 A GaAs-GaAlAs cw injection laser with a corrugated interface. The insert shows a scanning electron microscope photograph of the layered structure. The feedback is in third order ($l = 3$) and is provided by a corrugation with a period $\Lambda = 3\,\lambda_g/2 = 0.345\ \mu$m. The thin (0.2 μm) p-$Ga_{.83}Al_{.17}As$ layer provides a potential barrier that confines the injected electrons to the active (p-GaAs) layer, thus increasing the gain. (*Source:* Reference 20.)

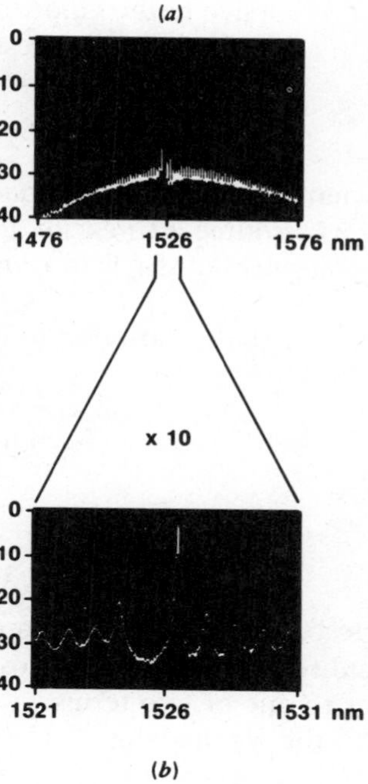

FIGURE 22.13 The spontaneous emission spectrum of a distributed feedback InP/GaInP laser showing (*a*) the longitudinal modes. (*b*) An expanded portion of (*a*) showing the forbidden gap and a few modes to either side of it. (*Source:* Reference 21.)

the optical fiber communication field. This is due to the fact that they can be made to oscillate in a single mode. (Our analysis has shown that two modes, one on each side of ω_0, have equal thresholds, but a slight asymmetry in the structure can favor one of these modes of the other.) This single mode feature is important in high data-rate long-distance communication in optical fiber using the preferred wavelengths of $\lambda = 1.33\ \mu$m and $1.55\ \mu$m. Oscillation at more than one wavelength severely limits the distance-data-rate product by causing pulse spreading (with distance) due to group velocity dispersion (see Section 22.10) as well as leading to excess noise that accompanies the hopping of the laser power between the modes. Figure 22.12 shows a GaAs-GaAlAs semiconductor laser with a corrugated interface (Reference 19).

In Figure 22.13, we show experimental data of the spectrum of the output of a $1.55\ \mu$m distributed feedback laser operated just below threshold so that the higher-order modes are comparable in their output power to that of the main ($m = 0, -1$) modes. Some unintentional residual asymmetry favors the $m = 0$ mode over the $m = -1$ one. The gap in the mode spectrum corresponds to the forbidden gap.

22.7 ELECTROOPTIC MODULATION AND MODE COUPLING IN DIELECTRIC WAVEGUIDES

As another important class of thin film devices, we consider the case of electrooptic modulation and mode coupling. To be specific, we consider the case where a forward-traveling TM mode $[E_x^{(\omega)} \neq 0]$ is applied to the guide in the presence of a dc field $E^{(0)}$. The effect, according to (16.0-1), is to produce a polarization

$$[P_{\text{pert}}(t)]_y \propto rE^{(0)}E_x^{(\omega)}e^{i\omega t} \tag{22.7-1}$$

Symbol r is an appropriate electrooptic coefficient (or a linear combination of such coefficients). This polarization, acting as a source, can excite, in accordance with (22.3-9), a TE wave $A_s^{(+)}$. The application of the field thus causes TM → TE power conversion.

If we use (16.0-1), the complex amplitude of the polarization produced by the TM mode in the presence of a dc field $\mathbf{E}^{(0)}$ is

$$P_y^{(\omega)}(\mathbf{r}) = \frac{\varepsilon^2 rE^{(0)}}{\varepsilon_0} E_x^{(\omega)}(\mathbf{r}) \tag{22.7-2}$$

where

$$\varepsilon^2 rE^{(0)} \equiv \varepsilon_i \varepsilon_j r_{ijk} l_{iy} l_{jx} E_k^{(0)} \tag{22.7-3}$$

Here, i, j, k refer to the crystalline principal dielectric axes, whereas x and y are the waveguide axes as in Figure 22.1. The l's are direction cosines. In most practical cases, the summation of (22.7-3) reduces to one or two terms.

Using (22.2-9), we write the input TM field in the lth mode as

$$E_x^{(l)}(\mathbf{r}, t) = \frac{\beta_l}{2\omega\varepsilon(x)} B_l \mathcal{H}_y^{(l)}(x) e^{i(\omega t - \beta_l z)} + \text{c.c.} \tag{22.7-4}$$

where $\mathcal{H}_y^{(l)}(x)$ is given by (22.2-10) and $|B_l|^2$ is the mode power per unit width in the y direction. The polarization (22.7-2) can thus be written as

$$P_y(\mathbf{r}, t) = \frac{\varepsilon^2 r(x, z)E^{(0)}}{2\omega\varepsilon_0\varepsilon(x)} \beta_l B_l \mathcal{H}_y^{(l)}(x) e^{i(\omega t - \beta_l z)} + \text{c.c.} \tag{22.7-5}$$

Substitution of (22.7-5) into the wave equation (22.3-9) leads to

$$\frac{dA_m^{(+)}}{dz} \exp(-i\beta_m^{\text{TE}} z) - \frac{dA_m^{(-)}}{dz} \exp(i\beta_m^{\text{TE}} z)$$
$$= -\frac{i}{4} \int_{-\infty}^{\infty} \frac{\varepsilon^2 r(x, z)E^{(0)}(x, z)}{\varepsilon(x)\varepsilon_0} \beta_l B_l \mathcal{H}_y^{(l)}(x) \mathcal{E}_y^{(m)}(x)\, dx \exp(-i\beta_l^{\text{TM}} z) \tag{22.7-6}$$

If $\beta_l^{\text{TM}} \approx \beta_m^{\text{TE}}$, the coupling excites only the A_m^+ wave, that is, it is codirectional. Dropping the plus and minus superscripts, we can rewrite (22.7-6) as

$$\frac{dA_m}{dz} = -i\kappa_{ml}(z) B_l e^{-i(\beta_m^{\text{TE}} - \beta_l^{\text{TM}})z} \tag{22.7-7}$$

$$\kappa_{ml} = \frac{\beta_l}{4} \int_{-\infty}^{\infty} \frac{\varepsilon^2 r(x, z)E^{(0)}(x, z)}{\varepsilon(x)\varepsilon_0} \mathcal{H}_y^{(l)}(x) \mathcal{E}_y^{(m)}(x)\, dx \tag{22.7-8}$$

Equation (22.7-8) is general enough to apply to a large variety of cases. The dependence of $E^{(0)}$ and $r(x, z)$ on x accounts for coupling by electrooptic material in the guiding or in the bounding layers. The z dependence allows for situations where $E^{(0)}$ or r depend on position. To be specific, we consider first the case where the guiding layer $-t < x < 0$ is uniformly electrooptic and where $E^{(0)}$ is uniform over the same region so that the integration is from $-t$ to 0. In that case, the overlap integral of (22.7-8) is maximum when the TE(m) and TM(l) modes are well confined and of the *same* order so that $l = m$. Under well-confined conditions $p,\ q \gg h$ and the expressions (22.2-3), (22.2-7) for $\mathscr{E}_y^{(m)}(x)$ and (22.2-10), (22.2-13) for $\mathscr{H}_y^{(m)}(x)$ in the guiding layer become

$$\mathscr{E}_y^{(m)}(x) \to \left(\frac{4\omega\mu}{t\beta_m^{\text{TE}}}\right)^{1/2} \sin\frac{m\pi x}{t}$$

$$\mathscr{H}_y^{(m)}(x) \to \left(\frac{4\omega\varepsilon_0 n_2^2}{t\beta_m^{\text{TM}}}\right)^{1/2} \sin\frac{m\pi x}{t}$$

where for well-confined mode $\beta_l^{\text{TM}} \simeq \beta_m^{\text{TE}} \equiv \beta = kn_2$. In this case, the overlap integral becomes

$$\int_{-t}^{0} \mathscr{H}_y^{(m)}(x)\mathscr{E}_y^{(m)}(\text{x})\, dx = \frac{4\omega\sqrt{\mu\varepsilon_2}}{t\beta}\int_{-t}^{0} \sin^2\frac{m\pi x}{t}\, dx = 2$$

and the coupling coefficient (22.7-8) achieves a maximum value of

$$\kappa \to \frac{n_2^3 k r E^{(0)}}{2} \qquad (22.7\text{-}9)$$

The coupling is thus described by

$$\frac{dA_m}{dz} = -i\kappa B_m e^{-i(\beta_m^{\text{TM}} - \beta_m^{\text{TE}})z}$$

and (22.7-10)

$$\frac{dB_m}{dz} = -i\kappa A_m e^{i(\beta_m^{\text{TM}} - \beta_m^{\text{TE}})z}$$

The second equation of (22.7-10) can be obtained by a process similar to that leading to the first equation or by invoking the conservation of total power (Reference 12), which shows that the above expression for dB_m/dz is consistent with

$$\frac{d}{dz}\left(|A_m|^2 + |B_m|^2\right) = 0$$

For the phase-matched condition $\beta_m^{\text{TM}} = \beta_m^{\text{TE}}$, the solution of (22.7-10) subject to $B_m(0) = B_0$, $A_m(0) = 0$ is

$$\begin{aligned} B_m(z) &= B_0 \cos(\kappa z) \\ A_m(z) &= -iB_0 \sin(\kappa z) \end{aligned} \qquad (22.7\text{-}11)$$

Using (22.7-9), we can show that the field length product $E^{(0)}L$ necessary to effect a complete TM $\leftrightarrow$ TE power transfer in a given distance is the same as that needed to go from "on" to "off" in the bulk modulator shown in Figure

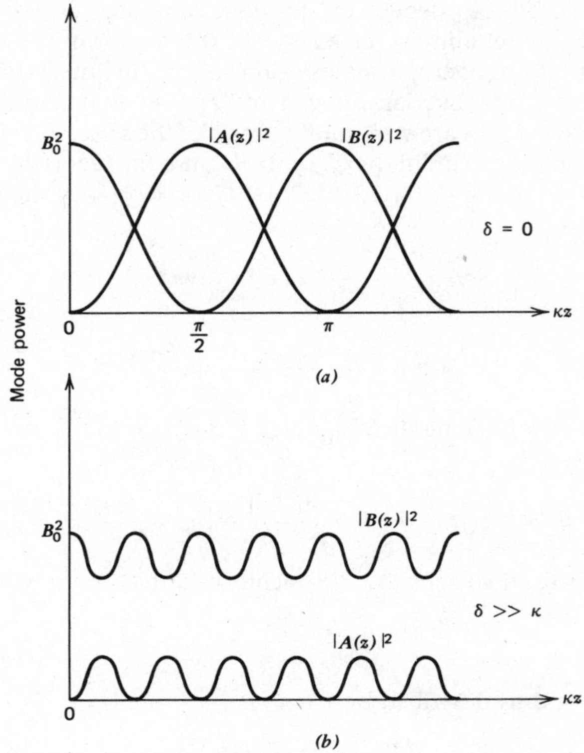

FIGURE 22.14 Power exchange between two coupled modes under (*a*) phase-matched conditions ($\beta_m^{TM} = \beta_m^{TE}$) as described by Eq. (22.7-11). (*b*) $\beta_m^{TM} \neq \beta_m^{TE}$, Eq. (22.7-12).

14.4. This result applies only in the limit of tight confinement. In general, the coupling coefficient κ is smaller than the value given by (22.7-9) and the $E^{(0)}L$ product needed to achieve a complete power transfer is larger.

When $\beta_m^{TM} \neq \beta_m^{TE}$, the solution of (22.7-10), subject to boundary conditions $B_m(0) = B_0$, $A_m(0) = 0$, is

$$B(z) = B_0 e^{i\delta z}\left\{\cos[(\kappa^2 + \delta^2)^{1/2}z] - i\frac{\delta}{(\kappa^2 + \delta^2)^{1/2}}\sin[(\kappa^2 + \delta^2)^{1/2}z]\right\}$$

$$A(z) = -iB_0 e^{-i\delta z}\frac{\kappa}{(\kappa^2 + \delta^2)^{1/2}}\sin[(\kappa^2 + \delta^2)^{1/2}z] \tag{22.7-12}$$

where

$$2\delta \equiv \beta_m^{TM} - \beta_m^{TE} \tag{22.7-13}$$

In contrast to the phase-matched case (22.7-11), the maximum fraction of the power that can be coupled from the input mode B to A is

$$\text{Fraction of power exchanged} = \frac{\kappa^2}{\kappa^2 + \delta^2} \tag{22.7-14}$$

and becomes negligible once $\delta \gg \kappa$.

A plot of the mode power for the phase-matched ($\delta = 0$) and $\delta \neq 0$ case is shown in Figure 22.14.

A deliberate periodic variation of $E^{(0)}(z)$ or $r(z)$ in this case with a period $2\pi/(\beta_m^{\text{TE}} - \beta_m^{\text{TM}})$ can be used, according to (22.7-7), to neutralize the mismatch factor $\exp[-i(\beta_m^{\text{TE}} - \beta_m^{\text{TM}})z]$ in (22.7-10), thus leading again to a phase-matched operation.

Example: GaAs Thin-Film Modulator at $\lambda = 1$ μm. To appreciate the order of magnitude of the coupling, consider a case where the guiding layer is GaAs and $\lambda = 1 \mu\text{m}$. In this case (see Table 14.2),

$$n_2 \simeq 3.5 \qquad n_2^3 r = 59 \times 10^{-12} \frac{\text{m}}{\text{V}}$$

Taking an applied field $E^{(0)} = 10^6$ V/m we obtain, from (22.7-9),

$$\kappa = 1.85 \text{ cm}^{-1}$$

$$l \equiv \frac{\pi}{2\kappa} = 0.85 \text{ cm}$$

for the coupling constant and the power-exchange distance, respectively.

22.8 DIRECTIONAL COUPLING—SUPERMODES

Exchange of power between guided modes of adjacent waveguides is known as directional coupling. Waveguide directional couplers perform a number of useful functions in thin-film devices, including power division, modulation, switching, frequency selection, and polarization selection.

We will treat the subject of directional coupling by two equivalent methods. The first one, which follows immediately, is based on the coupled mode approach in which the proximity of one guide to another is considered a dielectric perturbation on the latter and vice versa. The second approach taken up in Section 22.9 uses the concept and formalism of supermodes.

Consider the case of the two planar waveguides illustrated in Figure 22.15. The refractive index distributions for the two guides in the absence of coupling are given by $n_a(x)$ and $n_b(x)$. The transverse electric field distribution for a particular guided mode of waveguide a alone and a particular mode of waveguide b alone will be denoted by $\mathscr{E}_y^{(a)}(x)$ and $\mathscr{E}_y^{(b)}(x)$, and their propagation constants by β_a and β_b. The field in the coupled-guided structure with an index $n_c(x)$ (for propagation in the positive z direction) is approximated by

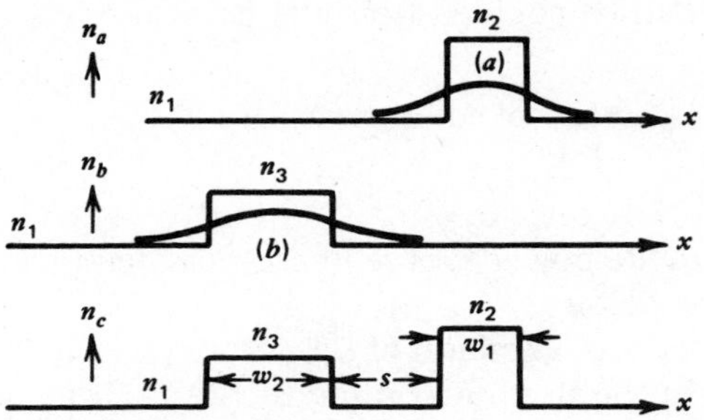

FIGURE 22.15 Spatial variation of the refractive index for uncoupled waveguides $n_a(x)$ and $n_b(x)$, and for a parallel waveguide structure $n_c(x)$.

the sum of the unperturbed fields

$$E_y = A(z)\mathscr{E}_y^{(a)}(x)e^{i(\omega t-\beta_a z)} + B(z)\mathscr{E}_y^{(b)}(x)e^{i(\omega t-\beta_b z)} \tag{22.8-1}$$

In the absence of coupling—that is, if the distance between guides a and b were infinite—$A(z)$ and $B(z)$ do not depend on z and will be independent of each other, since each of the two terms on the right side of (22.8-1) satisfies the wave equation (22.3-1) separately.

The perturbation polarization responsible for the coupling is calculated by substituting (22.8-1) into (22.3-3) and (22.3-9). The result is

$$P_{\text{pert}} = e^{i\omega t}\varepsilon_0[\mathscr{E}_y^{(a)}A(z)[n_c^2(x) - n_a^2(x)]e^{-i\beta_a z} + \mathscr{E}_y^{(b)}B(z)[n_c^2(x) - n_b^2(x)]e^{-i\beta_b z}] \tag{22.8-2}$$

where $n_c(x)$ is the index profile of the two-guide structure. Substituting (22.8-2) in (22.3-9) and integrating over x gives

$$\frac{dA}{dz} = -i\kappa_{ab}Be^{-i(\beta_a-\beta_b)z} - iM_aA$$
$$\frac{dB}{dz} = -i\kappa_{ba}Ae^{i(\beta_a-\beta_b)z} - iM_bB \tag{22.8-3}$$

$$\kappa_{\substack{ab\\ba}} = \frac{\omega\varepsilon_0}{4}\int_{-\infty}^{\infty}[n_c^2(x) - n_{\substack{b\\a}}^2(x)]\mathscr{E}_y^{(a)}\mathscr{E}_y^{(b)}\,dx \tag{22.8-4}$$

where

$$M_{(a,b)} = \frac{\omega\varepsilon_0}{4}\int_{-\infty}^{\infty}[n_c^2(x) - n_{(a,b)}^2(x)](\mathscr{E}_y^{(a,b)})^2\,dx \tag{22.8-5}$$

The terms M_a and M_b represent a small correction to the propagation constants β_a and β_b, respectively, because of the present of the second guide. Therefore, if we take the total field as

$$E_y = A(z)\mathscr{E}_y^{(a)}e^{i[\omega t-(\beta_a+M_a)z]} + B(z)\mathscr{E}_y^{(b)}e^{i[\omega t-(\beta_b+M_b)z]}$$

instead of (22.8-1), Eqs. (22.8-3) become

$$\frac{dA}{dz} = -i\kappa_{ab}Be^{-i2\delta z}$$
$$\frac{dB}{dz} = -i\kappa_{ba}Ae^{i2\delta z} \tag{22.8-6}$$

where

$$2\delta = (\beta_a + M_a) - (\beta_b + M_b)$$

The solution of (22.8-6) subject to a single input at guide $b[B(0) = B_0$, $A(0) = 0]$ is given by (22.7-12). In terms of powers $P_a = AA^*$ and $P_b = BB^*$ in the two guides, the solution in the case $\kappa_{ba} = \kappa_{ab}$ becomes

$$P_a(z) = P_0 \frac{\kappa^2}{\kappa^2 + \delta^2} \sin^2 [(\kappa^2 + \delta^2)^{1/2} z]$$
$$P_b(z) = P_0 - P_a(z) \tag{22.8-7}$$

where $P_0 = |B(0)|^2$ is the input power to guide b. Complete power transfer from b to a occurs in a distance $L = \pi/2\kappa$, provided $\delta = 0$ (that is, equal phase velocities in both modes). For $\delta \neq 0$, the maximum fraction of power that can be transferred is, from (22.8-7),

$$\frac{\kappa^2}{\kappa^2 + \delta^2} \tag{22.8-8}$$

The coupling constant κ is given by (22.8-4). It can be evaluated straightforwardly using the field expressions (22.2-3) in the case of TE modes. In the special case of identical waveguides $h_1 = h_2$ and $p_1 = p_2$ in Figure 22.15, one obtains

$$\kappa = \frac{2h^2 p e^{-ps}}{\beta(w + 2/p)(h^2 + p^2)} \tag{22.8-9}$$

The extension to channel waveguide couplers that are confined in the y, as well as in the x, direction is simple and is discussed in Reference 22. In the well-confined case, $w \gg 2/p$ and (22.8-9) simplifies to (Reference 23)

$$\kappa = \frac{2h^2 p e^{-ps}}{\beta w(h^2 + p^2)} \tag{22.8-10}$$

A typical value of κ obtained at $\lambda \sim 1\ \mu\text{m}$ with $w, s \sim 3\ \mu\text{m}$, and $\Delta n \sim 5 \times 10^{-3}$ is $\kappa \sim 5\ \text{cm}^{-1}$ so that coupling distances are of the order of magnitude of $\kappa^{-1} \approx 2$ mm.

A form of an electrooptic switch based on directional coupling is as follows.

The length L of the coupler is chosen so that for $\delta = 0$ (that is, synchronous case) $\kappa L = \pi/2$. From (22.8-7), it follows that all the input power to guide b exits from guide a at $z = L$. The switching is achieved by applying an electric field to guide a(or b) in such a way as to change its propagation constant until

$$\delta L = \tfrac{1}{2}(\beta_a - \beta_b)L = \frac{\sqrt{3}}{2}\pi \tag{22.8-11}$$

that is, $\delta = \sqrt{3}\kappa$. It follows from (22.8-7) that at this value of δ

$$P_a = 0 \qquad P_b = P_0$$

that is, the power reappears at the output of guide b. A control of δ can thus be used to achieve any division of the powers between the outputs of guides a and b.

In practice, a convenient way to control δ is to fabricate the directional coupler in an electrooptic crystal. In this case, according to (14.1-11), the application of an electric field E across one of the two waveguides will cause the index of refraction to change by

$$\Delta n \sim n^3 rE$$

where r is the appropriate electrooptic tensor element. The change Δn will give rise to a change in the propagation constant

$$\delta \sim \frac{\omega}{c}\Delta n \sim \frac{\omega}{c} n^3 rE$$

The control of the power output from both arms of a directional coupler by means of an applied voltage is illustrated in Figure 22.16. The electrode geometry for applying a field to the waveguide is illustrated in Figure 22.17.

One of the interesting applications for electrooptically switched directional couplers is in the area of very high-frequency ($>5 \times 10^9$ Hz) sampling and of multiplexing and demultiplexing of optical binary pulse trains. An example of the latter is demonstrated by Figure 22.17. Two independent, but synchronized, data pulse trains A and B are fed into legs a and b, respectively, of a directional coupler. The length of the coupling section satisfies the power transfer condition $\kappa L = \pi/2$. The phase mismatch δ between the two waveguides is controlled, as discussed above, by an electric field applied across one of the waveguides. This electric field is due to a microwave signal at a frequency ω_m. The resulting peak phase constant mismatch, which occurs at the

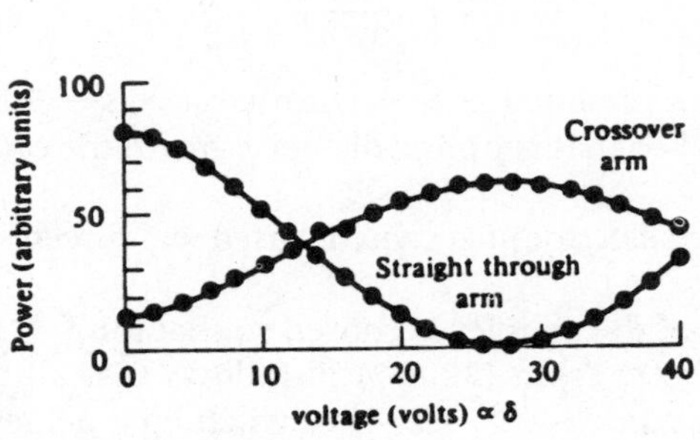

FIGURE 22.16

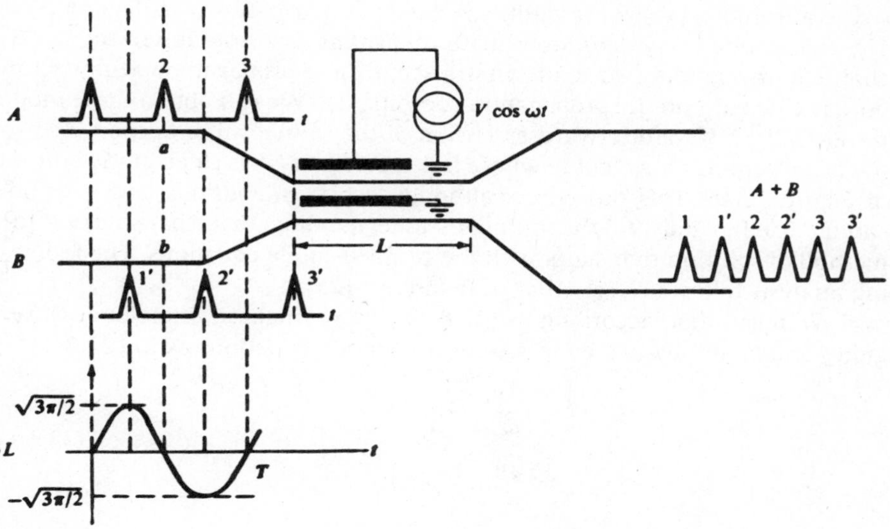

FIGURE 22.17

maxima and minima of the applied voltage, satisfies the condition (22.8-11)

$$|\delta_{\max}| = \frac{\sqrt{3}\,\pi}{2L} \tag{22.8-12}$$

so that the B pulses, which are synchronized to arrive during the extrema of the microwave signal, exit from arm b. Pulses A, on the other hand, arrive when the applied field, hence δ, is zero and, since $\kappa L = \pi/2$, cross over and exit from guide b. The result is that both pulse trains A and B are interleaved, or in the electrical engineering parlance, multiplexed in the output of guide b. The (combined) output from b can be fed into the input of a second directional coupler fed with a signal at $2\omega_m$ and multiplexed thereby with a second data train, and so on.

The device can, of course, be operated in reverse, right to left in the figure, and act as s demultiplexer for separating the dense bit train $A + B$ entering b into the individual trains A and B.

22.9 THE EIGENMODES OF A COUPLED WAVEGUIDE SYSTEM ("SUPERMODES")

In the preceding section we treated the important case of directional coupling between two parallel waveguides (Figure 22.15) by means of the coupled-mode formalism. The same problem may be approached from a different, and equivalent, point of view that is better suited for the treatment of certain important classes of experimental and device configurations. From this new point of view, we seek to obtain the propagating eigenmodes of the *complete* two-waveguide system (or multiwaveguide system) shown in Figure 22.1. The eigenmode is, by definition, that propagating field solution of the multichannel waveguiding structure that, except for a propagation delay factor, does not depend on the propagation coordinate z. We can obtain these mode solutions by a straightforward extension of the formalism of Section 22.1 to the more complex waveguide whose index profile $n_c(x)$ is given at the bottom of Figure 22.15. This procedure, although exact, is laborious and does not submit itself readily to the intuitive understanding that characterizes the method of solution that starts with the coupled-mode equations. The following analysis follows closely that of Reference [25].

We recall that, according to (22.8-6), the normalized, individual waveguide, mode amplitudes obey the coupled-mode equations as in (22.8-6)

$$\frac{dA}{dz} = \kappa B e^{-i2\delta z}$$
$$\frac{dB}{dz} = -\kappa^* A e^{i2\delta z} \tag{22.9-1}$$

These equations, subject to boundary conditions $A(0)$ and $B(0)$ at $z = 0$, specify as in (22.8-1) the total field in terms of the individual waveguides' fields.

$$\text{In guide } a \qquad a(x, z, t) = A(z)\mathscr{E}_y^m(x)e^{i[\omega t-(\beta_a+M_a)z]}$$
$$\text{In guide } b \qquad b(x, z, t) = B(z)\mathscr{E}_y^{(l)}(x)e^{i[\omega t-(\beta_b+M_b)z]} \tag{22.9-2}$$

with m and l denoting the transverse mode order. Since the individual mode field profiles $\mathscr{E}_y^{(m)}(x)$ and $\mathscr{E}_y^{(l)}(x)$ are known as well as β_a, β_b, M_a, M_b, and the frequency ω, *the total field is specified once the (complex) amplitudes $A(z)$ and $B(z)$ are given.* We thus may describe the field at z *uniquely* by means of

$$R(z) = \bar{A}(z)\mathscr{E}_y^{(m)}(x)e^{i(\omega t-(\beta_a+M_a)z]} + \bar{B}(z)\mathscr{E}_y^{(l)}(x)e^{i(\omega t-(\beta_b+M_b)z]} \tag{22.9-2a}$$

or equivalently by means of a "vector"

$$\mathbf{E}(z) \equiv \begin{vmatrix} B(z)e^{-i\beta_b' z} \\ A(z)e^{-i\beta_a' z} \end{vmatrix} \equiv \begin{vmatrix} E_1(z) \\ E_2(z) \end{vmatrix} \tag{22.9-3}$$

with

$$\beta'_{\substack{a\\b}} \equiv \beta_{\substack{a\\b}} + M_{\substack{a\\b}}$$

The evolution of $\mathbf{E}(z)$ is obtained from (22.9-1) as

$$\frac{d\mathbf{E}}{dz} = \tilde{\mathbf{C}}\mathbf{E} \tag{22.9-4}$$

with the matrix $\tilde{\mathbf{C}}$ given by

$$\tilde{\mathbf{C}} = \begin{vmatrix} -i\beta_b' & -\kappa^* \\ \kappa & -i\beta_a' \end{vmatrix} \tag{22.9-5}$$

Since an eigenmode depends on z only through a propagation phase factor $\exp(i\beta z)$, we postulate a solution

$$\mathbf{E}(z) = \mathbf{E}(0)e^{i\beta z} \tag{22.9-6}$$

Combining (22.9-4) and (22.9-6) results in

$$\tilde{\mathbf{C}}\mathbf{E} = -i\beta\mathbf{E} \tag{22.9-7}$$

This is a standard matrix algebra eigenvalue problem where $\mathbf{E}$ is the eigenvector and $-i\beta$ is the eigenvalue of the matrix $\mathbf{C}$. To determine $\mathbf{E}$ and β, we write out the two equations represented by (22.9-7)

$$\begin{aligned} -i(\beta_b' + \beta)E_1 - \kappa^* E_2 &= 0 \\ \kappa E_1 - i(\beta_a' + \beta)E_2 &= 0 \end{aligned} \tag{22.9-8}$$

The condition for nontrivial solutions for $\mathbf{E}_1$ and $\mathbf{E}_2$ is the vanishing of the determinant of the coefficients in (22.9-8). The solution of the resulting quadratic equation yields the eigenvalues

$$\beta_{1,2} = -\frac{\beta_a' + \beta_b'}{2} \pm \frac{1}{2}\sqrt{(\beta_a' - \beta_b')^2 + 4\kappa^2} = -\bar{\beta} \pm S \tag{22.9-9}$$

$$\bar{\beta} \equiv \frac{1}{2}(\beta_a' + \beta_b') \qquad S \equiv \sqrt{\delta^2 + \kappa^2} \qquad \kappa^2 \equiv \kappa\kappa^* \tag{22.9-10}$$

The two values β_1 and β_2 are substituted, one at a time, in (22.9-8) to obtain, to within an arbitrary constant, the corresponding eigenvectors. The result is

$$\mathbf{E}_1(z) = \begin{vmatrix} \dfrac{i\kappa^*}{\delta + S} \\ 1 \end{vmatrix} e^{-i(\bar{\beta} - S)z} \tag{22.9-11a}$$

$$\mathbf{E}_2(z) = \begin{vmatrix} \dfrac{i\kappa^*}{\delta - S} \\ 1 \end{vmatrix} e^{-i(\bar{\beta} + S)z} \tag{22.9-11b}$$

We note that, as expected, $\mathbf{E}_1 \cdot \mathbf{E}_2^* = 0$, i.e., the eigenmodes are normal. The mode norms $\mathbf{E}_{1,2} \cdot \mathbf{E}_{1,2}^* = 1 + |\kappa|^2/(\delta + S)^2$ are proportional to the respective (eigen) mode powers and are thus a constant. The two components $i\kappa^*/(\delta \pm S)$ and 1 of each eigenvector represent, respectively, the normalized amplitude of the individual waveguide modes $\mathscr{E}_y^{(l)}(x)$ in guide b and $\mathscr{E}_y^{(m)}(x)$ in guide a, which together make up the eigenmode of the two-waveguide system. The ratio of the power in waveguide b to that in a of these two "supermodes" is thus $|\kappa|^2/(\delta \pm S)^2$. In the limit $\kappa/\delta \to 0$, the "velocity mismatch" limit, $\mathbf{E}_1$ and $\mathbf{E}_2$ become

$$\begin{aligned} \mathbf{E}_{1\,(\kappa \ll \delta)} &\to \begin{vmatrix} 0 \\ 1 \end{vmatrix} e^{-i\beta_a' z} \\ \mathbf{E}_{2\,(\kappa \ll \delta)} &\to \begin{vmatrix} 1 \\ 0 \end{vmatrix} e^{-i\beta_b' z} \end{aligned} \tag{22.9-12}$$

to within a multiplicative constant, i.e., the super (eigenmodes) become the uncoupled single-waveguide modes.

Another important situation occurs when the two individual waveguide modes have the same phase velocity, i.e., $\delta = 0$. In this case,

$$\underset{(\delta=0)}{\mathbf{E}_1(z)} = \begin{vmatrix} i\dfrac{\kappa^*}{|\kappa|} \\ 1 \end{vmatrix} e^{-i(\beta_0-|\kappa|)z}$$

$$\beta_0 \equiv \beta_1 = \beta_2 \qquad (22.9\text{-}13)$$

$$\underset{(\delta=0)}{\mathbf{E}_2(z)} = \begin{vmatrix} -i\dfrac{\kappa^*}{|\kappa|} \\ 1 \end{vmatrix} e^{-i(\beta_0+|\kappa|)z}$$

The admixture is 50-50 and each waveguide carries half of the total power. In the case of identical waveguides at $\delta = 0$ and for $l = m$ (i.e., the same order modes), the coupling constant κ is a negative imaginary number [see Eqs. (22.8-3) and (22.8-4)] so that the two eigenvectors [of (22.9-13)] take the form

$$\underset{(\delta=0)}{\mathbf{E}_1(z)} = \begin{vmatrix} -1 \\ 1 \end{vmatrix} e^{-i(\beta_0-|\kappa|)z}$$

$$\qquad (22.9\text{-}14)$$

$$\underset{(\delta=0)}{\mathbf{E}_2(z)} = \begin{vmatrix} 1 \\ 1 \end{vmatrix} e^{-i(\beta_0+|\kappa|)z}$$

$\mathbf{E}_1(z)$ is thus the odd symmetric mode, whereas $\mathbf{E}_2(z)$ is even symmetric as depicted in Figure 22.18.

According to (22.9-1), the admixture—hence, the profile of the supermode—depends on the phase velocity mismatch parameter δ. A situation may exist in which the uncoupled dispersion curves of the individual guides may cross each other at some frequency ω_0 as shown in Figure 22.19. In the vicinity of ω_0, the supermode profile is thus a strong function of δ and, hence, of ω. If we approximate the mismatch parameter near ω_0 by $\delta =$ constant $(\omega - \omega_0)$, the supermode's dispersion curves $\gamma_1(\omega)$ and $\gamma_2(\omega)$ are as shown in the figure. Also shown are the (super)mode profiles at ω_0 and at a frequency ω where $\delta(\omega) \geq \kappa$.

Since the supermode (i.e., eigenmode) description of the waveguide problem is formally equivalent to one that is based on individual waveguide coupled modes, it is instructive to consider how we might, for example, describe the phenomenon of directional coupling [see (22.8-7)] by means of our new point of view. We consider the case $\delta = 0$ (phase-matching) and referring to Figure 22.18. We will assume that at $z = 0$ the power is fed into guide a (on the left) only. This boundary condition can be satisfied by expanding the total field at $z = 0$ as an equal admixture of the two supermodes (22.9-14) so that

$$\mathbf{E}_{\text{tot}}(0) = \mathbf{E}_1(0) + \mathbf{E}_2(0) = \begin{vmatrix} 0 \\ 2 \end{vmatrix} \qquad (22.9\text{-}15)$$

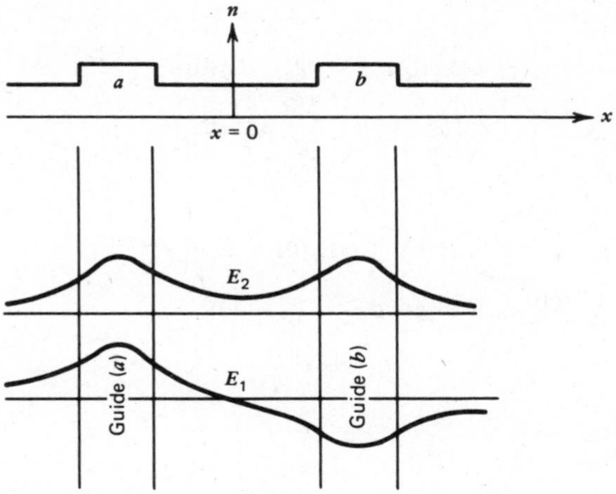

FIGURE 22.18 The transverse (x) field distribution of the two supermodes at the phase velocity matching ($\delta = 0$) condition of the parallel two-guide structure whose index of refraction profile is shown at the top.

It is also clear that if one were to add algebraically the fields of Figure 22.18, they would reinforce each other on the left and cancel each other on the right, leading to the column vector (22.9-15). Having established in (22.9-15) the proper admixture that satisfies the boundary condition at $z = 0$, we can determine the field at any z by simply reinserting the z dependence of each

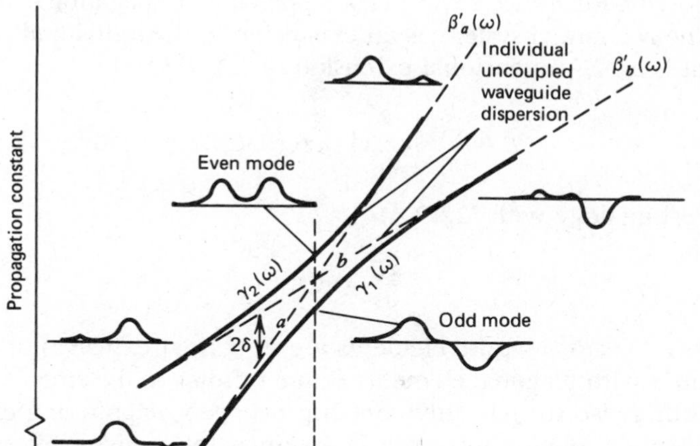

FIGURE 22.19 Some field distributions and the dispersion curves of the two lowest-order "supermodes" of a two-waveguide configuration near the phase-matching frequency (ω_0). The dashed curves correspond to the guide's dispersion at the absence of coupling.

supermode

$$\begin{aligned}\mathbf{E}_{\text{tot}}(z) &= \mathbf{E}_1(0)e^{-i(\beta_0-|\kappa|)z} + \mathbf{E}_2(0)e^{-i(\beta_0+|\kappa|)z} \\ &= \left|\begin{matrix}\dfrac{i\kappa^*}{\kappa} \\ 1\end{matrix}\right| e^{-i(\beta_0-|\kappa|)z} + \left|\begin{matrix}-\dfrac{\kappa^*}{|\kappa|} \\ 1\end{matrix}\right| e^{-i(\beta_0+|\kappa|)z} \\ &= e^{-i(\beta_0-|\kappa|)z}[\mathbf{E}_1(0) + \mathbf{E}_2(0)e^{-i2|\kappa|z}]\end{aligned} \tag{22.9-16}$$

At a distance z where

$$|\kappa|z = \frac{\pi}{2}$$

the total field becomes

$$\begin{aligned}\mathbf{E}_{\text{tot}}\left(z = \frac{2\pi}{|\kappa|}\right) &= e^{-i(\beta_0-|\kappa|)z}[\mathbf{E}_1(0) - \mathbf{E}_2(0)] \\ &= e^{-i(\beta_0-|\kappa|)z}\left|\begin{matrix}\dfrac{2i\kappa^*}{|\kappa|} \\ 0\end{matrix}\right|\end{aligned} \tag{22.9-17}$$

so that the power is completely in the right waveguide. This exchange of power between the two waveguides thus takes place with a spatial period $\Delta z = \pi/2|\kappa|$ as predicted by the coupled-mode solution (22.8-7). Here, however, we attribute the sloshing of power between guides to the interference term, $\exp(-i2|\kappa|z)$ in (22.9-16), between two supermodes.

The matrix formalism can be extended straightforwardly to the case of N modes following Reference 25. We can express any propagating field configuration in the N channel system as an expansion in the individual waveguide modes that is an N-dimensional extension of (22.9-2a)

$$E = \sum_{l=1}^{N} \mathscr{E}_l(x, y)A_l(z)\exp(i\beta_\ell z) \tag{22.9-18}$$

and in direct analogy with (22.9-4)

$$\frac{d\mathbf{E}}{dz} = i\tilde{\mathbf{C}}\mathbf{E} \tag{22.9-19}$$

where $\mathbf{E}$ is a "vector" whose elements are $E_l = A_l(z)\exp(i\beta_l z)$ and $\mathbf{C}$ is the $N \times N$ matrix with diagonal elements β_l and off-diagonal elements $\kappa_{l,l+1} \equiv \kappa_l$ and zero otherwise so that only coupling between neighboring channels is assumed. The coupling constant κ is assumed to be given by expressions similar to (22.8-4) derived for the case of a pair of coupled waveguides. The array supermodes are by definition the eigensolutions of (22.9-19), i.e., those "vectors" satisfying as in (22.9-6)

$$\mathbf{E}^\nu(z) = \mathbf{E}^\nu(0)\exp(i\beta_\nu z) \tag{22.9.20}$$

β_ν being the propagation constant of the supermode ν. Using (22.9-20) in (22.9-19) leads, in a manner analogous to (22.9-7), to

$$(\tilde{\mathbf{C}} - \beta_\nu \tilde{\mathbf{I}})\mathbf{E}^\nu = 0 \tag{22.9-21}$$

whose solution yields the N eigenvalues β_ν and the N supermodes $\mathbf{E}^\nu$, $\nu = 1, 2, \ldots, N$. Each such eigenmode consists of a unique *phase-locked* combination of the individual channel amplitudes E_l^ν propagating with a single propagation constant β_ν. An example of the supermodes in the case of $N = 4$ is shown in Figure 22.20. Also shown are the far-field patterns of the supermodes.

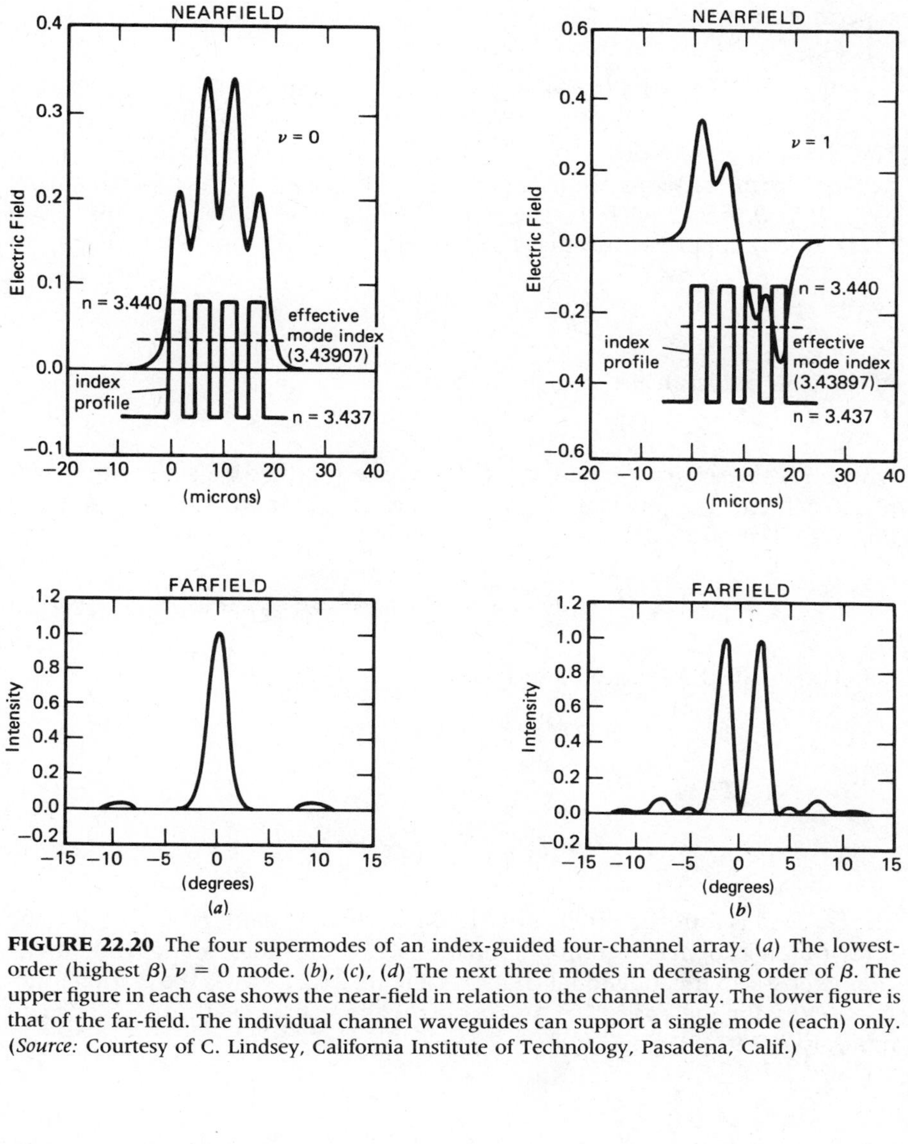

FIGURE 22.20 The four supermodes of an index-guided four-channel array. (*a*) The lowest-order (highest β) $\nu = 0$ mode. (*b*), (*c*), (*d*) The next three modes in decreasing order of β. The upper figure in each case shows the near-field in relation to the channel array. The lower figure is that of the far-field. The individual channel waveguides can support a single mode (each) only. (*Source:* Courtesy of C. Lindsey, California Institute of Technology, Pasadena, Calif.)

In the special case of identical channels with identical coupling $\kappa_l \equiv \kappa$, $\beta_l \equiv \beta$, the solution of (22.9-21) is

$$E_l^\nu = \sin\left(l\frac{\pi\nu}{N+1}\right) \quad l = 1, 2, \ldots N \tag{22.9-22}$$

$$\beta_\nu = \beta + 2\kappa\cos\left(l\frac{\pi\nu}{N+1}\right) \tag{22.9-23}$$

Note that the splitting between adjacent modes in β_ν is of the order of κ/N so that for $N \gg 1$, the propagation constants form a quasi-continuum spanning the range $\beta - 2\kappa \geq \beta \geq \beta + 2\kappa$.

From (22.9-18) and (22.9-20) we can write the near field of the νth supermode as

$$E^\nu(x, y, z) = \sum_{l=1}^{N} A_l^\nu \mathscr{E}_l(x, y)\exp(i\beta_\nu z) \tag{22.9-24}$$

The constant amplitudes A_l^ν are the solution of (22.9-21) for the νth mode. Let $E_l(\theta)$ be the far-constant field lth channel field $\mathscr{E}_l(x, y)$ (i.e., $A_l = 1$) all by itself with θ measured from the normal to the exit plane. It follows from (22.9-24) by superposition that the far-field of the νth supermode is

$$F^\nu(\theta) = \sum_{l=1}^{N} A_l^\nu E_l(\theta) \tag{22.9-25}$$

In the case of identical channels, we have

$$E_l(\theta) = E_{l+1}(\theta)e^{ik_0 S\sin\theta} = E_0(\theta)e^{ik_0 lS\sin\theta}$$

where S is the separation between two channels so that $k_0 S \sin\theta$ is the extra phase delay in the far-field between the fields from two adjacent channels. From the last two equations,

$$F^\nu(\theta) = E_0(\theta)\sum_{l=0}^{N-1} A_l^\nu e^{ik_0 lS\sin\theta} \tag{22.9-26}$$

The far-field intensity pattern is

$$|F^\nu(\theta)|^2 = |E_0(\theta)|^2 G^\nu(\theta) \tag{22.9-27}$$

with

$$G^\nu(\theta) \equiv \left|\sum_{l=0}^{N-1} A_l^\nu e^{ik_0 lS\sin\theta}\right|^2 \tag{22.9-28}$$

The far-field is thus the product of the single channel pattern $|E_0(\theta)|^2$ and the array (often known as "grating") function $G^\nu(\theta)$ that does not depend on any characteristic of the individual channels. Figure 22.21 shows the grating function $G^\nu(\theta)$ for the case $N = 5$. To obtain the actual far-field, we need to multiply (22.9-28) of $|E_0(\theta)|^2$.

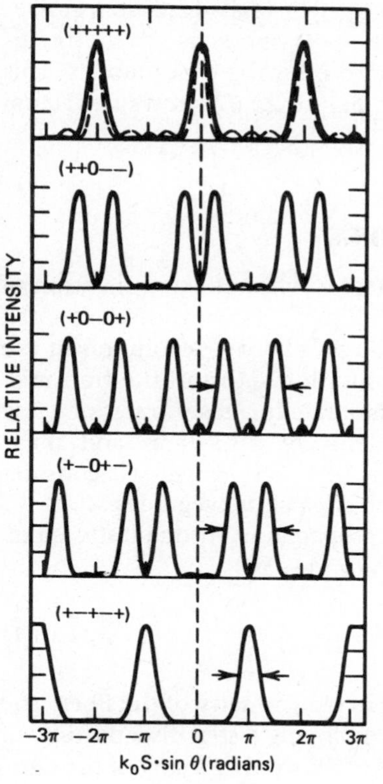

FIGURE 22.21 The grating function G for $N = 5$. The dashed curve corresponds to an array of five equal amplitude ($A_i = 1$) radiators. (*Source:* Reference 25.)

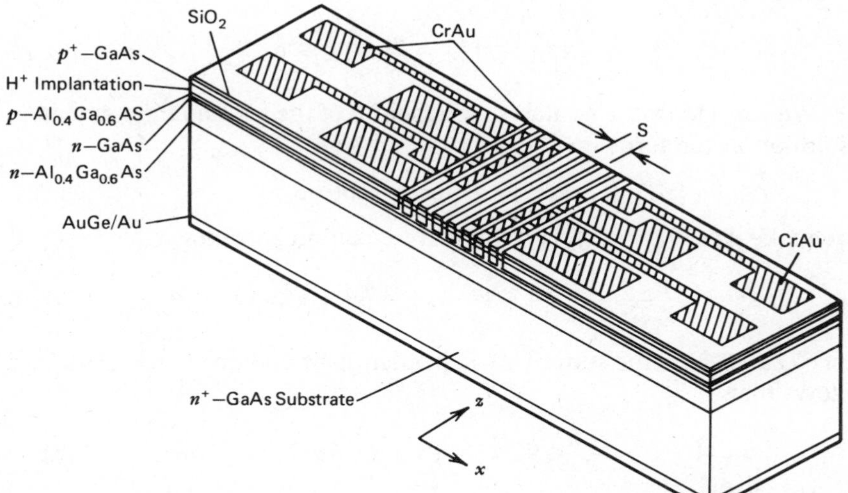

FIGURE 22.22. A sketch of the physical layout of a seven-element semiconductor laser array. The separation between individual channels is achieved by ion implantation. (*Source:* Reference 26.)

The truncation of $G^{\nu}(\theta)$ in (22.9-27) by $|E_0(\theta)|^2$ usually leads to a single-lobed far-field pattern for the fundamental (++++) mode as shown in Figure 22.20*a* and to a double-lobed far-field pattern for the other modes. This point is considered further in Problem 22.10. Figure 22.22 shows the actual structure of an array semiconduction laser.

22.10 PROPAGATION IN OPTICAL FIBERS

We would be remiss if we concluded this chapter without some mention of optical fibers. The successful development of silica fibers with losses as small as 0.2 db/km at 1.53 μm (Reference 27) brought about a revolution in the technology of data transmission and communication. Most of the transmission is of digital pulses. As the transmission data rate increases (rates of a few gigabits/s are already contemplated), problems of dispersion and pulse spreading in the fibers become important. In this section we will consider some basic issues related to pulse propagation in such waveguides.

We choose to consider the problem of propagation in quadratic index fibers, that is, fibers whose index of refraction varies as

$$n^2(r) = n^2\left(1 - \frac{n_2}{n} r^2\right) \tag{22.10-1}$$

where r is the distance from the cylindrical axis of symmetry of the fiber. The analysis of these fibers is simpler than those employing a step discontinuity at the core-cladding interface, but is still capable of illustrating many of the basic features common to all fibers.

The vector wave equation (22.1-1) becomes

$$\nabla^2\mathbf{E} - k^2\left(1 - \frac{n_2}{n} r^2\right)\mathbf{E} = 0 \tag{22.10-2}$$

We consider some (scalar) component E of the last equation and assume a solution in the form

$$E(x, y) = \psi(x, y)\exp(-i\beta z)$$

If we take $\psi(x, y) = f(x)g(y)$, the wave equation becomes

$$\frac{1}{f}\frac{\partial^2 f}{\partial x^2} + \frac{1}{g}\frac{\partial^2 g}{\partial y^2} + k^2 - k^2\frac{n_2}{n}(x^2 + y^2) - \beta^2 = 0 \tag{22.10-3}$$

Since (22.10-3) is the sum of a y-dependent part and an x-dependent part, it follows that

$$\frac{1}{f}\frac{d^2 f}{dx^2} + \left(k^2 - \beta^2 - k^2\frac{n_2}{n} x^2\right) = C \tag{22.10-4}$$

$$\frac{1}{g}\frac{d^2 g}{dy^2} - k^2\frac{n_2}{n} y^2 = -C \tag{22.10-5}$$

where C is some constant. Consider first (22.10-5). If we define a variable ξ by

$$\xi = \alpha y \qquad \alpha \equiv k^{1/2}\left(\frac{n_2}{n}\right)^{1/4} \tag{22.10-6}$$

(22.10-5) becomes

$$\frac{d^2 g}{d\xi^2} + \left(\frac{C}{\alpha^2} - \xi^2\right) g = 0 \tag{22.10-7}$$

This is a well-known differential equation and is identical to the Schrödinger equation (2.2-2) of the harmonic oscillator. The eigenvalue C/α^2 thus must satisfy

$$\frac{C}{\alpha^2} = (2m + 1) \qquad m = 1, 2, 3 \ldots \tag{22.10-8}$$

and corresponding to an integer m, the solution is given by

$$g_m(\xi) = H_m(\xi) e^{-\xi^2/2} \tag{22.10-9}$$

where H_m is the Hermite polynomial of order m.

We now repeat the procedure with (22.10-4). If we substitute

$$\zeta = \alpha x$$

it becomes

$$\frac{\partial^2 f}{\partial \zeta^2} + \left(\frac{k^2 - \beta^2 - C}{\alpha^2} - \zeta^2\right) f = 0$$

so that, as in (22.10-8),

$$\frac{k^2 - \beta^2 - C}{\alpha^2} = (2l + 1) \qquad l = 1, 2, 3, \ldots \tag{22.10-10}$$

and

$$f_l(\zeta) = H_l(\zeta) e^{-\zeta^2/2} \tag{22.10-11}$$

The total solution for ψ is thus

$$\psi(x, y) = H_1\left(\frac{\sqrt{2}x}{\omega}\right) H_m\left(\frac{\sqrt{2}y}{\omega}\right) e^{-(x^2+y^2)/\omega^2}$$

where the "spot size" ω is, according to (22.10-6),

$$\omega = \frac{\sqrt{2}}{\alpha} = \sqrt{\frac{2}{k}}\left(\frac{n}{n_2}\right)^{1/4} = \sqrt{\frac{\lambda}{\pi}}\left(\frac{1}{nn_2}\right)^{1/4} \tag{22.10-12}$$

The total (complex) field is

$$\begin{aligned} E_{l,m}(x, y, z) &= \psi_{l,m}(x, y) e^{-i\beta_{l,m} z} \\ &= E_0 H_l\left(\sqrt{2}\,\frac{x}{\omega}\right) H_m\left(\sqrt{2}\,\frac{y}{\omega}\right) \exp\left(-\frac{x^2 + y^2}{\omega^2}\right) \exp(-i\beta_{l,m} z) \end{aligned} \tag{22.10-13}$$

The propagation constant $\beta_{l,m}$ of the l, m mode is obtained from (22.10-8) and (22.10-10)

$$\beta_{l,m} = k\left[1 - \frac{2}{k}\sqrt{\frac{n_2}{n}}\,(l + m + 1)\right]^{1/2} \tag{22.10-14}$$

Two features of the mode solutions are noteworthy. (1) Unlike the homogeneous medium solution ($n_2 = 0$), the mode spot size ω is independent of z. This can be explained by the focusing action of the index variation ($n_2 > 0$), which counteracts the natural tendency of a confined beam to diffract (spread). In the case of an index of refraction that increases with $r (n_2 < 0)$, it follows from (22.10-12) and (22.10-13) that $\omega^2 < 0$ and no confined solutions exist. The index profile in this case leads to defocusing, thus reinforcing the diffraction of the beam. (2) The dependence of β on the mode indices l, m causes the different modes to have phase velocities $v_{l,m} = \omega/\beta_{l,m}$ as well as group velocities $(v_g)_{l,m} = d\omega/d\beta_{l,m}$ that depend on l and m.

Let us consider the modal dispersion (that is, the dependence on l and m) of the group velocity of mode l, m

$$(v_g)_{l,m} = \frac{d\omega}{d\beta_{l,m}} \tag{22.10-15}$$

If the index variation is small so that

$$\frac{1}{k}\sqrt{\frac{n_2}{n}}\,(l + m + 1) \ll 1 \tag{22.10-16}$$

we can approximate (22.10-14) as

$$\beta_{l,m} \cong k - \sqrt{\frac{n_2}{n}}\,(l + m + 1) - \frac{n_2}{2kn}\,(l + m + 1)^2 \tag{22.10-17}$$

so that the group velocity becomes

$$(v_g)_{l,m} = \frac{d\omega}{d\beta_{l,m}} = \frac{c/n}{\left[1 + \dfrac{(n_2/n)}{2k^2}\,(l + m + 1)^2\right]} \tag{22.10-18}$$

Pulse Spreading in Fibers

The main limitation on the distance-bandwidth product of low loss optical fibers is pulse spreading. If the pulse excites many modes in the fiber, then the excitation in a given mode (l, m) travels with the group pulse velocity $(v_g)_{l,m}$, resulting in a spread

$$\Delta\tau \sim L\left[\frac{1}{(v_g)_{l_{\max},m_{\max}}} - \frac{1}{(v_g)_{0,0}}\right] \tag{22.10-19}$$

where L is the length of the fiber and $l_{\max}$, $m_{\max}$ are the highest mode indices excited by the input pulse. Step index fibers, in a manner analogous to the slab waveguide considered in Section 22.1, can support only a finite number of propagating modes. Our idealized quadratic index fiber shows no such restraint since the index (22.10-1) can decrease, mathematically, without bound. In reality, the maximum change Δn is limited by the finite core radius, and this fiber too can support only a finite number of modes. If the number of propagating modes is reduced to one, then the modal spreading leading to (22.10-19) does not exist. In this case, the pulse broadening is due to group velocity dispersion, i.e., the dependence of the group velocity v_g on frequency. This problem was considered briefly in Section 6.10. We will expand on it here. We will first consider the broadening in a fiber of length L of a pulse of light

$$\begin{aligned} E(z = 0, t) &= e^{-\alpha t^2 + i\omega_0 t} \\ &= e^{i\omega_0 t} \int_{-\infty}^{\infty} F(\Omega) e^{i\Omega t} \, d\Omega \end{aligned} \tag{22.10-20}$$

formed by modulating an idealized perfectly monochromatic laser source oscillating at ω_0. $F(\Omega)$ is the Fourier transform of envelope function $e^{-\alpha t^2}$

$$\begin{aligned} F(\Omega) &= \frac{1}{2\pi} \int_{-\infty}^{\infty} e^{-\alpha t^2 - i\Omega t} \, dt \\ &= \frac{1}{\sqrt{4\pi\alpha}} e^{-\Omega^2/4\alpha} \end{aligned} \tag{22.10-21}$$

so that the full width at half maximum (FWHM) of the pulse spectrum $|F(\Omega)|^2$ is

$$(\Delta\Omega)_{\text{FWHM}} = \sqrt{8(\ln 2)\alpha} \tag{22.10-22}$$

whereas the initial pulse width (FWHM) is

$$\tau_0 = \left(\frac{2 \ln 2}{\alpha}\right)^{1/2} \tag{22.10-23}$$

The field at a distance z is obtained by multiplying each frequency component $(\omega_0 + \Omega)$ in (22.10-20) by $\exp[-i\beta(\omega_0 + \Omega)z]$. If we expand $\beta(\omega_0 + \Omega)$ near ω_0 as

$$\beta(\omega_0 + \Omega) = \beta(\omega_0) + \left.\frac{d\beta}{d\omega}\right|_{\omega_0} \Omega + \frac{1}{2} \left.\frac{d^2\beta}{d\omega^2}\right|_{\omega_0} \Omega^2 + \cdots$$

we obtain

$$E(z, t) = e^{i(\omega_0 t - \beta_0 z)} \int_{-\infty}^{\infty} d\Omega F(\Omega) \exp\left\{i\left[\Omega t - \frac{\Omega z}{v_g} - \frac{1}{2}\frac{\partial}{\partial\omega}\left(\frac{1}{v_g}\right)\Omega^2 z\right]\right\} \tag{22.10-24}$$

where

$$\beta_0 \equiv \beta(\omega_0), \qquad \frac{d\beta}{d\omega} = \frac{1}{v_g} = \frac{1}{\text{group velocity}}$$

The field envelope is given by the integral in (22.10-24)

$$\begin{aligned}\mathscr{E}(z,t) &= \int_{-\infty}^{\infty} d\Omega F(\Omega)\exp\left\{i\Omega\left[\left(t-\frac{z}{v_g}\right)-\frac{1}{2}\frac{d}{d\omega}\left(\frac{1}{v_g}\right)\Omega z\right]\right\}\\ &= \int_{-\infty}^{\infty} d\Omega F(\Omega)\exp\left\{i\Omega\left[\left(t-\frac{z}{v_g}\right)-a\Omega z\right]\right\}\end{aligned} \tag{22.10-25}$$

where

$$a \equiv \frac{1}{2}\frac{d}{d\omega}\left(\frac{1}{v_g}\right) = -\frac{1}{2v_g^2}\frac{dv_g}{d\omega} = \frac{1}{2}\frac{d^2\beta}{d\omega^2}\bigg|_{\omega=\omega_0}$$

After substituting for $F(\Omega)$ from (22.10-21), the last equation becomes

$$\mathscr{E}(z,t) = \sqrt{\frac{1}{4\pi\alpha}}\int_{-\infty}^{\infty}\exp\left\{-\left[\Omega^2\left(\frac{1}{4\alpha}+iaz\right)+i\left(t-\frac{z}{v_g}\right)\Omega\right]\right\}d\Omega \tag{22.10-26}$$

Carrying out the integration yields

$$\mathscr{E}(z,t) = \frac{1}{\sqrt{1+i4a\alpha z}}\exp\left[-\frac{(t-z/v_g)^2}{1/\alpha+16a^2z^2\alpha}\right]\exp\left[i\frac{4az(t-z/v_g)^2}{1/\alpha^2+16a^2z^2}\right] \tag{22.10-27}$$

The pulse duration τ at z can be taken as the separation between the two times when the pulse envelope squared (intensity) is smaller by a factor of $\frac{1}{2}$ from its peak value, that is,

$$\tau(z) = \sqrt{2\ln 2}\sqrt{\frac{1}{\alpha}+16a^2z^2\alpha} \tag{22.10-28}$$

The pulse width after propagating a distance L can thus be expressed as

$$\tau(L) = \tau_0\sqrt{1+\left(\frac{8aL\ln 2}{\tau_0^2}\right)^2} \tag{22.10-29}$$

At large distances such that $aL \gg \tau_0^2$, we obtain

$$\tau(L) \sim \frac{(8\ln 2)aL}{\tau_0} \tag{22.10-30}$$

If we use the definition of the factor a [see the line following (22.10-25)], the last expression becomes

$$\tau(L) = \frac{4\ln 2}{v_g^2}\frac{dv_g}{d\omega}\frac{L}{\tau_0} \tag{22.10-31}$$

The group velocity dispersion is often characterized by $D \equiv L^{-1}(dT/d\lambda)$, where T is the pulse transmission time through length L of the fiber. This definition is related to the second-order derivative of β with respect to ω as

$$D = -\frac{2\pi c}{\lambda^2}\left(\frac{d^2\beta}{d\omega^2}\right) = \frac{2\pi c}{\lambda^2 v_g^2}\frac{dv_g}{d\omega} \tag{22.10-32}$$

and is related to the parameter a used above by

$$D = -\frac{4\pi c}{\lambda^2}a \tag{22.10-33}$$

With this new definition, the pulse-width expression (22.10-29) can be written as

$$\tau(L) = \tau_0 \sqrt{1 + \left(\frac{2 \ln 2}{\pi c} \frac{DL\lambda^2}{\tau_0^2}\right)^2} \tag{22.10-34}$$

If DL is in units of picoseconds per nanometer, λ is in units of micrometers, and τ is in units of picoseconds, the pulse width can be written as

$$\tau(L) = \tau_0 \sqrt{1 + \left(\frac{1.47DL\lambda^2}{\tau_0^2}\right)^2} \tag{22.10-35}$$

If we combine (22.10-27) with (22.10-24), we find that the total field at z is

$$\begin{aligned} E(z, t) &= \mathscr{E}(z, t)e^{i(\omega_0 t - \beta_0 z)} \\ &= \frac{e^{-i\beta_0 z}}{\sqrt{1 + i4a\alpha z}} \exp\left\{i\left[\omega_0 t + \frac{4az(t - z/v_g)^2}{\alpha^{-2} + 16a^2z^2}\right] - \frac{(t - z/v_g)^2}{1/\alpha + 16a^2z^2\alpha}\right\} \end{aligned} \tag{22.10-36}$$

The oscillation phase is thus

$$\Phi(z, t) = \omega_0 t + \frac{4az(t - z/v_g)^2}{\alpha^{-2} + 16a^2z^2} - \beta_0 z \tag{22.10-37}$$

The local "frequency" $\omega(z, t)$ is then

$$\omega(z, t) = \frac{\partial\Phi}{\partial t} = \omega_0 + \frac{8az(t - z/v_g)}{\alpha^{-2} + 16a^2z^2} \tag{22.10-38}$$

and consists of the original frequency ω_0 and a linear frequency sweep (chirp), which is proportional to the group velocity dispersion term a. The chirp can be understood intuitively by the fact that, due to group velocity dispersion, different frequencies travel with different (group) velocity. According to (22.10-35), the broadening ratio $\tau(L)/\tau_0$ for a given laser pulse will be small (i.e., ~ 1), provided the group dispersion–length product DL is very small compared to $(\tau_0/\lambda)^2$. It is, of course, desirable to transmit light pulses at the condition when the group velocity dispersion is zero ($D = 0$).

The discussion leading to (22.10-35) applies to transform-limited optical pulses in which the spectral width of the pulse is due to the finite pulse duration. In the above example, we have from (22.10-22) and (22.10-23) that

$$(\Delta\Omega)_{\text{FWHM}} = \frac{4 ln 2}{\tau_0}$$

Most often—and especially in optical communication systems employing semiconductor lasers—the spectral width of the pulse is not transform-limited, but is due to the residual linewidth of the laser source. Let us denote that linewidth as $\Delta\omega$ so that for "long" distances the pulse width becomes

$$\begin{aligned} \Delta\tau &\sim \left|\frac{d}{d\omega}\left(\frac{L}{v_g}\right)\right| \Delta\omega \\ &= \frac{L}{v_g^2}\left|\frac{dv_g}{d\omega}\right| \Delta\omega \equiv LD\,\Delta\lambda \end{aligned} \tag{22.10-39}$$

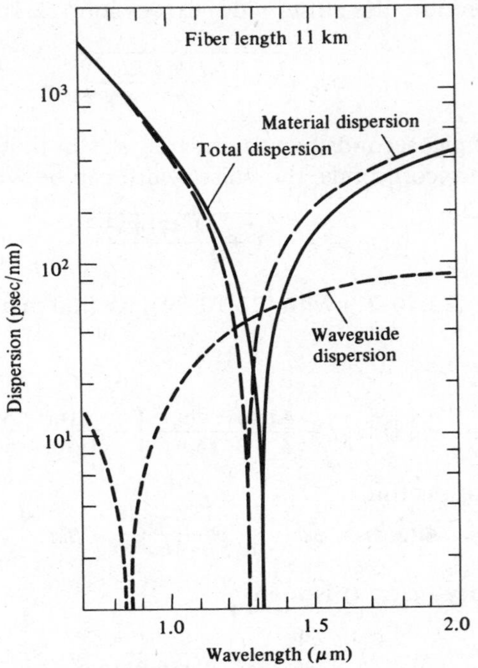

FIGURE 22.23 Pulse dispersion through a 11-km-long strand of single-mode graded-index (see Section 3.4) optical fibers. The "zero-dispersion" wavelength is 1.32 μm. The material dispersion has different signs in the wavelength regions below and above its "zero-point" at 1.27 μm. To obtain D in units of psec/(nm-km), we need divide the abcissa by the fiber length $L = 11$ km.

where $\Delta\lambda = -(\lambda^2/2\pi c)\,\Delta\omega$ is the spectral width expressed in wavelength and D is defined by (22.10-32).

The dependence of V_g (or D) on frequency is due to two main causes that are operative in *any* material waveguide. The first—waveguide dispersion—is due to the spatial confinement of the wave in the finite dimension waveguide and the fact that such confinement and, hence, the phase and group velocities of any mode depend on the frequency ω. The second—material dispersion—is the dependence of the index of refraction of the waveguide medium on ω. This mechanism will be operative even in nonguided propagation. In the special case of the quadratic index fiber we obtained in Eq. (22.10-18) in the case of a $l = 0$, $m = 0$ mode

$$v_g \equiv (v_g)_{0,0} = \frac{c/n}{\left(1 + \dfrac{n_2/n}{2k^2}\right)} \approx \frac{c}{n}\left(1 - \frac{n_2/n}{2k^2}\right)$$

so that

$$\frac{dv_g}{d\omega} \approx \frac{\lambda^2 c}{2\pi n^2} D = \frac{\partial v_g}{\partial \omega} + \frac{\partial v_g}{\partial n}\frac{\partial n}{\partial \omega}$$
$$\cong \frac{(n_2/n)}{k^3} - \frac{c}{n^2}\frac{\partial n}{\partial \omega} \tag{22.10-40}$$

where $k = \omega n/c = 2\pi n/\lambda$. The first term in (22.10-40) represents waveguide dispersion (the mode confinement is due to the quadratic index dependence represented by n_2, whereas the second one is the material index dispersion. In general, for any waveguide we can write

$$D = D_{\text{waveguide}} + D_{\text{material}} \tag{22.10-41}$$

As can be seen from (22.1-40), it is possible, in principle, in a given waveguide to find a wavelength where $D_{\text{waveguide}} = -D_{\text{material}}$ so that $D = 0$. In this case, the residual pulse broadening will be due to the much smaller neglected high-order terms in Taylor expansion of $\beta(\omega)$ in (22.10-14). Figure 22.23 shows a plot of $D_{\text{waveguide}}$, D_{material} and their sum of a single mode silica fiber. We note that $D = 0$ at $\lambda = 1.32$ μm and indeed this has become the wavelength of choice for optical fiber communication.

The loss spectrum of typical commercial silica fibers is shown in Figure 22.24. We note a minimum loss of ~0.2 db/km near $\lambda = 1.55$ μm. This implies that a fiber whose length is 150 km will cause the attenuation of the signal by 30 db (10^{-3}) that in many practical scenarios is a very acceptable loss. A major effort is now under way to develop fibers with "zero" dispersion near the minimum loss wavelength $\lambda = 1.55$ μm. The ability to transmit data rates of, say, at 10 gigabits/s, which according to Figure 22.23 should be possible, over distances of >100 km in a single-mode fiber is, indeed, a development deserving of the adjective "revolutionary."

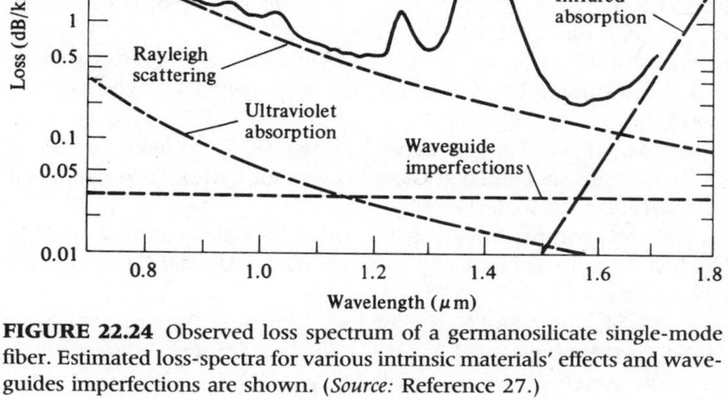

FIGURE 22.24 Observed loss spectrum of a germanosilicate single-mode fiber. Estimated loss-spectra for various intrinsic materials' effects and waveguides imperfections are shown. (*Source:* Reference 27.)

References

1. Yariv, A. and R. C. C. Leite, "Dielectric Waveguide Mode of Light Propagation in *p-n* Junctions," *Appl. Phys. Letters* **2,** 55 (1963).
2. Osterberg, H. and L. W. Smith, "Transmission of Optical Energy Along Surfaces," *J. Opt. Soc. Amer.* **54,** 1073 (1964).
3. Hall, D., A. Yariv, and E. Garmire, "Optical Guiding and Electrooptic Modulation in GaAs Epitaxial Layers," *Opt. Comm.* **1,** 403 (1970).
4. Shubert, R. and J. H. Harris, "Optical Surface Waves on Thin Films and Their Application to Integrated Data Processors," *IEEE Trans. Microwave Theory Tech.* (1968 Symp. Issue) **MTT-16,** 1048–1054 (1968). Also, see Miller, S. E., "Integrated Optics, an Introduction," *Bell Syst. Tech. J.* **48,** 2059 (1969).
5. Yariv, A., "The Beginning of Integrated Optoelectronic Circuits," *IEEE Trans. Electron. Devices* **ED-31,** 1656 (1984).
6. Goell, J. E., "A Circular Harmonic Computer Analysis for Rectangular Dielectric Waveguides," *Bell Syst. Tech. J.* **48,** 2133 (1968).
7. Marcatili, E. A. J., "Dielectric Rectangular Waveguide and Directional Couplers for Integrated Optics," *Bell Syst. Tech. J.* **48,** 2071 (1969).
8. Collins, R. E., *Field Theory of Guided Waves* (New York: McGraw-Hill, 1960), p. 470.
9. Yariv, A., *Introduction to Optical Electronics* (New York: Holt, Rinehart and Winston, 1971), p. 46.
10. Hammer, J. M., D. J. Channin, and M. T. Duffy, "High Speed Electrooptic Grating Modulators," RCA Technical Report, unpublished.
11. Tien, P. K., R. Ulrich, and R. J. Martin, "Modes of Propagating Light in Thin Deposited Semiconductor Films," *Appl. Phys. Letters* **14,** 291 (1969).
12. Yariv, A., "Coupled Mode Theory for Guided Wave Optics," *IEEE J. of Quant. Electr.* **QE-9,** 919 (1973).
13. Kuhn, L., M. L. Dakss, P. F. Heidrich, and B. A. Scott, "Deflection of Optical Guided Waves by a Surface Acoustic Wave," *Appl. Phys. Letters* **17,** 265 (1970).
14. Dixon, R. W., "The Photoelastic Properties of Selected Materials and Their Relevance to Acoustic Light Modulators and Scanners," *J. Appl. Phys.* **38,** 5149 (1967).
15. Stoll, H. and A. Yariv, "Coupled Mode Analysis of Periodic Dielectric Waveguides," *Opt. Commun.* **8,** 5 (1973).
16. Flanders, D. C., H. Kogelnik, R. V. Schmidt, and C. V. Shank, "Grating Filters for Thin Film Optical Waveguides," *Appl. Phys. Letters* **24,** 194 (1974).
17. Kogelnik, H. and C. V. Shank, "Coupled Wave Theory of Distributed Feedback Lasers," *J. Appl. Phys.* **43,** 2328 (1972).
18. Nakamura, M., A. Yariv, H. W. Yen, S. Somekh, and H. L. Garvin, "Optically Pumped GaAs Surface Laser with Corrugation Feedback," *Appl. Phys. Letters* **22,** 515 (1973).
19. Nakamura, M., H. W. Yen, A. Yariv, E. Garmire, S. Somekh, and H. L. Garvin, "Laser Oscillations in Epitaxial GaAs Waveguides with Corrugation Feedback," *Appl. Phys. Letters* **23,** 224 (1973).
20. Aiki, K., M. Nakamura, J. Umeda, A. Yariv, A. Katzir, and H. W. Yen, "GaAs-GaAlAs Distributed Feedback Laser with Separate Optical Carrier Confinement," *Appl. Phys. Letters* **27,** 145 (1975).
21. Sasaki, S., M. M. Choy, and N. K. Cheung, "Effects of Dynamic Spectral Behavior and Mode Partitioning of 1550 nm Distributed Feedback Lasers on Gigabit/sec Transmission Systems," *Electr. Letters* **24,** 26 (1987).

22. Somekh, S., Ph.D. Thesis, California Institute of Technology, Pasadena, Calif., 1973.
23. Marcatili, E. A. J., "Dielectric Rectangular Waveguides and Directional Couplers for Integrated Optics," *Bell Sys. Tech. J.* **48,** 2071 (1969).
24. Campbel, J. C., F. A. Blum, D. W. Shaw, and K. L. Lawley, "GaAs Electro-optic Directional Coupler Switch," *Appl. Phys. Letters* **27,** 202 (1975).
25. Kapon, E., J. Katz, and A. Yariv, "Supermode Analysis of Phase-Locked Arrays of Semiconductor Lasers," *Opt. Letters* **9,** 125 (1984).
26. Katz, J., E. Kapon, C. Lindsey, S. Margalit, V. Sohreter, and A. Yariv, "Phase-Locked Semiconductor Laser Array with Separate Contacts," *Appl. Phys. Letters* **43,** 521 (1983).
27. Miya, T., Y. Terunuma, T. Hosaka, and T. Miyashita, "Ultimate Low-Loss Single Mode Fiber at 1.55 μm," *Electr. Letters* **15,** 106 (1979).
28. Gambling, W. A., H. Matsumara, and C. M. Ragdale, "Zero Total Dispersion in Graded Index Single Mode Fibers," *Electr. Letters* **15,** 474 (1979).

Supplementary References

1. For a collection of reprints up to 1972 in integrated optics, see *Integrated Optics,* D. Marcuse, ed. (New York: IEEE Press, 1972).
2. Tien, P. K., "Light Waves in Thin Films and Integrated Optics," *Applied Optics* **10,** 2395 (1971).
3. Kaminow, I. P., *An Introduction to Electrooptic Devices,* (New York: Academic Press, 1974).
4. Arnaud, J. A., *Beam and Fiber Optics,* (New York: Academic Press, 1976).
5. Midwinter, J. E., *Optical Fibers for Transmission,* (New York: Wiley, 1979).
6. Yariv, A. and P. Yeh, *Optical Waves in Crystals/Propagation and Control of Laser Radiation,* (New York: Wiley-Interscience, 1984).
7. Special Issue on Integrated Optics, *IEEE J. Quant. Electr.* **QE-22** (June 1986).
8. Solimeno, S., B. Crosignani, and P. DiPorto, *Guiding, Diffraction and Confinement of Optical Radiation,* (New York: Academic Press, 1986).
9. Marcuse, D., *Light Transmission Optics* (New York: Van Nostrand, 1972).

Problems

22.1 Derive Eq. (22.2-7).

22.2 Show that the form of (22.7-10) is consistent with the conservation of the modes' power.

22.3 Derive the equations in (22.7-12).

22.4 Derive an expression for the modulation power of a transverse electrooptic waveguide modulator of length L and cross section $2\lambda \times 2\lambda$ (λ is the vacuum wavelength of the light). Compare to the bulk result (Reference 9, p. 325). Estimate the power requirement for a $LiNbO_3$ modulator at $\lambda = 1$ μm, $L = 5$ mm.

22.5 Derive the condition for distributed feedback laser oscillation for the case of gain perturbation, that is, $\gamma_1 \neq 0$, $n_1 = 0$. Compare with the result of Reference 17.

22.6 Calculate the threshold gain constant for a lossless distributed feedback in GaAs with $n_1 = 1$, $n_2 = 3.5$, $n_3 = 3.4$.

$$t = 3\ \mu\text{m}, \qquad a = 500\ \text{Å}, \qquad \lambda = 0.85\ \mu\text{m}$$

$$L = 200,\ 300,\ 500\ \mu\text{m}$$

Plot the gain versus L.

22.7 Write a short report on the theory and practice of launching waveguide modes by prism and grating couplers. Consult Supplementary References 1 and 2 for background material.

22.8 Consider the N-channel array described by (22.9-28). Assume that the width of each channel is equal to $S/2$ where S is the interchannel separation.
(a) Show that, for $\theta \ll 1$, the far-field pattern $G^{\nu}(\theta)$ is periodic in θ. What is the periodicity?
(b) Show that for the fundamental $(++++)$ mode the far-field consists of essentially one central lobe with an angular width $\Delta\theta \sim \lambda/NS$. Show that for the other modes the far-field has a double-lobed pattern.

22.9 Referring to Fig. 22.22 assume that the regions between the optical waveguides are highly absorbing. Explain qualitatively why the (desirable) $(++\cdots+)$ supermode has a smaller (modal) gain, at a given injection current, than the $(+-+-\cdots+)$ supermode.

22.10 (a) Plot the grating function of a six-channel in-phase $(+\cdots+)$ supermode assuming $A_l^{\nu} = 1$.
(b) Plot the far-field distribution function $|F^{\nu}(\theta)|^2$

$$F_{1(x)} = \begin{cases} \sin\left(\dfrac{x}{W}\pi\right) & -\dfrac{W}{2} \le x \le \dfrac{W}{2} \\ 0, & \text{elsewhere} \end{cases}$$

where W is the channel width and x is measured from the center of the channel (see Figure 22.22 for the definition of the x direction). Assume $k_0 S = 6\pi$ and (1) $W = 0.05S$, $0.2S$, $0.6S$.
(c) Describe semianalytically how the ratio W/S of the channel width to the channel separation affects the number of far-field lobes.

22.11 Consider the coupled mode equations [see, for example, Eq. (22.7-10)] for a lossless system

$$\frac{da}{dz} = \kappa_{ab} b e^{-i(\beta_b - \beta_a)z}$$

$$\frac{db}{dz} = \kappa_{ab} a e^{-i(\beta_a - \beta_b)z}$$

Derive the relationship between κ_{ab} and κ_{ba} in
(a) The case when modes a and b carry power in the same direction so that $|a(z)|^2 + |b(z)|^2 = \text{constant}$.
(b) The modes carry power in the opposite directions so that $|a(z)|^2 - |b(z)|^2 = \text{constant}$.

APPENDIX 1

The Kramers–Kronig Relations

Here we present the derivation of the Kramers–Kronig relations

$$\chi'(\omega) = \frac{1}{\pi}\,\text{P.V.}\int_{-\infty}^{+\infty} \frac{\chi''(\omega')}{\omega' - \omega}\,d\omega'$$

$$\chi''(\omega) = -\frac{1}{\pi}\,\text{P.V.}\int_{-\infty}^{+\infty} \frac{\chi'(\omega')}{\omega' - \omega}\,d\omega' \qquad \text{(A.1-1)}$$

that were given without proof in Section 8.1. These relations are valid when $\chi(\omega) = \chi'(\omega) - i\chi''(\omega)$ has no poles in the lower half-complex-ω-plane [when $\chi(\omega)$ has no singularities in the upper half-plane, similar relations, but with opposite signs, result]. For this case, we integrate the function $\chi(\omega')/(\omega' - \omega)$ over the contour shown in Figure A.1.

$$\int_{c'} \frac{\chi(\omega')}{\omega' - \omega}\,d\omega' + \int_{-R}^{\omega-\varepsilon} \frac{\chi(\omega')}{\omega' - \omega}\,d\omega' + \int_{\omega+\varepsilon}^{R} \frac{\chi(\omega')}{\omega' - \omega}\,d\omega' + \int_{c} \frac{\chi(\omega')}{\omega' - \omega}\,d\omega' = 0 \qquad \text{(A.1-2)}$$

where c' is the semicircle extending from $-R$ to R, whereas c is the semicircle around ω. The right side is zero since $\chi(\omega')/(\omega' - \omega)$ has no singularities inside the contour. We next take the limit of (A.1-2) as $R \to \infty$ and $\varepsilon \to 0$. The integral over c' vanishes for $\chi(\infty) = 0$ while the integral over c becomes

$$\lim_{\varepsilon\to 0}\int_{c} \frac{\chi(\omega')}{\omega' - \omega}\,d\omega' = \lim_{\varepsilon\to 0}\int_{\pi}^{2\pi} \frac{\chi(\omega + \varepsilon e^{i\phi})\,i\varepsilon e^{i\phi}}{\varepsilon^{i\phi}}\,d\phi = \pi i\chi(\omega)$$

where we took $\omega' = \omega + \varepsilon e^{i\phi}$ over c. The sum of the second and third integrals of (A.1-2) in the limit $\varepsilon \to 0$, $R \to \infty$, is, by definition, the principal value of the integral between $-\infty$ and ∞. The final result is

$$\chi(\omega) = \frac{i}{\pi}\,\text{P.V.}\int_{-\infty}^{+\infty} \frac{\chi(\omega')}{\omega' - \omega}\,d\omega' \qquad \text{(A.1-3)}$$

Taking $\chi(\omega) = \chi'(\omega) - i\chi''(\omega)$ and equating the real and imaginary parts of both sides of (A.1-3) yield Eqs. (A.1-1).

Another useful form results from the requirement that $\chi(-\omega) = \chi^*(\omega)$ so that $\chi''(\omega')$ is an odd and $\chi'(\omega')$ is an even function of ω'. We can multiply (A.1-1) by $(\omega' + \omega)/(\omega' + \omega)$ and obtain

$$\chi'(\omega) = \frac{2}{\pi}\,\text{P.V.}\int_{0}^{+\infty} \frac{\chi''(\omega')\omega'}{\omega'^2 - \omega^2}\,d\omega' \qquad \text{(A.1-4)}$$

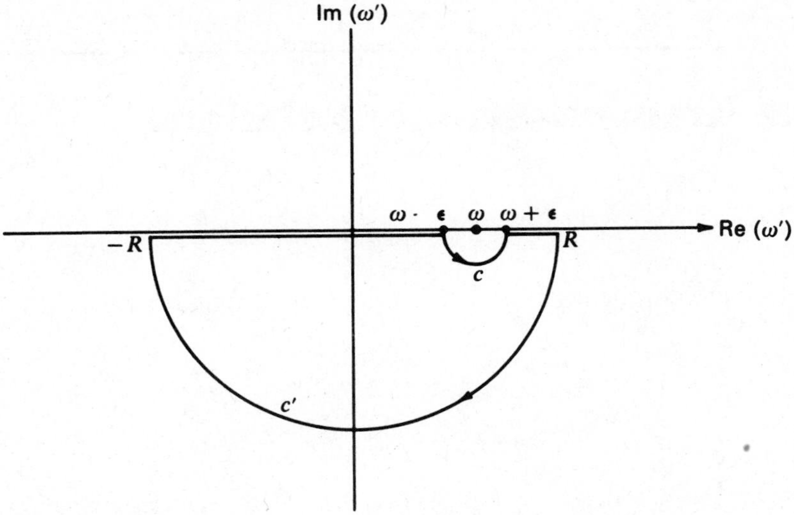

FIGURE A.1 The integration contour of (A.1-2) used to derive the Kramers–Kronig relations.

where the integral involving $\chi''(\omega')$ is zero, since $\chi''(\omega')$ is odd. In a similar fashion, we derive

$$\chi''(\omega) = \frac{2\omega}{\pi} \text{ P.V.} \int_0^{+\infty} \frac{\chi'(\omega')}{\omega'^2 - \omega^2} \, d\omega' \qquad \text{(A.1-5)}$$

The requirement that $\chi(\omega)$ have no poles in the lower half-plane is satisfied by *passive* linear systems. This is so because the Fourier transform $P(\omega)$ of the polarization $P(t)$ is equal in a linear system to the product $\chi(\omega)E(\omega)$ where $E(\omega)$ is the Fourier transform of $E(t)$. One consequence of this relation is that the natural frequencies of vibration are the poles of $\chi(\omega)$ and a pole $\omega_0 - i\omega_i$ in the lower half-plane corresponds to a solution $e^{i\omega_0 t}e^{\omega_i t}$. This represents an indefinite increase in energy with time, which is not possible in a passive linear system. The presence of poles in the lower half-plane can also be shown to violate the causality relation since it corresponds to a response of the polarization $P(t)$, which precedes the driving "force" $E(t)$.

APPENDIX 2

Solid Angle Associated with a Blackbody Mode

It is shown in Chapter 21 that, for an opening with a cross section of area A in the output screen of the amplifier, the minimum solid angle required for an essentially complete transmission of the signal power is $\boldsymbol{\Omega}_{\text{beam}} = \lambda^2/n^2 A$ and that the amount of noise radiated into this solid angle is equal to N_0. We will show below that this is exactly the solid angle "occupied" by a single blackbody mode. The blackbody radiation field is usually derived by quantizing the propagation vector k of a typical field component in a rectangular box of dimension a (along the x direction), b (along y), and c (along z).

We shall assume that the beam (or aperture) area is equal to the area of a side wall of the box given by $A = ab$, and that the propagation direction is in the z direction. The transverse (xy) quantization is thus given by (see Section 5.3)

$$\Delta k_x = \frac{2\pi}{a}$$
$$\Delta k_y = \frac{2\pi}{b} \tag{A.2-1}$$

Let us isolate, in k space, four neighboring modes whose k vectors lie

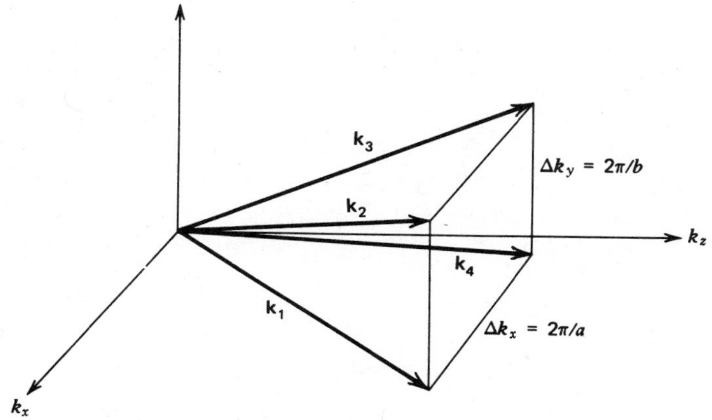

FIGURE A.2 Four adjacent blackbody modes in **k** space.

essentially in the z direction (i.e., $k \approx k_z$), as is shown in Figure A.2. The solid angle associated with a single blackbody mode is the solid angle subtended at the origin of **k** space by the area $\Delta k_x \, \Delta k_y$, or

$$\Omega_{\text{blackbody mode}} = \frac{\Delta k_x \, \Delta k_y}{k^2} = \frac{4\pi^2}{abk^2} = \frac{\lambda^2}{An^2} \tag{A.2-2}$$

where $\lambda \equiv 2\pi n/k$. This is the solid angle that contains blackbody noise power of $2N_0$ as shown in Section 21.1.

A P P E N D I X 3

The Spontaneous Emission Lifetime for a Vibrational-Rotational Transition in a Linear Molecule

A knowledge of the spontaneous emission lifetime is necessary in evaluating the induced transition rate (8.3-10) or, equivalently, the gain or loss due to a given population distribution (8.4-4). In Section 8.3 we obtained an expression (8.3-7) for the spontaneous emission lifetime of an electronic transition. In what follows, we consider the spontaneous emission lifetime for a rotational-vibrational transition.

To simplify the mathematical complexity, we consider a linear molecule. If we take the direction of the molecular axis as (θ, ϕ), the form of the Schrödinger equation becomes identical to (2.3-1), except that now m is a reduced mass, whereas r represents the vibrational coordinate. $V(r)$ is the total potential energy including the contribution from ground state electrons (Reference 1). The electronic coordinates are presented by $\mathbf{r}_e$.

The solution of the eigenfunctions can be taken directly from (2.3-9) as

$$u(\mathbf{r}_e, r, \theta, \phi) = u_r(\mathbf{r}_e)\, u_v(r)\, N_{jm} P_J^m(\cos\theta)\, e^{im\phi} \tag{A.3-1}$$

where we assumed that the electronic wavefunction adjusts instantaneously to the vibrational coordinate r that on the electronic time scale changes very "slowly." The solution of $u_v(r)$ depends on the potential energy function $V(r)$ and, in the following, is assumed known.

To derive the spontaneous lifetime for a transition $|v = 2, J, m\rangle \rightarrow |v = 1, J', m'\rangle$, we proceed in a manner similar to that of Section 8.3. We consider the case of an applied electromagnetic field with a z component and, assuming a dipolar interaction, take the perturbation Hamiltonian as in (8.3-1)

$$\mathcal{H}' = -\mu_z E_z = -\mu(r)\cos\theta E_z \tag{A.3-2}$$

where $\boldsymbol{\mu}(r) = \mathbf{a}_{\theta,\phi}\mu(r) = e\int |u_r(\mathbf{r}_e)|^2 \mathbf{r}_e\, d^3\mathbf{r}_e$. Here, $\mathbf{a}_{\theta,\phi}$ is a unit vector along the molecular axis of symmetry, $\mathbf{r}_e$ represents the electronic coordinates, and μ_z is the component of the electronic dipole moment along the field direction. We now expand the electronic dipole moment in a Taylor series

$$\mu(r) = \mu_0 + e'r + b_1 r^2 + \cdots \tag{A.3-3}$$

Ignoring the possibility of a permanent dipole, that is, take $\mu_0 = 0$, and neglecting terms with powers larger than 1, we have

$$\mathcal{H}' = -e'rE_z \cos\theta \tag{A.3-4}$$

where we took the dipole direction along (θ, ϕ). Note that e' is an effective charge and is defined by Eq. (A.3-3).

A comparison of (A.3-4) to (8.3-1) shows that the formalism of Section 8 applies here, provided we replace ey by $e'r\cos\theta$.

We thus obtain, from (8.3-7),

$$\underset{2,J,m\to 1,J',m'}{W_{\text{spont}}} = \frac{2n^3 e'^2\omega^3}{\varepsilon hc^3} |\langle 2Jm|\, r\cos\theta\, |1J'm'\rangle|^2 \tag{A.3-5}$$

We now proceed to evaluate the matrix element. Using (A.3-1) and

$$N_{Jm} = \frac{1}{4\pi}\left[\frac{(2J+1)(J-|m|)!}{(J+|m|)!}\right]^{1/2} \tag{A.3-6}$$

we obtain

$$r_z \equiv \langle 2Jm|\, r\cos\theta\, |1J'm'\rangle$$

$$= \int_0^\infty u_2^*(r)\,u_1(r)\,r\,dr \int_0^\pi\int_0^{2\pi} N_{Jm}N_{J'm'}P_J^m(\theta)\,P_{J'}^{m'}(\theta)\,e^{-i(m-m')\phi}\cos\theta\sin\theta\,d\theta\,d\phi$$

Using (A.3-6) and recognizing that $m = m'$ for a nonvanishing integral, we have

$$r_z = f_{12}\left[\frac{(2J+1)(J-|m|)!}{(J+|m|)!}\right]^{1/2}\left[\frac{(2J'+1)(J'-|m|)!}{(J'+|m|)!}\right]^{1/2}\int_{-1}^{1} vP_J^m(v)\,P_{J'}{}^m(v)\,dv \tag{A.3-7}$$

where $v = \cos\theta$ and

$$f_{12} = \frac{1}{8\pi}\int_0^\infty u_2^*(r)\,u_1(r)\,r\,dr \tag{A.3-8}$$

Using

$$vP_J^m(v) = \frac{1}{2J+1}\left[(J-|m|+1)P_{J+1}^m + (J+|m|)P_{J-1}^m\right] \tag{A.3-9}$$

and the relation

$$\int_{-1}^{1} P_K^m(v)\,P_J^m(v)\,dv = 2\delta_{KJ}\frac{(J+|m|)!}{(2J+1)(J-|m|)!} \tag{A.3-10}$$

we find that the integral in (A.3-7) vanishes unless

$$J' = J+1 \qquad (P \text{ transition})$$

or

$$J' = J-1 \qquad (R \text{ transition})$$

To be specific, we choose $J' = J + 1$ and obtain after substantial, but straightforward, simplification

$$\frac{r_z}{2f_{12}} = \frac{(J - |m| + 1)^{1/2}(J + |m| + 1)^{1/2}}{(2J + 1)^{1/2}(2J + 3)^{1/2}} \tag{A.3-11}$$

The discussion up to this point assumed that the molecule was initially in the state $|2, J, m\rangle$. Since there are $2J + 1$ degenerate levels with $m = J, J - 1, \ldots, -J$, a given molecule has an equal probability of being initially in any one of these levels. We thus need to average the absolute value squared of (A.3-11) over the $2J + 1$ possible initial m states.

$$\frac{\overline{r_z^2}}{4f_{12}^2} = \frac{1}{(2J + 1)^2(2J + 3)} \sum_{m=-J}^{J} [(J + 1)^2 - m^2] \tag{A.3-12}$$

If we use $\Sigma_{m=1}^{J} m^2 = \frac{1}{6}J(J + 1)(2J + 1)$, the summation in (A.3-12) is found to be equal to $\frac{1}{3}(2J + 1)(J + 1)(2J + 3)$ so that

$$\begin{aligned} \frac{\overline{r_z^2}}{4f_{12}^2} &= \frac{1}{3}\frac{J + 1}{2J + 1} \qquad J \rightarrow J + 1 \\ &= \frac{1}{3}\frac{J}{2J + 1} \qquad J \rightarrow J - 1 \end{aligned} \tag{A.3-13}$$

Substitution in (A.3-5) gives

$$W_{\substack{\text{spont} \\ 2,J\rightarrow 1,J\pm 1}} = \frac{8n^3e'^2f_{12}^2\omega^3}{3\varepsilon hc^3}\frac{(J + \frac{1}{2} \pm \frac{1}{2})}{(2J + 1)} = \frac{64\pi^3n^3e'^2f_{12}^2}{3\varepsilon h\lambda^3}\frac{(J + \frac{1}{2} \pm \frac{1}{2})}{(2J + 1)} \tag{A.3-14}$$

which is the form given in (10.6-8).

Reference

1. Townes, C. H. and A. L. Schawlow, *Microwave Spectroscopy* (New York: McGraw-Hill, 1955), p. 6.

APPENDIX 4

Quantum Mechanical Derivation of Nonlinear Optical Constants

In Section 16.3 we presented a classical derivation of the nonlinear optical constant d_{ijk} that relates, according to (16.1-4), the amplitude of the polarization at $\omega_j \pm \omega_k$ to the fields' product, $E_j E_k$. In the following, we will present a quantum mechanical derivation of this result.

The formalism employed is that of the density matrix as introduced in Section 3.15 and 3.16. The equation of motion of the density matrix ρ_{ij} is given by (3.16-5). We need to modify it to include the effects of collisions.

We use the result of Karplus and Schwinger (Reference 1) who showed that by assuming a "collision" time Γ^{-1} the equation of motion (3.16-5) becomes

$$\frac{\partial \rho}{\partial t} = -\frac{i}{\hbar}[\mathcal{H}, \rho] - \Gamma(\rho - \bar{\rho}) \tag{A.4-1}$$

where

$$\mathcal{H} = \mathcal{H}_0 + V(t)$$

and

$$\bar{\rho} = \exp(-\mathcal{H}_0/kT)/\mathrm{tr}[\exp(-\mathcal{H}_0/kT)] \tag{A.4-2}$$

is the thermal equilibrium value of the density matrix. It follows from (A.4-2) that in a representation in which $\mathcal{H}_0$ is diagonal, $\bar{\rho}$ is a diagonal matrix.

To allow for the possibility of different decay rates of the diagonal matrix elements (T_1 processes) as well as the off-diagonal elements (T_2 processes) as discussed in Section 8.1, we modify (A.4-1) to

$$\frac{\partial \rho_{ij}}{\partial t} = -\frac{i}{\hbar}[\mathcal{H}, \rho]_{ij} - \gamma_{ij}(\rho - \bar{\rho})_{ij} \tag{A.4-3}$$

so that γ_{ii} is the relaxation rate for the population in level i, whereas γ_{ij} is the rate at which ρ_{ij} $(i = j)$ relaxes to zero.

If we use $\mathcal{H} = \mathcal{H}_0 + V(t)$ where $V(t)$ is a perturbation, choose the representation where $(\mathcal{H}_0)_{ij} = E_i \delta_{ij}$, and define $\omega_{ij} = (E_i - E_j)/\hbar$, the equation of motion (A.4-3) becomes

$$\frac{\partial \rho_{ij}}{\partial t} = -i\omega_{ij}\rho_{ij} - \frac{i}{\hbar}[V, \rho]_{ij} - \gamma_{ij}(\rho - \bar{\rho})_{ij} \tag{A.4-4}$$

We will next obtain an integral perturbation solution for ρ. Let us introduce, as in Section 3.12, a "turning-on" parameter λ and replace V in (A.4-4) by λV. The density matrix is then expanded in a series with ascending powers of λ

$$\rho_{ij} = \rho_{ij}^{(0)} + \lambda \rho_{ij}^{(1)} + \lambda^2 \rho_{ij}^{(2)} + \cdots \tag{A.4-5}$$

The formal procedure consists of solving for ρ_{ij} by iteration up to some desired order and then putting $\lambda = 1$ in the final result. We substitute (A.4-5) in (A.4-4), replace V by λV and, after equating the same powers of λ on both sides, obtain $\rho_{ij}^{(0)} = \bar{\rho}_{ij}\delta_{ij}$ and

$$\frac{\partial \rho_{ij}^{(n)}}{\partial t} = -(i\omega_{ij} + \gamma_{ij})\rho_{ij}^{(n)} - \frac{i}{\hbar}[V, \rho^{(n-1)}]_{ij}$$

whose solution is

$$\rho_{ij}^{(n)}(t) = -\frac{i}{\hbar}\int_{-\infty}^{t} e^{i(\omega_{ij} - i\gamma_{ij})(t'-t)}[V(t'), \rho^{(n-1)}(t')]_{ij}\, dt' \tag{A.4-6}$$

This last result, in spite of its formal appearance, constitutes one of the most important theoretical tools for the study of photon-atom interactions (Reference 2).

To illustrate its importance, we will apply it to obtain the nonlinear polarization due to the mixing of the fields $E_3 \exp(i\omega_3 t)$ and $E_2 \exp(i\omega_2 t)$ in an atom (or molecule) such as shown in Figure A.4-1 to produce the difference frequency $\omega_1 = \omega_3 - \omega_2$. Let us assume that the photon energies $\hbar\omega_2$ and $\hbar\omega_3$ are nearly resonant with the transition energies $E_2 - E_1$ and $E_3 - E_1$, respectively, and keep, in the analysis, only those terms with nearly vanishing denominators (i.e., resonant). Assuming a dipolar interaction $V = -\boldsymbol{\mu} \cdot \mathbf{E}(t)$ where $\boldsymbol{\mu}$ is the dipole moment operator, $\mu_{11} = \mu_{22} = \mu_{33} = 0$, and taking

$$\mathbf{E}(t) = \frac{\mathbf{E}_3}{2} e^{i\omega_3 t} + \frac{\mathbf{E}_2}{2} e^{i\omega_2 t} + \text{c.c.} \tag{A.4-7}$$

we have

$$V = \begin{vmatrix} 0 & \left(-\mu_{12}\frac{E_2}{2} e^{i\omega_2 t} + \text{c.c.}\right) & \left(-\mu_{13}\frac{E_3}{2} e^{i\omega_3 t} + \text{c.c.}\right) \\ \left(-\mu_{12}\frac{E_2}{2} e^{i\omega_2 t} + \text{c.c.}\right) & 0 & 0 \\ \left(-\mu_{13}\frac{E_3}{2} e^{i\omega_3 t} + \text{c.c.}\right) & 0 & 0 \end{vmatrix} \tag{A.4-8}$$

where nonresonant terms have been omitted so that the only matrix elements V_{ij} in (A.4-8) are those where the field frequency is nearly equal to ω_{ij}.

The polarization is given by

$$\begin{aligned} P_\alpha &= N tr(\rho \mu_\alpha) \\ &= N tr\{[\rho^{(0)} + \rho^{(1)} + \cdots], \mu_\alpha\} \end{aligned} \tag{A.4-9}$$

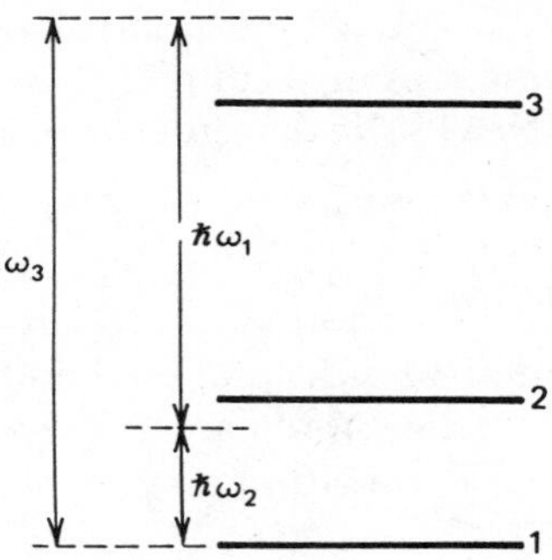

FIGURE A.4-1 The energy levels and photon energies involved in deriving the nonlinear polarization.

where N is the density of atoms in the states 1, 2, and 3, so that the polarization component at $\omega_1 = \omega_3 - \omega_2$ arises from the term $\rho^{(2)}$. Since the time dependence of $\rho^{(2)}$ is, according to (A.4-6), that of $[V(t'), \rho^{(1)}(t')]$, the desired mixing term will result from the product involving $V_{31}\exp(i\omega_3 t)\rho_{21}^{(1)}(t)$. Our first task is thus to obtain $\rho_{21}^{(1)}(t)$.

We use (A.4-8) in (A.4-6), taking $\rho_{ij}^{(0)} = \bar{\rho}_{ii}\delta_{ij}$. The commutator $[V, \rho^{(0)}]$ becomes

$$[V, \rho^{(0)}] = \begin{bmatrix} 0 & V_{12}(\bar{\rho}_{22} - \bar{\rho}_{11}) & -\bar{\rho}_{11}V_{13} \\ V_{12}(\bar{\rho}_{11} - \bar{\rho}_{22}) & 0 & 0 \\ V_{13}\bar{\rho}_{11} & 0 & 0 \end{bmatrix}$$

where we assumed that the thermal equilibrium population of level 3 is negligible and took $\bar{\rho}_{33} = 0$. Using the last result in (A.4-6) gives

$$\rho_{12}^{(1)}(t) = -\frac{i}{2\hbar}\int_{-\infty}^{t} e^{i(\omega_{12}-i\gamma_{12})(t'-t)}[-(\bar{\rho}_{22} - \bar{\rho}_{11})\mu_{12}E_2 e^{i\omega_2 t'} - (\bar{\rho}_{22} - \bar{\rho}_{11})\mu_{12}E_2^* e^{-i\omega_2 t'}]\,dt' \tag{A.4-10}$$

Since $\omega_{12} \simeq -\omega_2$, the first term in the square brackets dominates and

$$\rho_{12}^{(1)} \simeq \frac{i}{2\hbar}\frac{(\bar{\rho}_{22} - \bar{\rho}_{11})\mu_{12}E_2}{i(\omega_{12} + \omega_2 - i\gamma_{12})}e^{i\omega_2 t} \equiv \rho_{12A}^{(1)}e^{i\omega_2 t} \tag{A.4-11}$$

$$\rho_{21}^{(1)} = \rho_{12}^{(1)*}$$

The only other resonant matrix elements $\rho_{ij}^{(1)}$ are $\rho_{13}^{(1)}$ and $\rho_{31}^{(1)} = \rho_{13}^{(1)*}$.

$$\rho_{13}^{(1)} = -\frac{i}{2\hbar}\frac{\bar{\rho}_{11}\mu_{13}E_3}{[i(\omega_{13} + \omega_3) + \gamma_{13}]}e^{i\omega_3 t} \equiv \rho_{13A}^{(1)}e^{i\omega_3 t} \tag{A.4-12}$$

The second-order polarization component along the α direction is

$$P_\alpha^{(2)} = Ntr[\rho^{(2)}\mu_\alpha] \tag{A.4-13}$$

An inspection of (A.4-6) and the form of $\rho_{ij}^{(1)}$ reveals that the component of $P^{(2)}$ oscillating at $\omega_1 = \omega_3 - \omega_2$ is

$$P_\alpha^{(2)} = \rho_{23}^{(2)}(\mu_\alpha)_{32} + \rho_{32}^{(2)}(\mu_\alpha)_{23} \tag{A.4-14}$$

The resonant part of $\rho_{23}^{(2)}$ oscillating at ω_1 is

$$\rho_{32}^{(2)} = -\frac{i}{2\hbar} e^{-i(\omega_{32}-i\gamma_{32})t} \int_{-\infty}^{t} e^{i(\omega_{32}-i\gamma_{32})t'} \times [-\rho_{12A}^{(1)}\mu_{13}E_3^* e^{-i\omega_1 t'} + \rho_{31A}^{(1)}\mu_{12}E_2 e^{-i\omega_1 t'}]\, dt' \tag{A.4-15}$$

$$= \frac{i}{2\hbar}\left[\frac{\rho_{12A}^{(1)}\mu_{13}E_3^*}{i(\omega_{32}-\omega_1-i\gamma_{32})} - \frac{\rho_{31A}^{(1)}\mu_{12}E_2}{i(\omega_{32}-\omega_1-i\gamma_{32})}\right] e^{-i\omega_1 t}$$

that, if we use (A.4-11, 12), becomes

$$\rho_{32}^{(2)} = \frac{1}{4\hbar_2}\left\{\frac{\bar{\rho}_{11}\mu_{12}\mu_{13}}{[-i(\omega_3-\omega_{31})+\gamma_{13}][i(\omega_{32}-\omega_1)+\gamma_{32}]} - \frac{(\bar{\rho}_{22}-\bar{\rho}_{11})\mu_{12}\mu_{13}}{[i(\omega_2-\omega_{21})+\gamma_{12}][i(\omega_{32}-\omega_1)+\gamma_{32}]}\right\} E_3^* E_2 e^{-i\omega_1 t} \tag{A.4-16}$$

Substituting (A.4-16) in (A.4-14) leads to the desired result

$$P_\alpha^{(2)} = \frac{1}{4\hbar^2}\left\{\frac{N_1\mu_{12}\mu_{13}(\mu_\alpha)_{32}}{[\gamma_{13}+i(\omega_3-\omega_{31})][\gamma_{32}-i(\omega_{32}-\omega_1)]} - \frac{(N_2-N_1)\mu_{12}\mu_{13}(\mu_\alpha)_{32}}{[\gamma_{12}-i(\omega_2-\omega_{21})][\gamma_{32}-i(\omega_{32}-\omega_1)]}\right\} E_3 E_2^* e^{i\omega_1 t} + \text{c.c.} \tag{A.4-17}$$

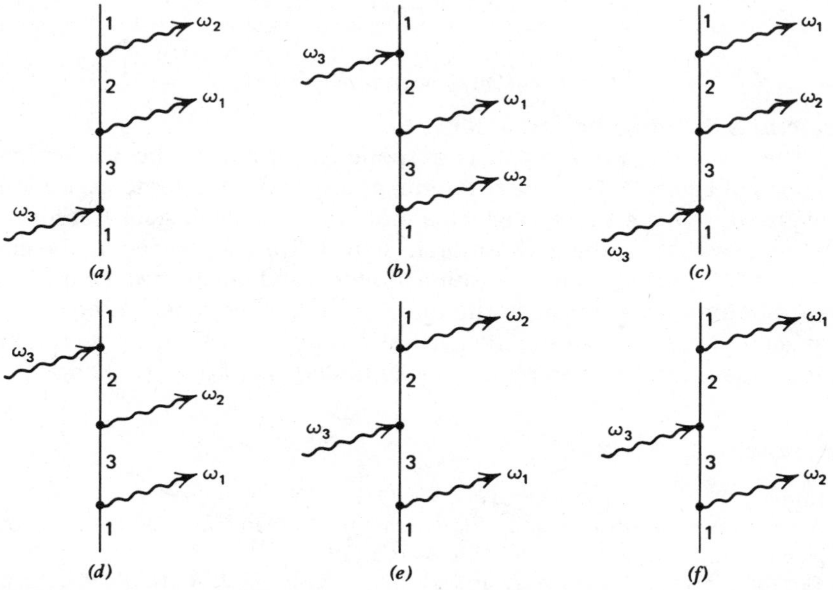

FIGURE A.4-2 The Feynman diagrams used to derive the six terms of the optical nonlinear constants $d_{\alpha 32}^{\omega_1=\omega_3-\omega_2}$. These diagrams correspond to the six permutations of the sequence of photon interactions in a process in which an atom makes a transition $|1\rangle \to |3\rangle \to |2\rangle \to |1\rangle$ while absorbing a photon at ω_3 and emitting photons at ω_1 and ω_2. Since $\omega_3 = \omega_1 + \omega_2$, energy is conserved. The amplitudes corresponding to these diagrams are given in (A.4-19).

In obtaining (A.4-17), we used $N_i = N\bar{\rho}_{ii}$ as the population density of level i and $\rho_{ij}^{(n)} = \rho_{ji}^{(n)*}$.

Using the definition (16.1-4) for the nonlinear susceptibility tensor $\tilde{\tilde{d}}$, we obtain, from (A.4-17),

$$d_{\alpha 32}^{\omega_1=\omega_3-\omega_2} = \frac{1}{2\hbar^2}\left\{\frac{N_1\mu_{12}\mu_{13}(\mu_\alpha)_{32}}{[\gamma_{13} + i(\omega_3 - \omega_{31})][\gamma_{32} - i(\omega_{32} - \omega_1)]} - \frac{(N_2 - N_1)\mu_{12}\mu_{13}(\mu_\alpha)_{32}}{[\gamma_{12} - i(\omega_2 - \omega_{21})][\gamma_{32} - i(\omega_{32} - \omega_1)]}\right\} \tag{A.4-18}$$

The nonlinear constants characterizing phenomena such as triple frequency mixing to produce radiation at $\omega_i \pm \omega_j \pm \omega_k$ and Raman scattering are obtained by carrying the perturbation calculation to third order in V (Reference 2).

If we operate far from resonance so that none of the factors in the denominators of (A.4-18) approaches zero, we can neglect the relaxation rates γ_{ij}. In this case, of course, we need to repeat the analysis and keep all the previously discarded "nonresonant" terms. The result, after considerable relabelling and regrouping of terms and taking $N_2 = N_3 = 0$, is (Reference 3)

$$\begin{aligned} d_{\alpha 32}^{\omega_1=\omega_3-\omega_2} = \frac{N_1}{2\hbar^2}\sum_{2,3}\Bigg[&\frac{(\mu_\alpha)_{32}(\mu_2)_{21}(\mu_3)_{13}}{(\omega_2 - \omega_{21})(\omega_3 - \omega_{31})} + \frac{(\mu_\alpha)_{23}(\mu_2)_{12}(\mu_3)_{31}}{(\omega_3 + \omega_{31})(\omega_2 + \omega_{21})} \\ &+ \frac{(\mu_\alpha)_{12}(\mu_2)_{23}(\mu_3)_{31}}{(\omega_1 - \omega_{21})(\omega_3 - \omega_{31})} + \frac{(\mu_\alpha)_{31}(\mu_2)_{23}(\mu_3)_{12}}{(\omega_3 - \omega_{21})(\omega_1 - \omega_{31})} \\ &+ \frac{(\mu_\alpha)_{31}(\mu_2)_{12}(\mu_3)_{23}}{(\omega_2 - \omega_{21})(\omega_1 + \omega_{31})} + \frac{(\mu_\alpha)_{12}(\mu_2)_{31}(\mu_3)_{23}}{(\omega_1 - \omega_{21})(\omega_2 + \omega_{31})}\Bigg] \end{aligned} \tag{A.4-19}$$

where μ_j is the projection of $\boldsymbol{\mu}$ along $\mathbf{E}_j$.

The six terms in (A.4-19) correspond, in order, to the six Feynman diagrams of Figure A.4-2. The first term, as an example, represents according to Figure A.4-2*a* a process in which a molecule initially in state 1 absorbs an ω_3 photon while making a transition to state 3. This is followed by the emission of a photon at ω_1 with a transition to state 2 and finally, by an emission of an ω_2 photon and a return to the initial state 1. Energy is conserved since $\omega_3 = \omega_1 + \omega_2$. The molecule thus acts as a catalyst breaking up a photon at ω_3 into one ω_1 and one ω_2 photon while undergoing no (net) change.

References

1. Karplus, R. and J. Schwinger, *Phys. Rev.* **73,** 1020 (1948).
2. An elegant use of this equation for deriving the Raman scattering cross section is given in C. L. Tang, *Phys. Rev.* **134,** A1166 (1964).
3. The six terms in (A.4-19) were first given by Armstrong, J. A., N. Bloembergen, J. Ducuing, and P. S. Pershan, *Phys. Rev.* **127,** 1918–1939 (1962), Equations 2.13 and 2.14. The correspondence is complete if in the cited reference we correct a misprint and interchange *j* and *j'* in (2.13) but not in (2.14).

APPENDIX 5

The Interaction of an Electron and an Electromagnetic Field

Consider the modification of the total Hamiltonian that is due to the interaction of an electron with an electromagnetic field. The resulting interaction Hamiltonian will play a central role in many theoretical developments in the remainder of the book.

Consider an electron that is subject to a static potential $V(\mathbf{r})$ and, simultaneously, to an electromagnetic field described by a vector potential $\mathbf{A}(\mathbf{r}, t)$. For a charge-free medium, the field $\mathbf{E}$ and $\mathbf{A}$ are related by $\mathbf{E} = -(\partial/\partial t)\mathbf{A}$. We shall choose the Coulomb gauge (see Section 5.1)

$$\nabla \cdot \mathbf{A} = 0 \tag{A.5-1}$$

The total Hamiltonian is

$$H = \frac{1}{2m}(\mathbf{p} - e\mathbf{A})^2 + V(\mathbf{r}) \tag{A.5-2}$$

where m is the electron mass and $\mathbf{p} = -i\hbar\nabla$ is the momentum operator. Since $\mathbf{A}$ in general depends on $\mathbf{r}$, $\mathbf{A}$ and $\mathbf{p}$ do not commute so that

$$H = \frac{p^2}{2m} - \frac{e}{2m}(\mathbf{p} \cdot \mathbf{A} + \mathbf{A} \cdot \mathbf{p}) + \frac{e}{2m}A^2 + V(\mathbf{r}) \tag{A.5-3}$$

Let us consider the term $\mathbf{p} \cdot \mathbf{A}$. Using the general result (see Problem 1.11)

$$[p_i, F(\mathbf{q}, t)] = -i\hbar \frac{\partial F}{\partial q_i}$$

we obtain

$$[p_i, A_i] = [p_i A_i - A_i P_i] = -i\hbar \frac{\partial A_i}{\partial q_i}$$

where summation over repeated indices is understood. The last result can be rewritten as

$$\mathbf{p} \cdot \mathbf{A} = \mathbf{A} \cdot \mathbf{p} - i\hbar\, \nabla \cdot \mathbf{A} = \mathbf{A} \cdot \mathbf{p}$$

since $\nabla \cdot \mathbf{A} = 0$. H thus takes the form

$$H = \frac{p^2}{2m} - \frac{e}{m}(\mathbf{A} \cdot \mathbf{p}) + \frac{e}{2m} A^2 + V(\mathbf{r}) \tag{A.5-4}$$

It will be left as an exercise to show that for allowed optical transitions

$$\frac{\left\langle i \left| \frac{e^2}{2m} A^2 \right| j \right\rangle}{\langle i | e/m(\mathbf{A} \cdot \mathbf{p}) | j \rangle} \sim \frac{E_{\text{external}}}{E_{\text{atomic}}} \sim \frac{E_{\text{ext}}}{10^{12}(V/m)}$$

Consequently, the term involving A^2 will be neglected in what follows. We must remember, however, that for applied optical fields that are sufficiently large or when the matrix elements of p are very small, the term A^2 is significant. It can, as an example, in first order account for a transition in which two photons are involved.

We write the total Hamiltonian including the electromagnetic field as

$$H = V(\mathbf{r}) + \frac{p^2}{2m} + \sum_{\mathbf{k},\lambda} \hbar\omega_k \left(a^\dagger_{\mathbf{k},\lambda} a_{\mathbf{k},\lambda} + \frac{1}{2} \right) - e\frac{\mathbf{A} \cdot \mathbf{p}}{m} \equiv H_0 + H' \tag{A.5-5}$$

We can view the first three terms on the right side of (A.5-5) as the unperturbed Hamiltonian H_0 and the last term, involving $\mathbf{A} \cdot \mathbf{p}$, as the interaction term that is responsible for transition between the eigenstates of H_0. Denoting these eigenfunctions, each of which is a product of an electronic wavefunction and a radiation field eigenfunction, as $|i\rangle$ we need consider matrix elements of the type

$$\langle j|H'|i\rangle = -\frac{e}{m} \sum_{\mathbf{k},\lambda} \sqrt{\frac{\hbar}{2\varepsilon V \omega_k}} \langle j | a^\dagger_{\mathbf{k},\lambda} \hat{\mathbf{e}}_{\mathbf{k},\lambda} e^{+i\mathbf{k}\cdot\mathbf{r}} \cdot \mathbf{p} + a_{\mathbf{k},\lambda} \hat{\mathbf{e}}_{\mathbf{k},\lambda} e^{-i\mathbf{k}\cdot\mathbf{r}} \cdot \mathbf{p} | i \rangle \tag{A.5-6}$$

In the optical regime $k = (2\pi/\lambda) \approx 10^7$ (m^{-1}), $r \approx$ atomic radius $\sim 10^{-10}$ m $\Rightarrow \mathbf{k} \cdot \mathbf{r} \ll 1$, and we can take $e^{i\mathbf{k}\cdot\mathbf{r}} = 1$. This is the so-called dipole approximation. It is obviously not valid in the case of hard x-rays where $k \sim 10^{11}$, but will be assumed throughout this book. Limiting our attention to one electromagnetic mode and taking the total eigenfunction as a product of an electronic part $|a\rangle$ and a radiation part $|n\rangle$, we obtain

$$\langle j|H'|i\rangle = -\frac{e}{m} \sqrt{\frac{\hbar}{2\varepsilon V \omega_k}} \langle \alpha, n | a^\dagger_{\mathbf{k},\lambda} \hat{\mathbf{e}}_{\mathbf{k}} \cdot \mathbf{p} + a_{\mathbf{k},\lambda} \hat{\mathbf{e}}_{\mathbf{k}} \cdot \mathbf{p} | \beta, l \rangle$$

which using $\langle \alpha | p | \beta \rangle = i\omega_{\alpha\beta} m \langle \alpha | r | \beta \rangle$ (see Problem 1.7), $\omega_{\alpha\beta} \equiv (E_\alpha - E_\beta)/\hbar$, and the conservation of energy relation $\omega_{\alpha\beta} = -\omega_k$ for emission and $\omega_{\alpha\beta} = \omega_k$ for absorption leads to

$$\langle j|H'|i\rangle = ie \sqrt{\frac{\hbar\omega_k}{2\varepsilon V}} \langle j | a^\dagger_{\mathbf{k},\lambda} \hat{\mathbf{e}}_{\mathbf{k},\lambda} \cdot \mathbf{r} - a_{\mathbf{k},\lambda} \hat{\mathbf{e}}_{\mathbf{k},\lambda} \cdot \mathbf{r} | i \rangle$$

Using (5.6-15), we can rewrite the last relation as

$$\langle j|H'|i\rangle = \langle j| - e\mathbf{r} \cdot \mathbf{E} |i\rangle \tag{A.5-7}$$

In the remainder of this book, we will take the interaction Hamiltonian in the form

$$H' = \boldsymbol{\mu} \cdot \mathbf{E} \tag{A.5-8}$$

where

$$\boldsymbol{\mu} = -e\mathbf{r} \tag{A.5-9}$$

is the dipole moment operator of an electron. A scalar version of (A.5-8) appropriate when **E** is linearly polarized was used in Eq. (8.1-1).

APPENDIX 6

The Derivation of the Spontaneous Emission Langevin Fluctuation "Power"

Here we will derive the scale constant Q appearing in Eq. 21.5-8

$$\langle \Delta_r(\lambda_1)\Delta_r(\lambda_2)\rangle = QD(\lambda_1 - \lambda_2) \tag{A.6-1}$$

where $D(x)$ is the "Delta" function of x. Our approach will be to calculate the optical energy in a single mode just below threshold. This energy is proportional to $\langle \Delta^2(t)\rangle \propto Q$. We will derive a second expression for this energy starting with elementary bookkeeping of spontaneous transitions. By equating the two results to each other, we will obtain an expression for Q. Below threshold we can neglect the induced polarization term $\ddot{P}$ in (21.4-5) so that the laser oscillator equation becomes

$$\ddot{E} + \frac{\dot{E}}{\tau_p} + \omega_n^2 E = \Delta e^{i\omega_m t}$$

which after Fourier transformation leads to

$$\tilde{E}(\Omega) = \frac{\tilde{\Delta}(\Omega - \omega_m)}{(\omega_n^2 - \Omega^2) + i\dfrac{\Omega}{\tau_p}} \tag{A.6-2}$$

Equation A.6-2 relates the Fourier transform $\tilde{E}$ of the resonator field to that of the Langevin driving force $\tilde{\Delta}$. Now the spectral density $W_f(\Omega)$ of any function $f(t)$ is proportional to the absolute value squared of its Fourier transform[1]. Thus by taking the absolute value (squared) of both sides of (A.6-2), we obtain for $\Omega \approx \omega_n$

$$\underset{\Omega \approx \omega_n}{W_E(\Omega)} = \frac{2W_{\Delta_r}^{(\Omega - \Omega_m)}}{4\omega_n^2(\omega_n - \Omega)^2 + \left(\dfrac{\Omega}{\tau_p}\right)^2} \tag{A.6-3}$$

In (A.6-3) we used (21.4-13) as well as (21.5-8) and assumed that the fluctuation powers of the in-phase (Δ_r) and quadrature component (Δ_i) are equal, $\langle \Delta_r^2\rangle = \langle \Delta_i^2\rangle$, so that $W_\Delta(\Omega) = 2W_{\Delta_r}(\Omega) = 2W_{\Delta_i}(\Omega)$. We can use the Wiener-Kintchine theorem (21.3-4) and the first of Eqs. 21.5-8 to obtain an expression for $W_{\Delta_r}(\Omega)$

$$W_{\Delta_r}(\Omega) = \frac{1}{\pi}\int_{-\infty}^{\infty} \langle \Delta_r(0)\Delta_r(\tau)\rangle e^{-i\Omega\tau} = \frac{Q}{\pi} \tag{A.6-4}$$

which corresponds to a "white spectrum. We can now rewrite (A.6-3) as

$$\underset{\Omega \approx \omega_n}{W_E(\Omega)} = \frac{2Q}{\pi\left[4\omega_n^2(\omega_n - \Omega)^2 + \left(\frac{\Omega}{\tau_p}\right)^2\right]} \tag{A.6-5}$$

According to (21.3-5)

$$\langle E^2(t)\rangle = \frac{1}{2}\int_{-\infty}^{\infty} W_E(\Omega)\, d\Omega$$

performing the indicated integration we obtain from (A.6-5)

$$\langle E^2(t)\rangle = \frac{Q\tau_p}{4\omega_n^2} \tag{A.6-6}$$

so that the energy per mode is

$$\text{Energy/mode} = \varepsilon\langle E^2(t)\rangle V$$

where we used $(\Omega/\tau_p)^2 \approx (\omega_n/\tau_p)^2$ and V is the mode volume.

We can also derive the mode energy below threshold from a power conservation argument by requiring that the energy fed into the (single) mode by spontaneous emission be equal to that dissipated. The total spontaneous emission power into all modes just below threshold is

$$P_{\text{spont}} = \frac{N_2\hbar\omega}{t_{\text{spont}}}$$

where N_2 is the (total) population of the upper laser transition level just below threshold. This power is divided among the p modes "on speaking terms" with the transition (Eq. 9.3-24)

$$p = \frac{2n^3\omega^2\Delta\nu V}{\pi c^3} \tag{A.6-7}$$

where $\Delta\nu$ is the spectral width of the transition. We thus have

$$P_{\text{spont/mode}} = \frac{N_2\hbar\omega}{t_{\text{spont}}p} \tag{A.6-8}$$

The last relation which we need in order to complete our derivation is that of the threshold inversion density. From the footnote on page 187 we have

$$(N_2 - N_1)_{\text{threshold}} = \frac{pt_{\text{spont}}}{(\tau_p/2)} \tag{A.6-9}$$

We can thus rewrite (A.6-8) as

$$P_{\text{spont/mode}} = \frac{N_2(N_2 - N_1)\hbar\omega}{(N_2 - N_1)t_{\text{spont}}p} = \frac{\varepsilon\langle E^2(t)\rangle V}{(\tau_p/2)}$$

where the cavity photon decay time constant is $t_c = \tau_p/2$. If we substitute (A.6-8) for p in the last equality, this leads to

$$\langle E^2(t)\rangle = \frac{\hbar\omega\left(\frac{N_2}{N_2 - N_1}\right)}{\varepsilon V} \tag{A.6-10}$$

and from (A.6-6)

$$Q = \frac{4\hbar\omega^3 \left(\dfrac{N_2}{N_2 - N_1}\right)}{\varepsilon V \tau_p} \tag{A.6-11}$$

as stated in (21.5-14).

The factor $[N_2/(N_2 - N_1)]$ taken at threshold represents the increase in N_2 and thus in spontaneous emission noise power $(N_2 V\hbar\omega)/t_{\text{spont}}$, which results when the lower laser level is populated $(N_1 \neq 0)$. It is called the inversion factor. In a semiconductor laser where the electron distribution obeys Fermi-Dirac statistics, we replace this factor, using (11.2-15), with

$$\left(\frac{N_2}{N_2 - N_1}\right) \rightarrow \frac{f_c(\hbar\omega)(1 - f_v(\hbar\omega))}{f_c(\hbar\omega) - f_v(\hbar\omega)} \tag{A.6-11}$$

where f_c and f_v are the Quasi-Fermi functions for the conduction and valence band, respectively, evaluated at the transition energy.

References

1. See, for example, A. Yariv, *Optical Electronics*, 3rd ed., Holt, Rinehart and Winston, New York, p. 311.

Index